DEVELOPMENTAL NEUROBIOLOGY
OF BREATHING

LUNG BIOLOGY IN HEALTH AND DISEASE

Executive Editor: **Claude Lenfant**

Director, National Heart, Lung, and Blood Institute
National Institutes of Health
Bethesda, Maryland

Volume 1   IMMUNOLOGIC AND INFECTIOUS REACTIONS IN THE LUNG,
            edited by Charles H. Kirkpatrick and Herbert Y. Reynolds

Volume 2   THE BIOCHEMICAL BASIS OF PULMONARY FUNCTION,
            edited by Ronald G. Crystal

Volume 3   BIOENGINEERING ASPECTS OF THE LUNG,
            edited by John B. West

Volume 4   METABOLIC FUNCTIONS OF THE LUNG,
            edited by Y. S. Bakhle and John R. Vane

Volume 5   RESPIRATORY DEFENSE MECHANISMS (in two parts),
            edited by Joseph D. Brain, Donald F. Proctor, and Lynne M. Reid

Volume 6   DEVELOPMENT OF THE LUNG,
            edited by W. Alan Hodson

Volume 7   LUNG WATER AND SOLUTE EXCHANGE,
            edited by Norman C. Staub

Volume 8   EXTRAPULMONARY MANIFESTATIONS OF RESPIRATORY DISEASE,
            edited by Eugene Debs Robin

Volume 9   CHRONIC OBSTRUCTIVE PULMONARY DISEASE,
            edited by Thomas L. Petty

Volume 10   PATHOGENESIS AND THERAPY OF LUNG CANCER,
            edited by Curtis C. Harris

Volume 11   GENETIC DETERMINANTS OF PULMONARY DISEASE,
            edited by Stephen D. Litwin

Volume 12   THE LUNG IN THE TRANSITION BETWEEN HEALTH AND DISEASE,
            edited by Peter T. Macklem and Solbert Permutt

Volume 13   EVOLUTION OF RESPIRATORY PROCESSES:
            A COMPARATIVE APPROACH,
            edited by Stephen C. Wood and Claude Lenfant

Volume 14   PULMONARY VASCULAR DISEASES,
edited by Kenneth M. Moser

Volume 15   PHYSIOLOGY AND PHARMACOLOGY OF THE AIRWAYS,
edited by Jay A. Nadel

Volume 16   DIAGNOSTIC TECHNIQUES IN PULMONARY DISEASE (in two parts),
edited by Marvin A. Sackner

Volume 17   REGULATION OF BREATHING (in two parts),
edited by Thomas F. Hornbein

Volume 18   OCCUPATIONAL LUNG DISEASES:
RESEARCH APPROACHES AND METHODS,
edited by Hans Weill and Margaret Turner-Warwick

Volume 19   IMMUNOPHARMACOLOGY OF THE LUNG,
edited by Harold H. Newball

Volume 20   SARCOIDOSIS AND OTHER GRANULOMATOUS
DISEASES OF THE LUNG,
edited by Barry L. Fanburg

Volume 21   SLEEP AND BREATHING,
edited by Nicholas A. Saunders and Colin E. Sullivan

Volume 22   *PNEUMOCYSTIS CARINII* PNEUMONIA:
PATHOGENESIS, DIAGNOSIS, AND TREATMENT,
edited by Lowell S. Young

Volume 23   PULMONARY NUCLEAR MEDICINE:
TECHNIQUES IN DIAGNOSIS OF LUNG DISEASE,
edited by Harold L. Atkins

Volume 24   ACUTE RESPIRATORY FAILURE,
edited by Warren M. Zapol and Konrad J. Falke

Volume 25   GAS MIXING AND DISTRIBUTION IN THE LUNG,
edited by Ludwig A. Engel and Manuel Paiva

Volume 26   HIGH-FREQUENCY VENTILATION IN INTENSIVE CARE
AND DURING SURGERY,
edited by Graziano Carlon and William S. Howland

Volume 27   PULMONARY DEVELOPMENT: TRANSITION
FROM INTRAUTERINE TO EXTRAUTERINE LIFE,
edited by George H. Nelson

Volume 28   CHRONIC OBSTRUCTIVE PULMONARY DISEASE, second edition, revised and expanded, *edited by Thomas L. Petty*

Volume 29   THE THORAX (in two parts), *edited by Charis Roussos and Peter T. Macklem*

Volume 30   THE PLEURA IN HEALTH AND DISEASE, *edited by Jacques Chrétien, Jean Bignon, and Albert Hirsch*

Volume 31   DRUG THERAPY FOR ASTHMA: RESEARCH AND CLINICAL PRACTICE, *edited by John W. Jenne and Shirley Murphy*

Volume 32   PULMONARY ENDOTHELIUM IN HEALTH AND DISEASE, *edited by Una S. Ryan*

Volume 33   THE AIRWAYS: NEURAL CONTROL IN HEALTH AND DISEASE, *edited by Michael A. Kaliner and Peter J. Barnes*

Volume 34   PATHOPHYSIOLOGY AND TREATMENT OF INHALATION INJURIES, *edited by Jacob Loke*

Volume 35   RESPIRATORY FUNCTION OF THE UPPER AIRWAY, *edited by Oommen P. Mathew and Giuseppe Sant'Ambrogio*

Volume 36   CHRONIC OBSTRUCTIVE PULMONARY DISEASE: A BEHAVIORAL PERSPECTIVE, *edited by A. John McSweeny and Igor Grant*

Volume 37   BIOLOGY OF LUNG CANCER: DIAGNOSIS AND TREATMENT, *edited by Steven T. Rosen, James L. Mulshine, Frank Cuttitta, and Paul G. Abrams*

Volume 38   PULMONARY VASCULAR PHYSIOLOGY AND PATHOPHYSIOLOGY, *edited by E. Kenneth Weir and John T. Reeves*

Volume 39   COMPARATIVE PULMONARY PHYSIOLOGY: CURRENT CONCEPTS, *edited by Stephen C. Wood*

Volume 40   RESPIRATORY PHYSIOLOGY: AN ANALYTICAL APPROACH, *edited by H. K. Chang and Manuel Paiva*

Volume 41   LUNG CELL BIOLOGY, *edited by Donald Massaro*

Volume 42   HEART-LUNG INTERACTIONS IN HEALTH AND DISEASE, *edited by Steven M. Scharf and Sharon S. Cassidy*

Volume 43 CLINICAL EPIDEMIOLOGY OF CHRONIC OBSTRUCTIVE PULMO-
NARY DISEASE
edited by Michael J. Hensley and Nicholas A. Saunders

Volume 44 SURGICAL PATHOLOGY OF LUNG NEOPLASMS
edited by Alberto M. Marchevsky

Volume 45 THE LUNG IN RHEUMATIC DISEASES
edited by Grant W. Cannon and Guy A. Zimmerman

Volume 46 DIAGNOSTIC IMAGING OF THE LUNG
edited by Charles E. Putman

Volume 47 MODELS OF LUNG DISEASE: MICROSCOPY AND STRUCTURAL
METHODS
edited by Joan Gil

Volume 48 ELECTRON MICROSCOPY OF THE LUNG
edited by Dean Schraufnagel

Volume 49 ASTHMA: ITS PATHOLOGY AND TREATMENT
*edited by Michael A. Kaliner, Peter J. Barnes,
and Carl G. A. Persson*

Volume 50 ADULT RESPIRATORY DISTRESS SYNDROME
edited by Warren M. Zapol and Francois Lemaire

Volume 51 LUNG DISEASE IN THE TROPICS
edited by Om P. Sharma

Volume 52 EXERCISE: PULMONARY PHYSIOLOGY AND PATHOPHYSIOLOGY
edited by Brian J. Whipp and Karlman Wasserman

Volume 53 DEVELOPMENTAL NEUROBIOLOGY OF BREATHING
edited by Gabriel G. Haddad and Jay P. Farber

Additional Volumes in Preparation

MEDIATORS OF PULMONARY INFLAMMATION
edited by Michael Bray and Wayne Anderson

AIRWAY EPITHELIUM IN HEALTH AND DISEASE
edited by Stephen G. Farmer and Douglas W.P. Hay

DEVELOPMENTAL NEUROBIOLOGY
OF BREATHING

Edited by

Gabriel G. Haddad

Yale University
School of Medicine
New Haven, Connecticut

Jay P. Farber

University of Oklahoma
Health Sciences Center
Oklahoma City, Oklahoma

MARCEL DEKKER, INC.   New York • Basel • Hong Kong

Library of Congress Cataloging--in--Publication Data

Developmental neurobiology of breathing/edited by Gabriel G. Haddad.
   Jay P. Farber
        p.   cm. – -- (Lung biology in health and disease; v. 53)
      Includes bibliographical references and indexes.
      ISBN 0-8247-8459-6 (alk. paper)
      1. Respiration. 2. Developmental neurology. I. Haddad, Gabriel
   G. II. Farber, Jay P. III. Series.
      [DNLM: 1. Anoxia. 2. Asphyxia. 3. Central Nervous System-
   --physiology. 4. Chemoreceptors-- --physiology. 5. Neural
   Transmission-- --physiology. 6. Respiration-- --physiology.
   7. Respiratory System-- –growth & development. W1 LU62 v. 53/WF
   102 D489]
   QP123.D48   1991
   599'.012-- –dc20
   DNLM/DLC
   for Library of Congress   91-16450
                                                                     CIP

This book is printed on acid-free paper.

MARCEL DEKKER, INC.
270 Madison Avenue, New York, New York 10016

Current printing (last digit):
10 9 8 7 6 5 4 3 2 1

PRINTED IN THE UNITED STATES OF AMERICA

To my parents, George and Ida, who always encouraged me
To my wife Karen and my children Chris, Diana, and Justin

GGH

To my wife Barbara and my daughter Sharon

JPF

and to Peter Getting
a friend, a colleague, and a teacher

INTRODUCTION

Volume 6 of the series of monographs Lung Biology in Health and Disease was entitled *Development of the Lung*. One of its chapters addressed the development of respiratory control mechanisms in the fetus and the newborn. The authors, D. E. Woodrum, R. D. Guthrie, and W. A. Hodson, concluded with this statement: "Study of the development of control of ventilation is clearly an intellectual challenge, being an integration of developmental anatomy, neurochemistry and physiology. Many intriguing questions can be posed." This volume was published in 1977!

Since then, neurobiology has truly come of age and an enormous amount of data has been collected and interpreted in medical terms in hopes of benefiting public health. The common use of disciplines—such as cell biology, molecular biology, and histochemistry, among others—has greatly contributed to the many advances in neurobiology that we have witnessed. One area that has very much expanded during the past two decades is the developmental neurobiology of breathing. Just ten years ago, V. Chernick stated, "Despite very significant advances in understanding during the past decade, it is quite clear that many gaps remain and in many areas more confusion than enlightenment still exists. Yet, the future looks bright; advances in knowledge continue at a rapid pace." And, indeed, this prediction has been true.

It is therefore not surprising that after reviewing the field, Drs. Haddad and Farber saw so much new material that they were able to conceive a monograph that would be timely and fill a gap in the series. There is no question that this monograph reports on the state of knowledge in developmental aspects of respiratory control, but it does much more. It identifies areas where more research and new research are needed. Further, it advocates the use of the newer methodologies.

I firmly believe that those who read this monograph will be stimulated by the unique and outstanding contributions of the authors. I want to express my deepest appreciation to all who participated in this volume. Not only does it add a lot to the series Lung Biology in Health and Disease, but more importantly, it adds to the field and provides a blueprint for future investigations.

Claude Lenfant, M.D.
Bethesda, Maryland

PREFACE

It is clear that the health sciences community has experienced a major revolution in the past two decades. This revolution has resulted from major advances in scientific technology, including those of molecular and cellular biology and genetics. Many examples come to mind. Thanks to major advances in the 1970s and 1980s that allowed us to study single ion channels and to isolate and identify specific genes, the major defect in cystic fibrosis has been discovered. Our increasing knowledge of growth factors has led us to better understand normal cellular proliferation as well as malignancies. Advances in the molecular genetics of cholesterol metabolism have reduced the incidence of heart disease. The neurosciences are also rapidly generating new knowledge and solving health problems. Tremendous strides have been made to provide answers for Huntington's disease, myasthenia gravis, multiple sclerosis, Alzheimer's disease, Duchenne's muscular dystrophy, schizophrenia, and seizure disorders. Neuroscientists are now so bold that they are attempting to understand human mentation, learning and memory, at the cellular and molecular level.

Technical advances at cellular and molecular levels in neurophysiology and neuroanatomy have increased our ability to study how breathing is regulated, but we still have major questions to answer. One of these questions relates to respiratory rhythmogenesis itself. The respiratory oscillator remains an enigma and we know little about the nature of its cellular composition, location, and function. Most neurobiologists have not been interested in the brainstem or other brain regions involved with respiratory regulation but rather in higher functions and learning. Although respiratory physiologists have made some progress in understanding respiratory control mechanisms, they have not really taken full advantage of the historical progress recently made in the neural sciences.

In order to advance our field of interest, it would undoubtedly be useful to excite young neuroscientists to work on some of the aspects of neural control of respiration. It would be equally beneficial for respiratory physiologists interested in neural control to study how model neural networks behave and to provide a mechanistic basis for understanding bursting patterns of neuronal activity. In brief, it is very important for neuroscientists and respiratory physiologists to exchange ideas; *one intent of this monograph is to provide a potential forum for such an exchange.* Actually, about one-third of our contributors can be considered as "outsiders," making their contributions essential to the basic idea of this exercise. Two to three decades ago, it was relatively difficult for scientists in different disciplines, such as biochemists, physiologists, geneticists, and biologists, to communicate—even when they were working on the same problem. Today, the revolution of cellular and molecular biology has erased borders between disciplines, and the methodologies that once separated investigators from each other are now responsible for bringing these same scientists together.

Another major goal of this volume is to provide a comprehensive basis for understanding the complex interactions that take place between synaptic input, cellular properties, and the oscillatory output of a neural network, especially in the maturing or developing nervous system. Emphasis is placed on drawing a parallel between the neonate and the mature subject in order to highlight some of the different strategies that the newborn uses when challenged with stress. It will become clear to the reader that we are only beginning to put together the complex interactions between peripheral input and central nervous system activity; in some cases it will be possible only to indicate potential avenues for future research. In designing this volume, we were less concerned with exhaustive coverage of previous work on developmental aspects of respiratory control than with highlighting the relationship of current knowledge to needed areas of inquiry. To accomplish this task, several chapters have been included that do not specifically deal with development of respiratory control; instead, these chapters consider aspects of neurobiology that are important for interpreting the developmental information.

This monograph is divided into five sections. The first part starts with general neurobiological concepts and approaches to central control of respiration. Thus Drs. Millhorn, Sahyoun, and their colleagues present some of the newer advances in neurobiological concepts and Drs. Dekin, Haddad, Martin, Farber, Lawson, and colleagues discuss some of our current knowledge and approaches in addressing the developing central nervous system as it relates to respiration. In Part II of this volume, the developmental role of neural inputs in shaping the central respiratory output is then considered, emphasizing the functions of lung and upper airway receptors as well as carotid chemoreceptors. Hence, Drs. Sant 'Ambrogio, Fisher, Farber, and Donnelly present a comprehensive view of how airway reflexes work at the cellular and integrated level, especially in early life. Drs. Biscoe, Mulligan, and colleagues discuss some of the exciting new work on the carotid chemo-reflex and the biophysical basis for O_2 chemotransduction, and Dr. Nattie reviews data on chemoreception by the brainstem. In Part III, cellular and integrated aspects of the respiratory motor output are discussed by Drs. Sieck, Bigland-Ritchie, Bazzy, Fisher, and their colleagues. These chapters include discussions on interactions between skeletal muscles (respiratory and limb) and CNS, development of structure and function, and how muscles respond to stress. Part IV of this volume deals with the neurophysiological basis for state-dependent respiratory behavior. Dr. Harper emphasizes the neurophy-siological alterations during sleep, and Drs. Stark and Eichenwald review the maturational aspects of respiration as a function of sleep state. Conse-quences of hypoxia and asphyxia are the central theme of Part V of this monograph. Drs. Haddad and Rosen present a general overview of hypoxia and its effects and present some newer data on hypoxia susceptibility and resistance. Drs. Chance and Nioka discuss the subcellular biochemical changes that occur during hypoxia, especially in the young, and Dr. Trippenbach presents data on the effect of hypoxia on the ionic environment in the cen-tral nervous system. Dr. Lagercrantz reviews the neuropharmacology of breathing during hypoxia and Dr. Thach and colleagues discuss asphyxia and autoresuscitation in early life.

It is our hope that this book will provide crucial background for re-searchers with interests in developmental aspects of respiratory control. Application of new cellular and molecular techniques promises to yield im-portant information within the next few years, regarding not only develop-ment, but also the basis of respiratory rhythmogenesis. If discussions in this book convince some scientists to cross disciplines in order to learn new meth-odologies and make progress in answering the many outstanding questions in this field, we will certainly be gratified.

We would like to thank all the contributors for their chapters and ex-press our special gratitude to Dr. Claude Lenfant for his continued encour-agement, perseverance, and help throughout this past year.

Gabriel G. Haddad
Jay P. Farber

CONTRIBUTORS

Douglas A. Bayliss, Ph.D.   Department of Physiology and Curriculum in Neurobiology, University of North Carolina, Chapel Hill, North Carolina

Alia R. Bazzy, M.D.   Assistant Professor, Department of Pediatrics, Section of Respiratory Medicine, Yale University School of Medicine, New Haven, Connecticut

Brenda Bigland-Ritchie, Ph.D., D.Sc.   Fellow, John B. Pierce Laboratory, Inc., New Haven, Connecticut

Tim J. Biscoe, F.R.C.P.   Judrell Professor of Physiology, Department of Physiology, University College London, London, England

Britton Chance, Ph.D.   Department of Biochemistry and Biophysics, University of Pennsylvania School of Medicine, Philadelphia, Pennsylvania

Michael S. Dekin, Ph.D.   Assistant Professor, T. H. Morgan School of Biological Sciences, University of Kentucky, Lexington, Kentucky

David F. Donnelly, Ph.D.   Research Scientist, Department of Pediatrics, Division of Respiratory Medicine, Yale University School of Medicine, New Haven, Connecticut

Michael R. Duchen, Ph.D., M.R.C.P.   Royal Society 1983 University Research Fellow and Lecturer in Physiology, Department of Physiology, University College London, London, England

Eric C. Eichenwald, M.D.   Instructor in Pediatrics, Joint Program in Neonatology, Harvard Medical School, The Children's Hospital, Boston, Massachusetts

Jay P. Farber, Ph.D. Professor, Department of Physiology and Biophysics, University of Oklahoma Health Sciences Center, Oklahoma City, Oklahoma

Jonathan D. Feldman, M.D. Fellow, Department of Pediatrics, Section of Critical Care and Applied Physiology, Yale University School of Medicine, New Haven, Connecticut

John T. Fisher, Ph.D. Associate Professor, Department of Physiology, and Assistant Professor, Department of Anaesthesiology, Queen's University, Kingston, Ontario, Canada

Mario Fournier, Ph.D.* Research Assistant Professor, Department of Biomedical Engineering, University of Southern California, Los Angeles, California

William M. Gershan, M.D. Director, Pediatric Pulmonology, Department of Pediatrics, Medical College of Wisconsin, Milwaukee, Wisconsin

Gabriel G. Haddad, M.D. Professor, Department of Pediatrics, and Director, Section of Respiratory Medicine, Department of Pediatrics, Yale University School of Medicine, New Haven, Connecticut

Ronald M. Harper, Ph.D. Professor, Department of Anatomy and Cell Biology, UCLA School of Medicine, Los Angeles, California

Raymond H. Ho, Ph.D. Associate Professor, Department of Cell Biology, Neurobiology, and Anatomy, The Ohio State University College of Medicine, Columbus, Ohio

Tomas Hökfelt, M.D. Professor, Department of Histology and Neurobiology, Karolinska Institute, Stockholm, Sweden

Margaret S. Jacobi, M.B., Ph.D. Division of Newborn Medicine, Department of Pediatrics, Washington University School of Medicine, St. Louis, Missouri

Hugo Lagercrantz, M.D., Ph.D. Professor, Neonatal Research Unit, Department of Pediatrics, Karolinska Hospital, Nobel Institute for Neurophysiology, Karolinska Institute, Stockholm, Sweden

Present affiliation: Department of Anesthesiology, Mayo Clinic and Mayo Medical School, Rochester, Minnesota

Edward E. Lawson, M.D.   Professor of Pediatrics, Department of Pediatrics, University of North Carolina at Chapel Hill, Chapel Hill, North Carolina

Harry LeVine III, Ph.D.   Research Investigator, Glaxo Research Laboratories, Research Triangle Park, North Carolina

George F. Martin, Ph.D.   Professor, Department of Cell Biology, Neurobiology, and Anatomy, The Ohio State University College of Medicine, Columbus, Ohio

Oommen P. Mathew, M.B.B.S., F.A.A.P.   Professor of Pediatrics and Physiology and Biophysics, Department of Perinatal Pediatrics, University of Texas Medical Branch, Galveston, Texas

Robert B. Mellins, M.D.   Professor of Pediatrics and Director, Pediatric Pulmonary Division, Department of Pediatrics, Columbia University College of Physicians and Surgeons, New York, New York

David E. Millhorn, Ph.D.   Associate Professor, Department of Physiology and Curriculum in Neurobiology, University of North Carolina, Chapel Hill, North Carolina

Eileen M. Mulligan, Ph.D.   Assistant Professor of Physiology, Department of Physiology, Temple University School of Medicine, Philadelphia, Pennsylvania

Eugene E. Nattie, M.D.   Professor of Physiology, Dartmouth Medical School, Hanover, New Hampshire

Shoko Nioka, M.D., Ph.D.   Department of Biochemistry and Biophysics, University of Pennsylvania School of Medicine, Philadelphia, Pennsylvania

Ronda R. Pindzola, Ph.D.   Research Associate, Department of Neurosciences, Case Western Reserve University School of Medicine, Cleveland, Ohio

Carol L. Rosen, M.D.   Assistant Professor, Section of Respiratory Medicine, Department of Pediatrics, Yale University School of Medicine, New Haven, Connecticut

Michael Runold   Neonatal Research Unit, Department of Pediatrics, Karolinska Hospital, Nobel Institute for Neurophysiology, Karolinska Institute, Stockholm, Sweden

Naji Sahyoun, M.D. Head, Section of Molecular and Cellular Neurobiology, Division of Cell Biology, Burroughs Wellcome Company, Research Triangle Park, North Carolina

Giuseppe Sant 'Ambrogio, M.D. Professor of Physiology and Biophysics, Department of Physiology and Biophysics, University of Texas Medical Branch, Galveston, Texas

Kim B. Seroogy, Ph.D.* Department of Physiology and Curriculum in Neurobiology, University of North Carolina, Chapel Hill, North Carolina

Gary C. Sieck, Ph.D.† Department of Biomedical Engineering, University of Southern California, Los Angeles, California

Ann R. Stark, M.D. Joint Program in Neonatology, Harvard Medical School, The Children's Hospital, Boston, Massachusetts

Caroline L. Szymeczek Department of Physiology and Curriculum in Neurobiology, University of North Carolina, Chapel Hill, North Carolina

Bradley T. Thach, M.D. Professor of Pediatrics, Division of Newborn Medicine, Department of Pediatrics, Washington University School of Medicine, St. Louis, Missouri

Teresa Trippenbach, M.D., Ph.D. Associate Professor, Department of Physiology, McGill University, Montreal, Quebec, Canada

Mark A. Waldron, Ph.D. Department of Physiology, Queen's University, Kingston, Ontario, Canada

Xiao Ming Xu Department of Cell Biology, Neurobiology, and Anatomy, The Ohio State University College of Medicine, Columbus, Ohio

---

*Present affiliation: Assistant Professor, Department of Anatomy and Neurobiology, University of Kentucky, Lexington, Kentucky

†*Present affiliation*: Professor of Physiology, Departments of Anesthesiology and Physiology and Biophysics, Mayo Clinic and Mayo Medical School, Rochester, Minnesota

CONTENTS

Introduction   *v*
Preface   *vii*
Contributors   *xi*

Clinical Observations of Respiratory Control:
Challenges for Basic Research   **1**

Robert B. Mellins

   I. Breathing Behavior   1
  II. Complexity and Redundancy   2
 III. Emotion and State   3
 IV. Structural Versus Functional Abnormalities   3
  V. The Importance of Respiratory Timing   4
 VI. Afferent Input to Respiration   4
VII. Neurochemicals, Hormones, and Growth Factors   5
VIII. Respiratory Arrhythmia   5
 IX. Rates of Maturation   6
  X. Conclusions   6
    References   7

PART ONE: GENERAL NEUROBIOLOGICAL APPROACHES
TO CENTRAL CONTROL OF RESPIRATION

1. Cellular, Molecular, and Developmental Aspects of Chemical
Synaptic Transmission   **11**

David E. Millhorn, Caroline L. Szymeczek, Douglas A. Bayliss,
Kim B. Seroogy, and Tomas Hökfelt

 I. Introduction   11
II. Cellular and Molecular Mechanisms of Synaptic Transmission   12

III. Regulation of Gene Expression by Synaptic and Hormonal
     Signals   31
IV. Ontogenesis of Neurotransmitter and Peptide Systems in the
    Brain   44
    References   54

2.   **Neuronal Signaling: Pathways and Protein Kinases**   **71**

     Harry LeVine III and Naji Sahyoun

  I. Introduction: Signaling Pathways   71
 II. Families of Signaling Molecules   73
III. Protein Kinase C   76
 IV. CaMKinase II   84
  V. Signal Termination by Phosphoprotein Phosphatases   93
     References   95

3.   **Comparative Neurobiology of Invertebrate Motor Networks:**
     Implications for the Control of Breathing in Mammals   **111**

     Michael S. Dekin

  I. Introduction   111
 II. A Primer on the Motor Organization of Invertebrates   112
III. Implications for the Control of Breathing   127
 IV. Summary   144
     References   145

4.   **Cellular and Membrane Properties of Brainstem Neurons**
     in Early Life   **155**

     Gabriel G. Haddad

   I. Introduction   155
  II. Synaptic Input Versus Membrane Properties: Historical
      Perspective   156
 III. Repetitive Firing Properties: Examples   157
  IV. Passive Electrophysiological Properties   157
   V. Active Membrane Properties and Ionic Currents   160
  VI. Morphology and Connectivity   168
 VII. Importance to Breathing   169
VIII. Conclusions and Implications for Future Research   171
      References   172

Contents   *xvii*

5. **The Development of Descending Spinal Pathways
    in the North American Opossum**   **177**

    *George F. Martin, Raymond H. Ho, Xiao Ming Xu, and
    Ronda R. Pindzola*

      I. Introduction   177
     II. The Development of Brainstem-Spinal Projections   178
    III. The Development of Corticospinal Projections   189
     IV. Summary and Conclusions   191
         Abbreviations   192
         References   193

6. **Neurophysiological Organization of Respiratory Neurons
    in Early Life**   **199**

    Jay P. Farber and Edward E. Lawson

      I. Introduction   199
     II. Studies Using Extracellular Recording and Related Techniques   200
    III. Studies Using Intracellular Recording Techniques: The
         Newborn Piglet Model   207
     IV. Future Directions/Problems   213
         References   213

**PART TWO: INPUT TO THE CNS: CELLULAR NEUROBIOLOGY
              AND INTEGRATED BEHAVIOR**

7. **Morphological and Neurophysiological Aspects of
    Airway and Pulmonary Receptors**   **219**

    *John T. Fisher, Oommen P. Mathew, and Giuseppe
    Sant'Ambrogio*

      I. Introduction   219
     II. Upper Airway Receptors   220
    III. Lower Airway Receptors   230
     IV. Conclusion   239
         References   239

8. **Development of Pulmonary and Chest Wall Reflexes
    Influencing Breathing**   **245**

    Jay P. Farber

      I. Introduction   245
     II. Reflex Responses Associated with Vagal Airway Receptors   247

III. Reflexes Associated with Chest Wall Receptors   261
IV. Implications for Future Research   262
References   263

9.   **Laryngeal Reflexes: Integrated and Cellular Aspects**   **271**

      David F. Donnelly

  I. Introduction   271
 II. Laryngeal Nerve Fibers   272
III. Effect and Mechanism of Action   274
IV. Maturational Changes   280
 V. Future Directions   282
References   283

10.   **How Do Arterial Chemoreceptors Work? Studies of Cells
      from the Carotid Body of the Rabbit**   **289**

      Michael R. Duchen and Tim J. Biscoe

    I. Introduction   289
   II. Cellular Elements and Identity of the Transducer   291
  III. The Evidence   292
   IV. Single-Cell Studies   293
    V. Intracellular Recordings from Type I Cells   295
   VI. Measurements of Intracellular Calcium   299
  VII. Other Recent Investigations   305
 VIII. The Centrifugal Pathway   307
   IX. The Transmitters   308
    X. Effects of Hypoxia on Neurons   309
   XI. Prospects for the Future   311
References   312

11.   **Discharge Properties of Carotid Bodies: Developmental
      Aspects**   **321**

      Eileen M. Mulligan

  I. Development of Carotid Body Chemoreceptors: Evidence from
     Ventilatory Studies   321
 II. Carotid Chemoreceptor Recording: Methodology   322
III. Carotid Chemoreceptor Afferent Nerve Responses in the Fetus 324
IV. Carotid Chemoreceptor Afferent Nerve Responses in the
     Neonate   325
 V. Summary   335
References   336

12. **Central Respiratory Chemoreceptors: Cellular Mechanisms
     and Developmental Aspects**   **341**

     Eugene E. Nattie

      I. Introduction   341
     II. Evidence for the Existence of Central Chemoreception   341
    III. Location of Central Chemoreceptors   342
     IV. An Expanded Role for Neurons Within the VLM   348
      V. Mechanisms of Central Chemoreception   354
     VI. Central Chemoreception During Development   358
    VII. Conclusions   361
         References   361

**PART THREE   RESPIRATORY MOTOR OUTPUT: CELLULAR
               FUNCTION AND INTEGRATED BEHAVIOR**

13. **Developmental Aspects of Diaphragm Muscle Cells:
     Structural and Functional Organization**   **375**

     Gary C. Sieck and Mario Fournier

      I. Introduction   375
     II. Motor Unit Physiological Properties   375
    III. Neural Control of Muscle Tension   381
     IV. Factors Contributing to Diaphragm Muscle Fatigue   389
      V. Diaphragm Fiber Type Proportions During Postnatal
         Development   407
     VI. Summary and Conclusions   415
         References   416

14. **Functional Aspects of Human Muscles and Their
     Central Nervous System Interactions: 429
     Implications for Muscle Failure**

     Brenda Bigland-Ritchie

      I. Introduction   429
     II. Problems in Fatigue Studies   431
    III. Contractile Properties of Human Motor Units   433
     IV. Potential Causes of Fatigue   436
      V. Fatigue Processes in Human Voluntary Contractions   439
     VI. Relation Between CNS Motor Drive and Muscle Contractile
         Properties   444
    VII. Fatigue of Respiratory Muscles   446
   VIII. Implications for Neonatal Respiratory Muscle Failure   448
     IX. Summary and Conclusions   448
         References   449

15. **Integrated Response of the Respiratory Muscles to Load**   **455**

Alia R. Bazzy and Jonathan D. Feldman

   I. Introduction   455
  II. Response to Acute Loads   456
 III. How Respiratory Muscle Training Modulates the Response
     of Respiratory Muscles to Acute Loads   466
 IV. The Developing Respiratory Muscles   472
     References   473

16. **Neural Control of Airway Smooth Muscle in the Newborn**   **483**

Mark A. Waldron and John T. Fisher

   I. Introduction   483
  II. Species Differences   484
 III. Autonomic Airway Innervation in the Adult:
     An Overview   485
 IV. Vagal Control of Airway Smooth Muscle   487
  V. Adrenergic/Sympathetic Control of Airway Smooth Muscle   502
 VI. Conclusion   506
     References   507

PART FOUR: NEUROPHYSIOLOGICAL BASIS FOR STATE-
                DEPENDENT RESPIRATORY BEHAVIOR

17. **State-Dependent Electrophysiological Changes in Central
    Nervous System Activity**   **521**

Ronald M. Harper

   I. Introduction   521
  II. Electrophysiologic Aspects of REM Sleep   523
 III. Descending "Arousal" Systems: What Do Descending
     Forebrain Systems Do for Respiratory Control?   528
 IV. Ascending Projection Systems   535
  V. Relationship of Slow-Wave Activity to Respiratory Patterning 538
 VI. Summary   540
     References   541

18. **Respiratory Motor Output: Effect of State and Maturation
    in Early Life**   **551**

Eric C. Eichenwald and Ann R. Stark

   I. Introduction   551
  II. Determination of State   552

III. Fetal Breathing Movements 554
IV. Newborn Period 558
V. Respiratory Muscle Output 567
VI. Metabolic Control: Effect of State 571
VII. Conclusions 574
    References 575

PART FIVE   **HYPOXIA AND ASPHYXIA: CELLULAR, INTEGRATED, AND PATHOPHYSIOLOGICAL ASPECTS**

19. **Ventilatory Response to Hypoxia: Integrated, Cellular, and Molecular Aspects**   **591**

    Gabriel G. Haddad and Carol L. Rosen

  I. Introduction 591
 II. Integrated Response 592
III. Cellular and Molecular Response 595
    References 607

20. **Energy Metabolism and Ionic Homeostasis During Hypoxic Stress in the Developing Mammalian Brain**   **615**

    Shoko Nioka and Britton Chance

   I. Introduction 615
  II. Energy Metabolism in Early Life 616
 III. Hypoxic Tolerance in the Developing Brain 626
 IV. Summary 635
    References 635

21. **Ionic Environment in the Central Nervous System and Effects of Hypoxia in Early Life**   **643**

    Teresa Trippenbach

   I. Introduction 643
  II. Anatomical Organization of Respiration-Related Neurons 644
 III. Ventilatory Effects of Hypoxia in Newborn Animals 645
 IV. Extracellular Ion Activities in the CNS of Mammals 646
  V. Extracellular Ion Activities During Low Oxygen Supply 650
    References 656

22.   **Hypoxia and Neuropharmacology of Breathing**   **661**

   Hugo Lagercrantz and Michael Runold

   I. Introduction   661
   II. Neuroactive Agents in the Regulation of Breathing   662
   III. Respiratory-Controlling Structures Affected by Hypoxia   663
   IV. Fetal Hypoxia   667
   V. Infant Hypoxia   668
   VI. Clinical Aspects   672
   VII. Conclusions   672
       References   673

23.   **Control of Breathing During Asphyxia and Autoresuscitation**   **681**

   *Bradley T. Thach, Margaret S. Jacobi, and William M.
   Gershan*

   I. Introduction   681
   II. The Pattern of Respiration During Acute Asphyxiation:
       Hyperpnea, Apnea, and Gasping   682
   III. The Role of Gasping in Spontaneous Recovery from Hypoxic
        Apnea   688
   IV. Summary   694
       References   695

Author Index   *701*
Subject Index   *753*

DEVELOPMENTAL NEUROBIOLOGY
OF BREATHING

Clinical Observations of Respiratory Control
Challenges for Basic Research

ROBERT B. MELLINS

Columbia University College of Physician and Surgeons
New York, New York

The great wall between body and behavior is being bored from within
(starting with the neuron) and from without (starting with behavior).
As the boring proceeds, we realize that the wall is not in nature but in
philosophy and theology.

Mario Bunge, 1989

I. Breathing Behavior

Living matter is characterized by cell growth, development, and decay. Development includes such phenomena as differentiation, remodeling, and
maturation. Decay includes aging. All of these activities require both the
sending and receiving of signals, and nowhere is this more critical than in
the growth and development of the nervous system. At its most basic, disease or pathophysiology interferes with the orderly progression of normal
growth and development. This book is focused on one specific aspect of
neurobiological function: that relating to breathing. As von Euler (1986,
1987) and before him Fenn have pointed out, breathing is a complex *behavior*

1

that depends not only on a series of reflexes including central and peripheral chemoreceptors, but also on mental activities and emotional states. While it is the integrated behavior itself, namely breathing, that will be of the greatest clinical interest to us, our ability to diagnose specific disorders and to devise therapy will depend on our understanding of the basic building blocks. Thus, of necessity, we will need to understand molecules, membranes, messengers, mediators, and modulators if we are to be able to address such respiratory abnormalities as apnea, aspiration, and respiratory failure. If the challenge of our predecessors was to marry physiology to clinical medicine, our challenge is even greater, for we must now bridge the gap between molecules and behavior. Until recently, the approach to understanding respiratory control has depended on studying the physiological responses to ablation-type experiments or the superimposition of stimuli (e.g., hypoxia or hypercapnia) or the administration of agonists and antagonists. Given the complexity of respiratory control, and the likelihood that any given stimulus or pharmacological agent may have different effects on a number of the components of respiratory control, it is not surprising that the results have sometimes seemed contradictory.

II. Complexity and Redundancy

From a teleoleogical point of view, one might ask why it is necessary for the control of breathing to be so complex. First, to maintain normal acid-base balance at least in the face of rapid changes in metabolism, the respiratory system has been given the responsibility to adjust pH rapidly, at least on a short-term basis, through the ability to blow off CO_2. Second, there is the need to preserve respiratory gas concentrations within relatively narrow limits across a wide spectrum of activities (rest and exercise), environments (altitude or hypoxia, heat and cold), states (awake and sleep), and positions (upright and supine). Third, some anatomical components of the respiratory control system are shared with other organ systems and their functions must be coordinated according to changing priorities. Thus, speech, swallowing, expulsive acts (cough, grunting, defecation, micturition) all share components of the upper airways. During sleep, breathing must be automatic, but during eating, speaking, and playing of wind instruments there must be the possibility to override the automaticity, depending on the importance or priority of the activity. Also, periodic sighs are necessary either as an expression of emotion (deep sigh of relief) or to preserve lung function by reversing microatelectasis. Given the widely varying and sometimes apparently conflicting needs of the body, it is not at all surprising that so many agonists and antagonists are involved in respiratory control with build-in redundancy.

The complexity of respiratory control and the fact that parts of the respiratory system are shared with the alimentary system may account for several conditions in which there is discoordination in swallowing with recurrent aspiration presenting either as chronic cough or recurrent pneumonia. This is most dramatically seen in patients with familial dysautonomia but is also present in patients with congenital tracheoesophageal fistula even after the fistula has been repaired. We have also seen this during the early months or years of life in several patients who have not shown any other gross neurological abnormalities (Mellins et al., 1970). Perhaps research at the cellular or molecular level will open up new approaches to these challenging clinical problems.

III. Emotion and State

The importance of emotion to respiratory control is revealed by several clinical examples, including (1) breath holding in childhood (temper tantrums), (2) alkalotic tetany following anxiety-provoked hyperventilation, and (3) laryngeal asthma (Christopher et al., 1983). The importance of state (both state of consciousness and sleep state) to respiratory control is clearly revealed by the need to support respiration during anesthesia as well as by the failure of adequate spontaneous ventilation during sleep but not wakefulness in patients with failure of automatic control of ventilation, primary pulmonary hypoventilation, or Ondine's curse (Mellins et al., 1970).

IV. Structural Versus Functional Abnormalities

Lesions of the medullary reticular formation and its outflow to the spinal cord have also resulted in abnormalities in the automatic control of ventilation, especially during sleep. Other conditions, including myelomeningoceles, hydrocephalus, and Arnold-Chiari malformation, have resulted in abnormal ventilatory pattern during sleep in infants (Davidson et al., 1986). Recently adolescents with myelomeningocele have been shown to have decreased hypercapnic ventilatory responses during wakefulness and sleep (Swaminathan et al., 1989). Although this book is focused on the biology of breathing, there are clinical examples that suggest that abnormalities in the development of the nervous system can have profound effects on other functions not generally believed to relate to breathing. Thus transient disease of the hypothalamus, including voracious appetite, has been reported in a child with primary pulmonary hypoventilation (Mellins et al., 1970). Perhaps even more intriguing has been the association of failure to automatic control of breathing, heart rate, gastrointestinal motility (Haddad et al., 1978). Since this was written we have learned of several other patients with the same constellation of abnormalities. This initial report is a striking example of the insights that can be derived from the thorough study of a single patient. The observation that the abnormalities were not constant but state related ($PaCO_2$ was above 80 mm Hg when asleep and normal when awake) suggested that the problem was not in "hard wiring." Also, abnormalities were noted in respiratory frequency with relatively normal tidal volume, suggesting independent mechanisms for controlling these two determinants of ventilation.

V. The Importance of Respiratory Timing

Thus far we have been discussing respiration in a global sense. However, it is well recognized that a given level of ventilation can be achieved by different combinations of tidal volume and frequency of breathing depending on physiological alterations (e.g., pregnancy) or pathophysiology (e.g., pulmonary fibrosis or kyphoscoliosis). The timing of respiratory events is critical as well. Impulses reach the tongue and other upper airway muscles a fraction of a second before the diaphragm, thus maintaining the patency of the upper airways. Two dramatic examples of the clinical relevance of this are the upper airway obstruction seen when (1) patients with primary pulmonary hypoventilation undergo electrophrenic pacing in the absence of a tracheostomy and (2) patients with neuromuscular disease are placed in cuirass-type negative-pressure ventilators. In both these circumstances the sequential innervation of the upper airway muscles and then the diaphragm is disturbed so that the collapsing force of the negative pressure within the upper airways is not opposed by the stimulation of the muscles in the upper airways.

VI. Afferent Input to Respiration

It has been well recognized that afferent input to the central nervous system can have profound effects on respiratory control. Afferent input can modify or modulate inherent cellular firing or rhythmicity, but how this comes about remains elusive. It is not clear, for example, whether the primary defect in apnea of prematurity is at the afferent or efferent levels or at the level of the central nervous system processing of signals. Decrease in sensation from the upper airways in familial dysautonomia is believed to cause dysphagia and predispose to pulmonary aspiration: Whether there are other sensory or afferent abnormalities in dysautonomia and whether these result in defects in respiratory control is not known. There are also powerful reflexes from the larynx and upper airways that are protective in nature, inhibiting respiration. Both maturation and the balance of stimulatory and inhibitory reflexes are likely to be very important since studies in lambs have demonstrated that with maturation, the central neural respiratory circuits become more responsive to stimulation by CO_2 and less responsive to inhibition by superior laryngeal nerve stimulation (Kurth et al., 1989).

VII. Neurochemicals, Hormones, and Growth Factors

Abnormalities or alterations in other systems can have profound effects on breathing as the result of changes in neuromediators, hormones, and growth factors. The use of progesterone, for example, as a respiratory stimulant derived from clinical observations that there was an increase in ventilation during the luteal phase of the menstrual cycle as well as during pregnancy. The number of neurochemicals involved in respiratory control during development is increasing rapidly and includes both inhibitory modulators [adenosine, gamma aminobutyric acid (GABA), opioids, and prostaglandins], stimulatory modulators (progesterone, substance P, and thyrotropic-releasing hormone), and modulators with effects on respiratory control that are less well characterized (serotonin and calecholamines) (Moss and Inman, 1989). These neurochemicals, hormones, and growth factors may play a major role in state (state of arousal) and adaptation, as well as in cell proliferation and differentiation.

VIII. Respiratory Arrhythmia

With respect to apnea, epidemiological and clinical studies have helped to define populations at risk, but even when coupled with studies of cardiorespiratory physiology and neuropharmacology, these have not elucidated the pathogenesis. Major progress in the treatment of cardiac arrhythmias depended not only on recognition of the role of pacemakers, but on more fundamental understanding of how electric impulses spread through the heart. By analogy, it seems likely that advances in treating respiratory arrhythmias, including apnea of prematurity, will be made when we have a better understanding of respiratory rhythm generation both at the neuronal level and as the impulses spread throughout the respiratory network. Although the cellular basis for the central pattern generator rhythm has not been elucidated, progress has been made in understanding the cellular properties of respiratory neurons, including the types of ionic conductances (Dekin and Haddad, 1990).

IX. Rates of Maturation

Given the possibility that there are differences in the maturation of systems sharing common anatomical or functional pathways, a more fundamental understanding of the cellular and subcellular development of these systems seems now to be a necessary step to building up a more comprehensive understanding of respiratory control. Nowhere is this more important than in the understanding of the dynamic aspects of breathing in early life, a period during which there is apparently an excessive number of neurons that are refined and axons that are pruned.

Two clinical areas have received increased recent attention, in part because of some evidence suggesting that maturation plays an important role. These are (1) the ventilatory response to hypoxia and (2) respiratory muscle fatigue or failure. Although the biphasic response to hypoxia in early life, i.e., an initial increase followed by a decrease below baseline, is well recognized, our understanding of the phenomenon remains incomplete. It is not clear whether the biphasic response results from changes in organ physiology with maturation, e.g., redistribution of blood flow away from relatively high metabolic areas or in alterations in cellular properties with maturation. With respect to respiratory muscle failure, initial studies had suggested that the diaphragm is more susceptible to failure under loaded conditions in early life, but recent work has questioned this conclusion. More fundamental work on the differences in muscle cell properties with age will be needed to resolve the question of whether there are differences in ease of fatigability and failure with maturation.

X. Conclusions

One conclusion to be derived from these clinical observations is that it will not only be important to understand function at the cellular and molecular levels, but it will also be important to understand the relative rates of growth or of maturation of the various systems involved in respiratory control. Indeed, it seems that disease or disorders of respiration are just as likely to result from abnormalities in the *rates* of maturation of different functions as they are from abnormalities in the functions themselves.

A second conclusion that seems inescapable is that major advances in our ability to diagnose and treat clinical disorders of respiratory control are more likely to emerge from cross fertilization of fields and the application of newer technologies to understand the cellular and molecular basis for respiratory rhythm. Since widely disparate fields now use simular techniques to explore mechanisms of disease at the cellular and molecular level, a common vocabulary is developing. Thus it ought not to be difficult to foster the necessary interdisciplinary research. To make this prediction a reality, however, we will need to foster more basic science training for the clinical investigator of the future.

References

Bunge, M. (1989). From neuron to mind. *News Physiol. Sci.* **4**:206-209.

Christopher, K. L. Raymond, P., Wood, R. P. II, Eckert, R. C., Blager, F. B., Raney, R. A., and Souhrada, J. F. (1983). *N. Engl. J. Med.* **308**: 1566-1570.

Davidson, W. S. L., Jacobs, J. A., and Gates, E. P. (1986). Abnormal ventilatory patterns during sleep in infants with myelomeningocele. *J. Pediatri.* **109**:631-634.

Dekin, M. S., and Haddad, G. G. (1990). Membrane and cellular properties in oscillating networks: Implications for respiration. *J. Appl. Physiol.* (in press).

Euler, C. von (1986). Brain-stem mechanisms for generation and control of the breathing pattern. In *Handbook of Physiology. The Respiratory System.* Vol. 2. Edited by N. S. Cherniack and J. G. Widdicombe. Bethesda, MD, American Physiological Society, pp. 1-67.

Euler, C. von (1987). Breathing behavior. In *Neurobiology of the Control of Breathing.* Edited by C. von Euler and H. Lagercrantz. New York, Raven Press, pp. 3-8.

Haddad, G. G., Mazza, N. M., Defendini, R. F., Blanc, W. A., Driscoll, J. M., Epstein, A., Epstein, R. A., and Mellins, R. B. (1978). Congenital failure of automatic control of ventilation, gastrointestinal motility and heart rate. *Medicine* **57**:517-526.

Kurth, C. D., Hutchison, A. A., Caton, D. C., and Davenport, P. W. (1989). Maturational and anesthetic effects on apneic thresholds in lambs. *J. Appl. Physiol.* **67**:643-647.

Mellins, R. B. (1986). Recurrent aspiration, chronic cough and reactive airways disease. In *III Congresso Nazionale di Bronchopneumologia Pediatrica.* Edited by Mario LaRosa. Catania, Acireale, pp. 104-109.

Mellins, R. B., Balfour, H. H., Turino, G. M., and Winters R. W. (1970). Failure of automatic control of ventilation (Ondines curse). *Medicine* **49**:487-504.

Moss, I. R., and Inman, J. G. (1989). Neurochemicals and respiratory control during development. *J. Appl. Physiol.* **67**:1-13.

Swaminathan, S., Paton, J. Y., Davidson Ward, S. L., Jacobs, R. A., Sargeant, C. W., and Keens, T. G. (1989). Abnormal control of ventilation in adolescents with myelodysplasia. *J. Pediatri.* **115**:898-903.

Part One

GENERAL NEUROBIOLOGICAL APPROACHES TO CENTRAL CONTROL OF RESPIRATION

1

Cellular, Molecular, and Developmental Aspects of Chemical Synaptic Transmission

DAVID E. MILLHORN,
CAROLINE L. SZYMECZEK,
DOUGLAS A. BAYLISS, and
KIM B. SEROOGY*

University of North Carolina
Chapel Hill, North Carolina

TOMAS HÖKFELT

Karolinska Institute
Stockholm, Sweden

I. Introduction

At the turn of the century, the renowned Spanish neuroanatomist Santiago Ramon y Cajal showed unequivocally, using the cell staining technique of Golgi, that the nervous system is composed of structurally independent cells rather than a syncytium of interconnecting fibrils (i.e., the reticular theory), as previously believed (Ramon y Cajal, 1908). From an historical perspective, this simple observation completely revolutionized the study of the nervous system. It was evident to Ramon y Cajal and other neuroscientists of the era that even the simplest reflex responses required the coordinated activation and deactivation of an immense number of independent nerve cells. This, of course, meant that neurons must possess the ability to communicate with each other and nonneural target tissue (e.g., striated and smooth muscle). Sherrington (1906) proposed that neurons communicate with each other at sites of close apposition or functional contact, which he termed the *synapse*.

Although it was a foregone conclusion by the end of the first decade of this century that neurons must be able to communicate with each other, there

---

*Present Affiliation: University of Kentucky, Lexington, Kentucky

was little insight as to how cell-to-cell communication in the nervous system might occur. Most neuroscientists of the time believed that signaling in the nervous system occurred via *electrical* connections. Support for electrical transmission was based, for the most part, on teleological considerations. It was widely believed that the speed of signaling in the nervous system, assessed from measurements of reflex latencies, could only be explained by electrical coupling; the idea of chemical transmission was dismissed by most workers as being much too slow. However, as early as 1905, Elliott presented evidence that chemical mediators were involved in synaptic transmission in sympathetic ganglia. Due largely to the work of such notable neuroscientists as Loewi, Langley, Dale, and Cannon, support for chemical transmission grew during the 1920s and 1930s. Debate on this topic endured until the late 1940s and early 1950s, when Eccles, a prominent neuroscientist and proponent of electrical transmission, presented electrophysiological evidence that both excitatory and inhibitory signaling is accomplished by release from the presynaptic terminal of special chemical compounds called *neurotransmitters* [for an in-depth account of this early work refer to the monograph written by J.C. Eccles (1957)].

Acceptance of chemical transmission as the primary mode of communication in the nervous system marked the birth of the modern era of neurobiology. Certainly, investigations of nervous system function during the last quarter century have been dominated by research concerning different aspects of chemical synaptic transmission. Much of this research has been directed at identification of chemical compounds involved in synaptic transmission and elucidation of intracellular signal transduction mechanisms that mediate many of the postsynaptic effects associated with neurotransmission. In this chapter we shall provide an account of cellular and molecular mechanisms associated with chemical transmission and signal transduction in the mature mammalian nervous system. We shall then discuss chemical transmission in the developing nervous system and emphasize the changing nature of the neuronal chemical phenotypes (i.e., transmitter and peptide content) during ontogeny.

II. Cellular and Molecular Mechanisms of Synaptic Transmission

Synaptic transmission is perhaps the most widely studied phenomenon in neurobiology today. As a result, there have been many new and exciting findings in recent years concerning the ultrastructure of the synapse and the cellular and molecular mechanisms involved in transmitter biosynthesis, transmitter release, receptor function, signal transduction, and the role of synaptic transmission in regulation of gene expression. In order to understand the role of chemical synaptic transmission in the developing animal, one must first understand its role in the mature animal. For this reason we believe it is important to begin with an account of cellular and molecular mechanisms involved in synaptic transmission and signal transduction in the mature animal. As a matter of convenience, we shall discuss first the basic structure and function of the presynaptic terminal and then turn our attention to the molecular events that occur postsynaptically following receptor activation.

A. The Presynaptic Neuron

In the simplest scenario, the presynaptic cell initiates the process of synaptic transmission by releasing neurotransmitter into the synaptic cleft. This type of arrangement is generally referred to as "axodendritic" or "axosomatic" transmission, depending on whether the presynaptic neuron releases transmitter onto the dendrites or soma of the postsynaptic cell. Although we use this simple model in this discussion, the reader should bear in mind that the presynaptic cell may also receive synaptic input (i.e., "axoaxonic" connections) and in a sense is also a postsynaptic cell. Presynaptic inhibition is often mediated by axoaxonic connections.

The synapse appears as electron-dense material on both the pre- and postsynaptic membranes when viewed through an electron microscope (Palade, 1954). The presynaptic membrane is separated from the postsynaptic membrane by a cleft (i.e., the synaptic cleft) that measures 100-300 angstroms (Palade, 1954; Gray, 1969; Bodian, 1972). This distinguishing feature of presynaptic terminal is the large number of clear and dense core spherical vesicles of various sizes (200-800 angstroms in diameter) that are clustered very close to the synaptic membrane (Palade, 1954; De Robertis and Bennett, 1954; Palay, 1954, 1956; De Robertis, 1956). There is now biochemical and morphological evidence that these vesicles actually contain transmitter molecules (Pelletier et al., 1981; Freid et al., 1985). Thus, the presynaptic element is morphologically well suited for storage and release of neurotransmitters.

Transmitter Release

Much of what is known today about transmitter release comes from early studies of the neuromuscular junction where acetylcholine (ACh) is the neurotransmitter. The classical electrophysiological studies of Katz and associates (see Katz, 1969) were instrumental in showing that transmitter release from the presynaptic terminal is not an "all or nothing" event but rather occurs in graded fashion. Intracellular recordings from postjunctional membrane revealed small, randomly occurring, spontaneous depolarizations of equal amplitude even in the absence of impulses on the motor nerve (Fatt and Katz, 1952; Del Castillo and Katz, 1954a; Katz, 1969). These small fluctuations in membrane potential are commonly referred to as *minature end-plate potentials* (MEPPs). An important observation was that the frequency, but not the amplitude, of MEPPs increased as the nerve was depolarized and decreased when the nerve was hyperpolarized (Del Castillo and Katz, 1954b; Liley, 1956). This observation led to formulation of the "quantal" theory of transmitter release, which holds that each individual MEPP results from release of an equal amount or "packet" of transmitter (Katz, 1969; Kuno, 1971). During stepwise depolarization of the axon terminal, an increasing number of quanta are released, which results in stepwise depolarization of the postsynaptic membrane.

It is now generally believed that the vesicles that were visualized previously in the presynaptic terminal are the morphological substrate for quantal release of transmitter. In support of this are findings from cell fractionation studies which showed that the presynaptic vesicles actually contain transmitter (Whitaker, 1971). Moreover, procedures that depolarize the nerve ending and cause transmitter release also cause depletion of vesicles in the presynaptic terminal (Jones and Kwanbunbumpen, 1970; Ceccarelli et al., 1972; Heuser and Reese, 1973). Electron micrographs of the terminal revealed that vesicles actually rupture in the vicinity of the membrane (Hubbard and Kwanbunbumpen, 1968).

Early electrophysiological studies showed that the divalent cation calcium (Ca^{2+}) is essential for transmitter release (Jenkinson, 1957; Blackman et al., 1963; Katz and Miledi, 1967a,b; Hubbard, 1970). The work of Katz and Miledi (1967a,b) was crucial for gaining insight concerning the role of Ca^{2+} in transmitter release. These workers discovered that Ca^{2+}, but not other divalent ions such as magnesium (Mg^{2+}), facilitated transmitter release when applied to the synapse just prior to depolarization of the presynaptic terminal. They hypothesized that Ca^{2+} enters the terminal from the extracellular fluid during depolarization and interacts with the vesicles to cause transmitter release. Support for this hypothesis came from the work of Blaustein (1971), who showed that presynaptic terminals actually accumulate radioactive Ca^{2+} during nerve stimulation. It is now known that Ca^{2+} enters the terminal via voltage-gated channels when the membrane is depolarized (see Levitan and Kaczmarek, 1987).

The exact role of Ca^{2+} in transmitter release remains unclear. However, it is generally believed that the increase in intracellular concentration of Ca^{2+} when the presynaptic terminal is depolarized causes the synaptic vesicles to fuse with the membrane and release their contents (i.e., transmitter) into the synaptic cleft. New light on this subject has recently been shed by the findings of Greengard and associates (1984, 1987, 1989). They report that a phosphoprotein called *synapsin I* is closely associated with synaptic vesicles and intricately involved in transmitter release. Synapsin I is a phosphoprotein that becomes phosphorylated in response to a variety of physiological signals, including depolarization of the axon terminal (Hemmings *et al.,* 1989). Phosphorylation of synapsin I requires increased Ca^{2+} concentration (Forn and Greengard, 1978) and activation of Ca^{2+}/calmodulin (CaM)- and cyclic AMP (cAMP)-dependent protein kinases (Greengard, 1987).

The most recent theory of transmitter release holds that the dephosphorylated form of synapsin I binds to synaptic vesicles and prevents them from fusing with the synaptic membrane. Once phosphorylated, synapsin I becomes less tightly bound, which allows the vesicle to fuse with the presynaptic membrane and discharge its contents into the cleft. Experimental support for this theory comes from the work of Llinas and colleagues (1985). They found that injection of dephosphorylated synapsin I directly into the presynaptic terminal of the squid giant axon caused a diminution in both the amplitude and rate of rise of postsynaptic potential. This finding indicates that dephosphorylated synapsin I inhibits transmitter release supposedly by preventing the vesicles from fusing with the membrane. In addition, these workers found that injection of Ca^{2+}/CaM-dependent protein kinase II into the presynaptic terminal led to a marked increase in the amplitude and rate of rise of the postsynaptic potential presumably by enhancing phosphorylation of synapsin I, and thus augmenting transmitter release. It appears that the influx of Ca^{2+} initiates a complex series of biochemical reactions that leads ultimately to transmitter release. It is likely that future research will elucidate additional key elements involved in regulation of transmitter release.

Identification of Chemical Messengers

During the last two decades there has been considerable effort to identify chemical compounds in neurons that might function as neurotransmitters. Here we use the term *chemical messenger* to refer to all compounds that might function as a neurotransmitter. Five minimal requirements have been established for acceptance of a compound as a neurotransmitter: (1) the compound is present in presynaptic nerve terminals, (2) the presumptive transmitter is released during depolarization of the presynaptic terminal, (3) exogenous application of the compound to the postsynaptic membrane mimics the effects caused by stimulation of the presynaptic fibers that contain the compound, (4) a mechanism exists to enzymatically degrade the substance in the synaptic cleft, and (5) antagonists affect the response to exogenously applied and endogenously released transmitter in a similar fashion.

At the start of the present decade a number of chemical compounds that had been identified in nerve cells appeared to meet most of these criteria; these include acetylcholine (ACh), the catecholamines (dopamine, norepinephrine, and epinephrine), 5-hydroxytryptamine (serotonin; 5-HT), histamine, and certain amino acids (glycine, GABA, glutamate, and aspartate). These small-molecular-weight compounds are now generally accepted as neurotransmitters and are referred to collectively as *classical neurotransmitters*. In addition, there is also growing evidence that certain purine compounds such as adenosine and its nucleotides (e.g., adenosine triphosphate, ATP) might also serve as neurotransmitters in the central nervous system (Phillis and Wu, 1981).

The use of sophisticated immunological, biochemical, and molecular techniques during the last decade has led to an unprecedented growth in the number of compounds being considered for status as a neurotransmitter. This phenomenal growth is due largely to inclusion of an entire new class of compounds, the *neuropeptides,* as potential neurotransmitter molecules (Snyder, 1980). To date, more than 30 peptides, ranging in size from a few to 40 amino acids, have been identified in neurons. Do peptides actually function as neurotransmitters? There is now evidence that some of the identified peptides are released from axon terminals, bind to postsynaptic receptors, and cause a change in the activity of the target cell (Otsuka and Takahashi, 1977; Pernow, 1983; Snyder, 1980; Bloom, 1988). However, it is also probable that some neuropeptides are not involved directly in cell-to-cell communication, but instead subserve some other function (e.g., a trophic factor).

It is interesting that many of the peptides that have been identified in the brain were discovered in the gut, the so-called gut-brain peptides. In addition, as we shall discuss later, the levels of some peptides in certain regions of the nervous system change during development. Thus, the role of a particular peptide may change during ontogeny. A list of peptides that have been identified in the mammalian nervous system is provided in Table 1.

Biosynthesis of Neurotransmitters and Peptides

Although the role of peptides in chemical neurotransmission remains unclear, it is entirely possible that some are involved in mediating signals across the synapse, whereas others might be involved in some other aspect of synaptic function (e.g., modulation of transmitter release). Even if a neuropeptide were involved directly in transmission of impulses from one cell to another, its role would most certainly be different from that of a classical neurotransmitter. This conclusion is based strictly on differences in the biosynthesis of classical neurotransmitters and peptides and the rate of replenishment of these compounds once released.

Table 1 Neuroactive Peptides Found in Mammalian Nervous Tissue
(Listed Alphabetically)

Adenocorticotropin	Luteinizing hormone
Angiotensin II	α-Melanocyte-stimulating
Bombesin	hormone
Bradykinin	Motilin
Calcitonin gene-related peptide	Neurokinin A (substance K)
Cholecystokinin (CCK)	Neuropeptide tyrosine (NPY)
Corticotropin-releasing hormone	Neurophysin
β-Endorphin	Neurotensin
(Leucine)-Eukephalin	Oxytocin
(Methionine)-Enkephalin	Prolactin
Galanin	Secretin
Gastrin	Sleep peptide
Glucagon	Somatostatin
Growth hormone	Substance P
Growth hormone-releasing hormone	Thyrotropin-releasing hormone (TRH)
Insulin	Vasoactive intestinal peptide (VIP)
	Vasopressin

Classical neurotransmitters are synthesized at the site of release from readily available substrate by specific enzymes (see Siegel et al., 1981) as depicted in Figure 1. These enzymes are synthesized in the cell body and transported to the terminal by axonal transport systems, where each enzyme repeatedly synthesizes transmitter. In most cases, the substrate that results from enzymatic breakdown of the transmitter following its release is taken up by the presynaptic terminal and used in the biosynthesis of new transmitter. This situation is ideally suited for fast and continuous transmission of impulses.

The story for peptides, on the other hand, is markedly different. Neuropeptides are the direct products of gene expression (see Fig. 1). Thus, neuropeptides are synthesized from specific messenger RNA (mRNA) in the soma and then transported via axonal transport systems to the presynaptic terminal. The distance the newly synthesized peptide must travel before it can be released is often considerable. For instance, peptides in premotor neurons

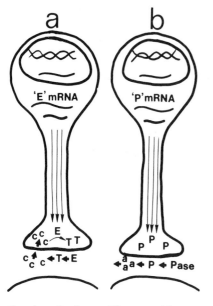

Figure 1 Biosynthesis of classical neurotransmitters (a) and peptides (b) is illustrated. Classical neurotransmitters are synthesized from substrate in the presynaptic terminal. Enzymes (E) that catalyze these reaction are products of gene expression. The transmitter is enzymatically degraded after its release by enzymes in the synaptic cleft. Some or all of the components (c) of this catabolic reaction are taken up by the terminal and used as substrate for production of new transmitter. Thus, transmitters can be rapidly resynthesized at the site of release. Peptides (P), on the other hand, are synthesized in the soma and transported to the terminal. Once released, they are degraded to amino acids (a) in the synaptic cleft by peptidases (Pase). Peptides are replaced by expression of new peptide in the cell body and transport to the site of release. Thus peptides are not ideal candidates for rapid, continuous synaptic transmission.

involved in regulation of respiration (i.e., bulbospinal projections) must be transported from the brainstem to various sites in the spinal cord. Peptides released into the synaptic cleft are degraded by peptidases and can only be replaced by transport of newly synthesized peptides, a process that may require many hours. It seems unlikely, therefore, that peptides mediate rapid and continuous signaling. Instead, peptides located in such pathways may modulate the effect of a classical neurotransmitter.

Coexistence of Multiple Chemical Messengers in Individual Neurons

The distributions of classical neurotransmitters and peptides have been mapped in the peripheral and central nervous systems using both immunohistochemical (Björklund et al., 1984-1985) and biochemical (Krieger et al., 1983; Palkovits and Brownstein, 1985) approaches. An important finding is that transmitters and peptides are not distributed randomly in the nervous system, but instead show precise distribution patterns that are well conserved within and, more often than not, among mammalian species. Moreover, brain areas or nuclei are not homogenous in terms of transmitter and peptide content, but contain a wide variety of these compounds. For example, the nucleus of the solitary tract (NTS), a major autonomic integrative area in the medulla oblongata, contains most of the classical neurotransmitters and most of the identified peptides (see Björkland et al., 1984-1985; Kalia et al., 1984; Leslie, 1985; Palkovits, 1985).

Findings from immunohistochemical mapping studies revealed that the distributions of many classical neurotransmitter and peptides often overlap in certain regions of the nervous system. This observation led to speculation that individual neurons might contain more than one transmitter and that Dale's law, i.e., that single neurons utilize only one neurotransmitter species, might be incorrect. First direct evidence for *coexistence* of multiple messengers in individual neurons in the central nervous system was reported by Hökfelt and colleagues (1978) and by Chan-Palay and associates (1978). Both groups presented evidence that individual neurons in the obscurus and pallidus regions of the medullary raphe nuclei contained the classical neurotransmitter serotonin and the peptide substance P.

During the decade since this important discovery, many different coexistence situations have been identified in the peripheral and central nervous systems (see Hökfelt et al., 1986, 1987, 1988; Millhorn and Hökfelt, 1988). The most frequently encountered coexistence situation involves a single neuron that contains a classical neurotransmitter and one or more peptides. A partial list of coexistence situations involving transmitters and peptides is given in Table 2. In addition, an increasing number of coexistence situations involving either two classical transmitters or two peptides (without a classical transmitter) have been reported (Hökfelt et al., 1987, 1988; Millhorn and Hökfelt, 1988; Millhorn et al., 1987a, b, 1988; Ceccatelli et al., 1989). Examples of transmitter-peptide, peptide-peptide, and classical transmitter-classical transmitter coexistence situations in the ventrolateral aspect of the medulla oblongata, an area involved in regulation of the respiratory and cardiovascular systems (Millhorn and Eldridge, 1986), are shown in Figures 2, 3, and 4, respectively.

In recent years, there has been much speculation concerning the physiological significance of coexistence of multiple messengers. A number of possibilities have been proposed (Hökfelt et al., 1981, 1987; Lundberg and Hökfelt, 1983; Millhorn and Hökfelt, 1988). Perhaps the most interesting are those in which a coexisting neuropeptide modulates the efficacy of the primary transmitter. The serotonin (5-HT)-containing neurons in the caudal raphe nuclei that also contain substance P (SP) and thyrotropin-releasing hormone (TRH) (Johansson et al., 1981) provide a model for understanding how coexisting peptides might modulate the postsynaptic effect of a classical neurotransmitter (Fig. 5) (Hökfelt et al., 1981; Millhorn et al., 1988, 1989). In this model, one of the coexisting peptides, SP, modulates the release of 5-HT, whereas the other peptide, TRH, potentiates the postsynaptic effect of 5-HT. Accordingly, 5-HT released into the synaptic cleft acts

Table 2 Selected Cases of Coexistence of Classical Transmitters and Peptides Mainly in the Mammalian CNS[a]

Classical transmitter	Peptide	Brain region (special)
DA	CCK	Ventral mesencephalon (rat, cat, mouse, monkey, human?)
	Neurotensin	Ventral mesencephalon (rat)
		Hypothalamic arcuate nucleus (rat)
NA	Enkephalin	Locus coeruleus (cat)
	NPY	Medulla oblongata (human, rat)
		Locus coeruleus (rat)
	Vasopressin	Locus coeruleus (rat)
Adrenaline	Neurotensin	Medulla oblongata (rat)
	NPY	Medulla oblongata (rat)
	Substance P	Medulla oblongata (rat)
	Neurotensin	Solitary tract nucleus (rat)
5-HT	Substance P	Medulla oblongata (rat, cat)
	TRH	Medulla oblongata (rat)
	Substance P + TRH	Medulla oblongata (rat)
	CCK	Medulla oblongata (rat)
	Enkephalin	Medulla oblongata, pons (cat)
		Area postrema (rat)
ACh	Enkephalin	Superior olive (guinea pig)
		Spinal cord (rat)
	Substance P	Pons (rat)
	VIP	Cortex (rat)
	Galanin	Basal forebrain (rat, monkey)
	CGRP	Medullary motor nuclei (rat)
GABA	Motilin (?)	Cerebellum (rat)
	Somatostatin	Thalamus (cat)
		Cortex, hippocampus (rat, cat, monkey)
	CCK	Cortex, hippocampal formation (cat, monkey, rat)
	NPY	Cortex (cat, monkey)
	Enkephalin	Retina (chicken)
		Ventral pallidum, hypothalamus (rat)
	Opioid peptide	Basal ganglia (rat)
	Substance P	Hypothalamus (rat)
	VIP	Hippocampal formation (rat)
Glycine	Neurotensin	Retina (turtle)

[a]For references for these coexistence situations see Hökfelt et al. (1987).
Source: Adapted from Hökfelt et al. (1987).

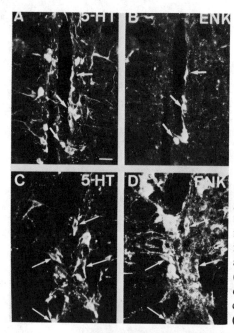

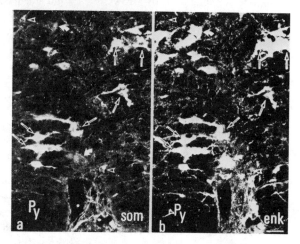

Figure 2 Immunofluorescence micrographs of the ventral medulla showing nucleus raphe obscurus (A, B) and nucleus raphe pallidus (C, D) after incubation with antibodies directed against 5-HT (A, C) and enkephalin (ENK; B, D). Individual cells that show coexistence of both compounds are designated by arrows. Bar = 50 μm. (From Millhorn et al., 1989a.)

Figure 3 Immunofluorescence micrographs of section of nucleus raphe magnus processed for immunohistochemistry for somatostatin (som) (a) and enkephalin (enk) (b). Numerous cells (arrows) contained both som- and enk-like immunoreactivity. Arrowheads point to som-positive cells that do not contain enk and curved arrows to enk-positive cells that lack som-like immunoreactivity. Bar = 50 μm. Py = pyramidal tract. (From Millhorn et al., 1987b.)

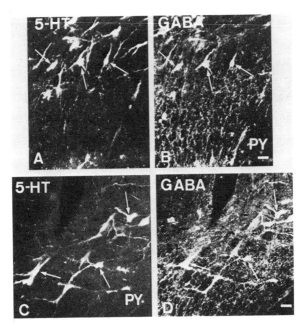

Figure 4 Immunofluorescence micrographs of the ventral medulla showing the lateral aspect of nucleus raphe magnus after incubation with antisera directed against 5-HT (A, C) and GABA (A, D). Individual cells that contained immunoreactivity for both neurotransmitters are designated by arrows. Bars = 50 μm. PY = pyramidal tract. (From Millhorn et al., 1988.)

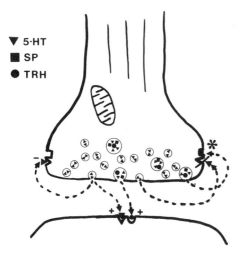

▼ 5-HT
■ SP
● TRH

Figure 5 Model of a synapse in which the presynaptic terminal contains serotonin (5-HT), substance P (SP), and thyrotropin-releasing hormone (TRH). Small vesicles contain only 5-HT, whereas large vesicles contain all three messengers. When 5-HT is released it diffuses to the postsynaptic terminal, where it depolarizes the membrane. 5-HT also diffuses to a site on the presynaptic terminal (autoreceptor) to inhibit further release of 5-HT (designated by the minus sign). When the presynaptic terminal is excited sufficiently, both small and large vesicles release their contents into the synaptic cleft. When this occurs the effects of 5-HT are potentiated by the excitatory action of TRH postsynaptically and by the action of SP presynaptically. It is hypothesized that SP interacts (*) with the 5-HT autoreceptor to enhance release of 5-HT. (Adapted from Hökfelt, et al., 1981.)

postsynaptically to alter the behavior of the target cell and presynaptically on 5-HT autoreceptors to inhibit its own release. When SP and TRH are co-released, SP inhibits the 5-HT autoreceptor, which permits sustained release of 5-HT. TRH, on the other hand, interacts with a postsynaptic receptor to enhance the behavioral effect of 5-HT on the target cell. There is experimental support for this model. For instance, it has been shown that addition of SP, but not TRH, to the synaptic cleft enhances release of 5-HT (Mitchell and Fleetwood-Walker, 1981; Iverfeldt et al., 1989). There is also evidence that TRH potentiates the depolarizing effect of 5-HT on motoneurons (Barbeau and Bedard, 1981).

This particular model might have functional relevance with respect to regulation of respiration. Recently, our laboratory reported that brief (10 min) electrical stimulation of nucleus raphe obscurus, a 5-HT-rich area in the medulla, causes long-lasting (hours) facilitation of phrenic nerve activity in cat (Millhorn, 1986) and rat (Erickson and Millhorn, 1990). This long-lasting response is prevented by pretreatment with a serotonin receptor antagonist. We proposed that the long-lasting nature of this response might be due to persistent release of 5-HT onto respiratory-related neurons. Results from immunohistochemical and axon tracing studies revealed a 5-HT, SP, and TRH pathway from nucleus raphe obscurus to nucleus tractus solitarii (NTS) (Thor and Helke, 1987; Erickson et al., 1989), a major integrative area for respiration, and to the phrenic motor nucleus in the spinal cord (Holtman et al., 1984). In addition, we now have evidence from microdialysis studies that stimulation (15-45 Hz) of raphe nucleus causes long-term release (>1 hr) of 5-HT in NTS and in the ventral horn of the cervical spinal cord (Brodin et al., in press). We are currently testing the hypothesis that co-released substance P mediates this enduring release of 5-HT.

How is differential release of coexisting messengers achieved? It is surprising that relatively few studies have been performed to attempt to answer this question. However, findings from several studies indicate that frequency of stimulation of the presynaptic terminal may be the crucial signal for differential release of a classical neurotransmitter and coexisting peptides (see Lundberg and Hökfelt, 1983). This is based on observations that small-molecular-weight classical transmitters are released during low-frequency stimulation and that both classical transmitter and coexisting peptides are released at higher stimulation frequencies. For example, Lundberg and colleagues (1982) found that low-frequency stimulation of the parasympathetic neurons that innervate the salivary gland led to release of acetylcholine (ACh), whereas high-frequency stimulation evoked release of both ACh and vasoactive intestinal peptide (VIP), a coexisting peptide.

The subcellular substrate for differential release of coexisting messengers might be differential storage of transmitter and peptides in presynaptic vesicles. There is biochemical and morphological evidence that small, clear vesicles contain exclusively the small-molecular-weight transmitter, whereas large, dense core vesicles contain both classical transmitter and peptide(s) (Lundberg et al., 1981; Pelletier et al., 1981; Freid et al., 1985). Thus, in this scenario the small, clear vesicles containing the classical neurotransmitter are released exclusively during low-frequency stimulation, whereas the large, dense core vesicles containing both the transmitter and peptide(s) are also released during high-frequency stimulation.

One of the most important challenges for the future is to elucidate the functional significance of coexistence. In our opinion, the most likely function for a coexisting peptide is to "modulate" the effects mediated by primary transmitters.

B. The Postsynaptic Neuron

Once released from the presynaptic terminal, the transmitter diffuses across the synaptic cleft and interacts with receptor molecules embedded in the membrane of the postsynaptic cell. The binding of a transmitter to its receptor initiates a cascade of intracellular events that leads ultimately to a change in the behavior and, in some cases, a change in the phenotype of the target cell. It is important to realize that the cellular effects evoked by the released transmitter (or peptide) are determined by the activated receptor, *not* the transmitter. If there were only one receptor species for each transmitter, this would not be such an important concept. However, this is not the case. Multiple receptor "subtypes" have been identified for most of the classical neurotransmitters and also for some neuropeptides (see Bloom, 1988). For example, seven different 5-HT receptor subtypes ($5\text{-HT}_{1a\text{-}e}$, 5-HT_2, and 5-HT_3) have been identified in the nervous system (see Conn and Sanders-Busch, 1987; Peroutka, 1988; Leonhardt et al., 1989). Activation of some of these subtypes cause depolarization of the postsynaptic neuron, whereas activation of others cause hyperpolarization. Thus, the same transmitter molecule can cause either depolarization or hyperpolarization depending on the receptor subtype activated.

Membrane receptors are separated into two broad classes: (1) receptors that mediate their effects *directly* by serving as ligand-gated ion channel (i.e., ionophore), and (2) receptors that mediate their effects *indirectly* via a complex molecular linkage with special signaling molecules located in the membrane and cytoplasm of the postsynaptic cell.

Ligand- or Transmitter-Gated Channels (Ionophores)

Receptors of this type have a transmitter-binding domain located on the extracellular surface of the membrane and hydrophobic membrane-spanning region that functions as an ion channel. At rest (inactivated), the receptor-channel complex does not permit passage of ions. Activation of the complex by transmitter leads to a conformational change in the membrane-spanning channel, which then allows passage of specific ions. Receptor-channel complexes are ideally suited for fast synaptic signaling. In recent years, several of these complexes have been cloned and their molecular structure and biophysical characteristics determined. Here we shall discuss briefly the molecular configuration and proposed function of two such receptor-channel complexes, the nicotinic ACh receptor and the family of glutamate receptor(s), both of which are involved in excitatory synaptic transmission in the nervous system.

Nicotinic Acetylcholine Receptor

Perhaps the most thoroughly characterized ligand-gated receptor-ionophore is the nicotinic ACh receptor (see Raftery et al., 1980; Changeux, 1981; Noda et al., 1983; Finer-Moore and Stroud, 1984; Karlin et al., 1984). Nicotinic ACh receptors are found at the neuromuscular junction, in sympathetic and parasympathetic ganglia, and at certain sites in the central nervous system. Findings from molecular cloning studies (Noda et al., 1983) revealed that the nicotinic receptor is a pentamer made from four subunits (α_2, β, γ, δ), each of which transverses the lipid bilayer membrane five times. These subunits are arranged in a rosette which defines a central channel that allows passage of certain positively charged ions when activated by ACh (Numa et al., 1983). The receptor is activated and the channel opened by the binding of two ACh molecules, one molecule to the extracellular domain of each of the α subunits (Numa et al., 1983). Because of a net negative charge on the interior surface of the channel, passage is restricted to positively charged ions, primarily Na^+ and K^+ (Lester, 1977). The rate of movement of ions through the ACh channel depends on their transmembrane concentration and electrochemical driving forces. At rest the driving force for movement of K^+ out of the cell is near zero, whereas there is a substantial electrochemical force for Na^+ to enter the cell. Thus, the result of activation of the ACh nicotinic receptor is a net influx of Na^+ into the cell, which results in rapid depolarization of the target cell.

L-Glutamate Receptors

In recent years there has been a virtual explosion of interest and information concerning excitatory amino acid transmission in the central nervous system. Much of this interest has focused on L-glutamate, which is generally considered to be a leading candidate as the transmitter at many excitatory synapses in the central nervous system. Findings from radioligand (Garthwaite, 1982; Sladeczek et al., 1985) and pharmacological (Fagg and Matus, 1984; Cotman and Iversen, 1987) studies revealed four different glutamate receptors. Three glutamate receptor subtypes are classified (and named) on the basis of the selective agonist action of N-methyl-D-aspartate (the NMDA receptor), quisqualate (the Q receptor), and kainate (the K receptor). The fourth glutamate receptor is classified by the antagonist action of L-2-amino-phosphonobutyric acid. The NMDA, Q, and K subtypes have been studied most extensively and are therefore the subject of the ensuing discussion.

Fast monosynaptic effects of L-glutamate are, for the most part, mediated by the Q and K subtypes. This conclusion is based on the findings that amino-5-phosphonovaleric acid (APV), a specific antagonist of NMDA receptor, administered to the spinal cord blocked polysynaptic, but not monosynaptic, responses to dorsal root stimulation (Davies and Watson, 1983). It is now generally believed that the Q and K glutamate receptors mediate rapid depolarizations by increasing permeability to certain cations, whereas the NMDA receptor mediates a much slower depolarization (McDermott and Dale, 1987). The diverse temporal properties of the different glutamate receptors indicate that they may be involved in a variety of neuronal functions.

One of the most exciting topics in neurobiology today is the NMDA receptor. The NMDA receptor has been implicated in a wide range of functions, including neuronal plasticity, learning and memory, burst firing, and generation of rhythmic activity (see Fagg et al., 1986). The NMDA receptor appears to be much more complex than the other glutamate receptor subtypes. For instance, binding sites for both L-glutamate (or NMDA) and glycine are found on the extracellular domain of the receptor (Bowery, 1987). It appears that the binding of glycine to its site potentiates the effects of glutamate (or NMDA) (Johnson and Ascher, 1987). In addition, conductance through the NMDA channel is inhibited in a voltage-dependent fashion by the binding of magnesium (Mg^{2+}) to a site on the inside of the channel (Nowak et al., 1984). At resting membrane potential, Mg^{2+} blocks the open channel and prevents passage of ions. As the membrane depolarizes, Mg^{2+} is excluded from the channel and positively charged ions are allowed to pass through the channel. The voltage-dependent regulation of ion flow through the channel by Mg^{2+} is responsible for the slow depolarization kinetics of the NMDA channel. A model of the NMDA receptor is shown in Figure 6.

Once glutamate is released from presynaptic terminals, it can activate any one or all of the receptor subtypes. Activation of the Q and/or K subtype results in rapid depolarization of the postsynaptic membrane. Thus, the Q and K receptors are excellent candidates for mediating fast excitatory monosynaptic transmission (e.g., primary sensory signaling from arterial chemoreceptors and baroreceptors). Activation of the NMDA receptor, on the other hand, is not accompanied by an immediate conductance of cations into the postsynaptic neuron. This does not occur until the membrane has

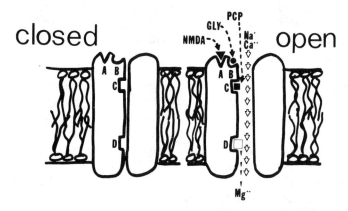

Figure 6 Model of NMDA receptor for L-glutamate showing the closed and activated (open) states. See text for description and function of the ligand-gated ionophore.

been depolarized so as to remove the Mg^{2+} block of the channel. The mechanism that mediates the initial depolarization needed to activate the NMDA receptor may vary from cell to cell, but could in some cases be mediated by either the Q or K receptors.

There are other receptor-channel complexes in the nervous system that may mediate either excitatory or inhibitory responses. For instance, the $GABA_A$ receptor is an ionophore that mediates inhibition (Mamalaki et al., 1987). Again, these receptors differ from other types of receptors in that they function as an ion channel once activated.

G Protein-Linked Receptors and Signal Transduction Molecules

The other major class of receptors found in the nervous system are those that mediate their action via a rather complex sequence of molecular events involving several intramembranous and cytoplasmic biochemical intermediaries. Because these receptors initiate their effect via a membrane-bound guanine nucleotide-binding protein, they are commonly referred to as "G protein-linked receptors." A number of transmitters, including the biogenic amines (dopamine, norepinephrine, epinephrine, serotonin, and histamine), ACh (via the muscarinic receptor), GABA (via the $GABA_B$ receptor), and certain peptides (e.g., substance P, neurokinin A), activate G protein-linked receptors. The sequence of events that begins with transmitter-receptor binding and terminates with protein phosphorylation is known as *signal transduction*. Here we shall give a brief account of the molecular mechanisms that mediate the cellular response to activation of G protein-linked receptors.

During the past several years many of the G protein-linked receptors have been cloned and their nucleotide and amino acid sequence determined (Kubo et al., 1986; Bonner et al., 1987; Dohlman et al., 1987; Kobilka et al., 1987; Julius et al., 1988). A prototypic G protein-linked receptor is presented in Figure 7. Thus far, all of the characterized G protein-linked receptors have been found to share a similar configuration; they are single molecules that consist of an extracellular domain (amino terminus), seven membrane-spanning regions, and a cytoplasmic domain (carboxy terminus). Although there is considerable difference in the nucleotide and amino acid sequences of the extracellular and cytoplasmic domains, the membrane-spanning regions of the different G protein-linked receptors are highly conserved. It is generally believed that the second membrane-spanning region is involved in ligand (transmitter) recognition and binding (Strader et al., 1987; Chung et al., 1988). The third cytoplasmic loop (amino acid sequence between the fifth and sixth membrane-spanning regions) interacts with G protein (Strader et al., 1987). In addition, the carboxy terminus contains potential sites (ser-

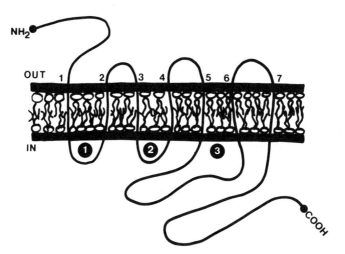

Figure 7 Prototypic model of a G protein-linked receptor. Distinguishing features are an extracellular amino (NH_2) terminus, seven transmembrane regions, and a cytoplasmic carboxy (COOH) terminus. The third cytoplasmic domain is believed to be involved in the molecular linkage with G protein. Transmitter binding is thought to occur on a site on the second membrane-spanning region. (Adapted from Julius et al., 1988.)

ine residues) that may serve as substrates for phosphorylation by second-messenger systems and thus provide a mechanism for receptor sensitization and desensitization (Sibley et al., 1987; Julius et al., 1988).

Transmitter binding to its receptor initiates a cascade of events that eventually leads to an alteration in the behavior of the target (postsynaptic) cell. As mentioned above, the "activated" receptor transmits a synaptic signal into the cell by means of a membrane-bound heterotrimeric *G protein* complex (see Gilman, 1987). Although a number of G proteins have been identified, they all appear to consist of a complex of three subunits (α, β, γ). The G protein is situated so that it can interact with receptors as well as molecules on the cytoplasmic side of the membrane.

The sequence of events involved in transmembrane signaling has been elucidated in recent years by Gilman and colleagues (see Gilman, 1987). Briefly, transmitter-receptor binding leads to a conformational change in the receptor molecule, which, in turn, causes activation of a G protein. The "activated" G protein binds to guanosine triphosphate (GTP) on the cytoplasmic side of the membrane. The binding of the G protein to GTP alters the configuration of the G protein, which allows it to "activate" adenylate cyclase, a membrane-bound amplifying enzyme that repeatedly catalyzes the formation of adenosine 3', 5'-monophosphate (cyclic AMP, cAMP) from adenosine triphosphate (ATP) (Sevilla et al., 1976; Birnbaumer and Iyengar, 1981). Activation of adenylate cyclase leads to accumulation of cAMP, an important *second messenger*. Once activated, signal transduction via this pathway does not terminate automatically, but requires disruption of the GTP-G protein complex by hydrolysis of GTP to GDP (Gilman, 1987). The signal transduction pathway is reactivated only after the next ligand-receptor binding event which causes dissocation of GDP from G protein. A diagram of a receptor/ G protein-linked signal transduction system is shown in Figure 8.

G proteins can either stimulate or inhibit the enzymatic activity of adenylate cyclase; G proteins that stimulate adenylate cyclase and cause accumulation of cAMP are termed G_s, whereas those that inhibit adenylate cyclase are referred to as G_i (Gilman, 1987). It is important to recognize that interaction of a single transmitter molecule with its receptor is enhanced many times over by the "amplifying" action of adenylate cyclase. For example, the binding of a single molecule of epinephrine to the β-adrenergic receptor leads to accumulation of many molecules of cAMP inside the target cell (Levitzki, 1988). This occurs because the G_s-adenylate cyclase complex remains intact and causes accumulation of cAMP until the complex is inactivated by hydrolysis of GTP to GDP (Gilman, 1987). The amplifying action

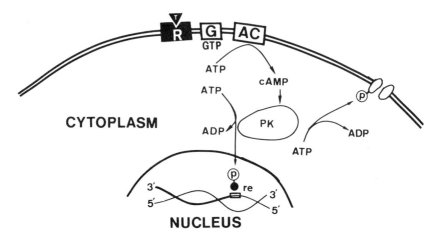

Figure 8 Model of signal transduction pathway that utilizes the cAMP second-messenger system. Signal transduction events that occur following binding of transmitter to its receptor (R) are described in the text. (From Millhorn et al., 1989a.)

of adenylate cyclase therefore enhances the cellular response to transmitter-receptor binding by causing accumulation of second-messenger molecules.

Accumulation of second messenger in the cell following transmitter-receptor binding activates a specific protein kinase that transfers a phosphate (e.g., from ATP or GTP) to a serine or threonine residue on an existing intracellular protein (see Greengard, 1976, 1978a,b, 1987). Phosphoproteins that result from this mechanism represent a final common pathway in signal transduction. As depicted in Figure 8, a transmitter may cause depolarization of the target cell by causing phosphorylation (via the above-described signal transduction pathway) of an ion channel protein which results in altered ion conductance through the channel (e.g., decreased conductance of K^+). In addition, phosphoproteins resulting from activation of a signal transduction pathway may serve as transcription factors that interact with chromatin to alter gene expression and, therefore, the phenotype of the target cell (Laufer and Changeux, 1989). Activity-dependent regulation of gene expression is discussed in the next section.

It is important for the reader to realize that some transmitters activate receptors and G protein signal transduction mechanisms that lead to an increase in second messengers other than cAMP. There are three other major second-messenger systems: (1) cGMP-dependent protein kinase, (2) inositol trisphosphate/diacylglycerol-protein kinase C, and (3) Ca^{2+}/calmodulin-dependent protein kinase. Several recent reviews on second-messenger-protein kinase systems are available (Berridge and Irvine, 1984; Nishizuka, 1986; Berridge, 1987; Blackshear et al., 1988; Greengard, 1987; Shulman, 1988; Hemmings et al., 1989).

III. Regulation of Gene Expression by Synaptic and Hormonal Signals

Modification of cell behavior subsequent to activation of a receptor and production of biochemical intermediaries involved in signal transduction are usually considered to last from seconds to minutes. However, there is now growing evidence that proteins phosphorylated via signal transduction pathways interact with the promoter region of certain genes and serve as regulators of transcription, i.e., transcription factors (see Nagamine and Reich, 1985; Yamamoto et al., 1988). Activity-dependent modification of gene expression might be the underlying mechanism for a number of neural phenomena, including learning, and memory, plasticity, and physiological adaptation might be explained by such a mechanism. Moreover, gene expression for neurotransmitters, peptides, and receptors might change during ontogeny and therefore mediate certain events during fetal and neonatal development (e.g., trophic events).

A basic scheme by which activation of signal transduction pathways alter gene expression is shown in Figure 8. In this particular example, activation of the cAMP-dependent protein kinase leads to phosphorylation of an existing nuclear protein, which then binds to a specific *cis*-acting sequence on the promoter region of the gene (i.e., a response element) to either enhance or repress its expression. This alteration in gene expression changes the phenotype of the target cell, which, in turn, may lead to long-term alterations in its behavior. Such a mechanism might be involved in mediating long-term changes in respiratory drive, i.e., the adaptive response (acclimatization) to hypoxia.

Our understanding of the mechanisms that regulate gene expression in eukaryotic cells has increased enormously during the last decade. As mentioned above, there is growing evidence that a number of genes, including some that are responsible for biosynthesis of molecules involved in synaptic transmission, are regulated to some extent by intermediaries associated with signal transduction (Nagamine and Reich, 1985). Perhaps the best example of this involves the gene that encodes for the neuropeptide somatostatin. The somatostatin gene has a specific sequence located on the promotor or regulatory region of the gene (i.e., 5 ' to the coding region) that binds a regulatory protein factor (the CREB protein) which is phosphorylated by cAMP-dependent protein kinase (Montminy et al., 1986; Montminy and Bilezikjian,

1987). The binding of the CREB protein to the regulatory *cis*-acting sequence on the promotor region, i.e., the cAMP response element (CRE), enhances expression of somatostatin. Thus, the somatostatin gene can be activated by an intermediary of signal transduction. There is also evidence that gene expression of vasoactive intestinal polypeptide (VIP), which is found in various regions of central and peripheral nervous systems, is regulated by proteins phosphorylated by cAMP-dependent kinase (Hayakawa et al., 1984) or the catalytic unit of cAMP (Riabowol et al., 1988).

A. Activity-Dependent Regulation of Gene Expression

Research on the neuromuscular junction provides evidence that synaptic activity might alter gene expression (Fontaine et al., 1987a; Laufer and Changeux, 1989). It has recently been shown that motor neurons contain calcitonin gene-related peptide (CGRP) as well as ACh (Fontaine et al., 1987b). It is believed that corelease of CGRP onto the motor end-plate causes increased biosynthesis of subunits for the ACh nicotinic receptor (Fontaine et al., 1987a). The action of CGRP on gene expression for the subunits of the nicotinic receptor appears to be mediated by a cAMP-dependent protein kinase (Fontaine et al., 1987a). Thus, release of CGRP onto the end-plate leads to increased biosynthesis of subunits of the nicotinic receptor. As yet, there is no information concerning the conditions (e.g., increased impulse frequency) during which CGRP is released and it is not known if there is a specific receptor mechanism on the target cell that mediates the action of CGRP. However, accumulation of cAMP in the muscle implies that CGRP acts via a G protein-linked receptor (Fontaine et al., 1987a).

There is also evidence that activity-dependent regulation of gene expression is involved in mediating certain long-lasting alterations in neuronal function. Perhaps the best illustration of this comes from studies of long-term sensitization of the gill- and siphon-withdrawal reflex in the marine mollusk *Aplysia* (see Goelet et al., 1986). Tactile stimulation of the siphon in *Aplysia* causes reflex withdrawal of the gill and subsequent enhancement of synaptic transmission between the sensory and motor components of this reflex. It also causes long-term facilitation of this reflex, which may last from hours to days. The short-term facilitation is mediated by release of transmitter (5-HT) and activation of a second-messenger system involving cAMP in the presynaptic terminal (Dale et al., 1988). Long-term facilitation (i.e., sensitization) is activated by the same transmitter/second-messenger system (Greenberg et al., 1987; Schacher et al., 1988; Scholz and Byrne, 1988; Castellucci et al., 1988), but also involves both transcriptional and translational mechanisms that lead to biosynthesis of new protein (Goelet et al., 1986). Thus, depending on the nature of the stimulus, synaptic activation of this signal transduction pathway leads to either short- or long-term enhancement of the gill-withdrawal reflex. Short-term facilitation may result from phorphorylation of an existing protein (e.g., a K^+ channel protein) that alters ion conductance and the state of depolarization of the cell. Long-term facilitation of the same neurons, on the other hand, involves activation of factors that alter gene expression.

Evidence for altered gene expression in mediating this enduring behavior comes from studies where pharmacological block of protein synthesis prevents long-term, but not short-term, facilitation (Goelet et al., 1986). In addition, several proteins have been identified on 2-D gels whose biosynthesis appeared to be enhanced substantially following activation of long-term facilitation of the gill-withdrawal reflex (Castellucci et al., 1988). Thus, findings from studies concerning a rather simple reflex in an invertebrate nervous system indicate that synaptic activation of target cells can lead to altered gene expression and persistent alterations in the behavior of the target cell.

What about the mammalian nervous system? Does activity-dependent alteration of gene expression mediate long-lasting behavioral changes in the mammalian nervous system? There is evidence that this is indeed the case. A good example is the widely studied phenomenon of *long-term potentiation* (LTP) in hippocampal neurons, which is thought to be the cellular substrate for memory and learning (see Brown et al., 1988). It is well documented that brief (seconds) repetitive stimulation of either the perforant pathway or Schaffer collaterals causes long-term (hours to days) enhancement of synaptic evoked potentials in neurons of the hippocampal CA1 region and dentate gyrus, respectively. LTP consists of two phases, or components: (1) the *induction* phase, and (2) the *maintenance* phase (Collingridge, 1987). There is now evidence that L-glutamate acting via the NMDA receptor (see discussion of this receptor above) is involved in the induction of LTP (Collingridge, 1987). It appears that activation of the NMDA receptor mediates an increase in intracellular Ca^{2+} which serves as a second messenger to activate specific protein kinases (e.g., Ca^{2+}/calmodulin protein kinase, protein kinase C) (see Akers et al., 1986; Brown et al., 1988; Lynch et al., 1988). In fact, protein kinase C activity was found to be increased by more than twofold in hippocampal cells at 1 h following induction of LTP (Akers et al., 1986).

Based on electrophysiological measurements, it is now known that both short- and long-term potentiation occur in hippocampal cells (see Brown et al., 1988). Short-term potentiation (minutes to hours) is likely to be mediated by covalent modification (e.g., phosphorylation) of ion channels which alters conductance of ions and renders the cell more excitable. LTP (hours to days), in contrast, may require a more permanent mechanism, such as altered gene expression. This is based on findings from experiments in which long-term potentiation in the hippocampus is prevented by pretreating the tissue with chemicals that block de novo protein synthesis (Duffy et al., 1981; Stanton and Sarvey, 1984).

In the above discussion, we have described mechanisms by which synaptic input might alter gene expression and thereby mediate enduring alterations in cell behavior. Might such mechanisms play a role in regulation of respiration and other homeostatic systems? We believe so. Our laboratory is investigating the possibility that activity-dependent regulation of gene expression is involved in mediating long-lasting changes in respiratory neural activity in response to certain stimuli. We reported recently that brief (10 min) activation of carotid body sensory afferents with either physiological (hypoxia) or electrical stimuli causes long-lasting facilitation (hours) of phrenic nerve activity in cat (Millhorn et al., 1980a,b) and rat (Erickson and Millhorn, 1990). Pretreatment of animals with antiserotonin drugs either prevents or significantly attenuates this persistent enhancement of phrenic activity (Millhorn et al., 1980b; Millhorn, 1986), which, of course, implicates the involvement of this classical neurotransmitter. In addition, we found that brief electrical stimulation of neurons in the caudal raphe complex, the primary source of 5-HT in the medulla, also leads to long-lasting facilitation of phrenic nerve activity of similar duration (Millhorn, 1986; Erickson and Millhorn, 1990).

There is evidence that long-lasting facilitation of respiration induced by these inputs involves both transmitter- and gene expression-mediated mechanisms. For instance, a short-term (up to 1 hr) enhancement of respiration might be mediated by a persistent release of 5-HT that involves the coexisting peptides, substance P and thyrotyropin-releasing hormone (see Fig. 5), whereas long-term (many hours to days) facilitation may involve altered gene expression. As described above, axon tracing studies show that neurons in the caudal raphe complex that contain 5-HT and coexisting peptides project to NTS (Thor and Helke, 1987; Erickson et al., 1989), a major integrative area for control of respiration, and the phrenic motonucleus in the spinal cord (Holtman et al., 1984). We found that electrical stimulation of neurons in raphe obscurus led to release of 5-HT both in NTS and in the ventral horn of the spinal cord that persisted for an hour or more following cessation of the stimulus (Brodin et al., in press). This persistent release of 5-HT is thought to be mediated by the action of a coexisting peptide (see above discussion). Thus, we believe that transmitter release is involved in facilitation of respiration that can last from minutes to hours. We believe, however, that longer-lasting facilitation of respiratory drive (hours to days) requires altered gene expression and de novo biosynthesis of protein. For example, synaptic input from carotid body primary afferents might activate signal transduction pathways that lead to altered gene expression in respiratory-related cells in the central nervous system.

What is the physiological significance of activity-dependent regulation of gene expression in control of breathing? At the present time, the answer to this question is left to speculation. One intriguing possibility is that genomic mechanisms are involved in mediating certain previously unexplained long-term adjustments in respiratory drive. Perhaps the best example of this is respiratory adaptation or acclimatization to hypoxia. In this example, the initial increase in ventilation during exposure to hypoxia is due to synaptic activation of neurons in the central nervous system by input from the arterial O_2 chemoreceptors (i.e., carotid bodies). During the next few hours to days, there is secondary "adaptive" increase in ventilation, referred to as *acclimatization*. The actual mechanism(s) that mediate ventilatory acclimatization during chronic hypoxia is not explained by traditional ventilatory stimulants (e.g., CO_2, hydrogen ion) and thus remains unknown. We propose that acclimatization might be the result of activity-dependent alteration in gene expression which leads to long-term enhancement in the excitability of the neurons that mediate this phenomenon.

In this scenario, synaptic input from the carotid body chemoreceptors activates neurons (perhaps second- or third-order neurons) in the brainstem via a specific receptor-signal transduction mechanism that leads to modification of some factor that alters gene expression. A protein (e.g., channel protein) that might result from altered gene expression might change the state of polarization of a population of respiratory-related cells, which, in turn, leads to a long-lasting enhancement of respiration without additional synaptic input. If altered gene expression were the mechanism responsible for ventilatory acclimatization to hypoxia, one would predict that ventilation would continue to be enhanced for some time after removal of the stimulus, i.e., return to normoxic conditions. Indeed, this is exactly what happens; subjects continue to hyperventilate for many days after return to normoxic conditions (see Dempsey and Forster, 1982).

We believe that activity-dependent regulation of gene expression offers a viable explanation to a much investigated, yet unexplained phenomenon. We believe that the long-lasting facilitation of respiration measured in laboratory animals following either carotid body afferent (Millhorn et al., 1980a,b; Erickson and Millhorn, 1990) or raphe simulation (Millhorn, 1986) might involve a neural substrate that is similar to (if not the same as) that responsible for mediating the process of acclimatization and thus might serve as a useful model to study the cellular and molecular mechanisms of such long-lasting phenomena.

In recent years, molecular techniques such as in situ hybridization have been developed which permit examination of gene expression (i.e., mRNA concentration) in individual neurons. Figure 9 shows cells in the nodose (Fig. 9A,B) and petrosal ganglia (Fig. 9C) that express mRNA (denoted by silver grains) for the peptides CGRP (Fig. 9A,B) and preprotackykinin A (Fig. 9C), the precursor for substance P and neurokinin A (Czyzyk-Krzeska et

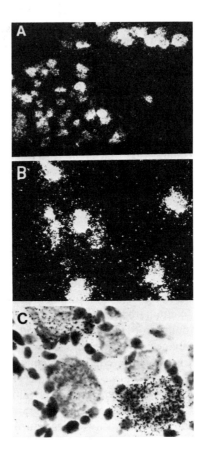

Figure 9 Dark (A, B) and light (C) field micrographs from experiments in which cells in the nodose (A, B) and petrosal (C) ganglia that express mRNA for CGRP (A, B) and preprotachykinin A (substance P/neurokinin A) (C) were identified with in situ hybridization.

al., 1990; Czyzyk-Krzeska et al., in press; Millhorn et al., in press). Studies using this and similar molecular approaches might allow identification and quantification of alterations in gene expression in respiratory-related neurons during different environmental pertubations. For example, does the mRNA concentration for these peptides in this chemoreceptor pathway increase during chronic hypoxia?

B. Ligand-Dependent Regulation of Gene Expression

Thus far, we have considered only synaptic activity in the process of cell-to-cell communication in the nervous system. In this section, we shall discuss briefly how the endocrine system influences neural function. Unlike transmitters that are produced and released from cells located near their targets, hormones are usually produced at some remote site and reach their targets via the circulation. We shall limit our discussion to a specific class of hormones, the gonadal steroids (i.e., progesterone, estrogen, and testosterone), some of which have been shown to have profound effects on respiration (see Bayliss et al., 1987, 1990).

As we shall see in the ensuing discussion, receptors for steroid hormones are markedly different from receptors that mediate responses to neurotransmitters. One major difference is that steroid receptors are located in the cytoplasm and/or nucleus rather than the membrane of the target cell. Another major difference is that the "activated" steroid receptor binds to DNA and functions as a *trans*-acting transcription factor. A generic model of the mechanism of action of a steroid hormone is shown in Figure 10. There is debate concerning the actual location of steroid receptors in the cell; some evidence suggests that the "unactivated" steroid receptors reside in cytoplasm, whereas other evidence suggests that they are actually located within the nucleus (see Blaustein, 1986; Blaustein and Olster, 1989, for review).

Regardless of the intracellular location of unactivated steroid receptor, the basic mechanism of action is the same. Briefly, steroid hormones enter the brain from the circulation and diffuse to target cells. Steroid hormones are lipid soluble and thus diffuse readily through the cell membrane. The

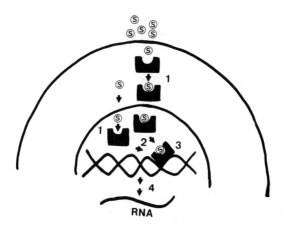

Figure 10 Generic model of mechanism of action of steroid receptor. Note that steroid hormone diffuses into cell and binds with its receptor in either the cytoplasm or nucleus (step 1). The "activated" receptor then translocates (step 2) to the DNA, where it regulates gene transcription (step 3) for the biosynthesis of new mRNA (step 4) and eventually protein.

hormone is not directed to specific cells, but instead diffuses into all cells. However, only those cells that contain the receptor are able to concentrate the hormone. In other words, the presence of the receptor determines which cells are targets. The hormone binds the receptor and the "activated" hormone-receptor complex is translocated to specific sites on the chromatin. Once attached to the DNA, the hormone-receptor complex acts as a transcription factor to alter the expression of certain genes (see O'Malley and Strott, 1973; O'Malley and Means, 1974; Chan and O'Malley, 1976; Yamamoto, 1985; Clark et al., 1985; Blaustein, 1986; Evans, 1988; Blaustein and Olster, 1989; Rories and Spelsberg, 1989).

In recent years considerable progress has been made in understanding the molecular mechanisms by which steroid hormones act to alter the behavior of target cells. Much of this progress is due to the recent cloning and elucidation of the molecular structure of gonadal steroid receptors (see Evans, 1988; Clark et al., 1985; Spelsberg et al., 1989, for review). Results from cloning studies revealed that the overall structures of the different classes of steroid receptors are remarkably similar. As a general rule, steroid receptors consist of a DNA-binding domain in the central core of the protein that is flanked on one side (COOH terminus) by a steroid-binding domain and on the other side (NH_2 terminus) by a regulatory domain. The nucleotide and amino acid sequence of the DNA binding domain and, to a lesser extent, the steroid binding is conserved among the different classes of steroid receptors. The amino acid and nucleotide sequence of the regulatory domain, on the other hand, is poorly conserved. It is now generally believed that the DNA-binding domain fits a zinc-finger motif that binds to chromatin. The steroid-binding domain, as the name implies, is involved in the binding of steroid hormone. The role of the regulatory domain remains unknown.

Behavioral Effects of Gonadal Steroid Hormones

Steroid receptor autoradiography studies in the early 1970s (Sar and Stumpf, 1973; Pfaff and Keiner, 1973) and immunohistochemical studies in the 1980s (Warembourg et al., 1986, 1989; Blaustein and Turcotte, 1989) revealed the existence of gonadal steroid receptors in discrete regions (primarily the hypothalamus and limbic system) of the central nervous system. First evidence that gonadal steroids might be involved in mediating central nervous system function came from studies on reproductive behavior which revealed that estrogen and progesterone acting via their receptors facilitate female sexual behavior in some species (see McEwen et al., 1987; Blaustein and Olster, 1989, for review). The brain area involved in this behavior is the hypothalamus (McEwen et al., 1979).

In recent years, the cellular and molecular mechanisms that mediate this behavior have become clearer. It is now known that estradiol released from the ovaries during the proestrous stage of the reproductive cycle acts on estrogen receptors in the hypothalamus to induce gene expression for progesterone receptor. Subsequently released progesterone acts via these newly synthesized receptors to induce gene expression for unknown factors (proteins) that mediate the behaviors associated with mating (see McEwen et al., 1987). Thus, the central dogma is that estrogen acting through its receptor induces gene expression for progesterone receptor, which, in turn, activates genes that mediate the behavioral response. The identity of the genes regulated by the "activated" progesterone receptor is unknown.

Progesterone and Facilitation of Respiration

We have detailed the cellular and molecular events involved in gonadal hormone-mediated sexual behavior because we believe a similar mechanism is involved in mediating the facilitatory effect of progesterone on respiration. It has been known since the turn of the century that women hyperventilate during the luteal phase of the menstrual cycle and during pregnancy (Magnus-Levy, 1904; Hasselbalch and Gammeltoft, 1915). Because the levels of circulating progesterone correlate positively with increased ventilation, progesterone has been proposed as the ventilatory stimulant in these conditions. This contention is supported by studies in which exogenously administered progesterone caused hyperventilation in normal male subjects (Skatrud et al., 1978; Kimura et al., 1984) and in patients with breathing disorders (Cullen et al., 1959; Tyler, 1960; Skatrud et al., 1980). Historically, studies concerning the respiratory effects of progesterone have been either clinical or correlational; little emphasis has been placed on determining either the site of action or the mechanism by which progesterone stimulates respiration (see Dempsey et al., 1986).

Our laboratory has recently performed studies to elucidate the mechanism(s) that mediate the facilitatory effect of progesterone on respiration (Bayliss et al., 1987, 1990). We found that progesterone (and synthethic progestins) act via its receptor to cause long-lasting (hours) facilitation of respiration (Bayliss et al., 1987); the response is blocked by pretreating animals (male and female cat) with RU486, a specific progesterone receptor antagonist (Baulieu, 1985). We also performed studies in ovariectomized cats to determine if facilitation of respiration by progesterone is estrogen dependent; i.e., does it require up-regulation of the progesterone receptor by estrogen? We found that this is indeed the case. Figure 11 shows results from experiments in which progesterone was given to an ovariectomized animal that was not "primed" with estrogen (Fig. 11a) and findings from an ovariectomized animal that was pretreated with estrogen for 3 days prior to administration of progesterone (Fig. 11b). It is clear from this example that progesterone has a strong facilitatory effect on phrenic nerve activity in the

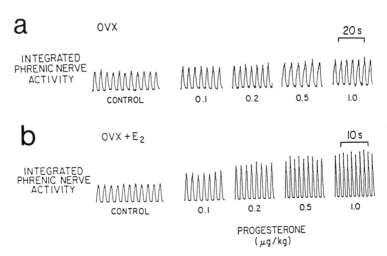

Figure 11 Effect of progesterone on phrenic nerve activity in a control animal (cat) that was ovariectomized (OVX) but not supplemented with estrogen (a) and an ovariectomized animal that was pretreated for several days with estrogen (OVX + E₂) (b). End-tidal P_{CO_2} and temperature were kept constant. Note that in the control animal sequential doses of progesterone had little if any effect on phrenic nerve activity. In the estrogen-primed animal, on the other hand, progesterone led to a dose-dependent increase in phrenic nerve activity. The response was measured 45 min after each dose (administered intravenously). Note that very small doses of progesterone were administered. These findings show clearly that progesterone-mediated facilitation of respiration requires up-regulation of its receptor with estrogen. (Adapted from Bayliss et al., 1990.)

estrogen-primed, but not the control, animal. We found this to be the case in all animals studied. Thus, similar to sexual behaviors mediated by progesterone (discussed above), the facilitatory effect of progesterone on respiration requires up-regulation of its receptor by estrogen (Bayliss et al., 1990).

Does facilitation of respiration by progesterone require gene expression and de novo protein synthesis? Indeed it does, and we found that the response was blocked or attenuated significantly by drugs that blocked either DNA transcription (actinomycin D, ACT D; Fig. 12a) or protein synthesis (anisomycin, ANI; Fig. 12b) (Bayliss et al., 1990). To our knowledge, this is first evidence for a respiratory response that requires altered gene expression. The nature of the genes and protein that are induced by progesterone and necessary for the respiratory effects is unknown.

Where are the neurons located that mediate this response? To answer this question, we measured the respiratory response to progesterone following

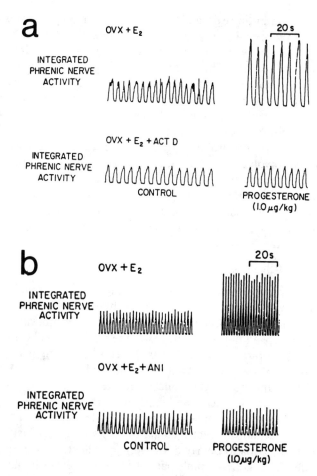

Figure 12 Effect of inhibitors of RNA, (actinomycin D) (a), and protein (anisomycin) (b), synthesis on the respiratory response to progesterone in ovariectomized and estrogen primed (OVX + E_2) cats. The effect on integrated phrenic activity of a single dose of progesterone (1.0 μg/kg) is shown. Measurements were made 150 min after administration of progesterone. In the control animals (OVX + E_2 (a and b) progesterone caused a striking increase in phrenic nerve activity. In contrast, progesterone caused only a slight increase in phrenic activity in animals in which RNA (OVX + E_2 + ACT D) (a) and protein (OVX + E_2 + ANI) (b) synthesis were inhibited. These findings show that gene expression and de novo protein synthesis are required in mediating the respiratory response to progesterone. (Adapted from Bayliss et al., 1990.)

ablation of certain brain regions. We found that the response was blocked when brain tissue rostral to the midbrain was removed surgically (i.e., decerebration), but not when the diencephalon (i.e., decortication) was left intact (Bayliss et al., 1990). Thus, it appears that the same region of the brain (i.e., the diencephalon, which includes the hypothalamus) that mediates progesterone-induced sexual behaviors also mediates the respiratory response to progesterone. The exact location of the cells within the diencephalon that mediate the response was not determined.

In summary, progesterone stimulates respiration via a hypothalamic mechanism that involves estrogen-induced progesterone receptors, gene expression, and protein synthesis. The functional significance of this mechanism is unclear, but it might be an important factor for increasing the partial-pressure of O_2 in maternal arterial blood, thus allowing better oxygenation of the fetus. Certainly, this mechanism illustrates how the endocrine and nervous system interact to regulate a vital homeostatic function.

Clinical and Experimental Considerations

An important practical aspect of understanding the cellular and molecular mechanism(s) by which progesterone, and perhaps other steroid hormones, affect respiration concerns the therapeutic use of these substances. For example, exogenously administered progesterone has been used with only limited success as a respiratory stimulant in patients (including neonates) with breathing disorders (see Sutton et al., 1974; Skatrud et al., 1980, 1981; Milerad et al., 1985). Inconsistencies concerning the ability of progesterone to stimulate respiration in patients and experimental animals (see Dempsey et al., 1986) is most likely attributed to a lack of understanding of the cellular and molecular mechanisms that mediate the response. This lack of understanding, in turn, leads to poorly designed experiments. For example, progesterone administered to a female patient or experimental animal at a time when there is an inadequate number of progesterone receptors (i.e., when the subject is not in the proper hormonal state) would yield an insignificant response. If, on the other hand, progesterone were administered at a time when there is estrogen-induced up-regulation of progesterone receptor levels (e.g., luteal phase of the menstrual cycle), progesterone would evoke a large respiratory response. Thus, the prevailing hormonal state (and sex of the patient or subject) must be considered when evaluating the effect of progesterone on respiration.

A number of other factors should also be considered when progesterone is used therapeutically as a respiratory stimulant in neonates. For instance, are progesterone receptors expressed in the neonate? If so, are these receptors responsive to progesterone? Are progesterone receptors in the neonate up-regulated by estrogen?

Another important consideration concerns the size of the dose administered. It is not unusual to read reports in which doses of 5 or more mg/kg body weight of progesterone were used. This dose far exceeds the saturation level for progesterone receptors. We found that a maximum response could be elicited when doses several orders of magnitude smaller (0.1-2.0 μg/kg, i.v.) were used (Bayliss et al., 1987). There have been recent reports that the respiratory response to progesterone is not dose-dependent. The reason for the inability to measure a dose response in these studies is quite obvious: the smallest doses used far exceed the dose needed to yield the maximum response. Thus, the receptors were saturated with the hormone even at the smallest dose. It is clear from the results from the ovariectomized estrogen-primed animal shown in Figure 11 that the respiratory response to progesterone is indeed dose-dependent.

C. Regulation of Gene Expression for Transmitter Enzymes and Neuropeptides by Steroid Hormones

There is accumulating evidence that steroid hormones can influence neurotransmitter and neuromodulator systems by exerting effects on key synthetic or degradative enzymes, receptors, and neuropeptides, often at the level of gene expression (see Harlan, 1988; Blaustein and Olster, 1989). In the hypothalamus, progesterone and estrogen receptors have been found in phenotypically identified neurons. For instance, a substantial number of neurons that concentrate either one or the other of these hormones also show immunoreactivity for tyrosine hydroxylase, the rate-limiting enzyme in the biosynthesis of catecholamines (Sar, 1984, 1988). GABAergic neurons in the preoptic area of the hypothalamus have been reported to contain receptors for estrogen (Flugge et al., 1986). Gonadal steroid receptors have also been found in hypothalamic neurons that are immunoreactive for a number of peptides, including, vasopressin (Sar and Stumpf, 1980), β-endorphin (Shivers et al., 1984; Jirikowski et al., 1986), met-enkephalin (Harlan, 1988), somatostatin (Sar and Stumpf, 1983), and substance P (Akesson and Micevych, 1988).

Can steroid hormones alter gene expression for transmitter and peptides? There is growing evidence which suggests that this might actually be the case. Nuclease footprinting experiments have been used successfully to identify specific nucleotide sequences (i.e., response elements) on the promotor regions of genes. These sequences are believed to be the binding sites of steroid receptors (see Yamamoto, 1985). Theoretically, any gene that contains a response element for a steroid receptor can be modulated by the hormone-"activated" receptor. First evidence that steroid hormones affect transmitter systems in the brain came from biochemical studies which showed that gonadal steroid hormones altered the turnover rates for a number of classical neurotransmitters, including dopamine, norepinephrine, serotonin, and acetylcholine (see McEwen et al., 1979). In addition, there is evidence that gene expression for certain peptides is either enhanced or repressed by steroid hormones. For example, gene expression for the decapeptide luteinizing hormone-releasing hormone (LHRH) is enhanced by estradiol (Rothfeld et al., 1987). In contrast, estrogen pretreatment led to a large decrease in expression of the proopiomelanocortin (POMC) gene, the precursor for β-endorphin (Wilcox and Roberts, 1985). It appears, therefore, that circulating gonadal hormones (and other steroid hormones) could modify or modulate neuronal function in certain neurons by altering gene expression for molecules that might be involved in synaptic transmission.

Alteration of gene expression by steroid hormones provides a mechanism by which the endocrine system can cause long-term modulation (i.e., biasing) of neuronal activity. Such a mechanism might be especially important during fetal and early postnatal development, a period when circulating levels of steroid hormone fluctuate widely. For a more in-depth review of the effects of gonadal steroid hormones on transmitter and peptide systems, the reader is referred to several recent reviews on this topic (McEwen et al., 1979; Harlan, 1988; Blaustein and Olster, 1989; Yamamoto, 1985).

IV. Ontogenesis of Neurotransmitter and Peptide Systems in the Brain

Historically, there has been considerable interest in elucidating the factors that regulate neuronal growth and development in the fetal and neonatal nervous system. The developing nervous system is extremely "plastic" in the sense that new synaptic connections are made and old ones rearranged, modified, or eliminated. This dynamic anatomical state leads to neuronal connections that allow the developing animal to express behaviors (e.g., motor patterns, social behaviors) that are crucial for survival and propagation of the species. For the past quarter century, there has been an intense effort to identify factors that regulate nerve growth, target recognition, and synaptic rearrangement during ontogeny (see Purves and Lichtman, 1985).

In addition to developmental changes involving establishment of synaptic connections, the chemical phenotype of neurons in many regions of the nervous system, including areas involved in regulation of respiration, undergoes change. Today, there is considerable interest and research to elucidate the role the transmitters and peptides in mediating certain functions. An idealistic approach to gain information of this type is to alter the phenotype of neurons and determine how this change affects their function. A major problem with this approach is that technology is not available to regulate expression of specific genes and thus manipulate the phenotype of the neurons of interest. However, nature provides such an opportunity; the chemical phenotype of neurons changes naturally during development. We refer to these alterations in phenotype during development as "programmed" gene expression. Much insight might be gained by studying these systems during development. For instance, the role of a chemical messenger might be elucidated by studying the relationship between its appearance and the onset of a particular physiological process or function during development of the nervous system.

In this section, we shall discuss the ontogeny of neurotransmitter and peptide systems. We shall not attempt to give an exhaustive review of this topic, but instead we shall provide examples of how the neurotransmitter and peptide phenotypes of neurons might change during fetal life and early postnatal development. Moreover, we shall not discuss factors that promote growth and guide neurons to their targets. For more information on this topic the reader is referred to the monograph by Purves and Lichtman (1985).

A. Ontogeny of Classical Neurotransmitters

The transmitter systems studied most extensively during development are the monoamines, dopamine, epinephrine, norepinephrine, and serotonin. In the mature animal, monoamines are found in cell bodies in discrete, but well-described nuclei, and in fibers throughout the central nervous system (see Björklund et al., 1984-1985). Catecholamines and serotonin are synthesized from aromatic amino acids via specific biochemical pathways involving several enzymes; the aromatic L-amino acid decarboxylase is the second enzyme in this pathway. This enzyme appears in the nervous system of rat embryos well before birth (gestational day 15-19) (Jaeger, 1986). Moreover, immunoreactivities for other key enzymes in the biosynthesis of catecholamines, including tyrosine hydroxylase (catalyzes conversion of tyrosine to L-DOPA) and phenylethanolamine *N*-methyltransferase (catalyzes conversion of epinephrine from norepinephrine), have been found in the dorsal and ventrolateral aspects of the medulla oblongata and in the hypothalamus prior to brith (Foster et al., 1985; Daikoku et al., 1986). Thus, enzymes involved in the biosynthesis of catecholamines are expressed well before birth, indicating that catecholamine neurotransmission might be functional in the fetus and neonate. Moreover, dopamine and norepinephrine have been found in respiratory-related brainstem nuclei in full-term fetuses and newborn rabbits (McNamara and Lawson, 1983).

There is also evidence from immunohistochemical studies that serotonin (Lidov and Mollover, 1982; Wallace and Lauder, 1983; Goto and Sano, 1984) and GABA (Lauder et al., 1986) phenotypes differentiate very early during embryonic life in rat. Fibers containing these transmitters are found throughout the neuraxis in the term fetus and neonate. There is also evidence that the amino acid transmitters glutamate (Campochiaro and Coyle, 1978; Wong and McGeer, 1981) and glycine (Johnston and Davies, 1974) are present in the newborn animal.

Immunohistochemical evidence that 5-HT- and GABA-immunoreactive neurons are present in nervous system of neonatal rat at birth is shown in Figure 13. The photomicrograph shows 5-HT- (Fig. 13A,B) and GABA- (Fig. 13C) immunoreactive neurons in the medullary raphe complex in a 0-day-old rat. In addition, we found that a number of individual neurons in this region contain both 5-HT and GABA (Fig. 13B,C; arrows), a coexistence pattern reported previously in the adult rat brain (see Fig. 4) (Millhorn et al., 1987a, 1988). This previously unreported observation is, to our knowledge, first evidence for coexistence of two classical neurotransmitters

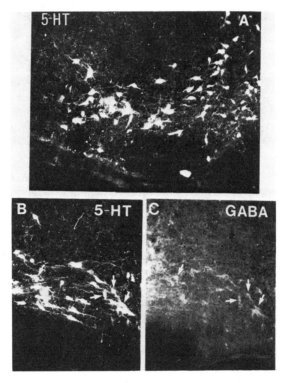

Figure 13 Immunofluorescence micrographs showing the distribution of serotonin (5-HT) (A, B) and GABA (C) in the region of nucleus raphe magnus in the medulla oblongata of a 0-day-old rat. Numerous individual perikarya contain both 5-HT (B) and GABA (C) as denoted by arrows.

in the neonate nervous system. There is, however, evidence that a classical neurotransmitter (catecholamines) and a neuropeptide (neuropeptide tyrosine, NPY) coexist in neurons of the ventrolateral medulla in embryonic rat (Foster et al., 1984).

Although the different classical neurotransmitter systems appear at different times during embryogenesis, most of these systems are present and possibly functional at the time of birth. The level of these compounds, however, may change during the early stages of postnatal development. There is need for more research to determine when these systems actually become functional, and more information is needed concerning coexistence of transmitters with other compounds (i.e., transmitters and peptides) during ontogeny.

B. Ontogeny of Neuropeptides

In recent years, a relatively large number of studies concerning the ontogeny of peptide systems in the brain have been performed. It appears that the pattern of development in these systems is more variable than that observed for classical neurotransmitters in the sense that some peptides are expressed before birth and others are not apparent until days to weeks after birth. This, of course, may relate directly to the function(s) that certain peptides may mediate (e.g., a trophic effect) during this critical time in life. Here we shall give a brief overview of some of the more interesting observations concerning ontogeny of peptide systems in the brain. We shall not attempt to chronicle all studies that have been performed on this interesting topic. We shall conclude with a discussion of recent findings from our laboratory in which we have studied gene expression for peptides in respiratory-related regions of the medulla oblongata during early postnatal development.

The two most widely studied peptide groups in the mammalian nervous system are the tachykinins and opiates. We shall use these two peptide families to illustrate how peptide systems change during ontogenesis. More than a dozen tackykinin peptides have been identified, but only three of these (substance P; neurokinin A, also known as substance K; and neurokinin B, also known as neuromedin K) have been identified in the mammalian nervous system (Maggio, 1985). A distinguishing feature of this family of peptides is the highly conserved carboxyl-terminal sequence Phe-X-Gly-Leu-Met-NH_2 (where X represents an amino acid in this sequence that often varies among the different tachykinins). The molecular cloning of the tackykinins expressed in mammalian brain revealed that substance P and neurokinin A are expressed by a single gene (preprotachykinin A, PPT-A) and neurokinin B is expressed by another gene (preprotachykinin B, PPT-B) (Kotani et al., 1986; Krause et al., 1987).

Substance P is perhaps the most extensively studied peptide and is believed to play a role in mediating a number of functions, including nociception, autonomic regulation, and somatic motor activity (see Pernow, 1983). Although substance P is found in cell bodies and fibers in numerous brain regions in the mature animal (Ljungdahl et al., 1978), there are reports that substance P levels in the spinal cord (Holzer et al., 1981; Senba et al., 1982b) and brain (Gilbert and Emson, 1979) decrease during postnatal development. In addition, Charlton and Helke (1986) reported the existence of a large number of diffusely organized binding sites for substance P (a measure of substance P receptor density) in the spinal cord of the newborn rat pup. It is interesting that although the total number of binding sites in the spinal cord decreased during the first few postnatal weeks, the number of sites became more concentrated in certain regions, i.e., the phrenic motor nucleus and preganglionic sympathetic neurons of the intermediolateral cell column. Thus, expression for substance P and its receptor appears to show an overall decrease during early postnatal development; however, concentration of substance P receptor in certain areas of the nervous system appears to occur during the same period. The physiological significance of these changes remains unclear, but the increase in substance P receptor density in the phrenic nucleus and intermediolateral cell column may indicate that this peptide plays an increasingly important role in modulating synaptic transmission at these sites during the later stages of development and in the mature animal.

Three classes of opioid peptides are found in the nervous system as enkephalins (met- and leu-, β-endorphin, and dynorphin. The biosynthesis of these peptides is rather complex and involves several distinct genes. The proenkephalin gene contains four copies of met-enkephalin, one copy of leu-enkepahlin, and one copy of met-enkephalin-Arg,-Gly,-Leu (Rosen et al., 1984). Dynorphin (and α-neo-endorphin) is derived from the prodynorphin gene (Civelli et al., 1985). β-Endorphin is derived from a single precursor molecule, proopiomelanocortin (POMC), that also gives rise to the nonopiate peptides, α-melanocyte-stimulating hormone (MSH), and adrenocorticotropic hormone (ACTH) (Roberts et al., 1979).

Findings from radioimmunoassay studies revealed that endorphin- and enkephalin-like peptides are present in rat brain as early as day 16 of gestation; thereafter, these compounds increase until birth (Bayon et al., 1979). Immunohistochemical studies were used to identify the brain regions that contain enkephalin during ontogeny. Enkephalin immunoreactivity was observed in perikarya of various regions of the medulla oblongata and pons by prenatal day 15 (Palmer et al., 1982). Moreover, enkephalin-positive fibers were found in late-gestational animals (Palmer et al., 1982). Pickel et al. (1981) reported that immunoreactivity for enkephalin is apparent in the caudal brainstem several days before it is evident in higher regions of fetal brain. Enkephalin immunoreactivity was observed in several brain nuclei in fetus, including nucleus tractus solitarii, substantia gelatinosa, spinal trigeminal nucleus, and caudate putamen (Pickel et al., 1981). It is also interesting that many enkephalin immunoreactive cells that were observed at prenatal day 15 were not found in the term fetus (Palmer et al., 1982), which indicates that this peptide may mediate a function in fetus that is not essential in the neonate. It is noteworthy that the enkephalin receptor is detectable at approximately the same time that immunoreactivity for the peptide is first observed (Clendeninn et al., 1976; Coyle and Pert, 1976).

After birth the overall number of enkephalin neurons in the brain appears to decrease slightly and the number of fibers that are immunoreactive for this peptide increases progressively until they reach the adult pattern several weeks after birth (Senba et al., 1982a). However, in several regions (e.g., trigeminal nucleus) both the number of cell bodies and fibers were found to increase after birth (Senba et al., 1982a). Certainly, enkephalin-containing neurons (Elde et al., 1976; Hökfelt et al., 1979; Finley et al., 1981) and opiate receptors (Atweh and Kuhar, 1977a,b) are found in many regions of the mature nervous system, including major respiratory-related regions of the brainstem.

A number of other peptides have been identified in the nervous systems of the embryo and reonate. For example, NPY, which is found in the fetal nervous system at day 13 of gestation, continues to be expressed throughout life (Foster et al., 1984). At birth, the neurotensin level in the forebrain was estimated to be only 10% of that found in the mature animal; the mature level is reached at the age of 5 weeks (Goedert et al., 1985). Somatostatin (SOM) immunoreactivity was found in the brainstem of fetal rat on gestational day 15 and appeared to increase until birth (Shiosaka et al., 1981). Thereafter, SOM levels decreased progressively during postnatal development; relatively few SOM immunoreactive neurons were found in brainstem structures in the mature animal. Other workers have found a considerable population of SOM cells in the medulla oblongata (including NTS, vagal motor nucleus, and the ventrolateral aspect of the medulla) in mature rat (Johansson et al., 1984; Millhorn et al., 1987b). However, SOM is found in some discrete areas in the brainstem at birth and then disappears gradually during the next few postnatal weeks, which indicates that this peptide might mediate a trophic function in these areas during early development.

This vignette on the ontogeny of peptides was presented to illustrate that peptide systems in the brain can change markedly during development. We did not intend to present an exhaustive review on this topic, but instead to present situations that emphasize the "plasticity" of peptide systems during this crucial period of life.

C. Gene Expression for Peptides During Development

Is there a molecular basis for alterations in peptide levels in the nervous system during ontogeny? What factors regulate gene expression for neuropeptides during development and are these factors present in all neurons? These are important questions since a given peptide may disappear from cells in one brain region and yet be present in other regions throughout life. Is the gene that expresses the precursor molecule for this peptide regulated differentially in different regions of the nervous system and could this explain regional changes in peptide phenotype during ontogeny? Certainly, answers to these questions would provide much-needed insight concerning neuronal plasticity during pre- and postnatal development.

We conclude this chapter with a brief discussion of recent work from our laboratory on regulation of gene expression for neuropeptides during development. Specifically, we shall discuss the ontogeny of somatostatin, which disappears from the hypoglossal (XII) nucleus, and thyrotropin-releasing hormone (TRH), which appears in the caudal raphe nuclei during the first month of life.

Somatostatin (SOM) immunoreactivity is found in fibers but not cell bodies in the hypoglossal nucleus of the adult rat (Johansson et al., 1984). Recently, however, we found that many perikarya in the caudal portion of this nucleus contain SOM (Seroogy et al., in preparation). An example of this is shown in Figure 14. Notice the SOM immunoreactive neurons in the dorsal vagal complex [vagal motor nucleus (X) and NTS] in both the newborn and adult indicating that SOM is present in this region throughout development. We also found (not shown) that calcitonin gene-related peptide (CGRP) is found in hypoglossal neurons throughout development and coexists with SOM in these neurons at birth. This observation is important in that it shows that the loss of SOM in this nucleus is not the result of natural cell death during development. If this were the case, CGRP would disappear as well. We tested the hypothesis that the decline and disappearance of SOM in the hypoglossal nucleus is related to decreased levels of preprosomatostatin (preproSOM) mRNA (Seroogy et al., in preparation). In situ hybridization was employed to determine if preproSOM mRNA concentration in the hypoglossal nucleus decreases during the first month of postnatal development. We found this to be the case. In fact, the decrease in preproSOM mRNA paralleled the disappearance of the peptide. This finding suggests that SOM may play an important role in hypoglossal motor neurons (these neurons innervate the tongue and upper airway) during early postnatal life that is no longer necessary in the mature animal.

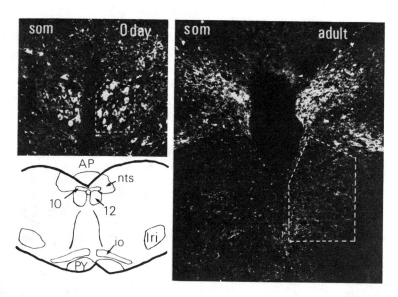

Figure 14 Immunofluorescence micrographs showing neurons in the hypoglossal nucleus (12; dashed arrow) of a 0-day-old rat that contain somatostatin (SOM) immunoreactivity. In contrast, SOM immunoreactivity is not found in cell bodies in this nucleus (dashed area) in the adult. However, SOM immunoreactivity is found in fibers of the hypoglossal nucleus and in cell bodies in nucleus tractus solitarii (nts, see diagram) and the vagal motor nucleus (10; see diagram). Thus, SOM is transiently expressed in cell bodies in the hypoglossal nucleus during ontogeny.

SOM mRNA is found in the dorsal vagal complex and other regions of the medulla throughout postnatal development and in the adult. It therefore appears that the preproSOM gene is repressed by some unknown factor that is present in the hypoglossal nucleus but not present in cells that express this gene in the mature animal. We are currently performing experiments to identify factors from nuclear protein extracts that might interact with the promotor region of the SOM gene to alter its expression. Previously, a cAMP-related protein (CREB protein) has been shown to interact with this gene to enhance its expression (Montminy et al., 1986; Montminy and Bilezikjian, 1987), but no factor has been identified that represses its expression. However, we have recently identified protein-DNA interactions at sites that do not coincide with the CREB protein binding site (Szymeczek et al., 1970). Experiments are underway to identify the DNA sequence that binds these proteins and to determine if these proteins regulate expression of the SOM gene.

Millhorn et al.

Many neurons in the caudal medullary raphe nuclei (obscurus and pallidus) of adult contain the tripeptide, thyrotropin-releasing hormone (TRH);
(see Palkovits and Brownstein, 1985). We found that mRNA for this peptide is either absent or present in very low concentration in raphe neurons at
birth and is found at progressively higher concentrations during the first 3-4
weeks after birth (Szymeczek et al., 1990). This progressive increase in TRH
mRNA in the medullary raphe complex is shown in autoradiogram from in situ

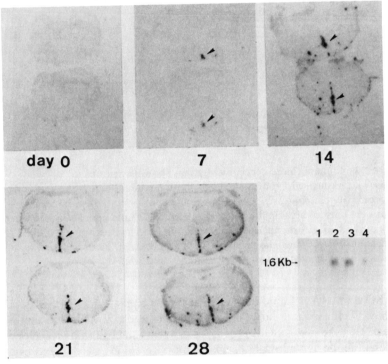

Figure 15 Autoradiographs from in situ hybridization and Northern analysis for
thyrotropin-releasing hormone (TRH) mRNA in nucleus raphe obscurus and pallidus
(arrowheads) in the medulla oblongata. The autoradiographs show that on day 0
(day of birth) there was no indication of TRH mRNA in the raphe complex. However, on days 7-21 there was progressive increase in TRH mRNA concentration in
raphe cells (dark region denoted by arrowheads). There appeared to be a slight decrease in TRH mRNA level from day 21 to day 28 as indicated by a reduction in the
intensity of the labeling (arrowheads). These findings were confirmed by results from
Northern analysis (lower right panel). A single hybridizing band of poly A[+] RNA
(1.6 Kb) which represents TRH mRNA was found to increase during the first postnatal
month. Note that the signal is barely visable at day 0 (lane 1) and is progressively larger
on day 7 (lane 2) and day 14 (lane 3). The signal decreases slightly between day 14 and
day 28 (lane 4).

after birth (Szymeczek et al., 1990). This progressive increase of TRH mRNA in the medullary raphe complex is shown in autoradiographs from in situ hybridization experiments (Fig. 15). Neurons that express TRH mRNA are found along the midline and at discrete areas near the ventral surface of the medulla (indicated by arrowheads). Note the progressive increase in the concentration of TRH mRNA in these regions from day 0 to day 21. The signal appears to decrease slightly between day 21 and day 28 (the adult level). This observation is confirmed by findings from Northern blot analysis (bottom right). Lanes 1-4 contain poly (A^+) RNA extracted from the medulla oblongata of neonatal rat at 0, 7, 14, and 28 days after birth, respectively. A hybridizing species that migrates as 1.6-kb RNA, appropriate for TRH mRNA, is evident. It is clear that the signal became progressively stronger in the 7 (lane 2)- and 14 (lane 3)-day-old neonate and then decreased by age 28 days (lane 4). This result is similar to that observed with in situ hybridization. Thus, this is an example of a peptide, in an area that has been implicated in regulation of respiration (Millhorn, 1986), whose levels increase after birth.

The significance of appearance of TRH postnatally is unknown. However, since TRH coexists with 5-HT and SP in neurons of the raphe complex (Johansson et al., 1981), and since neurons in the raphe complex that express TRH mRNA project to NTS (Millhorn et al., 1990b), it is conceivable that TRH plays an increasingly more important role in 5-HT transmission (see Fig. 5) and perhaps regulation of respiration as the animal matures.

V. Conclusions

Cell-to-cell communication in the nervous system is a complex event that involves many different transmitter and receptor types as well as intricate intracellular signal transduction pathways. In this chapter, we have provided a basic overview of the cellular and molecular mechanisms involved in synaptic transmission and signal transduction in the mammalian nervous system. A number of mechanisms, including memory and learning, sensitization, long-term potentiation, and physiological adaptation to certain environmental pertubations, are mediated either directly or indirectly by synaptic mechanisms. Perhaps one of the most interesting discoveries of the last decade is that synaptic activity via signal transduction pathways can alter gene expression and the phenotype of the target cell. This might be the underlying mechanism for a number of long-lasting respiratory phenomena, such as ventilatory adaptation (i.e., acclimatization) to chronic hypoxia.

Chemical synaptic transmission may change considerably during ontogeny and play an important factor in development of neuronal systems. Thus, certain synaptic elements (e.g., transmitters, peptides, receptors) may play an important trophic role during ontogeny. Certainly, the chemical phenotype (i.e., transmitter and peptide) content of neurons in various regions of the nervous system changes during fetal and neonatal development. Not only might these changes in phenotype have a functional significance in the developing nervous system, they may also offer an excellent opportunity to study transmitter/peptide-mediated responses. Additional research is needed to identify the factors that mediate phenotypic alterations in transmitter, peptide, and receptor content during ontogeny and to determine the functional significance of such changes.

Acknowledgment

Research from the authors' laboratories was supported by grants from the NIH (HL33831), the American Heart Association (88-1108), and the Swedish MRC (04X-2287). DEM is a Career Investigator of the American Lung Association. D.A.B. is supported by a fellowship from Glaxo Pharmaceutical Co. KBS is supported by an NRSA postdoctoral fellowship (NS08525).

References

Akers, R. F., Lovinger, D. M., Colley, P. A., Linden, D. J., and Routtenberg, A. (1986). Translocation of protein kinase C activity may mediate hippocampal long-term potentiation. *Science* **231**:587-589.

Akesson, T. R., and Micevych, P. E. (1988). Estradiol concentration by substance P immunoreactive neurons in the medial basal hypothalamus of the female rat. *J. Neurosci. Res.* **19**:412-419.

Atweh, S., and Kuhar, M. (1977a). Autoradiographic localization of opiate receptors in rat brain. I. Spinal cord and lower medulla. *Brain Res.* **124**: 53-67.

Atweh, S., and Kuhar, M. (1977b). Autoradiographic localization of opiate receptors in rat brain. II. The brain stem. *Brain Res.* **129**:1-12.

Barbeau, H., and Bedard, P. (1981). Similar motor effects of 5-HT and TRH in rats following chronic spinal transection and 5,7-dihydroxytryptamine injection. *Neuropharmacology* **20**:477-481.

Baulieu, E. E. (1985). RU486: An antiprogestin steroid with contragestational activity in women. In *The Antiprogestin Steroid RU486 and Fertility Control*. Edited by E. E. Baulieu and S. J. Segal. New York, Plenum Press.

Bayliss, D. A., Cidlowski, J. A., and Millhorn, D. E. (1990). The stimulation of respiration by progesterone in ovariectomized cat is mediated by an estrogen-dependent hypothalamic mechanism requiring gene expression. *Endocrinology* **126**.

Bayliss, D. A., Millhorn, D. E., Gallman, E. A., and Cidlowski, J. A. (1987). Progesterone stimulates respiration through a central nervous system steroid receptor-mediated mechanism in cat. *Proc. Natl. Acad. Sci. USA* **84**:7788-7792.

Bayon, A., Shoemaker, W. J., Bloom, F. E., Mauss, A., and Guillemin, R. (1979). Perinatal development of the endorphin- and enkephalin-containing systems in the rat brain. *Brain Res.* **179**:93-101.

Berridge, M. J. (1987). Inositol trisphosphate and diacylglycerol: Two interacting second messengers. *Annu. Rev. Biochem.* **56**:159-193.

Berridge, M. J., and Irvine, R. F. (1984). Inositol trisphosphate, a novel second messenger in cellular signal transduction. *Nature* **312**:315-321.

Birnbaumer, L., and Iyengar, R. (1981). Coupling of receptors to adenylate cyclase. In *Biochemistry of Cyslic Nucleotides.* Edited by J. A. Nathanson and J. W. Kebabian. New York, Springer-Verlag.

Björklund, A., Hökfelt, T., and Kuhar, M. (Eds.) (1984-1985). *Handbook of Chemical Neuroanatomy.* Vols. 2-4. Amsterdam, Elsevier.

Blackman, J. G., Ginsborg, B. L., and Ray, C. (1963). Spontaneous synaptic activity in synaptic ganglia of the frog. *J. Physiol. (London)* **167**: 389-401.

Blackshear, P. J., Narin, A. C., and Kuo, J. F. (1988). Protein kinases 1988: A current perspective. *FASEB J.* **2**:2957-2969.

Blaustein, J. D. (1986). Steroid receptors and hormone action in the brain. *Ann. NY Acad. Sci.* **474**:400-414.

Blaustein, J. D., and Olster, D. H. (1989). Gonadal steroid hormone receptors and social behaviors. In *Advances in Comparative and Environmental Physiology.* Edited by J. Balthazart. New York, Springer-Verlag.

Blaustein, J. D., and Turcotte, J. C. (1989). Estradiol-induced progestin receptor immunoreactivity is found only in estrogen receptor-immunoreactive cells in guinea pig brain. *Neuroendocrinology* **49**:454-461.

Blaustein, M. P. (1971). Preganglionic stimulation increases calcium uptake by sympathetic ganglia. *Science* **172**:391-393.

Bloom, F. E. (1988). Neurotransmitters: Past, present, and future directions. *FASEB J.* **2**:32-41.

Bodian, D. (1972). Neuron junctions: A revolutionary decade. *Anat. Rec.* **174**:73-82.

Bonner, T. I., Buckley, N. J., Young, A. C., and Brann, M. R. (1987). Identification of a family of muscarinic acetylcholine receptor genes. *Science* **237**:527-532.

Bowery, N. G. (1987). Glycine-binding sites and NMDA receptors in the brain. *Nature* **326**:338, 1987.

Brodin, E., Linderoth, B., Goiny, M., Yamamoto, Y., Gazelius, B., Millhorn, D. E., Hkfelt, T., and Ungerstedt, U. In vivo release of serotonin in cat nucleus tractus solitarii and cervical ventral horn induced by electrical stimulation of the medullary raphe nuclei. *Brain Res.* (in press).

Brown, T. H., Chapman, P. F., Kairiss, E. W., and Keenan, C. L. (1988). Long-term synaptic potentiation. *Science* **242**:724-728.

Campochiaro, P., and Coyle, J. T. (1978). Ontogenetic development of kainate neurotoxicity: correlates with glutamatergic innervation. *Proc. Natl. Acad. Sci. USA* **75**:2025-2029.

Castellucci, V. F., Kennedy, T. E., Kandel, E. R., and Goelet, P. (1988). A quantitative analysis of 2-D gels identifies proteins in which labeling is increased following long-term sensitization in *Aplysia*. *Neuron* **1**:321-328.

Ceccarelli, B., Hurlbut, W. P., and Mauro, A. (1972). Depletion of vesicles from frog neuromuscular junction by prolonged tetanic stimulation. *J. Cell Biol.* **54**:30-38.

Ceccatelli, S., Millhorn, D. E., Hökfelt, T., and Goldstein, M. (1989). Evidence for the occurrence of an enkephalin-like peptide in adrenaline and noradrenaline neurons of the rat medulla oblongata. *Exp. Brain Res.* **74**:631-640.

Chan, L., and O'Malley, B. W. (1976). Mechanisms of action of the sex steroid hormones. *N. Engl. J. Med.* **294**:1322-1328.

Changeux, J-P. (1981). The acetylcholine receptor: An "allosteric" membrane protein. *Harvey Lecture* **75**:85-254.

Changeux, J-P., Devillers-Thiery, A., Giraudat, J., Dennis, J., Heidmann, T., Revah, F., Mulle, C., Heidmann, O., Klarsfeld, A., Fontaine, B., Laufer, R., Nghiem, H., Kordeli, E., and Cartaud, J. (1987). The acetylcholine receptor: functional organization and evolution during synapse formation. In *Strategy and Prospects in Neuroscience*. Edited by O. Hayaishi. Tokyo, Japan Scientific Societies, pp. 29-76.

Chan-Palay, V., Jonsson, G., and Palay, S. (1978). Serotonin and substance P coexist in neurons of the rat's central nervous system. *Proc. Natl. Acad. Sci. USA* **75**:1582-1586.

Charlton, C. G., and Helke, C. J. (1986). Ontogeny of substance P receptors in rat spinal cord: Quantitative changes in receptor number and differential expression in specific loci. *Dev. Brain Res.* **29**:81-91.

Charlton, C. G., and Helke, C. J. (1987). Substance P-containing projections to the intermediolateral cell column: Identification with retrogradely transported rhodamine-labeled latex microspheres and immunohistochemistry. *Brain Res.* **418**:245-254.

Chung, F-Z., Wand, C. D., Potter, P. C., Venter, J. C., and Fraser, C. M. (1988). Site-directed mutagenesis and continuous expression of common beta-adrenergic receptors: Identification of a conserved aspartate residue involved in agonist binding and receptor activation. *J. Biochem.* **263**:4052-4055.

Civelli, O., Douglass, J., Goldstein, A., and Herbert, E. (1985). Sequence and expression of the rat prodynorphin gene. *Proc. Natl. Acad. Sci. USA* **82**:4291-4295.

Clark, J. H., Schrader, W. T., and O'Malley (1985). Mechanisms of steroid hormone action. In *William's Textbook of Endocrinology*. Edited by J. D. Wilson and D. W. Foster. Philadelphia, Saunders.

Clendeninn, N. J., Petraitis, M., and Simon, E. J. (1976). Ontological development of opiate receptors in rodent brain. *Brain Res.* 118:157-160.

Collingridge, G. (1987). The role of NMDA receptors in learning and memory. *Science* 230:604-605.

Conn, P. J., and Sanders-Busch, E. (1987). Central serotonin receptors: Effector systems, physiological roles and regulation. *Psychopharmacology* 92:267-277.

Cotman, C. W., and Iversen, L. L. (1987). Excitatory amino acids in the brain--focus on NMDA receptor. *Trends Neurosci.* 7:263-302.

Coyle, J. T., and Pert, C. B. (1976). Ontogenetic development of (^{3}H)-naloxone binding in rat brain. *Neuropharmacology* 15:555-560.

Cullen, J. H., Brum, V. C., and Redet, W. V. (1959). The respiratory effects of progesterone in severe pulmonary emphysema. *Am. J. Med.* 27:551-557.

Czyzyk-Krzeska, M. F., Bayliss, D. A., Seroogy, K., and Millhorn, D. E. Gene expression for peptides in neurons of the petrosal and nodose ganglia in rat. *Experimental Brain Res.* (in press).

Czyzyk-Krzeska, M. F., Seroogy, K. B., Bayliss, D. A., and Millhorn, D. E. (1990). Messenger RNAs for neuropeptides in primary sensory neurons of the petrosal ganglion in rat. In *Carotid Body Chemoreceptors and Chemoreflexes*. Edited by H. Acker, R. O'Regan, and A. Trzebski. New York, Plenum Press.

Daikoku, S., Kawano, H., Okamura, Y., Tokuzen, M., and Nagatsu, I. (1986). Ontogenesis of immunoreactive tyrosine hydroxylase-containing neurons in rat hypothalamus. *Dev. Brain Res.* 28:85-98.

Dale, N., Schacher, S., and Kandel, E. R. (1988). Long-term facilitation in *Aplysia* involves increase in transmitter release. *Science* 239:282-285.

Davies, J., and Watkins, J. C. (1983). Role of excitatory amino acid receptors in mono- and polysynaptic excitation in the cat spinal cord. *Exp. Brain Res.* 49:280-290.

Del Castillo, J., and Katz, B. (1954a). Statistical factors involved in neuromuscular facilitation and depression. *J. Physiol.* 124:574-584.

Del Castillo, J., and Katz, B. (1954b). Changes in endplate activity produced by presynaptic polarization. *J. Physiol.* 124:586-604.

Dempsey, J. A., and Forster, H. V. (1982). Mediation of ventilatory adaptations. *Physiol. Rev.* 62:262-346.

Dempsey, J. A., Olson, E. B., and Skatrud, J. B. (1986). Hormones and neurochemicals in the regulation of breathing. In *Handbook of Physiology*, Vol. 2. Edited by A. P. Fishman, N. S. Cherniack, and J. G. Widdicombe. Washington, DC, American Physiological Society.

De Robertis, E. D. P. (1956). Submicroscopic changes of the synapse after nerve section in the acoustic ganglion of the guinea pig. An electron microscope study. *J. Biochem. Biophys. Cytol. Suppl.* **2**:503-512.

De Robertis, E. D. P., and Bennett, H. S. (1954). Sub-microscopic vesicular component in the synapse. *Fed. Proc.* **13**:35.

Dixon, R., Sigal, I. S., Candelore, M. R., Register, R. B., Scattergood, W., Rands, E., and Strader, C. D. (1987). Structural features required for ligand binding to the beta-adrenergic receptor. *EMBO J.* **6**:3269-3275.

Dohlman, H. G., Caron, M. G., and Lefkowitz, R. J. (1987). A family of receptors coupled to guanine nucleotide regulatory proteins. *Biochemistry* **26**:2657-2664.

Eccles, J. C. (1957). *The Physiology of Nerve Cells*. Baltimore, Johns Hopkins Press.

Elde, R., Hökfelt, T., Johansson, O., and Terenius, L. (1976). Immunohistochemical studies using antibodies to leucine-enkephalin: Initial observations on the nervous system of the rat. *Neuroscience* **1**:349-351.

Erickson, J. T., and Millhorn, D. E. (1990). Long-lasting facilitation of phrenic nerve activity in rat is induced by stimulation of carotid bodies or caudal raphe nuclei. *Fed. Proc.* (in press).

Erickson, J. T., Millhorn, D. E., and Gorcs, T. (1989). Evidence for serotonin and substance P coexistence in fibers projecting from caudal midline structures to the nucleus tractus solitarii in cat. *Fed. Proc.* **3**:403.

Evans, R. M. (1988). The steroid and thyroid hormone superfamily. *Science* **240**:889-895.

Fagg, G. E., Foster, A. C., and Ganong, A. H. (1986). Excitatory amino acid synaptic mechanisms and neurological function. *Trends Neurosci.* **7**:357-363.

Fagg, G. E., and Matus, A. (1984). Selective association of N-methyl aspartate and quisqualate types of L-glutamate receptor with brain postsynaptic densities. *Proc. Natl. Acad. Sci. USA* **81**:6876-6880.

Fatt, P., and Katz, B. (1952). Spontaneous subthreshold activity at motor nerve endings. *J. Physiol.* **117**:109-128.

Finer-Moore, J., and Stroud, R. M. (1984). Amphipathic analysis and possible formation of the ion channel in an acetylcholine receptor. *Proc. Natl. Acad. Sci. USA.* **81**:155-159.

Flugge, G., Oertel, W. H., and Wuttke, W. (1986). Evidence for estrogen-receptive GABAergic neurons in the preoptic/anterior hypothalamic area of the rat brain. *Neuroendocrinology* **43**:1-5.

Fontaine, B., Klarsfeld, A., and Changeux, J. P. (1987a). Calcitonin gene-related peptide and muscle activity regulate acetylcholine receptor alpha-subunit mRNA levels by distinct intracellular pathways. *J. Cell Biol.* **105**:1337-1342.

Fontaine, B., Klarsfeld, A., Hökfelt, T., and Changeux, J. P. (1987b). Calcitonin gene-related peptide, a peptide present in spinal cord moto-neurons, increase the number of acetylcholine receptors in primary cultures of chick embryo myotubes. *Neurosci. Lett.* **71**:59-65.

Forn, J., and Greengard, P. (1978). Depolarizing agents and cyclic nucleotides regulate the phosphorylation of specific neuronal proteins in rat cerebral cortex slices. *Proc. Natl. Acad. Sci. USA* **75**:5195-5199.

Foster, G. A., Schultzberg, M., and Goldstein, M. (1984). Differential and independent manifestation within co-containing neurones of neuropeptide Y and tyrosine hydroxylase during ontogeny of the rat central nervous system. *Neurochem. Int.* **6**:761-771.

Foster, G. A., Schultzberg, M., Goldstein, M., and Hökfelt, T. (1985). Ontogeny of phenylethanolamine *N*-methyltransferase- and tyrosine hydroxylase-like immunoreactivity in presumptive adrenaline neurones of the foetal rat central nervous system. *J. Comp. Neurol.* **236**:348-381.

Freid, G., Terenius, L., Hökfelt, T., and Goldstein, M. (1985). Evidence for the differential localization of noradrenaline and neuropeptide Y (NPY) in neuronal storage vesicles isolated from rat vas deferens. *J. Neurosci.* **5**:450-458.

Garthwaite, J. (1982). Excitatory amino acid receptors and guanosine 3 ', 5 ' cyclic monophosphate in incubated slices of immature and adult rat cerebellum. *Neuroscience* **7**:2491-2497.

Gilbert, R. F. T., and Emson, P. C. (1979). Substance P in rat CNS and duodenum during development. *Brain Res.* **171**:166-170.

Gilman, A. G. (1987). G Proteins: Transducers of receptor generated signals. *Annu. Rev. Biochem.* **56**:615-649.

Goedert, M., Hunt, S. P., Mantyh, P. W., and Emson, P. C. (1985). The ontogenetic development of neurotensin-like immunoreactivity and neurotensin receptors in the cat striatum. *Dev. Brain Res.* **20**:127-131.

Goelet, P., Castellucci, V. F., Schacher, S., and Kandel, E. R. (1986). The long and the short of long-term memory—A molelcular framework. *Nature* 419-422.

Goto, M., and Sano, Y. (1984). Ontogenesis of the central serotonin neuron system of the rat—An immunohistochemical study. *Neurosci. Res.* **1**: 3-18.

Gray, E. G. (1969). Electron microscopy of excitable and inhibitory synapses: A brief review. *Prog. Brain Res.* **31**:141-155.

Greenberg, S. M., Castellucci, V. F., Bayley, H., and Schwartz, J. H. (1987). A molecular mechanism for long-term sensitization in *Aplysia*. *Nature* **329**:62-65.

Greengard, P. (1976). Possible role for cyclic nucleotides and phosphorylated membrane proteins in postsynaptic actions of neurotransmitters. *Nature* **260**:101-108.

Greengard, P. (1978a). *Cyclic Nucleotides, Phosphorylated Proteins and Neuronal Function.* New York, Raven Press.

Greengard, P. (1978b). Phosphorylated proteins as physiological effectors. *Science* **199**:146-152.

Greengard, P. (1987). Neuronal phosphoproteins—Mediators of signal transduction. *Mol. Neurobiol.* **1**:81-119.

Harlan, R. E. (1988). Regulation of neuropeptide gene expression by steroid hormones. *Mol. Neurosci.* **2**:183-200.

Hasselbalch, K. A. (1912). Ein Beitrag zur Respiration Physiolgia der Graviditat. *Skand. Arch. Physiol.* **27**:1-12.

Hasselbalch, K. A., and Gammeltoft, S. A. (1915). Die neuralitatisie gravideson des gravidesn organismus. *Biochem. Z.* **68**:206-264.

Hayakawa, Y., Obata, K. I., Itoh, N., Yanaihara, N., and Okamoto, H. (1984). Cyclic AMP regulation of provasoactive intestinal polypeptide/PHM-27 synthesis in human neuroblastoma cells. *J. Biol. Chem.* **259**: 9207-9211.

Hemmings, H., Narin, A. C., McGuiness, T. L., Huganir, R., and Greengard, P. (1989). Role of protein phosphorylation in neuronal signal transduction. *FASEB J.* **3**:1583-1592.

Heuser, J. E., and Reese, T. S. (1973). Evidence for the recycling of synaptic vesicle membrane during transmitter release at the frog neuromuscular junction. *J. Cell Biol.* **57**:315-344.

Hökfelt, T., Holets, V. R. Staines, W., Meister, B., Melander, T., Schalling, M., Schultzberg, M., Freedman, J., Björklund, A., Olson, L., Lindh, B., Elfvin, L. G., Lundberg, J., Samuelsson, B., Terenius, L., Post, C., Everitt, B., and Goldstein, M. (1986). Coexistence of neuronal messengers—An overview. *Prog. Brain Res.* **68**:33-70.

Hökfelt, T., Johansson, O., and Goldstein, M. (1981). Chemical anatomy of the brain. *Science* **225**:477-481.

Hökfelt, T., Ljungdahl, A., Steinbusch, H., Verhofstad, A., Nilsson, G., Brodin, E., Pernow, P., and Goldstein, M. (1978). Immunohistochemical evidence of substance P-like immunoreactivity in some 5-hydroxytryptamine-containing neurons in the rat central nervous system. *Neuroscience* **3**:517-538.

Hökfelt, T., Meister, B., Melander, T., Schalling, M., Staines, W., Millhorn, D., Seroogy, K., Tsuruo, Y., Holets, V., Ceccatelli, S., Villar, M., Ju, G., Freedman, J., Olson, L., Lindh, B., Bartfai, T., Fisone, C., Le Greves, P., Terenius, L., Post, C., Mollenholt, P., Dean, J., and Goldstein, M. (1988). Coexistence of multiple neuronal messengers: New aspects on chemical transmission. In *Fidia Foundation Neuroscience Award Lectures.* Vol 2. Edited by E. Costas. New York, Raven Press.

Hökfelt, T., Millhorn, D. E., Seroogy, K., Tsuruo, Y., Ceccatelli, S., Lindh, B., Meister, B., Melander, T., Schalling, M., and Terenius, L.

(1987). Coexistence of peptides eith classical neurotransmitters. *Experentia* **43**:768-779.

Hökfelt, T., Terinius, L., Kuypers, H. G. J. M., and Dann, O. (1979). Evidence for enkephalin immunoreactive neurons in the medulla oblongata projecting to the spinal cord. *Neurosci. Lett.* **14**:55-60.

Holtman, J. R., Norman, W. P., Skirboll, L., Dretchen, K. L., Cuello, C. Visser, T. J., Hökfelt, T., and Gillis, R. A. (1984). Evidence for 5-hydroxytryptamine, substance P and thyrotropin-releasing hormone in neurons innervating the phrenic motor nucleus. *J. Neurosci.* **4**:1064-1071.

Holzer, P., Emson, P. C., Holzbauer, M., Iversen, L. L., and Sharman, D. F. (1981). Tissue distribution of substance P in the domestic pig during postnatal development. *Dev. Rrain Res.* **1**:455-459.

Hubbard, J. I. (1970). Mechanisms of transmitter release. *Prog. Biophys. Md. Biol.* **21**:33-124.

Hubbard, J. I., and Kwanbunbumpen, S. (1968). Evidence for the vesicle hypothesis. *J. Physiol. (London)* **194**:407-420.

Iverfeldt, K., Serfözö, P., Diaz Arnesto, L., and Bartfai, T. (1989). Differential release of coexisting neurotransmitters: Frequency of the influx of substance P, thyrotropin releasing hormone and [3H] serotonin from tissue slices of rat ventral spinal cord. *Acta Physiol. Scand.* **137**: 63-71.

Jaeger, C. B. (1986). Aromatic L-amino acid decarboxylase in the rat brain: Immunocytochemical localization during prenatal development. *Neuroscience* **18**:121-150.

Jenkinson, D. H. (1957). The nature of the antagonism between calcium and magnesium ions at the neuromuscular junction. *J. Physiol. (London)* **138**:434-444.

Jirikowski, G. F., Merchenthaler, I., Rieger, G. E., and Stumpf, W. E. (1986). Estradiol target sites immunoreactive for β-endorphin in the arcuate nucleus of rat and mouse hypothalamus. *Neurosci. Lett.* **65**: 121-126.

Johansson, O., Hökfelt, T., and Elde, R. (1984). Immunohistochemical distribution of somatostatin-like immunoreactivity in the central nervous system of the adult rat. *Neuroscience* **13**:265-339.

Johansson, O., Hökfelt, T., Pernow, B., Jeffcoate, S. L., White, N., Steinbusch, H. W. M., Verhofstaad, A. A. J., Emerson, P., and Spindel, E. (1981). Immunohistochemical support for three putative transmitter in one neuron: coexistence of 5-hydroxytryptamine-, substance P- and thyrotropin releasing hormone-like immunoreactivity in medullary neurons projectiang to the spinal cord. *Neuroscience* **6**:1857-1881.

Jones, S. F., and Kwanbunbumpen, S. (1970). The effects of nerve stimulation and hemicholinium on synaptic vesicles at the mammalian neuromuscular junction. *J. Physiol.* **215**:31-50.

Johnson, J. W. and Ascher, P. (1987). Glycine potentiates the NMDA receptor in cultured mouse brain neurons. *Nature* 325:529-531.

Johnston, G. A. R., and Davies, L. P. (1974). Postnatal changes in high-affinity uptake of glycine and GABA in the rat central nervous system. *J. Neurochem.* 22:101-105.

Julius, D., McDermott, A. B., Axel, R., and Jessel, T. M. (1988). Molecular characterization of a functional cDNA encoding the serotonin lc receptor. *Science* 241:558-564.

Kalia, M., Fuxe, K., Hökfelt, T., Johansson, O., Lang, R., Ganten, D., Cuello, C., and Terenius, L. (1984). Distribution of neuropeptide immunoreactive nerve terminals within the subnuclei of the nucleus of the tractus solitarius of the rat. *J. Comp. Neurol.* 222:409-444.

Karlin, A., Holtzman, E., Yodh, N., Lobel, P., Wall, J., and Hainfeld, J. (1984). Arrangement of the subunits of the acetylcholine receptor of *Torpedo californica. J. Comp. Neurol.* 222:409-444.

Katz, B. (1969). *The Release of Transmitter Substance.* Liverpool, Liverpool University Press.

Katz, B., and Miledi, R. (1967a). The timing of calcium action during neuromuscular transmission. *J. Physiol. (London)* 189:535-544.

Katz, B., and Miledi, R. (1967b). A study of synaptic transmission in the absence of nerve impulses. *J. Physiol. (London)* 407-436.

Kimura, H., Hayashi, F., Yoshida, A., Watanabe, S., Hashizume, I., and Honda, Y. (1984). Augmentation of CO_2 drives by chlormadinone acetate, a synthetic progesterone. *J. Appl. Physiol.* 56:1627-1632.

Kobilka, B. K., Frielle, T., Collins, S., Yang-Feng, T., Kobilka, T. S., Francke, U., Lefkowitz, R. J., and Caron, M. G. (1987a). An intronless gene encoding a potential member of the family of receptors coupled to guanine nucleotide regulatory proteins. *Nature* 329:75-79.

Kobilka, B. K., Matsui, H., Kobilka, T. S., Yang-Feng, T. L., Francke, Y., Caron. M. G., Lefkowitz, R. J., and Regan, J. W. (1987b). Cloning, sequencing, and expression of the gene coding for the human platlet alpha 2-adrenergic receptor. *Science* 238:650-656.

Kotani, H., Hoshimaru, M., Nawa, H., and Nakanishi, S. (1986). Structure and gene organization of bovine neuromedin K precursor. *Proc. Natl. Acad. Sci. USA* 83:7074-7078.

Krause, J. E., Chirgwin, J. M., Carter, M. S., Xu, Z. S., and Hershey, A. D. (1987). Three rat preprotachykinin mRNAs encode the neuropeptides substance P and neurokinin A. *Proc. Natl. Acad. Sci. USA* 84: 881-885.

Krieger, D. T., Brownstein, M. J., and Martin, J. B. (Eds.) (1983). *Brain Peptides.* New York, Wiley.

Kubo, T., Fukuda, K., Mikami, A., Maeda, A., Takahashi, H., Mishina, M., Haga, T., Ichigama, A., Kangawa, K., Kojima, M., Matsuo, H., Hirose, T., and Numa, S. (1986). Cloning, sequence and expression of complementary DAN encoding the muscarinic acetylcholine receptor. *Nature* 323:411-416.

Kuno, M. (1971). Quantum aspects of central and ganglionic synaptic transmission in vertebrates. *Physiol. Rev.* 51:647-c678.

Lauder, J. M., Han, V. K. M., Henderson, P., Verdoorn, T., and Towle, A. C. (1986). Prenatal ontogeny of the GABAergic system in the rat brain: An immunocytochemical study. *Neuroscience* 19:465-493.

Laufer, R., and Changeux, J. P. (1989). Activity-dependent regulation of gene expression in muscle and neuronal cells. *Mol. Neurobiol.* 3:1-53.

Leonhardt, S., Herrick-Davis, K., and Titler, M. (1989). Detection of a novel serotonin receptor subtype, (5-HT$_{1E}$) in human brain: Interaction with a GTP-binding protein. *J. Neurochem.* 53:465-471.

Leslie, R. A. (1985). Neuroactive substances in the dorsal vagal complex of the medulla oblongata: Nucleus tractus solitarius, area postrema, and the dorsal motor nucleus of the vagus. *Neurochem. Int.* 7:191-212.

Lester, H. A. (1977). The response of acetylcholine. *Sci. Am.* 236:106-118.

Levitan, L., and Kaczmarek, L. K. (1987). Ion currents and ion channels: substrates for neuromodulation. In *Neuromodulation*. Edited by L. Kaczmarek and L. Levitan. New York, Oxford University Press, pp. 18-38.

Levitzki, A. (1988). From epinephrine to cyclic AMP. *Science* 241:800-806.

Lidov, H. G. W., and Mollover, M. E. (1982). Immunohistochemical study of the development of serotonergic neurons in the rat CNS. *Brain Res. Bull.* 9:559-604.

Liley, A. W. (1956). The effects of presynaptic polarization on the spontaneous activity at the mammalian neuromuscular junction. *J. Physiol.* 134:427-443.

Ljungdahl, A., Hökfelt, T., and Nilsson, G. (1978). Distribution of substance P-like immunoreactivity in the central nervous system of the rat. I. Cell bodies and nerve terminals. *Neuroscience* 3:861-943.

Llinas, R., McGuiness, T. L., Leonard, C. S., Sugimori, M., and Greengard, P. (1985). Intraterminal injection of synapsin I or calcium/calmodulin-dependent protein kinase II alters neurotransmitter release at the squid giant synapse. *Proc. Natl. Acad. Sci. USA* 82:3035-3039.

Lundberg, J., Fried, G., Fahrenkrug, J., Holmstedt, B., Hökfelt, T., Langercratz, H., Lundgren, G., and Anggard, A. (1981). Subcellular fractionation of cat submandibular gland: Comparative studies on the distribution of acetylcholine and vasoactive intestinal polypeptide. *Neuroscience* 6:1001-1010.

Lundberg, J., Hellund, B., Anggard, A., Fahrenkrug, J., Hökfelt, T., Tatemoto, T., and Bartfai, T. (1982). Co-storage of peptides and classical transmitter in neurons. In *Systemic Role of Regulatory Peptides*. Edited by S. R. Bloom, J. M. Polak, and E. Lindenlaub. Stuttgart, Schattsauer Co., pp. 93-119.

Lundberg, J., and Hökfelt, T. (1983). Coexistence of peptides and classical neurotransmitters. *Trends Neurosci.* **6**:325-333.

Lynch, G., Muller, D., Seubert, P., and Larson, J. (1988). Long-term potentiation: Persisting problems and recent results. *Brain Res. Bull.* **21**: 363-372.

Maggio, J. E. (1988). Tachykinins. *Annu. Rev. Neurosci.* **11**:13-28.

Mamalaki, C., Stephenson, F. A., and Barnard, E. A. (1987). The $GABA_A$/-bensodiazepine receptor is a heterotetramer of homologous alpha and beta subunits. *EMBO J.* **6**:561-565.

McDermott, A. B., and Dale, N. (1987). Receptors, ion channels and synaptic potentials underlying the integrative actions of excitatory amino acids. *Trends Neurosci.* **7**:280-284.

McEwen, B. S., Davis, P., Parsons, B., and Pfaff, D. W. (1979). The brain target for steroid hormone action. *Annu. Rev. Neurosci.* **2**:65-112.

McEwen, B. S., Jones, K. J., and Pfaff, D. W. (1987). Hormonal control of sexual behavior in the female rat: Molecular, cellular and neurochemical studies. *Biol. Reprod.* **36**:37-45.

McNamara, M. C., and Lawson, E. E. (1983). Ontogeny of biogenic amines in respiratory nuclei of the rabbit brainstem. *Dev. Brain Res.* **7**: 181-185.

Milerad, J., Lagercrantz, H., and Lofgren, O. (1985). Alveolar hypoventilation treated with medroxyprogesterone. *Arch. Dis. Childhood* **60**: 150-155.

Millhorn, D. E. (1986). Stimulation of raphe (obscurus) nucleus causes long-term potentiation of phrenic nerve activity in cat. *J. Physiol.* **381**:169-179.

Millhorn, D. E., Bayliss, D. A., and Erickson, J. T. (1990a). Thyrotropin releasing hormone (TRH) innervation of dorsal vagal complex is from caudal raphe nuclei but not paraventricular nucleus: A combined in situ hybridization and retrograde labeling study. *FASEB J* **4**:A881.

Millhorn, D. E., Bayliss, D. A., Erickson, J. T., Gallman, E. A. Szymeczek, C. L., Czyzyk-Krzeska, M., and Dean, J. B. (1989a). Cellular and molecular mechanisms of chemical synaptic transmission. *Am. J. Physiol.: Lung Mol. Cell Physiol.* **3**.

Millhorn, D. E., Czyzyk-Krzeska, M. F., Bayliss, D. A., and Seroogy, K. B. (1990b). Gene expression for neuropeptides in the ganglia of the vagus (nodose) and glossopharyngeal (petrosal) nerves. In *Cardiorespiratory and Motor Coordination*. Edited by H. P. Koepchen. Berlin, Springer-Verlag. (in press).

Millhorn, D. E., and Eldridge, F. L. (1986). The role of the ventrolateral medulla in regulation of respiratory and cardiovascular systems. *J. Appl. Physiol.* **61**:1249-1263.

Millhorn, D. E., Eldridge, F. L., and Waldrop, T. G. (1980a). Prolonged stimulation of respiration by a new central neural mechanism. *Resp. Physiol.* **41**:87-103.

Millhorn, D. E., Eldridge, F. L., and Waldrop, T. G. (1980b). Prolonged stimulation of respiration by endogenous central serotonin. *Resp. Physiol.* **42**:171-103.

Millhorn, D. E., and Hökfelt, T. (1988). Chemical messengers and their coexistence in individual neurons. *News Physiol. Sci.* **3**:1-5.

Millhorn, D. E., Hökfelt, T., Seroogy, K., Oertel, W., Verhofstad, A. A. J., and Wu J-Y. (1987a). Immunohistochemical evidence for colocalization of gamma-aminobutyric acid and serotonin in neurons of the ventral medulla oblongata projecting to the spinal cord. *Brain Res.* **410**: 179-185.

Millhorn, D. E., Hökfelt, T., Seroogy, K., and Verhofstad, A. A. J. (1988). Extent of colocalization of serotonin and GABA in neurons of the ventral medulla oblongata in rat. *Brain Res.* **461**:169-428.

Millhorn, D. E., Hökfelt, T., Terenius, L., Buchan, A., and Brown, J. C. (1987b). Somatostatin- and enkephalin-like immunoreactivities are frequently colocalized in neurons in the caudal brain stem of rat. *Exp. Brain Res.* **67**:420-428.

Millhorn, D. E., Hökfelt, T., Verhofstad, A. A. J., and Terenius, L. (1989b). Individual cells in the raphe nuclei of the medulla oblongata in rat that contain immunoreactivities for both serotonin and enkephalin project to the spinal cord. *Exp. Brain Res.* **75**:536-542.

Mitchell, R., and Fleetwood-Walker, S. (1981). Substance P, but not TRH, modulates the 5-HT autoreceptor in ventral lumbar spinal cord. *Eur. J. Pharmacol.* **76**:119-120.

Montminy, M. R., and Bilezikjian, L. M. (1987). Binding of a nuclear protein to the cyclic-AMP response element of the somatostatin gene. *Nature* **328**:175-178.

Montminy, M. R., Sevarino, K. A., Wagner, J. A., Mandel, G., and Goodman, R. H. (1986). Identification of a cyclic-AMP response element within the rat somatostatin gene. *Proc. Natl. Acad. Sci. USA* **83**:6682-6686.

Nagamine, Y., and Reich, E. (1985). Gene expression and cAMP. *Proc. Natl. Acad. Sci. USA* **82**:4606-4610.

Nestler, E. J., and Greengard, P. (1984). *Protein Phosphorylation in the Nervous System*. New York, Wiley.

Nishizuka, Y. (1986). Studies and perspectives of protein kinase C. *Science* **233**:305-312.

Noda, M., Takahashi, H., Tanabe, T., Toyosato, M., Kikyotani, S., Furutani, Y., Hirose, T., Takashima, H., Inayama, S., Miyata, T., and Numa, S. (1983). Structural homology of *Torpedo californica* acetylcholine receptor subunits. *Nature* **302**:528-532.

Nowak, L., Bregestoviski, P., Ascher, P., Herbet, A., and Prochiantz, A. (1984). Magnesium gates glutamate-activated channels in mouse central nervous system. *Nature* **307**:462-465.

Numa, S., Noda, M., Takahashi, H., Tanabe, T., Toyosato, M., Furutani, Y., and Kikyotani, S. (1983). Molecular structure of the nicotinic acetylcholine receptor. *Cold Spring Harbor Symp.* **49**:9-25.

O'Malley, B. W., and Means, A. R. (1974). Female steroid hormones and target cell nuclei. *Science* **183**:610-620.

O'Malley, B. W., and Strott, C. A. (1973). The mechanism of action of progesterone. In *Handbook of Physiology, Endocrinology of Female Reproductive System.* Vol. II. Edited by R. O. Greep and E. B. Astuved. Washington, DC, Am. Physiol. Soc.

Otsuka, M., and Takahashi, T. (1977). Putative peptide neurotransmitters. *Am. Rev. Pharmacol. Toxicol.* **17**:425-439.

Palade, G. E. (1954). Electron microscope observations of interneuronal and neuromuscular synapses. *Anat. Rec.* **118**:335-336.

Palay, S. L. (1954). Electron microscope study of the cytoplasm of neurones. *Anat. Rec.* **118**:336, 1954.

Palay, S. L. (1956). Synapses in the central nervous system. *J. Biochem. Biophys. Cytol. Suppl.* **2**:193-201.

Palkovits, M. (1985). Distribution of neuroactive substances in the dorsal vagal complex of the medulla oblongata. *Neurochem. Int.* **7**:213-220.

Palkovits, M., and Brownstein, S. (1985). Distribution of neuropeptides in the central nervous system using giochemical micromethods. In *Handbook of Chemical Neuroanatomy.* Edited by A. Björklund and T. Hökfelt. Amsterdam, Elsevier.

Palmer, M. R., Miller, R. J., Olson, L., and Seiger, A. (1982). Prenatal ontogeny of neurons with enkephalin-like immunoreactivity in the rat central nervous system: An immunohistochemical mapping investigation. *Med. Biol.* **60**:61-88.

Pelletier, G., Steinbusch, H. W. M., and Verhofstad, A. A. J. (1981). Immunoreactive substance P and serotonin present in the same dense-core vesicles. *Nature* **293**:71-72.

Pernow, B. (1983). Substance P. *Pharmacol. Rev.* **35**:85-141.

Peroutka, S. J. (1988). 5-Hydroxytryptamine receptor subtypes. *Annu. Rev. Neurosci.* **11**:45-60.

Pfaff, D. W., and Keiner, M. (1973). Atlas of estradiol-concentrating cells in the central nervous system of the female rat. *J. Comp. Neurol.* **151**:121-157.

Phillis, J. W., and Wu, P. H. (1981). The role of adenosine and its nucleotides in central synaptic transmission. *Prog. Neurobiol.* **16**:187-239.

Pickel, V. M., Beckley, S. C., Sumal, K. K., Joh, T. H., Reis, D. J., and Miller, R. J. (1981). Light and electron microscopic localization of enkephalin and tyrosine hydroxylase in neostriatum of fetal and adult rat brain. *Acta Histochem.* **105**(Suppl. 24):97-105.

Purves, D., and Lichtman, J. W. (1985). *Principles of Neural Development.* Sunderland, MA, Sinauer Associates.

Raftery, M. A., Hunkapillar, M. W., Strader, C. D., and Wood, L. E. (1980). Acetylcholine receptor: Complex of homologous subunits. *Science* **208**: 1454-1457.

Ramon y Cajal, S. (1908). Neuron theory or reticular theory. *Arch. Fisol.* **5**:1908.

Riabowol, K. T., Fink, J. S., Gilman, M. Z., Walsh, D. A., Goodman, R. H., and Feramisco, J. R. (1988). The catalytic subunit of cAMP-dependent protein kinase induces expression of genes containing cAMP-responsive enhancer elements. *Nature* **336**:83-86.

Roberts, J. L., Seeburg, P. H., Shine, J., Herbert, E., Baxter, J. D., and Goodman, H. M. (1979). Corticotropin and beta-endorphin: Construction and analysis of recombinant DNA complementary to mRNa for the common precursor. *Proc. Natl. Acad. Sci. USA* **76**:2153-2157.

Rories, C., and Spelsberg, T. C. (1989). Ovarian steroid action on gene expression: Mechanisms and models. *Annu. Rev. Physiol.* **51**:653-681.

Rosen, H., Douglass, J., and Herbert, E. (1984). Isolation and characterization of the rat proenkephalin gene. *J. Biol. Chem.* **259**:14309-14313.

Rothfeld, J. M., Hejtmancik, J. F., and Pfaff, D. W. (1987). Quantitation of LHRH mRNA within the female rat forebrain following estrogen treatment. *Anat. Rec.* **218**:117a.

Sar, M. (1984). Estradiol is concentrated in tyrosine hydroxylase containing neurons of the hypothalamus. *Science* **223**:938-940.

Sar, M. (1988). Distribution of progestin-concentrating cells in rat brain: colocalization of [^{3}H] ORG.2058, a synthetic progestin, and antibodies to tyrosine hydroxylase in hypothalamus by combined autoradiography and immunocytochemistry. *Endocrinology* **123**:1110-1118.

Sar, M., and Stumpf, W. E. (1973). Neurons of the hypothalamus concentrate [^{3}H]progesterone or its metabolites. *Science* **182**:1266-1271.

Sar, M., and Stumpf, W. E. (1980). Simultaneous localization of [^{3}H] estradiol and neurophysin I or arginine vasopressin in hypothalamic neurons demonstrated by a combined technique of dry-mount autoradiography and immunohistochemistry. *Neurosci. Lett.* **17**:179-184.

Sar, M., and Stumpf, W. E. (1983). Simultaneous localization of steroid hormones and neuropeptides in the brain by combined autoradiography and immunocytochemistry. *Methods Enzymol.* **103**:631-638.

Schacher, S., Castellucci, V. F., and Kandel, E. R. (1988). cAMP evokes long-term facilitation in *Aplysia* sensory neurons that requires new protein synthesis. *Science* **240**:1667-1669.

Scholtz, K., and Byrne, J. H. (1988). Intracellular injection of cAMP induces a long-term reduction of neuronal K$^+$ currents. *Science* **240**:1664-1666.

Senba, E., Shiosaka, S., Hara, Y., Inagaki, S., Kawai, Y., Takatsuki, K., Sakanaka, M., Iida, H., Takagi, H., Minagawa, H., and Tohyama, M. (1982a). Ontogeny of the leucine-enkephalin neuron system of the rat: Immunohistochemical analysis. I. Lower brainstem. *J. Comp. Neurol.* **205**:341-359.

Senba, E., Shiosaka, S., Hara, Y., Inagaki, S., Sakanaka, M., Takatsuzi, K., Kawai, Y., and Tohyama, M. (1982b). Ontogeny of peptidergic system in the rat spinal cord: Immunohistochemical analysis. *J. Comp. Neurol.* **208**:54-66.

Seroogy, K. B., Bayliss, D. B., Szymeczek, C. L., Hökfelt, T., and Millhorn, D. E. Transient expression of somatostatin mRNA and peptide in the hypoglossal nucleus of the neonatal rat. (in preparation).

Sevilla, N., Steer, M. L., and Levetzki, A. (1976). Synergistic activation of anenylate cyclase by guanylyl imidophosphate and epinephrine. *Biochemistry* **15**:3493-3499.

Sherrington, C. S. (1906). *The Integrative Action of the Nervous System.* New York, Scribners.

Shiosaka, S., Takatsuki, K., Sakanaka, M., Inagaki, S., Takagi, H., Senba, E., Kawai, Y., and Tohyama, M. (1981). Ontogeny of somatostatin-containing neuron system of the rat: Immunohistochemical observations. I. Lower brainstem. *J. Comp. Neurol.,* **203**:173-188.

Shivers, B. D., Harlan, R. E., and Pfaff, D. W. (1984). Hypothalamic immunoreactive prolactin neurons are targets for estrogenic action. *Soc. Neurosci.* **10**:156 (Abstract).

Shulman, H. (1988). The multifunctional Ca^{2+}/calmodulin-dependent protein kinase. In *Advances in Second Messenger and Phosphoprotein Research.* Edited by P. Greengard and G. A. Robinson. New York, Raven. Vol. 22.

Sibley, D. R., Benovic, J. L., Caron, M. G., and Lefkowitz, R. J. (1987). Regulation of transmembrane signaling by receptor phosphorylation. *Cell* **48**:913-922.

Siegel, G., Albers, R. W., Agranoff, B. W., and Katzman, R. (Eds.) (1981). *Basic Neurochemistry*, 3rd ed. Boston, Little, Brown.

Skatrud, J. B., Dempsey, J. A., Bhansali, P., and Irvin, C. (1980). Determinants of chronic carbon dioxide retention and its correction in humans. *J. Clin. Invest.* **65**:813-821.

Skatrud, J. B., Dempsey, J. A., and Kaiser, D. G. (1978). Ventilatory response to medroxyprogesterone acetate in normal subjects: Time course and mechanism. *J. Appl. Physiol.* **44**:939-944.

Skatrud, J. B., Dempsey, J. A., Iber, C., and Berssenbrugge, A. (1981). Correction of CO_2 retention during sleep in patients with chronic obstructive pulmonary disease. *Am. Rev. Respir. Dis.* **124**:260-268.

Sladeczek, F., Pin, J. P., Recasens, M., Bockaert, J., and Weiss, S. (1985). Glutamate stimulates inositol phosphate formation in striatal neurons. *Nature* **317**:717-719.

Snyder, S. (1980). Brain peptides as neurotransmitters. *Science* **209**:976-983.

Spelsberg, T. C., Rories, C., Rejman, J. J., Goldberger, A., Fink, K., Lau, C. K., Colvard, D. S., and Wiseman, G. (1989). Steroid action on gene expression: Possible roles of regulatory genes and nuclear acceptor sites. *Biol. Reprod.* **40**:54-69.

Stanton, P. K., and Sarvey, J. M. (1984). Blockade of long-term potentiation in rat hippocampal CA1 region by inhibitors of protein synthesis. *J. Neurosci.* **12**:3080-3088.

Strader, C. D., Sigal, I. S., Register, R. B., Candelore, M. R., Rands, E., and Dixon, R. A. (1987). Identification of residues required for ligand binding to the beta-adrenergic receptor. *Proc. Natl. Acad. Sci. USA* **84**:4384-4388.

Sutton, F. D., Zwillich, C. W., Creogh, E. E., Peirson, D. J., and Weil, J. V. (1974). *Ann. Intern. Med.* **83**:476-479.

Szymeczek, C. L., Bayliss, D. A., and Millhorn, D. E. (1990). Developmental changes in thyrotropin releasing hormone mRNA levels in rat raphe medullary neurons. *FASEB J.* **4**:A1109.

Thor, K. B., and Helke, C. J. (1987). Serotonin- and substance P-containing projections to the nucleus tractus solitarii of the rat. *J. Comp. Neurol.* **265**:275-293.

Tyler, J. M. (1960). The effect of progesterone on the respiration of patients with emphysema and hypercapnia. *J. Clin. Invest.* **39**:34-41.

Wallace, J. A., and Lauder, J. M. (1983). Development of the serotonergic system in rat embryo: An immunohistochemical study. *Brain Res. Bull.* **10**:459-479.

Warembourg, M., Logeat, F., and Milgrom, E. (1986). Immunocytochemical localization of progesterone receptor in the guinea pig central nervous system. *Brain Res.* **384**:121-131.

Warembourg, M., Jolivet, A., and Milgrom, E. (1989). Immunocytochemical evidence of the presence of estrogen and progesterone receptors in the same neurons of the guinea pig hypothalamus and preoptic area, *Brain Res.* **480**:1-15.

Whitaker, V. P. (1971). Subcellular localization of neurotransmitters. In *Advances in Cytopharmacology*. Edited by F. Clementi and B. Ceccarelli. New York, Raven Press.

Wilcox, J. N., and Roberts, J. L. (1985). Estrogen decreases rat hypothalamic pro-opiomelanocortin messenger ribonucleic acid levels. *Endocrinology* **117**:2392-2396.

Yamamoto, K. R. (1985). Steroid receptor regulated transcription of specific genes and gene networks. *Annu. Rev. Genet.* **19**:209-252.

Yamamoto, K. R., Gonzalez, G. A., Biggs, W. H., III, and Montminy, M. R. (1988). Phosphorylation-induced binding and transcriptional efficacy of nuclear factor CREB. *Nature* **334**:494-498.

Yoshikawa, K., Williams, C., and Sabol, S. L. (1984). Rat brain preproenkephalin mRNA. *J. Biol. Chem.* **259**:14301-14308.

2

Neuronal Signaling
Pathways and Protein Kinases

HARRY LEVINE III

Glaxo Research Laboratories
Research Triangle Park, North Carolina

NAJI SAHYOUN

Burroughs Wellcome Company
Research Triangle Park, North Carolina

I. Introduction: Signaling Pathways

Neuronal communication is largely mediated by a network of intracellular signaling pathways. Understanding the structure, function, and assembly of the molecular components of these pathways represents a major aspect of the multidisciplinary study of neurons and neuronal networks. Intracellular signaling pathways regulate a broad spectrum of overlapping neuronal functions, including neurotransmitter release, neurotransmitter action, ionic fluxes, neurite growth, gene expression, and long-term as well as short-term plasticity. This chapter will focus on general concepts emerging from investigations of the assembly and operation of signaling pathways and will employ these concepts in the illustrative discussion of two prominent protein kinases that participate in Ca^{2+}-dependent neuronal signaling. More detailed discussion of other components of signaling systems exceeds the scope of this chapter.

The diversity and versatility of intracellular signaling follow from the use of multicomponent pathways as well as multiple pathways. Stimulus-response coupling involves a minimum of three elements: a receptor, which is sensitive to various extracellular signals such as photons, hormones, neurotransmitters, growth factors, or cell adhesion molecules; an effector, which generates the intracellular response, such as nucleotide cyclases, phosphodiesterases, ion channels, or phospholipases; and a coupler, which mediates the transfer of molecular information between the former two components.

Receptor-effector systems display a variety of modes of structural organization. The three basic components can occur variously as separate molecular entities, as subunits of the same oligomeric complex, or as domains within the same polypeptide. Receptor-linked GTP-binding protein (G-protein) systems (Dohlman et al., 1987; Axelrod et al., 1988; Freissmuth et al., 1989), ligand-gated ion channels (Cotman et al., 1988; Stephenson, 1988; Sternweis and Pang, 1990), and tyrosine kinase-containing growth factor receptors (Carpenter, 1987; Yarden and Ullrich, 1988) exemplify the different organizational schemes, respectively. Receptors often contain additional subunits which serve in a regulatory or "accessory" capacity, a condition best illustrated by the antigen receptor of T lymphocytes (Weissman et al., 1989).

Multiple signaling pathways can be arranged serially in the manner of cascades, or in parallel, and may converge or diverge at different molecular levels. They are subject to regulation by feedback and feedforward reactions analogous to those described for metabolic pathways, and by "cross-talk" (Yoshimasa et al., 1987) among different pathways. The molecular composition and regulation of a signaling pathway can lead to amplification, attenuation, and/or oscillation of a signal.

Plasticity in signaling mechanisms derives from changes in the concentration and/or activity of the constituent molecules. Activity changes usually involve posttranslational modifications by the covalent addition of phosphate groups, fatty acids, carboxymethyl groups, or isoprenyl units. Changes in component concentration derive from alterations in the relative rates of synthesis or degradation of specific polypeptides either by posttranslational mechanisms or from alterations in gene expression.

Subcellular compartmentalization of components of signaling pathways is an important mechanism which allows propagation of signals throughout the cell and the spatiotemporal coordination of complex cellular responses (Rasmussen et al., 1986; Alkon and Rasmussen, 1988). A number of regulatory components undergo activity-dependent translocations between different subcellular compartments, including the plasma membrane, cytoplasm, and nucleus.

The discovery of cyclic AMP more than 30 years ago and the elucidation of the "second messenger" principle proved prototypic for the elaboration of more complex signaling pathways. The advent of powerful molecular tools has proven indispensable to the unraveling of this complexity. Their scope encompasses the use of affinity purification methods, monoclonal antibodies, protein microsequencing techniques, synthetic peptides, biological toxins, patch-clamp electrophysiological techniques, imaging techniques, cation-sensitive fluorescent indicators, and the multitude of advances in recombinant DNA technology which has led to a veritable explosion of information concerning the domain structure of relevant polypeptides, structure/function correspondance, and the relationship of various polypeptides to one another.

II. Families of Signaling Molecules

An improved understanding of transmembrane signaling mechanisms has resulted from application of the preceding methods. Receptors, G-proteins (couplers), and effectors occur as families and superfamilies which share specific structural and functional attributes. Moreover, the ability of receptor/coupler/effector molecules to bind their respective ligands and to interact with each other is determined by specific domains whose occurrence can be employed to predict the nature of the ensuing interactions. For example, study of the photon receptor rhodopsin and beta-adrenergic receptors (Oprian et al., 1987; Dixon et al., 1988) provided seminal information which could be applied to other receptors linked to G-proteins, including muscarinic cholinergic, serotonergic, and dopaminergic subtypes (Peroutka, 1989; Schofield et al., 1990). The hallmark of these receptor polypeptides is that they contain seven hydrophobic alpha-helical segments, which anchor the receptor in the membrane by passing through the membrane lipid bilayer, connected by loops on the cytoplasmic and extracellular sides of the membrane. Mutational analysis of beta-adrenergic receptors suggests that the ligand-binding site is generated by interactions among the transmembrane helices within or close to the membrane bilayer, and that the amino and carboxyl regions of the third cytoplasmic loop are important for G-protein interaction (Dixon et al., 1988). The amino acid sequence of the receptor reveals the presence of putative consensus phosphorylation sites for the cyclic AMP-dependent protein kinase and the beta-adrenergic receptor kinase, βARK, which are involved in receptor inactivation processes and in down-regulation of the receptor by internalization (Sibley et al., 1987; Klein et al., 1989). Variations in amino acids at particular positions in the protein sequence generate receptor subtypes that are pharmacologically distinct with respect to ligands and can interact with different species of G-protein.

   GTP-binding proteins themselves have been the focus of concentrated study since the original observation that GTP is required for hormonal activation of adenylate cyclase (Rodbell et al., 1971). The ability to bind GTP and to hydrolyze the γ-phosphate residue from GTP resides in the α-subunit of a heterotrimeric complex containing β and γ subunits as well. The existence of several distinct genes and of differential mRNA splice products underlies the multiplicity of α-subunit types, allowing G-proteins to couple to a variety of receptors, on one hand, and to different effectors, on the other. A number of physiological functions not known to be classically receptor-linked may also be coupled through G-proteins. These include processes such as protein sorting in the Golgi apparatus (Melancon et al., 1987) and exocytosis and thus possibly neurotransmitter release (Cockcroft et al., 1987). Amino acid sequence and mutational analysis have identified specific domains within the α-subunit mediating binding and hydrolysis of GTP, ADP-ribosylation by bacterial toxins, and binding to receptors and to effector molecules (Gilman, 1987; Casey and Gilman, 1988; Freissmuth et al., 1989; Ross, 1989).

Whereas the molecular weight of the various α-subunits is approximately 40 kDa, another group of small-molecular-weight (20-26 kDa) GTP-binding proteins has recently been described. Much of the interest in this area has been triggered by the discovery that the protooncogene product c-ras is a GTP-binding protein of this class. Other members of this family also appear to be protooncogene products (Burgoyne, 1989) whose normal function in mammalian cells remains elusive. Recent studies reveal that some of the small-molecular-weight G-proteins interact with a class of GTPase activating proteins called "GAP's" which may propagate and transduce the G-protein signal within the cell (Hall, 1990).

The analysis of certain receptors and their corresponding G-proteins in terms of submolecular domains and molecular families applies equally well to other components of transmembrane signaling systems. These include G-protein-linked effectors such as adenylate cyclase (Krupinski et al., 1990; Livingstone et al., 1984), phospholipase C (Rhee et al., 1989), and K^+ (Blatz and Magleby, 1987) and Ca^{2-} channels (Ellis et al., 1988; Tsien et al., 1988). Non-G-protein-linked systems follow the same pattern. Multisubunit, voltage-gated ion channels for Na^+, K^+, and Ca^{2+} (Jan and Jan, 1989; Miller, 1989; Stühmer et al., 1989; Unwin, 1989; Gordon et al., 1987) and/or ligand-gated ion channels which are regulated by kainate (Stevens, 1989), NMDA (Cotman et al., 1988), glycine (Grenningloh et al., 1990), GABA (Schofield, 1989; Stephenson, 1988), and nicotinic cholinergic agonists (Nef et al., 1988) contain receptor and effector elements as subunits within the same macromolecular complex.

Even tighter coupling between receptor and effector functions is achieved when they are present as domains within the same polypeptide chain. This condition characterizes certain growth factor receptors that feature a ligand-binding site and a protein tyrosine kinase domain within the same molecule

(Yarden and Ullrich, 1988; Carpenter, 1987; Hunter and Cooper, 1985). Growth factor receptors comprising a single polypeptide chain such as the EGF receptor are classified as type I, whereas type II receptors such as insulin and IGF-I contain two tyrosine kinase-bearing β subunits and two ligand-binding α subunits linked to each other and to the β subunits by extracellular disulfide bonds. A class of transmembrane phosphotyrosine phosphatases also contain a potential ligand-binding extracellular binding domain and an intracellular effector domain on the same polypeptide chain (Tonks and Charbonneau, 1989). Another example of the same scheme is the atrial natriuretic factor (ANF) receptor with its associated intracellular guanylate cyclase (Chinkers et al., 1989; Schulz et al., 1989).

Neurons as well as other cells also carry surface receptors responsible for specific cell-cell and cell-substratum adhesion. These interactions are particularly important both for cellular migration along defined surfaces and for the assembly of cellular networks. Receptors for cell-cell binding are exemplified by Ca^{2+}-dependent and Ca^{2+}-independent cell adhesion molecules (CAM's) (Edelman, 1985) while cell-substratum binding is partly mediated by α/β multisubunit integrins which bind to extracellular matrix proteins such as laminin and fibronectin (Buck and Horwitz, 1987). Both CAM's and integrins can be classified according to domains and into families. Recent evidence further suggests that CAM's and integrins serve as more than passive mechanical receptors but may also behave as transmembrane signaling elements which elicit intracellular responses.

The preceding abridged discussion indicates that signaling mechanisms can regulate a variety of prominent neuronal functions pertaining to ion channels, neurotransmitters, neuromodulators, hormones, growth factors, and cell adhesion molecules. While multiple transmembrane signaling pathways exist to accommodate a diversity of neuronal functions, analysis of the component signaling elements has revealed a unity that simplifies the assembly and operation of higher-order signaling systems. This "combinatorial" principle strives to generate maximal diversity and specificity by combining a limited number of shared building blocks into unique permutations. The concept appears to be equally applicable to other biological mechanisms of similar complexity, such as the regulation of gene expression (Herschman, 1989). Thus, unique signaling pathways can share one or more component polypeptides, and unique polypeptides can share one or more homologous domains. The manner in which the various polypeptides and domains are combined allows for different types of interactions among signaling elements. For example, a signaling system may contain three independent proteins, such as a β-adrenergic receptor, a Gs G-protein, and adenylate cyclase; alternatively, it may be organized into a multisubunit structure such as a ligand-gated channel; or it may be arranged within a single polypeptide chain such as the EGF receptor or the ANF-sensitive guanylate cyclase. Each type of molecular arrangement results in different types of regulatory interactions and different kinetics for signal propagation. Thus, the specific molecular organization of a signaling system as well as its combinatorial properties can provide the biochemical basis for the observed physiological and pharmacological characteristics of signaling pathways.

The principal function of several transmembrane signaling pathways is to generate an intracellular second messenger which can, in turn, regulate the activity of target intracellular signaling systems. The discovery of cyclic AMP and cyclic GMP and the concomitant formulation of the second-messenger hypothesis quickly led to the suggestion of a similar role for Ca^{2+} and to a search for alternative second messengers. This search was richly rewarded by the discovery of second-messenger roles for IP3 and diacylglycerol (Berridge, 1987, 1986; Nishizuka, 1984) and a possible role for other inositol phosphates and arachidonic acid and its metabolites.

Second-messenger-sensitive targets play a major role in determining the cellular response to extracellular stimuli. Two primary targets for second messengers appear to be ion channels and protein kinases. Purification, cloning, and reconstitution and functional analysis of the IP3 glycoprotein receptor of the endoplasmic reticulum into lipid vesicles reveal that it is an IP3-activated Ca^{2+} channel (Furuichi et al., 1989). In the rod visual photoreceptor, cyclic GMP acts by binding directly to and opening a cation channel, which has been shown to contain both a cyclic GMP binding site and an ion channel structure in the same polypeptide chain (Kaupp et al., 1989). Lipoxygenase metabolites of arachidonic acid such as 12-HPETE can directly activate K^+ channels in *Aplysia* sensory neurons (Buttner et al., 1989). However, protein kinases remain the major venue for the action of cyclic nucleotides, Ca^{2+}, and diacylglycerol. The isolation of a cyclic AMP-dependent protein kinase in 1968 (Walsh et al., 1968) was followed almost a decade later by the discovery of two major protein kinases that have proved to be especially relevant to neuronal signaling, namely, a Ca^{2+}/phospholipid-dependent protein kinase (protein kinase C) and Ca^{2+}/calmodulin-dependent protein kinase II (CaMKinase II). Because of the prominence and importance of Ca^{2+} in neuronal signaling, the abundance of these two protein kinases, and their impact on neuronal function, they will provide the focus for the remainder of this chapter. The conclusions arrived at from the study of these two enzymes serve to highlight several of the general features of signaling pathways discussed above.

III. Protein Kinase C

A protein kinase activity phosphorylating a lysine-rich histone fraction (H1) and generated by proteolysis with calpain, a calcium-dependent neutral protease, was described by Nishizuka and co-workers (Inoue et al., 1977). The proteolytic cleavage was subsequently found to have removed an inhibitory regulatory domain of the enzyme resulting in a chronically activated enzyme (Kishimoto et al., 1989). It was also found that this same kinase was activated by Ca^{2+} and acidic phospholipids, while further activation could be achieved with diglycerides, in particular those bearing an unsaturated fatty acid in the 2-position (Takai et al., 1979; Go et al., 1987). These properties immediately marked this protein kinase as a transducer integrating two signals, Ca^{2+} and diglycerides, themselves a product of hydrolytic cleavage of a phosphatidylinositol-4,5-bisphosphate pool responsive to hormonal stimuli. The discovery that a class of plant alkaloid tumor-promoting agents, phorbol esters, would substitute for diacylglycerol and permanently activate the enzyme provided immediate pharmacological relevance (Kikkawa et al., 1983; Niedel et al., 1983; Castagna et al., 1982).

A. Molecular Properties

Protein kinase C purified from brain yields a polypeptide preparation with a molecular weight around 80 kDa. However, this kinase is now known to be a family of proteins consisting of at least seven subspecies or isozymes (Kikkawa et al., 1989). Cloning technology has elucidated the structure and genetics of these highly similar, yet distinct enzymes. Four of the subtypes (I, IIa, IIb, III) have been isolated from brain by hydroxylapatite chromatography (Huang et al., 1986b) and correspond to the cDNA clones γ, β I/β II, and α, whose genes are localized on human chromosomes 19, 16, and 17, respectively (Coussens et al., 1986). The two forms of the beta isozyme arise from differential mRNA splicing. Three other forms of the enzyme (δ, ϵ, ζ) have been characterized more recently (Ono et al., 1988).

The major subtypes (α, βI, βII, and γ) are Ca^{2+}/phospholipid/diacylglycerol-dependent; the δ, ϵ, and ζ isozymes lack Ca^{2+} sensitivity, and the ζ isozyme fails to respond to diacylglycerol (Nishizuka, 1988). The difference in responsiveness to activators has been explained on the basis of the molecular data obtained by cDNA cloning and sequencing of the relevant isozymes (Nishizuka, 1988). The general model contains four constant (C1-C4) and five variable (V1-V5) regions. C1 and C2 are present in the N-terminal regulatory domain, while C3 and C4 are located in the C-terminal catalytic domain. C1, which features tandem cysteine-containing zinc-finger repeats, appears to be required for phorbol ester or diacylglycerol binding, while C2 may be required for Ca^{2+} sensitivity. The absence of C2 from the δ, ϵ, and ζ isozymes may underlie their Ca^{2+} insensitivity. Similarly, the absence of one of the cysteine-rich zinc-finger repeats from the ζ isozyme may account for its lack of dependence on diacylglycerol. The availability of the cloned isozymes from transfected cultured cells is facilitating the study of their biochemical properties (Knopf et al., 1986; Ohno et al., 1988). Immunological and nucleic acid probes derived from the cloning information are being used to distinguish the biology of the subtypes to probe their physiology.

B. Developmental Expression

Types I and II protein kinase C are detectable in rat brain in the late embryo, and levels increase steadily throughout postnatal development, stabilizing around 3 weeks postnatally, but declining in adulthood. By contrast, type I protein kinase C is barely detectable up to almost 2 weeks postnatally and then rapidly approaches the levels of the other two isozymes by the third week (Yoshida et al., 1988; Hashimoto et al., 1988a; Girard et al., 1988; Sposi et al., 1989.) This burst of expression of the "brain-specific" isozyme of protein kinase C is correlated with the period of synaptogenesis. In the cerebellum, this corresponds to the enlargement of the junctional processes of the parallel fiber network and the formation of connections on Purkinje cells (Yoshida et al., 1988).

C. Cell-Specific Expression

Brain protein kinase C is primarily localized in neurons. Most neurons express all three major isotypes, with type III located in the cytosol and types I and II associated with membranes (Worley et al., 1986; Huang et al., 1987; Huang et al., 1988). Protein kinase C is highly concentrated in the hippocampus, the cerebral cortex, and the amygdaloid complex. In the hippocampus, pyramidal granule cells contain all three isozymes with the hilus of the dentate gyrus expressing type III > I, II; fiber tracts have only small amounts of the enzyme (Huang et al., 1988). This particular localization is instructive when considering the putative role of protein kinase C in synaptic plasticity of long-term potentiation. The cerebellum displays a distinctive cellular and subcellular localization of the isozymes. Purkinje cell apical dendrites contain type I protein kinase C, which is lacking from the Purkinje cell body and basket or stellate neurons. Granule cells contain type II protein kinase C. Type III enzyme is also found in granule and Purkinje cells, as well as in basket and stellate neurons, but not in the Purkinje cell dendrites (Huang et al., 1988; Kose et al., 1988).

D. Compartmentalization

A subpopulation of protein kinase C exists in the soluble (cytosolic) phase under basal conditions. In response to increasing intracellular Ca^{2+}, it translocates to membranes where it can interact with the membrane phospholipids as well as the activating diacylglycerol (Wolf et al., 1985b; Kraft and Andersen,

1983). Phorbol esters or diacylglycerol synergizes this interaction. Many of the apparent in vivo substrates of protein kinase C are membrane-bound or cytoskeletal, which is compatible with the effects of Ca^{2+} and diacylglycerol on enzyme compartmentalization. At certain lipid surface pressures, brain protein kinase C can insert into the lipid bilayer, becoming permanently activated and an integral membrane protein, requiring detergent for solubilization (Bazzi and Nelsestuen, 1988a,b). This property of membrane insertion could account for some of the long-term effects of protein kinase C stimulatory signals.

In mature brain protein kinase C is enriched in presynaptic terminals (Girard et al., 1985); in fetal rat brain protein kinase C is highly concentrated in the growth cones of extending processes, where it appears to participate in signal transduction (Katz et al., 1985). Postsynaptic densities also contain a subtype of protein kinase C that is cytoskeletally associated (Wolf et al., 1986). The role of this protein kinase C is not known, but it is situated where it could readily participate in synaptic regulatory events. Cytochemical analysis provides evidence for perinuclear localization, particularly prominent on the nuclear envelope (Girard et al., 1985). Biochemical studies also detect binding sites for the enzyme on both the neuronal nuclear envelope and the chromatin and nuclear matrix fraction which is the nucleoskeletal scaffold for DNA (Misra and Sahyoun, 1987). The ζ-isozyme appears to be the predominant protein kinase C in neuronal nuclei (Hagiwara et al., 1990).

E. Mechanism of Activation

Protein kinase C displays the special property among protein kinases of responding simultaneously to two different signals, changes in free Ca^{2+} levels and the presence of diacylglycerol in the 1,2-sn configuration (Takai et al., 1979). Acidic phospholipids such as phosphatidylserine are required for activation in a reaction that involves a high degree of positive cooperativity with respect to Ca^{2+} and the lipid concentration (Hannun et al., 1985; Wolf et al., 1985a). There appear to be some differences in the rank order of effectiveness of the different phospholipids to serve as coactivators with Ca^{2+} among the isoforms of protein kinase C. Diacylglycerol provides an incremental increase in protein kinase C activity, with only the 1,2-sn configuration showing activity (Zhou et al., 1988). The state of the lipid bilayer is important in both the recognition of substrate and the activity of protein kinase C (Bazzi and Nelsestuen, 1987; Hannun et al., 1985).

A major biochemical difference among the different protein kinase C isotypes is the comparatively marked sensitivity of the type I isozyme to *cis*-unsaturated fatty acids (Murakami et al., 1986; Naor et al., 1988; Sekiguchi et al., 1987), such as arachidonic acid, a phospholipase A2 product and pre-cursor to the cycloxygenase and lipoxygenase pathways (Moncada and Higgs, 1988). Such sensitivity could represent a feedback mechanism at the synapse since these lipid products may be released from the postsynaptic to the pre-synaptic terminals which are rich in protein kinase C. Protein kinase C also responds to alternative activators such as phosphatidylinositol-4,5-bisphos-phate (Chauhan et al., 1989) or lysophospholipids linking protein kinase C regulation to the action of phospholipases A1 and A2 (Oishi et al., 1988).

Like many other protein kinases, protein kinase C undergoes autophos-phorylation, which accounts for modest effects on enzymatic activity (Han-nun and Bell, 1990; Mochly-Rosen and Koshland, 1987; Huang et al., 1986a). An increased dissociation rate of phosphorylated protein kinase C from membranes is observed (Wolf and Sahyoun, 1986) as well as an increased susceptibility to proteolysis (Kishimoto et al., 1989). Persistent stimulation of protein kinase C by physiological ligands or by phorbol esters can lead to proteolysis by calpain or other proteases. Such down-regulation can occur separately for different isotypes of protein kinase C within the same cell (Ase et al., 1988; Cooper et al., 1989).

The mechanism of activation of protein kinases has been hypothesized to reflect release from constitutive inhibition by an intramolecular peptide segment called a "pseudosubstrate" (Soderling, 1990; Kemp et al., 1989, 1987; House and Kemp, 1987). First observed in an inhibitory protein and in the regulatory subunit of cyclic AMP-dependent protein kinase (Scott et al., 1985), a pseudosubstrate "prototope" (Kemp et al., 1989) would interact with the active site of the kinase, forming a nonproductive complex. The pseudosubstate sequence in protein kinase C is constant among all isozymes with regulatory domains. Synthetic peptides were used to verify the hyp-thesis and to probe the specificity of the interaction. Activation would occur as a consequence of conformational changes in the regulatory domain caused by binding of phospholipid, Ca^{2+}, and diacylglycerol, removing the pseudo-substrate from the active site and allowing access to substrates. The selectivity of the interaction of the pseudosubstrate domain with the catalytic center of the interaction of the pseudosubstrate domain with the catalytic center of protein kinase C has allowed the design of more specific inhibitors for pro-tein kinase C to probe the physiological function of this kinase.

F. Substrates

A well-defined number of substrates are observed to be phosphorylated in cells upon treatment with phorbol esters. Some of the effects of protein ki-nase C activation appear to be mediated by quantitatively minor proteins poised in key regulatory positions in various processes, such as the activation of transcription factors, which may not be observed under normal labeling conditions. In the nervous system, several proteins are prominently phosphorylated by protein kinase C. We will consider three examples where function is altered by the phosphorylation.

87 kDal MARCKS Protein

An acidic, myristoylated, alanine-rich, membrane-associated 80-87-kDa protein is a substrate for protein kinase C originally described in synaptosomes (Albert et al., 1984; Wang et al., 1988; Wu et al., 1982) and subsequently found in growth cones of developing brain (Katz et al., 1985). Although its function in synaptic and growth cone physiology is unknown, synaptic stimulation under conditions that activate protein kinase C, but not calmodulin-dependent protein kinases, is associated with phosphorylation and translocation of MARCKS from the membrane to the cytosol (Wang et al., 1989). These events are dependent on extracellular Ca^{2+}; repolarization of the synaptosomes results in dephosphorylation and reassociation with the membrane. This protein has recently been cloned (Stumpo et al., 1989) and found to be a calmodulin-binding protein in the presence of Ca^{2+} (Graff et al., 1989). Phosphorylation by protein kinase C prevents calmodulin binding (Stumpo et al., 1989), thus presumably releasing calmodulin to interact with other effector proteins.

GAP43 Protein

Another acidic protein enriched in growth cones and found also in synaptosomes has been described variously as B-50, F1 protein, and P-57, as well as GAP-43. The name of this protein, growth-associated protein, reflects the enhancement of expression during neuronal regeneration and in areas of the brain undergoing neurite extension and axonal growth, specifically in the growth cones (Benowitz and Routtenberg, 1987). It is a major substrate for protein kinase C, which is localized in regions of neurons involved in membrane addition or extensive vesicle recycling, and may be associated with other proteins of the neuronal cytoskeleton (Gordon-Weeks, 1989). Both Ca^{2+}-dependent and phosphoinositide-dependent signaling pathways may be modulated by this protein (Van Hoff et al., 1989). Phosphatidylinositol phosphate kinase is apparently inhibited by B-50 phosphorylation, thereby potentially reducing the pool of phosphatidylinositol-4,5-bisphosphate available for hormone-dependent phospholipase C activity. P-57 binds calmodulin in the *absence,* but not in the presence, of Ca^{2+} (Cimler et al., 1985). Calmodulin binding is decreased by protein kinase C phosphorylation, which would increase the free pool of calmodulin available for other

Ca²⁺ signaling targets. Calcineurin, a Ca^{2+}/calmodulin-dependent protein phosphatase, reverses the phosphorylation, restoring high-affinity calmodulin binding.

The high concentration of protein kinase C in synaptosomes is suggestive of a role for the enzyme in modulating neurotransmitter release. Phorbol esters have been observed to facilitate the release of norepinephrine and dopamine from synaptosomes (Kaczmarek, 1987). Antibodies to B-50 interfere with neurotransmitter release, possibly by inhibiting vesicle recycling (Dekker et al., 1989). The mechanism by which GAP-43 phosphorylation potentiates neurotransmitter release remains unclear. A recent report also provides an intriguing connection between GAP-43 and the G-protein, Go (Strittmatter et al., 1990). Like GAP-43, this G-protein subtype was highly enriched in growth cones, and its ability to bind the GTP analog GTP-γ-S was enhanced by nanomolar concentrations of GAP-43. This activity of GAP-43 resides in the first 24 amino acids and may be a consequence of a sequence homology with receptor domains that interact with G-proteins. Thus, GAP-43 may serve as a convergence point for several signaling elements, including Go as well as calmodulin and protein kinase C. With several putative integrative roles ascribed to GAP43, it is not surprising that its regulation by protein kinase C is also implicated in plastic modifications of neuronal function such as long-term potentiation.

Adducin

A membrane cytoskeletal heterodimer found in erythrocytes (97 kDa and 103 kDa) and in brain (104 kDa and 107-110 kDa) mediates interactions between spectrin-actin complexes in the cytoskeleton. Ca^{2+}/calmodulin binds to adducin, leading to inhibition of its cytoskeletal interactions (Bennett et al., 1988). Adducin also binds acidic phospholipids and protein kinase C itself (Wolf and Sahyoun, 1986). Phosphorylation of adducin by protein kinase C inhibits both these interactions, thus modulating the association of the cytoskeleton with the membrane lipid bilayer. Regulation of the assembly of the membrane skeleton by protein kinase C through adducin could influence the redistribution of membrane components and remolding of cell shape.

Among neuronal targets of protein kinase C, ion channels also represent a prominent class of substrates. There is an extensive literature on the effects of phosphorylation by protein kinases on ion channels (Armstrong, 1989; Kaczmarek, 1987, 1988; Levitan, 1988; Tsien et al., 1988). A voltage-dependent Ca^{2+} channel where the channel protein itself is the target for protein kinase C has been described (Jahn et al., 1988). Similar evidence is available for the homologous neuronal Na⁺ channel and cyclic AMP-dependent protein kinase (Catterall, 1986). However, a detailed discussion of the effects of protein kinase C activators on various ion channel conductances is beyond the scope of this discussion.

G. Long-Term Potentiation

Central nervous system neurons sum the inputs of numerous presynaptic cells, which can be either excitatory or inhibitory depending on whether they bring the cell closer to or further from the threshold for generation of action potentials. Several types of both short- and long-term modulation of synaptic communication have been distinguished based on the time courses of their development and decay (Kennedy, 1989; Nicoll et al., 1988). High-frequency stimulation of certain pathways such as the perforant and mossy fiber hippocampal pathways leads to long-term potentiation (LTP), which can be divided into two phases: induction and maintenance. Several experimental approaches indicate that protein kinase C may be involved in the induction phase of LTP. Translocation of the kinase from cytosol to membranes accompanies high-frequency stimulation of the perforant path that induces LTP (Akers et al., 1986). Activation of protein kinase C by phorbol esters (Malinow et al., 1988) or inhibition by sphingosine or the intracellular injection of pseudosubstrate peptide (Malenka et al., 1989; Malinow et al., 1988) further suggests a role for the kinase in LTP induction. Maintenance of LTP may involve release of arachidonic acid from the postsynaptic terminals and its retrograde modulation of presynaptic function (Piomelli et al., 1987). One possible, though not exclusive, function of arachidonic acid may be activation of protein kinase C in presynaptic terminals. The mechanism by which protein kinases modulate synaptic transmission in LTP entails the phosphorylation of specific substrates. While substrates critical for LTP have not been identified, candidate targets include the synaptosome-enriched GAP-43 and MARCKS (Akers and and Routtenberg, 1987; Pfenninger, 1986) but may also involve ion channel subunits (Cotman et al., 1988). A role for protein kinase C has also been postulated in a standard learning paradigm. Protein kinase C has been shown histochemically to translocate from the cytosol to the membrane in regions of the brain known to be involved in associative learning (Olds et al., 1989). Longer-term effects (days) are accompanied by a redistribution of protein kinase C from the cell soma to the dendrites within the affected region. Concomitantly, gene expression and protein synthesis are altered, which presumably results from a signal conveyed to the nucleus (Goelet et al., 1986).

H. Genetic Modulation

Long-term modulation of cellular function extends to the nuclear level and ultimately to the regulation of gene transcription. The expression of several genes is altered by protein kinase C activation. Protein kinase C itself is found in neuronal nuclei, distributed between the nuclear envelope and chromatin (Hagiwara et al., 1990; Misra and Sahyoun, 1987). There it may have access to DNA-related enzymes such as topoisomerase II, which is phosphorylated and activated by protein kinase C (Sahyoun et al., 1986). Double-stranded DNA-binding proteins that interact with and regulate the activity of the promoters and enhancers of target genes (Maniatis et al., 1987; Wigender, 1988) are also putative targets for protein kinase C regulation. In fact, protein kinase C phosphorylates an inhibitory subunit of one such protein, NF-$\varkappa$B, releasing the DNA binding moiety that regulates a host of target genes (Baeuerle and Baltimore, 1988). A novel protein that binds NF-$\varkappa$-B-like sequences has been detected in brain (Korner et al., 1989) and may prove to be a neuronal target for protein kinase C regulation of gene expression.

The preceding discussion of protein kinase C illustrates a number of attributes that are germane to signaling pathways in general. These include the occurrence of a family of related polypeptides and of submolecular domains within each polypeptide, the interaction of the protein kinase with regulatory signals, the phosphorylation of specific substrates in different subcellular compartments, and the ensuing regulation of short-term and long-term neuronal function such as neurotransmitter release or ion channel conductance, on one hand, and gene expression, on the other.

IV. CaMKinase II

The multifunctional Ca^{2+}/calmodulin-dependent protein kinase (CaMKinase) type II belongs to a group of calmodulin-dependent protein kinases that also includes myosin light-chain kinase, phosphorylase kinase, CaMKinases I and III (Schulman, 1988), and the neuron-specific CaMKinase Gr, which is enriched in the cerebellar granule cells (Ohmstede et al., 1989). CaMKinase II has assumed special prominence because of its abundance in brain tissue, its localization to pre- and postsynaptic nerve endings, and its ability to phosphorylate several important substrates.

A. Molecular Properties

CaMKinase II was purified in the early 1980s as a soluble ~600-kDa molecular complex which is comprised of 10-12 α and β/β' subunits with monomeric molecular weights of 50 and 58-60 kDa, respectively. Subsequent recombinant DNA studies suggested that the α and β polypeptides are separate gene products whereas the β and β' polypeptides are generated from differential mRNA splicing of one message (Bennett and Kennedy, 1987; Lin et al., 1987).

These studies further disclosed that the α and β subunits are themselves members of a family of polypeptides that now includes γ and δ subunits. The different subunits (isotypes) of CaMKinase represent variations on a general structural theme. The N-terminal 314 amino acids are 80-90% homologous, depending on the isotype, as are the C-terminal 150 amino acids. Major deviations from sequence homology arise in a region just C-terminal to the calmodulin binding domain in which reside two insertions of a total of about 100 amino acids. However, it is the CaMKinase II-containing α and β subunits which are enriched in neurons while the δ and γ polypeptides display a broader peripheral tissue distribution (Tobimatsu and Fujisawa, 1989).

Molecular and biochemical studies of the α and β subunits indicate that each polypeptide can be divided into three regions which embody the functional characteristics of the enzyme (Soderling, 1990; Colbran et al., 1989a; Schulman and Lou, 1989; Schulman, 1988). The N-terminal half contains the catalytic domain with binding sites for ATP and the protein substrate. A protease-sensitive "hinge" connects the catalytic portion to a regulatory portion, which includes autophosphorylation sites, a calmodulin-binding domain, and a possible inhibitory internal peptide sequence (pseudosubstrate prototope). The regulatory region is flanked by an "association" domain which may specify the interactions of the enzyme subunits with each other and with other polypeptides thought to be components of the neuronal cytoskeleton, and may have a regulatory role as well.

B. Developmental Expression

CaMKinase II production during specific stages of brain development has acquired enhanced significance because of its correlation with the period of active synaptogenesis and subsequent synaptic maturation. The levels of CaMKinase II have been assessed by measuring autophosphorylation, ^{125}I-labeled calmodulin binding, subunit immunoreactivity, and mRNA levels. Growth cones from fetal rat brain are relatively devoid of CaMKinase II activity (Hyman and Pfenninger, 1985), which subsequently becomes enriched in mature synaptosomes. A major burst of enzyme production occurs between the first and fourth neonatal weeks in rat brain, which is reflected in increased enzyme levels in the soluble, cytoskeletal, and nuclear compartments (Rostas et al., 1988; Kelly et al., 1987; Weinberger and Rostas, 1986; Kelly and Vernon, 1985; Sahyoun et al., 1985b). Deposition of the enzyme onto postsynaptic densities continues after the period of synaptogenesis in a process that leads to synaptic consolidation and thickening of the postsynaptic density itself (Rostas et al., 1986). Increased CaMKinase II levels are also accompanied by a switch from β-subunit preponderance to α-subunit dominance in rat forebrain and by a shift from a distribution favoring the soluble compartment to one favoring the particulate-cytoskeletal fraction

(Dunkley et al., 1988; Weinberger and Rostas, 1986; Sahyoun et al., 1985b). Maintenance of the low levels of a α-subunit production in the cerebellum results in preponderance of the β-subunit in the mature cerebellum (Sahyoun et al., 1985b).

The general trend of increased CaMKinase II protein levels during development is corroborated by Northern blot analysis of mRNA levels; these studies reveal a commensurate increase in the 5.3-5.4-kb and 4.8-5.0-kb mRNA species which code for the α and β subunits, respectively (Lin et al., 1987; Bennett and Kennedy, 1987). The developmental profile of this protein suggests that the genetic mechanisms regulating the expression of CaMKinase II will shed light on other proteins involved in synaptic maturation. A 5'-regulatory region of the gene for the α subunit has been isolated and shown to interact with nuclear proteins from the forebrain and cerebellum; these appear to differ from analogous binding proteins of peripheral tissues, and a protein coded for by the protooncogene c-myb may play a role in regulating expression of CaMKinase II α subunit (Sunyer and Sahyoun, 1990).

Expression of CaMKinase II in primary neuronal cultures has shown that the forebrain type of the kinase was produced by hippocampal pyramidal cells, with the β subunit appearing after 2 days in culture and the α subunit after 4 days (Scholz et al., 1988). In cerebellar granule cell cultures the β subunit only is produced during an 8-9-day culture period (Fukunaga et al., 1989). The enzyme has also been purified from a neuroblastoma/glioma cell line demonstrating a differentiation-dependent increase in levels (Vallano and Beaman-Hall, 1989).

The study of CaMKinase II development has led to valuable insights into the mechanisms of synaptogenesis and neuronal maturation through knowledge of its subcellular distribution, regional expression of different enzyme subunits, genetic regulation, and production in cultured neurons. Ultimately, these studies represent an attempt to understand the mechanisms that influence the assembly, distribution, and modulation of a prominent Ca^{2+}-signaling pathway.

C. Regional and Subcellular Distribution

Differences in the regional distribution of CaMKinase II subunits became readily evident upon comparison of cerebellar and forebrain enzyme preparations. The latter showed a three- to fourfold preponderance of the α over the β subunit, whereas the former showed the inverse ratio (McGuinness et al., 1985; Miller and Kennedy, 1985). Subsequent studies of enzyme distribution employed immunochemical and molecular probes. The α subunit of the kinase was highly enriched in the hippocampus, where it accounted for 2% of the total protein, but was far less abundant in the pons/medulla, where it amounted to 0.1% of the total protein (Erondu and Kennedy, 1985). In general, CaMKinase II was enriched in neurons compared to glia and was particularly concentrated in the hippocampus, cerebral cortex, amygdala, and striatum (Ouimet et al., 1984). Immunohistochemical visualization revealed that the kinase was evident in somata, dendrites, and postsynaptic densities, with weak immunoreactivity in axons and nuclei (Erondu and Kennedy, 1985; Ouimet et al., 1984). The enzyme was also found in association with microtubules, microfilaments, and neurofilaments (Vallano et al., 1986). Subcellular fractionation (Weinberger and Rostas, 1986) revealed four notable compartments containing CaMKinase II: soluble, synaptosomal, cytoskeletal, and nuclear. The soluble enzyme was comprised of a mixture of a population of symmetrical oligomers of α/β subunits and one of elongated oligomers containing tubulin as well (LeVine and Sahyoun, 1988). The synaptosomal enzyme was present in presynaptic terminals but was exceptionally concentrated in postsynaptic densities (Kennedy et al., 1983). The presence of CaMKinase II in pre- and postsynaptic endings renders it eminently suitable for participation in synaptic communication. In particular, the identification of CaMKinase II as the major protein component of isolated postsynaptic densities (Kelly et al., 1984; Kennedy et al., 1983) proved a landmark observation which has elicited much speculation about its role in postsynaptic Ca^{2+}-mediated signaling events. The cytoskeletal enzyme was biochemically isolated as the major component of the membrane-associated cytoskeleton (Sahyoun et al., 1985a) and in conjunction with cold stable microtubules and neurofilaments (Vallano et al., 1986). The nuclear enzyme was found associated with the nuclear equivalent of the cytoskeleton called the nuclear matrix (Sahyoun et al., 1984a,b).

CaMKinase II subunits are present in critical neuronal subcellular compartments and are capable of associating not only with themselves, but with other polypeptides as well, to generate structural and functional diversity and to allow participation in Ca^{2+} signaling at multiple subcellular levels. Several considerations combine to suggest a role for CaMKinase II in neuronal Ca^{2+} signaling. These include occurrence of a family of related polypeptides with selective tissue distribution, functional submolecular domains, critical regional distribution, and subcellular localization. Particularly relevant are the regulation of enzyme expression during neuronal development, its mechanism of activation, and the phosphorylation of prominent neuronal substrates. Such criteria can also be used to assess the role of other protein kinases in signaling mechanisms.

D. Mechanism of Activation

CaMKinase II, like several other calmodulin-activated proteins, is in a tonically inhibited state through the interaction of a regulatory domain residing between residues 281 and 309 of the α-subunit with the active site of the enzyme. Kinetic analysis of the inhibition suggests that the domain does not compete with peptide or protein substrates but with the ATP binding site, which remains inaccessible in the absence of calmodulin (Soderling, 1990; Colbran et al., 1989b). The binding of Ca^{2+}-calmodulin, which interacts with residues 296-309, releases this inhibition. Ca^{2+} binding to calmodulin causes a conformational change in that molecule, exposing a hydrophobic surface for interaction with an amphipathic helix in the regulatory domain of CaMKinase II. The cytoskeletal form of the kinase allows the functional distinction of calmodulin binding from enzyme activation. Calmodulin binding to the particulate enzyme was measured with a radiolabeled ligand while enzyme activity was examined by autophosphorylation or phosphorylation of exogenous substrate (LeVine et al., 1985b). The cytoskeletal enzyme represents a supramolecular structure, possibly containing >1000 subunits. Such a molecular organization is conducive to intersubunit interactions that would mediate information transfer from one subunit to another. Subunit-subunit interactions may account for the observation that calmodulin binding to 10% of the subunits seems to result in activation of the entire molecular complex. Moreover, presumed interaction among subunits may underlie the ability of calmodulin to activate the cytoskeletal enzyme with positive cooperativity. This cooperativity is manifest as a nonlinear relationship between calmodulin binding and enzyme activation in which a minimum "threshold" of calmodulin binding has to be achieved before substantial enzyme activation can occur. Such a threshold could be achieved in a cell containing 10-20 μM calmodulin when the free $[Ca^{2+}]$ is increased from 100 to 300 nM; however, the soluble CaMKinase II, which contains only 10-12 subunits, does not exhibit cooperativity in its activation by calmodulin (LeVine et al., 1986).

Activated enzyme undergoes intramolecular autophosphorylation in two phases, which generate CaMKinase states with distinguishable characteristics and important regulatory features. Phase I autophosphorylation requiring the presence of Ca^{2+}-calmodulin, occurs on threonine 286 of the α subunit (threonine 287 of the β subunit) and leads to enzyme activation in the absence of Ca^{2+}-calmodulin (Miller et al., 1988). Generation of this "autonomous" state of enzyme activity allows the enzyme to remain active even after the Ca^{2+}-signal is withdrawn and has led to the characterization of CaMKinase II as a "molecular switch" (Kennedy, 1988; Miller and Kennedy, 1986). The mechanism by which phase I autophosphorylation leads to enzyme activation seems to involve the negative charge of the covalently attached phosphate group and can be mimicked by introducing the negatively charged amino acid aspartate at the position of threonine 286 (Fong et al., 1989). Moreover, activation by autophosphorylation seems to involve subunit-subunit interactions because only a fraction of the total subunits of the soluble enzyme must be autophosphorylated in order to achieve apparent activation of all subunits (Miller and Kennedy, 1986). Phase II is observed as a "burst" of autophosphorylation which requires removal of Ca^{2+}-calmodulin and involves multiple sites, which may include threonine 305/306 and serine 314 of the α subunit (Miller et al., 1988). The functional consequences of autophosphorylation of the threonine residues manifest as reduced calmodulin binding and enzyme activity (Colbran and Soderling, 1990; Colbran et al., 1989a; LeVine et al., 1985b). Experimental verification of in vivo phosphorylation and activation of CaMKinase II has been difficult for technical reasons. Nevertheless, phase I autophosphorylation of CaMKinase II has been observed in intact synaptosomes (Wang et al., 1988; Gorelick et al., 1988), cultured cerebellar granule cells (Fukunaga et al., 1989), and hippocampal slices (Yip and Kelly, 1989). The in vivo autophosphorylation of CaMKinase II appears to be transient, indicating that phosphorylation phosphates can readily dephosphorylate CaMKinase II. Such phosphatase activities have indeed been detected in cerebellar extracts, synaptosomes, and postsynaptic densities and may include phosphatases type 2A and type 1 (LeVine et al., 1985a; Shields et al., 1985).

Whereas the modulation of protein kinase C activity appears to be predominantly governed by enzyme translocation and proteolysis, it appears that the modulation of CaMKinase II activity results mainly from autophosphorylation. Nevertheless, it is possible that translocation and proteolysis do play some role in regulating CaMKinase II activity. Autophosphorylation of the enzyme in an invertebrate system is accompanied by translocation between a particulate and a soluble compartment (Saitoh and Schwartz, 1985). Moreover, a proteolytic fragment of CaMKinase II is a 30-kDa soluble, fully activated monomer that can be derived from either the soluble or cytoskeletal form of the enzyme (LeVine and Sahyoun, 1987). It is devoid of calmodulin binding, does not associate with other proteins, does not autophosphorylate detectably, and does not retain phosphate from previously autophosphorylated holoenzyme. Kwiatkowski and King (1989) suggest that autophosphorylation of the holoenzyme directs the cleavage event. Under certain circumstances, particularly in synaptic plasticity, proteolytic activation and release of the enzyme could be physiologically important.

Finally, it is worth noting that unraveling of the mechanism of activation of CaMKinase II as well as protein kinase C has inspired the development of models for molecular memory. Translocation and persistent membrane association may represent a major pathway for molecular memory involving protein kinase C. However, autophosphorylation-dephosphorylation, subunit-subunit interaction, and cooperativity appear to be the major ingredients of molecular memory models based on CaMKinase II. Such a molecular memory with graded storage as a function of the oligomer structure and the observed phosphorylation-dephosphorylation cycle has been proposed as a basis for long-term synaptic plasticity (Lisman, 1989; Lisman and Goldring, 1988).

E. Regulation of Synaptic Communication by CaMKinase II Phosphorylation of Synapsin I

The study of cyclic AMP and Ca^{2+}-dependent phosphorylation in isolated nerve terminals has led to exceptional insights into the regulation of neurotransmitter release. Early studies revealed that cyclic AMP stimulated the phosphorylation of a protein designated protein I in a synaptic membrane fraction (Ueda and Greengard, 1977). Purified protein I is represented by two highly basic polypeptides Ia and Ib with molecular weights of 86 and 80 kDa, respectively (Ueda and Greengard, 1977). Biochemical and immuno-histochemical examination confirmed that this protein, now called synapsin I, was a substrate for both cyclic AMP and CaMKinases, and was specifi-cally expressed in the central and peripheral nervous systems where it localized to the cytoplasmic surface of small smooth synaptic vesicles (DeCamilli et al., 1983a, b; Huttner et al., 1983). Eventual cDNA cloning and sequencing of synapsin I disclosed the occurrence of a family of four polypeptides: synapsins Ia, Ib, IIa, and IIb, which represent differential splicing products of two separate genes (Sudhoff et al., 1989). Amino acid sequence compari-sons among the four synapsins defined the boundaries of several domains in each polypeptide. The N-terminal domains A, B, and C are shared by all synapsins, but the C-terminal domains D-I are represented in different com-binations for each polypeptide. Domain A, which is contained within the N-terminal "head" region, contains the serine residue (site I) phosphorylated by cyclic AMP-dependent protein kinase and CaMKinase I; domain D, which is unique to synapsins Ia and Ib, contains two other serine residues (sites II and III) in the "tail" region of the molecule which are phosphorylated by CaMKinase II (Huttner et al., 1981). The head and tail regions are conveni-ently separated from each other by *Staphylococcus aureus* V8 protease digestion (Kennedy and Greengard, 1981). The tail region of synapsins Ia and Ib is sensitive to bacterial collagenase digestion due to the high pro-line content of the D domain (Huttner et al., 1981). Immunohistochemical visualization shows that synapsin Ib is widely distributed in most synaptic endings, while the other synapsins are differentially expressed in various types of nerve endings (Sudhoff et al., 1989).

The functional role of synapsin I derives from its specific molecular associations and their sensitivity to phosphorylation. It is these modulated associations which figure prominently in the regulation of neurotransmitter release. Synapsin I is capable of multiple protein-protein interactions with calmodulin, spectrin, microtubules, microfilaments, and synaptic vesicles themselves (Baines and Bennett, 1986; Bähler and Greengard, 1987; Baines, 1987; Schiebler et al., 1986). The latter two types of interactions have provided a basis for explanation of the regulation of neurotransmitter release as well as organelle movement in neurons in general (McGuinness et al., 1989). Synapsin I binds to isolated synaptic vesicles in a saturable, high-affinity reaction not mimicked by other basic proteins (Schiebler et al., 1986). In vitro phosphorylation by cyclic AMP-dependent protein kinase had a minimal effect on the affinity of synapsin I for synaptic vesicles, whereas phosphorylation by CaMKinase II caused a fivefold decrease in the affinity, thus favoring dissociation of synapsin I from the vesicle membrane (Schiebler et al., 1986). These results suggested that the ability of synapsin I to bind to synaptic vesicles and to cross-link them is weakened by the action of CaM-Kinase II. Synapsin I not only interacts with synaptic vesicles, but possesses, in addition, two actin binding sites and induces bundling of actin polymers. The bundling reaction is reduced following phosphorylation by cyclic AMP-dependent protein kinase and is completely abolished by CaMKinase II phosphorylation (Bähler and Greengard, 1987; Petrucci and Morrow, 1987). Synapsin I itself is translocated into the soluble phase of synaptosomes following activation by depolarization (Sihra et al., 1989). Consequently, synapsin I can mediate the formation of a presynaptic matrix in which synaptic vesicles are bound to actin polymers, while the phosphorylation of synapsin I by CaMKinase II favors the release of synaptic vesicles, rendering them available for fusion with the presynaptic membrane, liberating neurotransmitter by exocytosis.

This hypothesis was tested by injecting synapsin I and CaMKinase II into the squid giant synapse (Llinas et al., 1985). As expected, neurotransmitted release was inhibited by unphosphorylated synapsin I and was augmented by CaMKinase II, but not by cyclic AMP-dependent protein kinase. More recently, cortical synaptosomes were loaded with the "autonomous," Ca^{2+}-dependent CaMKinase II stably activated by thiophosphorylation (Nichols et al., 1990). The autonomous enzyme potentiated the release of endogenous glutamate and preloaded ^{3}H-norepinephrine following K^+-induced depolarization or treatment with the ionophore ionomycin in the presence of extracellular Ca^{2+}. Conversely, the introduction of a CaMKinase II peptide inhibitor, whose sequence was derived from the regulatory domain of the kinase, inhibited neurotransmitter release (Nichols et al., 1990). These studies represent the culmination of a research effort that focused on synaptosomal protein kinases and their substrates, providing strong support for the utility of this molecular approach in unraveling the underlying mechanisms of neurotransmitter release. Moreover, the effects of CaMKinase II on neuronal and synaptic function extend to the phosphorylation of other substrates, which have been considered elsewhere (Shulman, 1988; Nairn et al., 1985), including the neurotransmitter-synthesizing enzymes tyrosine hydroxylase and tryptophan monooxygenese, the microtubule-associated proteins MAP-2 and tau, and possibly ion channels as well.

CaMKinase II phosphorylation of synapsin I is an example of a form of short-term modulation of synaptic efficacy; however, this kinase has also been implicated in the extended synaptic modification of the experimental paradigm of long-term potentiation (LTP) (Kennedy, 1989). The exceptional enrichment of the enzyme in the postsynaptic density, its obligomeric structure and cooperative activation, and its regulation by autophosphorylation-dephosphorylation have rendered CaMKinase II an attractive candidate for long-term synaptic modulation. More direct evidence for participation of the kinase in the postsynaptic induction of hippocampal LTP was obtained following the injection of specific peptide inhibitors and anticalmodulin drugs (Malinow et al., 1989; Malenka et al., 1989). The detailed molecular mechanisms of CaMKinase II participation in LTP, however, remain more obscure than the more immediate effects on neurotransmitter release.

F. Genetic Modulation

Phosphorylation of neurotransmitter-synthesizing enzymes, synapsin I, and induction of LTP constitute key loci for the regulation of synaptic communication by CaMKinase II. A logical further site for long-term regulatory effects of CaMKinase II is the neuronal nucleus. Biochemical studies with purified neuronal nuclei show that this kinase is a major nuclear target for Ca^{2+} signaling (Sahyoun et al., 1984a,b). It is associated with the nuclear matrix and selectively phosphorylates a chromatin protein thought to be involved in the regulation of transcription of RNA from DNA (Sahyoun et al., 1984a). Hence, activation of nuclear CaMKinase II may prove to be another limb of neuronal CaMKinase II signaling. Increased neuronal activity (Morgan and Curran, 1989) and intracellular neuronal Ca^{2+} have been linked to the regulation of gene expression, such as induction of *c-fos* (Dragunow and Robertson, 1987; Morgan et al., 1987; Greenberg et al., 1986). In view of these findings, it is tempting to speculate on the role of nuclear CaMKinases in the regulation of neuronal gene expression. However, it remains to be seen whether CaMKinase II actually regulates gene expression, and whether the products of such neuronal genes are synaptic components whose levels control synaptic communication. Suffice it to say here that Ca^{2+}-sensitive protein kinases, which include protein kinase C, CaMKinase II, and CaMKinase-Gr (Ohmstede et al., 1989), appear to be associated with neuronal nuclei and may well mediate long-term Ca^{2+}-signaling reactions at the nuclear level.

The analysis of CaMKinase II structure, regional distribution, developmental expression, mechanism of activation, subcellular localization, substrates, and synaptic regulation provides a wealth of observations on short-term and long-term Ca^{2+} signaling. These and similar studies of other components of signaling pathways have begun to define the principles governing the mechanisms of cellular stimulus-response coupling, and information transfer, storage, propagation, and modulation in the nervous system.

V. Signal Termination by Phosphoprotein Phosphatases

Signaling pathways are ultimately controlled by their integrated on-off mechanisms and the modulation thereof. Reversal of protein phosphorylation naturally entails the action of phosphoprotein phosphatases (PP-ases) whose properties have been lucidly presented in the reviews cited below. Some of these properties will be mentioned briefly here insofar as they pertain to the molecular structure, subcellular distribution, and activity regulation of other components of signaling pathways.

Two classes of PP-ases are currently recognized: those which act on phosphorylated serine or threonine residues and those which dephosphorylate proteins containing phosphotyrosine (PTP-ases). The former class (Cohen, 1989, Cohen and Cohen, 1989) contains four major types of enzymes: PP-1, PP-2A, PP-2B, and PP-2C. The first three enzymes and their isotypes are members of the same gene family and contain homologous catalytic domains whereas isotypes of PP-2C appear to derive from a distinct gene family. PP2B, also known as calcineurin (Klee et al., 1979), is enriched in brain, contains a Ca^{2+}-binding subunit that is homologous to calmodulin, and binds to authentic calmodulin as well. These properties allow Ca^{2+} as well as Ca^{2+}-calmodulin to activate the enzyme. An autoinhibitory domain has also been demonstrated in calcineurin (Hashimoto et al., 1990) and the phosphatase can be regulated by phosphorylation by CaMKinase II (Hashimoto et al., 1988b). Interestingly, some of the substrates of cyclic AMP-dependent protein kinase, such as inhibitor I and DARRP-32, are dephosphorylated by calcineurin, which permits cyclic AMP, and Ca^{2+}-dependent signaling pathways to interact at the level of substrate phosphorylation-dephosphorylation.

The catalytic subunit of PP-1 is inhibited by the heat-stable proteins 1 and 2, and its subcellular distribution appears to be determined by its association with what have been called "targeting" subunits, such as the glycogen-associated "G"-subunit. Interestingly, both the activity and distribution of PP-1 are regulated by cyclic AMP-dependent phosphorylation and Ca^{2+}-dependent dephosphorylation. Moreover, PP-1 as well as PP-2A may have an antagonistic role to protein kinase C because they reverse the phosphorylation of protein kinase C substrates. Thus, both protein kinase C activators such as phorbol esters and the PP-1/PP-2A inhibitor okadaic acid augment the phosphorylation of similar substrates and behave as tumor promoters. In fact, the discovery that the hydrophobic, complex fatty acid okadaic acid is a selective PP-1/PP-2A inhibitor provides a powerful tool for investigation of the in situ action of these enzymes (Cohen et al., 1990; Bialojan and Takai, 1988).

The discovery of protein tyrosine kinases (PTK's) led to a search for PP-ases (PTP-ases) that would act specifically on PTK substrates (reviewed in Tonks and Charbonneau, 1989; Hunter, 1989). Multiple PTP-ase activities can be detected in several tissues and as many as seven have been chromatographically resolved from brain extracts. The purification and sequencing of a soluble, 38,000-kDa placental PTP-ase (PTP-ase 1B) proved to be a milestone achievement which facilitated the detection of homologous PTP-ases. These include a soluble T-lymphocyte enzyme and the transmembrane polypeptide CD45, which is restricted to cells of hematopoietic lineage, and LAR, which has a broader tissue distribution. The membrane-spanning property of CD45 and LAR immediately raised the question of their potential regulation of extracellular ligands in a manner analogous to the activation of transmembrane PTK's by specific growth factors. This speculation was given additional impetus by the observation that the extracellular domain of LAR contains tandem sequences homologous to the cell adhesion molecule, N-CAM, suggesting that LAR activity may be influenced by cell-cell interactions. The domain structure of CD45 and LAR reveals other interesting features, such as the presence of a duplicated, cytoplasmic catalytic domain, fibronectin-type sequences, and a potential protein kinase C phosphorylation site in the cytoplasmic portion of CD45 which is adjacent to the plasma membrane. The discovery of novel PTP-ases, the regulation of their activity, and their joint action with PTK's on specific substrates are likely to represent valuable contributions to the role of tyrosine phosphorylation in cellular signaling. Such contributions will probably lead to a better understanding of the role of PTK's, PTP-ases, and their substrates in neuronal development and communication.

The study of PP-ases, like the study of other proteins that participate in cellular signaling, has benefited from analyzing their domain structure, subunit composition, "family affiliation," subcellular distribution, activation-inactivation mechanisms, and interaction with other components of the signaling "apparatus." These considerations are rapidly evolving into an integrated framework for the understanding of the assembly and operation of individual signaling pathways and of signaling networks. The application of this molecular approach to the study of many aspects of neuronal development and communication has proven to be quite rewarding.

References

Akers, R. F., Lovinger, D. M., Colley, P. A., Linden, D. J., and Routtenberg, A. (1986). Translocation of protein kinase C activity may mediate hippocampal long-term potentiation. *Science* **231**:587-589.

Akers, R. F., and Routtenberg, A. (1987). Calcium-promoted translocation of protein kinase C to synaptic membranes: Relation to the phosphorylation of an endogenous substrate. (Protein F1) involved in synaptic plasticity. *J. Neurosci.* **7**:3976-3983.

Albert, K. A., Wu, W. C-S., Nairn, A. C., and Greengard, P. (1984). Inhibition by calmodulin of calcium/phospholipid-dependent protein phosphorylation. *Proc. Natl. Acad. Sci. USA* **81**:3622-3625.

Alkon, D. L., and Rasmussen, H. (1988). A spatial-temporal model of cell activation. *Science* **239**:998-1005.

Armstrong, D. L. (1989). Calcium channel regulation by calcineurin, a Ca^{2+}-activated phosphatase in mammalian brain. *Trends Neurosci.* **12**:117-122.

Ase, K., Berry, N., Kikkawa, U., Kishimoto, A., and Nishizuka, Y. (1988). Differential down-regulation of protein kinase C subspecies in KM3 cells. *FEBS Lett.* **236**:396-400.

Axelrod, J., Burch, R. M., and Jelsema, C. L. (1988). Receptor-mediated activation of phospholipase A2 via GTP-binding proteins: Arachidonic acid and its metabolites as second messengers. *Trends Neurosci.* **11**:117-123.

Baeuerle, P. A., and Baltimore, D. (1988). IϰB: A specific inhibitor of the NF-ϰB transcription factor. *Science* **242**:540-546.

Bähler, M., and Greengard, P. (1987). Synapsin I bundles F-actin in a phosphorylation-dependent manner. *Nature* **326**:704-707.

Baines, A. J. (1987). Synapsin I and the cytoskeleton. *Nature* **326**:646.

Baines, A. J., and Bennett, V. (1986). Synapsin I is a microtubule-bundling protein. *Nature* **319**:145-147.

Bazzi, M. D., and Nelsestuen, G. L. (1987). Role of substrate in determining the phospholipid specificity of protein kinase C activation. *Biochemistry* **26**:5002-5008.

Bazzi, M. D., and Nelsestuen, G. L. (1988a). Association of protein kinase C with phospholipid monolayers: Two-stage irreversible binding. *Biochemistry* **27**:6776-6783.

Bazzi, M. D., and Nelsestuen, G. L. (1988b). Properties of membrane-inserted protein kinase C. *Biochemistry* **27**:7589-7593.

Bennett, M. K., and Kennedy, M. B. (1987). Deduced primary structure of the beta subunit of brain type II Ca^{2+}/calmodulin-dependent protein kinase determined by molecular cloning. *Proc. Natl. Acad. Sci. USA* **84**:1794-1798.

Bennett, V., Gardner, K., and Steiner, J. P. (1988). Brain adducin: A protein kinase C substrate that may mediate site-directed assembly of the spectrin-actin junction. *J. Biol. Chem.* **263**:5860-5869.

Benowitz, L. I., and Routtenberg, A. (1987). A membrane phosphoprotein associated with neural development, axonal regeneration, phospholipid metabolism, and synaptic plasticity. *Trends Neurosci.* **10**:527-532.

Berridge, M. J. (1986). Inositol phosphates as second messengers. *Receptor Biochem. Methodol.* **7**:25-45.

Berridge, M. J. (1987). Inositol triphosphate and diacylglycerol: Two interacting second messengers. *Annu. Rev. Biochem.* **56**:159-193.

Bialojan, C., and Takai, A. (1988). Effects of okadaic acid on isometric tension and myosin phosphorylation of chemically skinned guinea pig *Taenia coli. Biochem. J.* **256**:283-290.

Blatz, A. L., and Magleby, K. L. (1987). Calcium-activated potassium channels. *Trends Neurosci.* **10**:463-467.

Buck, C. A., and Horwitz, A. F. (1987). Integrin, a transmembrane glycoprotein complex mediating cell-substratum adhesion. *J. Cell Sci.* **7** (suppl.): 231-250.

Burgoyne, R. D. (1989). Small GTP-binding proteins. *Trends Biochem. Sci.* **14**:394-396.

Buttner, N., Siegelbaum, S. A., and Volterra, A. (1989). Direct modulation of *Aplysia* S-K$^+$ channels by a 12-lipoxygenase metabolite of arachidonic acid. *Nature* **342**:553-555.

Carpenter, G. (1987). Receptors for epidermal growth factor and other polypeptide mitogens. *Annu. Rev. Biochem.* **56**:881-914.

Casey, P. J., and Gilman, A. G. (1988). G. protein involvement in receptor-effector coupling. *J. Biol. Chem.* **263**:2577-2580.

Castagna, M., Takai, Y., Kaibuchi, K., Sano, K., Kikkawa, U., and Nishizuka, Y. (1982). Direct activation of calcium-activated, phospholipid-dependent protein kinase by tumor-promoting phorbol esters. *J. Biol. Chem.* **257**:7847-7851.

Catterall, W. A. (1986). Molecular properties of voltage-sensitive sodium channels. *Annu. Rev. Biochem.* **55**:953-985.

Chauhan, A., Chauhan, V. P. S., Deshmukh, D. S., and Brockerhoff, H. (1989). Phosphatidylinositol 4,5-bisphosphate competitively inhibits phorbol ester binding to protein kinase C. *Biochemistry* **28**:4952-4956.

Chinkers, M., Garbers, D. L., Chang, M-S., Lowe, D. G., Chin, H., Goeddel, D. V., and Shulz, S. (1989). A membrane form of guanylate cyclase is an atrial natriuretic peptide receptor. *Nature* **338**:78-83.

Cimler, B. M., Andreasen, T. J., Andreasen, K. I., and Storm, D. R. (1985). P-57 is a neural-specific calmodulin-binding protein. *J. Biol. Chem.* **260**:10784-10788.

Cockcroft, S., Howell, T. W., and Gomperts, B. D. (1987). Two G-proteins act in series to control stimulus-secretion coupling in mast cells: Use of neomycin to distinguish between G-proteins controlling polyphospho-inositide phosphodiesterase and exocytosis. *J. Cell Biol.* **105**:2745-2750.

Cohen, P. (1989). The structure and regulation of protein phosphatases. *Annu. Rev. Biochem.* **58**:453-508.

Cohen, P., and Cohen, P. T. W. (1989). Protein phosphatases come of age. *J. Biol. Chem.* **264**:21435-21438.

Cohen, P., Holmes, F. B., and Tsukitani, Y. (1990). Okadaic acid: A new probe for the study of cellular regulation. *Trends Biochem. Sci.* **15**:98-102.

Colbran, R. J., Schworer, C. M., Hashimoto, Y., Fong, Y-L., Rich, D. P., Smith, M. K., and Soderling, T. R. (1989a). Calcium/calmodulin-dependent protein kinase II. *Biochem J.* **258**:313-325.

Colbran, R. J., Smith, M. K., Schworer, C. M., Fong, Y-L., and Soderling, T. R. (1989b). Regulatory domain of calcium/calmodulin-dependent kinase II: Mechanism of inhibition and regulation of phosphorylation. *J. Biol. Chem.* **264**:4800-4804.

Colbran, R. J., and Soderling, T. R. (1990). Calcium/calmodulin-independent autophosphorylation sites of calcium/calmodulin-dependent protein kinase II. Studies of the effect of phosphorylation of threonine 305/306 and serine 314 on calmodulin binding using synthetic peptides. *J. Biol. Chem.* **265**:11213-11219.

Cooper, D. R., Watson, J. E., Acevedo-Duncan, M., Pollet, R. J., Standaert, M. L., and Farese, R. V. (1989). Retention of specific protein kinase C isozymes following chronic phorbol ester treatment in BC3H-1 myocytes. *Biochem. Biophys. Res. Commun.* **161**:327-334.

Cotman, C. W., and D. T., Monaghan, Ganong, A. H. (1988). Excitatory amino acid neurotransmission: NMDA receptors and Hebb-type synaptic plasticity. *Annu. Rev. Neurosci.* **11**:61-80.

Coussens, L., Parker, P. J., Rhee, L., Yang-Feng, T. L., Chen, E., Waterfield, M. D., Francke, U., and Ullrich, A. (1986). Multiple distinct forms of bovine and human protein kinase C suggest diversity in cellular signaling pathways. *Science* **233**:859-866.

DeCamilli, P., Cameron, R., and Greengard, P. (1983a). Synapsin I (protein I), a nerve terminal-specific phosphoprotein. I. Its general distribution in synapses of the central and peripheral nervous system demonstrated by immunofluorescence in frozen and plastic sections. *J. Cell Biol.* **96**:1337-1354.

DeCamilli, P., Harris, S. M., Jr., Huttner, W. B., and Greengard, P. (1983b). Synapsin I (protein I), a nerve terminal specific phosphoprotein. II. Its specific association with synaptic vesicles demonstrated by immunocytochemistry in agarose-embedded synaptosomes. *J. Cell Biol.* **96**: 1355-1373.

Dekker, L. V., DeGraan, P. N. E., Oestreicher, A. B., Versteeg, D. H. G., and Gispen, W. H. (1989). Inhibition of noradrenaline release by antibodies to B-50 (GAP-43). *Nature* **342**:74-76.

Dixon, R. A. F., Sigal, I. S., and Strader, C. D. (1988). Structure-function analysis of the beta-adrenergic receptor. *Cold Spring Harbor Symp. Quant. Biol.* **53**:487-497.

Dohlman, H. G., Caron, M. G., and Lefkowitz, R. J. (1987). A family of receptors coupled to guanine nucleotide regulatory proteins. *Biochemistry* **26**:2657-2664.

Dragunow, M., and Robertson, H. A. (1987). Kindling stimulation induces c-*fos* protein(s) in granule cells of the rat dentate gyrus. *Nature* **329**: 441-442.

Dunkley, P. R., Jarvie, P. E., and Rostas, J. A. P. (1988). Distribution of calmodulin- and cyclic AMP-stimulated protein kinases in synaptosomes. *J. Neurochem.* **51**:57-68.

Edelman, G. M. (1985). Cell adhesion and the molecular processes of morphogenesis. *Annu. Rev. Biochem.* **54**:135-169.

Ellis, S. B., Williams, M. E., Ways, N. R., Brenner, R., Sharp, A. H., Leung, A. T., Campbell, K. P., McKenna, E., Koch, W. J., Hui, A., Schwartz, A., and Harpold, M. M. (1988). Sequence and expression of mRNAs encoding the $\alpha 1$ and $\alpha 2$ subunits of a DHP-sensitive calcium channel. *Science* **241**:1661-1664.

Erondu, N. E., and Kennedy, M. B. (1985). Regional distribution of type II Ca^{2+}/calmodulin-dependent protein kinase in rat brain. *J. Neurosci.* **5**:3270-3277.

Fong, Y-L., Taylor, W. L., Means, A. R., and Soderling, T. R. (1989). Studies of the regulatory mechanism of Ca^{2+}/calmodulin-dependent protein kinase II. Mutation of threonine 286 to alanine and to aspartate. *J. Biol. Chem.* **264**:16759-16763.

Freissmuth, M., Casey, P. J., and Gilman, A. G. (1989). G proteins control diverse pathways of transmembrane signaling. *FASEB J.* **3**:2125-2131.

Fukunaga, K., Rich, D. P., and Soderling, T. R. (1989). Generation of the Ca^{2+}-independent form of Ca^{2+}/calmodulin-dependent protein kinase II in cerebellar granule cells. *J. Biol. Chem.* **264**:21830-21836.

Furuichi, T., Yoshikawa, S., Miyawaki, A., Wada, K., Maeda, N., and Mikoshiba, K. (1989). Primary structure and functional expression of the inositol 1,4,5-triphosphate-binding protein P400. *Nature* **342**:32-38.

Gilman, A. G. (1987). G-proteins: Transducers of receptor-generated signals. *Annu. Rev. Biochem.* **56**:615-649.

Girard, P. R., Mazzei, G. J., Wood, J. G., and Kuo, J. F. (1985). Polyclonal antibodies to phospholipid/Ca^{2+}-dependent protein kinase and immunocytochemical localization of the enzyme in rat brain. *Proc. Natl. Acad. Sci. USA* **82**:3030-3034.

Girard, P. R., Wood, J. G., Fresch, J. E., and Kuo, J. F. (1988). Immunocytochemical localization of protein kinase C in developing brain tissue and in primary neuronal cultures. *Dev. Biol.* **126**:98-107.

Go, M., Sekiguichi, K., Nomura, H., Kikkawa, U., and Nishizuka, Y. (1987). Further studies on the specificity of diacylglycerol for protein kinase C activation. *Biochem. Biophys. Res. Commun.* **144**:598-605.

Goelet, P., Castellucci, V. F., Schacher, S., and Kandel, E. R. (1986). The long and the short of long-term memory—A molecular framework. *Nature* **322**:419-422.

Gordon, D., Merrick, D., Auld, V., Dunn, R., Goldin, A. L., Davidson, N., and Catterall, W. A. (1987). Tissue-specific expression of the RI and RII sodium channel subtypes. *Proc. Natl. Acad. Sci. USA* **84**:8682-8686.

Gordon-Weeks, P. R. (1989). GAP-43—What does it do in the growth cone? *Trends Neurosci.* **12**:363-365.

Gorelick, F. S., Wang, J. K. T., Lai, Y., Nairn, A. C., and Greengard, P. (1988). Autophosphorylation and activation of Ca^{2+}/calmodulin-dependent kinase II in intact nerve terminals. *J. Biol. Chem.* **263**:17209-17212.

Graff, J. M., Young, T. N., Johnson, J. D., and Blackshear, P. J. (1989). Phosphorylation-regulated calmodulin binding to a prominent cellular substrate for protein kinase C. *J. Biol. Chem.* **264**:21818-21823.

Greenberg, M. E., Ziff, E. B., and Greene, L. A. (1986). Stimulation of neuronal acetylcholine receptors induces rapid gene transcription. *Science* **234**:80-83.

Grenningloh, G., Schmieden, V., Schofield, P. R., Seeburg, P. H., Siddique, T., Mohandas, T. K., Becker, C. M., and Betz, H. (1990). Alpha subunit variants of the human glycine receptor: Primary structures, functional expression, and chromosomal localization of the corresponding genes. *EMBO J.* **9**:771-776.

Hagiwara, M., Uchida, C., Usuda, N., Nagata, T., and Hidaka, H. (1990). ζ-Related protein kinase C in nuclei of nerve cells. *Biochem. Biophys. Res. Commun.* **168**:161-168.

Hall, A. (1990). The cellular functions of small GTP-binding proteins. *Science* **249**:635-640.

Hannun, Y. A., and Bell, R. M. (1990). Rat brain protein kinase C: Kinetic analysis of substrate dependence, allosteric regulation, and autophosphorylation. *J. Biol. Chem.* **265**:2962-2972.

Hannun, Y. A., Loomis, C. R., and Bell, R. M. (1985). Activation of protein kinase C by Triton X-100 mixed micelles containing diacylglycerol and phosphatidylserine. *J. Biol. Chem.* **260**:10039-10043.

Hashimoto, T., Ase, K., Sawamura, S., Kikkawa, U., Saito, N., Tanaka, C., and Nishizuka, Y. (1988a). Postnatal development of a brain-specific subspecies of protein kinase C in rat. *J. Neurosci.* **8**:1678-1683.

Hashimoto, Y., King, M. M., and Soderling, T. R. (1988b). Regulatory interactions of calmodulin-binding proteins: Phosphorylation of calcineurin by autophosphorylated Ca^{2+}/calmodulin-dependent protein kinase II. *Proc. Natl. Acad. Sci. USA* **85**:7001-7005.

Hashimoto, Y., Perrino, B. A., and Soderling, T. R. (1990). Identification of an autoinhibitory domain in calcineurin. *J. Biol. Chem.* **265**:1924-1927.

Herschman, H. R. (1989). Extracellular signals, transcriptional responses, and cellular specificity. *Trends Biochem. Sci.* **14**:455-458.

House, C., and Kemp, B. E. (1987). Protein kinase C contains a pseudosubstrate prototope in its regulatory domain. *Science* **238**:1726-1728.

Huang, F. L., Yoshida, Y., Nakabayashi, H., and Huang, K-P. (1987). Differential distribution of protein kinase C isozymes in the various regions of brain. *J. Biol. Chem.* **262**:15714-15720.

Huang, F. L., Yoshida, Y., Nakabayashi, H., Young, W. S., III, and Huang, K-P. (1988). Immunocytochemical localization of protein kinase C isozymes in rat brain. *J. Neurosci.* **8**:4734-4744.

Huang, K-P., Chan, K-F. J., Singh, T. J., Nakabayashi, H., and Huang, F. L. (1986a). Autophosphorylation of rat brain Ca^{2+}-activated and phospholipid-dependent protein kinase. *J. Biol. Chem.* **261**:12134-12140.

Huang, K-P., Nakabayashi, H., and Huang, F. L. (1986b). Isozymic forms of rat brain Ca^{2+}-activated and phospholipid-dependent protein kinase. *Proc. Natl. Acad. Sci. USA* **83**:8535-8539.

Hunter, T., (1989). Protein-tyrosine phosphatases: The other side of the coin. *Cell* **58**:1013-1016.

Hunter, T., and Cooper, J. A. (1985). Protein-tyrosine kinases. *Annu. Rev. Biochem.* **54**:897-930.

Huttner, W. B., DeGennaro, L. J., and Greengard, P. (1981). Differential phosphorylation of multiple sites in purified protein I by cyclic AMP-dependent and calcium-dependent protein kinases. *J. Biol. Chem.* **256**:1482-1488.

Huttner, W. B., Schiebler, W., Greengard, P., and DeCamilli, P. (1983). Synapsin I (protein I), a nerve terminal-specific phosphoprotein. III. Its association with synaptic vesicles studied in a highly purified synaptic vesicle preparation. *J. Cell Biol.* **96**:1374-1388.

Hyman, C., and Pfenninger, K. H. (1985). Intracellular regulators of neuronal sprouting: Calmodulin-binding proteins of nerve growth cones. *J. Cell Biol.* **101**:1153-1160.

Inoue, M., Kishimoto, A., Takai, Y., and Nishizuka, Y. (1977). Studies on a cyclic nucleotide-independent protein kinase and its proenzyme in mammalian tissues. II. Proenzyme and its activation by calcium-dependent protease from brain. *J. Biol. Chem.* **252**:7610-7616.

Jahn, H., Nastainczyk, W., Rohrkasten, A., Schneider, T., and Hofmann, F. (1988). Site-specific phosphorylation of the purified receptor for calcium channel blockers by cAMP and cGMP-dependent kinase, protein kinase C, calmodulin-dependent protein kinase II and casein kinase II. *Eur. J. Biochem.* **178**:535-542.

Jan, L. Y., and Jan, Y. N. (1989). Voltage-sensitive ion channels. *Cell* **56**: 13-25.

Kaczmarek, L. K. (1987). The role of protein kinase C in the regulation of ion channels and neurotransmitter release. *Trends Neurosci.* **10**:30-34.

Kaczmarek, L. K. (1988). The regulation of neuronal calcium and potassium channels by protein phosphorylation. *Adv. Second Messengers Phosphoprotein Res.* **22**:113-138.

Katz, F., Ellis, L., and Pfenninger, K. H. (1985). Nerve growth cones isolated from fetal rat brain. Calcium-dependent protein phosphorylation. *J. Neurosci.* **5**:1402-1411.

Kaupp, U. B., Niidome, T., Tanabe, T., Terada, S., Bonigk, W., Stuhmer, W., Cook, N. J., Kangawa, K., Matsuo, H., Hirose, T., Miyata, T., and Numa, S. (1989). Primary structure and functional expression from complementary DNA of the rod photoreceptor cyclic GMP-gated channel. *Nature* **342**:762-766.

Kelly, P. T., McGuinness, T. L., and Greengard, P. (1984). Evidence that the major postsynaptic density protein is a component of a Ca^{2+}/calmodulin-dependent protein kinase. *Proc. Natl. Acad. Sci. USA* **81**: 945-949.

Kelly, P. T., Shields, S., Conway, K., Yip, R., and Burgin, K. (1987). Developmental changes in calmodulin-kinase II activity at brain synaptic junctions: Alterations in holoenzyme composition. *J. Neurochem.* **49**: 1927-1940.

Kelly, P. T., and Vernon, P. (1985). Changes in subcellular distribution of calmodulin kinase II during brain development. *Dev. Brain Res.* **18**: 211-224.

Kemp, B. E., Pearson, R. B., Guerriero, V., Jr., Bagchi, I. C., and Means, A. R. (1987). The calmodulin binding domain of chicken smooth muscle myosin light chain kinase contains a pseudosubstrate sequence. *J. Biol. Chem.* **262**:2542-2548.

Kemp, B. E., Pearson, R. B., House, C., Robinson, P. J., and Means, A. R. (1989). Regulation of protein kinases by pseudosubstrate prototopes. *Cellular Signalling* **1**:303-311.

Kennedy, M. B. (1988). Synaptic memory molecules. *Nature* 335:770-772.

Kennedy, M. B. (1989). Regulation of synaptic transmission in the central nervous system. *Cell* 59:777-787.

Kennedy, M. B., Bennett, M. K., and Erondu, N. E. (1983). Biochemical and immunochemical evidence that the "major postsynaptic density protein" is a subunit of a calmodulin-dependent protein kinase. *Proc. Natl. Acad. Sci. USA* 80:7357-7361.

Kennedy, M. B., and Greengard, P. (1981). Two calcium/calmodulin-dependent protein kinases, which are highly concentrated in brain, phosphorylate protein I at distinct sites. *Proc. Natl. Acad. Sci. USA* 78:1293-1297.

Kikkawa, N., Kishimoto, A., and Nishizuka, Y. (1989). The protein kinase C family: Heterogeneity and its implications. *Annu. Rev. Biochem.* 58:31-44.

Kikkawa, U., Takai, Y., Tanaka, Y., Miyake, R., and Nishizuka, Y. (1983). Protein kinase C as a possible receptor protein of tumor promoting phorbol esters. *J. Biol. Chem.* 258:11442-11445.

Kishimoto, A., Mikawa, K., Hashimoto, K., Yasuda, I., Tanaka, S-I., Tominaga, M., Kuroda, T., and Nishizuka, Y. (1989). Limited proteolysis of protein kinase C subspecies by calcium-dependent neutral protease (calpain). *J. Biol. Chem.* 264:4088-4092.

Klee, C. B., Crouch, T. H., and Krinks, M. H. (1979). Calcineurin: A calcium and calmodulin-binding protein of the nervous system. 76:6270-6273.

Klein, W. L., Sullivan, J., Skorupa, A., and Aguilar, J. S. (1989). Plasticity of neuronal receptors. *FASEB J.* 3:2132-2140.

Knopf, J. L., Lee, M-H., Sultzman, L. A., Kriz, R. W., Loomis, C. R., Hewick, R. M., and Bell, R. M. (1986). Cloning and expression of multiple protein kinase C cDNAs. *Cell* 46:491-502.

Korner, M., Rattner, A., Mauxion, F., Sen, R., and Citri, Y. (1989). A brain-specific transcription factor. *Neuron* 3:563-572.

Kose, A., Saito, N., Ito, H., Kikkawa, U., Nishizuka, Y., and Tanaka, C. (1988). Electron microscopic localization of type I protein kinase C in rat Purkinje cells. *J. Neurosci.* 8:4262-4268.

Kraft, A. S., and Andersen, W. B. (1983). Phorbol esters increase the amount of Ca^{2+} phospholipid-dependent protein kinase associated with plasma membrane. *Nature* 301:621-623.

Krupinski, J., Coussen, F., Bakalyar, H. A., Tang, W.-J., Feinstein, P. G., Orth, K., Slaughter, C., Reed, R. R., and Gilman, A. G. (1990). Adenylyl cyclase amino acid sequence: Possible channel- or transporter-like structure. *Science* 244:1558-1564.

Kwiatkowski, A. P., and King, M. M. (1989). Autophosphorylation of the type II calmodulin-dependent protein kinase is essential for formation of a proteolytic fragment with catalytic activity: Implications for long-term synaptic potentiation. *Biochemistry* 28:5380-5385.

LeVine, H., III, and Sahyoun, N. E. (1987). Characterization of a soluble Mr-30 000 catalytic fragment of the neuronal calmodulin-dependent protein kinase II. *Eur. J. Biochem.* **168**:481-486.

LeVine, H., III, and Sahyoun, N. E. (1988). Two types of brain calmodulin-dependent protein kinase II: Morphological, biochemical, and immuno-chemical properties. *Brain Res.* **439**:47-55.

LeVine, H., III, Sahyoun, N., and Cuatrecasas, P. (1985a). Endogenous dephosphorylation of synaptosomal calmodulin-dependent protein kinase type II. *Biochem. Biophys. Res. Commun.* **131**:1212-1218.

LeVine, H., III, Sahyoun, N., and Cuatrecasas, P. (1985b). Calmodulin binding to the cytoskeletal calmodulin-dependent protein kinase is regulated by autophosphorylation. *Proc. Natl. Acad. Sci. USA* **82**:287-291.

LeVine, H., III, Sahyoun, N., and Cuatrecasas, P. (1986). Binding of cal-modulin to the neuronal cytoskeletal protein kinase type II coopera-tively stimulates autophosphorylation. *Proc. Natl. Acad. Sci. USA* **83**:2253-2257.

Levitan, I. B. (1988). Modulation of ion channels in neurons and other cells. *Annu. Rev. Neurosci.* **11**:119-136.

Lin, C. R., Kapiloff, M. S., Durgesian, S., Takemoto, K., Russo, A. F., Hanson, P., Schulman, H., and Rosenfeld, M. G. (1987). Molecular cloning of a brain-specific calcium/calmodulin-dependent protein ki-nase. *Proc. Natl. Acad. Sci. USA* **84**:5962-5966.

Lisman, J. E. (1989). A mechanism for the Hebb and the anti-Hebb pro-cesses underlying learning and memory. *Proc. Natl. Acad. Sci. USA* **86**:9574-9578.

Lisman, J. E., and Goldring, R. A. (1988). Feasibility of long-term storage of graded information by the Ca^{2+}/calmodulin-dependent protein ki-nase molecule of the postsynaptic density. *Proc. Natl. Acad. Sci. USA* **85**:5320-5324.

Livingstone, M. S., Sziber, P. P., and Quinn, W. G. (1984). Loss of calcium/calmodulin responsiveness in adenylate cyclase of rutabaga, a *Droso-phila* learning mutant. *Cell* **37**:205-215.

Llinas, R., McGuiness, T. L., Leonard, C. S., Sugimori, M., and Green-gard, P. (1985). Intraterminal injection of synapsin I or calcium/calmod-ulin-dependent protein kinase II alters neurotransmitter release at the squid giant axon. *Proc. Natl. Acad. Sci. USA* **82**:3035-3039.

Malenka, R. C., Kauer, J. A., Perkel, D. J., Mauk, M. D., Kelly, P. T., Nicoll, R. A., and Waxham, M. N. (1989). An essential role for post-synaptic calmodulin and protein kinase activity in long-term potentia-tion. *Nature* **340**:554-557.

Malenka, R. C., Madison, D. V., and Nicoll, R. A. (1986). Potentiation of synaptic transmission in the hippocampus by phorbol esters. *Nature* **321**:175-177.

Malinow, R., Madison, D. V., and Tsien, R. W. (1988). Persistent protein kinase activity underlying long-term potentiation. *Nature* **335**:820-824.

Malinow, R., Schulman, H., and Tsien, R. W. (1989). Inhibition of postsynaptic PKC or CaMKII blocks induction but not expression of LTP. *Science* **245**:862-866.

Maniatis, T., Goodbourn, S., and Fischer, J. A. (1987). Regulation of inducible and tissue-specific gene expression. *Science* **236**:1237-1245.

McGuinness, T. L., Brady, S. T., Gruner, J. A., Sugimori, M., Llinas, R., and Greengard, P. (1989). Phosphorylation-dependent inhibition by synapsin I of organelle movement in squid axoplasm. *J. Neurosci.* **9**: 4138-4149.

McGuinness, T. L., Lai, Y., and Greengard, P. (1985). Ca^{2+}/calmodulin-dependent protein kinase II. Isozymic forms from rat forebrain and cerebellum. *J. Biol. Chem.* **260**:1696-1704.

Melancon, P., Glick, B. S., Malhotra, V., Weidman, P. J,. Serafini, T., Gleason, M. L., Orci, L., and Rothman, J. E. (1987). Involvement of GTP-binding "G" proteins in transport through the Golgi stack. *Cell* **51**:1053-1082.

Miller, C. (1989). Genetic manipulation of ion channels: A new approach to structure and mechanism. *Neuron* **2**:1195-1205.

Miller, S. G., and Kennedy, M. B. (1985). Distinct forebrain and cerebellar isozymes of type II Ca^{2+}/calmodulin-dependent protein kinase associate differently with the postsynaptic density fraction. *J. Biol. Chem.* **260**:9039-9046.

Miller, S. G., and Kennedy, M. B. (1986). Regulation of brain type II Ca^{2+}/calmodulin-dependent protein kinase by autophosphorylation: A Ca^{2+}-triggered molecular switch. *Cell* **44**:861-870.

Miller, S. G., Patton, B. L., and Kennedy, M. B. (1988). Sequences of autophosphorylation sites in neuronal type II CaM kinase that control Ca^{2+}-independent activity. *Neuron* **1**:593-604.

Misra, U., and Sahyoun, N. (1987). Protein kinase C binding to isolated nuclei and its activation by a Ca^{2+}/phospholipid-independent mechanism. *Biochem. Biophys. Res. Commun.* **145**:760-768.

Mochly-Rosen, D., and Koshland, D. E., Jr. (1987). Domain structure and phosphorylation of protein kinase C. *J. Biol. Chem.* **262**:2291-2297.

Moncada, S., and Higgs, E. A. (1988). Metabolism of arachidonic acid. *Ann. NY Acad. Sci.* **522**:454-463.

Morgan, J. I., Cohen, D. R., Hempstead, J. L., and Curran, T. (1987). Mapping patterns of c-*fos* expression in the central nervous system after seizure. *Science* **237**:192-197.

Morgan, J. I., and Curran, T. (1989). Stimulus-transcription coupling in neurons: Role of cellular immediate-early genes. *Trends Neurosci.* **12**: 459-462.

Murakami, K., Chan, S. Y., and Routtenberg, A. (1986). Protein kinase C activation by *cis*-fatty acid in the absence of Ca^{2+} and phospholipids. *J. Biol. Chem.* **261**:15424-15429.

Naor, Z., Shearman, M. S., Kishimoto, A., and Nishizuka, Y. (1988). Calcium-independent activation of hypothalamic type I protein kinase C by unsaturated fatty acids. *Mol. Endocrinol.* **2**:1043-1048.

Nairn, A. C., Hemmings, H. C., Jr., and Greengard, P. (1985). Protein kinases in the brain. *Annu. Rev. Biochem.* **54**:931-976.

Nef, P., Oneyser, C., Alliod, C., Couturier, S., and Ballivet, M. (1988). Genes expressed in the brain define three distinct neuronal nicotinic acetylcholine receptors. *EMBO J.* **7**:595-601.

Nichols, R. A., Sihra, T. S., Czernik, A. J., Nairn, A. C., and Greengard, P. (1990). Calcium/calmodulin-dependent protein kinase II increases glutamate and noradrenaline release from synaptosomes. *Nature* **343**:647-651.

Nicoll, R. H., Kauer, J. A., and Malenka, R. C. (1988). The current excitement in long-term potentiation. *Neuron* **1**:97-103.

Niedel, J. E., Kuhn, L. J., and Vandenbark, G. R. (1983). Phorbol ester-receptor copurifies with protein kinase C. *Proc. Natl. Acad. Sci. USA* **80**:36-40.

Nishizuka, Y. (1984). The role of protein kinase C in cell surface signal transduction and tumor promotion. *Nature* **308**:693-698.

Nishizuka, Y. (1988). The molecular heterogeneity of protein kinase C and its implications for cellular regulation. *Nature* **334**:661-665.

Ohmstede, C-A., Jensen, K. F., and Sahyoun, N. E. (1989). Ca^{2+}/calmodulin-dependent protein kinase enriched in cerebellar granule cells: Identification of a novel neuronal calmodulin-dependent protein kinase. *J. Biol. Chem.* **264**:5866-5875.

Ohno, S., Akita, Y., Konuo, Y., Imajoh, S., and Suzuki, K. (1988). A novel phorbol ester receptor/protein kinase, nPKC, distantly related to the protein kinase C family. *Cell* **53**:731-741.

Oishi, K., Raynor, R. L., Charp, P. A., and Kino, J. F. (1988). Regulation of protein kinase C by lysophospholipids. *J. Biol. Chem.* **263**:6865-6871.

Olds, J. L., Anderson, M. L., McPhie, D. L., Staten, L. D., and Alkon, D. L. (1989). Imaging of memory-specific changes in the distribution of protein kinase C in the hippocampus. *Science* **245**:866-869.

Ono, Y., Fugii, T., Ogita, K., Kikkawa, U., Igarashi, K., and Nishizuka, Y. (1988). The structure, expression, and properties of additional members of the protein kinase C family. *J. Biol. Chem.* **263**:6927-6932.

Oprian, D., Sakmar, T., Nakayama, T., Chen, H-B., Franke, R., and Khorana, H. G. (1987). Molecular biological studies of the visual pigment, rhodopsin. *Discussions Neurosci.* **4**:20-28.

Ouimet, C. C., McGuinness, T. L., and Greengard, P. (1984). Immunochemical localization of calcium/calmodulin-dependent protein kinase II in rat brain. *Proc. Natl. Acad. Sci. USA* **81**:5604-5608.

Peroutka, S. J. (1989). 5-Hydroxytryptamine receptor subtypes: Molecular, biochemical, and physiological characterization. *Trends Neurosci.* **11**:496-500.

Petrucci, T. C., and Morrow J. S. (1987). Synapsin I: An actin-bundling protein under phosphorylation control. *J. Cell Biol.* **105**:1355-1363.

Pfenninger, K. H. (1986). Of nerve growth cones, leukocytes and memory: Second messenger systems and growth-related proteins. *Trends Neurosci.* **9**:562-565.

Piomelli, D., Shapiro, E., Feinmark, S. J., and Schwartz, J. H. (1987). Metabolites of arachidonic acid in the nervous system of *Aplysia*: Possible mediators of synaptic modulation. *J. Neurosci.* **7**:3675-3686.

Rasmussen, H., Apfeldorf, W., Barrett, P., Takuwa, N., Zawalich, W., Kreutter, D., Park, S., and Takuwa, Y. (1986). Inositol lipids: Integration of cellular signaling systems. *Receptor Biochem. Methodol.* **7**:109-147.

Rhee, S. G., Suh, P. G., and Lee, S. Y. (1989). Studies of inositol phospholipid-specific phospholipase C. *Science* **244**:546-550.

Rodbell, M., Birnbaumer, L., Pohl, S. L., and Krans, H. M. J. (1971). The glucagon sensitive adenyl cyclase system in plasma membranes of rat liver. *J. Biol. Chem.* **246**:1877-1882.

Ross, E. M. (1989). Signal sorting and amplification through G-protein-coupled receptors. *Neuron* **3**:141-152.

Rostas, J. A. P., Seccombe, M., and Weinberger, R. (1988). Two developmentally regulated isozymes of calmodulin-stimulated protein kinase II in rat forebrain. *J. Neurochem.* **50**:945-953.

Rostas, J. A. P., Weinberger, R. P., and Dunkley, P. R. (1986). Multiple pools and multiple forms of calmodulin-stimulated protein kinase during development: Relationship to postsynaptic densities. *Prog. Brain Res.* **69**:355-371.

Sahyoun, N., LeVine, H., III, Bronson, D., and Cuatrecasas, P. (1984a). Calmodulin-dependent protein kinase in neuronal nuclei. *J. Biol. Chem.* **259**:9341-9344.

Sahyoun, N., LeVine, H., III, Bronson, D., Greenstein, F., and Cuatrecasas, P. (1985a). Cytoskeletal calmodulin-dependent protein kinase: Characterization, solubilization, and purification from rat brain. *J. Biol. Chem.* **260**:1230-1237.

Sahyoun, N., LeVine, H., III, Burgess, S. K., Blanchard, S., Chang, K-J., and Cuatrecasas, P. (1985b). Early postnatal development of calmodulin-dependent protein kinase II in rat brain. *Biochem. Biophys. Res. Commun.* **132**:878-884.

Sahyoun, N., LeVine, H., III, and Cuatrecasas, P. (1984b). The Ca^{2+}-calmodulin-dependent protein kinase from the neuronal nuclear matrix and the postsynaptic density are structurally related. *Proc. Natl. Acad. Sci. USA* **81**:4311-4315.

Sahyoun, N., Wolf, M., Besterman, J., Hsieh, T-S., Sanders, M., LeVine, H., III, Chang, K-J., and Cuatrecasas, P. (1986). Protein kinase C phosphorylates topoisomerase II: Topoisomerase activation, and its possible role in phorbol ester-induced differentiation in HL-60 cells. *Proc. Natl. Acad. Sci. USA* **83**:1603-1607.

Saitoh, T., and Schwartz, J. H. (1985). Phosphorylation-dependent subcellular translocation of a Ca^{2+}/calmodulin-dependent protein kinase produces autonomous enzyme in *Aplysia* neurons. *J. Cell. Biol.* **100**: 835-842.

Schiebler, W., Jahn, R., Doucet, J-P., Rothlein, J., and Greengard, P. (1986). Characterization of synapsin I binding to small synaptic vesicles. *J. Biol. Chem.* **261**:8383-8390.

Schofield, P. R. (1989). The $GABA_A$ receptor: Molecular biology reveals a complex picture. *Trends Pharmacol. Sci.* **10**:476-478.

Schofield, P. R., Shivers, B. D., and Seeburg, P. H. (1990). The role of receptor subtype diversity in the CNS. *Trends Neurosci.* **13**:8-11.

Scholz, W. K., Baitinger, C., Schulman, H., and Kelly, P. T. (1988). Developmental changes in Ca^{2+}/calmodulin-dependent protein kinase II in cultures of hippocampal pyramidal neurons and astrocytes. *J. Neurosci.* **8**:1039-1051.

Schulman, H. (1988). The multifunctional Ca^{2+}/calmodulin-dependent protein kinase. *Adv. Sec. Messengers Phosphoprotein Res.* **22**:39-112.

Schulman, H., and Lou, L. L. (1989). Multifunctional Ca^{2+}/calmodulin-dependent protein kinase: Domain structure and regulation. *Trends Biochem. Sci.* **14**:62-66.

Schulz, S., Chinkers, M., and Garbers, D. L. (1989). The guanylate cyclase/receptor family of proteins. *FASEB J.* **3**:2026-2035.

Scott, J. D., Fischer, E. H., DeMaille, J. G., and Krebs, E. G. (1985). Identification of an inhibitory region of the heat-stable protein inhibitor of the cAMP-dependent protein kinase. *Proc. Natl. Acad. Sci. U.S.A.* **82**: 4379-4383.

Sekiguchi, K., Tsukuda, M., Ogita, K., Kikkawa, U., and Nishizuka, Y. (1987). Three distinct forms of rat brain protein kinase C.: Differential response to unsaturated fatty acids. *Biochem. Biophys. Res. Commun.* **145**:797-802.

Shields, S. M., Ingebritsen, T. S., and Kelly, P. T. (1985). Identification of protein phosphatase I in synaptic junctions: Dephosphorylation of endogenous calmodulin-dependent kinase II and synapse-enriched phosphoproteins. *J. Neurosci.* **5**:3414-3422.

Sibley, D. R., Benovic, J. C., Caron, M. G., and Lefkowitz, R. J. (1987). Regulation of transmembrane signaling by receptor phosphorylation. *Cell* **48**:913-922.

Sihra, T. S., Wang, J. K. T., Gorelick, F. S., and Greengard, P. (1989). Translocation of synapsin I in response to depolarization of isolated nerve terminals. *Proc. Natl. Acad. Sci. USA* **86**:8108-8112.

Soderling, T. R. (1990). Protein kinases: Regulation by autoinhibitory domains. *J. Biol. Chem.* **265**:1823-1826.

Sposi, N. M., Bottevo, L., Cossu, G., Russo, G., Testa, U., and Peschle, C. (1989). Expression of protein kinase C genes during ontogenic development of the central nervous system. *Mol. Cell. Biol.* **9**:2284-2288.

Stephenson, F. A. (1988). Understanding the $GABA_A$ receptor: A chemically-gated ion channel. *Biochem. J.* **249**:21-32.

Sternweis, P. C., and Pang, I-H. (1990). The G protein-channel connection. *Trends Neurosci.* **13**:122-126.

Stevens, C. F. (1989). A finger on brain receptors. *Nature* **342**:620-621.

Strittmatter, S. M., Valenzuela, D., Kennedy, T. E., Neer, E. J., and Fishman, M. C. (1990). Go is a major growth cone protein subject to regulation by GAP-43. *Nature* **344**:836-841.

Stühmer, W., Ruppersberg, J. P., Schröter, K. H., Sakmann, B., Stocker, M., Giese, W. P., Perschka, M., Baumann, A., and Pongs, O. (1989). Molecular basis of functional diversity of voltage gated potassium channels in mammalian brain. *EMBO J.* **8**:3235-3244.

Stumpo, D. J., Graff, J. M., Albert, K. A., Geengard, P., and Blackshear, P. J. (1989). Molecular cloning, characterization and expression of a cDNA encoding the "80- to 87-kDa" myristoylated alanine-rich C kinase substrate: A major cellular substrate for protein kinase C. *Proc. Natl. Acd. Sci. USA* **86**:4012-4016.

Sudhoff, T. C., Czernik, A. J., Kao, H-T., Takei, K., Johnston, P.A., Horiuchi, A., Kanazir, S. D., Wagner, M. A., Perin, M. S., DeCamilli, P., and Greengard, P. (1989). Synapsins: Mosaics of shared and individual domains in a family of synaptic vesicle phosphoproteins. *Science* **245**:1474-1480.

Sunyer, T., and Sahyoun, N. (1990). Sequence analysis and DNA-protein interactions within the 5'-flanking region of the Ca^{2+}/calmodulin-dependent protein kinase II alpha subunit gene. *Proc. Natl. Acad. Sci. USA* **87**:278-282.

Takai, Y., Kishimoto, A., Kikkawa, U., Mori, T., and Nishizuka, Y. (1979). Unsaturated diacylglycerol as a possible messenger for the activation of calcium-activated, phospholipid-dependent protein kinase system. *Biochem. Biophys. Res. Commun.* **91**:1218-1224.

Tobimatsu, T., and Fujisawa, H. (1989). Tissue-specific expression of four types of rat calmodulin-dependent protein kinase II mRNAs. *J. Biol. Chem.* **264**:17907-17912.

Tonks, N. K., and Charbonneau, H. (1989). Protein tyrosine dephosphorylation and signal transduction. *Trends Biochem. Sci.* **14**:497-500.

Tsien, R. W., Lipscombe, D., Madison, D. V., Bley, K. R., and Fox, A. P. (1988). Multiple types of neuronal calcium channels and their selective modulation. *Trends Neurosci.* **11**:431-438.

Ueda, T., and Greengard, P. (1977). Adenosine 3 ':5 '-monophosphate-regulated phosphoprotein system of neuronal membranes. I. Solubilization, purification, and some properties of an endogenous phosphoprotein. *J. Biol. Chem.* **252**:5155-5163.

Unwin, N. (1989). The structure of ion channels in membranes in excitable cells. *Neuron* **3**:665-676.

Vallano, M. L., and Beaman-Hall, C. M. (1989). Differentiation increases type II calmodulin-dependent protein kinase in the neuroblastoma/glioma cell line 108CC15 (NG108-15). *J. Neurosci.* **9**:539-547.

Vallano, M. L., Goldenring, J. R., Lasher, R. S., and DeLonzo, R. I. (1986) Association of calcium/calmodulin-dependent protein kinase with cytoskeletal preparations: Phosphorylation of tubulin, neurofilaments, and microtubule-associated proteins. *Ann. NY Acad. Sci.* **466**:357-374.

Van Hoff, C. O. M., DeGraan, P. N. E., Oestreicher, A. B., and Gispen, W. H. (1989). B-50 phosphorylation and polyphosphoinositide metabolism in nerve growth cone membranes. *J. Neurosci.* **8**:1789-1795.

Walsh, D. A., Perkins, J. P., and Krebs, E. G. (1968). An adenosine 3 ', 5 '-monophosphate-dependent protein kinase from rabbit skeletal muscle. *J. Biol. Chem.* **243**:3763-3774.

Wang, J. K. T., Walaas, S. I., and Greengard, P. (1988). Protein phosphorylation in nerve terminals: Comparison of calcium/calmodulin-dependent and calcium/diacylglycerol-dependent systems. *J. Neurosci.* **8**: 281-288.

Wang, J. K. T., Walaas, S. I., Sihra, T. S., Aderem, A., and Greengard, P. (1989). Phosphorylation and associated translocation of the 87-kDa protein, a major protein kinase C substrate in isolated nerve terminals. *Proc. Natl. Acad. Sci. USA* **86**:2253-2256.

Weinberger, R. P., and Rostas, J. A. P. (1986). Subcellular distribution of a calmodulin-dependent protein kinase activity in rat cerebral cortex during development. *Dev. Brain Res.* **29**:37-50.

Weissman, A. M., Bonifacino, J. S., Klausner, R. D., Samalson, L. E., and O'Shea, J. J. (1989). T Cell antigen receptor: Structure, assembly, and function. *Year Immunol. 1988* **4**:74-93.

Wigender, E. (1988). Compilation of transcription regulating proteins. *Nucleic Acids Res.* **16**:1879-1902.

Wolf, M., Burgess, S., Misra, U. K., and Sahyoun, N. (1986). Postsynaptic densities contain a subtype of protein kinase C. *Biochem. Biophys. Res. Commun.* **140**:691-698.

Wolf, M., Cuatrecasas, P., and Sahyoun, N. (1985a). Interaction of protein kinase C with membranes is regulated by Ca^{2+}, phorbol esters, and ATP. *J. Biol. Chem.* **260**:15718-15722.

Wolf, M., LeVine, H., III. May, W. S., Cuatrecasas, P., and Sahyoun, N. (1985b). A model for intracellular translocation of protein kinase C involving synergism between Ca^{2+} and phorbol esters. *Nature* **317**:546-549.

Wolf, M., and Sahyoun, N. (1986). Protein kinase C and phosphatidylserine bind to Mr 110,000/115,000 polypeptides enriched in cytoskeletal and postsynaptic density preparations. *J. Biol. Chem.* **261**:13327-13332.

Worley, P. F., Baraban, J. M., and Snyder, S. H. (1986). Heterogeneous localization of protein kinase C in rat brain: Autoradiographic analysis of phorbol ester receptor binding. *J. Neurosci.* **6**:199-207.

Wu, W. C., Walaas, S. I., Nairn, A. C., and Greengard, P. (1982). Calcium/phospholipid regulates phosphorylation of a Mr "87k" substrate protein in brain synaptosomes. *Proc. Natl. Acad. Sci. USA* **79**:5249-5253.

Yarden, Y., and Ullrich, A. (1988). Growth factor receptor tyrosine kinases. *Annu. Rev. Biochem.* **57**:443-478.

Yip, R. K., and Kelly, P. T. (1989). In situ protein phosphorylation in hippocampal tissue slices. *J. Neurosci.* **9**:3618-3630.

Yoshida, Y., Huang, F. L., Nakabayashi, H., and Huang, K-P. (1988). Tissue distribution and developmental expression of protein kinase C isozymes. *J. Biol. Chem.* **263**:9868-9873.

Yoshimasa, T., Sibley, D. R., Bouvier, M., Lefkowitz, R. J., and Caron, M. G. (1987). Cross-talk between cellular signaling pathways suggested by phorbol-ester-induced adenylate cyclase phosphorylation. *Nature* **327**:67-70.

Zhou, Q., Raynor, R. L., Wood, Jr., M. G., Menger, F. M., and Kuo, J. F. (1988). Structure-activity relationship of synthetic branched-chain distearoylglycerol (distearin) as protein kinase C activators. *Biochemistry* **27**:7361-7365.

3

Comparative Neurobiology of Invertebrate Motor Networks

Implications for the Control of Breathing in Mammals

MICHAEL S. DEKIN

University of Kentucky
Lexington, Kentucky

I. Introduction

During the past decade, the study of the control of breathing in mammals has witnessed several technical innovations allowing the cellular neurobiology of respiratory neurons to be described. Among these, the use of in vitro brainstem slices has provided a greater appreciation of the integrative capabilities of respiratory neurons (Dekin and Getting, 1984, 1987a,b; Champagnat et al., 1986; Dekin et al., 1987a,b; see also this volume, Chapter 7). The simple view that these cells possess only the classic sodium and potassium conductances responsible for the action potential has evolved to include an array of time- and voltage-dependent conductances that can participate in sculpturing their repetitive firing activity. The expression of these ionic conductances underlies such membrane excitability changes as spike frequency adaptation (Dekin and Getting, 1987a,b), delayed excitation (Dekin and

This work was supported in part by NIH Grants HL39929, HL40369, and HL-02314, and a University of Kentucky Biomedical Research Support Grant.

Getting, 1984, 1987a), and postburst hyperpolarization (Dekin, 1988a, 1988b). An equally diverse chemical neuroanatomy of the central respiratory circuit has been identified (for an example, see this volume, Chapter 22). The richness of potential neurotransmitter and neuromodulatory substances within the respiratory circuit suggests that multiple chemical mechanisms exist for fine tuning the activity of respiratory muscles. Our understanding of how these neurochemicals interact with putative respiratory neurons at the cellular level has also benefited from the use of brainstem slices. For instance, both substance P and thyrotropin-releasing hormone have been localized within nerve terminals in the dorsal respiratory group (DRG) (Hök-felt et al., 1975a,b, 1980; Maley and Elde, 1981) and are known to alter rhythmic breathing movements (Yamamoto and Lagercrantz, 1985, 1987; Hedner et al., 1983; Mueller et al., 1985). Using the in vitro brainstem slice, it has been shown that both these putative respiratory neurotransmitters alter the membrane properties of bulbospinal neurons in the region of the DRG (Dekin et al., 1985; Dekin 1988a, also see Section III.C).

Progress in elucidating the cellular complexity of the central respiratory network has been paralleled by similar studies on "simple" invertebrate neural networks. Although invertebrate systems contain far fewer neuronal elements, they too exhibit an extensive repertoire of intrinsic (voltage- and time-dependent conductances) and extrinsic (synaptic and modulatory) membrane properties. Because of their experimental accessibility, however, the role of cellular properties in these simple neural networks has been well characterized. As more has been learned about the cellular sophistication of invertebrate motor circuits, our conceptual views of how these circuits are organized have been challenged. This reevaluation has led to a greater appreciation for the adaptive flexibility of invertebrate rhythmic motor behaviors, which makes comparisons with similar behaviors in vertebrates more appealing. In light of these developments, it is both timely and informative to reconsider invertebrate motor systems for what they may tell us about the control of breathing in mammals. As a starting point, this chapter will review current hypotheses concerning the organization of invertebrate motor behaviors and consider their applicability to the control of breathing. Cellular mechanisms underlying this organization will then be examined and compared with the emerging body of data on the cellular properties of respiratory neurons.

II. A Primer on the Motor Organization of Invertebrates

Rhythmic motor behaviors in both vertebrates and invertebrates have long attracted the attention of neurophysiologists. In the early part of this century, however, the experimental emphasis was clearly on vertebrates. In large part, this stemmed from the pioneering work of Sherrington (1906) on spinal reflexes and locomotor activity in cats and dogs. Sherrington was able to demonstrate that rhythmic limb movements could still be elicited when the spinal cord was isolated from higher CNS levels. A popular interpretation of these results was that a series of "chain reflexes" were responsible for the movement. That is, proprioceptive sensory feedback originating from one segment of the limb movement cycle was used to organize the next segment of the cycle. This process would continue for as long as appropriate sensory feedback was generated, and so the idea of "peripheral control" of rhythmic movements gained acceptance. Later, Graham Brown (1911) demonstrated that the same limb movements could be observed when the spinal cord was isolated from higher centers as well as all sensory input from the limbs. Brown argued that this was evidence for a "central control" mechanism residing in the spinal cord itself.

Although Sherrington and Brown appeared to reach a consensus opinion that rhythmic motor movements were central in origin but highly regulated by sensory feedback (see Liddell and Sherrington, 1924), some investigators continued to ardently support the peripheral control model well into the 1950s and 1960s (reviewed by Delcomyn, 1980). At about that time invertebrate CNS preparations were gaining popularity as experimental tools. One of the main advantages of these preparations is that they contain relatively few neurons which are generally large by vertebrate standards. This allows the use of intracellular recording electrodes for both electrophysiological and anatomical studies. These neurons can also be identified from one preparation to another by their morphology and location. Equally important is the ability to maintain the isolated invertebrate CNS in vitro for long periods of time. Combined, these properties made the invertebrates excellent models in which to test the competing hypotheses of peripheral and central control of rhythmic motor movements. Indeed, the fact that it is now generally accepted that all rhythmic motor behaviors are controlled by a defined group of central neurons which initiates and coordinates the activity of muscles involved in the behavior can be largely attributed to the persuasive evidence that has accumulated from work with invertebrates over the last 30 years.

A simplified model for the organization of invertebrate rhythmic motor circuits is shown in Figure 1. Although many studies contributed to this model, two papers in particular stand out as laying a firm foundation for its acceptance. One was the work of Donald Wilson (1961) with locusts in which he demonstrated that the removal of peripheral (sensory) timing cues failed to suppress the basic motor pattern underlying flight behavior. These data provided unequivocal evidence for a *central pattern generator* (CPG), which coordinates the proper spatiotemporal sequence of motor activity during a

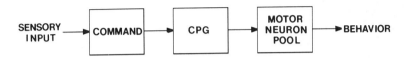

Figure 1 Functional organization of rhythmic motor circuits. Sensory inputs other than that acting on the command function are not shown.

rhythmic movement. CPGs provide two types of instructive input to a moto-neuron pool: (1) a rhythmic drive, which determines both cycle period and the phase relationship between the activation of agonist and antagonist muscles, and (2) a patterned drive within a particular movement phase, which determines how a muscle(s) will be activated. Explicit criteria have been proposed for defining neurons that are part of the CPG (Kristan et al., 1977). These criteria exploit the fact that rhythm generation is a CPG function. As suggested by Getting et al. (1980): "If a neuron is involved directly in the generation of the motor pattern, then an imposed change in its normal burst time should reset the burst timing at both the interneuronal and motoneuronal levels."

The other observation that helped to mold our current concept of how motor circuits are organized was that of Wiersma and Ikeda (1964), who found that single, identified interneurons could trigger or release a CPG to express a motor program. These interneurons were called *command neurons.* Command neurons are thought to be permissive in the sense that they merely signal when a particular motor behavior is to be activated (Kennedy, 1973). As defined by Kupferman and Weiss (1976), a command neuron must be both "necessary and sufficient" for the expression of a motor behavior. It would not, however, actually participate in coordinating the sequence of motor events underlying the behavior. Thus, unlike CPG neurons, the pattern(s) of spike activity in the command neuron and its associated sensory inputs were thought to lack instructive value.

While a CPG does not require phasic sensory feedback in order to generate a properly coordinated motor program, sensory input is necessary if a motor behavior is to respond appropriately to changes in an animal's internal and external environment. The CPG might best be thought of as a "primal" motor pattern that is embellished by sensory feedback. For example, going back to Wilson's work with locust flight behavior in which the importance of a CPG was emphasized, subsequent work from his laboratory demonstrated the need for sensory feedback in adjusting wing beat frequency (Wilson and Gettrup, 1963; Wilson and Wyman, 1965). More recently, it has been shown that the motor rhythm generated by the flight CPG can actually be reset by phasic sensory feedback (Bacon and Mohl, 1983). These data suggest that afferent inputs have direct access to a CPG. A complete rhythmic motor circuit, therefore, contains not only elements for initiating and maintaining the rhythmic activity in muscles, but also sensory input for modulating that activity as it progresses.

A. Central Pattern Generators

The CPGs for a wide variety of behaviors in invertebrates have now been studied. These include swimming, walking, flight, feeding, gut motility, and the neurogenic heartbeat in crustacea and leech. In all these studies, the cellular basis of the CPG oscillator has dominated the interest of investigators. Three types of oscillator networks can be defined. There are those whose basic rhythm arises from the ensemble synaptic interactions between members of the CPG. This is called an *emergent* network since no single member of the CPG population can, by itself, express rhythmic activity. Rather, rhythmicity is a function of the monosynaptic connections between these cells. In contrast, rhythmicity in some circuits is expressed as a consequence of cellular pacemakers within the CPG network. Such cells, called *endogenous bursters*, possess intrinsic voltage- and time-dependent ionic conductances which allow them to generate bursts of action potentials in the absence of an extrinsic influence (Alving, 1968; Meech, 1979). Finally, a third type of oscillator employs pacemaker neurons whose expression of endogenous bursting activity is under neurochemical control (Barker and Smith, 1976). That is, these neurons do not exhibit the voltage- and time-dependent ionic conductances necessary for autorhythmic activity unless an appropriate chemical signal is received from some outside source.* Because such cells depend on an extrinsic input to organize their bursting activity, they are referred to as *conditional bursters*. For purposes of clarity, I have drawn a distinction between endogenous and conditional bursters. It is clear, however, that some networks that employ endogenously active cells, such as the pyloric circuit in the lobster stomatogastric ganglion, also possess neurochemical mechanisms for modulating that activity (Flamm and Harris-Warrick, 1986a,b; see also Section II.A). Physiologically, therefore, the separation of conditional and endogenous bursting oscillators may be inappropriate.

Oscillatory activity is but one facet of a CPG; coordination and shaping of the output to muscles is also important. Even networks employing endogenous oscillators still depend heavily on synaptic interactions (both electrical and chemical) and intrinsic membrane properties (in addition to their autorhythmic activity) to process their output. In fact, there are instances

---

*This would include both neurotransmitters released from presynaptic terminals and humoral agents released into the blood.

where endogenous oscillator networks can continue to express a rhythm, albeit greatly altered, when the pacemaker cell is removed (Miller and Selverston, 1982b). Other networks, such as that controlling the heartbeat in the leech, have clearly separated a *timing* component of the CPG, which consists of four bilateral pairs of autoactive heart interneurons called HN(1-4), and a *patterning* component consisting of three additional pairs of interneurons called HN (5-7) (Peterson and Calabrese, 1982; Calabrese and Peterson, 1983). Although endogenous bursting properties play an important role in the oscillatory behavior of this network (Calabrese, 1979), it is the synaptic connectivity between all 14 interneurons which determines the final output to the heart motoneurons. The separation of a CPG into separate timing and patterning components raises an important caveat for the criteria used to identify CPG neurons (Kristan et al., 1977; Getting et al., 1980). That is, cells within the patterning component of the CPG would not meet the resetting test of the rhythm. In this case, a demonstration that such cells are an integral connection between the oscillator and the motoneuron pool is sufficient to consider them part of the CPG.

As pointed out by Peter Getting in a recent review of invertebrate CPGs (see Getting, 1988), "It appears that no two pattern generators are alike. Each seems to be a unique entity with its own set of peculiarities and properties." Indeed, the diversity of known CPGs throughout the animal kingdom suggests that the rules for assembling these circuits lie not in the form of their final configuration, but in the evolutionary conservation of a common set of synaptic and biophysical "building blocks." The heterogeneity of CPGs arises as a consequence of how these building blocks are combined in a circuit. Cellular building blocks can be separated into intrinsic and extrinsic components. Intrinsic components include those properties residing within a cell, such as voltage-dependent ionic conductances and intracellular biochemical pathways. Extrinsic components include the synaptic and neuromodulatory inputs to the cell which regulate the expression of intrinsic properties. In the following section, the *Tritonia* swim network will be used to illustrate some of the building blocks that make up a CPG.

Cellular Building Blocks

One of the best-studied invertebrate motor networks is escape swimming behavior in the mollusc *Tritonia diomedea*. This system was first described by Dennis Willows (1967; see also Willows and Hoyle, 1969) and later characterized at the cellular level in the laboratory of Peter Getting (see Getting, 1983c, Getting and Dekin, 1985b, for review). Escape swimming in *Tritonia* consists of a brief series of alternating dorsal and ventral flexions (Willows and Hoyle, 1969). The purpose of these movements is to lift the animal off the ocean floor and into the current so that it can elude predators. The network consists of three groups of identified premotor interneurons (Getting, 1977; Taghert and Willows, 1978; Getting et al., 1980) and one class of un-identified cell, called the I cell, which is postulated to exist on the basis of functional interactions (Getting and Dekin, 1985a). The identified neurons are cerebral cell 2 (C2), the dorsal swim interneurons (DSI), and the ventral swim interneurons (VSI). Within the VSI population, there are two subclasses of cells, VSI-A and VSI-B. Of these, VSI-B appears to the most important in terms of network oscillation (Getting, 1983b). The neural correlate of swimming behavior (Fig. 2) is characterized by each cell type firing a burst of action potentials during either the dorsal or ventral flexion phase of a swim. The DSI fire coactively and alternate with burst activity in the VSI. The bursts in C2 overlap with those in the DSI, but also span a short gap between the DSI and VSI bursts. One swim cycle is measured from the onset of a burst in DSI to the onset of the next DSI burst (Lennard et al., 1980). Note that as the swim progresses, there is a gradual increase in cycle period.

Oscillation

Since no single cell in the *Tritonia* swim CPG is capable of endogenous bursting activity, this network is an example of an emergent oscillator (Getting, 1983a). The basic synaptic connectivity of the swim CPG is shown in Figure 3.

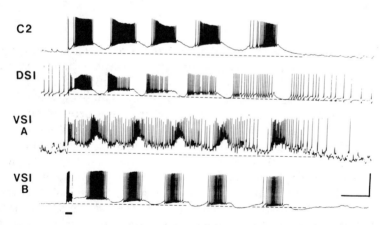

Figure 2 Simultaneous intracellular recordings from the four groups of identified *Tritonia* swim interneurons. The swim was initiated by an electrical stimulation of a peripheral nerve (bar). The dashed line represents the resting potentials for each cell. Note that during the swim a prolonged ramp-like depolarization is observed in all cell types. Calibrations: vertical, 50 mV for C2, DSI, and VSI-B; 25 mV for VSI-A; time scale, 5 s. (Adapted from Getting and Dekin, 1985a).

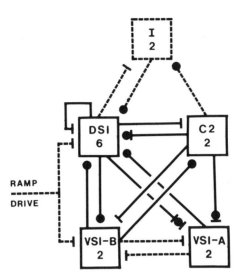

Figure 3 Monosynaptic connections between interneurons in the *Tritonia* swim CPG. Dashed lines are postulated connections. Each cell group is shown as a labeled box with the number of neurons in the group indicated. Excitatory connections are shown as bars, inhibitory connections as filled circles. Mixed symbols denote multifunctional synapses with the temporal sequence of action given by the order of symbols going from the presynaptic to postsynaptic neuron. (Adapted from Getting and Dekin, 1985a.)

The basis for rhythmic activity in this network is reciprocal synaptic inhibition between the DSI and VSI, paralleled by a delayed synaptic excitation via C2 from DSI to VSI (Getting, 1983a,b). The purpose of reciprocal inhibition is to prevent the coactivation of antagonist muscles, while the role of the delayed synaptic excitation is to switch phases between dorsal and ventral flexions. Reciprocal inhibition between the DSI and VSI is mediated by fast IPSPs (Getting, 1983b). The delay in synaptic excitation is more complex. The majority of the excitatory synaptic drive from the DSI to the VSI is filtered through C2, which receives direct monosynaptic excitation from the DSI (Getting, 1981). In VSI-A, C2 makes a multifunctional synapse consisting of an initial IPSP followed by an EPSP (Getting et al., 1980; Getting, 1981, 1983a). This multifunctional synapse ensures that the excitation that originated with DSI activity is effectively delayed before reaching VSI-A. In VSI-B, C2 makes only monosynaptic EPSPs. VSI-B, however, possesses an intrinsic membrane property which prevents the effects of excitation from being immediately expressed (Getting, 1983b). This intrinsic mechanism is the A-current first described by Connor and Stevens (1971). A-current is a fast, transient outward current carried by potassium ions.* This current turns on with membrane depolarization, but then quickly inactivates. Hyperpolarization of the membrane potential removes this inactivation. In VSI-B, the resting membrane potential is sufficient to remove A-current inactivation but not to turn it on. When the barrage of EPSPs from C2 arrive and depolarize the membrane, A-current is activated. This causes VSI-B to initially display a period of reduced excitability during which no action potentials fire. This period of reduced excitability may last for seconds, and like the multifunctional synapse in VSI-A, it effectively delays the transfer of excitation from part of the network to the other. Once the VSI start to fire, the DSI and C2 become silent because of inhibitory feedback from the VSI (see Fig. 2). As the activity in C2 is suppressed, the excitatory drive to the VSI begins to diminish. This reduction in synaptic drive is further enhanced by a time-dependent decrease in spike firing rate known as *spike frequency* adaptation. Spike frequency adaptation in these cells is caused by the expression of a slow, calcium-activated potassium current (Hume and Getting, 1982). This current turns on as a consequence of calcium crossing the membrane through voltage-dependent calcium channels that are activated by depolarization (Meech, 1978; Adams et al., 1980). When the activity in the VSI is sufficiently reduced, the DSI are released from their reciprocal inhibition and the next cycle of the swim starts.

Initiation and Termination

The network described so far would produce an alternating pattern of dorsal and ventral flexions during a swim provided the DSI were firing at rates appropriate to activate C2. Swimming behavior, however, is episodic and the network is normally inactive. A swim is only initiated when a tonic excitatory sensory input, called the ramp drive (Lennard et al., 1980; Getting and Dekin, 1985a), is directed to the DSI and VSI cell populations (see Fig. 2). When this happens, the synaptic interactions between members of the DSI population are altered so that swimming can occur. The synaptic connectivity of the DSI consists of monosynaptic excitation via both fast EPSPs (Lennard et al., 1980; Getting, 1981) and electrical synapses (Getting, 1981) in parallel with polysynaptic inhibition (Getting and Dekin, 1985a). In the quiescent state, the DSI interactions are dominated by the polysynaptic inhibitory pathway (Getting and Dekin, 1985a). While this inhibition can be directly demonstrated, its source is unknown. Hence, the I cell in Figure 3 is postulated to exist on functional grounds. The DSI are thought to excite

---

*By definition, an inward current increases the positive charge inside of the cell (depolarization) while an outward current reduces the positive charge inside the cell (hyperpolarization).

the I cell, which, in turn, inhibits the DSI and masks their excitatory connections. The DSI also excite C2, which is thought to inhibit the I cell. If the C2 are active, therefore, the DSI interactions will switch to mutual excitation because their polysynaptic inhibition is removed. While this connection is active during swimming, it is not functional at rest. The reason for this is that the C2 have a high spike threshold. When coupled with the low firing rate of the DSI at rest, this high threshold prevents the C2 from generating action potentials.

Swim activity begins when the extrinsic ramp drive causes a large synchronous depolarization in the DSI population which raises C2 above threshold. When the C2 fire, the monosynaptic excitatory connections between the DSI enter into a positive feedback phase. Swimming then ensues because of the network interactions described above. Swim activity will continue for as long as the net gain of the DSI interactions remain excitatory. This usually entails 2-20 swim cycles. Several factors contribute to the termination of the swim. Unlike the delayed transfer of excitation from DSI to VSI, there is no reciprocal delayed excitatory pathway. That is, VSI activity does not contribute to eventual excitation in the DSI. If it did, the network would be self-sustaining and swimming behavior would be expressed continuously. The DSI also possess spike frequency adaptation (Hume and Getting, 1982). With time, this adaptation accumulates over many swim cycles to reduce the overall firing rate of the DSI at a given level of depolarization. In addition, C2 makes a multifunctional synapse back onto the DSI consisting of an initial short EPSP and a longer IPSP (Getting, 1981). The time course of the IPSP (10-15 s) extends beyond an individual swim cycle, and so, over the course of the entire swim these IPSPs summate. And finally, the magnitude of the extrinsic ramp drive declines with time (Lennard et al., 1980; Getting and Dekin, 1985a). During the course of swimming activity, therefore, a net reduction in the membrane excitability of the DSI accrues and this makes it more difficult to depolarize C2 to threshold. As a result, the time interval between the beginning of activity in the DSI and when C2 reaches threshold gradually increases (see Fig. 2). This accounts for the increase in cycle period as swimming progresses. At some point, the firing rate of the DSI is insufficient to initiate C2 activity and the I cell again dominates. This signals the end of the swim.

The CPG underlying escape swimming behavior in *Tritonia* demonstrates the importance of both intrinsic and extrinsic membrane properties. Without intrinsic mechanisms such as the A-current in VSI-B and the high threshold of C2, coordinated activity could not occur. Similarly, the importance of extrinsic mechanisms is illustrated by the reciprocal inhibitory pathways between antagonists and the multifunctional synapses made by C2. The list of cellular mechanisms that form the building blocks of the *Tritonia* swim

CPG is by no means exhaustive. Many other mechanisms exist and their use by various motor circuits has been amply documented (see Getting, 1988, for review). The diversity of potential cellular mechanisms makes it impossible to point to a particular CPG and predict which ones will be important without a direct demonstration of their presence. The *Tritonia* swim CPG clearly demonstrates, however, that a knowledge of such mechanisms is an absolute requirement for understanding how a motor circuit works.

Plasticity

The *Tritonia* swim CPG raises several questions with regard to the organization of motor systems as depicted in Figure 1 (see Section II.A). For both C2 and DSI, tests of their ability to reset the swim rhythm using membrane potential perturbations have been successful (Getting et al., 1980; Getting, 1983b). Thus, these cells satisfy the criteria for a CPG neuron. Both cells, however, also appear to satisfy the criteria for command neurons. That is, they are necessary and sufficient for escape swimming behavior to occur. Tonic depolarization of C2 in a quiescent preparation initiates swim activity because of its ability to inhibit the I cell and change the sign of the DSI interactions (Getting, 1977; Lennard et al., 1980; Getting and Dekin, 1985a,b). Likewise, tonic hyperpolarization of the C2 prevents swimming from occurring (Getting and Dekin, 1985a). The same is true for the DSI (Getting and Dekin, 1985a,b). In this case, the artificially imposed depolarization overrides the effects of the I neuron, raises C2 to threshold and initiates swimming. The apparent hybrid command/CPG function of these cells argues against a strict hierarchial organization of motor systems where each level of the circuit (i.e., command, CPG, etc.) is assigned to specific types or groups of neurons. Such a convergence of function has also been observed in other systems, such as the pyloric circuit in the lobster stomatogastric ganglion (Selverston et al., 1977) and the swimmeret circuit in crayfish (Heitler and Mulloney, 1978). In both these networks motoneurons are part of the CPG.

Another issue raised by the *Tritonia* network is the definition of a command neuron and the plasticity of CPGs. Recall that the original observations of Wiersma and Ideda (1964) implied that the command function and its associated sensory input was permissive in that it simply released a preexisting network to be expressed. For the *Tritonia* swim network, this is not the case. The command function is instructive in that it organizes the swim CPG by reversing the sign of the DSI-DSI interactions. This also suggests that CPGs are not hardwired and that the swim interneurons can be organized into more than one functional configuration. In considering this possibility, Getting and Dekin (1985b) employed the term "polymorphic network" to emphasize the difference between an anatomically defined network consisting of monosynaptic connections and a motor circuit consisting of a

functional subset of these connections. For example, in a quiescent preparation the mutual excitation between members of the DSI population, the reciprocal inhibitory pathway between the DSI and VSI, and the excitatory
connections from C2 to VSI are not functional. That the swim interneurons
can still participate in other motor circuits while in this configuration is illustrated by the fact that a general reflexive withdrawal of appendages (i.e., gills,
rhinophores, etc.) occurs when a sensory input of short duration is directed
to the DSI and VSI (Getting et al., 1980; Lennard et al., 1980). During the
reflexive withdrawal, which last 1-2 s, the DSI and VSI fire coactively to control the appropriate flexion motoneurons for this motor act. Except for a
single spike at the very beginning of the withdrawal, C2 remains silent. It
seems from these observations that the quality of the initial sensory stimulus
does possess instructive value in determining which synaptic connections will
be functional within the *Tritonia* swim network. A long-duration stimulus
organizes one set of connections appropriate for swimming while a short-
duration stimulus organizes another set of connections for reflexive with-
drawals.

The finding that the command function and its associated sensory inputs have instructive value, in that they provide the inputs necessary to as-
semble specific circuits from a larger collection of potential monosynaptic
connections, is complemented by work in the pyloric system of the lobster
Panulirus interruptus. Here, central mechanisms for organizing multiple
circuits within a given network have been identified. The pyloric CPG is
located in the stomatogastric ganglion and functions to move and filter food
in the foregut. Several factors make this system uniquely suited for studying
the effects of neuromodulatory substances and CPG plasticity. First, like the
Tritonia swim system, the cellular circuitry for the pyloric rhythm has been
extensively characterized (Eisen and Marder, 1982; Maynard and Selverston, 1975; Miller and Selverston, 1982a,b). Second, the neurotransmitters
(Lingle, 1980; Marder, 1976; Marder and Eisen, 1984a,b) and intrinsic membrane properties (Russell and Hartline, 1978, 1982; Miller and Selverston,
1982a) used by the pyloric CPG are known. And finally, neuromodulatory
inputs from other parts of the lobster CNS have been identified (Russell and
Hartline, 1978; Claiborne and Selverston, 1982).

The pyloric system is an example of an oscillator circuit that employs
an endogenous bursting neuron called the anterior burster (AB) cell. AB is
the only interneuron in this CPG (Miller and Selverston, 1982a). The other
13 cells are all motoneurons and include two pyloric dilator cells (PD), the
lateral pyloric cell (LP), the ventricular dilator cell (VD), the inferior cardiac cell (IC), and eight pyloric cells (PY). Normally, the activity of this circuit is modulated and enhanced by inputs from the commissural and esophageal ganglia. These extrinsic inputs can induce membrane properties such as
conditional bursting activity (Miller and Selverston, 1982a,b; Marder and

Eisen, 1984b; Marder et al., 1986; Flamm and Harris-Warrick, 1986b) and plateau potentials (Russell and Hartline, 1978). They can also alter the synaptic drive to various cells in the circuit (Eisen and Marder, 1984). With the connectives from the commissural and esophageal ganglia left intact, all 14 cells within the pyloric circuit participate in generating the pyloric rhythm (Fig. 4A) (Russell, 1979). AB, PD, and VD are coupled by electrical synapses. All other connections within the circuit are inhibitory chemical synapses. Cycle period is determined primarily by the autorhythmic activity of AB,

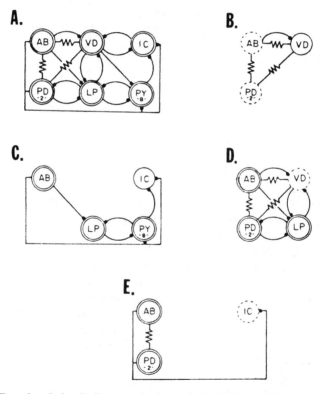

Figure 4   Functional circuit diagrams for the pyloric CPG are shown when inputs from the commissural and esophageal ganglia are present (A) or not present (B). C, D, and E represent the actions of 10^{-4} dopamine, 10^{-4} octopamine, and 10^{-5} serotonin, respectively (inputs from the commissural and esophageal ganglia are not present). The strength of activity in a given cell is illustrated by the circle type: double circle, strong activity; single circle, moderate activity; dashed circle, weak activity. Synaptic connections: resistor symbol, electrical synapse; filled circle, inhibitory synapse. For each circuit, inactive neurons are omitted. (Adapted from Flamm and Harris-Warrick, 1986a.)

which also entrains the activity in PD (see Selverston et al., 1983). Inhibitory synaptic connections, particularly those from the AB-PD pair to the rest of the network, and the reciprocal connections between the LP-IC and PY-VD cell pairs are important in defining phase relationships. One cycle of activity consists of an isolated AB-PD burst, followed by an LP-IC burst, and finally a PY-VD burst.

To determine the extent to which inputs from outside the stomatogastric ganglion contribute to generating and shaping the pyloric rhythm, the activity of the nonmodulated CPG must be observed. To do this, a sucrose block can be applied to the stomatogastric nerve, which connects modulatory inputs from the commissural and esophageal ganglia to the stomatogastric ganglion (Russell, 1979; Flamm and Harris-Warrick, 1986a). Under these conditions, endogenous bursting activity in the AB cell is greatly attenuated and the pyloric rhythm becomes much slower (Miller and Selverston, 1982b; Flamm and Harris-Warrick, 1986a). The circuit for this "basal" pyloric rhythm is shown in Figure 4B. Only the AB, PD, and VD cells are active. Using this isolated circuit, Flamm and Harris-Warrick (1986a,b) performed an elegant series of experiments to determine how the pyloric CPG was reorganized by several identified neuromodulatory inputs. These included serotonin (Beltz et al., 1984), dopamine (Barker et al., 1979; Kushner and Barker, 1983), and octopamine (Barker et al., 1979). Each amine assembles a slightly different circuit configuration, which, in turn, produces a modified pyloric rhythm in terms of cycle period and phase relationships (Fig. 4C-E). In all cases, the pyloric rhythm is enhanced owing to the amplification of bursting activity in the AB cell.

The particular cycle period associated with each amine can be attributed to a distinctive form and frequency of the bursting pacemaker potential (BPP) underlying AB cell bursting activity. Tetrodotoxin abolishes the BPP induced by serotonin and octopamine, while reducing extracellular Ca^{2+} abolishes the BPP induced by dopamine. These observations indicate that the AB cell can choose between multiple ionic mechanisms in generating its BPP and that each mechanism conveys unique information with regard to cycle period. Each drug also acts on other cells in the pyloric CPG. For example, dopamine induces tonic excitation in the LP, PY, and IC cells and tonic inhibition in the PD and VD cells.* By selectively changing the excitability in

---

*Not only do neurochemicals selectively affect subsets of cells, they may also affect cells within this subset differently (see also discussion in Marder and Eisen, 1984b). For instance, serotonin and octopamine both have the same action on the AB cell. In the VD and LP cells, however, serotonin induces tonic excitation while octopamine inhibits activity. Thus, in contrast to widely held views, the action of a neurochemical on a particular cell has little or no predictive value on how it will affect other cells within or outside the circuit.

subsets of cells in the CPG, the synaptic drive to other CPG cells will be altered (see Eisen and Marder, 1984). In large part, this accounts for the different phase relationships observed for each drug since these are primarily determined by reciprocal inhibitory connections within the network (see Selverston et al., 1983).

Chemical induction and modulation of CPGs is not limited to the pyloric rhythm of stomatogastric ganglion, nor is it solely an invertebrate phenomenon. In vertebrates, chemical activation of CPGs has long been a useful experimental tool. For instance, fictive swimming activity in the isolated lamprey spinal cord is typically induced by bath application of the excitatory amino acid D-glutamate (Poon, 1980; Cohen and Wallen, 1980), which is thought to activate *N*-methyl-D-aspartate receptors (Brodin et al., 1985; Brodin and Grillner, 1985). In the cat, locomotor activity can be induced by L-dopa (Jankowska et al., 1967; Grillner, 1969). L-dopa is a catecholamine precursor, and the actual neurotransmitter involved in activating this locomotor activity is thought to be norepinephrine (Forssberg and Grillner, 1973; Jordan, 1983). Even rhythmic breathing movements in mammals appear to be subject to this global type of neurochemical control. If breathing movements are suppressed by alcohol or barbiturate narcosis, exogenous application of thyrotropin-releasing hormone to the brainstem reestablishes a rhythmic breathing pattern (Breese et al., 1974, 1981; Kraemer et al., 1976; Andry and Horita, 1977).

B. Redefining the Command Function

Clearly, chemical activation of CPGs is a widespread phenomenon in the animal kingdom. The question arises, however, can the model depicted in Figure 1 account for these drug actions. Are they, as initially suggested by Grillner and Shik (1973), the neurochemical correlate of the command function? The data for the aminergic modulation of the pyloric rhythm argue against this since a rhythm can be induced by each individual amine in the absence of the others. Thus, they may be *sufficient* for the motor pattern to occur but they are not *necessary*. This is also the case for chemically induced locomotion in cats, where Steeves et al. (1980) demonstrated that pharmacological depletion or norepinephrine by 6-hydroxydopamine and α-methyltyrosine failed to block locomotion induced by electrical stimulation in the mesencephalic locomotor region.

A similar problem arises in studies of positioning behavior in crayfish and lobsters (reviewed by Larimer, 1988). Here, many neurons are found that are sufficient to evoke motor activity when electrically stimulated at high frequency (Jellies and Larimer, 1985, 1986). Hyperpolarization of these individual cells, however, does not prevent the behavior from occurring. An important observation made in these studies is that during the naturally induced behavior, the physiological spike frequency in such cells was always less than that required to evoke the behavior by artificial electrical stimulation. This has led to the idea of "command elements" which are subsets of neurons within a larger command system (Jellies and Larimer, 1986; Larimer, 1988). Each individual element is sufficient (if stimulated at artificially high levels) but not necessary for activating a behavior. Physiologically, it would be the coactivation of command elements which reach some threshold to activate the CPG for a behavior. The overall form of the command message can change since different subsets of command elements can combine to reach this threshold. Each specific form of the command function, therefore, determines a particular configuration of the CPG. Like the *Tritonia* swim system, this model implies that both the command function and its associated sensory inputs have instructive value.

The model presented above is similar to an earlier one suggested by Davis (1976) in which the command function is viewed as being distributed among many neuronal elements, subsets of which act in "consensus" to form the actual trigger for the behavior. The ability to organize subsets of control elements to produce slightly altered forms of a motor rhythm has been further characterized by Kien (1983) for locust walking behavior. The CPG controlling walking in locusts is located in the thoracic ganglia. When it is isolated from the rest of the nervous system, spontaneous walking does not occur although local tactile stimulation may activate the behavior. In this case, walking only proceeds for as long as the local sensory input is maintained. This is in contrast to the intact preparation where microstimulation in the rostral parts of the CNS initiates walking behavior that far outlasts the duration of the stimulus. It was also found that specific sites in the rostral CNS initiated different types of walking in terms of direction, speed, step size, and so forth. This led to the hypothesis that it is the "across fiber pattern" that determines the final form of the command function. As suggested by Kien (1983), ". . . walking is normally initiated by many fibers acting in consensus and the type of walking produced normally depends on which combination or pattern of recommendations to the walking generators is active. This pattern of activity across a large number of labeled input lines has been called the across fiber pattern." Across fiber pattern refers back to an idea first postulated by Pfaffman et al. (1976) for the processing of gustatory information. Using recordings from individual chorda tympani fibers, he demonstrated that each fiber had a preferred taste modality (i.e., salt, sweet, bitter, or sour) but that other modalities were also picked up. Likewise, fibers that did not prefer a specific taste modality still exhibited a small amount of activity in response to it. On the basis of these observations, Pfaffman et al. suggested that it was not the activity in a specific fiber (or labeled line) that conveyed taste quality, but the pattern of activity across all fibers in the pathway. As Kien points out, application of this concept to the initiation of motor programs resolves many problems with the original definition of a command function. First, the ensemble information carried by the command function would clearly be instructive. As we have seen for many systems (e.g., *Tritonia* swimming, lobster pyloric rhythm), this imparts plasticity to the CPG itself. Second, information carried by primary sensory afferents that directly acts on the CPG can now be included in the command function. As discussed above, such input has been demonstrated for locust walking, *Tritonia* swimming, and locust flight. In the intact animal, sensory input would be expected to work in conjunction with central inputs in determining the final form of the behavior.

Because these ideas depart so dramatically from the original concept of a command neuron, Kien (1983) has suggested the term *recommendation function* be used rather than command function. This seems reasonable in that it more accurately portrays the role of these inputs. With this in mind, it is equally appropriate to relabel the individual components of the recommendation function *control* rather than command elements. A formal definition of the recommendation function is that it represents the integrated activity of numerous individual control elements (including sensory inputs) which have organized themselves in such a fashion so as to specify a particular configuration for a CPG. Because the recommendation function is dynamic in the sense that it can instruct a given motor circuit how to adjust its pattern of activity, it provides the animal with the needed flexibility to deal with its environment.

III. Implications for the Control of Breathing

In the preceding sections we have seen that current views of invertebrate motor system organization are quite different from those envisioned even 10 years ago. Two areas have undergone a basic conceptual reorganization. First, the CPG is no longer viewed as a static hard-wired circuit. Rather, it possesses an inherent plasticity. It has also been shown that in some cases the CPG can be further divided into functional components (i.e., oscillator and patterning functions). The second area to have undergone major revision is the concept of a command function. The more contemporary idea of a recommendation function takes full advantage of the inherent plasticity of CPGs and also serves to organize the role of central and peripheral inputs to the CPG under the common theme of control elements. The underlying premise in all this work has been that the cellular sophistication of CPG neurons, in terms of their intrinsic and extrinsic membrane properties, allows for far more flexibility in the organization of motor circuits than was previously believed possible. In the following sections, these ideas will be applied to the central respiratory circuit, and potential cellular substrates for their sites of action will be discussed.

A. Oscillation and Pattern Formation in the Respiratory CPG

It has been known for some time that the minimal neuronal circuitry required to produce a rhythmic discharge pattern in the phrenic motor nerve (i.e., the respiratory CPG) is located in the lower brainstem (Pitts et al., 1939; Wang et al., 1957). In spite of this, we still have little insight into either the identity of those neurons which generate the basic respiratory rhythm or the mechanism used for oscillation. While there is some evidence for endogenous (or possibly conditional) bursting activity underlying the respiratory rhythm (Feldman and Cleland, 1982; Smith and Feldman, 1987), alternative models have also been presented (see Richter et al., 1986).* At present, the data are insufficient to firmly establish a particular oscillatory mechanism. This is largely due to the fact that most evidence for or against a particular model is indirect. Recent work with the isolated neonatal rat brainstem-spinal cord preparation (Suzue, 1984; Smith and Feldman, 1987), however, may provide an opportunity to finally address this question directly. This in vitro preparation still displays a respiratory rhythm and is mechanically stable enough for routine intracellular recordings to be made (Smith and Feldman, 1988). It should be possible, therefore, to identify oscillator neurons using the resetting criteria discussed above (see Section II) as well as test for the presence of endogenous (or conditional) bursting mechanisms.

Although little is known about the basic mechanism of oscillation in the respiratory CPG, another aspect of this circuit is far better understood. This involves the role of the dorsal and ventral respiratory groups (DRG and VRG) in shaping the output of the respiratory rhythm generator. As first suggested by Smith and Feldman (1987; see also Feldman et al., 1988), the respiratory CPG can likely be split into two components where burst timing and patterning are carried out by distinct groups of neurons. This is similar to the situation in the leech heart CPG where one group of endogenous bursting neurons controls cycle period and another group of neurons determines the shape of the motor outflow (Peterson and Calabrese, 1982; Calabrese and Peterson, 1983). The best evidence for a similar organization

---

*The possibility must also be considered that the respiratory rhythm employs different oscillator mechanisms at different times. There is some precedence for this in the pyloric rhythm of the lobster stomatogastric ganglion. Miller and Selverston (1982b) demonstrated that an altered pyloric rhythm could still be generated after removal of the autorhythmic AB cell.

in the respiratory CPG comes from sequential microlesioning studies of the DRG and VRG (Speck and Feldman, 1982). These experiments demonstrated that the gradual destruction of these areas reduced the amplitude of the phrenic nerve discharge but did not alter respiratory timing. This strongly suggests that the DRG and VRG are interposed between the respiratory oscillator and the phrenic motoneuron pool. Thus, they would be well placed to act as a premotor integrating centers (and part of the CPG) where modifications to the basal respiratory rhythm can be made.

The DRG, in particular, is suited for such a role. Most respiratory neurons in the DRG fire during inspiration (reviewed by Cohen, 1979; Feldman, 1986) and appear to be bulbospinal interneurons (Euler et al., 1973; Berger, 1977). Cross-correlation studies comparing the activity of DRG inspiratory cells and contralateral phrenic nerve activity suggest that as many as 80% of the inspiratory neurons in the DRG make monosynaptic connections with phrenic motoneurons (Cohen et al., 1974; Hilaire and Monteau, 1976). These conclusions are strengthened by anatomical studies that have shown that DRG neurons project to spinal cord levels where respiratory motoneurons are located (Loewy and Burton, 1978). The DRG also appears to be involved in the processing of central and peripheral inputs affecting ventilation. For instance, the nucleus of the tractus solitarius (NTS), within which the DRG is located, is the major sensory nucleus of the vagus nerve which carries information from sensory receptors within both the lungs and airways (Kalia and Mesulam, 1980) as well as peripheral chemoreceptors (Kirkwood et al., 1979; Donoghue et al., 1984). Central inputs to the NTS include the cerebral cortex (Mettler, 1935a,b; Torvik, 1956), hypothalamus (Bereiter et al., 1980), raphe nucleus (Palkovits et al., 1986), and pontine respiratory centers (Saper and Loewy, 1980). It is apparent from these observations that the DRG is strategically located to integrate the basal respiratory rhythm with a large number of central and peripheral inputs. As will be seen below (see Section III.C), the cellular properties of neurons in the NTS provide further evidence for a role of the DRG in sculpturing the final form of the respiratory CPG output.

B. Respiratory Drive: An Example of the Recommendation Function

A fundamental issue in the control of breathing is the source of the respiratory drive. Many inputs to the central respiratory circuit are recognized (reviewed by Feldman, 1986). Among these, central chemosensitivity toward CO_2/H^+ levels is continuously active. As Feldman (1986) points out, however, the continuous monitoring of CO_2/H^+ by the central chemoreceptors does not mean that this input is the most dominant influence on breathing. Episodic factors such as wakefulness (Hugelin, 1980), exercise (Eldridge et al., 1985; Eldridge and Millhorn, 1986), and phonation (Bunn and Mead, 1971) also make significant contributions to this drive and can cause dramatic changes in breathing pattern. When a particular input is active, it may alter the degree to which other inputs are expressed. For instance, during speech the ventilatory response to CO_2 is muted compared with that in quiet breathing (Bunn and Mead, 1971). This implies that the effects of chemosensitive inputs on the central respiratory circuit are transiently being overridden by phonation inputs. These observations suggest that the respiratory drive is composed of a variable number of inputs, and that a particular pattern of breathing depends on which inputs are active as well as the degree to which their effects are amplified or curtailed.

Although the individual effects of different types of inputs on breathing are well established, no conceptual framework exists for organizing their overall impact on the respiratory CPG. The contemporary interpretation of the recommendation function may provide such a framework. The individual control elements in this system would be comprised of the various factors discussed above (i.e., central chemosensation, wakefulness, etc.). Primary sensory afferents with access to the respiratory CPG may also be defined as control elements, provided they are sufficient to maintain rhythmic breathing movements. This would include the peripheral chemoreceptors but would exclude mechanoreceptors, which serve as a moment-to-moment adjustment to breathing movements. The respiratory recommendation function would represent the summated activity of all these control elements. In addition to providing a mechanism for maintaining regular breathing movements, the respiratory recommendation must also configure the respiratory CPG in such a way as to generate a pattern of breathing movements that is appropriate for the immediate needs of the animal. To do this, each control element would make its own unique contribution to the message carried by the respiratory drive. Within this message, individual control elements might preferentially act on certain aspects of the respiratory CPG. It is the resulting "across fiber pattern," however, that actually determines the final configuration of the respiratory CPG.

The action of a given recommendation on the respiratory CPG may range from changes in gain (e.g., during hypoxia) to modifications in circuit organization (e.g., sleep—see Feldman, 1986, for review). Because of the wide range of potential actions initiated by the respiratory recommendation, it is possible that individual control elements may sometimes contribute conflicting messages. Some mechanism, therefore, must exist for sorting out the pertinent information. Given the diverse origin of the various control elements affecting breathing, it is difficult to imagine how they could interact among themselves to sort out conflicting aspects of the recommendation. The two options that are left then are (1) a pre-CPG integrating area where the respiratory recommendation is modified to eliminate conflicting signals, or (2) an intrinsic CPG ability to intelligently decipher the recommendation message. Currently, there is no compelling anatomical evidence for an intervening structure between the respiratory CPG and its control elements.* One can, however, envision cellular mechanisms within the CPG that impart the ability to selectively respond to the respiratory recommendation. For instance, if one aspect of a particular recommendation called for increasing the expression of an A-current while another aspect of the message depolarized the cell into a range where this current undergoes steady-state inactivation, the former message would be negated (see Section III.C). Thus, there is an intrinsic hierarchical organization of membrane properties that could be the basis for a decision-making process within the CPG itself. Even if an anatomically defined pre-CPG integrating area(s) is eventually defined, one must ultimately look to cellular mechanisms for sorting out conflicting aspects of the respiratory recommendation.

A potential source of confusion in this analysis is the application of a recommendation function to a continuously expressed behavior such as breathing. The original rationale for a recommendation function stemmed from work with episodic behaviors, such as *Tritonia* swimming and locust walking, where the recommendation was needed for initiation of the behavior. This does not appear to be the case for the respiratory CPG. The hypothesis that the recommendation function is also instructive, however, does have significance for rhythmic breathing movements. This can be readily seen by considering the pyloric rhythm in the lobster stomatogastric ganglion. Here too, an essentially continuous rhythm is generated by a well-defined CPG in the absence of any extrinsic inputs (see Section II.B). It is unlikely that this basal rhythm provides a physiologically relevant motor output since it is both weak and subject to failure (i.e., spontaneously turns off) (see Flamm and Harris-Warrick, 1986a). It is only with the added embellishment provided by extrinsic inputs from the commissural and esophageal ganglia that the pyloric rhythm is able to meet the needs of the animal. Thus, while the respiratory CPG may be able to inherently generate a motor rhythm, it still depends on the recommendation function and its associated control elements to regularize the rhythm.

---

*Some afferents from peripheral chemoreceptors terminate in the dorsal medial part of the NTS and thus do not directly innervate DRG neurons (Berger, 1980; Donoghue et al., 1984). Whether other control elements terminate in this area is not known at present. It is possible, however, that areas such as this may take on increasing importance in the future as pre-CPG integrating areas.

C. Dorsal Respiratory Group Neurons

As stated above, there is compelling evidence suggesting that the DRG is an integral part of the "shaping" component of the respiratory CPG. In fulfilling this role, the DRG must be able to interpret the respiratory recommendation and convey this message to respiratory motoneurons. In invertebrate systems like the pyloric system of the lobster, it was demonstrated that a detailed knowledge of the cellular properties of CPG neurons was necessary in order to understand how the recommendation function is played out on the CPG. Until recently, such information was unavailable for respiratory CPG neurons. In the mid-1960s however, Yamamoto and McIlwain (1966) introduced the "thin section" brain slice preparation for studying the cellular properties of hippocampal neurons. Since then, the brain slice technique has been applied to almost every region of the mammalian brain, including brainstem respiratory areas. The advantages of this technique are quite similar to the original rationale for using invertebrate preparations to study rhythmic motor behaviors (see Section II). They include: (1) mechanical stability allowing intracellular recordings for current clamp, voltage clamp, and intracellular injection of dyes and pharmacological agents, (2) long-term viability permitting complicated experimental protocols to be carried out on single cells, and (3) complete control over the extracellular environment so that ionic and pharmacological manipulations can be done. There are also disadvantages to this technique. First, neural circuits are sufficiently disrupted so that the integrated activity of a network cannot usually be studied. This is the case for the respiratory CPG. Second, there is no way to uniquely identify respiratory neurons in the slice using currently accepted criteria (see Cohen, 1979). That is, respiratory neurons are generally identified by comparing their activity to the phrenic (or hypoglossal) nerve discharge.* This, of course, is not possible in the slice.

With these caveats in mind, one might question whether the brain slice technique can be successfully applied to the study of the control of breathing. These concerns are largely offset by the fact that respiratory areas in the brainstem have been extensively studied in situ and their neuroanatomy is well characterized. One is able, therefore, to use anatomical criteria such

---

*Identification of respiratory neurons involves comparing the phasic firing activity of a cell with the phrenic nerve discharge. Neurons characterized in this manner are called respiratory-related units (RRUs). An advantage of this method is that it allows further classification of respiratory neurons into subgroups such as early inspiratory, late inspiratory, expiratory, etc. (see Fig. 5 of Feldman, 1986). A major disadvantage of this classification scheme is that it is biased against potential respiratory neurons which fire tonically.

as location and connectivity (see below) to demonstrate that neurons studied in the slice are from respiratory centers such as the DRG and VRG. In the final analysis, however, it is the type of questions that are asked using the brain slice technique that determines if the data will be relevant to the control of breathing. With regard to the DRG, a primary question to ask is how many different types of neurons are located in this region? Can different classes of neurons be identified using cellular criteria such as morphology and membrane properties? And, if so, do these neuronal classes share any properties with DRG respiratory neurons characterized in situ? Using an in vitro brainstem slice preparation from adult guinea pigs, these questions have been successfully addressed.* While brief reviews of some of this work appeared (Dekin and Getting, 1987b; Dekin et al., 1987b), the following discussion will focus on the potential sites of interaction between the cellular properties of DRG neurons and the respiratory recommendation message.

Cellular Organization

In guinea pigs, respiratory-related units in the medulla have been mapped in situ using extracellular recording techniques (Richerson and Getting, 1983; see also Fig. 1 of Dekin et al., 1985). A high density of inspiratory units are found within the ventral parts of nucleus tractus solitarius (ventral NTS) from approximately the level of the obex to 1.2 mm rostral. This area is believed to be the DRG of the guinea pig. Consistent with this idea is the finding that many neurons in the vNTS are bulbospinal and project to the region of the phrenic motor nucleus (Johnson et al., 1986; Dekin et al., 1987a). When ventral NTS neurons are studied with intracellular microelectrodes in the in vitro slice preparation, they do not exhibit rhythmic bursting patterns suggestive of respiratory-related activity (Dekin and Getting, 1984; Champagnat et al., 1986; Dekin et al., 1987a). They often do, however, fire nonrhythmically at low frequency. By studying the morphology of ventral NTS cells and their repetitive firing properties during depolarizing stimuli, three classes of ventral NTS cells are able to be distinguished. These are termed types I, II, and III (Dekin et al., 1987a). The use of the notation I-III emphasizes that these cells have not been identified as respiratory neurons, and that the purpose of studies such as this is to characterize the *basic cellular organization* or a respiratory area.

---

*While most studies on the control of breathing have been done using cats, the guinea pig was chosen for slice studies because of its size, availability, and the fact that obtaining brain slices from guinea pigs was a demonstrated technology. Prior to these studies, the DRG in the guinea pig was mapped using extracellular recordings in situ (Richerson and Getting, 1983). The location and properties of RRUs in the guinea pig are similar to those described for the cat.

Figure 5 illustrates the repetitive firing activity and morphology of the three classes of ventral NTS neurons. To study their repetitive firing properties, two pulse paradigms can be used. In the first (Fig. 5, upper panels), the cell is simply depolarized from its resting level (i.e., −60 to 50 mV). Type I neurons respond with a brief burst of action potentials, which rapidly declines to a lower steady-state level of activity. This time-dependent decline in spike activity is called spike frequency adaptation (see Section II.A). Type II neurons also exhibit spike frequency adaptation, but to a lesser degree. Finally, type III neurons display little or no adaptation during the course of their depolarization. In the second test (Fig. 5, lower panels), this same depolarization is preceded by a membrane hyperpolarization to levels more negative than −70 mV. This alters the repetitive firing activity in types I and II neurons but not type III. In type I cells, the initial period of high-frequency spike activity is attenuated. In type II neurons there is an actual delay between the onset of the depolarization and the beginning of spike activity. This delay can be several hundred msec in duration and is termed delayed excitation (Dekin and getting, 1984; see also Getting, 1983b; Byrne, 1980). The ionic mechanisms underlying spike frequency adaptation and the delay in excitation are similar to those already described for the *Tritonia* swimming CPG.

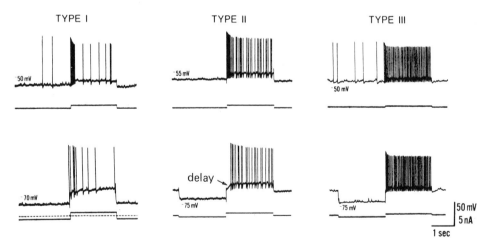

Figure 5 Intracellular recordings illustrating the repetitive firing properties in the three classes of ventral NTS neurons. Upper traces are voltage, lower traces are current injected through the microelectrode. Dashed line for type I neuron indicates the zero current level at a resting potential of −50 mV. In the lower panels, neurons were either tonically hyperpolarized (type I) or given a 2-s hyperpolarizing prepulse (types II and III). (Adapted from Dekin and Getting, 1987a.)

Each of these cell types is also associated with a unique cell body morphology as studied with intracellular injections of the fluorescent dye Lucifer yellow (Dekin et al., 1987a). In addition, retrograde transport of dyes injected into the phrenic motor nucleus demonstrates that types I and II neurons give rise to bulbospinal connections while type III neurons appear to be higher-order interneurons (see Johnson et al., 1986; Dekin et al., 1987a). An important consequence of the unique repetitive firing properties in ventral NTS neurons is that cells from each class can be routinely identified from one preparation to another.

The cellular organization of the ventral NTS raises an interesting point concerning the DRG. That is, several different types of respiratory neurons have been classified in this respiratory area, including cells with augmenting and decrementing discharges, as well as those which begin to fire at different times after the phrenic nerve discharge begins (see Cohen, 1979; Feldman, 1986, for review). In fact, there are many more functionally defined respiratory neurons in the DRG than there are neuronal classes in the ventral NTS. Thus, individual classes of ventral NTS neuron must give rise to more than one type of respiratory neuron. This implies that extrinsic inputs to NTS cells are important in defining particular types of respiratory neurons. A simple example of this is the observation that no single class of ventral NTS cell can exhibit an augmenting discharge, suggesting that it is the shape of the synaptic drive itself which determines this property in respiratory neurons.

Other characteristics of respiratory neurons may involve more complex interactions between both intrinsic and extrinsic properties (see section on "Pattern Formation in the DRG" below). The observation that extrinsic inputs are important in determining the patterns of activity displayed by respiratory neurons suggests that these patterns are not static. That is, they should change as the inputs to a neuron are altered. Recent studies employing chronic recording techniques (see Chang et al., 1988; Chang and Harper, 1989) in awake behaving guinea pigs have demonstrated that the discharge patterns of respiratory neurons do in fact change (Chang and Foster, 1986). Thus, the notion of a static respiratory neuron discharge pattern is likely an artifact of recording conditions where extrinsic inputs are minimized (i.e., anesthesia, decerebration, etc.). To explain the diversity and plasticity of the firing patterns displayed by respiratory neurons in the DRG, Dekin and Getting (1987b) suggested the term *pattern-forming units* (see also Dekin et al., 1987a). Pattern-forming units are defined as subsets of ventral NTS neurons (i.e., types I, II, or III) that receive a unique combination of synaptic drive and modulatory inputs (i.e., from the respiratory recommendation function). Because both synaptic drive and modulatory inputs can affect which intrinsic properties of a cell are expressed, and to what degree, they will have a profound effect on the inherent pattern-forming capabilities of neurons within the DRG. This implies that the ability of discrete components of the dorsal respiratory group to form patterned activity can be independently controlled without necessarily altering the overall syanptic connectivity of the respriatory CPG. A detailed knowledge of the intrinsic properties of ventral NTS neurons, therefore, is needed to understand how this sculpturing process occurs.

Intrinsic Membrane Properties

In addition to the traditional sodium and potassium conductances responsible for the action potential, ventral NTS neurons possess several other ionic conductances that shape their repetitive firing activity. These include calcium currents (Champagnat et al., 1986; Dekin and Getting, 1987a; Dekin, 1988b), calcium-activated potassium currents (Dekin and Getting, 1984, 1987a; Champagnat et al., 1986; Dekin, 1988a,b), the A-current (Dekin and Getting, 1984, 1987a; Champagnat et al., 1986), and an inward rectifier current (Dekin et al., 1987a; Dekin, 1989). Table 1 list these currents and their action on each class of ventral NTS neuron. The effect of a given current may not be the same in each cell type. One reason for this is that the time- and voltage-dependent properties of a current may vary from one cell class to another (Dekin and Getting, 1987a). In addition, the magnitude of a given current is not necessarily the same in all cell classes. For instance, the size of the calcium-activated potassium current that mediates spike frequency adaptation in ventral NTS neurons (see Table 1) follows the sequence type I > type II > type III* (Dekin and Getting, 1987a). As a result of these differences, the expression of various membrane currents is altered and this allows each cell type to display a different set of repetitive firing properties.

This point is illustrated by the interactions between the A-current and calcium current in types I and II neurons. Recall that the A-current is activated by membrane potential depolarization, but then rapidly inactivates (see Section II.A). Hyperpolarization of the membrane potential is needed to remove this inactivation. In type I neurons, the A-current activates at membrane potentials around − 65 mV, while in type II neurons this current is not expressed until the membrane potential is depolarized to about − 55 mV (Dekin and Getting, 1984, 1987a; Dekin et al., 1987b). A similar disparity exists for the voltage dependence of the removal of A-current inactivation. In type I neurons, significant A-current inactivation is only removed when the cell is hyperpolarized to membrane potential levels of about

---

*The small amount of spike frequency adaptation in type III cells is likely due to a rundown of the voltage-dependent sodium current with repetitive firing rather than the activation of a calcium-activated potassium current.

Table 1 Ionic Conductances

Currents	Type I	Type II	Type III
Sodium	Rising phase of spike	(Same as type I)	(Same as type I)
Delayed rectifier	Falling phase of spike	(Same as type I)	(Same as type I)
A-current	Prevents high-frequency burst of spikes	Causes long delay in spike firing	Absent[a]
Low-threshold calcium	Contributes to initial high-frequency burst of spikes	Absent	Absent
High-threshold calcium	Spike broadening	Spike broadening	Absent[a]
Calcium-activated potassium	Spike frequency adaptation and termination of initial burst of spikes	Spike frequency adaptation	Absent[a]
Inward rectifier[b]	Prevents cell from becoming too hyper-polarized	(Same as type I)	(Same as type I)

[a]These currents may be present in negligible amounts, but do not appear to contribute to the repetitive firing properties of these cells.
[b]The inward rectifier current in these cells is only active at membrane potentials more negative than -70 mV (Dekin et al., 1987a). It could, however, compete with expiration phase inhibitory synaptic drive and affect the amount of A-current inactivation that is removed (see Dekin, 1989).

-70 mV. In type II neurons, A-current inactivation is removed when the membrane potential becomes more negative than -60 mV. Thus, the voltage dependence of the activation and removal of inactivation for this current is shifted by approximately -10 mV in type I cells with respect to type II. There are also differences in the calcium currents expressed by these two classes of neurons (Dekin, 1988b). Both cell types exhibit a high-threshold calcium current, which turns on at membrane potential levels more positive than -35 mV. This current is normally activated during an action potential and contributes to a gradual broadening of the spike during repetitive firing (see Dekin and Getting, 1987a). Type I neurons also possess a second, low-threshold calcium current, which is expressed at membrane potentials between -60 and -50 mV. This current is absent in type II neurons (Dekin, 1988b; also see Dekin and Getting, 1987a).

These properties of the A-current and calcium current are the basis for the repetitive firing patterns illustrated in Figure 5. The upper panels in this figure represent the repetitive firing activity in ventral NTS cells when the A-current is undergoing steady-state inactivation. During depolarization then, only the voltage-dependent calcium currents in these cells would be expressed. An important effect of the low-threshold calcium current in type I neurons may be to enhance the underlying depolarizing drive (in this case, positive current injection through the microelectrode) and increase the initial spike firing rate. Along with the high-threshold calcium currents, however, this low-threshold current will raise intracellular calcium levels. This turns on the calcium-activated potassium current responsible for spike frequency adaptation. In terms of raising intracellular calcium levels, the effects of the low- and high-threshold calcium currents in type I neurons will be additive. Thus, the total accumulation of intracellular calcium in type I neurons will be larger and occur more quickly than in type II cells. This additive effect accounts for the differences in spike frequency adaptation between these two classes of cells (Dekin and Getting, 1987a).

The lower panels in Figure 5 represent the effects of removing A-current inactivation so that this current can be expressed during subsequent depolarization. In type I neurons the outward A-current will compete with the low-threshold inward calcium current for control of the membrane potential. This coactivation of an inward and outward current causes a reduction in the initial spike firing rate. The ability of A-current to cause a delay in spike firing in type II neurons can be explained by the absence of a low-threshold calcium current. Acting alone, the A-current is able to dominate the membrane potential and transiently prevent spike firing (also see Section II.A). As this current inactivates, its ability to compete with the depolarizing drive diminishes and spike activity begins. Based on this analysis, type I neurons should be able to display a delay in excitation due to A-current if their low-threshold calcium current is blocked. This prediction has been tested experimentally by blocking the calcium current with cobalt (Fig. 6) (see also Dekin and Getting, 1987a). In the upper panel, a type I neuron is depolarized from -75 mV. As stated above, the initial low-frequency firing activity reflects the coactivation of the A-current and calcium current. In the lower panel, the same depolarization is given in the presence of 3 mM cobalt. In this case, the calcium current is blocked and the A-current is activated alone causing a long delay (approximately 800 msec) before spike activity begins. This experiment

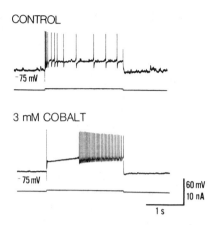

CONTROL

3 mM COBALT

Figure 6 Repetitive firing activity in a type I neuron depolarized from −75 mV before (upper panel) and following (lower panel) bath application of cobalt. Upper traces in each panel are voltage, lower traces are current injected through the micro-electrode. (Adapted from Dekin and Getting, 1987a.)

also illustrates one way in which extrinsic inputs could modulate intrinsic membrane properties (i.e., by either increasing or decreasing the expression of a current) and thereby greatly alter the repetitive firing activity of neurons.

Pattern Formation in the DRG

The characteristics of the A-current in type I and II neurons provide a direct example of how intrinsic membrane properties can interact with extrinsic synaptic (and modulatory) inputs. The removal of A-current inactivation depends on the level of membrane hyperpolarization, and as the membrane is made more negative more inactivation is removed. Thus, for a given level of depolarization the amount of A-current expressed will depend on the membrane potential level prior to depolarization. This *history-dependent* expression of A-current has important implications for the activity of respiratory neurons in the DRG. The synaptic drive to these cells is believed to consist of an alternating pattern of inhibition during expiration and excitation during inspiration (see Fig. 5 of Richter, 1982). Intracellular recordings from respiratory neurons in situ have shown that membrane potential levels as negative as − 80 mV can be reached during the inhibitory phase of their synaptic drive (Richter, 1982). In type II neurons, the removal of A-current inactivation occurs at membrane potential levels between − 60 and − 90 mV, which is within the range of membrane potential levels observed for DRG inspiratory neurons during expiration. The activation of A-current in type

II neurons causes a delay in spike firing during subsequent depolarization (i.e., delayed excitation). A similar delay is seen in a class of inspiratory DRG neurons which begin to fire after the start of inspiration (i.e., late inspiratory neurons; see Cohen and Feldman, 1984; see also Fig. 5 of Richter, 1982).

Late inspiratory neurons exhibit a range of delays (up to 600 msec) between the beginning of inspiration and when their spike activity begins. The history-dependent expression of A-current in type II neurons may provide a cellular basis for the activity of these late inspiratory neurons (Dekin and Getting, 1984, 1987a). This point is illustrated graphically in Figure 7, where the magnitude of delayed excitation in type II neurons was modulated by altering the level of membrane potential hyperpolarization prior to depolarization. The inset shows the pulse paradigm for this experiment. The delay (D) was measured as the time interval between the onset of depolarization and the occurrence of the first spike. The amplitude of a 2-s hyperpolarizing prepulse (V) was varied between − 40 and − 110 mV and then followed by a 2-s depolarizing step to a constant level. No delay was observed for hyperpolarizing prepulses between − 40 and − 50 mV. At − 55 mV, a small delay was observed. The magnitude of the delay increased as the prepulse became more negative and appeared to reach a plateau at about − 95 mV. This experiment illustrates that a range of delays similar to those observed in late inspiratory neurons can be generated in a homogeneous cell population simply by manipulating the level of membrane potential hyperpolarization prior

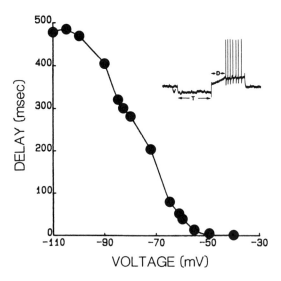

Figure 7 Dependence of delayed excitation in type II neurons on prepulse voltage. See text for details. (Adapted from Dekin and Getting, 1987a.)

to depolarization. It is possible, therefore, that modulation of expiratory phase inhibition is one mechanisms for altering the activity of respiratory neurons in vivo.

In type I neurons, the effect of membrane hyperpolarization on the removal of A-current inactivation occurs over a more limited range than in type II neurons. In these cells, inactivation starts to be removed at $-70\,mV$ and is complete at around $-80\,mV$. Thus, the removal of inactivation in type I neurons approaches an on-off situation. If these cells are depolarized from membrane potentials more positive than $-70\,mV$, they can produce initial firing rates up to 90 spikes/s (Dekin et al., 1987a). If they are depolarized from membrane potential levels more negative than $-70\,mV$, however, their peak firing rate decreases by approximately 75%. One hypothesis for the role of A-current in these cells is to maintain a subpopulation of type I neurons in a relatively quiescent state (i.e., by reducing their ability to generate high-frequency bursts of spikes) and thereby modulate the number of physiologically active respiratory neurons in the DRG (Dekin and Getting, 1987a). Since type I neurons are bulbospinal with projections to the phrenic motor nucleus, this could be a mechanism for adjusting the amplitude of the synaptic drive to the phrenic motor nucleus. Again, this could be accomplished by adjusting the level of expiratory phase inhibition, in this case around a membrane potential level of $-70\,mV$.

Chemical modulation of the intrinsic properties of ventral NTS neurons has also been described (Champagnat et al., 1983; Dekin et al., 1985; Dekin, 1988a). An example of such an action is the effect of thyrotropin-releasing hormone (TRH) on ventral NTS cells. There is evidence that TRH is a neurotransmitter (or modulator) involved in the control of breathing. It has been localized within nerve terminals in the NTS (Hökfelt et al., 1975b) and possesses a potent analeptic effect on the respiratory CPG. This effect is manifest as a large increase in the minute ventilation of normal animals (Hedner et al., 1983; Mueller et al., 1985) or a reversal of respiratory depression in animals poisoned with barbiturates (Kraemer et al., 1976; Andry and Horita, 1977; Breese et al., 1981) and alcohol (Breese et al., 1974). These data suggest that TRH plays a critical role in adjusting the excitatory drive to phrenic motoneurons. The increase in minute ventilation caused by TRH is usually associated with an increase in respiratory rate. In vagotomized animals, however, the increase in minute ventilation is attributable to a direct increase in tidal volume (Mueller et al., 1985). Information from lung stretch receptors, which limit lung inflation via the Breuer-Hering reflex, is carried in the vagus nerve. When activated, this reflex tends to mask large increases in tidal volume caused by TRH and perhaps transform them into an increase in rate. Thus, the action of TRH in animals lacking stretch receptor feedback may be a more accurate reflection of how this drug actually affects the respiratory CPG.

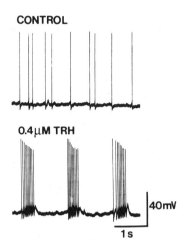

Figure 8 Activity in a type II neurons before (upper panel) and following (lower panel) bath application of TRH. Activity in the lower panel was observed approximately 30 min following the introduction of the drug into the bath. In both cases, the neuron was held at −45 mV by injecting positive current through the microelectrode. (Adapted from Dekin et al., 1985. Copyright 1989 by the AAAS.)

In the brainstem slice preparation, TRH acts directly on the membrane properties of some type I and II neurons (Dekin et al., 1985). At low concentrations (< 10 μM), TRH transforms these neurons into conditional bursters (Fig. 8). The initial step in this transformation is development of a depolarizing afterpotential (DAP) (Fig. 9). With time, the size of the DAP increases, eventually giving rise to bursting activity. These data suggest that TRH induces the expression of a new membrane current. At higher concentrations (< 10 μM), a second effect of TRH predominates. In this case, spike frequency adaptation is reduced, suggesting that TRH is blocking a calcium-activated potassium current (Dekin, unpublished observation). Although these actions of TRH in vitro clearly demonstrate a potential rhythm-generating capability in some ventral NTS neurons, such activity may not normally be expressed in vivo.* Indeed, the same membrane currents underlying en-

---

*It is possible, however, that the ability of exogenously applied TRH to reverse respiratory depression during barbiturate and alcohol narcosis reflects a rhythm-generating capability of neurons in the DRG. In this situation one would have to assume that the normal respiratory oscillator could no longer entrain the activity of DRG neurons, thus freeing them to express their endogenous bursting capabilities. Such an action would be similar to the ability of exogenous L-dopa to induce locomotion in cats and D-glutamate to induce swimming behavior in lamprey (see Section II.A).

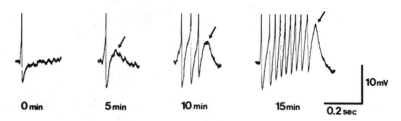

Figure 9 Time course of the development of the depolarizing afterpotential (DAP) and bursting activity. High gain records of spike activity at different times following exposure to TRH are shown for the same neuron as in Figure 8 (the tops of the spikes are truncated). Before exposure to TRH (0 min), the action potential undershoot gradually returns to baseline. At 5 min, a DAP (arrow) appeared. The size of the DAP increased with time and was associated with the development of bursting activity. (Adapted from Dekin et al., 1985. Copyright 1989 by the AAAS.)

dogenous bursting activity could normally function to maintain or modulate the subthreshold excitability in premotor respiratory neurons. For example, these membrane currents could affect the efficacy of the respiratory oscillator to depolarize premotor respiratory neurons. In this case, TRH would be modulating the gain of the respiratory CPG.

In addition to demonstrating how extrinsic inputs can modify intrinsic membrane properties, the actions of TRH on ventral NTS cells may also provide an example of how such inputs act to regulate cells at the submembrane level. It has been suggested in other systems that some types of endogenous bursting activity are due to the induction of calcium-activated inward currents (Lewis, 1984; Swandulla and Lux, 1985). If this were true, the action of TRH presents a paradox since at low concentrations it would be causing the expression of a calcium-activated inward current, while at high concentrations it would be blocking the expression of calcium-activated potassium currents. In the GH3 pituitary cell line, TRH has been shown to have a dual effect on intracellular calcium regulation (Drummond, 1985). At low TRH concentrations, internal calcium stores are mobilized. At higher TRH concentrations, however, intracellular calcium is sequestered. These effects are mediated by a second-messenger system involving inositol 1,4,5-triphosphate and protein kinase C, coupled with an elegant negative feedback mechanism, which buffers the cell from excessive internal calcium loads. In essence, the sequestering of internal calcium protects the cell from overstimulation by TRH. Although a similar mechanism has yet to be demonstrated in ventral NTS cells, in vitro preparations such as the brainstem slice (and possibly dissociated slices in which single channel recordings can be made; Kay and Wong, 1986) should eventually allow mechanisms such as this to be tested directly.

IV. Summary

The emphasis of this chapter has been on the conservation of mechanisms underlying rhythmic motor behaviors in a wide variety of animals. Some of the "invertebrate" mechanisms discussed, such as splitting the CGP into separate timing and patterning components, have immediate applicability for the respiratory CPG. Others, such as the recommendation function, suggest new ways in which to organize inputs that affect the activity of the respiratory CPG. In all cases, cellular properties and their inherent plasticity have provided a common theme in which to interpret these organizational concepts.

The current focus of attention on the adaptive flexibility of CPGs will certainly continue to dominate the interests of vertebrate and invertebrate neurobiologists alike. In the future, it is likely that this trend will lead us to search for mechanisms whereby different CPGs communicate and perhaps share information. This would have important implications for matching the activity of the cardiovascular and respiratory systems (see Koepchen, 1983), as well as coupling the respiratory CPG to other activities such as locomotion (Bramble and Carrier, 1983). The polymorphic network suggested by Getting and Dekin (1985b) for the *Tritonia* swim network may provide a starting point for addressing this issue at the cellular level (see Section II.B). Supporting evidence for the concept of polymorphic network has recently been provided by studies in the lobster stomatogastric ganglion, where the VD cell has been shown to switch from the pyloric CPG to the cardiac sac CPG following mechanosensory stimulation (Hooper and Moulins, 1989).* There remains little question, therefore, that CPG neurons possess the capability of participating in multiple neural circuits and that recommendation messages play a central role in determining which circuit a neuron will join. For the *Tritonia* system, it has also been demonstrated that neural circuits can be coupled since a reflexive withdrawal always precedes escape swimming behavior (Willows and Hoyle, 1969). It remains to be seen, however, whether individual CPG neurons can remain in "functional" contact with more than one circuit at a time.

Acknowledgments

I thank Chris and Sandy McNair for technical help in the preparation of this manuscript, and Drs. Dexter Speck and Kiisa Nishikawa for many insightful discussions on the nature of CPGs and command neurons.

---

*The cardiac sac CPG provides rhythmic dilations of the anterior foregut in the lobster. Because mechanosensory inputs are sufficient to gate this behavior (Stein, 1978), they can be viewed as a recommendation message.

References

Adams, D. J., Smith, S. J., and Thompson, S. H. (1980). Ionic currents in molluscan soma. *Annu. Rev. Neurosci.* **3**:141-167.

Alving, B. O. (1968). Spontaneous activity in isolated somata of *Aplysia* pacemaker neurons. *J. Gen. Physiol.* **51**:29-45.

Andry, D., and Horita, A. (1977). Thyrotropin-releasing hormone: Physiological concomitants of behavioral excitation. *Pharmacol. Biochem. Behav.* **6**:55-59.

Bacon, J. P., and Mohl, B. (1983). The tritocerebral commissure giant (TCG) wind-sensitive interneurone in the locust. I. Its activity in straight flight. *J. Comp. Physiol.* **150**:439-452.

Barker, D. L., Kushner, P. D., and Hooper, N. K. (1979). Synthesis of dopamine and octopamine in the crustacean stomatogastric nervous system. *Brain Res.* **161**:99-113.

Barker, J. L., and Smith, T. G. (1976). Peptide regulation of neuronal membrane properties. *Brain Res.* **103**:167-170.

Beltz, B., Eisen, J. S., Flamm, R., Harris-Warrick, R. M., Hooper, S. L., and Marder, E. (1984). Serotonergic innervation and modulation of the stomatogastric ganglion of three decapod crustaceans (*Panulirus interruptus, Homarus americanus,* and *Cancer irroratus*). *J. Exp. Biol.* **109**:35-54.

Bereiter, D., Berthoud, H. R., and Jeanrenaud, B. (1980). Hypothalamic input to brain stem neurons responsive to oropharyngeal stimulation. *Exp. Brain Res.* **39**:33-39.

Berger, A. J. (1977). Dorsal respiratory group neurons in the medulla of the cat: spinal projections, responses to lung inflation and superior laryngeal nerve stimulation. *Brain Res.* **135**:231-254.

Berger, A. J. (1980). The distribution of the cat's carotid sinus nerve afferent and efferent cell bodies using the horseradish peroxidase technique. *Brain Res.* **190**:309-320.

Bramble, D. M., and Carrier, D. R. (1983). Running and breathing in mammals. *Science* **219**:251-256.

Breese, G. R., Cott, J. M., Cooper, B. R., Prange, A. J., and Lipton, M. A. (1974). Antagonism of ethanol narcosis by thyrotropin-releasing hormone. *Life Sci.* **14**:1053-1063.

Breese, G. R., Mueller, R. A., Mailman, R. B., and Frye, G. D. (1981). Effects of TRH on central nervous system function. In *The Role of Peptides and Amino Acids as Neurotransmitters.* Edited by J. B. Lombardini and A. D. Kenny. New York, Alan R. Liss, pp. 99-116.

Brodin, L., and Grillner, S. (1985). The role of putative excitatory amino acid neurotransmitters in the initiation of locomotion in the lamprey spinal cord. I. The effects of excitatory amino acid antagonists. *Brain Res.* **360**:139-148.

Brodin, L., Grillner, S., and Rovainen, C. M. (1985). NMDA, kainate and qusiqualate receptors and the generation of fictive locomotion in the lamprey spinal cord. *Brain Res.* **325**:302-306.

Brown, G. (1911). The intrinsic factors in the act of progression in the mammal. *Proc. Roy. Soc. Ser. B* **84**:308-319.

Bunn, J. C., and Mead, J. (1971). Control of ventilation during speech. *J. Appl. Physiol.* **31**:870-872.

Byrne, J. H. (1980). Analysis of ionic conductance mechanisms in motor cells mediating inking behavior in *Aplysia californica*. *J. Neurophysiol.* **43**:630-650.

Calabrese, R. L. (1979). The roles of endogenous membrane properties and synaptic interaction in generating the heartbeat rhythm of the leech, *Hirudo medicinalis*. *J. Exp. Biol.* **82**:163-176.

Calabrese, R. L., and Peterson, E. L. (1983). Neural control of heartbeat in the leech, *Hirudo medicinalis*. In *Neural Origin of Rhythmic Movements*. Edited by A. Roberts and B. L. Roberts, Soc. Exp. Biol. Symp. 37. Cambridge, UK, Cambridge University Press, pp. 195-221.

Champagnat, J., Denavit-Saubie, M., and Siggins, G. R. (1983). Rhythmic neuronal activities of the nucleus of the tractus solitarius in vitro. *Brain Res.* **280**:155-159.

Champagnat, J., Jacquin, T., and Richter, D. (1986). Voltage dependent currents in neurons of the nuclei of the solitary tract of rat brainstem slices. *Pfluegers Arch.* **406**:272-379.

Chang, F-C. T. and Foster, R. E. (1986). Medullary inspiratory-related unit discharge patterns in awake, behaving guinea pigs. *Soc. Neurosci. Abstr.* **12**:305.

Chang, F-C. T., and Harper, R. M. (1989). A procedure for chronic recording of diaphragmatic electromyographic activity. *Brain Res. Bull.* **22**: 561-563.

Chang, F-C. T., Scott, T. R., and Harper, R. M. (1988). Methods of single unit recording from medullary neural substrates in awake behaving guinea pigs. *Brain Res. Bull.* **21**:749-756.

Claiborne, B. J., and Selverston, A. I. (1982). Localization of stomatogastric neuron cell bodies in lobster brain. *J. Comp. Physiol.* **154**:27-32.

Cohen, A. H., and Wallen, P. (1980). The neuronal correlates of locomotion in fish: Fictive swimming induced in an in vitro preparation of the lamprey spinal cord. *Exp. Brain. Res.* **41**:11-18.

Cohen, M. I. (1979). Neurogenesis of respiratory rhythm in the mammal. *Physiol. Rev.* **59**:1105-1173.

Cohen, M. I., and Feldman, J. L. (1984). Discharge properties of dorsal medullary inspiratory neurons: Relation to pulmonary afferent and phrenic efferent discharge. *J. Neurophysiol.* **51**:753-776.

Cohen, M. I., Piercey, M. F., Gootman, P. M., and Wolotsky, P. (1974). Synaptic connections between medullary inspiratory neurons and phrenic motoneurons as revealed by cross-correlation. *Brain Res.* **81**:319-324.

Connor, J. A., and Stevens, C. F. (1971). Voltage clamp studies of a transient outward current in gastropod neural somata. *J. Physiol. (Lond.)* **213**:21-30.

Davis, W. J. (1976). Organizational concepts in the central motor networks of invertebrates. *Adv. Behav. Biol.* **18**:265-292.

Dekin, M. S. (1988a). Substance P modulates the membrane excitability of neurons in the ventral part of the nucleus tractus solitarius of guinea pigs. *Fed. Proc.* **2**(4):A508.

Dekin, M. S. (1988b). Interaction between the A-current and calcium current in bulbospinal neurons in the ventral part of the nucleus tractus solitarius. *Soc. Neurosci. Abstr.* **14**(2):936.

Dekin, M. S. (1989). Inward rectification in bulbospinal neurons located in the ventral part of the nucleus tractus solitarius of the guinea pig. *Soc. Neurosci. Abstr.* **15**(2):878.

Dekin, M. S. and Getting, P. A. (1984). Firing pattern of neurons in the nucleus tractus solitarius: Modulation by membrane hyperpolarization. *Brain Res.* **324**:180-184.

Dekin, M. S., and Getting, P. A. (1987a). In vitro characterization of neurons in the ventral part of the nucleus tractus solitarius. II. Ionic mechanisms responsible for repetitive firing activity. *J. Neurophysiol.* **58**(1): 215-229.

Dekin, M. S., and Getting, P. A. (1987b). Pattern formation in ventral NTS neurons: Role of intrinsic membrane properties. In *Neurobiology of the Control of Breathing.* Edited by C. von Euler and H. Lagercrantz. New York, Raven Press, pp. 209-216.

Dekin, M. S., Johnson, S. M., and Getting, P. A. (1987a). In vitro characterization of neurons in the ventral part of the nucleus tractus solitarius. I. Identification of neuronal types and repetitive firing properties. *J. Neurophysiol.* **58**(1):195-214.

Dekin, M. S., Johnson, S. M., Richerson, G. B., and Getting, P. A. (1987b). Biophysical properties of bulbospinal neurons in the ventral part of the nucleus tractus solitarius in guinea pigs. In *Respiratory Muscles and Their Neuromotor Control.* Edited by G. C. Sieck, S. G. Gandevia, and W. E. Cameron. New York, Alan R. Liss, pp. 17-26.

Dekin, M. S., Richerson, G. B., and Getting, P. A. (1985). Thyrotropin-releasing hormone induces rhythmic bursting in neurons of the nucleus tractus solitarius. *Science* **229**:67-69.

Delcomyn, F. (1980). Neural basis of rhythmic behavior in animals. *Science* **210**:492-498.

Donoghue, S., Felder, R. B., Jordan, D., and Spyer, K. M. (1984). The central projections of carotid baroreceptors and chemoreceptors in the cat: A neurophysiological study. *J. Physiol. (Lond.)* **347**:397-409.

Drummond, A. H. (1985). Bidriectional control of cytosolic free calcium by thyrotropin-releasing hormone in pituitary cells. *Nature (Lond.)* **315**:752-755.

Eisen, J. S., and Marder, E. (1982). Mechanisms underlying pattern generation in lobster stomatogastric ganglion as determined by selective inactivation of identified neurons. III. Synaptic interactions of electrically coupled pyloric neurons. *J. Neurophysiol.* **48**:1392-1415.

Eisen, J. S. and Marder, E. (1984). A mechanism for production of phase shifts in pattern generators. *J. Neurophysiol.,* **51**(6):1375-1393.

Eldridge, F. L. and Millhorn, D. E. (1986). Hypothalamic central command and exercise hyperpnea. In *Neurobiology of the Control of Breathing.* Edited by C. von Euler and H. Lagercrantz. New York, Raven Press, pp. 35-43.

Eldridge, F. L., Millhorn, D. E., Kiley, J. P., and Waldrop, T. G. (1985). Stimulation by central command of locomotion, respiration and circulation during exercise. *Respir. Physiol.* **59**:313-337.

Euler, C. von., Hatward, J. M., Marttila, J., and Wyman, R. J. (1973). Respiratory neurons of the ventrolateral nucleus of the solitary tract of cat: Vagal input, spinal connections, and morphological identification. *Brain Res.* **61**:1-22.

Feldman, J. L. (1986). Neurophysiology of respiration in mammals. In *Handbook of Physiology*; *Section 1*: *The Nervous System*; *Vol. IV.* Edited by F. E. Bloom. Washington, D.C., American Physiological Society, pp. 463-524.

Feldman, J. L. and Cleland, C. L. (1982). Possible roles of pacemaker neurons in mammalian respiratory rhythmogenesis. In *Cellular Pacemakers*, Vol. II. Edited by D. O. Carpenter, New York, Wiley, pp. 104-128.

Feldman, J. L., Smith, J. C., McCrimmon, D. R., Ellenberger, H. H., and Speck, D. F. (1988). Generation of respiratory pattern in mammals. In *Neural Control of Rhythmic Movements in Mammals.* Edited by A. H. Cohen, S. Rossignol, and S. Grillner. New York, Wiley, pp. 73-100.

Flamm, R. E., and Harris-Warrick, R. M. (1986a). Aminergic modulation in lobster stomatogastric ganglion. I. Effects on motor output pattern and activity of neurons within the pyloric circuit. *J. Neurophysiol.* **55**: 847-865.

Flamm, R. E., and Harris-Warrick, R. M. (1986b). Aminergic modulation in lobster stomatogastric ganglion. II. Target neurons of dopamine, octopamine, and serotonin within the pyloric circuit. *J. Neurophysiol.* **55**: 866-881.

Forssberg, H., and Grillner, S. (1973). The locomotion of the acute spinal cat injected with clonidine i.v.. *Brain Res.* **50**:184-186.

Getting, P. A. (1977). Neural organization of escape swimming in *Tritonia*. *J. Comp. Physiol.* **121**:325-342.

Getting, P. A. (1981). Mechanisms of pattern generation underlying swimming in *Tritonia*. I. Neuronal network formed by monosynaptic connections. *J. Neurophysiol.* **46**:65-79.

Getting, P. A. (1983a). Mechanisms of pattern generation underlying swimming in *Tritonia*. II. Network reconstruction. *J. Neurophysiol.* **49**:1017-1035.

Getting, P. A. (1983b). Mechanisms of pattern generation underlying swimming in *Tritonia*. III. Intrinsic and synaptic mechanisms for delayed excitation. *J. Neurophysiol.* **49**:1036-1050.

Getting, P. A. (1983c). Neural control of swimming in *Tritonia*. In *Neural Origin of Rhythmic Movements*. Edited by A. Roberts and B. L. Roberts, Soc. Exp. Biol. Symp. 37. Cambridge, UK, Cambridge University Press, pp. 89-128.

Getting, P. A. (1988). Comparative analysis of invertebrate central pattern generators. In *Neural Control of Rhythmic Movements in Vertebrates*. Edited by A. H. Cohen, S. Rossignol, and S. Grillner. New York, Wiley, pp. 101-128.

Getting, P. A., and Dekin, M. S. (1985a). Mechanisms of pattern generation underlying swimming in *Tritonia*. IV. Gating of a central pattern generator. *J. Neurophysiol.* **53**:466-480.

Getting, P. A., and Dekin, M. S. (1985b). *Tritonia* swimming: A model for integration within rhythmic motor systems. In *Model Neural Networks and Behavior*. Edited by A. I. Selverston. New York, Plenum Press, pp. 3-20.

Getting, P. A., Lennard, P. R., and Hume, R. I. (1980). Central pattern generator mediating swimming in *Tritonia*. I. Identification and synaptic interactions. *J. Neurophysiol.* **44**:151-164.

Grillner, S. (1969). Supraspinal and segmental control of static and dynamic gamma-motoneurons in the cat. *Acta. Physiol. Scand. Suppl.* **327**:1-34.

Grillner, S., and Shik, M. L. (1973). On the descending control of the lumbosacral spinal cord from the "mesencephalic locomotor region." *Acta. Physiol. Scand.* **87**:320-333.

Hedner, J., Hedner, T., Wessberg, P., Lundberg, D., and Jonason, J. (1983). Effects of TRH and TRH analogues on the central regulation of breathing in the rat. *Acta Physiol. Scand.* **117**:427-437.

Heitler, W. J., and Mulloney, B. (1978). Crayfish motor neurons are an integral part of the swimmeret central oscillator. *Soc. Neurosci. Abstr.* **4**:381.

Hilaire, G., and Monteau, R. (1976). Connexions entre les neurones inspiratoires bulbaires et les motoneurones phréniques et intercostaux. *J. Physiol. (Paris)* **72**:987-1000.

Hökfelt, T., Fuxe, K., Johansson, O., Jeffcoate, S., and White, N. (1975a). Distribution of thyrotropin-releasing hormone (TRH) in the central nervous system as revealed with immunohistochemistry. *Eur. J. Pharmacol.* **34**:389-392.

Hökfelt, T., Fuxe, K., Johansson, O., Jeffcoate, S., and White, N. (1975b). Thyrotropin-releasing hormone (TRH) containing nerve terminals in certain brain stem nuclei and in the spinal cord. *Neurosci. Lett.* **1**:133-139.

Hökfelt, T., Johansson, O., Ljungdahl, A., Lundberg, J. M., and Schulzberg, M. (1980). Peptidergic neurones. *Nature (London)* **284**:515-521.

Hooper, S. L., and Moulins, M. (1989). Switching of neuron from one network to another by sensory-induced changes in membrane properties. *Science* **244**:1587-1589.

Hugelin, A. (1980). Does the respiratory rhythm originate from a reticular oscillator in the waking state? In *The Reticular Formation Revisited.* Edited by J. A. Hobson and M. A. B. Brazier. New York, Raven Press, pp. 261-274.

Hume, R. I., and Getting, P. A. (1982). Motor organization of *Tritonia* swimming. III. Contribution of intrinsic membrane properties to flexion neuron burst formation. *J. Neurophysiol.* **47**:91-102.

Jankowska, E., Jukes, M. G. M., Lund, S., and Lundberg, A. (1967). The effect of L-dopa on the spinal cord. 5. Reciprocal organization of pathways transmitting excitatory action to alpha motoneurons of flexors and extensors. *Acta Physiol. Scand.* **70**:369-388.

Jellies, J., and Larimer, J. L. (1985). Synaptic interactions between neurons involved in the production of abdominal posture in crayfish. *J. Comp. Physiol.* **156**:861-873.

Jellies, J., and Larimer, J. L. (1986). Activity of crayfish abdominal-positioning interneurones during spontaneous and sensory-evoked movements. *J. Exp. Biol.* **120**:173-188.

Johnson, S. M., Dekin, M. S., and Getting, P. A. (1986). Spatial distribution of phrenic motor neurons and bulbospinal neurons in the adult guinea pig. *Soc. Neurosci. Abstr.* **12**:493.

Jordan, L. M. (1983). Factors determining motoneuron rhythmicity during fictive locomotion. In *Origin of Rhythmic Movements.* Edited by A. Roberts and B. L. Roberts. Soc. Exp. Biol. Symp. 37. Cambridge, UK, Cambridge University Press, pp. 423-444.

Kalia, M., and Mesulam, M. M. (1980). Brainsteam projections of sensory and motor components of the vagus complex in the cat. II. Laryngeal, tracheobronchial, pulmonary, cardiac, and gastrointestinal branches. *J. Comp. Neurol.* **193**:467-508.

Kay, A. R., and Wong, K. S. (1986). Isolation of neurons suitable for patch clamping from adult mammalian central nervous system. *J. Neurosci. Methods* **16**:227-239.

Kennedy, D. (1973). Control of motor output. In *Control of Posture and Locomotion, Advances in Behavioral Biology, Vol. 7.* Edited by R. B. Stein, K. G. Pearson, R. S. Smith, and J. B. Redford. New York, Plenum Press, pp. 429-436.

Kien, J. (1983). The initiation and maintenance of walking in the locust: An alternative to the command concept. *Proc. Roy. Soc. Lond. Ser. B.* **219**: 137-174.

Kirkwood, P. A., Nisimaru, N., and Sears, T. A. (1979). Monosynaptic excitation of bulbospinal neurones by chemoreceptor afferents in the carotid sinus nerve. *J. Physiol. (London)* **293**:35P-36P.

Koepchen, H. P. (1983). Respiratory and cardiovascular "centres": Functional entirety or separate structures? In *Central Neurone Environment.* Edited by M. E. Schlafke, H. P. Koepchen, and W. R. See. Berlin, Springer-Verlag, pp. 221-237.

Kraemer, G. W., Mueller, R. A., Breese, G. R., Prange, A. J., Lewis, J. K., Morrison, H., and McKinney, W. T. (1976). Thyrotropin releasing hormone: Antagonism of pentobarbital narcosis in the monkey. *Pharmacol. Biochem. Behav.* **4**:709-712.

Kristan, W. B., Burrows, M., Elsner, N., Grillner, S., Huber, F., Jankowska, E., Pearson, K. G., Sears, T. A., and Stent, G. S. (1977). Neural control of movement. In *Function and Formation of Neural Systems.* Edited by G. S. Stent. Berlin, Dahlem Konferenzen, pp. 329-354.

Kupferman, I., and Weiss, K. R. (1976). The command neuron concept. *Behav. Brain. Sci.* **1**:3-39.

Kushner, P. D., and Barker, D. L. (1983). A neurochemical description of the dopaminergic innervation of the stomatogastric ganglion of the spiny lobster. *J. Neurobiol.* **14**:17-28.

Larimer, J. L. (1988). The command hypothesis: a new view using and old example. *TINS* **11**:506-510.

Lennard, P. R., Getting, P. A., and Hume, R. I. (1980). Central pattern generator mediating swimming in *Tritonia*. II. Initiation, maintenance, and termination. *J. Neurophysiol.* **44**:165-173.

Lewis, D. V. (1984). Spike aftercurrents in R15 of *Aplysia*: Their relationship to slow inward current and calcium influx. *J. Neurophysiol.* **51**: 387-403.

Liddell, E. G. T., and Sherrington, C. S. (1924). Reflexes in response to stretch (myotatic reflexes). *Proc. Roy. Soc. Ser. B* **96**:212-242.

Lingle, C. J. (1980). Sensitivity of decapod foregut muscles to acetylcholine and glutamate. *J. Comp. Physiol.* **138**:187-199.

Loewy, A. D., and Burton, H. (1978). Nuclei of the solitary tract: Efferent projections to the lower brainstem and spinal cord of the cat. *J. Comp. Neurol.* **181**:421-450.

Maley, B., and Elde, R. (1981). Immunohistochemical localization of putative neurotransmitters within the feline nucleus tractus solitarius. *Neusoscience* 7:2469-2490.

Marder, E. (1976). Cholinergic motor neurons in the stomatogastric system of the lobster. *J. Physiol. (London)* 257:63-86.

Marder, E., and Eisen, J. S. (1984a). Transmitter identification of pyloric neurons: Electrically coupled neurons use different transmitters. *J. Neurophysiol.* 51:1345-1361.

Marder, E., and Eisen, J. S. (1984b). Electrically coupled pacemaker neurons respond differently to the same physiological inputs and neurotransmitters. *J. Neurophysiol.* 51:1362-1374.

Marder, E., Hooper, S. L., and Siwicki, K. (1986). Modulatory action and distribution of the neuropeptide Proctolin in the crustacean stomatogastric nervous system. *J. Comp. Physiol.* 243:454-467.

Maynard, D. M., and Selverston, A. I. (1975). Organization of the stomatogastric ganglion of the spiny lobster. IV. The pyloric system. *J. Comp. Physiol.* 100:161-182.

Meech, R. W. (1978). Calcium-dependent potassium activation in nervous tissues. *Ann. Rev. Biophys. Bioeng.* 7:1-18.

Meech, R. W. (1979). Membrane potential oscillations in mollucan "burster" neurons. *J. Exp. Biol.* 81:93-112.

Mettler, F. A. (1935a). Corticofugal fiber connections of the cortex of *Macaca mulatta*. The frontal region. *J. Comp. Neurol.* 61:509-542.

Mettler, F. A. (1935b). Corticofugal fiber connections of the cortex of *Macaca mulatta*. The parietal region. *J. Comp. Neurol.* 62:263-292.

Miller, J. P., and Selverston, A. I. (1982a). Mechanisms underlying pattern generation in lobster stomatogastric ganglion as determined by selective inactivation of identified neurons. II. Oscillatory properties of pyloric neurons. *J. Neurophysiol.* 48:1378-1391.

Miller, J. P., and Selverston, A. I. (1982b). Mechanisms underlying pattern generation in lobster stomatogastric ganglion as determined by selective inactivation of identified neurons. IV. Network properties of pyloric system. *J. Neurophysiol.* 48:1416-1432.

Mueller, R. A., Towle, A. C., and Breese, G. R. (1985). The role of vagal afferents and carbon dioxide in the respiratory response to thyrotropin-releasing hormone. *Reg. Peptides* 10:157-166.

Palkovits, M., Mezey, E., Eskay, R. L., and Brownstein, M. J. (1986). Innervation of the nucleus of the solitary tract and the dorsal vagal nucleus by thyrotropin-releasing hormone-containing raphe neurons. *Brain Res.* 373:246-251.

Peterson, E. L., and Calabrese, R. L. (1982). Dynamic analysis of a rhythmic neural circuit in the leech *Hirudo medicinalis*. *J. Neurophysiol.* 47:256-271.

Pfaffman, C., Frank, M., Bartoshik, L. M., and Snell, T. C. (1976). Coding gustatory information in the squirrel monkey chorda tympani. In *Progress in Psychobiology and Physiological Psychology.* Edited by J. M. Sprague and A. N. Epstein. New York, Academic Press, pp. 1-27.

Pitts, R., Magoun, H. W., and Ranson, S. W. (1939). The origin of respiratory rhythmicity. *Am. J. Physiol.* **127**:654-670.

Poon, M. (1980). Induction of swimming in lamprey by L-DOPA and amino acids. *J. Comp. Physiol.* **136**:337-344.

Richerson, G. B., and Getting, P. A. (1983). Characteristics of respiratory neurons in the guinea pig. *Soc. Neurosci. Abstr.* **9**:1163.

Richter, D. W. (1982). Generation and maintenance of the respiratory rhythm. *J. Exp. Biol.* **100**:93-107.

Richter, D. W., Ballantyne, D., and Remmers, J. E. (1986). How is the respiratory rhythm generated? A model. *News Physiol. Sci.* **1**:109-112.

Russell, D. F. (1979). CNS control of central pattern generators in the lobster stomatogastric ganglion. *Brain Res.* **177**:598-602.

Russell, D. F., and Hartline, D. K. (1978). Bursting neural networks: A re-examination. *Science* **200**:453-456.

Russell, D. F., and Hartline, D. K. (1982). Slow active potentials and bursting motor patterns in pyloric network of the lobster, *Panulirus interruptus. J. Neurophysiol.* **48**:914-932.

Saper, C. B., and Loewy, A. D. (1980). Efferent connections of the parabrachial nucleus in the rat. *Brain Res.* **197**:291-317.

Selverston, A. I., Miller, J. P., and Wadepuhl, M. (1983). Cooperative mechanisms for the production of rhythmic movements. In *Origin of Rhythmic Movements.* Edited by A. Roberts and B. L. Roberts. Soc. Exp. Biol. Symp. 37. Cambridge, UK, Cambridge University Press, pp. 55-87.

Selverston, A. I., Russell, D. F., Miller, J. P., and King, D. G. (1977). The stomatogastric nervous system: Structure and function of a small neural network. *Prog. Neurobiol.* **7**:215-290.

Sherrington, C. S. (1906). *Integrative Action of the Nervous System.* New Haven, CT, Yale University Press.

Smith, J. C., and Feldman, J. L. (1987). Central respiratory pattern generation studied in an in vitro mammalian brainstem-spinal cord preparation. In *Respiratory Muscles and Their Neuromotor Control.* Edited by G. S. Sieck, S. C. Gandevia, and W. E. Cameron. New York, Alan R. Liss, pp. 27-36.

Smith, J. C., and Feldman, J. L. (1988). Discharge patterns of medullary respiratory neurons in mammalian brainstem in vitro. *Soc. Neurosci. Abstr.* **14**:1060.

Speck, D. F., and Feldman, J. L. (1982). Effects of microstimulation and microlesion in dorsal and ventral respiratory groups in medulla of cat on respiratory outflow. *J. Neurosci.* **2**:744-757.

Steeves, J. D., Schmidt, B. J., Skovgaard, B. J., and Jordan, L. M. (1980). The effect of noradrenalin and 5-hydroxytryptamine depletion on locomotion in the cat. *Brain Res.* **185**:349-362.

Stein, P. S. G. (1978). Motor systems, with specific reference to the control of locomotion. *Annu. Rev. Neurosci.* **1**:61-82.

Suzue, T. (1984). Respiratory rhythm generation in the in vitro brainstem-spinal cord preparation of the neonatal rat. *J. Physiol.* **354**:173-183.

Swandulla, D., and Lux, H. D. (1985). Activation of a nonspecific cation conductance by intracellular Ca^{2+} elevation in bursting pacemaker neurons of Helix pomatia. *J. Neurophysiol.* **54**:1430-1443.

Taghert, P. H. and Willows, A. O. D. (1978). Control of a fixed action pattern by single, central neurons in the marine mollusk, *Tritonia diomedea. J. Comp. Physiol.* **123**:253-259.

Torvik, A. (1956). Afferent connections to the sensory trigeminal nuclei, the nucleus of the solitary tract and adjacent structures. An experimental study in the rat. *J. Comp. Neurol.* **106**:51-132.

Wang, S. C., Ngai, S. H., and Frumin, M. J. (1957). Organization of central respiratory mechanisms in the brainstem of the cat: Genesis of normal respiratory rhythmicity. *Am. J. Physiol.* **190**:333-342.

Wiersma, C. A. G., and Ikeda, K. (1964). Interneurons commanding swimmeret movements in the crayfish *Procambarus clarkii* (Girard). *Comp. Biochem. Physiol.* **212**:509-525.

Willows, A. O. D. (1967). Behavioral acts elicited by stimulation of single, identifiable brain cells. *Science* **157**:570-574.

Willows, A. O. D., and Hoyle, G. (1969). Neuronal network triggering of fixed action pattern. *Science* **166**:1549-1551.

Wilson, D. M. (1961). The central nervous control of flight in a locust. *J. Exp. Biol.* **38**:471-490.

Wilson, D. M., and Gettrup, E. (1963). A stretch reflex controlling wingbeat frequency in grasshoppers. *J. Exp. Biol.* **40**:171-185.

Wilson, D. M., and Wyman, J. (1965). Motor output patterns during random and rhythmic stimulation of locust thoracic ganglia. *Biophys. J.* **5**:121-143.

Yamamoto, Y., and Lagercrantz, H. (1985). Some effects of Substance P on central respiratory control in rabbit pups. *Acta Physiol. Scand.* **124**:449-455.

Yamamoto, Y., and Lagercrantz, H. (1987). Substance P: A putative mediator of the hypoxic drive. In *Neurobiology of the Control of Breathing.* Edited by C. von Euler and H. Lagercrantz. New York, Raven Press, pp. 97-100.

Yamamoto, C., and McIlwain, H. (1966). Electrical activities in thin sections from the mammalian brain maintained in chemically-defined media in vitro. *J. Neurochem.* **13**:1333-1343.

4

Cellular and Membrane Properties of Brainstem Neurons in Early Life

GABRIEL G. HADDAD

Yale University School of Medicine
New Haven, Connecticut

I. Introduction

There is little doubt that the young mammal, during prenatal or early post-natal life, responds differently to stimuli than the more mature infant or adult (Haddad et al., 1982; Haddad and Mellins, 1984; Haddad, 1986). For example, whereas the fetus shuts off all breathing movements or efforts when exposed to a decrease in arterial PO_2, the adult hyperventilates for as long as the stimulus is maintained (Haddad et al., 1983; Haddad and Mellins, 1984). The newborn has an in-between response since only the initial phase consists of hyperventilation and this is followed by a fall in ventilation (Rigatto, 1979; Haddad and Mellins, 1984). Respiratory responses to CO_2, laryngeal and carotid afferent stimuli, and thermal alterations also depend on age in early life (Rigatto, 1979; Boggs and Bartlett, 1983; Johnson et al., 1983; Donnelly and Haddad, 1986, Johnson, 1988; Donnelly and Haddad, 1990).

Possible explanations for the maturational differences in reflexic behavior during development include peripheral and central elements of each reflex. For instance, the difference in laryngeal-induced apnea between the newborn and the older animal could be attributed, in part, to maturational changes occurring at the laryngeal receptor level itself, to changes in central nervous system "responsiveness," or to differences in the ability to integrate incoming information at the central level.

Whether central neurons in the immature subject are capable of integrating synaptic input in the same fashion as neurons in the adult is not well understood and is presently under study. Since the neuronal ability to integrate depends to a large degree on inherent cellular and membrane properties, the study of such properties in mature and immature tissue becomes crucial. Recent advances in neurobiological techniques have made it possible to study the cellular and membrane properties of central neurons (Llinas, 1988). Several groups of investigators have now shown that the morphological and electrophysiological properties of central neurons mature. The purpose of this chapter is therefore (1) to present evidence that emphasizes the importance of these properties for integration and thus determination of a neuronal system's output in general, and (2) to review the available evidence for the role of these properties in the neuronal function of the immature and mature brainstem. I will focus mostly on the electrophysiological properties since these have not been examined until recently and a review of present knowledge is timely. I will draw on information obtained not only from brainstem neuronal populations, but also from neurons in other parts of the central nervous system.

II. Synaptic Input Versus Membrane Properties: Historical Perspective

Earlier pioneering electrophysiological investigations and later work by Hodgkin and Huxley (Hodgkin and Huxley, 1952a, b, c) influenced the formulation of principles of excitability in central neurons in a major way. It was thought, for example, that the responsiveness of spinal motoneurons to synaptic input followed "faithfully" from connectivity and that spinal neurons acted as "followers." The excitability of central neurons was therefore assumed to be based only on inward Na^+ and outward K^+ conductances similar to those described in peripheral nerves or the squid giant axon.

The importance of invertebrate neurobiological work regarding cellular and network properties is readily illustrated historically at this juncture. Numerous reports appeared in the 1960s and 1970s showing that invertebrate neurons have inherent membrane properties and conductances that determined, to a large extent, their spiking output. One such example is the discovery of a fast outward current, or A-current, in gastropods by Connor and Stevens (1971). This current (see below) turned out later to be almost ubiquitous (Rogawski, 1985; Gean and Shinick-Gallagher, 1989). Mammalian neurons such as those in cat (Barrett and Crill, 1972), guinea pig (Gustafson et al., 1982), mouse, and rat (Segal et al., 1984) are all endowed with A-current in various brain regions. It is clear now that, besides the two conductances that were originally described, many other ion channels are selective for certain ions, are voltage-activated or receptor-operated, and play an important role in mammalian neuronal excitability and brain function (Llinas, 1988; Nicoll, 1988; Jan and Jan, 1989).

III. Repetitive Firing Properties: Examples

Central neurons generally exhibit a variety of repetitive firing patterns. Examples are shown in Figure 1 and include (A) a nonactive, silent neuron at resting membrane potential (e.g., hypoglossal neuron) (Haddad et al., in press), (B) a rhythmically beating neuron, which can vary in rhythmicity and frequency of firing (e.g., vagal and nucleus tractus solitarius neurons) (Haddad and Getting, 1989, unpublished observations from our laboratory), and (C) a bursting neuron (endogenous or conditional) (see this volume, Chapter 3; Dekin et al., 1985). The repetitive firing patterns of such neurons are determined by synaptic input and active and passive membrane properties. Hence, any change in these factors due to maturation or development in early life would induce alterations in firing patterns.

IV. Passive Electrophysiological Properties

Passive properties, which include both resistive and capacitive elements, play a role in determining neuronal integration and firing pattern. Neuronal integration is influenced by the time and space (or length) constants of individual neurons. Neuronal membrane time constant is generally much longer than the duration of synaptic currents and hence will alter the voltage change in the postsynaptic cell in such a way that the rise and the decline of excitatory or inhibitory postsynaptic potentials are slower. Temporal summation, therefore, would be enhanced in neurons with relatively long, rather than short, time constants. Temporal summation is defined as the process that leads to the addition of synaptic potentials when synaptic currents arrive consecutively and overlap in time on *the same site* of the postsynaptic membrane. Similarly, neurons with a long space constant are more effective than those with a short space constant since the former neurons would allow depolarization to reach the trigger zone with less decrement in postsynaptic potentials than the latter. Spatial summation is the process that leads to addition of synaptic potentials when *spatially discrete* synaptic currents ar-

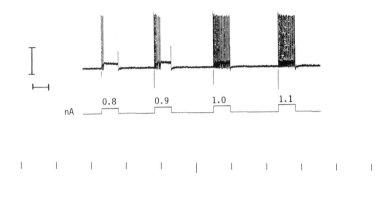

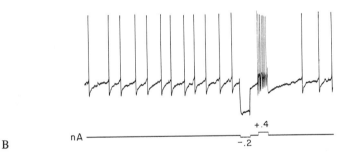

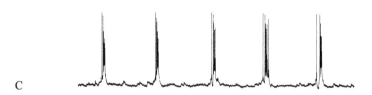

Figure 1 Three different patterns of neuronal firing. (A) *Silent* hypoglossal adult neuron; it fires only when current is imposed. (B) *Autoactive* or spontaneously firing DMNX cell; note how it hyperpolarizes or fires with negative or positive current injection, respectively. (C) A spontaneously *bursting* v-NTS neuron in a 5-day-old brainstem slice. Calibrations: In A, vertical bar = 40 mv; horizontal bar = 2 s. In B, tic marks correspond to seconds; action potential height = 85 mV. In C, tic marks correspond to 100 msec and action potential height = 75 mV.

rive on the postsynaptic membrane. In the final analysis, temporal and spatial summation influence neuronal excitability by modulating the membrane potential at the trigger zone.

A number of studies have demonstrated that membrane input resistance (R_N) decreases with age (Fulton and Walton, 1986; McCormick and Prince, 1987; Smith et al., 1988; Ziskind-Conhaim, 1988; Haddad et al., in press; Cameron et al., submitted). In neonatal rats, Fulton and Walton (1986) demonstrated that lumbar spinal motoneurons had high R_N compared to the adult. At 10 days postnatally, R_N dropped to less than a third of what it was at 3-5 days of age. Similarly, R_N in phrenic motoneurons in the kitten decreases by threefold between 2 and 14 weeks of age (Cameron et al., submitted). Hypoglossal (HYP) neurons also show a marked drop in R_N when comparing newborn to adult rats (Haddad et al., in press). These studies suggest that the increase in soma size is probably a major determinant of the decrease in R_N with age. However, it is important to note that studies on neurons in the ventral region of the tractus solitarius (v-NTS) did not show a major difference in R_N when rats in the first 2 weeks of life were compared to adult rats (Haddad and Getting, 1989). Although neonatal v-NTS neurons are smaller than their adult counterparts (Haddad and Getting, 1989 the similarity in R_N suggests that soma size is not the only determinant. In fact, neuronal membrane properties as well as dendritic dimensions can also be important (Gustafson and Pinter, 1984; Grogan and Tyc-Dumont, 1989).

Whether the age-related decrease in R_N influences the time or the space constant is not fully determined. We do not have enough information about neuronal membrane capacitance as a function of age, and studies to date are conflicting (Fulton and Walton, 1986; McCormick and Prince, 1987; Ziskind-Conhaim, 1988). For example, studies on spinal neurons in the neonatal rat have shown that, although R_N decreases substantially with age, the time constant does not, thus implicating a change in membrane capacitance (Fulton and Walton, 1986). On the other hand, studies on neocortical cells from animals aged 1-30 days postnatally demonstrated clearly that there was a change in the time constant from 35 msec to about 15 msec along with a decrease in R_N (McCormick and Prince, 1987). In addition, since the space constant is determined as $\sqrt{R_N/R_a}$ (R_a being axial resistance) and since dendrites grow markedly postnatally, it is difficult to know what this ratio would be in neonatal compared to adult rats for specific neuronal species, and information is lacking in this area.

In those neurons which have a larger R_N in early life, neuronal recruitment in newborns will be different than in adults. This is because recruitment depends, to a degree, on the magnitude of voltage change at a given synaptic current, which is bound to be larger in the newborn, all else being equal. Actually, the rheobase (defined as injected current necessary to induce firing) for hypoglossal (Haddad et al., in press) and phrenic motoneurons (Cameron et al., submitted) has been found to be smaller in newborns than in adults.

V. Active Membrane Properties and Ionic Currents

A. Action Potential Waveform

Elegant studies by Spitzer have demonstrated how an action potential develops in amphibian (*Xenopus*) neurons (Spitzer and Lamborghini, 1976; Spitzer, 1981). Intracellular records from cultured neurons obtained from *Xenopus* at various stages of development showed that action potentials were much wider in the neural tube stage than in subsequent stages of the young larva (Spitzer, 1981). This difference was related to the fact that impulse generation was much more dependent on the inward current carried by Ca^{2+} ions rather by the current carried by Na^+ ions in early life. As maturation progressed, there was a transition from a Ca^{2+}-generated to a Na^+-generated spike. Recent studies have demonstrated such a transition from Ca^{2+} currents to Na^+ currents in acutely dissociated neocortical neurons in the rat and mouse (MacDermott and Westbrook, 1986; Huguenard et al., 1988). Indeed, using a variety of approaches, it is well known now that Na conductance increases with maturation in postnatal life (Bader et al., 1983; Baumgold et al., 1983; Couraud et al., 1986).

We have shown that the action potentials recorded from vagal (dorsal motor, DMNX) or HYP neurons in a brainstem slice preparation are much wider ($\times 2$) in the newborn than in the adult rat (Haddad et al., in press; Haddad, unpublished observations). Although we did not perform experiments to examine ionic currents, these data suggest that a similar process of transition from calcium-based action potential to a Na^+ one takes place also in brainstem neurons.

The rate of rise (V_{max}) of action potentials has been measured in some cell populations and was found to be slower in immature neocortical cells than in similar cells from adults (Gardette et al., 1985; MacDermott and Westbrook, 1986; McCormick and Prince, 1987). The V_{max} was typically 3-4$\times$ steeper for the adult action potential. Since V_{max} has been related to membrane conductance for Na (G_{Na}^+), previous investigators have concluded that G_{Na}^+ increases with maturation (MacDermott and Westbrook, 1986; McCormick and Prince, 1987; Huguenard et al., 1988). This has been consistent with the increase in Na^+ channels with neuronal development found in studies using patch clamp techniques (Baumgold et al., 1983; Couraud et al., 1986).

Recently, hypothalamic neurons from embryonic rat brains have been shown to develop in culture (Petroski and Geller, 1989). Na channels develop seemingly after embryonic day 12. Furthermore, Ca^{2+} currents could be demonstrated using flow cytometry and Indo-1 at a time when inward Na^+ channels are absent using whole cell patch (O'Connell et al., 1989). K^+ channels (delayed rectifier) mature at an earlier date but the A-current shows maturation after embryonic day 14-15. Interestingly, as has been shown with cortical cells, only a small proportion of hypothalamic neurons fire early in life, and when they do fire, they are slower than mature cells (Ling et al., 1989).

Action potentials from HYP neurons were also characterized by an afterpotential hyperpolarization that was larger in magnitude and slower in time constant in the newborn than in the adult. Interestingly, adult HYP neurons exhibited little widening in action potential ($< 10\%$) with repetitive firing while the newborn showed much more widening (70% with currents of 1.5-2.0 nA). Based on data generated in molluscan neurons by Aldrich et al. (1979), this observation on HYP cells also suggests that a larger inward Ca^{2+} current exists in the newborn than in the adult. The characteristics of the action potentials in the neonate, including width, would promote a slower firing frequency with stimulation. This would be consistent with and suitable for the needs of neonatal skeletal muscles since these are notoriously slow muscles, as their sarcoplasmic reticulum is underdeveloped and sequestration and release of intracellular calcium is slow. Once again, the neuronal electrophysiological properties are well adapted to the contractile properties of the skeletal muscles they innervate.

B. Membrane Potential and Cellular Properties

Cellular properties vary widely from one neuron to another, especially when they belong to different populations. One of the emerging concepts in neurobiology is that neurons are varied not only in terms of their inherent properties, but also in terms of the developmental schedule of these specialized properties. These properties seem to follow different maturational timetables. There are many examples in the literature and below are some data pertaining to brainstem neurons and those in other regions.

v-NTS Neurons

Recently, neurons in the ventral region of the tractus solitarius (v-NTS) have been extensively studied in the guinea pig (Dekin and Getting, 1984, 1987; Dekin et al., 1987) and in the rat (Haddad and Getting, 1989). Adult rat v-NTS neurons were categorized into two groups (type A and type B) based on their electrophysiological properties (Haddad and Getting, 1989) but guinea pig neurons were categorized into three types (Dekin et al., 1987). Interestingly, the two types of neurons in the rat-vNTS were endowed with all the cellular properties that the three types of neurons in the guinea pig had (Dekin et al., 1987). Since guinea pig neuronal properties are described in Chapter 3 of this volume, we describe below the adult and newborn firing and cellular properties of the rat v-NTS neurons. Cellular properties are important since they impart on neurons their firing characteristics and ability to respond to excitatory and inhibitory stimuli. Examples of these properties include firing threshold, spike frequency adaptation, postburst hyperpolarization, postinhibitory rebound, delayed excitation, plateau potentials, and bursting.

An example of the location of v-NTS neurons impaled is shown in Figure 2. Upon depolarization, such cells immediately fire at a high rate, reaching maximal or peak frequencies of 75-100 spikes/s. The first interspike interval is always the shortest. As the injected current is increased, both peak and steady-state firing rate increase. The persistent large difference between peak and steady-state curves obtained as a function of injected currents indicates that there is adaptation in firing. These cells, therefore, exhibit fre-

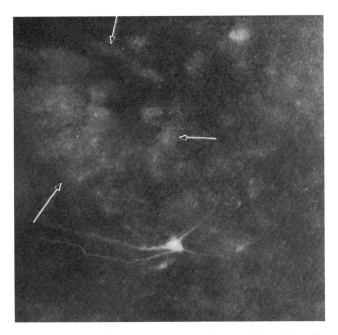

Figure 2 Lucifer-filled adult NTS neuron located ventrally and laterally to the tractus solitarius (arrows).

quency adaptation (SFA). The mean SFA index (defined as steady-state frequency divided by peak frequency × 100) for type A v-NTS cells was 40% at an injected current of 0.5 nA. The time constant for adaptation, determined by exponential fit, measured 100-280 msec for all cells tested.

When type A neurons are depolarized from membrane potentials more negative than resting, the pattern of activity changes remarkably. First, the onset of firing is delayed from the onset of depolarization (Fig. 3). Delayed excitation (DE) is defined as the time between the onset of depolarization and the time of occurrence of the first action potential. During the delay, the membrane potential gradually depolarizes in a "ramp" fashion to reach spiking threshold (see arrow, Fig. 3a). DE is modulated by both the duration and magnitude of the hyperpolarizing prepulse (Fig. 3). The maximum DE observed in any v-NTS neuron is approximately 900 msec. A second effect of hyperpolarizing prepulses is to decrease firing rate during a subsequent depolarization. Firing frequency is generally lower throughout when depolarization is preceded by hyperpolarization (Fig. 3).

Upon depolarization, type B neurons fire repetitively with an SFA index of about 48%. Exponential fit of the time course of adaptation yields a time constant of 2-14 msec, thus indicating that adaptation in type B neurons is much faster than in type A neurons. The frequency-current plot of type B v-NTS cells shows that both peak and steady-state frequencies increase to very high levels. Type B v-NTS cells have the highest firing rate so far obtained from brainstem neurons. Peak frequencies of type A cells are lower than peak and even steady frequencies of type B cells at all currents tested.

Type B cells do not show delayed excitation, even when cells are hyperpolarized to very negative levels (-110 mV). With hyperpolarizing prepulses, however, type B cells do exhibit a lower firing frequency throughout subsequent depolarizations, with the most marked decrease being at the beginning of the depolarizing pulse.

Some type B cells have two additional biophysical properties that are not found in type A cells. After a hyperpolarizing pulse, cells generate a slow depolarizing waveform with action potentials or postinhibitory rebound. These neurons also display depolarizing afterpotentials, a property usually seen in bursting neurons (Thompson and Smith, 1976; Adams et al., 1980; Dekin et al., 1985).

In contrast to the two populations in the adult v-NTS, neonatal v-NTS neurons are homogeneous and could only be grouped in one such population. When depolarized from rest, cells typically showed an immediate increase in firing rate. Spike frequency decreases in time throughout depolarization but the firing rate declines gradually and relatively little throughout the period of stimulation. Both peak and steady-state frequencies increase monotonically with applied current and the difference between them is small

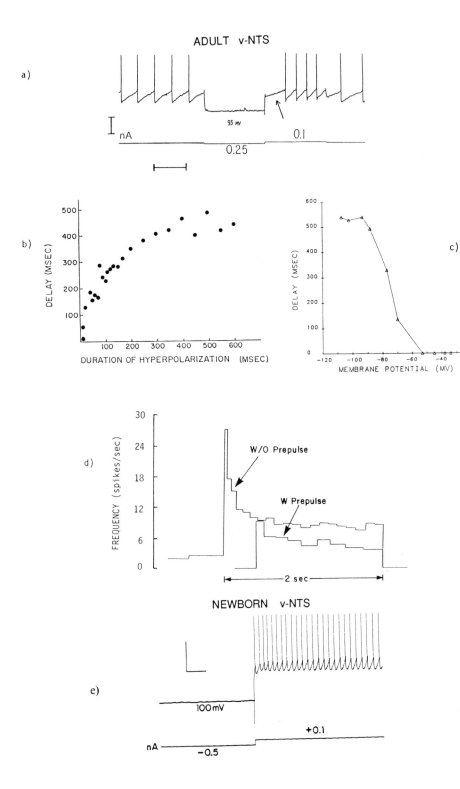

ADULT v-NTS

a)

93 mV

nA   0.1

0.25

b)

DELAY (MSEC)

500

400

300

200

100

100 200 300 400 500 600

DURATION OF HYPERPOLARIZATION (MSEC)

c)

DELAY (MSEC)

600

500

400

300

200

100

0

-120 -100 -80 -60 -40 -20

MEMBRANE POTENTIAL (MV)

d)

FREQUENCY (spikes/sec)

30

24

18

12

6

0

W/O Prepulse

W Prepulse

2 sec

NEWBORN v-NTS

e)

100 mV

nA

-0.5   +0.1

compared to the difference for adult cells, indicating that spike frequency adaptation develops postnatally. For example, mean SFA index was about 60% when neonatal v-NTS neurons were depolarized with 0.4-0.5 nA pulses; this is significantly different from mean SFA index for type A or B v-NTS neurons in the adult. Exponential fit of the frequency-time profile for neonatal v-NTS cells showed that the time constant for adaptation was between 300 and 750 msec, indicating that adaptation occurred at a much slower rate in the newborn than in either adult type A or B neurons.

When depolarized from membrane potentials that were more negative than rest, most neonatal v-NTS neurons (>90%) showed *no silent period* between the onset of the depolarizing pulse and the time of occurrence of the first spike, or delayed excitation (Fig. 3). Neonatal neurons tested for postinhibitory rebound did not exhibit this phenomenon, unlike some type B cells.

Thus, v-NTS neurons in the newborn showed a set of different electrophysiological properties than those in the adult rat. For example, the nature and magnitude of the spike frequency adaptation in neonatal cells are different from those in adult neurons. Delayed excitation is a characteristic of type A v-NTS cells in the adult rat, but DE occurs only in a minority of newborn cells and is of much smaller magnitude. Finally, neonatal neurons, unlike some adult cells, do not show postinhibitory rebound.

Since certain cellular properties develop with postnatal age, and since these properties are based on the presence of certain ionic membrane properties, we raise the question as to which membrane properties mature with age. Several hypotheses can be formulated. For example, we hypothesize that an A-current will need to be "up-regulated" in the neonatal cells. Type B and especially type A cells show evidence of this outward, fast inactivating K^+ current (Haddad and Getting, 1989), but only a minority of the neonatal cells show such a property. Another hypothesis is that a Ca^{2+}-activated K^+ current needs to be further developed in neonatal cells since adult but not neonatal cells show the major spike frequency adaptation upon depolarization.

Figure 3 Adult v-NTS neuron showing delayed excitation (arrow in a) (type A cell). Delayed excitation depends on duration of hyperpolarization (b) and on the magnitude of membrane potential reached with the polarization (c). The frequency of firing is changed throughout depolarization when the latter is preceded by a hyperpolarizing prepulse (d). The majority of newborn v-NTS cells (e) did not show any evidence of delayed excitation (see text). Calibration in a: vertical = 40 mV; horizontal = 1 s. Calibration for newborn v-NTS: vertical = 35 mV; horizontal = 100 msec. (From Haddad and Getting, 1989.)

Other Brainstem Neurons

We have also studied other brainstem neurons such as DMNX and HYP neuron. These neurons have different properties and their developmental schedule is different from that of v-NTS cells.

The firing threshold determines the level of excitation needed for a cell to fire, and HYP neurons, in the slice preparation, are generally resting at a membrane potential that is not close to firing threshold. HYP neurons have a resting membrane potential (V_m) of approximately -80 mV and neurons are generally silent unless invaded by sufficient synaptic current (Haddad et al., in press). On the other hand, DMNX neurons have a much lower V_m (-55 to -60 mV) and are almost always autoactive at rest (Haddad, unpublished observations). Both types of neurons in the v-NTS (type A and type B) are always beating, though at different rates and have different responses to imposed synaptic current, as shown above (Haddad and Getting, 1989).

Although there is a tendency toward a higher (more negative) V_m in the mature as compared to the immature neuron, studies have not been consistent (Fulton and Walton, 1986; McCormick and Prince, 1987; Haddad et al., in press). On average, neocortical and brainstem (HYP) neurons show an increase in V_m with maturation (McCormick and Prince, 1987; Haddad et al., submitted), whereas spinal neurons do not (Fulton and Walton, 1986). The change in V_m with age should not be surprising since conductances, channel densities, and leak currents may change with time in early life as discussed above.

In response to hyperpolarizing currents, adult HYP cells display a "sag" which is prominent with higher current pulses (Haddad et al., in press). This sag is a time-dependent decrease in membrane potential during hyperpolarization and is secondary to inward rectification. Although some newborn cells display sag, it is a much rarer phenomenon in newborns than in adults. This sag phenomenon has been seen in other central neurons such as in hippocampal neurons (Halliwell and Adams, 1982). Although the mechanism for this inward rectification is not clear, ionic currents, such as Iq, which is sensitive to cesium, have been implicated (Halliwell and Adams, 1982).

The SFA index of adult HYP neurons was 70-90%, with most of the decline in firing rate occuring between the first and second interspike intervals. Moreover, it was only at relatively high current intensities (1.5-2.5 nA) that the SFA index reached about 50%. In addition, the slope of the peak frequency of firing as a function of current was increased sharply at higher currents. Attempts to elicit delayed excitation were not successful; no delay in spike generation was observed in any neuron, even with prepulse hyperpolarizations leading to membrane potentials of -120 mV. Neonatal neurons showed also little adaptation (SFA = 80-90%), even at high currents, and no delayed excitation.

DMNX neurons respond very differently from HYP neurons. DMNX cells have a much larger and longer afterhyperpolarization than HYP neurons. SFA in DMNX cells is much larger than that in HYP cells and comparable to the SFA in type A adult -vNTS cells. All adult DMNX cells show delayed excitation, though smaller than that exhibited by type A v-NTS neurons. Interestingly, and in contrast to neonatal v-NTS cells, neonatal DMNX cells show major SFA as well as *delayed excitation*. In fact, every neuron we have impaled so far in the DMNX area in rats as young as 3-5 postnatal days showed delayed excitation.

Thus, spike frequency adaptation is considerably less in HYP than in vagal motoneurons or v-NTS neurons (Haddad and Getting, 1989; Haddad et al., in press). Since accommodation is minimal in HYP neurons beyond the first interspike interval, the importance of dynamic membrane characteristics that alter the discharge frequency is also limited. Because spike frequency adaptation depends on certain ionic channels and currents, it is likely from our studies (Haddad et al., in press) and those of others (Dekin and Getting, 1987) that HYP neurons have little or no Ca^{2+}-dependent K^+ channels and that Ca^{2+} influx is minimal during action potential genera-

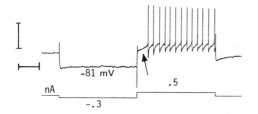

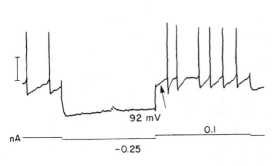

Figure 4 Adult DMNX (A) and newborn DMNX (B) neurons showed delayed excitation at the beginning of depolarization (arrows) following a hyperpolarizing prepulse. Calibration: vertical bar = 50 mV in A and 40 mV in B; horizontal bar = 500 msec in A. Prepulse in B = 2 s. (From Haddad and Getting, 1989.)

tion. This speculation is supported by the shape of the action potential and its duration in HYP neurons. Significant calcium currents generally prolong the action potential duration with a marked hump on the repolarization phase (Aldrich et al., 1979) or have the effect of widening action potentials during the inactivation of the delayed rectifier (e.g., during spike trains), and this is not seen in adult HYP cells. The delayed rectifier is an outward K^+ current responsible for the immediate repolarization phase of action potentials.

Differences between neonatal v-NTS and neonatal DMNX neurons are very interesting and demonstrate that different neurons can follow different developmental timetables. This is probably best exemplified by two cellular properties: frequency adaptation and delayed excitation. DMNX cells as early as in 3-day-old rats adapt and exhibit delayed excitation like their adult counterparts. On the other hand, neonatal v-NTS cells (as late as 12 days of age) show much less adaptation than adult cells and only a minority ($< 10\%$) have delayed excitation.

VI. Morphology and Connectivity

The rat is probably one of the most frequently studied animals with respect to neuronal morphology. It is unique since it is very immature at birth (Fish and Winick, 1969), unlike other animals such as the guinea pig (Dobbing and Sands, 1970). Investigators interested in development resort to the rat (or to the opossum, which is even less mature at birth than the rat) (Lent, 1981) since it is more likely that developmental trends, if present, would be detected in the rat than in other species.

The morphology and ultrastructure of neocortical neurons have been well studied during maturation (Eayrs and Goodhead, 1959). Neocortical neurons are packed at birth throughout the cortex with little evidence of segregation or laminae formation. With age, there is more separation between neurons and the ratio of cell/gray decreases markedly in the first 3 weeks of life (Eayrs and Goodhead, 1959). Cell/gray ratio is taken as an index of the inverse of distance between neurons. Dendrites and axons also develop postnatally. For example, there are very few dendrites on neocortical cells before 6 days of age but these grow fast after 12 days. It also seems that the number of primary dendrites is low at birth, and by 3 weeks of age their number reaches adult levels. Furthermore, the mean number of dendrites increases distally with increasing distance from the center of the perikaryon as a function of age, indicating that dendrites are extending. This continues to occur in neocortical neurons even beyond the first month of life in rats.

Brainstem and spinal neurons have recently been examined morphologically (Fulton and Walton, 1986; Cameron et al., submitted; Brozanski et al., in press). Hypoglossal cell soma in kittens increases in the first 2 months of life from about 25 to 40 μ in major axis and from about 18 to 24 μ in minor axis (Brozanski et al., in press). Interestingly, unlike neocortical cells, the number of primary dendrites in kitten hypoglossal neurons does not change with age. This is probably not species related since we have seen this also in the rat (Haddad, unpublished observations). Rat hypoglossal neurons, though they increase in soma size and extend their dendrites with age (Fig. 5), have the same number of primary dendrites at birth as in adults (about four to six). Work done on phrenic motoneurons in the cat also shows that these neurons continue to increase soma surface area until 8-10 weeks postnatally (Cameron et al., submitted).

Given the morphological changes that occur postnatally in neurons, a number of questions can be raised. For example, when do synapses form in early postnatal life, especially in brainstem nuclei? How "strong" are the synaptic connections already formed? What are the functional implications for neuronal performance and integration before and during synapses formation? What are the mechanisms involved in synapse formation? Clearly, these questions cannot be addressed in this chapter or in this volume, and an enormous amount of work is ongoing in a number of laboratories. However, suffice it to state that there is a developmental *decrease* in the amount of current needed to induce postsynaptic potentials; i.e., more current is needed to induce postsynaptic potentials in the young than in the more mature system (Prince and Gutnick, 1972). In addition, these synaptic potentials cannot be maintained with repetitive stimulation (Prince and Gutnick, 1972). These data clearly indicate that synaptic connections are immature at birth and that they mature postnatally. This immaturity can be attributed to anatomical connectivity or to mechanisms of neurotransmitter release and action.

VII. Importance to Breathing

The question we raise is how do the cellular properties we have described influence rhythmicity and respiratory cycling? Furthermore, we ask whether the maturational changes are relevant to the known differences in respiratory system behavior between mature and young subjects (Haddad et al., 1982, 1983; Haddad and Mellins, 1984; Haddad, 1986). Studies on neural networks in invertebrates (see Chapter 3) and vertebrates (Llinas, 1988; Grillner et al., 1988) have already shown that intrinsic cellular properties of neurons in these networks are crucial in influencing the overall system's

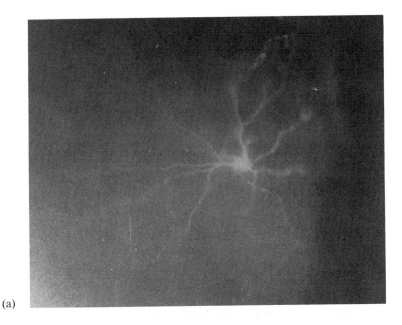

(a)

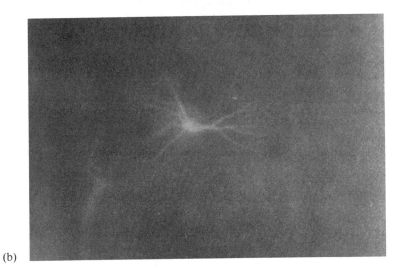

(b)

Figure 5 Lucifer-filled, 10-day-old hypoglossal neuron (a) and a 3-day-old hypoglossal neuron (b). Note difference in dendritic outgrowth.

output. It is difficult at present to ascertain how these properties influence breathing in specific terms. However, we speculate, for example, that a calcium-activated K^+ current, which has been described in brainstem neurons (Dekin and Getting, 1987), plays an important role in the decline of depolarization rate during inspiration. The A-current, also present in respiratory-related brainstem neurons, may be important in delaying neuronal firing and therefore in network timing such as phase switching.

Although it is more difficult to answer the question regarding maturation since our findings and those of others are rather recent, it is tempting to hypothesize that the slow firing activity and the few spikes generated per breath in respiration-related neurons in young versus adult opossums (Farber, 1988, 1989) have to be secondary to both the inherent properties of these neurons and, importantly, to the immaturity of synaptic connectivity. How this impacts on the overall function of breathing cannot be answered with certainty at present, and additional studies have to be performed.

VIII. Conclusions and Implications for Future Research

It is our current belief that understanding respiratory rhythmicity will depend not only on a system's approach, but also on the study of properties of individual neurons or neuronal groups that form the network system and on their synaptic interactions. With the advent of modern neurobiological and molecular techniques, answering questions such as those formulated in this chapter is no longer a major feat.

Development and maturation in early life is especially interesting and challenging. Since development may not proceed uniformly for various cellular and membrane properties, as we have seen in this chapter, another layer of complexity is added. However, it is the kinetic nature of developmental biology that will teach us how cell function is regulated and how assemblies of cells differentiate to reach maturity. The challenge for the future is to try to understand cellular and organ functions in terms of basic mechanisms and, more importantly, to try to understand abnormal function in terms of basic defects. Is apnea of prematurity a disease of synaptic immaturity? If so what are the molecular mechanisms responsible? Is it a condition related to a defect or an undeveloped neurotransmitter release system? Or is it a condition related to cellular differentiation and development of certain properties that provide cells with ability to better integrate incoming information? If so, what are the molecular mechanisms that induce such differentiation? These are only a few questions, and many such examples are awaiting new discoveries and therapy.

References

Adams, D. J., Smith, S. J., and Thompson, S. H. (1980). Ionic currents in molluscan soma. *Brain Res.* **250**:71-92.

Aldrich, R. W., Getting, P. A., and Thompson, S. H. (1979). Mechanism of frequency-dependent broadening of molluscan neurons soma spikes. *J. Physiol. (London)* **291**:531-544.

Bader, C. R., Bertrand, D., Dupin, E., and Kato, A. C. (1983). Development of electrical membrane properties in cultured avian neural crest. *Nature* **305**:808-810.

Barrett, J. N., and Crill, W. E. (1972). Voltage clamp analysis of conductances underlying cat motoneuron action potentials. *Fed. Proc.* **31**: 305.

Baumgold, J., Zimmerman, I., and Bambrick, L. (1983). Appearance of [³H] saxitoxin binding sites in developing rat brain. *Dev. Brain Res.* **9**: 405-407.

Boggs, D. F., and Bartlett, Jr., D. (1983). Chemical specificity of a laryngeal apneic reflex in puppies. *J. Appl. Physiol.* **53**:455-462.

Brozanski, B. S., Guthrie, R. D., Volk, E. A., and Cameron, W. E. Postnatal growth of genioglossal motoneurons. *Pediatr. Pulm.* (in press).

Cameron, W. E., Brozanski, B. S., and Guthrie, R. D. Postnatal development of phrenic motoneurons in the cat. *Dev. Brain Res.* (submitted).

Cameron, W. E., Jodkowski, J. S., Fang, He, and Guthrie, R. D. Electrophysiological properties of developing phrenic motoneurons in the cat. *J. Neurophysiol.* (submitted).

Connor, J. A., and Stevens, C. F. (1971). Voltage clamp studies of a transient outward current in gastropod neural somata. *J. Physiol.* **213**:21-30.

Couraud, F., Martin-Moutot, N., Koulakoff, A., and Berwald-Netter, Y. (1986). Neurotoxin-sensitive sodium channels in neurons developing in vivo and in vitro. *J. Neurosci.* **6**:192-198.

Dekin, M. S., and Getting, P. A. (1984). Firing pattern of neurons in the nucleus tractus solitarius: Modulation by membrane hyperpolarization. *Brain Res.* **324**:180-184.

Dekin, M. S., and Getting, P. A. (1987). In-vitro characterization of neurons in the ventral part of the nucleus tractus solitarius in guinea pigs. II. Ionic basis for repetitive firing properties. *J. Neurophysiol.* **58**: 215-229.

Dekin, M. S., Johnson, S., and Getting, P. A. (1987). In-vitro characterization of neurons in the ventral part of the nucleus tractus solitarius in guinea pigs. I. Identification of neuronal types and repetitive firing properties. *J. Neurophysiol.* **58**:195-214.

Dekin, M. S., Richardson, G. B., and Getting, P. A. (1985). Thyrotropin-releasing hormone induces rhythmic bursting in neurons of the nucleus tractus solitarius. *Science* **229**:67-69.

Dobbing, J., and Sands, J. (1970). Growth and development of the brain and spinal cord of the guinea pig. *Brain Res.* **17**:115-123.

Donnelly, D. F., and Haddad, G. G. (1986). Respiratory changes induced by prolonged laryngeal stimulation in awake piglets. *J. Appl. Physiol.* **61**: 1018-1024.

Donnelly, D. F., and Haddad, G. G. (1990). Prolonged apnea and impaired survival in piglets following sinus and aortic nerve section. *J. Appl. Physiol.* **68(3)**:1048-1052.

Eayrs, J. T., and Goodhead, B. (1959). Postnatal development of the cerebral cortex in the rat. *J. Anat. (London)* **93**:385-402.

Farber, J. P. (1988). Medullary inspiratory activity during opossum development. *Am. J. Physiol.* R578-R584.

Farber, J. P. (1989). Medullary expiratory activity during opossum development. *Am. J. Physiol.* **24**:R578-R584.

Fish, I., and Winick, M. (1969). Cellular growth in various regions of the developing rat brain. *Pediatr. Res.* **3**:407-412.

Fulton, B. P., and Walton, K. (1986). Electrophysiological properties of neonatal rat motoneurons studied in-vitro. *J. Physiol. (London)* **370**: 651-678.

Gardette, R., Debono, M., Dupont, J. L., and Crepel, F. (1985). Electrophysiological studies on the postnatal development of intracerebellar nuclei neurons in rat cerebellar slices maintained in vitro. II. Membrane conductances. *Dev. Brain Res.* **20**:97-106.

Gean, P. W., and Shinick-Gallagher, P. (1989). The transient potassium current, the A-current, is involved in spike frequency adaptation in rat amygdala neurons. *Brain Res.* **480**:160-169.

Grillner, S., Buchanan, J. T., Wallen, P., and Brodin, L. (1988). Neural control of locomotion in lower vertebrates: From behavior to ionic mechanisms. In *Neural Control of Rhythmic Movements in Vertebrates.* Edited by A. V. Cohen, S. Rossignol, and S. Grillner. New York, Wiley Interscience, pp. 1-40.

Grogan, P., and Tyc-Dumont, S. (1989). How do dendrites process neural information? *News Physiol. Sci.* **4**:127-130.

Gustafson, B., Galvan, M., Grafe, P., and Wigstrom, H. (1982). A transient outward current in a mammalian central neuron blocked by 4-aminophysidine. *Nature* **299**:252-254.

Gustafson, B., and Pinter, M. J. (1984). Relations among passive electrical properties of lumbar x-motoneurons of the cat. *J. Physiol. (London)* **356**:401-431.

Haddad, G. G. (1986). Control of breathing in children. In *Breathing Disorders of Sleep*. Edited by N. H. Edelman and T. Santiago. New York, Churchill Livingtone, pp. 57-80.

Haddad, G. G., Donnelly, D. F., and Getting, P. A. Repetitive firing properties of hypoglossal motoneurons in-vitro: Intracellular studies in adult and neonatal rats. *J. Appl. Physiol.* (in press).

Haddad, G. G., Gandhi, M. R., and Mellins, R. B. (1982). Maturation of ventilatory response to hypoxia in puppies during sleep. *J. Appl. Physiol.* **52**:309-314.

Haddad, G. G., and Getting, P. A. (1989). Repetitive firing properties of neurons in the ventral region of nucleus tractus solitarius. In-vitro studies in adult and neonatal rat. *J. Neurophysiol.* **62**:1213-1224.

Haddad, G. G., and Mellins, R. B. (1984). Hypoxia and respiratory control in early life. *Annu. Rev. Physiol.* **46**:629-643.

Haddad, G. G., Schaeffer, J. I., and Bazzy, A. R. (1983). Postnatal maturation of the ventilatory response to hypoxia. In *Hypoxia, Exercise and Altitude*. Edited by J. R. Sutton, C. S. Houston, and N. L. Jones. New York, Alan R. Liss, pp. 39-50.

Halliwell, J. V., and Adams, P. R. (1982). Voltage clamp analysis of muscarinic excitation in hippocampal neurons. *Brain Res.* **250**:71-92.

Hodgkin, A. L., and Huxley, A. F. (1952a). Currents carried by sodium and potassium ions through the membrane of the giant axon of *Loligo. J. Physiol. (London)* **116**:449-472.

Hodgkin, A. L., and Huxley, A. F. (1952b). The components of membrane conductance in the giant axon of *Loligo. J. Physiol. (London)* **116**:473-496.

Hodgkin, A. L., Huxley, A. F. (1952c). A quantitative description of membrane current and its application to conduction and excitation in nerve. *J. Physiol. (London)* **117**:500-544.

Huguenard, J. R., Hamill, O. P., and Prince, D. A. (1988). Developmental changes in Na^+ conductances in rat neocortical neurons: Appearance of a slowly inactivating component. *J. Neurophysiol.* **59**:778-795.

Jan, L. Y., and Jan Y. N. (1989). Voltage-sensitive ion channels. *Cell* **56**: 13-25.

Johnson, P. (1988). Airway reflexes and the control of breathing in postnatal life. In *The Sudden Infant Death Syndrome: Cardiac and Respiratory Mechanisms and Interventions*. Edited by P. J. Schwartz, D. P. Southall and M. Valdes-Dapena. New York, New York Academy of Sciences, pp. 262-275.

Johnson, P., Fewell, J. E., Fedorko, L. M., and Wollner, J. C. (1983). The vagal control of breathing in postnatal life: Implications for sleep related to respiratory failure. In *The Sudden Infant Death Syndrome*. Edited by J. T. Tildon, L. M. Roeder, and A. Steinschneider. New York, Academic Press, pp. 467-490.

Lent, R. (1981). The brain of baby opposums. *Trends Neurosci.* **4**:84-87.

Ling, D. S. F., Petroski, R. E., Chon, W., and Geller, H. M. (1989). Development of spontaneous activity of hypothalamic neurons in dissociated culture. *Soc. Neurosci. Abstr.* **15**:1021.

Llinas, R. R. (1988). The intrinsic electrophysiological properties of mammalian neurons: Insights into central nervous system function. *Science* **242**:1654-1663.

MacDermott, A. B., and Westbrook, G. L. (1986). Early development of voltage-dependent sodium currents in cultured mouse spinal cord neurons. *Dev. Biol.* **113**:317-326.

McCormick, D. A., and Prince, D. A. (1987). Post-natal development of electrophysiological properties of rat cerebral cortical pyramidal neurones. *J. Physiol.* **393**:743-762.

Nicoll, R. A. (1988). The coupling of neurotransmitter receptors to ion channels in the brain. *Science* **241**:545-551.

O'Connell, S. M., Grierson, J. P., Petroski, R. E., and Geller, H. M. (1989). Analysis of calcium fluxes in dissociated rat neurons during embryonic development. *Soc. Neurosci. Abstr.* **15**:1021.

Petroski, R. E., and Geller, H. M. (1989). Development of voltage-gated ion channels in differentiating hypothalamic neurons. *Soc. Neurosci. Abstr.* **15**:1021.

Prince, D. A., and Gutnick, M. J. (1972). Neuronal activities in epileptogenic foci of immature cortex. *Brain Res.* **45**:455-468.

Rigatto, H. (1979). A critical analysis of the development of peripheral and central respiratory chemosensitivity during the neonatal period. In *Central Nervous Control Mechanisms in Breathing*. Edited by C. von Euler and H. Lagercrantz. Oxford, Pergamon Press, pp. 137-148.

Rogawski, M. A. (1985). The A-current: How ubiquitous a feature of excitable cells is it? *Trends Neurosci.* **8**:214-219.

Segal, M., Rogawski, M. A., and Barker, J. L. (1984). A transient potassium conductance regulates the excitability of cultured hippocampal and spinal neurons. *J. Neurosci.* **4**:604-609.

Smith, J. C., Liu, G., and Feldman, J. L. (1988). Intracellular recording from phrenic motoneurons receiving respiratory drive in-vitro. *Neurosci. Lett.* **88**:27-32.

Spitzer, N. C. (1981). Development of membrane properties in vertebrates. *Trends Neurosci.* **4**:169-172.

Spitzer, N. C., and Lamborghini, J. E. (1976). The development of the action potential mechanism of amphibian neurons isolated in culture. *Proc. Natl. Acad. Sci. USA* **73**:1641-1645.

Thompson, S. H., and Smith, S. J. (1976). Depolarizing afterpotentials and burst production in molluscan neurons. *J. Neurophysiol.* **39**:153-161.

Ziskind-Conhaim, L. (1988). Electrical properties of motoneurons in the spinal cord of rat embryos. *Dev. Biol.* **128**:21-29.

5

The Development of Descending Spinal Pathways in the North American Opossum

GEORGE F. MARTIN
RAYMOND H. HO, and
XIAO MING XU

The Ohio State University
College of Medicine
Columbus, Ohio

RONDA R. PINDZOLA

Case Western Reserve University
School of Medicine
Cleveland, Ohio

I. Introduction

It is well known that descending spinal pathways modulate locomotor activity in adult animals (see Kuypers, 1982, for review), but their role during development is not known. The first limb movements are spontaneous, intermittent, uncoordinated, and presumably produced by pattern generators within the spinal cord (see Hamburger, 1973, for review). Although brainstem axons grow into the spinal cord prior to the development of limb movements (the opossum, Martin et al., 1978; the frog, Forehand and Farel, 1982; the chick, Okado and Oppenheim, 1985), it is not clear that they influence limb movements. Spontaneous and reflexive movements of the hindlimbs are not affected noticeably by transection of the thoracic cord in newborn rats (Stelzner et al., 1975; Stelzner, 1982) or pouch-young opossums operated on prior to postnatal day 30 (Martin et al., 1978). The subsequent development of spinal shock after transection of the spinal cord may be due to maturation of descending spinal pathways (Stelzner, 1982) and/or their postsynaptic targets. It is possible, however, that brainstem axons influence the early development of limb movements, but their influence is not detected by lesion studies. There is evidence for monoaminergic modulation of spontaneous limb movements in the newborn rat (Commissiong, 1983) and most monoaminergic axons in the spinal cord originate within the brainstem. Cortical axons do not innervate the spinal cord (Cabana and Martin, 1984, 1985) or brainstem (Cabana and Martin, 1986a) until well after spontaneous limb movements begin, so they cannot influence them.

In order to study the role of descending spinal pathways on the development of motor activity, including breathing, it is important to know their developmental history. In the following account we will describe selected observations made in our laboratory on the development of descending spinal pathways in the North American opossum, *Didelphis virginiana*. Opossums are useful for developmental studies because they are born in a very immature state, 12-13 days after conception (Hartman, 1952; McCrady, 1938) and are available for approximately 90 days in an external pouch where they can be manipulated experimentally. In addition, the development of the opossum's nervous system occurs over a relatively extended period of time, providing a relatively wide window through which to view developmental events.

II. The Development of Brainstem-Spinal Projections

A. The Origin of Brainstem-Spinal Pathways at Different Stages of Development

Using the retrograde transport of horseradish peroxidase (HRP) and the fluorescent marker Nuclear yellow, we have shown that neurons within the pontine reticular formation and the presumptive locus coeruleus project to the thoracic spinal cord by postnatal day (PND) 3 (Cabana and Martin, 1984). More recently, we have used the retrograde transport of the fluorescent marker Fast blue (FB) to determine if spinal axons originate from additional areas at the same age. The results from one of the cases injected at PND3 are plotted in Figure 1, where it can be seen that neurons were labeled in the bulbar raphe (Figs. 1D-F and 2A), the presumptive locus coeruleus (Figs. 1D and 2B), the reticular formation (Figs. 1C-G and 2C), the vestibular nuclei (Figs. 1D, E and 2D), the presumptive interstitial nucleus of Cajal and related areas (Fig. 1B, C), and the hypothalamus (Fig. 1A). In fact, neurons could be labeled in most of the areas labeled by comparable injections in adult animals (Crutcher et al., 1978). A notable exception was the red nucleus. Lumbar injections of FB did not label rubral neurons with certainty until PND11. At PND3, and for some time thereafter, labeling density was greater in the brainstem than in adult animals and neurons were labeled in parts of the locus coeruleus that were not labeled in adult animals.

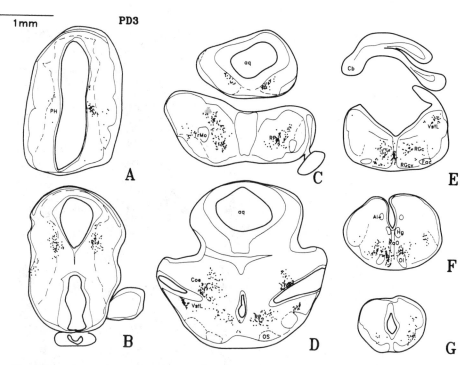

Figure 1 Plot of the labeled neurons (dots) present in representative sections of the brainstem after an injection of Fast blue into the caudal spinal cord of a postnatal day 3 opossum. The sections are arranged from rostral (A) to caudal (G).

It might be argued that lack of rubral labeling between PND3 and 11 reflects transport failure rather than lack of rubral axons in the caudal spinal cord or that some of the neurons labeled in the presumptive interstitial nucleus of Cajal were destined for the red nucleus. For these and other reasons, we made injections of wheat germ agglutinin conjugated to horseradish peroxidase (WGA-HRP) into the red nucleus in order to orthogradely label rubral axons in an age-graded series of animals. Details of methodology can be found in previous communications (Cabana and Martin, 1986b; Martin et al., 1988). At PND7 (left side, Fig. 3), rubral axons (Martin and Dom, 1970) were labeled contralaterally in the dorsal part of the lateral funiculus (arrows) at cervical and thoracic levels, but not in the lumbosacral cord. The injection included the interstitial nucleus of Cajal and the rostral metencephalon, probably accounting for the axonal labeling present in other funiculi and at lumbar levels. At PND13, rubral axons were labeled at lumbar

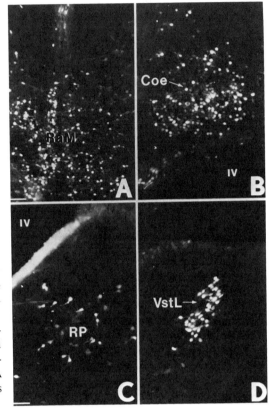

Figure 2 Fluorescence photomicrographs of Fast blue-labeled neurons in the bulbar raphe and adjacent reticular formation (A), the presumptive locus coeruleus (B), the pontine reticular formation (C), and the lateral vestibular nucleus (D) from the case plotted in Figure 1. The bar in A represents 30μm for both A and B, that in C represents 25 μm for C and D.

levels (Fig. 3) where they were limited to the marginal zone and by PND24 they had grown into most of the areas they occupy in adult animals.

B. The Development of Monoaminergic Projections to the Spinal Cord

In adult animals, monoaminergic axons innervate somatic motor, autonomic, and somatosensory areas of the spinal cord (Bowker et al., 1982; Martin et al., 1982; Westlund et al., 1982; Björklund and Skagerberg, 1982). Monoaminergic projections to spinal motoneurons have been hypothesized to provide motivational influence during movement (Kuypers, 1982), and it is well known that they influence autonomic (e.g., Coote and MacLeod, 1974; Coote et al., 1981) and somatosensory (Basbaum and Fields, 1978) functions at spinal levels. Although it is possible that monoaminergic axons

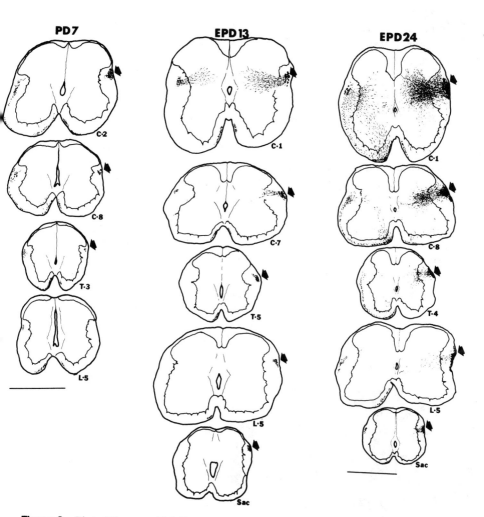

Figure 3 Plot of the axonal labeling produced at cervical (C-1, C-2, C-7, or C-8), thoracic (T-3, T-4, or T-5), lumbar (L-5), and sacral (Sac) levels of the spinal cord by rubral injections of wheat germ agglutinin conjugated to HRP at three different postnatal ages. The arrows indicate labeled axons contralateral to the injection which are presumed to be rubral in origin. The bar in each case represents 0.5 mm and that for the postnatal day (PD) 7 case can be used for those injected at estimated PD (EPD) 13. (Reproduced with permission from Cabana and Martin, 1986b.)

serve similar functions during development, it is equally possible that they influence the maturation of their postsynaptic targets (e.g., Lauder and Krebs, 1978).

Serotoninergic Projections

When the spinal cords of newborn opossums were processed for serotonin (5-HT) by the indirect antibody peroxidase-antiperoxidase (PAP) technique of Sternberger et al. (1970), evidence for 5-HT-like immunoreactive (LI) axons was found at most spinal levels (DiTirro et al., 1983). Such axons were present primarily in the marginal zone (Fig. 4A), although a few were found in the intermediate zone at cervical levels. By PND5, 5-HT-LI axons were present in most areas of the intermediate (mantle) zone they occupy in adult animals, particularly at cervical and thoracic levels, except for superficial areas of the dorsal horn [presumptive laminae I and II of Rexed (1952)]. By PND15, immunoreactive axons were abundant within presumptive laminae IV-X, and in lamina IX some of them approximated the somata of presumed motoneurons. It was not until approximately PND50, however, that immunoreactive axons were identified with certainty within laminae I and II. At that

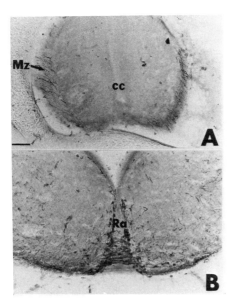

Figure 4 Photomicrographs of serotonin-like immunoreactivity in the rostral cervical cord (A) and brainstem (B) of a newborn opossum. In A, immunoreactive axons can be seen in the marginal zone (Mz), and in B, immunoreactive neurons are obvious in the raphe (Ra). The bar in A represents 80 μm for A and B.

age the intraspinal distribution of 5-HT-LI axons was comparable, at least qualitatively, to that in adult opossums (Martin et al., 1982).

At birth, 5-HT-LI neurons were present in most of the brainstem areas that contain them in adult animals (Martin et al., 1985b); a photomicrograph of such neurons is shown in Figure 4B. Since few 5-HT-LI neurons are found within the spinal cord at birth (DiTirro et al., 1983), it is likely that most of the 5-HT-LI axons at the same level originate within the brainstem as in the adult animal (Martin et al., 1982). In order to identify the neurons giving rise to such axons at later stages of development, we used the retrograde transport of FB together with immunofluorescence. The positions of neurons showing evidence for FB and 5-HT (double labeled) in an animal injected at PND11 are indicated by stars in Figure 5. Such neurons were present within those areas of the bulbar raphe and reticular formation expected from the results of comparable experiments in adult opossums. Examples of FB-labeled neurons (Fig. 6A), neurons showing 5-HT-LI (Fig. 6B), and double-labeled neurons (arrows, Fig. 6A, B) are provided from an animal injected at PD30.

Catecholaminergic Projections

Using antibodies generated against tyrosine hydroxylase (TH) and PAP immunohistochemistry (Sternberger et al., 1970), TH-LI axons were localized in the spinal cord of newborn opossums (Fig. 7A). Since tyrosine hydroxylase is an enzyme in the biosynthetic pathway for catecholamines, we have assumed that TH-LI axons are catecholaminergic. Such axons were found at most spinal levels at birth, where they were limited primarily to lateral parts of the marginal zone. By PND3, an occasional immunostained axon extended into the presumptive intermediolateral cell column and some were found near the central canal at cervical and lumbar levels.

In PND8 animals, TH-LI axons were present in the ventral and dorsal horns at all spinal levels as well as in the intermediolateral cell column at thoracic levels (Fig. 7B), but they were not found in superficial areas of the dorsal horn (presumptive laminae I and II). By PND15, an occasional immunostained axon was found in presumptive lamina II and, possibly, lamina I at cervical, thoracic, and lumbar levels. By PND23, such axons were present at all levels and by PND44, the distribution of TH-LI axons resembled that in adult opossums (Pindzola et al., 1988).

At birth, TH-LI neurons were found in all areas of the hypothalamus and brainstem that contribute to TH-LI axons in the spinal cord of adult opossums (Pindzola et al., 1988). A photomicrograph of TH-LI neurons in the presumptive locus coeruleus at PND1 is provided in Figure 7C. Since

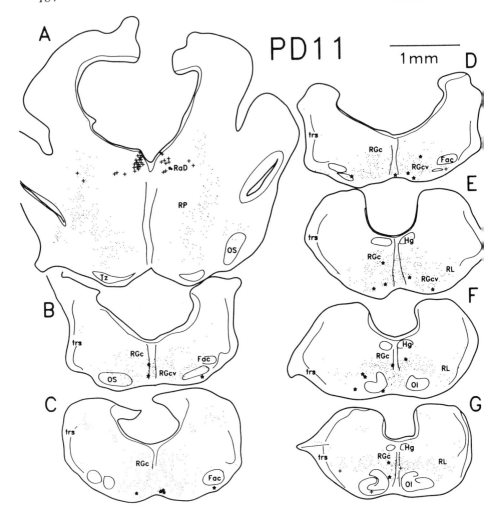

Figure 5 Plot of neurons labeled by Fast blue alone (dots), neurons that immuno-stained for serotonin (crosses), and neurons that contained both Fast blue and serotonin-like immunoreactivity (stars) in a case subjected to a lumbar injection of Fast blue at PD11 and immunofluorescence for serotonin. The sections are arranged from rostral (A) to caudal (G).

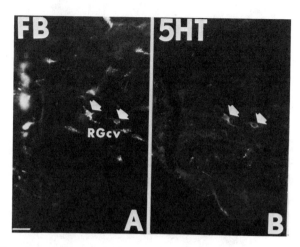

Figure 6   Fluorescence photomicrographs of Fast Blue-labeled neurons in the nucleus reticularis gigantocellularis, pars ventralis (RGcv), after an injection of Fast blue into the lumbar cord at PD30 (A) and, in the same section, neurons that show serotonin-like immunoreactivity (B). Neurons that contain both Fast blue and serotonin are indicated by arrows. The bar indicates 30 μm for A and B.

only a few TH-LI neurons were found in the spinal cord at birth, we have assumed that most axons of the same phenotype originated supraspinally.

Using the retrograde transport of FB in conjunction with immuno-fluorescence, the origin of TH-LI axons in the lumbar cord was determined at later stages of development. After injections of FB into the lumbar cord at PND5, neurons containing FB and TH-LI (stars, Fig. 8) were present in those areas of the hypothalamus and brainstem which contained them in adult animals (Pindzola et al., 1988). According to the atlas of Oswaldo-Cruz and Rocha-Miranda (1968), such areas included the nucleus paraventricularis hypothalami (Fig. 8A), the area hypothalamica lateralis, the presumptive nucleus coeruleus (Fig. 8C, D), and the ventrolateral medulla (Fig. 8E). Examples of FB-labeled and TH-LI neurons in the locus coeruleus (Fig. 9A, B) and ventrolateral medulla (Fig. 9C, D) are provided from an animal injected at PND20. Double-labeled neurons are indicated by arrows. At later stages of development, double-labeled neurons were present in areas where they were few to absent in adult opossums. Such areas included the dorsal part of the locus coeruleus and the paralemniscal area medial to the ventral nucleus of the lateral lemniscus.

C. The Development of Selected Peptidergic Projections to the Spinal Cord

It is clear that axons that contain enkephalins and substance P are abundant in the spinal cord of adult animals and that some of them originate within

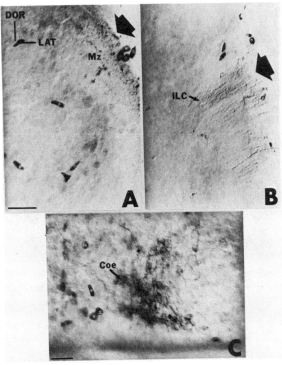

Figure 7 Photomicrographs of tyrosine hydroxylase-like immunoreactive axons (arrows) in the marginal zone (Mz) of the lumbar cord at birth (A) and at PND8 (B). The presumptive intermediolateral cell column is identified in B. A photomicrograph of tyrosine hydroxylase immunoreactive neurons in the presumptive locus coeruleus at postnatal day 1 is shown in C. Dorsal (DOR) and lateral (LAT) are shown in A for all figures. The bar in A represents 25 μm and can be used for A and B, whereas that in C represents 80 μm. Immunoreactive blood cells containing endogenous peroxidase activity can also be seen.

the brainstem (see review by Hökfelt et al., 1982). It is also clear that these peptides colocalize with monoamines in some spinally projecting neurons of the brainstem (e.g., Marson, 1989; Millhorn et al., 1989). Enkephalins inhibit nociceptive neurons of the dorsal horn (Duggan et al., 1977; Randic and Miletic, 1978) and apparently influence autonomic functions at spinal levels (Krukoff et al., 1985; Romagnano and Hamill, 1984). Substance P has been shown to increase the excitability of motoneurons (Nicoll, 1978; Otsuka and Yanagasiwa, 1980; White, 1985). Although it is possible that axons that contain these peptides are involved in similar functions during development, they may also influence the maturation of their targets (e.g., Zagon and McLaughlin, 1986).

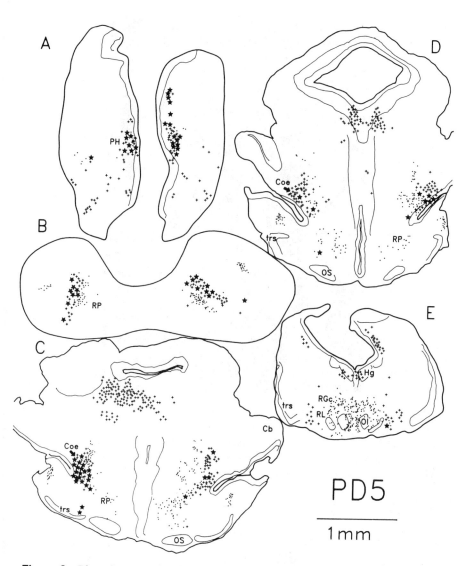

Figure 8 Plot of neurons labeled by Fast blue alone (dots), neurons that immuno-stained for tyrosine hydroxylase (crosses), and neurons that contained both Fast blue and tyrosine hydroxylase immunoreactivity (stars) in a case subjected to a lumbar injection of Fast blue at PD5 and immunofluorescence for tyrosine hydroxylase. The sections are arranged from rostral (A) to caudal (E).

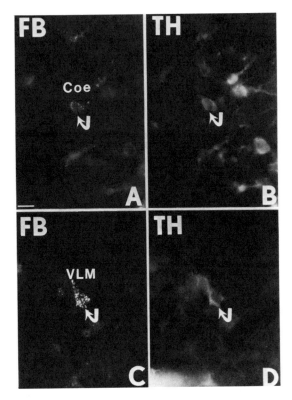

Figure 9 Photomicrographs of neurons labeled by Fast blue at PND20 in the presumptive locus coeruleus (Coe) and ventrolateral medulla (VLM) are shown in A and C. Neurons that immunostained for tyrosine hydroxylase (TH)-like immunoreactivity in the same sections are shown in B and D. Neurons that contained both Fast blue and serotonin are indicated by arrows. The bar in A represents 15 μm for all figures.

Enkephalin (ENK)-LI axons are present in the spinal cord of newborn opossums, where they are limited primarily to the dorsolateral marginal zone (DiTirro et al., 1983). By PND5, such axons are more abundant and an occasional one is found in the intermediate zone. As early as PND15, ENK-LI axons are scattered in deeper regions of the dorsal horn, the ventral horn, and the intermediolateral cell column. By PND30, they can be found within presumptive laminae I and II of the dorsal horn, thus completing the pattern present in adult opossums (DiTirro et al., 1981).

We tentatively conclude that most, if not all, ENK-LI axons in the spinal cord of newborn opossums originate within the brainstem. ENK-LI neurons were not found in the spinal cord until approximately PND30 (DiTirro et al., 1983), but they have been demonstrated in the brainstem at PND7 (Walker and King, 1989), even without colchicine treatment. In adult opossums, some of the ENK-LI axons in the spinal cord originate within the locus coeruleus; the locus coeruleus, pars alpha; the nuclei raphe magnus, pallidus, and obscurus; and the medullary reticular formation (Cassini et al., 1989).

Substance P-LI axons are also present in the spinal cord of newborn opossums, where, like ENK-LI axons, they are located primarily in the dorsolateral marginal zone (DiTirro et al., 1983). By PND5, such axons are present in the intermediate zone, where they are most abundant in presumptive laminae I and II. By PND30, SP-LI axons are more numerous within superficial areas of the dorsal horn and they are found within the presumptive intermediolateral cell column and lamina X. This general pattern of immunostaining persists, with variations in density, until maturity.

It is possible that most of the SP-LI axons in the spinal cord of newborn opossums originate within the brainstem. Relatively few SP-LI neurons are present in the spinal cord of developing (or adult) opossums (DiTirro et al., 1981, 1983), and in adult opossums, some SP-LI axons in the spinal cord originate within the locus coeruleus, pars alpha; the nuclei raphe magnus, pallidus, and obscurus; and the medullary reticular formation (Cassini et al., 1989). Some ENK and SP-LI bulbospinal neurons also contain 5-HT (Reddy et al., 1989).

III. The Development of Corticospinal Projections

In the opossum, corticospinal axons only project to cervical and rostral thoracic levels of the cord (Martin and Fisher, 1968; Martin and Cabana, 1985a). In order to determine when cortical axons reach spinal levels, we made injections of HRP and fluorescent markers into the cervical cord at different stages of development. In those experiments cortical neurons were not labeled until approximately 30 days after birth (Cabana and Martin, 1985). At that age labeled neurons were limited to a small area caudal to the orbital sulcus (Fig. 10C) in a region which by position corresponded to part of the primary somatosensory/motor cortex (Lende, 1963a,b). During later stages of development, neurons were labeled over a wider area of cortex than in adult animals (compare Fig. 10B and A).

In order to establish that the lack of cortical labeling prior to PND30 was due to absence of cortical axons, rather than transport failure or technical problems, we also studied the development of corticospinal axons using orthograde transport techniques (Cabana and Martin, 1985). In those experiments, multiple injections of WGA-HRP were made at different stages of development in order to fill the presumptive somatosensory/motor cortex.

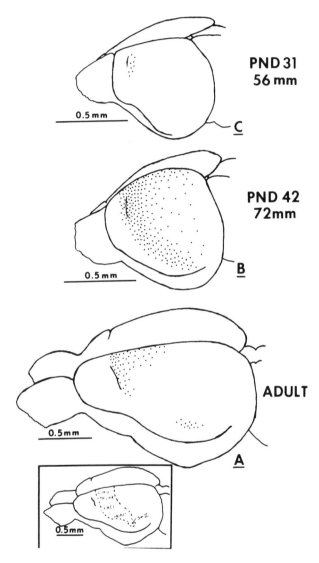

Figure 10   Diagram of the cortical labeling produced by cervical injections of re-trograde tracers in the adult opossum (A) and at two stages of development (B and C). The insert shows the somatosensory/motor areas in the adult opossum for reference. (Reproduced, with modification, by permission from Cabana and Martin, 1985.)

Although cortical axons could be labeled in the mesencephalon by PND10, and as far caudally as the motor facial nucleus by PND17 (Cabana and Martin, 1986a), they were not labeled in the spinal cord until approximately PND28 (Cabana and Martin, 1985). When cortical axons first enter the cord, they are limited to those portions of the dorsal and lateral funiculi which contain them in adult animals. Cortical axons do not grow into the gray matter until several days later and do not reach their caudal extent, the fifth thoracic segment of the cord, until approximately PND38. There are periods during development when the density of corticospinal innervation exceeds that in adult opossums and when cortical axons can be found in areas of the white and gray matter that do not contain them in adult animals.

IV. Summary and Conclusions

Brainstem axons, including those which transport monoamines and probably peptides, grow into the spinal cord very early in development and they originate primarily from areas that innervate the spinal cord in adult animals. Since the density of retrogradely labeled neurons in the brainstem after spinal injections of retrograde markers appears greater during some stages of development than in adult animals, it is possible that transient projections exist. Evidence for transient projections is most compelling in the locus coeruleus, where neurons can be labeled during development in areas that are not labeled by comparable injections in adult animals. Evidence for transient coeruleospinal projections has also been provided in the rat (Chen and Stanfield, 1987), and in that species, the transient projection is lost by axonal retraction or degeneration, not by degeneration of coeruleospinal neurons.

It appears that brainstem-spinal axons grow into the spinal cord asynchronously. For example, rubral axons reach lumbar levels of the cord after axons from most areas of the brainstem. Evidence for relatively late growth of rubral axons into the spinal cord has also been published for the chicken (Okado and Oppenheim, 1985) and clawed toad (ten Donkelaar and de Boer-van Huizen, 1982). It is possible that developmental sequences in the growth of brainstem-spinal axons are dictated by the timing of final mitosis and/or the maturation rate of postmitotic neurons. Since the interstitial nucleus of Cajal and the hypothalamus innervate the spinal cord before the red nucleus, clear rostral-to-caudal gradients do not exist.

Cortical axons do not grow into the spinal cord until well after axons from the brainstem have reached most of their adult targets and they do not innervate brainstem areas that project spinalward until relatively late (Cabana and Martin, 1986a). Evidence for spinal projections from cortical areas that do not give rise to them in adult animals has been reported for the opossum (Cabana and Martin, 1985), mouse (Crandall et al., 1985), rat (Ivy and Killackey, 1982; Stanfield et al., 1982), and rabbit (Distel and Hollander, 1980) and explanations for that phenomenon have been offered (e.g., Stanfield, 1984; O'Leary and Stanfield, 1989). In the rat, cortical neurons lose their transient spinal projections by selective elimination of axonal collaterals rather than cell death (O'Leary and Stanfield, 1985). Evidence for growth of cortical axons into areas of the spinal cord they do not supply in adult animals has been provided for the opossum (Cabana and Martin, 1985) and kitten (Theriault and Tatton, 1989). The role of transient projections is not certain.

Opossums are able to breathe at birth, but it is not known whether respiration is regulated at brainstem levels. It might be argued that newborn opossums must regulate breathing by chemoreceptor control over brainstem mechanisms in order to accommodate changes that occur in PCO_2 and PO_2 when the mother tightens or relaxes the pouch sphincter (Farber and Tenney, 1971). Since bulbospinal axons have grown to lumbar levels by PND3, they probably reach rostral cervical levels by birth. If so, they might influence respiration. Brainstem modulation of breathing has been documented by the end of the first postnatal week in opossums (Farber, 1972; Farber et al., 1972) and the bulbar areas that control respiration have been identified at PND21 (Farber, 1988). Because the developmental history of descending spinal pathways is becoming known in the opossum, it may make a good model for further studies on the development of brainstem control over respiration.

Abbreviations

Al	nucleus alaris
aq	cerebral aqueduct
Cb	cerebellum
CC	central canal
Coe	presumptive locus coeruleus
Fac	facial nucleus
FB	Fast blue
Hg	hypoglossal nucleus
5-HT	5-hydroxytryptamine (serotonin)
iv	fourth ventricle
ILC	intermediolateral cell column
Mz	marginal zone
Ol	inferior olivary nucleus
OS	superior olivary complex

PD   postnatal day
PH   presumptive paraventricular nucleus of the hypothalamus
Ra   raphe nucleus
RaM   nucleus raphe magnus
RaD   nucleus raphe dorsalis
RGc   nucleus reticularis gigantocellularis
RGcv   nucleus reticularis gigantocellularis, pars ventralis
RL   nucleus reticularis lateralis
RP   nucleus reticularis pontis
TH   tyrosine hydroxylase
TrMo   motor trigeminal nucleus
trs   spinal trigeminal tract
VLM   ventrolateral medulla
VstL   lateral vestibular nucleus

References

Basbaum, A. I., and Fields, H. L. (1978). Endogenous pain control mechanisms: Review and hypothesis. *Ann Neurol.* **4**:451-462.

Björklund, A., and Skagerberg, G. (1982). Descending monoaminergic projections to the spinal cord. In Fernström Foundation Series. *Brainstem Control of Spinal Mechanisms.* Amsterdam, Elsevier Biomedical Press.

Bowker, R. M., Westlund, K. N., Sullivan, M. C., and Coulter, J. D. (1982). Organization of descending serotonergic projections to the spinal cord. In: Progress in Brain Research. Vol. 57. *Descending Pathways to the Spinal Cord.* Edited by H.G.J.M. Kuypers and G. F. Martin. Amsterdam, Elsevier Biomedical Press.

Cabana, T., and Martin, G. F. (1984). Developmental sequences in the origin of descending spinal pathways. Studies using retrograde transport techniques in the North American opossum (*Didelphis virginiana*). *Dev. Brain Res.* **15**:247-263.

Cabana, T., and Martin, G. F. (1985). Corticospinal development in the North American opossum: Evidence for a sequence in the growth of cortical axons in the spinal cord and for transient projections. *Dev. Brain Res.* **23**:69-80.

Cabana, T., and Martin, G. F. (1986a). The development of projections from somatic motor-sensory areas of neocortex to the diencephalon and brainstem in the North American opossum. *J. Comp. Neurol.* **251**:506-516.

Cabana, T., and Martin, G. F. (1986b). The development of the rubrospinal tract. An experimental study using the orthograde transport of WGA-HRP in the North American opossum. *Dev. Brain Res.* **30**:1-11.

Cassini, P., Ho, R. H., and Martin, G. F. (1989). The brainstem origin of enkephalin and substance-P-like immunoreactive axons in the spinal cord of the North American opossum. *Brain Behav. Evolution* **34**:212-222.

Chen, K. S., and Stanfield, B. B. (1987). Evidence that selective elimination during postnatal development results in a restriction in the distribution of locus coeruleus neurons which project to the spinal cord in rats. *Brain Res.* **410**:154-158.

Coghill, G. E. (1938). Early movements of the opossum with special reference to the walking gait. *Proc. Soc. Exp. Biol. Med.* **39**:31-35.

Commissiong, J. W. (1983). Development of catecholaminergic nerves in the spinal cord of the rat. *Brain Res.* **264**:197-208.

Coote, J. H., and MacLeod, V. H. (1974). The influence of bulbospinal monoaminergic pathways on sympathetic nerve activity. *J. Physiol. (London)* **241**:453-475.

Coote, J. H., MacLeod, V. H., Fleetwood-Walker, S., and Gilbey, M. P. (1981). The response of individual sympathetic preganglionic neurones to microelectrophoretically applied endogenous monoamines. *Brain Res.* **215**:135-145.

Crandall, J. E., Whitcomb, J. M., and Caviness, V. S., Jr. (1985). Development of the spino-medullary projection from the mouse barrel field. *J. Comp. Neurol.* **239**:205-215.

Crutcher, K. A., Humbertson, A. O., and Martin, G. F. (1978). The origin of brainstem-spinal pathways in the North American opossum (*Didelphis virginiana*). Studies using the horseradish peroxidase method. *J. Comp. Neurol.* **179**:160-194.

Distel, H., and Hollander, H. (1980). Autoradiographic tracing of developing subcortical projections of the occipital region in fetal rabbits. *J. Comp. Neurol.* **192**:505-518.

DiTirro, F. J., Ho, R. H., and Martin, G. F. (1981). Immunohistochemical localization of substance-P, somatostatin and methionine-enkephalin in the spinal cord and dorsal root ganglia of the North American opossum, *Didelphis virginiana. J. Comp. Neurol.* **198**:351-363.

DiTirro, F. J., Martin, G. F., Ho, R. H. (1983). A developmental study of substance-P, somatostatin, enkephalin and serotonin immunoreactive elements in the spinal cord of the North American opossum. *J. Comp. Neurol.* **213**:241-261.

Duggan, A. W., Hill, J. G., and Headley, P. M. (1977). Enkephalins and dorsal horn neurons of the cat: Effects on responses to noxious and innocuous skin stimuli. *Br. J. Pharmacol.* **61**:399-408.

Farber, J. P. (1972). Development of pulmonary reflexes and pattern of breathing in the Virginia opossum. *Respir. Physiol.* **14**:278-286.

Farber, J. P. (1988). Medullary inspiratory activity during opossum development. *Am. J. Physiol.* **254** (*Regulatory Integrative Comp. Physiol.* 23):R578-R584.

Farber, J. P., Hultgren, H. N., and Tenney, S. M. (1972). Development of the chemical control of breathing in the Virginia opossum. *Respir. Physiol.* **14**:267-277.

Farber, J. P., and Tenney, S. T. (1971). The pouch gas of the Virginia opossum. *Respir. Physiol.* **11**:335-345.

Forehand, C. J., and Farel, P. B. (1982). Spinal cord development in anuran larvae. II. Ascending and descending pathways. *J. Comp. Neurol.* **209**: 495-508.

Hamburger, F. (1973). Anatomical and physiological basis of embryonic motility in birds and mammals. In *Behavioral Embryology.* Vol. 1. *Studies of the Development of Behavior and the Nervous System.* Edited by G. Gottlieb. New York, Academic Press.

Hartman, C. G. (1952). *Possums.* Austin, University of Texas Press.

Hökfelt, T., Skirboll, L., Dalsgaard, C.-J., Johansson, O., Lundberg, J. M., Norell, G., and Jancso, G. (1982). Peptide neurons in the spinal cord with special reference to descending systems. In Fernström Foundation Series. Vol. 1. *Brain Stem Control of Spinal Mechanisms.* Edited by B. Sjölund and A. Björklund. Amsterdam, Elsevier Biomedical Press.

Ivy, G. O., and Killackey, H. P. (1982). Ontogenetic changes in the projections of neocortical neurons. *J. Neurosci.* **2**:735-743.

Krukoff, T. L., Ciriello, J., and Calaresu, F. R. (1985). Segmental distribution of peptide- and 5-HT-like immunoreactivity in nerve terminals of the thoracolumbar sympathetic nuclei of the cat. *J. Comp. Neurol.* **240**: 103-116.

Kuypers, H. G. J. M. (1982). A new look at the organization of the motor system. In Progress in Brain Research. Vol. 57. *Descending Pathways to the Spinal Cord.* Edited by H. G. J. M. Kuypers and G. F. Martin. Amsterdam, Elsevier Biomedical Press.

Lauder, J. M., and Krebs, H. (1978). Serotonin as a differentiation signal in early neurogenesis. *Dev. Neurosci.* **1**:15-30.

Lende, R. A. (1963a). Sensory representation in the cerebral cortex of the opossum (*Didelphys virginiana*). *J. Comp. Neurol.* **121**:395-403.

Lende, R. A. (1963b). Motor representation in the cerebral cortex of the opossum (*Didelphys virginiana*). *J. Comp. Neurol.* **121**:405-415.

Marson L. (1989). Evidence of colocalization of substance-P and 5-hydroxytryptamine in spinally projecting neurons from the cat medulla oblongata. *Neurosci. Lett.* **96**:54-59.

Martin G. F., Beals, J. K., Culberson, J. L., Dom, R., Goode, G., and Humbertson, A. O. (1978). Observations on the development of brainstem-spinal systems of the North American opossum. *J. Comp. Neurol.* **181**:271-289.

Martin, G. F., and Cabana, T. (1985a). Cortical projections to superficial laminae of the dorsal horn and to the ventral horn of the spinal cord in the North American opossum. Studies using the orthograde transport of WGA-HRP. *Brain Res.* **337**:188-192.

Martin, G. F., Cabana, T., DiTirro, F. J., Ho, R. H., and Humbertson, A. O. (1982). Reticular and raphe projections to the spinal cord of the North American opossum. Evidence for connectional heterogeneity. In Progress in Brain Research. Vol. 57. *Descending Pathways to the Spinal Cord.* Edited by H. G. J. M. Kuypers and G. F. Martin. Amsterdam, Elsevier Biomedical Press.

Martin, G. F., Cabana, T., and Hazlett, J. C. (1988). The development of selected rubral connections in the North American opossum. *Behav. Brain Res.* **28**:21-28.

Martin, G. F., De Lorenzo, G., Ho, R. H., Humbertson, A. O., and Waltzer, R. (1985b). Serotonergic innervation of the forebrain of the North American opossum. *Brain Behav. Evolution* **26**:196-228.

Martin, G. F., and Dom, R. (1970). The rubrospinal tract of the opossum, *Didelphis virginiana. J. Comp. Neurol.* **138**:19-30.

Martin, G. F., and Fisher, A. M. (1968). A further evaluation of the origin, the course and the termination of the opossum corticospinal tract. *J. Neurol. Sci.* **7**:177-187.

McCrady, E. (1938). *The Embryology of the Opossum.* Amer. Anat. Memoirs, No. 16. Philadelphia, Wistar Institute of Anatomy and Biology.

Millhorn, D. E., Hökfelt, T., Verhofstad, A. A. J., and Terenius, L. (1989). Individual cells in the raphe nuclei for both serotonin and enkephalin project to the spinal cord. *Exp. Brain Res.* **75**:536-542.

Nicoll, R. A. (1978). The action of thyrotropin-releasing hormone, substance P and related peptides on frog motoneurons. *J. Pharmacol. Exp. Ther.* **207**:817-824.

Okado, N., and Oppenheim, R. (1985). The onset and development of descending pathways to the spinal cord in the chick embryo. *J. Comp. Neurol.* **232**:143-161.

O'Leary, D. D. M., and Stanfield, B. B. (1985). Occipital cortical neurons with transient pyramidal tract axons extend and maintain collaterals to subcortical but not intracortical targets. *Brain Res.* **336**:326-333.

O'Leary, D. D. M., and Stanfield, B. B. (1989). Selective elimination of axons extended by developing cortical neurons is dependent on regional locale: Experiments utilizing fetal cortical transplants. *J. Neurosci.* **9**:2230-2246.

Oswaldo-Cruz, E., and Rocha-Miranda, C. E. (1968). *The Brain of the Opossum* (Didelphis marsupialis). Rio de Janeiro, Instituto de Biofisica, Universidade Federal de Rio de Janeiro.

Otsuka, M., and Yanagisawa, M. (1980). The effects of substance P and baclofen on motor neurons of isolated spinal cord of the newborn rat. *J. Exp. Biol.* **89**:201-214.

Pindzola, R. R., Ho, R. H., and Martin, G. F. (1988). Catecholaminergic innervation of the spinal cord in the North American opossum, *Didelphis virginiana. Brain Behav. Evolution* **32**:281-292.

Randic, M., and Miletic, V. (1978). Depressant actions of methionine-enkephalin and somatostatin in cat dorsal horn neurons activated by noxious stimuli. *Brain Res.* **152**:196-202.

Reddy, V. K., Cassini, P., Ho, R. H., and Martin, G. F. (1990). The origins and terminations of bulbospinal axons that contain serotonin and either enkephalin or substance-P in the North American opossum. *J. Comp. Neurol.* **294**:96-108.

Rexed, B. (1952). The cytoarchitectonic organization of the spinal cord in the cat. *J. Comp. Neurol.* **96**:415-495.

Romagnano, M. A., and Hamill, R. W. (1984). Spinal sympathetic pathway: An enkephalin ladder. *Science* **225**:737-739.

Sawchenko, P. E., and Swanson, L. W. (1981). A method for tracing biochemically defined pathways in the central nervous system using combined fluorescence retrograde transport and immunohistochemical techniques. *Brain Res.* **210**:31-51.

Stanfield, B. B. (1984). Postnatal reorganization of cortical projections: the role of collateral elimination. *Trends Neurosci.* **7**:37-41.

Stanfield, B. B., O'Leary, D. D. M., and Fricks, C. (1982). Selective collateral elimination in early postnatal development restricts cortical distribution of rat pyramidal tract neurons. *Nature* **289**:371-373.

Stelzner, D. J. (1982). The role of descending systems in maintaining intrinsic spinal function: A developmental approach. In Fernström Foundation Series. Vol. 1. *Brain Stem Control of Spinal Mechanisms.* Edited by B. Sjölund and A. Björklund. Amsterdam, Elsevier Biomedical Press.

Stelzner, D. J., Ershler, W. B., and Weber, E. D. (1975). Effects of spinal transection in neonatal and weanling rats: Survival of function. *Exp. Neurol.* **46**:156-177.

Sternberger, L. A., Hardy, P. H., Cuculis, J. J., and Meyer, H. G. (1970). The unlabeled antibody enzyme method of immunohistochemistry. Preparation and properties of soluble antigen-antibody complex (horseradish peroxidase-anti-horseradish peroxidase) and its use in identification of spirochetes. *J. Histochem. Cytochem.* **18**:315-333.

ten Donkelaar, H. J., and de Boer-van-Huizen, R. (1982). Observations on the development of descending pathways from the brain stem to the spinal cord in the clawed toad *Xenopus laevis. Anat. Embryol.* **163**: 461-473.

Theriault, E., and Tatton, W. G. (1989). Postnatal redistribution of pericruciate motor cortical projections within the kitten spinal cord. *Dev. Brain Res.* **45**:219-237.

Walker, J. J., and King, J. S. (1989). Ontogenesis of enkephalinergic afferent systems in the opossum cerebellum. *Dev. Brain Res.* **48**:35-58.

Westlund, K. N., Bowker, R. M., Ziegler, M. G., and Coulter, J. D. (1982). Descending noradrenergic projections and their spinal terminations. In Progress in Brain Research. Vol. 57. *Descending Pathways to the Spinal Cord.* Edited by H. G. J. M. Kuypers and G. F. Martin. Amsterdam, Elsevier Biomedical Press.

White, S. R. (1985). A comparison of the effects of serotonin, substance P and thyrotropin-releasing hormone on excitability of rat spinal motoneurons in vivo. *Brain Res.* **335**:63-70.

Zagon, I. S., and McLaughlin, P. J. (1986). Opioid-induced modulation of cerebral and hippocampal development: Histological and morphometric studies. *Brain Res. Bull.* **28**:233-246.

6

Neurophysiological Organization of Respiratory Neurons in Early Life

JAY P. FARBER

University of Oklahoma
Health Sciences Center
Oklahoma City, Oklahoma

EDWARD E. LAWSON

University of North Carolina at Chapel Hill
Chapel Hill, North Carolina

I. Introduction

The aim of this chapter is to discuss various means to study the cellular organization of the respiratory control network in developing mammals and to consider maturational influences on generation of respiratory rhythm and pattern of respiratory muscle activation. We will focus on studies using either extracellular or intracellular recording techniques from the brainstem of intact animals. Remarkably few studies are available that have utilized single neuronal recording techniques to study the developmental physiology of the respiratory control circuitry within the braimstem of neonates or fetuses. Therefore, the few relatively comprehensive studies that have been performed will be discussed in some detail. The task of examining respiratory rhythmogenesis within the context of development is made more difficult because of the lack of an accepted model in adult mammals; however, where possible, data from the newborn will be evaluated with respect to current ideas regarding neural circuitry underlying respiratory rhythm. The reader is referred to Chapters 3 and 4 in this volume for further discussion of respiratory rhythm generation by the mammalian brainstem.

II. Studies Using Extracellular Recording and Related Techniques

To date, only the opossum has been studied longitudinally with respect to development of respiration-related unit activity (RRUs) in the medulla. There are, in addition, published studies that consider RRUs in the fetal sheep (Ioffe et al., 1989), and recordings have been made in the newborn cat (Lucier et al., 1979) and pig (Sica et al., 1987). The opossum and, indeed, all marsupial mammals are particularly intriguing with respect to the development of respiratory control because their development at birth is consistent with fetuses of placental mammals, yet the animals must breathe to support gas exchange. The opossum can easily be removed from the mother's pouch for study, so that evaluations can be made from a neurophysiologically immature stage to the fully developed adult. As a point of reference, the opossum at 50-60 (postnatal) days of age has about the same physical appearance and motor coordination as a newborn cat or rabbit. Myelination of the vagus nerve in a 50-day-old opossum is comparable to that of a newborn cat or rabbit (Krous et al., 1985). Extracellular unit activity in the medulla, in phase with various portions of the respiratory cycle, has been obtained in opossums as young as approximately 15 days of age; these results (Farber, 1988, 1989) are summarized below.

A. Location of RRUs in Opossums: Relationships to Other Species

Considering all types of RRUs in immature opossums, locations within the medulla lie between the region of nucleus ambiguus and the hypoglossal nucleus/nucleus tractus solitarius. The most widely studied adult mammal, the cat, demonstrates a clear separation of RRUs in the region of nucleus tractus solitarius (the dorsal respiratory group, DRG) from RRUs located in a region just lateral to nucleus ambiguus (the ventral respiratory group, VRG). The ventral respiratory group of cells extends in the rostrocaudal direction from behind the nucleus ambiguus up to the retrofacial nucleus. Results from immature opossums do not show demarcation into a dorsal and ventral respiratory group. In adult opossums, inspiratory neurons in the region of (most often medial to) nucleus ambiguus have been obtained; both rostral and caudal portions of nucleus ambiguus are associated with expiratory neurons in adult opossums. Most rostral expiratory cells examined (92%) did not have either vagal or bulbospinal projections; the majority of caudal medullary expiratory cells examined (85%) were bulbospinal. Other portions of the medulla of adult opossums have not been systematically explored for RRUs. Distribution of RRUs in adult rabbits seems generally comparable to that observed in the young opossum (Fallert and Wassmeyer, 1977); the adult rat has most of its medullary RRUs in the region of nucleus ambiguus (Ezure et al., 1988).

In piglets, concentrations of units have been described as localized in a DRG (Sica et al., 1987) and a VRG (Lawson et al., 1989b). Compared to the cat, the piglet VRG neurons seem to be diffusely distributed, but definitive localization techniques, such as intracellular dye injection, have not yet been routinely performed. While the stereotactic localization of neurons appears diffuse when comparing piglets of different ages, we have developed the impression that respiratory neurons form a rather compact column within individual animals (Lawson, unpublished observations). The diameter of the column seems not to exceed 300 μm. A definite rostral-caudal organization of VRG respiratory neurons has been noted using stereotactic techniques in the piglet (Lawson et al., 1989b), but the rostral reaches of the medullary VRG neurons have not been explored systematically. The VRG neurons are more likely to be postinspiratory or inspiratory when found in the portion of the medulla at about the level of the obex. In contrast, RRUs are almost exclusively expiratory when localized 2 or more mm caudal to the obex; many of these are bulbospinal as determined by antidromic invasion following spinal cord stimulation.

It has been suggested that other medullary regions besides the dorsal and ventral respiratory group in cats [e.g., nucleus paragigantocellularis (von Euler, 1986)] might importantly influence respiratory output. In fact, given that lesions in dorsal and ventral respiratory groups do not eliminate respiratory rhythmicity (Speck and Feldman, 1982; Huang and St. John, 1988), it is also possible that higher brainstem regions, particularly the pons, are critical for establishing a "normal" respiratory rhythm (Huang and St. John, 1988). While the vagotomized opossum has very few RRUs in the rostrolateral pons (unpublished results), in comparison to numbers obtained in the cat, it will be shown that activation of this region using an excitatory amino acid can strongly influence breathing patterns at all ages.

B. Discharge Properties of Inspiratory Neurons in the Developing Opossum

The most striking feature of inspiration in the young opossum is the relatively brief interval of diaphragm activation (approx. 100 msec) as compared to the adult animal (about 0.4-0.8 s, obtained in REM and NREM sleep; Farber and Marlow, 1976) and a burst rather than the ramp-like increase in inspiratory motor activity. A relatively brief inspiratory effort is a characteristic shared by the young opossum with other immature mammals. In young opossums, anesthetized with 5-s butyl-5-ethyl-2-thio-barbituric acid (Inactin), units with an inspiratory phased discharge typically produced few action potentials within the brief inspiratory interval (Fig. 1). In animals between approximately 15 and 60 days of age, the number of spikes per inspiration averaged only 1.3 for those units whose discharge was totally confined to the interval of diaphragm discharge and 3.3 spikes/breath for those units whose discharge commenced before the start of diaphragm electromyogram (EMG) activity. Peak interspike interval, where available, averaged 59 msec. These values represent a small fraction of the spikes generated by inspiratory neurons in adult animals. In animals up to about 60 days of age 44% of successfully tested inspiratory cells were observed to have bulbospinal projections. In animals younger than 60 days of age, conduction velocities of all bulbospinal cells were in the unmyelinated range (0.3-1.8 m/sec). This lack of myelination and slow conduction velocity in bulbospinal inspiratory neurons suggests that those cells whose discharge commenced before inspiration were activating inspiratory spinal motor neurons. Non-bulbospinal inspiratory modulated cells, whose discharge began before the diaphragm, could be laryngeal motor units since the posterior cricoarytenoid muscles began to discharge before the diaphragm in these young animals. In addition to the preceding, neurons were recorded whose discharge could be described as late expiratory or preinspiratory; other cells spiked at the end of inspiration and beginning of expiration. These latter cell types were observed much less often than the previously described units.

The present developmental data from inspiratory neurons may have particularly important implications with respect to regulation of inspiration using peripheral feedback mechanisms. With the few spikes per breath in both bulbospinal and nonprojecting medullary neurons, results of feedback would be expected to produce relatively coarse changes in timing or ampli-

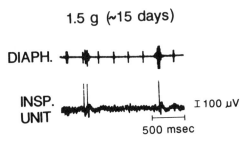

1.5 g (~15 days)

DIAPH.

INSP.
UNIT

I 100 μV

500 msec

Figure 1 Medullary unit activity in phase with inspiration from a 2-3-week-old opossum suckling. Top trace shows raw diaphragm EMG. Bottom trace shows unit activity obtained with a glass microelectrode.

tude of bursts. Fine modulation of the inspired breath would be better achieved with a high firing frequency which could then be more easily regulated. In the case of feedback from the lungs which limits duration of inspiration in adult animals, it is not clear that pulmonary afferent discharge is adequate to influence the inspired breath of very young opossums. That is, pulmonary afferents at an early age are unmyelinated and are expected to produce very few action potentials per inspiration when the lungs are inflated during a normal respiratory cycle (Farber et al., 1984); most afferent action potentials will occur relatively late in the inspired breath as transpulmonary pressure is increased. However, action potentials reaching the medulla relatively late in the breath will have limited possibility to influence inspiratory duration. This is a result of slow conduction velocities in unmyelinated fibers and the brief (ca. 100 msec) inspired breath. For example, consider the situation where conduction velocities of a pulmonary afferent, bulbospinal neuron, and phrenic motor neuron are all 0.5 mm/s (a value recorded in several instances from bulbospinal cells), and the total pathway is measured to be about 20 mm between pulmonary receptor and diaphragm motor end-plate in a 5-g animal. Under those conditions, there will be a 40-msec conduction delay (excluding brainstem synaptic events) between the initiation of an afferent action potential and possible termination of the inspired breath as measured at the diaphragm. Since most pulmonary afferent activity is expected relatively late in the breath, the most effective signal for inspiratory termination may reach brainstem control circuitry near the time of inspiratory off-switching or after inspiratory off-switching by the medulla; such input would be ineffective. With slow conduction time in the efferent as well as afferent pathway (about 26 msec of the total 40-msec conduction delay in the above example), the diaphragm is expected to discharge well after bulbospinal inspiratory neurons have shut down. This scenario seems to be born out by the lack of inspiratory prolongation during end-expiratory airway occlusions or after vagotomy in opossums 2-4 weeks of age (discussed in Chapter 8). One other possibility, however, is that such afferent impulses are gated out entirely from medullary respiratory control circuitry in young opossums.

The relatively long interspike interval observed in inspiratory neurons, if relatively synchronized in the motor outflow, or if very few motor units were activated, would be expected to produce widely spaced bursts of motor output. It is of interest, in this connection, that high-frequency oscillations in the phrenic motor activity of newborn dogs and cats are slow compared with the adult and have a dominant frequency that differs little from the value obtained using the long interspike interval of medullary cells in young opossums (Suthers et al., 1977). Additionally, for an inspiratory muscle where discharge can be prolonged in the very young opossum (e.g., activity of the posterior cricoarytenoid muscles during lung inflation), the bursts of activity are slow enough for the larynx to rapidly open and close with a fluttering pattern. It may be asked whether the avoidance of flutter in respiratory muscles is a potential reason for limiting the duration of the inspired breath in immature animals.

C. Discharge Properties of Expiratory Neurons in the Developing Opossum

Single-unit medullary neural activity with a clear expiration-phased discharge was difficult to obtain in spontaneously breathing, anesthetized, immature opossums. Very often units appeared to have an inconsistent expiratory modulation. Since breathing against a positive airway pressure (PPB) activates abdominal expiratory muscles in animals after the 4th postnatal week, it was reasoned that exploring the medulla under such conditions should result in the rhythmic activation of a larger number of expiratory units. This prediction was borne out experimentally in that 44% of expiratory units obtained with positive pressure showed no apparent rhythmic discharge when the load was removed. In most cases discharge was eliminated within the first two unloaded breaths. In the illustrated example (Fig. 2), firing pattern off-load was very erratic, with clear expiratory activity being noted in some sequences of breaths but not others. In most other cells the number of spikes per breath was greatly reduced off-load. Firing rates of expiratory cells, when breathing against a positive airway pressure, increased as a function of age (body weight); there was also a decrease in expiratory duration when breathing against a positive airway pressure as the animals matured. Results from adult opossums showed a much reduced incidence of cells that lost rhythmicity when pressure breathing stopped (less than 10% of tested cells). On the basis of a relatively high incidence of bulbospinal projections from expiratory neurons obtained in young opossums, it could be inferred that most were caudal medullary (see Section I.A). Typically, caudal medullary expiratory cells in adult animals showed an augmenting discharge which reached a plateau (with a delay between the end of inspiration and the start of spike activity); rostral cells more often discharged at about the same rate through their discharge period (e.g., discharge stepped up to a plateau value), while other rostral cells had either a decrementing discharge or an augmenting discharge. In most cases where the number of spikes was sufficient to make a judgment, the discharge of units in the young opossums reached a plateau; in the illustrated example (Fig. 2) the discharge appears to be decrementing. Only 2 of 46 units examined in young opossums could be designated as early expiratory.

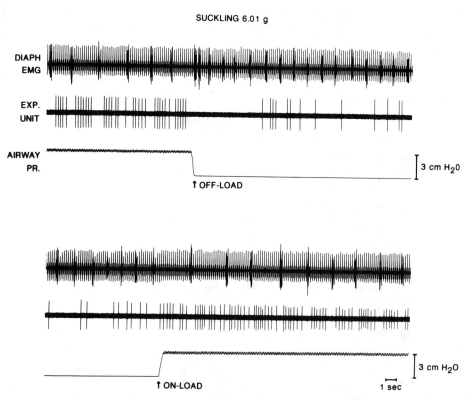

Figure 2 Medullary unit activity in phase with expiration in a suckling opossum when breathing against a positive airway pressure and under unloaded conditions. For upper and lower panels, top traces show raw diaphragm EMG, middle traces show expiratory unit activity from the medulla, and bottom traces show airway pressure. In the upper panel load was removed at arrow; in the lower panel load was reinstated at arrow.

These results suggest that in the immature opossum a substantial number of expiratory cells are not rhythmically active and are unlikely to be part of the respiratory rhythm generating circuitry of the brainstem. Interestingly, the location of most of these cells is caudal medullary; other investigators using adult cats have made similar conclusions about caudal medullary expiratory neurons based on several different techniques (Arita et al., 1987; Budzinska et al., 1985; Huang and St. John, 1988). In the young opossum, some expiratory cells might increase their activity as breathing is stimulated by the more asphyxiant gas composition (on average) in the mother's pouch

(Farber and Tenney, 1971); however, the amount of rebreathing of pouch air is expected to be influenced by the behavior of the mother, the size of the pouch, and a young animal's position in the pouch. The lack of discharge or low rates of discharge of expiratory cells in the youngest opossums may account, in part, for the lack of expiratory muscle activation during PPB in animals younger than 3-4 weeks of age (Farber, 1985).

D. Interpretation of Results from Opossums with Respect to Models of Respiratory Rhythmogenesis

In terms of analyzing respiratory rhythmogenesis in the young opossum, one recent model of interest consisted of a neural network simulation using pre-inspiratory, inspiratory, late inspiratory, postinspiratory, and expiratory cells (Ogilvie et al., 1989; also see Richter et al., 1986). Individual unit discharge patterns observed in the medulla of immature opossums are consistent with all these cell types. Inspiratory and expiratory units (during PPB) were most commonly observed, probably because of their relationship to motor output. Postinspiratory behavior seems clearly present in the decrementing discharge pattern of rostral medullary units in adult opossum as well in early expiratory cells; the illustrated expiratory cell from the young opossum (Fig. 2) might also illustrate such behavior. Despite the presence of the appropriate cell types, the above data cannot exclude either a pacemaker mechanism or other crucial sites for respiratory rhythmogenesis. With respect to the latter, a recent investigation in our laboratory, summarized below, illustrates the possibility that, at least under some conditions, strong influences of pontine mechanisms on respiratory rhythm can occur in very young opossums (Farber, 1990).

E. Medullary RRUs in Other Immature Mammals

Using the piglet, Sica et al. (1987) found that inspiratory neurons from the solitary tract region projected less frequently to the spinal cord than was the case for the adult cat; it was speculated that efferent projections might increase as a function of development, but differences related to the use of different species cannot be discounted. Ioffe and co-workers (1983, 1989) recorded RRUs in fetal sheep. While units could be identified that fired during various phases of the respiratory cycle, these cells often had an intermittent bursting type of discharge even when the fetus stopped breathing (e.g., in slow-wave-stage sleep). These data were interpreted on the basis of subthreshold or alternate inputs to these RRUs. Recent studies by Orem (1987) in adult cats also suggest possible behavioral modulation of some types of RRUs. Neither inspiratory nor expiratory units in sheep showed tonic activity in the absence of fetal breathing. Effects of superior laryngeal nerve stimulation on RRUs in newborn kittens will be considered in relationship to intracellular data to be presented in Section III.

F. Effects of Excitatory Amino Acids
in the Rostral Pons of Opossums

Rostral pontine injection of glutamate, a putative excitatory amino acid neurotransmitter, influenced breathing pattern in the youngest (ca. 2 weeks of age) animals tested. Injected volumes for the smallest animals were about 100 pl and represented only a few picomoles of amino acid. Effects on breathing pattern ranged from excitatory to inhibitory to mixed responses; but the highest concentrations (1 M) of glutamate typically caused apneas (17 of 20 animals). In contrast to the preceding, adults most often responded to the same concentration of glutamate (with injection volumes corrected to account for increases in volume of the pons) by increasing the respiratory motor output. Injection sites were found in the region of the parabrachial nucleus as well as in the Koelliker-Fuse, trigeminal, and lateral lemniscus regions. These data (Farber, 1990) suggest that input from pons to medulla can more easily disrupt breathing rhythm in immature animals. Because some of the active sites are associated with the trigeminal region, it is possible that a diving type of response was being activated; it can be speculated that such pontine inputs might also be involved with the frequent apneas occurring in premature infants. The fact that smaller amounts of glutamate altered rather than interrupted breathing pattern is consistent with the possibility that normal breathing pattern could be modulated by the rostral pons. In the newborn cat, inspiratory apneusis was not present after vagotomy and rostral pontine section (Duron and Marlot, 1980); instead, apnea with activation of expiratory muscles was observed. It is also of interest that the rostrolateral pons is necessary for the expression of a biphasic respiratory response to hypoxia in newborn lambs (Gluckman and Johnston, 1987). Clearly, the role of pontine mechanisms in the development of respiratory control awaits further elucidation.

III. Studies Using Intracellular Recording Techniques:
The Newborn Piglet Model

Both intracellular and extracellular recording techniques have been successfully utilized to record respiratory neurons in developing piglets. Intracellular recording techniques have an advantage over extracellular techniques because inhibitory and excitatory changes in membrane potential may be readily observed. Further, intracellular techniques do not depend on generation of action potentials in order to determine activity patterns and response characteristics of the neuron. Several animal models were considered before deciding to utilize piglets for intracellular neuronal recording. Already established models for developmental studies of other aspects of respiratory control were considered, including; rabbit, cat, dog, rat, and sheep. Piglets were chosen because of size, cost, and availability. Piglets are relatively inexpensive, yet they are raised in controlled conditions primarily for agricultural purposes. Also, piglets are bred such that litters are born throughout the year on a controlled basis with known gestation and nutrition throughout the pregnancy. However, the size (0.5-2.0 kg) at birth is the primary factor determining the preference for piglets. This relatively large size allows extensive surgery and instrumentation as well as paralysis with controlled ventilation, resulting in complete identification of recorded neurons and the possibility of relatively long, stable recordings. The principal disadvantage of piglets is their advanced stage of maturation at birth (especially compared to marsupials and rodents). A secondary limitation in using swine is that longitudinal studies are limited by rapid postnatal growth. Theoretically, this disadvantage may be overcome partially by use of miniswine rather than the more common strains of swine available from agriculture sources. However, longitudinal studies have not been performed using miniswine as cost and availability are limiting factors. Another disadvantage of swine is the lack of existing detailed description of the anatomical regions of the brain and brainstem. Hence, conclusions correlating stereotactic location and adjacent nuclear groupings are somewhat tentative. This problem is actually rather minor as the brainstem nuclear groupings appear very similar to those of the cat and comparisons are relatively easy.

A. Triphasic Pattern of Membrane Potential Oscillation During the Respiratory Cycle in Newborn Piglets

We have recorded a few neurons in a dorsal and medial location presumably corresponding to the DRG, but most of the experiments have concentrated on recording from neuronal units located 2.5-3.5 mm lateral to the midline and from +1 mm to −4 mm rostral-caudal in relation to the obex (Lawson et al., 1989). The latter location most likely represents neurons of the ventral respiratory group, as they were most frequently recorded 2.7-3.5 mm below the dorsal surface of the brainstem. Intracellular records of piglet brainstem neurons were considered acceptable for analysis when a stable maximal negative membrane potential exceeded −40 mV. Such recordings were considered especially reliable when spontaneous action potentials exceeded 30 mV in spike height. For all such VRG neurons recorded in the newborn piglet three distinct levels of membrane potential were distinguishable. Figure 3 demonstrates characteristics of the membrane potential for each of the three major classes of neuron observed in the ventral respiratory group of piglets.

Inspiratory neurons, usually recorded within ±1 mm of the obex (in the rostral-caudal axis), depolarized during the period of respiratory cycle associated with a ramp increase in neural activity of the phrenic nerve. Often these cells demonstrated a quick-onset depolarization followed by ramp depolarization of the membrane potential or a transient rapid rate of action

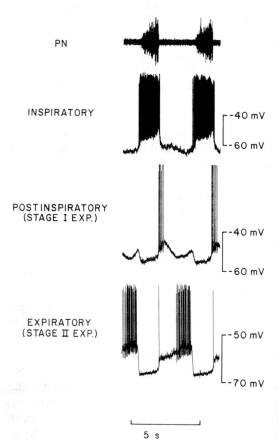

PN

INSPIRATORY

-40 mV

-60 mV

POSTINSPIRATORY
(STAGE I EXP.)

-40 mV

-60 mV

EXPIRATORY
(STAGE II EXP.)

-50 mV

-70 mV

5 s

Figure 3 Intracellular recordings obtained from the three major cell types identified in the VRG of piglet medulla. The top line represents phrenic neuronal activity obtained while simultaneously recording the inspiratory neuron demonstrated on the second line. The third and forth lines represent postinspiratory and expiratory neurons, respectively. The latter neurons were recorded from different animals than the inspiratory cell but the respiratory rate was similar, and they are shown together to demonstrate the timing relationships between all three groups of neurons in relation to phrenic activity. Membrane potential for each neuron is represented on the right of each neuron recording and the bottom horizontal line represents the time scale (5 s) for all recordings.

potential discharge followed by a more gradual, ramp-like increase in the action potential discharge rate. These ramp inspiratory neurons abruptly stopped discharging at the end of inspiration. The membrane potential then rapidly repolarized to one level, which changed to a second level after a short period. This biphasic expiratory period, demonstrated in Figure 3, was observed in all neurons having ramp inspiratory activity. In adult cats these two levels of membrane potential have been associated with two distinct waves of chloride-mediated inhibitory postsynaptic potentials (Ballantyne and Richter, 1986). One of the waves has been associated with postinspiratory phrenic nerve activity, which is regularly observed in anesthetized cats, but not in anesthetized piglets. The other wave occurs during the latter period of expiration and is termed stage II expiration or late expiration by different investigators.

Examples of two other types of inspiratory neurons have been observed. Several early inspiratory neurons have been identified. These RRUs are characterized by rapid attainment of maximal discharge rate or maximal depolarization early in inspiration followed by continuing decline throughout the remainder of the inspiratory period. One hypothesis regarding the function of these cells is that they inhibit the ramp inspiratory neurons and the decline in activity of the early inspiratory neurons results in the increase of activity in the ramp inspiratory neurons due to disinhibition (Richter et al., 1986). Direct evidence for inspiratory declining chloride-mediated synaptic inhibition of the ramp inspiratory neurons has been demonstrated in the adult cat (Ballantyne and Richter, 1984). The other type of inspiratory neuron is characterized by ramp depolarization during inspiration and continued depolarization during the postinspiratory period. Most of these neurons are bulbospinal and presumably are the units that activate the portion of the phrenic activity which remains during the postinspiratory period.

Postinspiratory (PI) neurons have been identified, in the same general region as the inspiratory units, that are maximally depolarized during the period immediately following the cessation of inspiration. In newborn piglets these cells remain depolarized for only a short interval (0.1-0.3 s) followed by gradual repolarization until the end of stage II expiration. An example of postinspiratory neuronal activity is demonstrated in Figure 3. During the inspiratory period these cells are maximally hyperpolarized. Only a few of these neurons have been shown to be antidromically activated following spinal cord stimulation, indicating that most postinspiratory neurons are probably interneurons with processes limited to the brainstem. Richter and co-workers (1986) speculated that since the activation of these neurons is associated with a simultaneous strong wave of inhibitory synaptic input to inspiratory neurons immediately upon cessation of inspiration, then the PI neurons may represent the network neurons responsible for the "irreversible off-switch." This concept was originally proposed by von Euler (1983) as an essential component of the respiratory network. It is clear that these neurons produce a strong inhibitory influence on inspiratory units and probably also shape the ramp depolarization of expiratory neurons by inhibitory synaptic potentials.

A third major neuronal activation pattern is observed in cells having their principal activation during the latter phase of expiration (defined as stage II expiration by some and late expiration by others). One such expiratory cell is demonstrated in Figure 3. These cells are strongly inhibited during inspiration, as proven by chloride reversal of this phase (Ballantyne and Richter, 1986). A sharp depolarization follows cessation of inspiration, followed in turn by a ramp increase in membrane potential to a late plateau ending at the onset of the next inspiratory phase. As demonstrated in Figure

3, the initial depolarization of the neuron following inspiratory inhibition may result in action potentials. This phenomen probably represents post-inhibitory excitation, a cellular property of these neurons, rather than synaptic excitation. As in the adult cat (Ballantyne and Richter, 1986), the ramp depolarization of these neurons is probably dependent on summation of excitatory and inhibitory postsynaptic potentials. The function of these cells is likely to be one of setting a slower respiratory frequency, as a two-phase oscillation between inspiration and postinspiration has been shown in rapid shallow breathing (Fig. 4) but a two-phase oscillation between expiratory neurons and either of the other two types has not been demonstrated. Further, it is clear that the most caudal of these neurons, which are almost exclusively bulbospinal in adults but perhaps less so in the newborn, mediate function of expiratory intercostal muscles.

B. Effect of Activation of Laryngeal Reflexes on Bulbar Respiratory Neurons

Lucier and co-workers (1979) were interested in age-related changes in reflex sensitivity to superior laryngeal nerve (SLN) stimulation. Recording from the region of tractus solitarius in kittens and cats, they observed that short latency excitatory inputs were rarer and the percentage of RRUs suppressed by SLN stimulation was greater in the kitten. It was suggested that inputs to RRUs changed as a function of development and might account for the more profound apneas caused by SLN electrical stimulation or physiological activation of laryngeal receptors in newborns.

Recently (Czyzyk-Krzeska and Lawson, 1990), we have shown that induction of apnea by laryngeal stimulation results in loss of phasic cycling in all three types of neurons described above. Further, inspiratory and expiratory neurons appear to be held at membrane potentials roughly equivalent to that of the postinspiratory phase in preceding control breaths. In contrast, postinspiratory neurons and vagal motor neurons were uniformly activated by laryngeal receptor stimulation. These results indicate that apnea induced by upper-airway stimulation in newborns is an active, synaptically mediated, process involving loss of phasic phrenic activity due to persistent activation of one phase of the respiratory cycle. That postinspiratory neurons are actively excited during apnea suggests further understanding of these neurons may indicate means to control apnea. Further, recovery from postinspiratory-induced apneas is preceded by expiratory activity before activation of inspiratory neurons or phrenic activity. Hence, the mechanism maintaining an apnea event appears different from that resulting in rapid, shallow breathing when postinspiratory activity alternates with inspiratory activity.

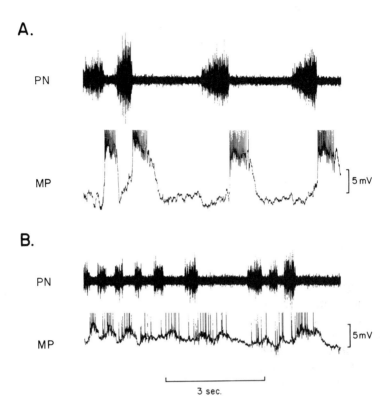

Figure 4   Representation of change from triphasic to biphasic respiratory pattern in nonantidromically activated postinspiratory neuron recorded from a 14-day-old piglet. A represents the three phase pattern observed during control periods. The first couplet of breaths, which may represent the functional equivalent of a sigh, also occurs without obtaining a stage II expiratory phase before onset of the second inspiration. B represents rapid, shallow breathing induced by retrograde insufflation of the larynx with ammonia-saturated gas. Many of the rapid breaths show alternating periods of inspiratory inhibition and postinspiratory activation without interspersed periods of expiratory inhibition. Neuronal membrane potentials consistent with stage II expiratory phases were present when the respiratory period abruptly lengthened, as illustrated by two instances in the latter half of this panel. Relative membrane potential is illustrated on the right and the time period is indicated by the bottom horizontal bar. PN, phrenic nerve activity; MP, membrane potential of the neuron. Action potentials are clipped to display absolute membrane potential changes. A minimal cardiac motion artifact is present in the membrane potential recording of A.

IV. Future Directions/Problems

To a great extent, understanding the pontomedullary organization of respiratory control circuitry during development will depend on obtaining an accepted model for respiratory rhythmogenesis in adult mammals. We especially need to determine whether control circuitry differs in any qualitative manner between neonates and adults; studies to date suggest such differences exist, but these data need to be integrated into more rigorous models for the generation of respiratory rhythm. It is clear that respiratory pattern generation is dependent on interactions among a complex network of neurons. Most of the neurons have been identified, and it is clear that synaptic and membrane properties of the neurons interact to fashion the basic respiratory pattern. Complicating factors in these investigations will include species difference and different maturities at birth. These differences in maturity may be reflected in the neural circuitry as alterations in numbers of involved cells, myelination of axons, cellular membrane properties, quantitative influences of excitatory or inhibitory central and peripheral reflexes, localization of specificity of central and peripheral afferent synaptic boutons, synthesis and degradation of neurotransmitters, and so forth. Control of breathing movements in fetuses is a special case where behavioral state becomes a primary determinant; it is important to understand how these behavioral mechanisms are expressed at birth and/or whether their persistence could impact on breathing in the newborn.

References

Arita, H. N., Kogo, N., and Koshiya, N. (1987). Morphological and physiological properties of caudal medullary expiratory neurons of the cat. *Brain Res.* **402**:258-266.

Ballantyne, D., and Richter, D. W. (1984). Post-synaptic inhibition of bulbar inspiratory neurones in the cat. *J. Physiol. (London)* **348**:67-87.

Ballantyne, D., and Richter, D. W. (1986). The non-uniform character of expiratory synaptic activity in expiratory bulbospinal neurones of the cat. *J. Physiol. (London)* **370**:433-456.

Budzinska, K., von Euler, C., Kao, F. F., Pantaleo, T., and Yamamoto, Y. (1985). Release of expiratory muscle activity by graded focal cold block in the medulla. *Acta Physiol. Scand.* **401**:258-266.

Czyzyk-Krzeska, M. F. and Lawson, E. E. (1990) Synaptic events in ventral respiratory neurons during apnea induced by superior laryngeal nerve stimulation in neonatal pig. *J. Physiol.* (London) in press.

Duron, B., and Marlot, D. (1980). Nervous control of breathing during postnatal development in the kitten. *Sleep* **3**:323-330.

Euler, C. von (1983). On the central pattern generator for basic breathing rhythmicity. *J. Appl. Physiol.* **55**:1647-1659.

Euler, C. von (1986). Brain stem mechanisms for generation and control of breathing pattern. In *Handbook of Physiology*. Section 3: The Respiratory System; Vol. 2: Control of Breathing: Edited by N. S. Cherniak, and J. G. Widdicombe. Bethesda, MD, American Physiological Society, pp. 1-67.

Ezure, K., Manabe, M., and Yamada, H. (1988). Distribution of medullary respiratory neurons. *Brain Res.* **455**:262-270.

Fallert, M., and Wassmeyer, B. (1977). Spacial distribution of various types of bulbar respiratory neurons in the rabbit. *Exp. Brain Res.* **30**: 339-351.

Farber, J. P. (1985). Motor responses to positive pressure breathing in the developing opossum. *J. Appl. Physiol.* **58**:1489-1495.

Farber, J. P. (1988). Medullary inspiratory activity during opossum development. *Am. J. Physiol.* R578-R584.

Farber, J. P. (1989). Medullary expiratory activity during opossum development. *J. Appl. Physiol.* **66**.

Farber, J. P. (1990). Effects on breathing of rostral pons glutamate injection during opossum development. *J. Appl. Physiol.* **69**:189-195.

Farber, J. P., Fisher, J. T., and Sant'Ambrogio, G. (1984). Airway receptor activity in the developing opossum. *Am. J. Physiol.* R753-R758.

Gluckman, P. D., and Johnston, B. M. (1987). Lesions in the upper lateral pons abolish the hypoxic depression of breathing in fetal lambs. *J. Physiol. (London)* **382**:373-383.

Huang, Q., and St. John, W. (1988). Respiratory neural activities after caudal-to-rostral ablation of medullary regions. *J. Appl. Physiol.* 1405-1411.

Ioffe, S., Jansen, A. H., and Chernick, V. (1983). Chronic extracellular recording of fetal medullary neuronal activity. *J. Appl. Physiol.* **55**: 1361-1364.

Ioffe, S., Jansen, A. H., and Chernick, V. (1989). Analysis of relation of respiratory neuronal firing and fetal diaphragmatic activity. *XVII Proc. Int. Union Physiol. Soc.,* Helsinki, Abstr. #P3386, p. 291.

Krous, H. F., Jordan, J., Wen, J., and Farber, J. P. (1985). Developmental morphology of the opossum vagus nerve. *Dev. Brain Res.* **20**:155-159.

Lawson, E. E., Richter, D. W., and Bischoff, A. (1989). Intracellular recordings of respiratory neurons in the lateral medulla of piglets. *J. Appl. Physiol.* **66**:983-988.

Lucier, G. E., Storey, A. T., and Sessle, B. J. (1979). Effects of upper respiratory tract stimuli on neonatal respiration: Reflex and single unit analysis in the kitten. *Biol. Neonate* **35**:82-89.

Ogilvie, M., Pack, A. I., and Richter, D. W. (1989). A model of respiratory rhythmogenesis. *FASEB J.* **3**:A251.

Orem, J. (1987). Inspiratory neurons that are activated when inspiration is inhibited behaviorally. *Neurosci. Lett.* 282-286.

Richter, D. W., Ballantyne, D., and Remmers, J. E. (1986). How is the respiratory rhythm generated? A model. *News Physiol. Sci.* **1** (June):109-112.

Sica, A. L., Donnelly, D. F., Steele, A. M., and Gandhi, M. R. (1987). Discharge properties of dorsal medullary inspiratory neurons in newborn pigs. *Brain Res.* **408**:222-226.

Speck, D. F., and Feldman, J. L. (1982). The effects of microstimulation and microlesions in the ventral and dorsal respiratory groups in the medulla of cat. *J. Neurosci.* **2**:744-757.

Suthers, G. K., Henderson-Smart, D. J., and Read, D. J. C. (1977). Postnatal changes in the rate of high frequency bursts of inspiratory activity in cats and dogs. *Brain Res.* **132**:537-540.

Part Two

**INPUT TO THE CNS: CELLULAR NEUROBIOLOGY
AND INTEGRATED BEHAVIOR**

7

Morphological and Neurophysiological Aspects of Airway and Pulmonary Receptors

JOHN T. FISHER

Queen's University
Kingston, Ontario, Canada

**OOMMEN P. MATHEW and
GIUSEPPE SANT 'AMBROGIO**

University of Texas Medical Branch
Galveston, Texas

I. Introduction

The purpose of this chapter is to compare the morphological and neurophysiological properties of upper and lower airway receptors of neonates and adults. As will become apparent, not much information is yet available on the developmental aspects of airway receptors, and this is an important factor in our present inability to explain several of the marked differences in the control of breathing between newborns and adults (see Chapters 8 and 9). Does the potency of feedback mechanisms remain the same during development? Are protective and defensive responses equally effective? Are the mechanisms subserving upper airway patency well developed at birth? Are the actions of chest wall muscles (intercostals, scaleni, diaphragm) properly regulated to stabilize the highly compliant rib cage of the newborn? Receptor properties are a necessary ingredient to explain such differences in the control mechanisms, and this review, admittedly based on insufficient data, aims at providing such information.

The research of the authors is supported by grants from NIH (HL-20122 and HL-32921), MRC Canada, and the Ontario Thoracic Society.

II. Upper Airway Receptors

A. Afferent Innervation

The afferent supply to the upper airway is mostly provided by the trigeminal, glossopharyngeal, and vagus nerves (Fig. 1). Nerve fibers conveying non-olfactory modalities of the nasal mucosa are part of the ophthalmic and maxillary divisions of the trigeminal nerve. The cell bodies of these fibers are located in the trigeminal ganglion. The ophthalmic branch gives origin to the anterior and posterior ethmoidal nerves that reach the upper and anterior regions of the lateral wall of the nasal cavity and the septum. The maxillary branch supplies fibers to the inferior and middle turbinates and the corresponding area of the septum, the lower anterior portion of the septum and the incisive canal, the anterior part of the floor of the nose and the inferior meatus. Recent data from cat's ethmoidal nerves (Biedenbach et al., 1975) report the presence of 1200-1500 myelinated fibers together with 600-700 Remak's bundles containing an unknown number of nonmyelinated fibers. From what is known for other nerves the number of nonmyelinated fibers should exceed that of the myelinated fibers. The histogram of the diameters of the myelinated fibers shows a unimodal distribution with a predominance of small axons of 2-6 μm. Several studies have demonstrated the presence of nonmyelinated fibers, presumed to be afferents terminating as free endings, in the nasal epithelium (Cauna, 1982). No evidence of any type of differentiated nervous receptors has been found. Nerve fibers containing substance P are abundant within and under the nasal epithelium and in the trigeminal ganglion (Uddman and Sundler, 1986). Pretreatment with capsaicin depletes these fibers of substance P and causes a loss of responses to various irritants. Other neuropeptides, such as calcitonin gene-related peptide (CGRP), galanin, and neurokinins, have been demonstrated in the nose. These data are consistent with the general distribution pattern of sensory neuropeptides (Uddman and Sundler, 1986). These neuropeptides in association with the afferent nerves of the nose are presumably involved in axon reflexes, local reflexes through the trigeminal ganglion, as well as central reflexes.

The nongustatory innervation of the pharynx is provided by the lingual branch of the mandibular division of the trigeminal nerve that supplies the anterior two-thirds of the tongue and the pharyngoesophageal branch of the glossopharyngeal nerve that supplies the posterior third of the tongue and part of the pharynx through the pharyngeal plexus (Fig. 1). The lingual nerve of the cat contains only a few nonmyelinated fibers; the myelinated fibers have a unimodal distribution of fiber size centered around a diameter of 2-6 μm (Biedenbach et al., 1975). No comparable study is available for the pharyngoesophageal branch of the glossopharyngeal nerve. Receptors

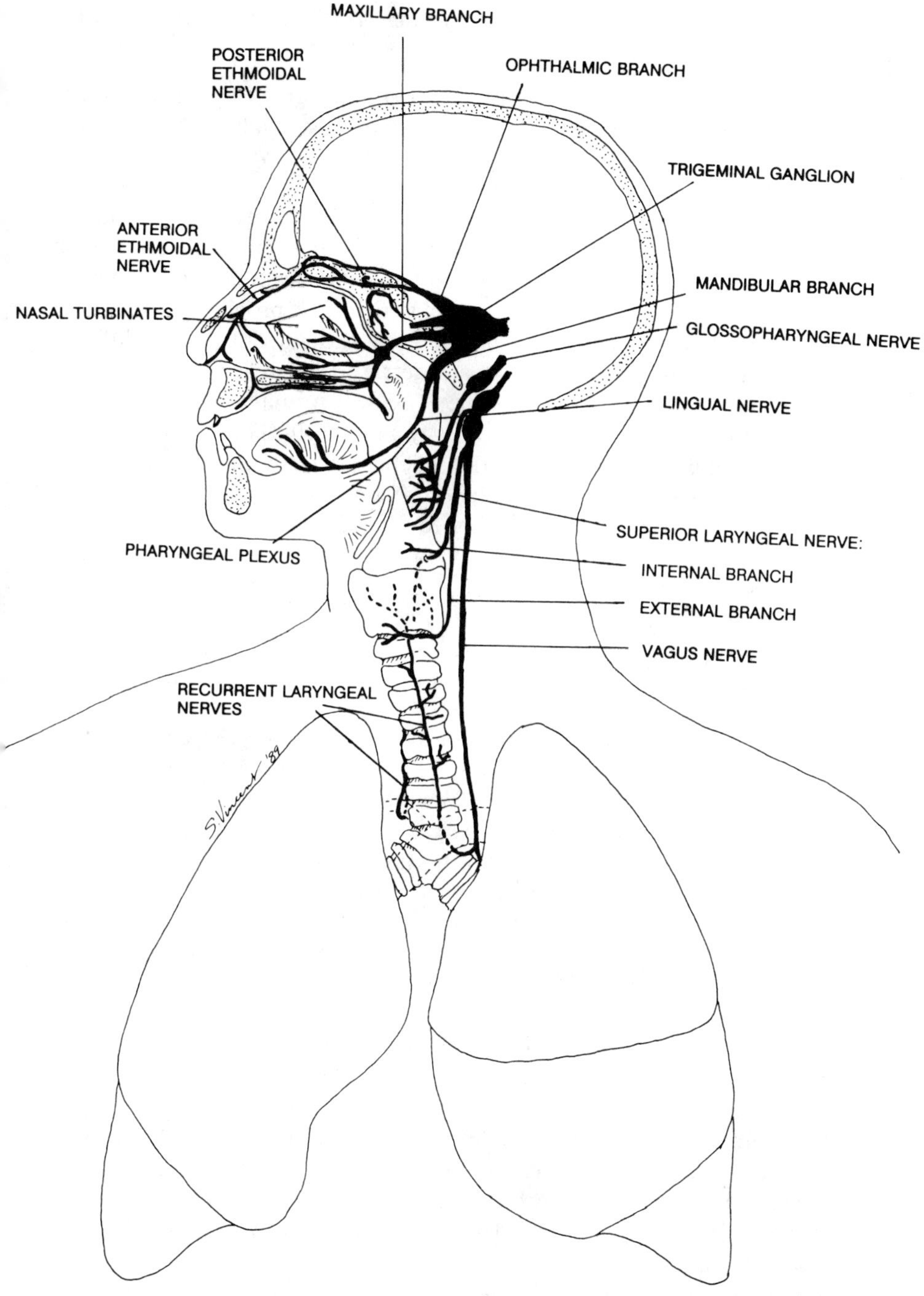

Figure 1 Schematic representation of the afferent innervation of the upper and lower airways.

with free endings have been identified under and within the squamous cells of the epithelium of the epipharynx. These endings could represent either the rapidly or the slowly adapting (Nail et al., 1969; Hwang et al., 1984) receptors having fibers in the pharyngoesophageal branch of the glossopharyngeal nerve. The muscles of the pharynx presumably have an afferent innervation similar to that of other skeletal muscles: muscle spindles, Golgi tendon organs, and so forth, but at present no morphological data are available.

The afferent supply to the laryngeal structures is provided by the vagus nerve through its laryngeal branches (Fig. 1). The superior laryngeal (SLN) is the most important afferent source, whereas the recurrent laryngeal nerve (RLN) and the external branch of the SLN, which supply the motor innervation to the laryngeal muscles, have only a minor role. The SLN has a considerable number of fibers in relation to the relatively small structure that it innervates. Despite the high density of afferent supply, the SLN has a relatively modest number of nonmyelinated fibers, at least in the cat (DuBois and Foley, 1936; Miller and Loizzi, 1974). The myelinated fibers have a unimodal distribution of their diameter frequency spectrum that is skewed toward the smaller sizes (Miller and Loizzi, 1974). The myelinated afferent fibers contained in the external branch of the SLN and the RLN are also of small diameter. The cell bodies of most of the laryngeal afferents are in the nodose ganglion.

Whereas no maturational data are available for the other upper airway nerves, some morphological studies have been performed on the SLN in newborn kittens (Fig. 2). At birth only 22% of the total fibers (efferent and afferent) are myelinated; this increases to 48% at 2 months (Miller and Dunmire, 1976). The estimated myelination of this nerve in the adult cat is about 90% (Miller and Loizzi, 1974). This rate is myelination is very similar to that reported in the same species for the vagus nerve (Marlot and Duron, 1979a). As for other nerves, the maturational changes include an increasing diameter of the fibers accompanied by an increase in conduction velocity (Miller and Loizzi, 1974; Sasaki, 1979).

Depending on their location, laryngeal receptors have been classified as "mucosal," "articular," and "muscular" (Wyke and Kirchner, 1976). A prevalence of undifferentiated, free-endings-type receptors has been found in all three locations. More complex "corpuscular" receptors have been described in submucosal layer and in the fibrous capsules of the intercartilagineous joints. Structures resembling muscle spindles and Golgi tendon organs have been demonstrated within the intrinsic laryngeal muscles of some species; studies of fiber caliber spectra of laryngeal nerve of humans and cats are also consistent with the presence of muscle proprioceptors (Wyke and Kirchner, 1976). Although we do not have data specifically related to receptor types in the newborn, some maturational inferences on muscle proprioceptors can be derived from the studies of Sasaki et al. (1977) and Sasaki

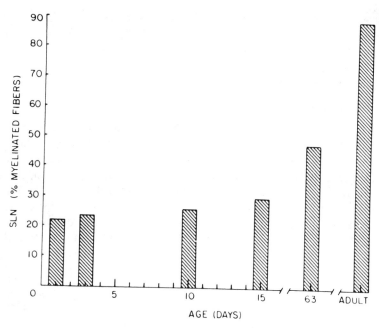

Figure 2 Postnatal maturation of the myelinated fibers in the internal branch of the superior laryngeal nerve (SLN) of the cat. Data for the kitten are from Miller and Dunmire (1976) and for the adult cat from Miller and Loizzi (1974). (From Mortola and Fisher, 1988.)

(1979) on developmental changes in the diameter and the myelination of the recurrent nerve of dogs. At birth most of the fibers are nonmyelinated, and it is only at 1 year of age that the RLN acquires the adult composition. Concurrent with these changes in myelination there is a progressive increase in the caliber of the myelinated fibers, which reach their adult values at about 8 weeks. Although specific information related to the afferent fibers is lacking, it is reasonable to speculate that these maturational changes in nerve composition and size reflect some functional limitation during development.

With regard to the C-fiber component of the afferent innervation to the larynx, major differences apparently exist among species. For instance, as previously mentioned, the cat's SLN has a scant supply of C-fibers (Miller and Loizzi, 1974), which, however, are more numerous in the rat (Rosenberg et al., 1989). The presence of substance P immunoreactivity does not seem to constitute a reliable index of C-fiber innervation and function as indicated by its presence in the laryngeal mucosa of both rats and dogs (Lundberg et al., 1983; Shin et al., 1987), species that differ markedly in their response to inhaled capsaicin (Palecek et al., 1989). Comparative data on laryngeal C-fiber function in the newborn are still lacking.

B. Functional Characteristics

Trigeminal Afferents

Changes in afferent activity have been documented in the trigeminal nerve in response to respiratory-related stimuli such as airflow and transmural pressure as well as irritant stimuli such as cigarette smoke (But and Klimova-Cherkasova, 1967; Glebovsky and Bayev, 1984; Tsubone, 1989a,b; Dawson, 1962; Ulrich et al., 1972). Single- and multifiber recordings from the ethmoidal branch of the trigeminal nerve have clearly demonstrated the presence of cold receptors responding to the cooling effect of airflow, as illustrated in Figure 3 (Tsubone, 1989a). Moreover, receptors stimulated mostly by

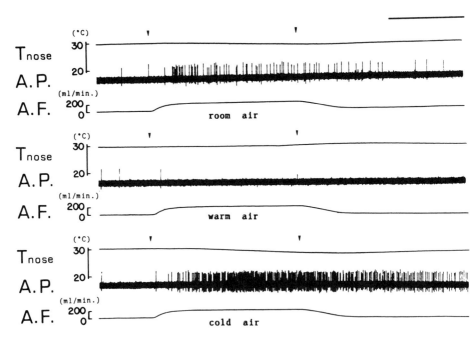

Figure 3 Effects of temperature on the activity of nasal "flow" receptors during constant airflow. T nose = intranasal temperature (°C), A.P. = action potential. A.F. = nasal airflow (200 ml/min). The horizontal line at the upper right corner shows 5 s. From top to bottom panels: room air, warm air, and cold air were applied to the nose between the arrows. (From Tsubone, 1989a.)

negative but also by positive pressure have been documented (Tsubone, 1989b). The presence of similar receptors in the newborn is yet to be documented. However, their presence can be presumed considering the reflexes elicited from this region.

The lingual nerve that innervates the anterior two-thirds of the tongue has afferent fibers representing tactile, nociceptive, and thermoceptive (warm and cold) modalities (Zotterman, 1936).

Glossopharyngeal Afferents

There is scant data on glossopharyngeal afferents even in adult animals. Since these studies were conducted in paralyzed, artificially ventilated animals, it is difficult to be certain about the presence of respiratory modulation during spontaneous breathing. Most of the endings studied were silent and responded with a rapidly adapting discharge to mechanical stimuli (Nail et al., 1969). Some of them exhibited a rapidly adapting discharge with airflow and pharyngeal distension as well. Hwang et al. (1984) documented the presence of receptors with fibers in the glossopharyngeal nerve which respond to negative and/or positive pressures with a slowly adapting discharge. Even though an involvement of these endings in the aspiration reflex and pharyngeal airway maintenance has been suggested, their role in the newborn is yet to be determined.

Laryngeal Afferents

Laryngeal endings have been studied extensively in both adults and newborns compared to trigeminal and glossopharyngeal afferents.

Respiratory-Related Stimuli

During each breathing cycle the larynx is subjected to changes in airflow, transmural pressure, and phasic contraction of its intrinsic laryngeal muscles. When the afferent activity emerging from the larynx is monitored by recording the whole nerve activity of the superior laryngeal nerve, a clear respiratory modulation is seen in both newborns and adults (Mathew et al., 1984b). Three types of laryngeal endings have been identified in both newborns and adults (Sant 'Ambrogio et al., 1983; Fisher et al., 1985). These studies in spontaneously breathing anesthetized dogs evaluated the behavior of laryngeal endings during four experimental conditions: (1) tracheostomy breathing, (2) upper airway breathing, (3) occlusion of the trachea, and (4) occlusion of the upper airway. These experimental conditions allowed us to identify the primary stimulus responsible for the activation of these receptors as schematically shown in Figure 4. For example, if the primary stimulus for activation of a receptor is airflow, this ending will be active only during

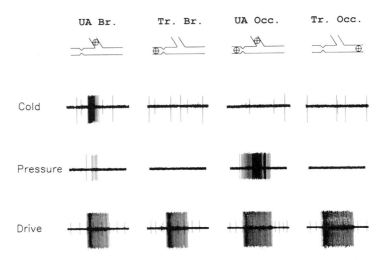

Figure 4 Behavior of laryngeal cold, pressure, and drive receptors during four experimental conditions. UA. Br. = upper airway breathing; Tr. Br. = tracheostomy breathing; UA. Occ. = upper airway occlusion; Tr. Occ. = tracheal occlusion. See text for details.

upper airway breathing (the only condition in which there is airflow through the upper airway) and not during the other three experimental conditions. Similarly, if the receptor is stimulated by pressure changes in the upper airway, the receptors will be active during both upper airway breathing and occlusion with maximum activation during upper airway occlusion, whereas a receptor stimulated by the active and passive movements of the larynx is expected to show maximal activity during airway occlusion but may be active during the other conditions as well.

The types of receptors responding to airflow through the larynx were initially designated as flow receptors. Subsequently it was shown that these receptors in fact respond to a decrease in laryngeal temperature (Sant 'Ambrogio et al., 1985a). The same flow at body temperature fails to stimulate these endings. Laryngeal endings that respond primarily to transmural pressure changes across the larynx are classified as pressure receptors. The majority of these endings are stimulated by negative pressure, some are stimulated by positive pressure, and a few respond to both (Mathew et al., 1984a). Receptors that respond to active (contraction of intrinsic laryngeal muscles) and passive (transmitted tracheal movements due to the contraction of thoracic inspiratory muscles) movements of the larynx are designated "drive" receptors (Sant 'Ambrogio et al., 1985b). All drive receptors invariably show some degree of pressure sensitivity, while the majority of pressure receptors exhibit some drive component. An example of a negative-pressure receptor with drive component in a newborn puppy is illustrated in Figure 5. In both adult and developing dogs most of the drive receptors studied showed an inspiratory modulation. It is likely that the number of expiratory modulated

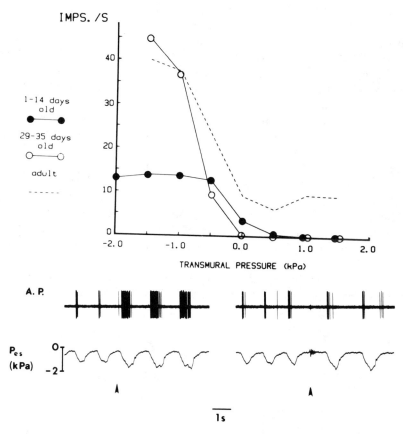

Figure 5 Upper panel: Relationship between transmural pressure and steady-state discharge frequency (imps/s) for negative-pressure receptors. Bottom panel: Behavior of a negative-pressure receptor during four experimental maneuvers. AP, action potentials; P_{es}, esophageal pressure. Left: Two breaths through upper airway followed (at arrowhead) by efforts against an occluded upper airway during which receptor activity increases markedly. Right: Following three breaths through tracheostomy (at arrowhead), trachea is occluded with no increase in receptor activity. Note that a residual inspiratory modulation is present during tracheal breathing and tracheal occlusion reflecting activity of inspiratory muscles of upper airway (drive component). (From Fisher et al., 1985.)

receptors was underestimated in these studies, since the level of anesthesia used depresses the activity of laryngeal expiratory muscles. Nevertheless, the relative distribution of laryngeal flow, pressure, and drive receptors is similar in puppies and adult dogs (Fig. 6). However, the mean discharge frequency of pressure receptors appears to be lower in newborn puppies when compared to values obtained in adults (Fig. 5).

Even though the behavior of flow, pressure, and drive receptors has not been studied extensively in newborns, discharge characteristics of these endings have been fairly well studied in adults. Temperature at which the conduction of these fibers is blocked indicates that all three types of receptors have myelinated fibers (Widdicombe et al., 1988). Similarly, inspiratory modulation of drive receptors is due primarily to the phasic contraction of intrinsic laryngeal muscles such as the PCA and transmitted tracheal movements (Sant 'Ambrogio et al., 1985b). Paralysis of the laryngeal muscles not only reduces or abolishes the respiratory modulation of most of the drive receptors, but also reduces the response of these endings to pressure changes (Mathew et al., 1988). That the activity of drive receptors is influenced by laryngeal muscle contraction does not necessarily imply a location in these muscles. Laryngeal cooling, in addition to stimulating the cold receptors, inhibits the respiratory modulation of pressure and drive receptors to a great extent (Sant 'Ambrogio et al., 1986a). This inhibitory effect of cooling is far greater than that observed for tracheal stretch receptors ($Q_{10} = 2.5$ vs. 1.2; Sant 'Ambrogio F.B., et al., 1986).

The three types of laryngeal receptors are presumed to be involved in several reflex responses originating from the larynx. For instance, it is clear

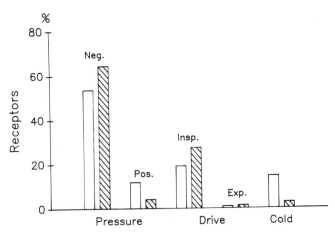

Figure 6   Relative occurrence of pressure, drive, and cold receptors in puppies (hatched columns) and adult dogs (open columns). Neg. = negative; Pos. = positive; Insp. = inspiratory; Exp. = expiratory.

that pressure-sensitive receptors of the larynx primarily account for the changes in breathing pattern observed with the application of negative pressure in the isolated upper airway in newborns and adults (Fisher et al., 1985). The application of negative pressure may induce apnea in newborn puppies, but not in adult dogs. These reflexes are discussed and reviewed in detail in Chapter 9. What is not clear is which of the different types of pressure-sensitive receptors mediate the above responses. Paralysis of the laryngeal muscles does not alter the response of pure pressure receptors as well as the response of drive receptors inhibited by negative pressure, whereas it reduces markedly the response of endings stimulated by drive and negative pressure (Mathew et al., 1988). Therefore, the latter type does not appear to play a major role in the reflex response to upper airway pressure changes. Although some of the pressure receptors have a deeper location, the superficially located endings appear to be primarily responsible for the reflex effects since the responses to pressure changes are abolished within minutes by topically applied lidocaine (Mathew et al., 1982).

Laryngeal cooling induces hardly any changes in breathing pattern in adult dogs, while a significant alteration of breathing pattern occurs in newborn puppies (Mathew et al., 1986, 1989). The relative contribution of the stimulation of cold receptors versus the inhibition of mechanoreceptors in the above response is not clear. In addition, laryngeal cooling has been shown to induce bronchospasm in adult cats (Jammes et al., 1983), nonmyelinated afferents in the superior laryngeal nerve are presumed to mediate this response (Jammes et al., 1987).

Chemical and Irritant Stimuli

The response of laryngeal endings to water has been studied extensively in newborn and adults of several species (Boushey et al., 1974; Harding et al., 1978; Storey and Johnson, 1975). Receptors from adult and newborn dogs exhibited similar responses to water and other solutions and their stimulation was shown to depend on the lack of chloride or other permeant anions (Boggs and Bartlett, 1982). Moreover, these water-responsive endings were not taste buds but were presumed to represent a distinct group of myelinated fibers with free endings located within the laryngeal mucosa (Boushey et al., 1974; Harding et al., 1978). Harding et al., (1978) showed that the water-responsive laryngeal endings do not belong to a homogeneous category but exhibit differences in latency and duration of response, mechanosensitivity, and so forth. Recently we have confirmed the presence of two distinct types of responses (Fig. 7) to water; a short-latency, short-duration response and a long-latency, long-duration response (Anderson et al., 1989). The short-latency, short-duration response depends on the lack of chloride ions, whereas the long-latency, long-duration response depends on the hypoosmolality of the solutions tested (Anderson et al., 1989). We have also documented

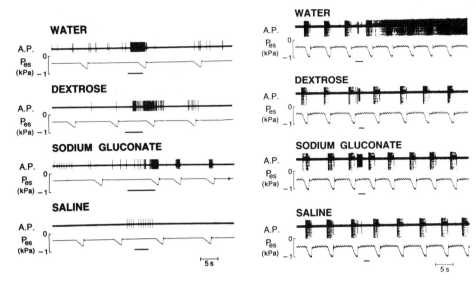

Figure 7 Effect of water and other test solutions on a randomly firing (left panel) and a respiratory modulated (right panel) laryngeal receptor. In each panel: A.P. = action potentials; P_{es} = esophageal pressure in kiloPascal (kPa). Applications of the test solutions are indicated by thick lines. (From Anderson et al., 1990.)

that the receptors that respond to hypoosmolality are generally respiratory modulated, whereas those stimulated by the lack of chloride ions are either silent or randomly active. Despite the similarities in receptor behavior of adult and newborn animals, the reflex responses are markedly greater in the newborn (Boggs and Bartlett, 1982).

Both receptor and reflex studies in adults clearly show that CO_2-sensitive endings are present in the larynx (Boushey et al., 1974; Boushey and Richardson, 1973). However, no such evidence exists in the newborn. Similarly, receptors responding to irritants such as smoke are well documented in adults (Lee et al., 1987), but similar studies have not been performed in newborns.

III. Lower Airway Receptors

A. Afferent Innervation

The vagus nerve supplies most of the afferent innervation to the tracheobronchial tree (Fig. 1). The extrathoracic trachea receives afferent fibers through the SLN and RLN nerves as well as from the pararecurrent laryngeal nerve (Fig. 1). The SLN is more important for the more cranial segments and the RLN and the pararecurrent for the lower segments. The intrathoracic trachea is innervated through fibers coming mostly from the RLN on the left side and directly from the main trunk of the thoracic vagus on the right side. It is worth noting that the trachea has a bilateral innervation while the bronchi receive their innervation almost exclusively through the ipsilateral vagus nerve.

As is the case in other peripheral nerves, the vagus nerve does not have many myelinated fibers at birth. The vagus nerve of the opossum, a particularly immature animal, does not have any myelination up to 20 days after birth (Krous et al., 1985). Kittens and rabbit pups have only 10-20% of their adult values of myelination at birth (Marlot and Duron, 1979a; DeNeef et al., 1982). Figure 8 shows the rate of myelination of the vagus nerve for three species. Other maturational changes in the vagus nerve include an increase in diameter of both nonmyelinated and myelinated fibers accompanied by an increase in conduction velocity. The increase in conduction velocity would be partially compensated by the parallel lengthening of the nerve in such a way that conduction time to the central nervous system remains largely unaffected (Hursh, 1939). However, recent data indicate that conduction times are longer in the newborn (Gregory and Proske, 1985).

To what extent the above-described developmental changes affect receptor function is not clear. We know, however, that activity generally similar to that of adult animals can be recorded from slowly adapting stretch receptors (SARs) and rapidly adapting receptors (RARs) of immature animals. Many of the fibers carrying this activity are likely to be nonmyelinated.

B. Functional Characteristics

Three major types of receptors have been identified in the adult. SARs have a distinct respiratory modulation with a smoothly increasing discharge during lung inflation. They adapt slowly to a maintained inflation (Sant' Ambrogio, 1982). RARs have a respiratory modulation that exhibits an irregular discharge pattern typically occurring at peak flow during inspiration (or occasionally during expiration). Both RARs and SARs have myelinated axons, although the conduction velocity of SARs is slightly higher than that of RARs (Sant 'Ambrogio, 1982). Both receptor types increase their discharge when lung compliance is low (Sant 'Ambrogio, 1982). C-fiber endings are nonmyelinated axons that do not typically have a respiratory modulation (Coleridge and Coleridge, 1984). They respond to injected capsaicin and have been subdivided into two types on the basis of their response to capsaicin injected into the right or left heart (Coleridge and Coleridge, 1984).

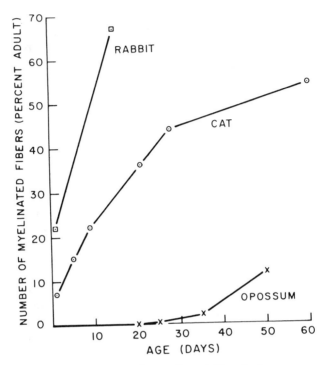

Figure 8 Postnatal maturation of the myelinated fibers in the vagus nerve of three species of mammals expressed as a percent of the adult value. Data for the cat are from Marlot and Duron (1979a); for the rabbit, from DeNeef et al., 1982; and for the opossum, from Krous et al. (1985). (From Fisher and Sant 'Ambrogio, 1985.)

Farber (Chapter 8, this volume) has summarized experimental evidence which shows that reflexes dependent on airway receptors are active in the newborn. However, observations regarding reflex "strength" do not necessarily reflect receptor behavior. For example, mechanical factors such as the lower lung compliance in the newborn would necessitate larger transpulmonary pressures for an inflation that is proportionally similar to that in the adult (Fisher and Mortola, 1980). Thus, receptors sensing stimuli related to lung inflation, other factors being equal, would be expected to discharge more than the adult. In contrast, the airways are more distensible in the newborn than the adult (Bhutani et al., 1981), and this too may affect the quality and quantity of the stimulus presented to the receptor ending. Without an analysis of receptor properties such as discharge characteristics during normal lung inflation, receptor frequency response curves to physiological stimuli, or receptor location, it is difficult to predict or interpret different reflex responses in the newborn.

Slowly Adapting Receptors

Slowly adapting receptors have been localized by probing the airways with catheters having inflatable cuffs (causing receptors to discharge when the airway is distended) or by gentle punctate stimulation of the lung with glass rods. On this basis they appear to be preferentially located within the airways and approximately 30% of the receptors are located within the trachea in the newborn dog (Fig. 9; Fisher and Sant 'Ambrogio, 1982a). The distribution is similar to that in the adult and indicates that at birth the anatomically complete tracheobronchial tree already provides afferent feedback from SARs located in strategic large airways. Figure 10 illustrates SAR discharge in a 20-day-old opossum during mechanical ventilation and during a maintained inflation. Two conclusions can be drawn from these recordings: First, the discharge frequencies are relatively low, an observation first made by Schwieler (1968) and confirmed later in several species (Marlot and Duron, 1979b; Fisher and Sant 'Ambrogio, 1982a; Marlot et al., 1982; Farber et al., 1984). Second, the receptor discharge is well maintained during lung inflation, in contrast to muscle spindles, which in the newborn cannot sustain a prolonged activation (Skoglund, 1960; Gregory and Proske, 1985). In order to compare discharge frequencies of SARs in the newborn and adult, the responses to maintained transpulmonary pressures have been measured (Fisher and

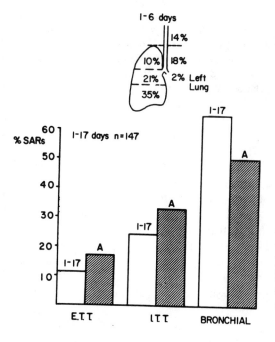

Figure 9 Distribution of slowly adapting stretch receptors (SARs) in the tracheobronchial tree of 1-17-day-old and adult dogs. The distribution appears to be similar in the newborn and adult animals, with a higher concentration of SARs in the larger airways. The inset shows the distribution of 81 SARs in 1-6-day-old puppies. E.T.T. = extrathoracic trachea; I.T.T. = intrathoracic trachea. The subdivision of the lung parenchyma by the broken line represents the upper, middle, and lower pulmonary lobes. (From Fisher and Sant 'Ambrogio, 1982a.)

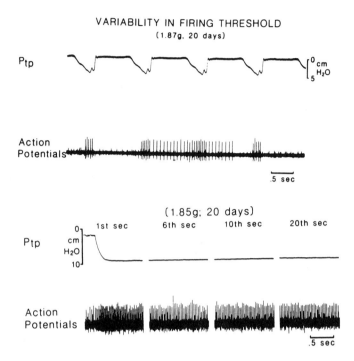

Figure 10 Upper panel: Slowly adapting receptor (SAR) discharge during ventilatory cycles. Polygraph records show transpulmonary pressure (P_{tp}) swings and action potentials from SAR of 20-day-old opossum during artificial ventilation from 0-cm H_2O end-expiratory airway pressure. Lower panel: Adaptation of slowly adapting receptor (SAR). Transpulmonary pressure (P_{tp}) and action potentials are shown for SAR during long inflation in 20-day-old opossum. Records are from the 1st, 6th, 10th, and 20th s of maintained inflation. Vagus nerve is unmyelinated at this age. (From Farber et al., 1984.)

Sant 'Ambrogio, 1982a; Farber et al., 1984). Figure 11 illustrates the response seen in opossums ranging in age from 20 days to adult and is typical of that found in other species such as the dog (Fisher and Sant 'Ambrogio, 1982a) or cat (Marlot et al., 1982). Considerable postnatal maturation occurs and discharge frequencies, at any given transpulmonary pressure, are significantly reduced in the 20-day-old opossum (20-50% of the adult values). Two other factors also contribute to a reduced SAR activity in the newborn: The first is a lower functional residual capacity compared to the adult, which results in a lower transpulmonary pressure stimulus applied to the receptor.

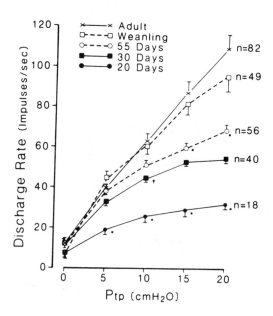

Figure 11 Receptor firing rate for slowly adapting receptors (SARs) during static inflations. Mean static firing rates for SARs are plotted as function of transpulmonary pressure (P_{tp}) for opossums of different age. Bars accompanying each datum point are ± SE. (From Farber et al., 1984.)

Second, the transpulmonary pressure threshold necessary to activate SARs is higher in the newborn than the adult (Fig. 12; Fisher and Sant'Ambrogio, 1982a). It is not clear which mechanisms are responsible for these differences, but possible candidates include a simpler receptor structure in the newborn that results in less stimulation, reduced capability to initiate a generator potential, or an altered smooth muscle/receptor coupling. Experiments that employ in vitro recordings and tension measurements (Bradley and Scheurmier, 1977) may help to shed light on possible mechanisms, since the applied stimulus would be closer to the "natural" stimulus of tension or distortion than transpulmonary pressure.

There are two mechanisms that appear to increase the amount of SAR feedback at any given transpulmonary pressure. First, in some species, the newborn appears to increase functional residual capacity above the passive value by the use of laryngeal adductors and postinspiratory muscle activity (Henderson-Smart and Read, 1979; England and Stogryn, 1986; Mortola et al., 1982). This would enhance receptor discharge by increasing transpulmonary pressure and is a mechanism that does not appear to be used in the adult (Harding et al.,

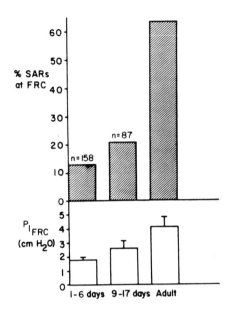

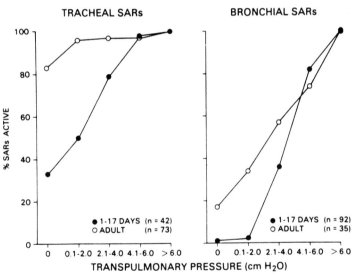

Figure 12 Upper panel: Percentage of SARs active at FRC and the elastic recoil pressure at FRC (P'_{FRC} mean ± SD in newborn (1-6 days), 9-17-day-old, and adult dogs. Note the increase of P'_{FRC} with age, which is paralleled by a concomitant increase of the number of SARs active at FRC. Lower panels: Transpulmonary (P_{tp}) threshold (abscissa) required to activate (ordinate, percentage of receptors active) tracheal (left panel) and bronchial (right panel) SARs. The P_{tp} threshold of 1-6- and 9-17-day-old puppies were the same and have been combined. The P_{tp} threshold required for SAR activity is higher in the 1-17-day-old dog compared to the adult and there are considerably fewer SARs active at low P_{tp} in the puppy. (From Fisher and Sant 'Ambrogio, 1982a)

1986). Second, efferent activation of airway smooth muscle increases the activity of the receptor, a mechanism that has been well described in the adult (Bartlett et al., 1976; Sant 'Ambrogio et al., 1986b; Sant 'Ambrogio, F.B., et al., 1988). Carbon dioxide or hypoxia can increase the activity of tracheal SARs due to a reflex activation of airway smooth muscle both in newborns and in adults. However, the response appears to be more easily elicited (especially for hypoxia) in the adult (Fisher and Sant 'Ambrogio, 1982b; Fisher et al., 1983). The muscarinic antagonist atropine eliminates the response indicating that the increase in receptor activity is solely due to bronchomotor tone. The reduced response to the newborn reflects little or no activation of airway smooth muscle during hypoxia (Fisher et al., 1987; Waldron and Fisher, 1988). Smooth muscle activation of bronchial SARs is presumably similar to that seen for tracheal receptors although recordings confirming this hypothesis are not available for the adult or newborn.

Rapidly Adapting Receptors

Recordings from rapidly adapting receptors have been obtained in newborn dogs and developing opossum. In the adult, RARs are less frequently encountered than SARs during vagal single-fiber recordings (Sant 'Ambrogio, 1982); however, in the newborn the relative number of RARs is further reduced compared to SARs (Fig. 13). Thus, in the newborn puppy, RARs account for 4% of the respiratory modulated receptors versus 15% in the adult dog (Fisher and Sant 'Ambrogio, 1982a), while in the 20-day-old opossum the value is 5% versus 18% for the adult (Farber et al., 1984). Studies in the newborn cat have not reported RAR activity (Schwieler, 1968; Marlot and Duron, 1979b; Marlot et al., 1982). It is possible that RARs are less frequently encountered in the newborn because axons of RARs are smaller than SARs and therefore more difficult to dissect for recording. However, in the 20-day-old opossum the vagus nerve is entirely unmyelinated and both types of receptors are found. It is also likely that axons recorded in the puppy are of larger diameter than those of the 20-day-old opossum. Therefore it seems likely that RARs are well within reach of single-fiber recording techniques. Clearly, data regarding the conduction velocities of SARs and RARs in the newborn will be important in resolving this issue.

C-Fiber Afferents

There are no direct recordings of C-fiber afferents in the newborn and conclusions regarding their activity are based on reflex studies. A study by Kalia (1976) examined the reflex response in the kitten to right heart injections of phenyldiguanide and found that the youngest animals failed to respond with

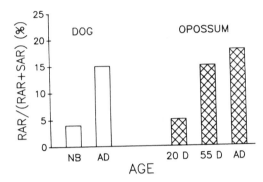

Figure 13 Relative occurrence of rapidly adapting receptors expressed as a percent of slowly adapting receptors encountered in dogs and opossums during development. NB = newborn; AD = adult; D = postnatal days.

the classical chemoreflex triad of apnea, bradycardia, and hypotension. Trippenbach (1988) reported that lactic acid causes a response in the newborn rabbit, although she suggested the response was attenuated compared to the adult since larger per kilogram doses were required. In the newborn dog C-fiber afferents appear to be active since right heart injection of capsaicin causes a reflex increase in pulmonary resistance that is eliminated by atropine (Fisher, unpublished observations). In addition to the nonphysiological stimuli used in the above studies, the profound effects on the pulmonary circulation at birth may be expected to affect C-fiber afferents due to alterations in pulmonary vascular and interstitial pressures. Single-fiber recordings should provide new and interesting insights into the behavior and role of these afferents in the newborn.

Antidromic stimulation of C-fiber afferents causes an increase in vascular permeability in the airways and leads to a "neurogenic inflammation" in the adult. Administration of capsaicin to neonatal rats causes the failure of substance P-containing afferents to develop, and adult animals that receive this treatment as neonates do not exhibit increases in airway vascular permeability in response to smoke (Lundberg and Saria, 1983). Although capsaicin administration eliminates the inflammatory response, it does not eliminate the effectiveness of vagal efferents or afferents. In fact, vagal stimulation still causes a bradycardia and atropine-sensitive bronchoconstriction (Lundberg et al., 1988) and the Hering-Breuer inflation reflex is still intact (L. Y. Lee, personal communication) following capsaicin treatment. However, to our knowledge there are no studies of the afferent activity of SARs and RARs in adult animals that have received neonatal capsaicin treatment. If these receptors are functional and exhibit discharge characteristics similar to those of control animals, it would imply that axons of SARs and RARs that are unmyelinated at birth have already acquired the specific properties that make them resistant to capsaicin treatment, while C-fibers that will be unmyelinated at maturity have specific properties that make them susceptible to capsaicin treatment.

Sympathetic Afferents

Sympathetic afferent fibers have been recorded in the adult (Kostreva et al., 1975). However, no data are available with respect to these fibers in the newborn.

IV. Conclusion

It is difficult to provide a unifying summary of the unique or special properties of airway and pulmonary receptors in the newborn. Throughout this chapter phrases such as "their presence in the newborn has not been documented", "no such evidence exists in the newborn", and "have yet to be determined in the newborn" are abundant. A recurring theme that stands out among each receptor group is the lack of information on morphological and neurophysiological properties of C-fiber endings in the newborn. Nevertheless, the receptors that have been studied provide precise quantitative feedback of physical stimuli that reflects the act of breathing. The discharge characteristics of the receptors are different from those of the adult, but the mechanisms responsible for these differences remain to be elucidated. Studies of sensory receptors in the anesthetized animal together with the use of in vitro techniques, and recordings of afferent activity in the unanesthetized animal, will enhance our ability to test hypotheses and explain these differences.

Several chapters in this volume appropriately reflect the surge in our knowledge and interest in molecular mechanisms underlying physiological responses. This information will contribute heavily to the design of in vivo studies and the formation of new hypotheses of receptor mechanisms. However, it is clear that much remains to be learned of the mechanisms responsible for the properties of sensory receptors in the newborn from the use of "conventional" morphological and neurophysiological techniques.

Acknowledgment

The authors thank Ms. Lynette Durant for the expert secretarial assistance.

References

Anderson, J. W., Sant 'Ambrogio, F. B., Mathew, O. P., and Sant 'Ambrogio, G. (1990). Water-responsive laryngeal receptors in the dog are not specialized endings. *Respir. Physiol.* **79**:33-44.

Bartlett, D., Jr., Jeffery, P., Sant 'Ambrogio, G., and Wise, J. C. M. (1976). Location of stretch receptors in the trachea and bronchi of the dog. *J. Physiol. (London)* **258**:409-420.

Bhutani, V. K., Rubenstein, S. D., and Shaffer, T. H. (1981). Pressure-volume relationships of tracheae in fetal, newborn, and adult rabbits. *Respir. Physiol.* **43**:221-231.

Biedenbach, M. A., Beuerman, R. W., and Brown, A. C. (1975). Graphic digitizer analysis of axon spectra in ethmoidal and lingual branches of the trigeminal nerve. *Cell Tissue Res.* **157**:341-352.

Boggs, D. F., and Bartlett, D. (1982). Chemical specificity of a laryngeal apneic reflex in puppies. *J. Appl. Physiol.* **53**:455-462.

Boushey, H. A., and Richardson, P. S. (1973). The reflex effects of intra-laryngeal carbon dioxide on the pattern of breathing. *J. Physiol. (London)* **228**:181-191.

Boushey, H. A., Richardson, P. S., Widdicombe, J. G., and Wise, J. C. M. (1974). The response of laryngeal afferent fibres to mechanical and chemical stimuli. *J. Physiol. (London)* **240**:153.

Bradley, G. W., and Scheurmier, N. (1977). The transduction properties of tracheal stretch receptors in vitro. *Respir. Physiol.* **31**:356-375.

But, V. I., and Klimova-Cherkasova, I. V. (1967). Afferentiation from upper respiratory tract. *Bull. Exp. Med.* **64**:13-16.

Cauna, N. (1982). Blood and nerve supply of the nasal lining. In *The Nose: Upper Airway Physiology and the Atmospheric Environment*. Edited by D. F. Proctor and I. B. Andersen. Amsterdam, Elsevier Biomedical Press, pp. 45-69.

Coleridge, J. C. G., and Coleridge, H. M. (1984). Afferent vagal C-fiber innervation of the lungs and airways and its functional significance. *Rev. Physiol. Biochem. Pharmacol.* **99**:1-110.

Dawson, W. W. (1962). Chemical stimulation of peripheral trigeminal nerve. *Nature* **196**:341-345.

DeNeef, K. J., Jansen, J. R. C., and Versprille, A. (1982). Developmental morphometry and physiology of the rabbit vagus nerve. *Dev. Brain Res.* **4**:265-274.

DuBois, F. S., and Foley, J. O. (1936). Experimental studies on the vagus and spinal accessory nerves in the cat. *Anat. Rec.* **64**:285-307.

England, S. J., and Stogryn, H. F. (1986). Influence of the upper airway on breathing pattern and expiratory time constant in unanesthetized dog pups. *Respir. Physiol.* **66**:181-192.

Farber, J. P., Fisher, J. T., and Sant 'Ambrogio, G. (1984). Airway receptor activity in the developing opossum. *Am. J. Physiol.* **246**:R752-R758.

Fisher, J. T., Mathew, O. P., Sant 'Ambrogio, F. B., and Sant 'Ambrogio, G. (1985). Reflex effects and receptor responses to upper airway pressure and flow stimuli in developing puppies. *J. Appl. Physiol.* **58**:258-264.

Fisher, J. T., and Mortola, J. P. (1980). Statics of the respiratory system in newborn mammals. *Respir. Physiol.* 41:155-172.

Fisher, J. T., and Sant 'Ambrogio, G. (1982a). Location and discharge properties of respiratory vagal afferents in the newborn dog. *Respir. Physiol.* 50:209-220.

Fisher, J. T., and Sant 'Ambrogio, G. (1982b). Effects of inhaled CO_2 on airway stretch receptors in the newborn dog. *J. Appl. Physiol.* 53:1461-1465.

Fisher, J. T. and Sant 'Ambrogio, G. (1985). Airway and lung receptors and their reflex effects in the newborn. *Pediatric Pulmonol.* 1:112-126.

Fisher, J. T., Sant 'Ambrogio, F. B., and Sant 'Ambrogio, G. (1983). Stimulation of tracheal slowly adapting stretch receptors by hypercapnia and hypoxia. *Respir. Physiol.* 53:325-339.

Fisher, J. T., Waldron, M. A., and Armstrong, C. J. (1987). Effects of hypoxia on lung mechanics in the newborn cat. *Can. J. Physiol. Pharmacol.* 65:1234-1238.

Glebovsky, V. D., and Bayev, A. V. (1984). Stimulation of nasal cavity mucosal trigeminal receptors with respiratory airflows. *Sechenov Physiol. J. USSR* 70:1534-1541.

Gregory, J. E., and Proske, U. (1985). Responses of muscle receptors in the kitten. *J. Physiol. (London)* 366:27-45.

Harding, R., England, S. J., Stradling, J. R., Kozar, L. F., and Phillipson, E. A. (1986). Respiratory activity of laryngeal muscles in awake and sleeping dogs. *Respir. Physiol.* 66:315-326.

Harding, R., Johnson, P., and McClelland, M. E. (1978). Liquid-sensitive laryngeal receptors in the developing sheep, cat and monkey. *J. Physiol. (London)* 277:409-422.

Henderson-Smart, D. J, and Read, D. J. C. (1979). Reduced lung volume during behavioral active sleep in the newborn. *J. Appl. Physiol.* 46:1081-1085.

Hursh, J. B. (1939). The properties of growing nerve fibers. *Am. J. Physiol.* 127:140-145.

Hwang, J. C., St. John, W. M., and Bartlett, D., Jr. (1984). Receptors corresponding to changes in upper airway pressure. *Respir. Physiol.* 55:355-366.

Jammes, Y., Barthelemy, P., and Delpierre, S. (1983). Respiratory effects of cold air breathing in anesthetized cats. *Respir. Physiol.* 54:41-54.

Jammes, Y., Nail, B., Mei, N., and Grimaud, Ch. (1987). Laryngeal afferents activated by phenyldiguanide and their response to cold air or helium-oxygen. *Respir. Physiol.* 67:379-389.

Kalia, M. (1976). Visceral and somatic reflexes produced by J. pulmonary receptors in newborn kittens. *J. Appl. Physiol.* 41:1-6.

Kostreva, D. R., Zuperku, E. J., Hess, G. L., Coon, R. L., and Kampine, J. P. (1975). Pulmonary afferent activity recorded from sympathetic nerve. *J. Appl. Physiol.* **39**:37-40.

Krous, H. F., Jordan, J., Wen, J., and Farber, J. P. (1985). Developmental morphometry of the vagus nerve in the opossum. *Dev. Brain Res.* **20**: 155-159.

Lee, L-Y., Sant'Ambrogio, F. B., Mathew, O. P., and Sant'Ambrogio, G. (1987). Acute effect of cigarette smoke on laryngeal receptors. *J. Appl. Physiol.* **62**:1575-1581.

Lundberg, J. M., Brodin, E., and Saria, A. (1983). Effects and distribution of vagal capsaicin-sensitive substance P neurons with special references to the trachea and lungs. *Acta Physiol. Scand.* **119**:243-252.

Lundberg, J. M., Lundbad, L., Anggard, A., Marting, C.-R., Theodorsson-Northeim, E., Stjarne, P., and Hokfelt, T. G. (1988). Bioactive peptides in capsaicin-sensitive afferents of the airways. In *The Airways*. Edited by M. A. Kaliner and P. J. Barnes. New York, Marcel Dekker, pp. 417-455.

Lundberg, J. M., and Saria, A. (1983). Capsaicin-induced desensitization of airway mucosa to cigarette smoke, mechanical and chemical irritants. *Nature* **302**:251-253.

Marlot, D., and Duron, B. (1979a). Postnatal maturation of phrenic, vagus and intercostal nerves in the kitten. *Biol. Neonate* **36**:264-272.

Marlot, D., and Duron, B. (1979b). Postnatal development of vagal control of breathing in the kitten. *J. Physiol. (Paris)* **75**:891-900.

Marlot, D., Mortola, J. P., and Duron, B. (1982). Functional localization of pulmonary stretch receptors in the tracheobronchial tree of the newborn kitten. *Can. J. Physiol. Pharmacol.* **60**:1073-1077.

Mathew, O. P., Abu-Osba, Y. K., and Thach, B. T. (1982). Genioglossus muscle responses to upper airway pressure changes: Afferent pathways. *J. Appl. Physiol.* **52**:445-450.

Mathew, O. P., Anderson, J. W., Sant'Ambrogio, F. B., and Sant'Ambrogio, G. (1989). Laryngeal cooling and breathing pattern in newborn puppies. *Physiologist.* **32**:228 (Abstract).

Mathew, O. P., Sant'Ambrogio, G., Fisher, J. T., and Sant'Ambrogio, F. B. (1984a). Laryngeal pressure receptors. *Respir. Physiol.* **57**:113-122.

Mathew, O. P., Sant'Ambrogio, G., Fisher, J. T., and Sant'Ambrogio, F. B. (1984b). Respiratory afferent activity in the superior laryngeal nerves. *Respir. Physiol.* **58**:41-50.

Mathew, O. P., Sant'Ambrogio, F. B., and Sant'Ambrogio, G. (1986). Effects of cooling on laryngeal reflexes in the dog. *Respir. Physiol.* **66**:61-70.

Mathew, O. P., Sant 'Ambrogio, F. B., and Sant 'Ambrogio, G. (1988). Laryngeal paralysis on receptor and reflex responses to negative pressure in the upper airway. *Respir. Physiol.* **74**:25-34.

Miller, A. J., and Dunmire, C. R. (1976). Characterization of the postnatal development of superior laryngeal nerve fibers in the postnatal kitten. *J. Neurobiol.* **7**:483-494.

Miller, A. J., and Loizzi, R. (1974). Anatomical and functional differentiation of superior laryngeal nerve fibers affecting swallowing and respiration. *Exp. Neurol.* **42**:369-387.

Mortola, J. P. and Fisher, J. T. (1988). Upper airway reflexes in newborns. In *Respiratory Function of the Upper Airway.* Edited by O. P. Mathew and G. Sant 'Ambrogio. New York, Marcel Dekker, pp. 303-357.

Mortola, J. P., Fisher, J. T., Smith, B., Fox, G., and Weeks, S. (1982). Dynamics of breathing in infants. *J. Appl. Physiol.* **52**:1209-1215.

Nail, B. S., Sterling, G. M., and Widdicombe, J. G. (1969). Epipharyngeal receptors responding to mechanical stimulation. *J. Physiol.* **204**:91-98.

Palecek, F., Mathew, O. P., Sant 'Ambrogio, F. B., and Sant 'Ambrogio, G. (1990) Cardiorespiratory response to inhaled laryngeal irritants. *Inhal. Toxicol.* **2**:93-104.

Rosenberg, S. I., Malgren, L. T., and Woo, P. (1989). Age-related changes in the internal branch of the rat superior laryngeal nerve. *Arch. Otolaryngol. Head Neck Surg.* **115**:78-86.

Sant 'Ambrogio, F. B., Sant 'Ambrogio, G., and Mathew, O. P. (1986). Effect of airway cooling on tracheal stretch receptors. *Respir. Physiol.* **66**:205-214.

Sant 'Ambrogio, F. B., Sant 'Ambrogio, G., Mathew, O. P., and Tsubone, H. (1988). Contraction of trachealis muscle and activity of trachea stretch receptors. *Respir. Physiol.* **71**:343-354.

Sant 'Ambrogio, G. (1982). Information arising from the tracheobronchial tree of mammals. *Physiol. Rev.* **62**:531-569.

Sant 'Ambrogio, G., Brambilla-Sant 'Ambrogio, F., and Mathew, O. P. (1986a). Effect of cold air on laryngeal mechanoreceptors in the dog. *Respir. Physiol.* **64**:45-56.

Sant 'Ambrogio, G., Mathew, O. P., Fisher, J. T., and Sant 'Ambrogio, F. B. (1983). Laryngeal receptors responding to transmural pressure, airflow and local muscle activity. *Respir. Physiol.* **54**:317-330.

Sant 'Ambrogio, G., Mathew, O. P., and Sant 'Ambrogio, F. B. (1985b). Role of intrinsic muscle and tracheal motion in modulating laryngeal receptors. *Respir. Physiol.* **61**:289-300.

Sant 'Ambrogio, G., Mathew, O. P., Sant 'Ambrogio, F. B., and Fisher, J. T. (1985a). Laryngeal cold receptors. *Respir. Physiol.* **59**:35-44.

Sant'Ambrogio, G., Sant'Ambrogio, F. B., and Mathew, O. P. (1986b). Kontrakcia trachealneho svalu a aktivita receptorov rozpatia trachey. *Bratisl. lek. Listy.* **85**:392-400.

Sasaki, C. T. (1979). Development of laryngeal function: Etiologic significance in the sudden infant death syndrome. *Laryngoscope* **89**:1964-1982.

Sasaki, C. T., Suzuki, M., and Horiuchi, M. (1977). Postnatal development of laryngeal reflexes in the dog. *Arch. Otolaryngol.* **103**:138-143.

Schwieler, G. H. (1968). Respiratory regulation during postnatal development in cats and rabbits and some of its morphological substrate. *Acta Physiol. Scand.* **304** (Suppl.):1-123.

Shin, T., Wada, S., and Maeyama, T. (1987). Substance P immunoreactive nerve fibers of the canine laryngeal mucosa. In *Neurolaryngology.* Boston, Toronto, San Diego, College-Hill, Little, Brown, pp. 65-74.

Skoglund, S. (1960). The activity of muscle receptors in the kitten. *Acta Physiol. Scand.* **50**:203-221.

Storey, A. T., and Johnson, P. (1975). Laryngeal water receptors initiating apnea in the lamb. *Exp. Neurol.* **47**:42-55.

Trippenbach, T. (1988). The pulmonary chemoreflex in newborn rabbits. *Physiologist* **31**:A6.

Tsubone, H. (1989a). Nasal "flow" receptors of the rat. *Respir. Physiol.* **75**:51-64.

Tsubone, H. (1989b). Nasal "pressure" receptors. *Proceedings of XXXI I.U.P.S. Congress,* Helsinki, Finland, July 9-14, Abstract.

Uddman, R., and Sundler, F. (1986). Innervation of the upper airway. *Clin. Chest Med.* **7**:201-209.

Ulrich, C. E., Haddock, M. P., and Alarie, Y. (1972). Airborne chemical irritants, role of the trigeminal nerve. *Arch. Environ. Health* **24**:37-42.

Waldron, M. A., and Fisher, J. T. (1988). Differential effects of CO_2 and hypoxia on bronchomotor tone in the newborn dog. *Respir. Physiol.* **72**:271-282.

Widdicombe, J., Sant'Ambrogio, G., and Mathew, O. P. (1988). Nerve receptors of the upper airway. In *Respiratory Function of the Upper Airway.* Edited by O. P. Mathew and G. Sant'Ambrogio. New York, Marcel Dekker, pp. 193-231.

Wyke, B. D., and Kirchner, J. A. (1976). Neurology of the larynx. In *Scientific Foundations of Otolaryngology.* Edited by R. Hinchcliffe and D. Harrison. London, Wm. Heinemann, pp. 546-574.

Zotterman, Y. (1936). Specific action potentials in the lingual nerve of cat. *Skand Arch. Physiol.* **75**:105-119.

8

Development of Pulmonary and Chest Wall Reflexes Influencing Breathing

JAY P. FARBER

University of Oklahoma Health Sciences Center
Oklahoma City, Oklahoma

I. Introduction

It has been very well established that reflexes originating from receptors in the lungs and chest wall can exert powerful effects on breathing pattern as well as individual breaths in adult mammals. Consideration of these reflexes from the standpoint of development is complicated by the fact that receptor function shows quantitative and sometimes qualitative differences between newborns and adults; further, the different levels of neuroanatomical and functional maturation within the central nervous system among the newborns of different species might well be expected to influence the expression of reflexes. For similar reasons, reflex expression may differ in the fetus or premature newborn in comparison to a neonate delivered after a normal gestational period.

In the present developmental examination of pulmonary and chest wall respiratory reflexes, events during inspiration and expiration will usually be treated separately; this is convenient yet artificial since information from specific receptors often influences more than one portion of the respiratory cycle. Also, some issues relating to reflex regulation of the postinspiratory period (stage 1 expiration) are considered in Chapter 6 in this volume. Although sympathetic afferents from the lungs have been identified (Kostreva et al., 1975), developmental information concerning possible effects of such sympathetic afferent pathways on breathing is virtually nonexistent and will not be considered in this chapter. There have been excellent reviews devoted to the development of lung and chest wall afferents and their reflexes (e.g., Fisher and Sant 'Ambrogio, 1985; Trippenbach, 1981). An important goal of the present chapter is, therefore, to examine how some of the more recent studies influence current ideas of reflex maturation.

A. Some Specific Considerations with Respect to Development

The respiratory reflexes of interest represent: transduction of sensory information (from lungs or chest wall) to afferent action potentials, processing of these action potentials by the central nervous system, and production of an efferent response in the form of changes in timing of breaths or output of respiratory muscles. Development can impose added complexity. For example, a recurring theme in this chapter will be that respiratory control circuitry of an immature mammal receives fewer action potentials from airway receptors (Farber et al., 1984; Fisher and Sant 'Ambrogio, 1982) during inspiration or expiration but converts that smaller input into effects on breathing pattern that may be comparable to or greater than those observed in adults. To analyze this developmental effect we would like to know how afferent signals impinge on brainstem neurons and/or brainstem circuits regulating either rhythm or bursts of motor activity; this requires a detailed understanding of excitatory and inhibitory interactions onto such cells and developmental influences on those interactions. Unfortunately, the ability to quantitatively assess processing of both lung and chest wall afferent inputs is greatly impaired by the relatively rudimentary knowledge of developmental processes in central nervous system (see Chapter 6). Another problem imposed by development is that changes in lung and respiratory system mechanics would be expected to influence the transduction of ventilatory events by lung and chest wall mechanoreceptors as well as the final expression of neuromuscular response (Davis and Bureau, 1987; Grunstein and Tanaka, 1985; Mortola et al., 1985).

In all but a few instances, instead of discussing development of central nervous system processing, it will be necessary to pursue the more modest goal of evaluating developmental changes in expression of lung and chest wall reflexes. Such expression relies on measurement of respiratory timing and/or muscle recruitment. Results at different ages will often be normalized to elements of the ongoing breathing pattern. When data are normalized, the presumed effectiveness of a given reflex could change with respect to development, but the absolute response might be unchanged or even be altered in the opposite direction. For example, if end-expiratory airway occlusion increased inspiratory duration (Ti) by 0.3 s in both a young animal whose unoccluded Ti was 0.3 s and an adult animal whose Ti was 0.6 s, normalization of the occluded value with respect to unoccluded Ti suggests that the reflex is stronger in the young animal. However, since absolute prolongation of Ti during occlusion is the same in both cases, these data could alternatively be interpreted to mean that effects on timing are unchanged. We will attempt in the following discussion to at least make note of the fact that conclusions are based on normalized or unnormalized responses.

II. Reflex Responses Associated with Vagal Airway Receptors

A. Effects During Inspiration

Reflex Shortening of the Inspired Breath by Lung Volume-Dependent Pathways

The best-studied vagal reflex influencing the inspired breath is the shortening of Ti caused by normal inspiratory increases in lung volume. The response has been characterized as being time dependent, whereby larger increases in volume are necessary to terminate the breath during the early portion of inspiration as opposed to later in inspiration (Clark and von Euler, 1972). Lung volume dependent termination of inspiration can be reversed by decreasing lung volume as inspiratory motor activity starts to "switch off" (Younes et al., 1978). Evidence strongly suggests that slowly adapting airway receptors (SARs) are important mediators of lung volume-dependent shortening of Ti in adult animals (Citterio et al., 1985; Clark and von Euler, 1972; Davies et al., 1978), but other airway receptors may also play a role (Citterio et al., 1985; Davies et al., 1978). While this reflex was initially characterized as not influencing the rate of ramp increase of inspiratory motor activity (phrenic neurogram or diaphragm electromyogram, EMG) both inhibitory and facilitatory influences on motor discharge have since been noted in some preparations (Citterio et al., 1985; DiMarco et al., 1981). Characteristics of this reflex have been reviewed in detail by von Euler (1986).

In terms of evaluating this reflex pathway in animals of different ages, three types of methodology have been employed: (1) airway occlusion at the end-expiratory volume; (2) vagotomy; and (3) progressive airway loading. All these techniques have provided insight into maturation of reflex pathways; however, some clear limitations in interpretation have also become apparent.

Airway Occlusion at the End-Expiratory Volume

An often-used method to evaluate the effect of airway receptors on Ti is to occlude the airway at end-expiration. Stopping airflow in the end-expiratory position results in prolongation of Ti in adult mammals; presumably, this is caused by the removal of phasic (lung volume dependent) input to brainstem processes which terminate the inspired breath. This test must be performed during anesthesia or sleep to eliminate conscious influences on the inspired breath; in the case of sleep, it is assumed that only the behavioral change of arousal is likely to alter the time course of the breath. It has been inferred that in the conscious human vagally mediated reflex shortening of the inspired breath is very weak or nonexistent, except with large inspired volumes (Clark and von Euler, 1972; Guz et al., 1970), but results of Polacheck et al. (1980) in anesthetized subjects clearly show increases in Ti accompanying end-expiratory airway occlusion.

With respect to human development, it was recently demonstrated that 2-29-month-old infants/children showed clear prolongation of Ti during sleep (induced with a sedative in the majority of subjects) when end-expiratory occlusion was performed on the lower airway (e.g., through a tracheal cannula). Measurements of occluded Ti using airway pressures or diaphragm EMG in this limited population were unable to demonstrate age-related changes in responses normalized to unoccluded Ti (Witte and Carlo, 1987). In addition, no changes in the rate of rise of inspiratory motor activity could be observed during occlusions. Several studies have confirmed that newborn human infants exhibit prolongation of the inspired breath during end-expiratory airway occlusion (Fisher et al., 1982; Gerhardt and Bancalari, 1981; Kosch et al., 1986; Olinksy et al., 1974; Taeusch et al., 1976; Thach et al., 1978); state of wakefulness has not always been rigorously determined in studies of infants, but at least by appearance most measurements were made during non-REM sleep. Finer et al. (1976) could not demonstrate Ti prolongation with end-expiratory occlusion in newborns during REM sleep. In premature infants, shortening rather than lengthening of Ti will often be observed with end-expiratory occlusion (Gerhardt and Bancalari, 1981; Knill and Bryan, 1976; Thach et al., 1978); it has been suggested that shortening of Ti is mediated by receptors in the chest wall (see Section III).

One other possible complication is that most studies of airway occlusion in human infants have included the upper airway, whereas receptors in the lungs, probably excluding the trachea (Citterio et al., 1985), are responsible for lung volume-dependent shortening of Ti. Recently, Thach and co-workers (1989a) showed that upper airway negative pressure would, in itself, cause a reduction in Ti if applied during inspiration in sleeping infants. With respect to the effects of pressures generated in the upper airway during end-expiratory occlusion, the results are inconsistent; Ti was not changed among subjects, but increases or decreases in Ti were obtained in a few individuals (Thach et al., 1989b). Clearly, the presence of other active reflex pathways makes it difficult to quantitatively evaluate the contribution of SARs, or other vagal afferents from the lungs, to control of the normal inspired breath. It is not surprising, therefore, that different studies suggest possible strengthening or weakening of this reflex pathway (in terms of percent prolongation of Ti) from premature to normal-term infants (Gerhardt and Bancalari, 1981; Olinsky et al., 1974; Thach et al., 1978). Another issue in infants is the influence of postnatal versus in utero development on regulation of the inspired breath; once again, the answer has differed among investigators when end-expiratory airway occlusion has been used as a probe. Thach et al. (1978) suggested that lung volume-dependent control of the inspired breath increased as a function of the first several postnatal days in either premature or term infants, but other studies have not obtained comparable effects (Fisher et al., 1982; Taeusch et al., 1976).

Results from other species have added some mechanistic insights, but have been unable to resolve some of the uncertainties concerning the interpretation of end-expiratory airway occlusion in human neonates. For example, while it is possible to use SO_2 to inhibit SARs in adult rabbits, the procedure is ineffective in newborns (Mortola et al., 1984); thus, the specific effect of this important group of receptors in the immature animal could not be assessed. Nevertheless, unanesthetized tracheotomized newborn rabbits, judged to be in non-REM sleep, showed inspiratory prolongation during end-expiratory airway occlusion; when normalized for unoccluded Ti and a tidal volume of 0 ml, prolongation of inspiration appeared to differ little as a function of age (Wyszogrodski et al., 1978). In contrast, Trippenbach et al. (1981) noted that the tracheotomized, anesthetized newborn kitten often had unanticipated responses to end-expiratory airway occlusions, including a decrease in Ti or, more often, the triggering of a deep breath. Depressed rate of rise of inspiratory motor activity was also observed. Prolongation of Ti did not occur as a uniform response until after the 2nd postnatal week. The presence of most of these early responses depended on an intact vagal innervation, suggesting that other types of airway receptors besides SARs could be stimulated during a normal breath or during the occlusion maneuver. Another possible interpretation of the preceding result is that afferent information from SARs in the kitten yielded different responses in younger versus older animals.

A probable example of age dependence in the central processing of information from SARs during inspiration is seen in young opossums. End-expiratory airway occlusion in 2-4-week-old anesthetized opossum sucklings most often produced no change in Ti, but increases or decreases in Ti were occasionally observed. The latter effects could have been due in part to irregularities in control breathing pattern (Farber, unpublished results). These marsupial mammals are particularly immature at birth and take many weeks to attain the overall development of many placental newborns. Ti is very brief (ca. 100 msec) in these young opossums and the rate of information transfer (action potentials) to medullary neurons by SARs is very slow compared to the duration of the inspired breath. Further, bulbospinal inspiratory neurons typically fire very few times during a breath (Farber, 1988; Farber et al., 1984). Under these conditions phasic increases in receptor activity accompanying increases in lung volume will have relatively little possibility to act on the inspired breath (see Chapter 6, this volume). Another potential example of age-dependent processing of lung afferent information is seen in premature human infants. Ti is not typically reduced in premature infants compared with normal-term infants (Gerhardt and Bancalari, 1981; Olinsky et al., 1974); however, premature infants with significant apneas had a lower Ti (Gerhardt and Bancalari, 1984) and more often had paradoxical responses accompanying end-expiratory airway occlusion. One possible explanation is that processing of phasic information from airway receptors was limited in the premature infant with significant apneas, and this allowed paradoxical responses to become apparent during airway occlusion.

Vagotomy

Complete loss of afferent information from the vagus nerves (e.g., vagotomy) would be expected to unmask all vagal effects on the inspired breath. It is well established that Ti in adult animals is typically increased by vagotomy, indicating that vagal afferents participate in the "switching off" of inspiration. The loss of lung volume-dependent activity from SARs as well as activity from other pulmonary receptors (see below) is likely to participate in this response. The effects of bilateral vagotomy in newborns are complicated by the fact that, in some species (rat, Fedorko et al., 1988; cat, Duron and Marlot, 1980; Schwieler, 1968; monkey, LaFramboise et al., 1985), duration of expiration also becomes exceedingly long (so long that the animal's viability can be compromised). Loss of vagal input under those conditions would appear to influence basic respiratory rhythmogenesis by the brainstem, thereby obscuring the interpretation of changes in Ti. Such findings are consistent with data by Sullivan et al. (1978) in adult dogs, where during sleep, loss of vagal and peripheral chemoreceptor input severely decreased breathing frequency. Not all immature mammals demonstrate untoward lengthening of expiration with bilateral vagotomy; the newborn rabbit and piglet show a moderate slowing of breathing with loss of vagal input (Clement et al., 1986; Mortola et al., 1984). Figure 1 illustrates the response to bilateral vagotomy in a very young (approx. 25 postnatal days) anesthetized

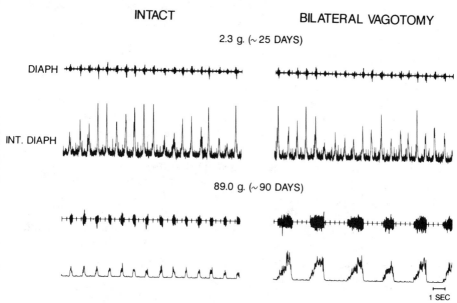

Figure 1 Effects of bilateral vagotomy on breathing pattern in Inactin-anesthe-tized suckling opossums of different ages. (Top) An animal about 25 days of age; (bottom) an older suckling near weaning age. (Upper traces) Raw diaphragm EMG; (lower traces) the moving average of the diaphragm EMG. Left-hand records were taken just before vagotomy; right-hand records were obtained after bilateral vagotomy. For further explanation, see text.

opossum in comparison to an animal near weaning age (90 postnatal days). Essentially no effect on breathing pattern is seen in the younger animal, while the older animal shows pronounced prolongation of Ti as well as duration of expiration. Again the suggestion can be made that, in a sufficiently immature preparation, the ability of the vagal afferent traffic to influence brainstem regulation of the inspired breath is limited.

Progressive Airway Loading

Grunstein et al. (1973) demonstrated in adult cats that elastic loading of the inspired breath decreased tidal volume while Ti was increased; the effect on Ti was due to decreased vagal afferent activity resulting from the de-creased tidal volume. Conversely, as tidal volume increased during CO_2 inhalation, increases in vagal feedback caused Ti to be reduced. Vagotomy prolonged the inspired breath to the value obtained with an infinite load (end-expiratory airway occlusion), and there were no further effects of air-way loading or CO_2 inhalation on Ti. The latter results implied that timing of inspiration through central nervous system mechanisms remained constant. Taeusch et al. (1976) observed small effects of elastic loading on inspiratory duration in infants, although complete end-expiratory occlusion did lengthen inspiration. Boychuk et al. (1977) noted increases in Ti in preterm infants which were more prominent during resistive as opposed to elastic loads. More recently, Kosch et al. (1986) carefully examined effects on inspired volume and timing of elastic and resistive loading in normal-term infants. Responses were compared with those obtained during total airway occlusion. These investigators used mechanical events to determine Ti, but EMG activity of the diaphragm was also used as a more direct assessment of neural activity. These authors suspected that mechanically derived and "neurally derived" assessments of Ti would differ between conditions of elastic and resistive load, an effect predicted by Miserocchi and Milic-Emili (1976). In fact, using the EMG measurement of Ti (Kosch et al., 1986), inspiration was prolonged as inspiratory volume was decreased during load, but values of Ti obtained through mechanical assessment during elastic loading of breaths were minimally changed. It was also shown that a more consistent effect on Ti prolongation was obtained when volume at peak EMG activity was used as the index of inspired volume. Most recently this same group examined effects of inspiratory loading on Ti in premature infants (Fox et al., 1988) and could find no differences between these infants and normal-term infants when data were normalized to the unloaded inspired breath.

Grunstein et al. (1973) noted that the relationship between Ti and inspired volume (VI) during airway loading and CO_2 inhalation could be linearized by plotting VI with respect to $1/Ti$; under those conditions the slope of $VI/(1/Ti)$ would show the effectiveness of vagal feedback in reducing Ti. It is uncertain whether such an assessment over a wide range of inspired volumes would work for human neonates, especially if chest wall reflexes become more prominent with increased loads. Further, this extrapolation could not be used for infants when breathing was stimulated and inspired volume was increased during CO_2 inhalation. Values of Ti should have decreased due to increased vagal afferent activity as VI became larger (Clark and von Euler, 1972; Grunstein et al., 1973), but infants showed no systematic change in Ti during CO_2 inhalation (Haddad et al., 1980; Taeusch et al., 1976; Thach et al., 1978). A caution to the preceding is that it is unknown whether Ti, independent of vagal feedback, is actually constant in human infants during CO_2 inhalation. Further, although there is evidence of vagally mediated shortening of Ti using end-expiratory occlusion in anesthetized human adults (Polacheck et al., 1980), Ti was not shortened during CO_2 inhalation (Gautier et al., 1981). The most effective use of airway loading would seem to be when changes in Ti can be measured under much less extreme airway loading conditions than airway occlusion. It is then possible that effects on Ti can be assessed with minimal distortion of the chest wall; this should lessen or eliminate contamination of phasic vagal pathways by chest wall reflexes. With minimal loading, the slope of the relationship, VI/(1/Ti) might provide a reasonable index of vagally mediated shortening of the inspired breath. Comparisons between different ages could be made by normalizing both volumes and Ti (e.g., Fox et al., 1988); in that case the "effectiveness" of the reflex at different ages would be obtained.

In summary, there is little doubt that lung volume-dependent shortening of Ti is present in unanesthetized human infants breathing with normal tidal volumes. In contrast, the presence of such reflex regulation in the adult human is problematical. For quantitation of reflex potency in premature and term infants, moderate levels of (resistive or elastic) airway loading would seem to be most applicable since effects attributable to other reflex pathways are minimized. The best current data suggest that there is little change in the effectiveness of this reflex in preterm and term infants; also, the reflex response seems little changed over the first several postnatal months. Animal studies also show minimal postnatal change in effectiveness of this reflex, once the response is clearly established. It is possible, however, that the sufficiently immature (or premature) newborn would be unable to effectively utilize this reflex pathway.

Other Pulmonary Afferent Mechanisms Contributing to Timing of the Inspired Breath

After SARs are presumably blocked by SO_2 in adult rabbits, vagotomy or end-expiratory occlusion causes further prolongation of Ti. This suggests that rapidly adapting airway receptors (RARs) and possibly C-fiber lung receptors could contribute to the vagally mediated shortening of Ti (Citterio et al., 1985; Davies et al., 1978). Results using partial vagal cooling (Fishman et al., 1973) and examination of RAR discharge using inflation at different airflows (Pack and Delaney, 1983) can be interpreted similarly. Unfortunately, conclusions based on these kinds of data cannot be rigorously defended because of the possibility of either incomplete receptor blockade or simultaneous effects on more than one type of lung receptor. There is essentially no developmental information concerning the specific effects of RARs or C-fiber receptors on eupneic breathing.

Pathways Regulating Spontaneous Deep Breaths

Effects on normal inspiration, likely attributable to RARs, must also include a discussion of spontaneous deep breaths (augmented breaths). In adult mammals, an intact vagal innervation is usually required for spontaneous deep breaths to occur (Barlett, 1971; Glogowska et al., 1972); chemo-receptor input also appears to influence their occurrence. Investigators have inferred that RARs, in particular, participate in this reflex (see Glogowska et al., 1972); an important part of this rationale is that RARs are activated by decreases in lung volume and/or lung compliance in the form of regional atelectasis. Spontaneous deep breaths have been observed in human and other newborns (Cross et al., 1960; Duron and Marlot, 1980; Thach and Taeusch, 1976). By the end of the first postnatal week, the opossum clearly demonstrated spontaneous deep breaths (Farber, 1972). This finding suggests one possible instance of pulmonary reflex mediation of breathing pattern in a very immature nervous system whose pulmonary afferent pathways are objectively in an undeveloped state (Farber et al., 1984). It is also of interest that lung inflation in a 1-week-old opossum typically elicited an augmented breath (Farber, 1972); augmented breaths can also be elicited with lung inflation in human newborns (Cross et al., 1960). These responses provide further support for the proposition that spontaneous deep breaths in newborns result from an airway reflex. Interestingly, the triggering of an augmented breath with lung inflation in newborns is similar to the paradoxical reflex of Head. The latter is obtained in adult animals by partially cooling the vagus nerves before inflating the lungs and may involve a qualitative change in the firing patterns of SARs (Paintal, 1966) as well as activation of RARs.

A specific role of spontaneous deep breaths in newborn placental mammals might be to inflate the lungs with air at birth (Cross et al., 1960). Thach and Taeusch (1976) noted that incidence of spontaneous deep breaths decreased in infants from the 1st to 5th postnatal day; it was suggested that spontaneous deep breaths in infants would be helpful in increasing homogeneity of ventilation as well as reestablishing pulmonary surfactant activity. The prominent presence of spontaneous deep breaths in the young opossum suggests that periodic large-volume inflation is of importance whether the lungs have already reached the alveolar stage at birth, as for many placental mammals, or are still at a very primitive bronchial stage, as for the opossum (Krause and Leeson, 1973).

B. Effects During Expiration

Prolongation of the Expired Breath—Inflation Reflex

Vagally mediated prolongation of expiratory duration (Te) during lung inflation is readily demonstrable in most anesthetized adult mammals (Widdicombe, 1961; actual index used was prolongation of the breath); other investigations have shown similar responses in unanesthetized dogs (Phillipson et al., 1971) and opossums (Farber and Marlow, 1976). Depending on the pattern of lung inflation, it is possible for either stage 1 expiration (post-inspiratory interval) or stage 2 expiration to be prolonged (Remmers et al., 1986). It has been inferred for many years that activity of SARs was responsible for this reflex. The best demonstration of the role of SARs was obtained using SO_2 inhalation in adult rabbits (which inhibits SAR firing); prolongation of Te with lung inflation was eliminated after SO_2 inhalation (Davies et al., 1978). In adult humans, inflation within the normal tidal range does not usually produce expiratory prolongation during sleep or wakefulness (Hamilton et al., 1988), although anesthetized subjects do show such an effect (Gautier et al., 1981). Increases in Te could be observed in newborn humans during lung inflation (Bodegard et al., 1969; Cross et al., 1960).

Quantitative comparisons have been made of prolongation of Te by lung inflation in animals of different ages. An important issue is how best to normalize the data. Transpulmonary pressure is clearly better for normalization than is lung volume from newborn to adult, but equal transpulmonary pressures do not account for differences in receptor discharge. Yet another approach is to normalize data to the tidal breath (e.g., occluding the breath at the end-inspiratory level). A major difficulty for comparison of responses is that prolongation of Te during lung inflation depends on status of respiratory drive as well as firing rate of receptors (Younes et al., 1974). For these reasons, it is necessary to carefully compare experimental conditions when relating various developmental studies to one another (Gaultier and Mortola, 1981; Marlot and Duron, 1979; Smejkal et al., 1985). In humans we have the relatively straightforward situation where prolongation of Te with lung inflation in the tidal range (using end-inspiratory occlusion) can be demonstrated in the unanesthetized neonate (Bodegard et al., 1969) and not in adults. Prolongation of Te due to lung inflation decreases over the first postnatal days in term infants (Bodegard et al., 1969; Cross et al., 1960), but may also be weaker in premature infants (Bodegard et al., 1969).

Expiratory muscles can be recruited during lung inflation through a vagal afferent mechanism. These muscles tend to be recruited toward the latter portion of the expired breath, but increases in respiratory drive promote earlier activation. Expiratory muscle activation during lung inflation has been demonstrated in both adult and immature animals (e.g., Farber, 1982, 1983; Younes et al., 1974).

With respect to breath-by-breath respiratory regulation, studies of Knox (1973) inferred that increased airway receptor firing due to increases in lung volume would cause increases in Te by transient as well as integrative processing of afferent information by the brainstem. Indeed, when SARs were blocked in the barbiturate-anesthetized adult rabbit, Te was shortened (Davies et al., 1978; Mortola et al., 1984); however, SO_2 blockade during urethane anesthesia resulted in prolongation of Te. Breathing rate with urethane anesthesia was very high and tidal volume was low in relation to barbiturate anesthesia. The pattern of SAR input would, in consequence, be different between the two anesthetics, and the changed relationship between SAR-mediated inspiratory inhibition and expiration prolongation might account for the different effects of withdrawing SAR input. Phillipson et al. (1971) showed that unilateral vagotomy in awake dogs was sufficient to block prolongation of Te by lung inflation, but did not influence breathing pattern; these data cannot be extrapolated to exclude a role for the lung volume-related reflex in regulation of Te since unilateral vagotomy also reduces inputs from RARs and lung C-fibers. In newborn human infants, Kosch et al. (1985) demonstrated that increases in Te occurred with increased single-breath elastic or resistive expiratory loads. Resistive loads produced greater volume feedback than elastic loads for the same end-expiratory volume, and Te was greater during resistive loads. This result supports the concept of an integrative mechanism for vagal control of Te, as proposed by Knox (1973).

In summary, using lung inflation within the range of resting tidal volume, a vagally mediated prolongation of the expired breath is readily observed in unanesthetized human infants; adults, on the other hand, need to be anesthetized to exhibit a comparable response. Although this reflex is typically active in most adult mammals and has been demonstrated in immature preparations, there are important problems of interpretation when comparing reflex strength between immature animals and adults. Results from studies using animals suggest that the reflex can prolong Te during normal tidal breathing; the normal increase in lung volume with each breath probably prolongs Te in human infants.

Other Influences on Expiration— Tonic Vagal Input

Whatever the modulation of normal Te in newborns by the inflation reflex, it is important to restate that complete removal of vagal input produces great prolongation of Te in at least some newborns (Duron and Marlot, 1980; Fedorko et al., 1988; LaFramboise et al., 1985; Schwieler, 1968). Since the previously discussed inflation reflex is likely to prolong Te, the excessive length of Te after vagotomy, in these species, suggests that the loss of nonphasic (e.g., tonic) information carried by the vagi can be much more important for the regulation of the expired breath in some newborns. On the other hand, the immature opossum shows minimal effect of vagotomy on Te (Fig. 1), although opossums of the same age demonstrate a clear prolongation of Te accompanying lung inflation (Farber, 1972). Another well-recognized tonic vagal influence on Te is caused by maintained changes in lung volume. This effect was very well illustrated by Bartoli et al. (1973), using anesthetized dogs with cardiopulmonary bypass, where lung volume could be changed independent of breathing pattern. Increases in lung volume were associated with prolongation of Te. The more complicated situation of maintained elevations in end-expiratory volume, combined with inflation of the lungs with each breath, is considered below.

Breathing with Increases in End-Expiratory Volume

Responses to increases in the end-expiratory volume [e.g., using positive pressure breathing (PPB) in anesthetized adult animals] may include either a decrease or increase in breathing frequency—largely caused by changes in expiratory duration; this is often accompanied by activation of abdominal and other expiratory muscles (Bishop and Bachofen, 1972a; Kelson et al., 1977). The awake guinea pig showed slowing of breathing with prolongation of Te at the initiation of PPB, but breathing rate increased as PPB was maintained (Gillespie et al., 1979). On the other hand, in unanesthetized adult humans, breathing frequency was little influenced by PPB (Gillespie et al., 1979); it was inferred that the weak reflex response to lung inflation in humans prevented the initial slowing of breathing. Effects on timing of breaths and activation of expiratory muscles accompanying increases in the end-expiratory volume require an intact vagal innervation; in the rabbit it can be demonstrated using SO_2 that SARs are involved (Davies et al., 1980).

Breathing rate was not consistently changed during maintained PPB in unanesthetized immature opossums, although prolongation of the expired breath was typically observed at the onset of load (Farber, 1983). Newborn anesthetized rabbits showed a prolongation of Te during PPB which depended on an intact vagal innervation (Trippenbach et al., 1985). Increase in tonic activity from SARs was thought responsible for the effect. In human term and preterm infants, where a relatively strong inflation reflex is demonstrable, expiratory prolongation was observed when end-expiratory volume was raised (Martin et al., 1978; Stark and Frantz, 1979). Results using end-expiratory airway occlusion suggested that tonic rather than phasic increases in airway receptor activity were responsible for most of the response.

Activation of expiratory muscles during PPB has been studied in detail during development of the opossum. Expiratory muscles began to be recruited in unanesthetized animals at about the 20th postnatal day, where they were progressively activated over several loaded breaths. Older opossum sucklings fully recruited these muscles during the first loaded breath. Expiratory intercostal muscles were also recruited during PPB; these muscles were recruited later in development than abdominal muscles (Farber, 1986). Further, muscles from the lower interspaces were recruited much more often and at a lower load than muscles of upper interspaces. Because of the probable dependence of expiratory muscle recruitment during PPB on discharge of SARs (Davies et al., 1980; Farber, 1982), it is possible that developmental responses are in part related to the particularly low discharge rates of such receptors in very young opossums (Farber et al., 1984). Immaturity of central mechanisms might also be a factor in the lack of recruitment of expiratory muscles in very young animals. Medullary expiratory neurons were often below threshold for activation during eupneic breathing (Farber, 1989); these cells were recruited during PPB but discharge rates were very low in comparison with older animals. Young suckling opossums, under anesthesia, could show paradoxical responses to the initiation of PPB. At an age when recruitment of abdominal muscles during PPB was not yet prevalent (near the end of the 1st postnatal month in the anesthetized animal), instances were found when ongoing expiratory abdominal muscle activity was completely suppressed at the onset of PPB (Farber, 1985). For adult cats (and older suckling opossums), lung inflation decreased the slope of any ongoing expiratory EMG activity, but because of the increased Te, the final level of EMG activity was often increased (Cohen et al., 1985). One potential mechanism for the strong suppression of abdominal muscle activity in the young opossum is the lack of development of the dendritic tree in spinal neurons. Inhibitory synapses tend to be closer to the cell soma (Burke and Rudomin, 1977), and it is possible that because of neuroanatomical constraints (a dominance of synapses near the cell soma in the immature spinal motor neuron), inhibitory interactions will dominate.

Positive pressure breathing can stabilize irregular respiratory rhythm, notably respiratory pauses and apneas, in premature human infants (Kattwinkel, 1977) as well as newborn sheep (Jonzon and Sedin, 1982). Possibly, the lack of activity in medullary expiratory neuronal circuitry, as noted for the young opossum, contributes to this effect because PPB is required to bring many of these neurons above firing threshold (Farber, 1989). However, rhythmic breathing occurs in the opossum even when these expiratory cells are not firing regularly. Another possibility is that repeated lung inflations, in the absence of PPB, do not always yield stable discharge of SARs; this effect has been observed in the sheep fetus and in young opossums (Farber et al., 1984; Ponte and Purves, 1973). Significant breath-by-breath changes in pulmonary afferent activity could have strong effects on respiratory rhythm. PPB will bring more receptors decisively above the firing threshold and would be expected to produce a more stable pattern of airway receptor firing.

In summary, when Te is prolonged during PPB or similar maneuvers, the response appears to depend on tonic increases in lung afferent activity. Behavioral state or other reflexes might mask the prolongation of Te in some preparations. In humans, only infants exhibit prolongation of Te when breathing at elevated lung volumes; this effect may be related to the relatively strong lung volume-dependent expiration-prolonging reflex in infants as opposed to adults. The ability of PPB to stabilize breathing patterns of immature infants may be, in part, dependent on a more consistent pattern of airway receptor discharge accompanying elevation in lung volume.

Recruitment of Expiratory Muscles with Chemostimulation of Breathing

In anesthetized adult cats, CO_2 inhalation produces a relatively weak recruitment of abdominal muscles, which is abolished by vagotomy (Bishop and Bachofen, 1972b). On the other hand, during CO_2 inhalation in anesthetized dogs, vagotomy eliminated the recruitment of expiratory muscles in a minority of cases (Kelson et al., 1977). In adult humans, expiratory muscles can be activated during hyperventilation due to hypoxic hypercapnia (Campbell and Green, 1955); the potential role of vagal mechanisms is unknown. With respect to development, during normoxic or hypoxic hypercapnia in anesthetized immature opossum, unilateral vagotomy greatly reduced the expiratory recruitment of abdominal muscles while minimally influencing the breathing pattern (Farber, 1983); however, intact vagi were not required for abdominal muscle recruitment in piglets (Watchko et al., 1990). Further, it has recently been verified that human infants recruit abdominal muscles during CO_2 inhalation; activation of abdominal muscles was greater in quiet versus active sleep (Praud et al., 1989). During hypoxic hypercapnia, expiratory muscles had little effect on respiratory effort in adult human subjects, except at relatively high levels of ventilation (Campbell and Green, 1955; Grimby et al., 1976). While other vagal reflexes are relatively active in human neonates, it is unknown whether vagally mediated recruitment of expiratory muscles during ventilatory chemostimulation has an important effect on respiratory efforts.

C. Reflex Responses with Important Inspiratory and Expiratory Components

Responses to Lung Deflation

Lung deflation in mature mammals often results in increased breathing frequency with increased activation of inspiratory muscles (see Coleridge and Coleridge, 1986), although prolonged inspiratory efforts can occur in some preparations (e.g., see below). Responses to lung deflation may involve changes in inputs from both SARs and RARs (Coleridge and Coleridge, 1986; Knox, 1973; Mills et al., 1970); more specifically, data using SO_2 blockade of SARs in rabbits infer that the tachypneic response to lung deflation involves SARs (Davies et al., 1978). In decerebrate or anesthetized kittens breathing through tracheal cannulas, it was found that lung deflation caused increased inspiratory efforts while Te was increased; older kittens and adult cats responded with prolonged inspiratory efforts (Marlot and Duron, 1979). Similarly, intact unanesthetized young opossums responded to deflation of the lungs and upper airway by increasing Te and increasing excitation of the diaphragm EMG, but increases in breathing frequency became prominent as the animals matured to the general developmental stage of a newborn placental mammal, such as the cat (Farber, 1972; Farber and Marlow, 1978). Strong increases in breathing rate were also observed with lung deflation in adult opossums breathing through a tracheostomy during paradoxical and slow-wave sleep (Farber and Marlow, 1976). In contrast, other studies have shown clear decreases in Te accompanying lung deflation in newborn cats and monkeys (Trippenbach, 1981). A similar result was obtained in anesthetized newborn rabbits, along with excitation of the diaphragm EMG and reduction in Ti (Trippenbach et al., 1985). It was suggested by Trippenbach et al. (1985) that vagal cardiac receptors might influence the breathing pattern of newborn rabbits during lung deflation. For the two studies from the kitten that are in conflict with respect to effects on Te, differences in anesthesia has been suggested as one reason for the discrepancy (Trippenbach, 1981). The prolongation of Te in the very young opossum during lung deflation may, on the other hand, be due to the immaturity of that preparation, but an upper airway reflex cannot be excluded.

One potential function of the responses to lung deflation as well as inflation in the young animal is the defense of the end-expiratory volume. The lack of an increased rate of breathing accompanying lung deflation seen in some studies might compromise the ability of a newborn to maintain the end-expiratory volume above the passive functional residual capacity (Olinsky et al., 1974). Other studies suggest, however, that an increase in breathing rate may not be required for raising the end-expiratory volume above the passively determined values. For example, human infants demonstrate a tonic activation of inspiratory muscles which can serve to increase the end-expiratory volume (Lopes et al., 1981). Upper airway braking of expiration can also be important (Noble et al., 1988; Stark et al., 1987). Strategies for maintaining the end-expiratory volume, and their effectiveness, appear to vary in infants during different conscious states (Stark et al., 1987).

Responses to Pulmonary Irritation

Important developmental effects have been noted in the responses to lung irritants. These responses are likely to involve RARs and/or C-fiber receptors (Mills et al., 1970). Defense reactions to mechanical airway irritation, presumably dependent on RARs, showed qualitative differences between term and premature infants (Fleming et al., 1978). Term infants uniformly responded by increasing inspiratory efforts, but premature infants typically responded with slowing of breathing or apnea, with or without initial increases in inspiratory efforts. Behavioral state was not noted in that study. From the standpoint of receptor stimulation, there appear to be fewer functional RARs in immature animals (Farber et al., 1984; Fisher et al., 1982); however, central processing of afferent information could also be important. In newborn anesthetized rabbits, intravenous injection of histamine, a procedure that should have strongly stimulated RARs, was increasingly effective over the first postnatal week in shortening Ti and causing an increase in breathing rate; the diaphragm EMG was strongly stimulated by this maneuver in the youngest animals (Trippenbach and Kelly, 1988). These effects depended on an intact vagal innervation. In kittens younger than 7 days of age, right atrial injection of phenyl diguanide did not induce either the rapid/shallow breaths or apnea typically associated with pulmonary C-fiber stimulation in adult animals (Kalia, 1976). After the first postnatal week, these C-fiber responses were present.

III. Reflexes Associated with Chest Wall Receptors

As mentioned previously, end-expiratory airway occlusion in infants does not always result in the expected prolongation of Ti. Instead, this occlusion maneuver in premature infants often elicits a shortening of the inspired breath. Given the fact that shortening of the inspired breath also occurs with chest wall afferent stimulation, some investigators postulated that the shortened Ti in infants was due to activation of reflexes from the chest wall. The presumed mechanism is that distortion of the highly compliant chest wall of neonates activates receptors in costal interspaces.

An inhibitory intercostal to phrenic supraspinal reflex in adult animals (Remmers, 1970) was thought to be mediated by muscle spindle receptors, but recent data indicate that major effects arise instead from tendon organ receptors. Specifically, stimulation of tendon organs from intercostal spaces can inhibit brainstem inspiratory neurons and result in early termination of the inspired breath; similar stimulation during expiration excites laryngeal constrictor muscles and inhibits activity of medullary expiratory neurons with probable projections to abdominal muscles (Bolser et al., 1987; Shannon et al., 1987).

Studies of newborn kittens indicate that various muscle and cutaneous receptors are functional; however, discharge appears to adapt more rapidly and peak firing rates are reduced (Ekholm, 1967; Skoglund, 1960). The extent of supraspinal mediation of phrenic nerve/diaphragm activity by reflexes originating from the chest wall in the newborn is not well established.

Studies by Trippenbach (1985) indicated that electrical stimulation of intercostal nerves, during either inspiration or expiration in newborn anesthetized vagotomized kittens, resulted in a shortening of Ti and prolongation of Te.

Reductions in Ti observed in infants during airway occlusion may not necessarily involve chest wall reflexes. For example, kittens may sometimes show a vagally dependent shortening of the inspired breath, among other paradoxical responses, during end-expiratory airway occlusion (Trippenbach et al., 1981). Further, in a newborn infant with a cervical cord transection, end-expiratory airway occlusion nevertheless produced a shortening of the inspired breath (Thach et al., 1980); this reaffirms that pathways other than chest wall reflexes can be responsible under some conditions for premature termination of inspiration.

With respect to load compensation through chest wall reflexes, Schwieler (1968), using end-expiratory occlusion in decerebrate, vagotomized, newborn kitten or rabbit, failed to show an augmentation of inspiratory intercostal activity until several days after birth. Further, dorsal root rhizotomy had no effect on ongoing inspiratory intercostal activity in the youngest animals. These data suggested negligible chest wall reflex load compensation in the newborns of these species. More recently, Trippenbach and Kelly (1983) noted increases in inspiratory intercostal nerve discharge in newborn, anesthetized, vagotomized kittens during end-expiratory occlusion; the increased intercostal activity was largely related to a small prolongation of Ti in their preparation. However, in several cases rate of rise of intercostal activity was augmented, suggesting the possible presence of chest wall reflex load compensation.

In summary, our knowledge of the developmental role of chest wall reflexes in respiratory control is inadequate. In human neonates, shortening of Ti with maneuvers such as end-expiratory airway occlusion, as well as other events that distort the chest wall, cannot necessarily be ascribed to chest wall afferents; vagal or upper airway reflexes may instead play a role. Nevertheless, animal studies do indicate that some chest wall reflexes are active at birth; the physiological circumstances of their activation needs to be better defined.

IV. Implications for Future Research

While it seems clear that vagal and possibly chest wall afferent pathways are capable of influencing the inspired and expired breath in immature mammals, our qualitative and quantitative understanding of the specific neural pathways must be improved. Obviously, important insight will be gained as the cellular level function of the central nervous system respiratory controller becomes understood within the context of development, especially in terms of response to peripheral inputs. However, even the more conservative goal of dissecting various reflexes in terms of gross effects on timing and respiratory muscle activation will require very sophisticated experiments to sort out the effects of each type of receptor and its elicited response(s).

References

Bartlett, D. (1971). Origin and regulation of spontaneous deep breaths. *Respir. Physiol.* **12**:230-238.

Bartoli, A., Bystrzycka, E., Guz, A., Jain, S. K., and Trenchard, D. (1973). Studies of the pulmonary vagal control of central rhythm in the absence of breathing movements. *J. Physiol. (London)* **230**:449-465.

Bishop, B., and Bachofen, H. (1972a). Vagal control of ventilation and respiratory muscles during elevated pressures in the cat. *J. Appl. Physiol.* **32**: 103-112.

Bishop, B., and Bachofen, H. (1972b). Comparison of neural control of the diaphragm and abdominal muscle activated in the cat. *J. Appl. Physiol.* **32**:798-805.

Bodegard, G., Schwieler, G. H., Skoglund, S., and Zetterstrom, R. (1969). Control of respiration in newborn babies. I. The development of the Hering-Breuer inflation reflex. *Acta Paeditr. Scand.* **58**:567-571.

Bolser, D. C., Lindsey, B. G., and Shannon, R. (1987) Medullary inspiratory activity: influence of intercostal tendon organ and muscle spindle endings. *J. Appl. Physiol.* **62**:1046-1056.

Boychuk, R. B., Seshia, M. M., and Rigatto, H. (1977). The immediate ventilatory response to added inspiratory elastic and resistive loads in preterm infants. *Pediatr. Res.* **11**:276-279.

Burke, R. E., and Rudomin, P. (1977). Spinal neurons and synapses. In *Handbook of Physiology. The Nervous System.* Edited by E. R. Kandel. sec. 1. Vol. 1. Bethesda, MD, American Physiological Society, pp. 877-944.

Campbell, E. J. M., and Green, J. H. (1955). The behaviour of abdominal muscles and the intra-abdominal pressure during quiet breathing and increased pulmonary ventilation. A study in man. *J. Physiol. (London)* **127**:423-426.

Citterio, G., Piccoli, S., and Agostoni, E. (1985). Breathing pattern and diaphragm EMG after SO_2 in rabbit intra- or extrathoracic airways. *Respir. Physiol.* **59**:169-183.

Clark, F. J., and Von Euler, C. (1972). On the regulation of rate and depth of breathing. *J. Physiol. (London)* **222**:267-295.

Clement, M. G., Mortola, J. P., Albertini, M., and Aguggini, G. (1986). Effects of vagotomy on respiratory mechanics in newborn and adult pigs. *J. Appl. Physiol.* **60**:1992-1999.

Cohen, M. I., Feldman, J. L., and Sommer, D. (1985). Caudal medullary expiratory neurone and internal intercostal nerve discharges in the cat: Effects of lung inflation. *J. Physiol. (London)* **368**:147-178.

Coleridge, H. M., and Coleridge, J. C. G. (1986). Reflexes evoked from tracheobronchial tree and lungs. In *Handbook of Physiology. The Respiratory System*. Sec. 3. Vol. 2. Edited by N. S. Cherniak and J. G. Widdicombe. Bethesda, MD, American Physiological Society, pp. 395-429.

Cross, K. W., Klaus, M., Tooley, W. H., and Weisser, K. (1960). The response of the new-born baby to inflation of the lungs. *J. Physiol. (London)* **151**:551-565.

Davies, A., Dixon, M., Callanan, D., Huszczuk, A., Widdicombe, J. G., and Wise, J. C. M. (1978). Lung reflexes in rabbits during pulmonary stretch receptor block by sulphur dioxide. *Respir. Physiol.* **34**:83-101.

Davies, A., Sant 'Ambrogio, F. B., and Sant 'Ambrogio, G. (1980). Control of postural changes of end expiratory volume (FRC) by airways slowly adapting mechanoreceptors. *Respir. Physiol.* **41**:211-216.

Davis, G. M., and Bureau, M. A. (1987). Pulmonary and chest wall mechanics in the control of respiration in the newborn. *Clin. Perinatol.* **14**:551-579.

DiMarco, A. F., Euler, C. von, Romaniuk, J. R., and Yamamoto, Y. (1981) Positive feedback facilitation of external intercostal and phrenic inspiratory activity by pulmonary stretch receptors. *Acta Physiol. Scand.* **113**:375-386.

Duron, B., and Marlot, D. (1980). Nervous control of breathing during postnatal development in the kitten. *Sleep* **3**:323-330.

Ekholm, J. (1967). Postnatal changes in cutaneous reflexes and in the discharge pattern of cutaneous and articular sense organs. *Acta Physiol. Scand.* (Suppl. 297).

Euler, C. von. (1986). Brain stem mechanisms for generation and control of breathing pattern. In *Handbook of Physiology. The Respiratory System*. Sec. 3. vol 2. Edited by N. S. Cherniak, and J. G. Widdicombe. Bethesda, MD, American Physiological Society, pp. 1-67.

Farber, J. P. (1972). Development of pulmonary reflexes and pattern of breathing in the Virginia opossum. *Respir. Physiol.* **14**:278-286.

Farber, J. P. (1982). Pulmonary receptor discharge and expiratory muscle activity. *Respir. Physiol.* **47**:219-229.

Farber, J. P. (1983). Expiratory motor responses in the suckling opossum. *J. Appl. Physiol.* **54**:919-925.

Farber, J. P. (1985). Motor responses to positive-pressure breathing in the developing opossum. *J. Appl. Physiol.* **58**:1489-1495.

Farber, J. P. (1986). Differential recruitment of expiratory muscles during opossum development. *J. Appl. Physiol.* **60**:841-845.

Farber, J. P. (1988). Medullary inspiratory activity during opossum development. *Am. J. Physiol.* **254**:R578-R584.

Farber, J. P. (1989). Medullary expiratory activity during opossum development. *J. Appl. Physiol.* **66**:1606-1612.

Farber, J. P., Fisher, J. T., and Sant 'Ambrogio, G. (1984). Airway receptor activity in the developing opossum. *Am. J. Physiol.* **246**:R753-R758.

Farber, J. P., and Marlow, T. A. (1976). Pulmonary reflexes and breathing pattern during sleep in the opossum. *Respir. Physiol.* **27**:73-86.

Farber, J. P., and Marlow, T. A. (1978). An obstructive apnea in the suckling opossum. *Respir. Physiol.* **34**:295-305.

Fedorko, L., Kelly, E. N., and England, S. J. (1988). Importance of vagal afferents in determining ventilation in newborn rats. *J. Appl. Physiol.* **65**:1033-1039.

Finer, N. N., Abroms, I. F., and Taeusch, H. W. Jr. (1976). Ventilation and sleep states in newborn infants. *J. Pediatr.* **89**:100-108.

Fisher, J. T., Mortola, J. P., Smith, J. B., Fox G. S., and Weeks, S. (1982) Respiration in newborns. Development of the control of breathing. *Am. Rev. Respir. Dis.* **125**:650-657.

Fisher, J. T., and Sant 'Ambrogio, G. (1982). Location and discharge properties of respiratory vagal afferents in the newborn dog. *Respir. Physiol.* **50**:209-220.

Fisher, J. T., and Sant 'Ambrogio, G. (1985). Airway and lung receptors and their reflex effects in the newborn. *Pediatr. Pulmonol.* **1**:112-126.

Fishman, N. H., Phillipson, E. A., and Nadel, J. A. (1973). Effect of differential vagal cold blockade on breathing pattern in conscious dogs. *J. Appl. Physiol.* **34**:754-758.

Fleming, P. J., Bryan, A. C., and Bryan, M. H. (1978). Functional immaturity of pulmonary irritant receptors and apnea in newborn preterm infants. *Pediatrics* **61**:515-518.

Fox, R. E., Kosch, P. C., Feldman, H. A., and Stark, A. R. (1988). Control of inspiratory duration in premature infants. *J. Appl. Physiol.* **64**:2597-2604.

Gaultier, C., and Mortola, J. P. (1981). Hering-Breuer inflation reflex in young and adult mammals. *Can. J. Physiol. Pharmacol.* **59**:1017-1021.

Gautier, H., Bonora, M., and Gaudy, J. H. (1981) Breuer-Hering inflation reflex and breathing pattern in anesthetized humans and cats. *J. Appl. Physiol.* **51**:1162-1168.

Gerhardt, T., and Bancalari, E. (1981). Maturational changes of reflexes influencing inspiratory timing in newborns. *J. Appl. Physiol.* **50**:1282-1285.

Gerhardt, T., and Bancalari, E. (1984). Apnea of prematurity. II. Respiratory reflexes. *Pediatrics* **74**:63-66.

Gillespie, J. R., Bruce, E., Alexander, J., and Mead, J. (1979). Breathing responses of unanesthetized man and guinea pigs to increased transrespiratory pressure. *J. Appl. Physiol.* **47**:119-125.

Glogowska, M., Richardson, P. S., Widdicombe, J. G., and Winning, A. J. (1972). The role of the vagus nerves, peripheral chemoreceptors and other afferent pathways in the genesis of augmented breaths in cats and rabbits. *Respir. Physiol.* **16**:179-196.

Grimby, G., Goldman, M., and Mead, J. (1976). Respiratory muscle action inferred from rib cage and abdominal V-P partitioning. *J. Appl. Physiol.* **41**:739-751.

Grunstein, M. M., and Tanaka, D. (1985). Ontogeny of respiratory control and pulmonary mechanics in newborn rabbits. *J. Appl. Physiol.* **59**: 1477-1486.

Grunstein, M. M., Younes, M., and Milic-Emili, J. (1973). Control of tidal volume and respiratory frequency in anesthetized cats. *J. Appl. Physiol.* **35**:463-476.

Guz, A., Noble, M. I. M., Eisele, J. H., and Trenchard, D. (1970). The role of vagal inflation reflexes in man and other animals. In *Breathing Hering-Breuer Centenary Symposium*. Edited by R. Porter. London, Churchill, pp. 17-40.

Haddad, G. G., Leistner, H. L., Epstein, R. A., Epstein, M. A. F., Grodin, W. K., and Mellins, R. B. (1980). CO_2-induced changes in ventilation and ventilatory pattern in normal sleeping infants. *J. Appl. Physiol.* **48**:684-688.

Hamilton, R. D., Winning, A. J., Horner, R. L., and Guz, A. (1988). The effect of lung inflation on breathing in man during wakefulness and sleep. *Respir. Physiol.* **73**:145-154.

Jonzon, A., and Sedin, G. (1982) A low continuous positive airway pressure induces regular breathing and increased inspiratory activity in newborn lambs. *Acta. Physiol. Scand.* **115**:415-419.

Kalia, M. (1976). Visceral and somatic reflexes produced by J pulmonary receptors in newborn kittens. *J. Appl. Physiol.* **41**:1-6.

Kattwinkel, J. (1977). Neonatal apnea: pathogenesis and therapy. *J. Pediatr.* **90**:342-347.

Kelson, S. G., Altose, M. D., and Cherniak, N. S. (1977). Interaction of lung volume and chemical drive on respiratory muscle EMG and respiratory timing. *J. Appl. Physiol.* **42**:287-294.

Knill, R., and Bryan, A. C. (1976). An intercostal-phrenic inhibitory reflex in human newborn infants. *J. Appl. Physiol.* **40**:352-356.

Knox, C. K. (1973). Characteristics of inflation and deflation reflexes during expiration in the cat. *J. Neurophysiol.* **36**:284-295.

Kosch, P. C., Davenport, P. W., Wozniak, J. A., and Stark, A. R. (1985). Reflex control of expiratory duration in newborn infants. *J. Appl. Physiol.* **58**:575-581.

Kosch, P. C., Davenport, P. W., Wozniak, J., and Stark, A. (1986). Reflex control of inspiratory duration in newborn infants. *J. Appl. Physiol.* **60**:2007-2014.

Kostreva, D. R., Zuperku, E. J., Hess, G. L., Coon, R. L., and Kampine, J. P. (1975). Pulmonary afferent activity recorded from sympathetic nerves. *J. Appl. Physiol.* **39**:37-40.

Krause, W. J., and Leeson, C. R. (1973). The postnatal development of the respiratory system of the opossum. I. Sight and scanning electron microscopy. *Am. J. Anat.* **137**:337-356.

LaFramboise, W. A., Woodrum, D. E., and Guthrie, R. D. (1985). Influence of vagal activity on the neonatal ventilatory response to hypoxemia. *Pediatr. Res.* **19**:903-907.

Lopes, J., Muller, N. L., Bryan, M. H., and Bryan, A. C. (1981). Importance of inspiratory muscle tone in maintenance of FRC in the newborn. *J. Appl. Physiol.* **51**:830-834.

Marlot, D., and Duron, B. (1979). Postnatal development of vagal control of breathing in the kitten. *J. Physiol. (Paris)* **75**:891-900.

Martin, R. J., Okken, A., Katona, P. G., and Klaus, M. H. (1978). Effect of lung volume on expiratory time in the newborn infant. *J. Appl. Physiol.* **45**:18-23.

Mills, J. E., Sellick, H., and Widdicombe, J. G. (1970). Epithelial irritant receptors in the lung. In *Breathing: Hering-Breuer Centenary Symposium.* Edited by R. Porter. London, Churchill, pp. 77-92.

Miserocchi, G., and Milic-Emili, J. (1976). Effect of mechanical factors on the relation between rate and depth of breathing in cats. *J. Appl. Physiol.* **41**:277-284.

Mortola, J. P., Fisher, J. T., and Sant 'Ambrogio, G. (1984). Vagal control of breathing pattern and respiratory mechanics in the adult and newborn rabbit. *Eur. J. Physiol.* **401**:281-286.

Mortola, J. P., Magnante, D., and Saetta, M. (1985). Expiratory pattern of newborn mammals. *J. Appl. Physiol.* **58**:528-533.

Noble, L. M., Carlo, W. A., Miller, M. J., DiFiore, J. M., and Martin, R. J. (1987). Transient changes in expiratory time during hypercapnia in premature infants. *J. Appl. Physiol.* **62**:1010-1013.

Olinsky, A., Bryan, M. H., and Bryan, A. C. (1974) Influence of lung inflation on respiratory control in neonates. *J. Appl. Physiol.* **36**:426-429.

Pack, A. I., and DeLaney, R. G. (1983). Response of pulmonary rapidly adapting receptors during lung inflation. *J. Appl. Physiol.* **55**:955-963.

Paintal, A. S. (1966). Re-evaluation of respiratory reflexes. *Q. J. Exp. Physiol.* **51**:151-163.

Phillipson, E. A., Hickey, R. F., Graf, P. D., and Nadel, J. A. (1971). Hering-Breuer inflation reflex and regulation of breathing in conscious dogs. *J. Appl. Physiol.* **31**:746-750.

Polacheck, J., Strong, R., Arens, J., Davies, C., Metcalf, I., and Younes, M. (1980) Phasic vagal influence on inspiratory motor output in anesthetized human subjects. *J. Appl. Physiol.* **49**:609-619.

Ponte, J., and Purves, M. J. (1973). Types of afferent nervous activity which may be measured in the vagus nerve of the sheep foetus. *J. Physiol. (London)* **229**:51-76.

Praud, J. P., Egreteau, L., Benlabed, M., Curzi, L., and Gaultier, Cl. (1989). The effect of rebreathing CO_2 on ventilation and respiratory muscle electromyography in infants. *FASEB J.* **3**:A1157 (Abstract).

Remmers, J. E. (1970) Inhibition of inspiratory activity by intercostal muscle afferents. *Respir. Physiol.* **10**:358-383.

Remmers, J. E., Richter, D. W., Ballantyne, D., Bainton, C. R., and Klein, J. P. (1986). Reflex prolongation of stage 1 of expiration. *Pflugers Arch.* **407**:190-198.

Schwieler, G. H. (1968). Respiratory regulation during postnatal development in cats and rabbits and some of its morphological substrate. *Acta Physiol. Scand.* (Suppl. 304).

Shannon, R., Bolser, D. C., and Lindsey, B. G. (1987). Medullary expiratory activity: Influence of intercostal tendon organs and muscle spindle endings. *J. Appl. Physiol.* **62**:1057-1062.

Skoglund, S. (1960). The activity of muscle receptors in the kitten. *Acta Physiol. Scand.* **50**:203-221.

Smejkal, V., Palacek, F., and Frydrychova, M. (1985). Developpement postnatal du reflexe de Breuer-Hering chez le rat. *J. Physiol. (Paris)* **80**:173-176.

Stark, A. R., Cohlan, B. A., Waggener, T. B., Frantz, I.D., III, and Kosch, P. C. (1987). Regulation of end-expiratory lung volume during sleep in premature infants. *J. Appl. Physiol.* **62**:1117-1123.

Stark, A. R., and Frantz, I. D., III. (1979). Prolonged expiratory duration with elevated lung volume in newborn infants. *Pediatr. Res.* **13**:261-264.

Sullivan, C. E., Kozar, L. F., Murphy, E., and Phillipson, E. A. (1978). Primary role of respiratory afferents in sustaining breathing rhythm. *J. Appl. Physiol.* **45**:11-17.

Taeusch, H. W. Jr., Carson, S., Frantz, I. D., and Milic-Emili, J. (1976). Respiratory regulation after elastic loading and CO_2 rebreathing in normal term infants. *J. Pediatr.* **88**:102-111.

Thach, B. T., Abroms, I. F., Frantz, I. D., III, Sotrel, A., Bruce, E. N., and Goldman, M. D. (1980). Intercostal muscle reflexes and sleep breathing patterns in the human infant. *J. Appl. Physiol.* **48**:139-146.

Thach, B. T., Frantz, I. D. III, Adler, S. M., and Taeusch, H. W., Jr. (1978). Maturation of reflexes influencing inspiratory duration in human infants. *J. Appl. Physiol.* **45**:203-211.

Thach, B. T., Menon, A. P., and Schefft, G. L. (1989a). Effects of negative upper airway pressure on pattern of breathing in sleeping infants. *J. Appl. Physiol.* **66**:1599-1605.

Thach, B. T., Schefft, G. L., Pickens, D. L., and Menon, A. P. (1989b). Influence of the upper airway negative pressure reflex on response to airway occlusion in sleeping infants. *J. Appl. Physiol.* **67**:749-755.

Thach, B. T., and Taeusch, H. W., Jr. (1976). Sighing in human newborn infants: Role of inflation-augmenting reflex. *J. Appl. Physiol.* **41**:502-507.

Trippenbach, T. (1981). Laryngeal, vagal and intercostal reflexes during the early postnatal period. *J. Dev. Physiol.* **3**:133-159.

Trippenbach, T. (1985). Chest wall reflexes in newborns. *Bull. Eur. Physiopathol. Respir.* **21**:115-122.

Trippenbach, T., and Kelly, G. (1983). Phrenic activity and intercostal muscle EMG during inspiratory loading in newborn kittens. *J. Appl. Physiol.* **54**:496-501.

Trippenbach, T., and Kelly, G. (1988). Respiratory effects of cigarette smoke, dust, and histamine in newborn rabbits. *J. Appl. Physiol.* **64**: 837-845.

Trippenbach, T., Kelly, G., and Marlot, D. (1985). Effect of tonic vagal input on breathing pattern in newborn rabbits. *J. Appl. Physiol.* **59**: 223-228.

Trippenbach, T., Zinman, R., and Mozes, R. (1981). Effects of airway occlusion at functional residual capacity in pentobarbital-anesthetized kittens. *J. Appl. Physiol.* **51**:143-147.

Watchko, J. F., O'Day, T. L., Brozanski, B. S., and Guthrie, R. D. (1990). Expiratory abdominal muscle activity during ventilatory chemostimulation in piglets. *J. Appl. Physiol.* **68**:1343-1349.

Widdicombe, J. G. (1961). Respiratory reflexes in man and other mammalian species. *Clin. Sci.* **21**:163-170.

Witte, M. K., and Carlo, W. A. (1987) Prolongation of inspiration during lower airway occlusion in children. *J. Appl. Physiol.* **62**:1860-1864.

Wyszogrodski, I., Thach, B. T., and Milic-Emili, J. (1978). Maturation of respiratory control in unanesthetized newborn rabbits. *J. Appl. Physiol.* **44**:304-310.

Younes, M. K., Remmers, J. E., and Baker, J. (1978). Characteristics of inspiratory inhibition by phasic volume feedback in cats. *J. Appl. Physiol.* **45**:80-86.

Younes, M., Vaillancourt, P., and Milic-Emili, J. (1974). Interaction between chemical factors and duration of apnea following lung inflation. *J. Appl. Physiol.* **36**:190-201.

9

Laryngeal Reflexes
Integrated and Cellular Aspects

DAVID F. DONNELLY

Yale University School of Medicine
New Haven, Connecticut

I. Introduction

The peripheral input to the respiratory system from laryngeal afferent fibers is unique compared to other respiratory afferents in several important respects. (1) On the basis of density of innervation the larynx exceeds that of all other respiratory segments. Approximately one-third of all vagal afferent fibers innervate the larynx, leaving two-thirds for the remainder of the tracheal/bronchial tree, lungs, heart, and gut. (2) Effects on respiratory neuronal activity caused by stimulation of laryngeal afferents generally occur at shorter latency than with other respiratory afferents, especially when compared to the long latency effects of pulmonary stretch receptor (Iscoe et al., 1979) and carotid chemoreceptors (Berger and Mitchell, 1976). (3) The duration of apnea by laryngeal stimulation and the change of this response with development are especially pronounced. In this chapter, we will consider these unique characteristics in greater detail and speculate on possible mechanisms for the reflex effects.

II. Laryngeal Nerve Fibers

A. Fiber Trajectory

Superior laryngeal nerve (SLN) afferent fibers innervate the laryngeal mucosa, internal laryngeal musculature, and epiglottis (Yoshida et al., 1986). The modal specificity of the afferent fibers has been addressed in Chapter 8 of this volume, but for our purposes, we will consider primarily three types of receptors: (1) those that respond to airway pressure (Storey, 1968), (2) chemoreceptors (also called water receptors) (Harding et al., 1978), and (3) cold or drive receptors (Fisher and Sant 'Ambrogio, 1985; Sant 'Ambrogio et al., 1985). Afferent nerve fibers from these end-organs comprise most of the superior laryngeal nerve, which joins the vagus nerve at the level of the nodose ganglia and runs with the vagus to termination sites in the brainstem and pons. Efferent fibers in the SLN innervate the cricothyroid muscle and provide efferent sympathetic and parasympathetic control to airway vessels (Lucier et al., 1986).

Upon entering the brainstem in the vagus, most of the fibers course caudally and comprise the lateral aspect of the solitary tract (Hanamori and Smith, 1986). Early studies that identified the termination sites using field-potential mapping suggested that termination sites occurred throughout the brainstem, particular in the NTS, dorsal raphe, and nucleus ambiguus (Biscoe, 1970a,b). Recent studies using anterograde transport of horseradish peroxidase (HRP) showed the termination sites to be more restricted. Reaction product consistent with nerve termination sites is localized ipsilaterally in the rostral medial nucleus tractus solitarius (NTS), interstitial NTS, intermediate NTS, lateral NTS, and commissural NTS, bilaterally (Nomura and Mizuno, 1983; Goding et al., 1987; Hanamori and Smith, 1986; Lucier et al., 1986) (Fig. 1). Afferents also terminate in the spinal trigeminal nucleus of the rat (Hamilton and Norgren, 1984) and hamster (Hanamori and Smith, 1986), but not in cats (Lucier et al., 1986) (Fig. 1). A bilateral sensory projection to nucleus ambiguus is reported for piglets (Goding et al., 1987). In addition, a rostral projection to the principal trigeminal nucleus of the pons has been observed in the lamb, but not in other species (Car et al., 1975) (Fig. 1).

Efferent fibers coursing in the SLN include muscle efferents to laryngeal constrictor muscles (cricothyroid muscle), parasympathetic efferents, and sympathetic fibers (Lucier et al., 1986). Cell bodies of origin reside in the nucleus ambiguous (NA) (Hamilton and Norgren, 1984) and rostral aspects of the dorsal motor nucleus of the vagus (DMNX) (Bieger and Hopkins, 1987; Nomura and Mizuno, 1983) and NTS (Hanamori and Smith, 1986).

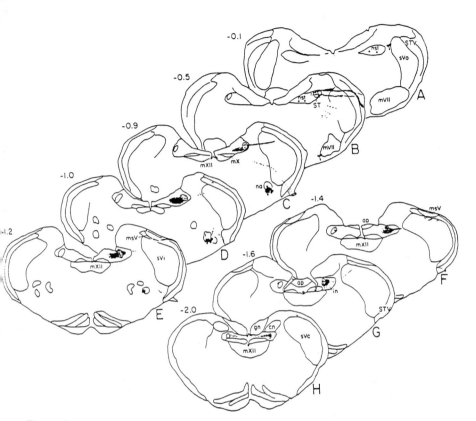

Figure 1 Camera lucida drawings of frontal sections through the brainstem showing the distribution of reaction product following incubation of the SLN with HRP. Section A is most rostral and section H is most caudal. Small dots represent regions of presumptive axon terminals, irregular short lines represent afferent and efferent fibers, and filled circles represent retrogradely labeled cell bodies. ap, Aprea postrema; cn, cuneate nucleus; gn, gracile nucleus; in, interstitial NTS; mX, dorsal motor nucleus of the vagus; mXII, hypoglossal nucleus; na, nucleus ambituus, ST, solitary tract.

III. Effect and Mechanism of Action

A. Respiratory and Autonomic Responses

Based on their location in and around the airway, it would be expected that laryngeal afferents would exert prominent effects on breathing. Many of the afferents are activated with normal respiratory movements and could potentially regulate breath-to-breath control of tidal volume and frequency. For instance, laryngeal "drive" receptors (or cold receptors) are activated with each inspiration (Jammes et al., 1987; Sant 'Ambrogio et al., 1985). However, bilateral section of the SLN in unanesthetized animals causes little change in breathing pattern (Citterio et al., 1985). Thus, the importance of SLN afferents in normal breath-to-breath control is probably small.

The major effects of SLN stimulation are found during airway challenge, with either chemical stimulation (Johnson et al., 1975; Kovar et al., 1979), pressure stimulation (Sullivan et al., 1978), or direct nerve stimulation (Donnelly and Haddad, 1986a,b; Lawson, 1981, 1982; Downing and Lee, 1975). A series of reflex maneuvers occurs which protects the lungs against the offending stimulus. If the stimulus is presented during inspiration (as it is with aspiration), then inspiratory drive is rapidly terminated (Iscoe et al., 1979) and the larynx is constricted (Sherry and Megirian, 1974), both of which protect the lungs against further progress of the offending substance. In addition, arousal, coughing, and swallowing are evoked in unanesthetized animals, maneuvers that remove the offending stimulus to the mouth or esophagus and away from the airway (Sullivan et al., 1978).

Cardiovascular changes are also evoked. Blood pressure increases, but heart rate decreases (Grogaard et al., 1980, 1982). Blood flow is decreased to the skin, gut, and kidneys but preserved to the heart and brain (Grogaard, 1982). Approximately 60% of these cardiovascular reflex effects are secondary to the respiratory (and arterial blood gas) changes since the magnitude of cardiovascular response may be reduced by artificial ventilation during the stimulus period (Grogaard, 1982). Conversely, superimposed hypoxia enhances the duration of apnea as well as depth of bradycardia (Wennergren et al., 1989).

These respiratory and cardiovascular changes may be of special interest to humans and in particular to human infants. Inspiratory inhibition and laryngeal constriction is present in the young (Perkett and Vaughan, 1982; Davies et al., 1988; Pickens et al., 1988; Wennergren et al., 1989) and adults (Ogura and Lam, 1953). However, infants show poorer coordination in swallowing and coughing than adults and thus may not clear the offending stimulus as readily. In addition, certain sleep states and depressant drugs, such as alcohol (Grogaard et al., 1985), may prolong the apnea and reduce the evoked swallowing, coughing (Sullivan et al., 1978), or arousal (Phillipson et al., 1978). Other factors may potentiate the duration of reflex apnea, for instance, low glucose (Johnson et al., 1975), low hematocrit (Lee and Downing, 1980), or an upper airway infection (Sessle and Lucier, 1983). Genetics also plays an important role. For instance, prolonged reflex apnea is never observed in Dorset Horn lambs but is often observed in other species (Johnson et al., 1975). Thus, given a set of genetic and environmental factors, the possibility exists of a "window of vulnerability" for a period of protracted apnea with consequent desaturation and hypoxic depression of neural systems. This laryngeal chemoreflex has been postulated as a possible mechanism for sudden death or the sudden infant death syndrome (SIDS, crib death) (French et al., 1972). This question of developmental changes will be taken up below.

B. Effect on Respiratory Muscle Drive

Since laryngeal afferents have a profound respiratory effect and they terminate in subnuclei of the NTS (near a major center of respiratory upper motoneurons), the anatomical substrate exists for a near-direct effect on respiratory neurons. Experiments that address this postulate are of two general types: (1) experiments in which a single electrical shock is delivered to the SLN, setting up a unified barrage on many SLN fibers and (2) experiments that employ stimulus trains to the SLN or physiological stimulation of the airway. The first protocol has the advantage that timing is well controlled by the experimenter and the stimulus may be delivered at discrete times within the respiratory cycle. Since the stimulus duration is short (one action potential), the latency to effect may be measured well. The second protocol more nearly reproduces the type of stimulation likely to be found under physiological conditions.

The effect on phrenic neural activity of a single SLN stimulus delivered during inspiration (I) is a brief excitation of the phrenic nerve contralateral to the stimulated SLN (latency 4.6 msec, Fig. 2) followed by a prolonged inhibition lasting 20-40 msec (latency 5.6 msec. Fig. 2) (Berger, 1977, 1978; Donnelly et al., 1989a,b). The brief contralateral excitation during I is observed as a rapid decrease in intracellular potential, but may also be elicited at identical latency if the stimulus is presented during expiration (E) (Berger, 1978). The phrenic nerve ipsilateral to the stimulated SLN shows only the inhibitory period; it occurs at identical latency to the inhibition of contralateral phrenic activity (Berger, 1977; Donnelly et al., 1989b) and this response is also not dependent on the phase of the respiratory cycle (Bellingham et al., 1989). The inhibitory effect on phrenic neurons is likely due to a postsynaptic effect and not simply a disfacilitation since the intracellular voltage change may be reversed with chloride iontophoresis (Bellingham et al., 1989). Stimulus trains or chemical stimulation of the laryngeal mucosa results in maintained suppression of phrenic activity (Lucier et al., 1979).

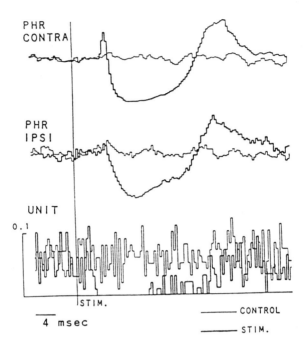

Figure 2 Poststimulus histograms (heavy lines) and control (nonstimulus) histograms (light lines) of phrenic activity contralateral to the SLN stimulus (upper traces), ipsilateral to the stimulus (middle traces), and activity of the dorsal respiratory group, I-augmenting unit (lower traces). This unit was ipsilateral to the stimulated SLN. Vertical scale for unit histograms: probability of discharge per bin. Stimuli were delivered at 400 msec delay from the start of I. Following the SLN stimulus, phrenic activity showed a brief excitation followed by depression. In this case the unit was not excited, was inhibited appreciably before phrenic inhibition, and resumed firing roughly in parallel with the return of phrenic activity. Average of 100 stimuli and 100 control trials.

Thus, the phrenic response has two patterns that need to be explained: (1) a brief excitatory response to the contralateral phrenic, and (2) a prolonged bilateral inhibitory period.

For the purposes of considering SLN effects, we may simplify the elements of the respiratory pattern generator. Within the brainstem, two regions have a high density of respiratory-related spiking activity: the dorsal respiratory group (DRG) in anatomical association with the NTS and the ventral respiratory group (VRG) in association with nucleus ambiguus (Vibert et al., 1976). The most caudal and rostral portions of VRG provide neural drive to expiratory intercostal motoneurons as well as inhibiting inspiratory neurons in the brainstem and spinal cord. DRG inspiratory neurons primarily excite phrenic motoneurons. VRG inspiratory neurons excite phrenic motoneurons and inspiratory intercostal and laryngeal motoneurons. This simplified model, while sufficient for our purposes, does not do justice to the current state of knowledge concerning the respiratory pattern generator. The interested reader may find some current and excellent reviews on this subject (Cohen, 1979; Feldman, 1988).

C. Effect on DRG Neurons

Since SLN afferents primarily terminate in subnuclei of the NTS, it is likely that SLN afferents exert short latency effects on these cells. Extracellular and intracellular recordings bear this prediction out. Fifty to ninety percent of DRG-I neurons show an excitatory response to a single SLN stimulus delivered during I (Donnelly et al., 1989b; Berger, 1977; Richter et al., 1986a). This excitatory response may be large enough to cause excitation during E (Berger, 1978), i.e., during a period when I neurons are receiving active inhibition. Recently, this excitation has been confirmed using the elegant techniques of intracellular recording, which allow greater resolution of latency (Richter et al., 1986a). Based on measurement of conduction delay and latency to synaptic effect, it is well substantiated that many SLN afferents monosynaptically excite DRG-I neurons. Since many of these I neurons are, in turn, excitatory to phrenic motoneurons (Cohen et al.,1 974), we may account for the contralateral excitatory phrenic response by postulating a two-neuron path from SLN to DRG-I neurons to phrenic motoneurons (Fig. 3). Consistent with this postulate is the observation that lesions to the ipsilateral DRG abolish the contralateral phrenic excitation in response to SLN stimulation (McCrimmon et al., 1987).

The neural substrate for DRG-I unit inhibition is more difficult to postulate. Extracellular recording has demonstrated that spiking activity of most NTS neurons is silenced for a period about the duration of phrenic silence (ca. 20-40 msec) (Donnelly et al., 1989a,b). However, using analysis of extracellular recordings, it is impossible to differentiate a direct postsynaptic inhibition from a loss of excitation (disfacilitation) to the DRG neuron. There is other evidence for both effects.

In approximately 20% of intracellular recordings of DRG-I neurons, in vivo, stimulation resulted in a short-latency inhibition, which could be reversed with chloride iontophoresis (Richter et al., 1986b). This evidence strongly supports the hypothesis that some I neurons are inhibited by SLN afferents, either directly or by an oligosynaptic pathway. However, extracellular recordings show that spiking activity in most DRG I neurons is inhibited. What is mediating the inhibition?

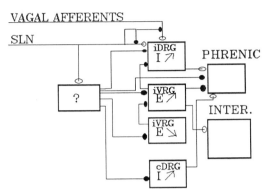

Figure 3 Schematic diagram of proposed neuronal interaction in response to SLN stimulation. Excitatory interactions are represented in open circles and inhibitory interactions are in closed circles. SLN afferents cause a monosynaptic excitation of DRG-I neurons but also presynaptically inhibit other vagal afferents. Short-latency inhibition of DRG-I neurons and phrenic neurons is mediated by an unidentified group of interneurons (?). iDRG, Inspiratory augmenting dorsal respiratory group cells ipsilateral to the stimulated SLN; cDRG, inspiratory augmenting dorsal respiratory group cells contralateral to the stimulated SLN; iVRG E^, expiratory augmenting ventral respiratory group neurons ipsilateral to the stimulus; iVRG E >, expiratory decrementing ventral respiratory group neurons ipsilateral to the SLN stimulus; PHRENIC, phrenic motoneurons; INTER, internal intercostal motoneurons.

One likely source of inhibition is through presynaptic inhibition of excitatory inputs on DRG cells. In this form, inhibition is mediated by a disfacilitation (or reduction) of excitatory drive. In support of this postulate, SLN afferents are established to have presynaptic effects on other SLN afferents as well as other vagal afferents, and the duration of the presynaptic interaction is approximately the same length as the observed duration of phrenic inhibition (Richter et al., 1986a; Rudomin, 1967; Sessle, 1973).

Even with the postulated roles for pre- and postsynaptic effects on DRG-I neurons, the model is incomplete. Following massive lesions within and around the DRG, phrenic I activity continued to be inhibited by SLN afferent stimulation (McCrimmon et al., 1987). This result implies that the inhibitory effects of SLN afferents may be mediated by an as yet unidentified group of interneurons which inhibit phrenic neurons, ipsilaterally and contralaterally.

D. Effect on VRG Neurons

The ventral respiratory group, in association with nucleus ambiguus, is composed of both inspiratory and expiratory neurons. Many inspiratory neurons are upper motoneurons to the phrenic pool and laryngeal motoneurons.

Rostral expiratory neurons function, at least in part, to inhibit inspiratory neurons of the VRG, DRG, and phrenic nerve (Merrill et al., 1983; Fedorko and Merrill, 1984). Caudal expiratory neurons provide upper motor neuron drive to intercostal and abdominal muscles.

The VRG contains two types of E neurons which may be important in mediating the SLN respiratory effects. E neurons, located in the caudal VRG, are an important source of excitation to internal intercostal motoneurons. Since these muscles are activated during laryngeal stimulation (Boushey et al., 1972) along with phrenic inhibition, it was postulated that this activation may be mediated by an increase in VRG expiratory neuron activity. Surprisingly, the effect of SLN afferents is inhibitory on caudal E neurons, i.e., opposite to that predicted (Jodkowski and Berger, 1988). Thus, the source of intercostal excitation remains unknown.

Rostral E neurons in VRG may also have an important role in mediating the laryngeal reflex. This conjecture is based on the critical role played by these neurons in inhibiting I neurons during E. In particular, these neurons project to and directly inhibit DRG-I neurons, bilaterally, and phrenic motoneurons. Thus, the hypothesis that phrenic and brainstem I unit inhibition may be caused by excitation of rostral E neurons would be logical. However, this hypothesis is not supported by experiments which demonstrated that rostral E neurons were inhibited by laryngeal stimulation (Pantaleo and Corda, 1985).

A third group of VRG E neurons may also play a critical role in mediating the laryngeal reflex. These are the E decrementing (also called post-I) neurons, which are hypothesized to inhibit I and late E neurons (Fig. 3) (Richter et al., 1986a; Remmers et al., 1986). Thus, brainstem I-unit inhibition by SLN afferents may be mediated by excitation of this neuron group and subsequent inhibition of I-unit activity. In support of this hypothesis, laryngeal expiratory motoneurons (which have a postinspiratory discharge pattern) are greatly stimulated by SLN stimulation (Lucier et al., 1978). However, recent results from 12 intracellular recordings demonstrated just the opposite effect; i.e., in 5 of 12 postinspiratory neurons tested, SLN stimulation caused an inhibition (Bianchi et al., 1988).

Although exhaustive studies have not been conducted on all classes of DRG and VRG neurons, the weight of evidence suggests that another, as yet unidentified, neuron group is involved in mediating the inhibition of brainstem I neurons and phrenic motoneurons. This speculation is primarily based on the demonstration that active inhibition occurs in brainstem and phrenic neurons during SLN stimulation. DRG neurons probably do not mediate this inhibition since it occurs following massive lesion in the area of the DRG (McCrimmon et al., 1987). VRG neurons probably do not med-

mediate the inhibition because the likely candidates for mediating the inhibition (rostral E and postinspiratory neurons) appear to be inhibited by SLN afferents. Thus, these inhibitory neurons are, as yet, unidentified.

IV. Maturational Changes

Laryngeal afferents undergo maturational changes in morphology and in particular in the reflexes that they cause.

A. Nerve Fibers

As in other areas of the nervous system, the laryngeal afferent nerves are unmyelinated or only lightly myelinated in the newborn and heavily myelinated in the adult (Miller and Dunmire, 1976). Little information is currently available on whether these axons innervate the same type of receptors as in the adult or whether the central projection of these axons changes with development. If a developmental change in projection site does, in fact, occur, then this may help explain the dramatic maturational changes in reflex effects (see below). The hypothesis is supported by one study in piglets which suggested that the caudal extent of the afferent termination site around NTS decreased with development, and a bilateral sensory projection to NA regressed to a unilateral projection (Goding et al., 1987).

B. Respiratory and Autonomic Response

The respiratory effects of laryngeal stimulation may be especially profound in the newborn compared to the adult. Experiments on anesthetized dogs (Boggs and Bartlett, 1983), cats (Lucier et al., 1979), and pigs (Long and Lawson, 1981) have strongly suggested that the duration of apnea by laryngeal stimulation decreases with age (Fig. 4). Although this developmental course seems clear, some caution should be exercised in interpreting the magnitude of respiratory inhibition in the presence of anesthesia. The duration of laryngeal-induced apnea is profoundly enhanced by the presence of increasing levels of anesthesia (Donnelly and Haddad, 1986b). Even with the same absolute level of anesthesia, differential sensitivity of newborn versus adult to anesthesia can complicate the interpretation of results.

Experiments on unanesthetized animals or babies may help alleviate these concerns regarding anesthetic. Although these experiments too have complicating factors—especially the ability of swallowing to remove the stimulus at any time—these experiments on unanesthetized subjects also suggest a developmental reduction in the magnitude of laryngeal reflex-induced apnea. Following placement of fluid on the larynx, some lambs

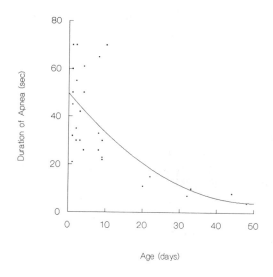

Figure 4 Duration of water-induced apnea as a function of age. At time zero, 10 ml of water was administered to the larynx of an anesthetized puppy. The ordinate represents the time period between onset of apnea and the first breath. The abscissa is the puppy's age. The data was fit to a quadratic regression curve. (Adapted from Boggs and Bartlett, 1983.)

would manifest a dramatic and prolonged respiratory pause, which resulted in the collapse of the animal and the need to resuscitate (Johnson et al., 1975; Marchal et al., 1982). This type of prolonged respiratory pause was not observed in the older (>7 day) animal. Also, in human preterm infants, a central apnea lasting 40 s in duration was produced by installation of 0.1 ml of fluid in the larynx (Davies et al., 1988). The duration may actually be longer but the apnea was terminated by intervention out of concern for the well-being of the infant. These prolonged respiratory pauses were never observed in the older infant.

Although the weight of evidence strongly suggests that the sensitivity of the respiratory system to SLN inhibition decreases with development, the cause for this maturation is not immediately clear. As mentioned above, maturational changes in the termination site of laryngeal sensory nerves may determine the magnitude of the reflex inhibition. Alternatively, the reduction in apnea duration may be due to maturational changes not directly related to the laryngeal pathway. For instance, results from intracellular recording, in vitro, and with synaptic blockade, demonstrated that hypoxia, itself, caused an enhanced excitability in some mature, but not immature, neurons (Haddad and Donnelly, 1990). Thus, apnea duration may be determined by changes in excitability caused by ensuing hypoxia and not changes in the "hardware" mediating the laryngeal reflex inhibition. Similarly, the maturation of swallowing and coughing reflexes may more readily remove the offending stimulus.

Possible Relationship to Sudden Infant Death Syndrome

Although the etiology of SIDS remains a mystery, the motivation for much of the experimental work dealing with the laryngeal reflex has stemmed from its parallels with SIDS, or crib death. The ability of a physiologically evoked reflex response to lead to the demise of an animal is unique within the cardio-respiratory system, and this collapse has been demonstrated in unanesthetized (Johnson et al., 1975; Marchal et al., 1982) as well as anesthetized (Lee et al., 1977; Lee and Downing, 1980; French et al., 1972; Sutton et al., 1978) animals. Development plays a strong influence in both cases, although the time factor is not identical. The incidence of SIDS peaks at 2 months and decreases thereafter. Laryngeal-induced apnea in human infants and lambs is greatest in newborns or premature babies and decreases thereafter. In cats, the apnea duration increases after birth and then decreases (Lucier et al., 1979). Upper airway infections are correlated with SIDS and are also known to prolong the duration of laryngeal-induced apnea (Sessle and Lucier, 1983). Genetic predisposition may play a role in SIDS and in determining the duration of laryngeal-induced apnea (Johnson et al., 1975). Whether this speculative linkage actually exists awaits further study.

V. Future Directions

Even after years of careful investigation, the role of laryngeal afferents in day-to-day life is enigmatic. In one sense, the laryngeal afferents may serve a protective function to signal that a foreign substance is in the airway and to evoke maneuvers to clear the offending stimulus. However, recent evidence suggests that laryngeal fibers may respond to natural airway secretions and to invoke swallowing to clear the continuous movement of fluid away from the lungs and airways (Davies et al., 1988). This reflex effect may be occurring hundreds of times an hour, or viewed another way, the airway is challenged hundreds of times an hour. We know from studies in lambs that circumstances of species, age, conscious state, temperature, and sugar level may predispose a laryngeal reflex response resulting in death. Why do some animals collapse during an airway challenge and some do not, and why is the observation inconsistent in animals that do collapse? The answer at present is unclear, but is important as we attempt to understand why some infants suffer unexpected and sudden death.

On a cellular level, the mechanism of laryngeal reflex respiratory inhibition is enigmatic and filled with paradoxes. The response to continuous SLN stimulation is a central apnea, but the observed response of many inspiratory upper motoneurons is a monosynaptic excitation, rather than an inhibition. How does SLN afferent activity cause phrenic inhibition? Which second-order neurons are involved in this response? Clearly, these questions need to be answered before we can understand the neuronal basis for reflex apnea.

References

Bellingham, M. C., Lipski, J., and Voss, M. D. (1989). Synaptic inhibition of phrenic motoneurons evoked by stimulation of the superior laryngeal nerve. *Brain Res.* **486**:391-395.

Berger, A. J. (1977). Dorsal respiratory group neurons in the medulla of the cat: Spinal projections, response to lung inflation and superior laryngeal nerve stimulation. *Brain Res.* **135**:231-254.

Berger, A. J. (1978). Respiratory gating of phrenic motoneuron responses to superior laryngeal nerve stimulation. *Brain Res.* **157**:381-384.

Berger, A. J., and Mitchell, R. A. (1976). Lateralized phrenic nerve response to stimulating respiratory afferents in the cat. *Am. J. Physiol.* **230**: 1314-1320.

Bianchi, A. L., Grelot, L., Iscoe, S., and Remmers, J. E. (1988). Electrophysiological properties of rostral medullary respiratory neurones in the cat: an intracellular study. *J. Physiol. (London)* **407**:293-310.

Bieger, D., and Hopkins, D. A. (1987). Viscerotopic representation of the upper alimentary tract in the medulla oblongata in the rat: The nucleus ambiguus. *J. Comp. Neurol.* **262**:546-562.

Biscoe, T. J., and Sampson, S. R. (1970a). Field potentials evoked in the brainstem of the cat by stimulation of the carotid sinus, glossopharyngeal, aortic and superior laryngeal nerves. *J. Physiol. (London)* **209**: 341-358.

Boggs, D. F., and Bartlett, D., Jr. (1983). Chemical specificity of a laryngeal apneic reflex in puppies. *J. Appl. Physiol.* **53**:455-462.

Bowes, G., Woolf, G. M., Sullivan, C. E., and Phillipson, E. A. (1980). Effect of sleep fragmentation on ventilatory and arousal responses of sleeping dogs to respiratory stimuli. *Am. Rev. Respir. Dis.* **122**:899-908.

Car, A., Jean, A., and Roman, C. (1975). A pontine primary relay for ascending projections of the superior laryngeal nerve. *Exp. Brain Res.* **22**:197-210.

Citterio, G., Mortola, J. P., and Agostoni, E. (1985). Reflex effects on breathing of negative pressure and SO₂ in upper airways. *Respir. Physiol.* **62**:203-215.

Cohen, M. I. (1979). Neurogenesis of respiratory rhythm in the mammal. *Physiol. Rev.* **59**:1105-1173.

Cohen, M. I., Piercey, M. F., Gootman, P. M., and Wolotsky, P. (1974). Synaptic connections between medullary inspiratory neurons and phrenic motoneurons as revealed by cross correlation. *Brain Res.* **81**:319-324.

Davies, A. M., Koenig, J. S., and Thach, B. T. (1988). Upper airway chemoreflex responses to saline and water in preterm infants. *J. Appl. Physiol.* **64**:1412-1420.

Donnelly, D. F., and Haddad, G. G. (1986a). Respiratory changes induced by prolonged laryngeal stimulation in awake piglets. *J. Appl. Physiol.* **61**:1018-1024.

Donnelly, D. F., and Haddad, G. G. (1986b). Effect of graded anesthesia on laryngeal-induced central apnea. *Respir. Physiol.* **66**:235-245.

Donnelly, D. F., Sica, A. L., Cohen, M. I., and Zhang, H. (1989a). Effects of contralateral superior laryngeal nerve stimulation on dorsal medullary inspiratory neurons. *Brain Res.* **505**:149-152.

Donnelly, D. F., Sica, A. L., Cohen, M. I., and Zhang, H. (1989b). Dorsal medullary inspiratory neurons: effects of superior laryngeal afferent stimulation. *Brain Res.* 243-252.

Downing, S. E., and Lee, J. C. (1975). Laryngeal chemosensitivity: Possible mechanism for sudden infant death. *Pediatrics* **55**:640-649.

Euler, C. von, Hayward, J. N., and Marttila, I. (1973). Respiratory neuones of the ventrolateral nucleus of the solitary tract of the cat: Vagal input, spinal connections and morphological identification. *Brain Res.* **51**:1-22.

Fagenholz, S. A., Lee, J. C., and Downing, S. E. (1979). Laryngeal reflex apnea in the chemodenervated newborn piglet. *Am. J. Physiol.* **237**:R10-R14.

Fedorko, L., and Merrill, E. G. (1984). Axonal projections from the rostral expiratory neurones of the Botzinger complex to medulla and spinal cord in the cat. *J. Physiol. (London)* **350**:487-496.

Feldman, J. L. (1988). In *Handbook of Physiology.* Bethesda, MD, American Physiological Society.

Fisher, J. T., and Sant 'Ambrogio, G. (1985). Airway and lung receptors and their reflex effects in the newborn. *Pediatr. Pulmon.* **1**:112-126.

French, J. W., Morgan, B. C., and Guntheroth, W. G. (1972). Infant monkeys—A model for crib death. *Am. J. Dis. Child.* **123**:480-484.

Goding, G. S., Richardson, M. A., and Trachy, R. E. (1987). Laryngeal chemoreflex: Anatomic and physiologic study by use of the superior laryngeal nerve in the piglet. *Otolaryngol. Head Neck Surg.* **97**:28-38.

Grogaard, J., Lindstrom, D. P., Stahlman, M. T., Marchal, F., and Sundell, H. (1982). The cardiovascular response to laryngeal water administration in young lambs. *J. Dev. Physiol.* **4**:353-370.

Grogaard, J., Sundell, H., and Stahlman, M. T. (1980). The cardiovascular response to laryngeal chemoreflex stimulation in newborn lambs. *Pediatr. Res.* **14**:445 (Abstract).

Grogaard, J., van den Abbeele, A., and Sundell, H. (1985). Effect of alcohol on apnea reflexes in young lambs. *J. Appl. Physiol.* **59**:420-425.

Haddad, G. G., and Donnelly, D. F. (1990). O_2 deprivation induces a major depolarization in brainstem neurons in the adult but not in the neonatal rat. *J. Physiol. (London)* (in press).

Hamilton, R. B., and Norgren, R. (1984). Central projections gustagory nerves in the rat. *J. Comp. Neurol.* **222**:560-577.

Hanamori, T., and Smith, D. V. (1986). Central projections of the hamster superior laryngeal nerve. *Brain Res. Bull.* **16**:271-279.

Harding, R. P., Johnson, P., Johnson, B. E., McClelland, M. F., and Wilkinson, A. R. (1976). Cardiovascular changes in new-born lambs during apnea induced by stimulation of laryngeal receptors with water. *J. Physiol. (London)* **256**:35P-36P.

Harding, R. P., Johnson, P., and McClelland, M. E. (1978). Liquid-sensitive laryngeal receptors in the developing sheep, cat and monkey. *J. Physiol. (London)* **277**:409-422.

Iscoe, S., Feldman, J. L., and Cohen, M. I. (1979). Properties in inspiratory termination by superior laryngeal and vagal stimulation. *Respir. Physiol.* **36**:353-366.

Jammes, Y., Nail, B., Mei, N., and Grimaud, C. (1987). Laryngeal afferents activated by phenyldiguanide and their response to cold air or helium-oxygen. *Respir. Physiol.* **67**:379-389.

Jodkowski, J. S., and Berger, A. J. (1988). Influences from laryngeal afferents on expiratory bulbospinal neurons and motoneurons. *J. Appl. Physiol.* **64**:1337-1345.

Johnson, P., Dawes, G. S., and Robinson, J. S. (1972). Maintenance of breathing in newborn lambs. *Arch. Dis. Child* **47**:151 (Abstract).

Johnson, P., Salisbury, D., and Storey, A. T. (1975). Apnea induced by stimulation of sensory receptors in the larynx. In *Development of Upper Respiratory Anatomy and Function: Implications for Sudden Infant Death*. Edited by J. F. Bosma and J. Showacre. Washington, DC, U.S. Government Printing Office, pp. 160-178.

Kovar, I., Selstam, U., Catterton, W. Z., Stahlman, M. T., and Sundell, H. W. (1979). Laryngeal chemoreflex in newborn lambs: respiratory and swallowing responses to salt, acids and sugars. *Pediatr. Res.* **13**:1144-1149.

Lawson, E. E. (1981). Prolonged central respiratory inhibition following reflex-induced apnea. *J. Appl. Physiol.* **50**:874-879.

Lawson, E. E. (1982). Recovery from central apnea: effect of stimulus duration and end-tidal CO_2 partial pressure. *J. Appl. Physiol.* **53**:105-109.

Lee, J. C., and Downing, S. E. (1980). Laryngeal reflex inhibition of breathing in piglets: Influences of anemia and catecholamine depletion. *Am. J. Physiol.* **239**:R25-R30.

Lee, J. C., Stoll, B., and Downing, S. E. (1977). Properties of the laryngeal chemoreflex in neonatal piglets. *Am. J. Physiol.* **233**:R30-R36.

Long, W. A., and Lawson, E. E. (1981). Maturation of the superior laryngeal nerve inhibitory effect. *Am. Rev. Respir. Dis.* **123**:166.

Lucier, G. E., Daynes, J., and Sessle, B. J. (1978). Laryngeal reflex regulation: Peripheral and central neural analyses. *Exp. Neurol.* **62**:200-213.

Lucier, G. E., Egizii, R., and Dostrovsky, J. O. (1986). Projections of the internal branch of the superior laryngeal nerve of the cat. *Brain Res. Bull.* **16**:713-721.

Lucier, G. E., Storey, A. T., and Sessle, B. J. (1979). Effects of upper respiratory tract stimuli on neonatal respiration: Reflex and single neuron analyses in the kitten. *Biol. Neonate* **35**:82-89.

Marchal, F., Corke, B. C., and Sundell, H. (1982). Reflex apnea from laryngeal chemo-stimulation in the sleeping premature newborn lamb. *Pediatr. Res.* **16**:621-627.

Marchal, F., Crance, J. P., and Arnould, P. (1986). Ventilatory and waking responses to laryngeal stimulation in sleeping mature lambs. *Respir. Physiol.* **63**:31-41.

McCrimmon, D. R., Speck, D. F., and Feldman, J. L. (1987). Role of the ventrolateral region of the nucleus of the tractus solitarius in processing respiratory afferent input from vagus and superior laryngeal nerves. *Exp. Brain Res.* **67**:449-459.

Merrill, E. G., Lipski, J. Kubin, L., and Fedorko, L. (1983). Origin of the expiratory inhibition of nucleus tractus solitarius inspiratory neurones. *Brain Res.* **263**:43-50.

Miller, A. J., and Dunmire, C. R. (1976). Characterization of the postnatal development of superior laryngeal nerve fibers in the postnatal kitten. *J. Neurobiol.* **7**:483-494.

Nomura, S., and Mizuno, N. (1983). Central distribution of efferent and afferent components of the cervical branches of the vagus nerve. *Anat. Embryol.* **166**:1-18.

Ogura, J. H., and Lam, R. L. (1953). Anatomical and physiological correlations on stimulating human superior laryngeal nerve. *Laryngoscope* **63**:947-959.

Pantaleo, T., and Corda, M. (1985). Expiration-related neurons in the region of the retrofacial nucleus: Vagal and laryngeal inhibitory influences. *Brain Res.* **359**:343-346.

Perkett, E. A., and Vaughan, R. L. (1982). Evidence for a laryngeal chemo-reflex in some human preterm infants. *Acta Paediatr. Scand.* **71**:969-972.

Phillipson, E. A., Sullivan, C. E., Read, D. J. C., Murphy, E., and Kozar, L. F. (1978). Ventilatory and waking responses to hypoxia in sleeping dogs. *J. Appl. Physiol.* **44**:512-520.

Pickens, D. L., Schefft, G., and Thach, B. T. (1988). Prolonged apnea associated with upper airway protective reflexes in apnea of prematurity. *Am. Rev. Respir. Dis.* **137**:113-118.

Remmers, J. E., Richter, D. W., Ballantyne, D., Bainton, C. R., and Klein, J. P. (1986). Reflex prolongation of stage I of expiration. *Pflugers Arch.* **407**:190-198.

Richter, D. W., Ballantyne, D., and Remmers, J. E. (1986b). How is the respiratory rhythm generated? A model. *News Physiol. Sci.* **1**:109-112.

Richter, D. W., Jordan, D., Ballantyne, D., Meesmann, M., and Spyer, K. M. (1986a). Presynaptic depolarization in myelinated vagal afferent fibres terminating in the nucleus of the tractus solitarius in the cat. *Pflugers Arch* **406**:12-19.

Rudomin, P. (1967). Presynaptic inhibition induced by vagal afferent volleys. *J. Neurophysiol.* **30**:964-981.

Sant 'Ambrogio, G., Mathew, O. P., Sant 'Ambrogio, F. B., and Fisher, J. T. (1985). Laryngeal cold receptors. *Respir. Physiol.* **59**:35-44.

Sessle, B. J. (1973) Presynaptic excitability changes induced in single laryngeal primary afferent fibers. *Brain Res.* **53**:333-342.

Sessle, B. J., and Lucier, G. E. (1983). Functional aspects of the upper respiratory tract and larynx: A review. In *Sudden Infant Death Syndrome.* New York, Academic Press, pp. 501-529.

Sherry, J. H., and Megirian, D. (1974). Spontaneous and reflexly evoked activity in pharyngeal, laryngeal and phrenic motoneurons of cat. *Exp. Neurol.* **42**:17-27.

Storey, A. T. (1968). A functional analysis of sensory units innervating epiglottis and larynx. *Exp. Neurol.* **20**:366-383.

Sullivan, C. E., Kozar, L. F., Murphy, E., and Phillipson, E. (1973). Primary role of respiratory afferents in sustaining breathing rhythm. *J. Appl. Physiol.* **45**:11-17.

Sullivan, C. E., Murphy, E., Kozar, L. F., and Phillipson, E. (1978). Waking and ventilatory responses to laryngeal stimulation in sleeping dogs. *J. Appl. Physiol.* **45**:681-689.

Sutton, D., Taylor, E. M., and Lindeman, R. C. (1978). Prolonged apnea in infant monkeys resulting from stimulation of the superior laryngeal nerve. *Pediatrics* **61**:519-527.

Vibert, J. F., Bertrand, F. Denavit-Saubié, M., and Hugelin, A. (1976). Three dimensional representation of bulbos-pontine respiratory networks architecture from unit density maps. *Brain Res.* **114**:227-244.

Wennergren, G., Hertzberg, T., Milerad, J., Bjure, J., and Lagercrantz, H. (1989). Hypoxia reinforces laryngeal reflex bradycardia in infants. *Acta Paediatr. Scand.* **78**:11-17.

Yoshida, Y., Tanaka, Y., Mitsumasu, T., Hirano, M., and Kanaseki, T. (1986). Peripheral course and intramucosal distribution of the laryngeal sensory nerve fibers of cats. *Brain Res. Bull.* **17**:95-105.

10

How Do Arterial Chemoreceptors Work?
Studies of Cells from the Carotid Body of the Rabbit

MICHAEL R. DUCHEN and **TIM J. BISCOE**

University College London
London, England

I. Introduction

Profound changes in the activity of excitable cells occur when the delivery of oxygen to the tissues of the body is reduced in disease, at high altitude, or during normal physiological activities at sea level. Rapid changes in available oxygen lead to hyperventilation, confusional states, and loss of consciousness. Other physiological modifications of bodily functions follow in the longer term. Prolonged, severe energy deprivation naturally enough leads to the rundown of all energy-dependent processes and ultimately to cell death. There must also be changes in sensitivity to hypoxia at or around the time of birth since we know that the fetus has a low arterial oxygen partial pressure yet is not being driven to hyperventilation. Some of these homeostatic effects are mediated by actions on specific sensory receptors, such as the carotid and aortic bodies, and others probably by less specific actions on excitable tissues.

The carotid and aortic bodies contain sensors that detect changes in arterial oxygen, carbon dioxide, and hydrogen ion concentrations. They are placed at the bifurcation of the carotid artery and around the arch of the aorta, respectively, ideally located to monitor the oxygen tension in blood en route to the brain. A fall in PaO_2 increases the frequency of action potentials propagated in their afferent nerves. The response of the receptor may be modulated by a wide variety of other events, including changes in pH or of $PaCO_2$, by application of chemicals like cyanide, by changes in activity in sympathetic nerves or a centrifugal pathway in the sinus nerve, and by mechanical stimulation. Moreover, some of these modulators of chemoreceptor activity may have multiplicative effects rather than simply adding together. In addition to the increased respiratory effort following excitation of the carotid body, lesser effects on other systems are seen. Much of the evidence for these responses and their mode of operation is to be found in Volume 17 of this series.

A central question is, how does the transduction mechanism work? How can it be, for example, that an increase in sinus nerve chemosensory discharge can be demonstrated as the arterial oxygen partial pressure changes from such a high level as 100 mm Hg to, say, 80 mm Hg partial pressure? At present, there is really no comprehensive description of the transduction process, despite years of research and considerable speculation. Since the 1920s the carotid bodies have been the more extensively studied because they are more easily accessible; the aortic bodies have the advantage that their afferent nerve responses may be recorded in the neck (vagus nerve) without exposing the sensory structures to air or other fluids. There is no reason to suppose that the transduction mechanism in these two groups of receptors differs, but the mechanism remains unresolved.

One of the major problems in the past has been that in nearly all previous studies whole animals, or an in vitro preparation of the intact structure, have been used to address these questions. Attempts to identify the *cellular* events that generate the output of this receptor are then made difficult by the complexity of the response of the whole animal or the whole structure to altered PaO_2. Changes in hormone output, alterations in blood or perfusate flow, alterations in the local concentrations of ions, and accumulation of transmitters and metabolites in the extracellular space, all confound attempts to disentangle the contributions made to the sensory output by the responses to hypoxia of cells within the sensor. While the totality of these changes ultimately determines the behavior of the whole complex, we must understand the events taking place in single cells if we are to comprehend the performance and function of the whole. Our objective, then, is to understand how transduction is effected in terms of the basic cell biology of the constituent cells. This requires, first, a definition of the relevant properties of the cells that may be involved and then a synthesis of those properties to allow description of the functioning of the whole complex.

II. Cellular Elements and Identity of the Transducer

Those cells believed to be primarily concerned with chemotransduction are the type I cells, the type II cells, and the associated nerve endings. The dominant nerve endings are those seen most clearly under the electron microscope juxtaposed to the type I cells. Their axons, in the case of the carotid body, run in the sinus nerve, a tributary of the glossopharyngeal nerve. There are also nerve endings associated with smooth muscle, and there may well be others (see, for example, Biscoe and Stehbens, 1966; McDonald, 1981; Pallot, 1987).

There are essentially three different hypotheses regarding the nature of the transducer.

First, the most widely accepted, and the textbook account, is that the type I cells are the site of transduction, an interpretation based on the work of De Castro (1926, 1928). In principle, this account conforms to the traditional opinion that receptor structures consist of a receptor cell, the transducer, which is associated with a sensory nerve ending (Davis, 1961). In recent years this view has received consistent support from Eyzaguirre and his group. Furthermore, transmission from the receptor cell to the nerve ending is, in this view, mediated by a chemical transmitter. The mechanism coupling an increased transmitter release and hypoxia is unknown.

One proposed mechanism to account for the sensitivity of the structure to PO_2, suggests that within the carotid body there is, in addition to the usual high-affinity mitochondrial cytochrome oxidase (aa3), a low-affinity cytochrome *aa3* (Mills and Jöbis, 1970, 1972). Their evidence was derived from spectrophotometric and fluorometric measurements on intact carotid bodies perfused with hemoglobin-free solutions. They found a compound having light absorption characteristics appropriate to aa3 but falling into two types, the high-affinity and the low-affinity forms. Mills and Joöbis concluded that the low-affinity cytochrome is involved with oxygen sensing and the high-affinity compound with all other aspects of cell economy, such as preservation of nerve function.

The notion that two cytochromes having different affinities are present conforms with the subsequent two-cytochrome model discussed by Nair et al. (1986). Their interpretation is that both cytochromes are necessary for normal chemoreceptor function, the low-affinity aa3 being important above 20 mm Hg PO_2 and the high-affinity aa3 at lower values of PO_2. They therefore differ somewhat from Mills and Jöbis in the attribution of significance for chemoreceptor function of the two cytochromes.

Spectrophotometric studies have also been published by Acker et al. (1985) and Acker and Eyzaguirre (1987), but it is not altogether clear that these authors excluded hemoglobin spectra from their records since their control, which utilized red blood cells rather than hemoglobin alone, may have been inappropriate. Keilin and Hartree (1941) showed that hemoglobin absorption at different wavelengths may be severely suppressed when hemoglobin in red blood cells rather than hemoglobin alone is examined. They explained the phenomenon as due to the properties of the surfaces separating hemoglobin from the surrounding medium.

The two-cytochrome model can be seen as supplying a mechanism for the first hypothesis. But the steps in the argument between detection of PO_2 and signal initiation in the sinus nerve are missing. Very recently, several attempts to account for these events have been published, and these will be discussed more fully below.

The other hypotheses receive far less support in the literature. The second is the small-nerve-ending hypothesis. This states that the transducers are very small nerve endings which respond to changes in PaO_2 because of the oxygen dependence of the sodium pump and because they have a very high surface-area-to-volume ratio. Therefore, they are more than usually dependent on the sodium pump to sustain their membrane potential. The argument requires that ATP should fall, or ADP rise, leading to inhibition of the pump and depolarization (i.e., excitation) of the terminals. The role of the type I cell, in this hypothesis, lies with an efferent control system where a transmitter is released and suppresses chemoreceptor activity (Biscoe, 1971).

The third hypothesis is that in the carotid body one cell type is sensitive to changes in the local PaO_2, the other responsible for the high rate of oxygen consumption. The sensor cell then undergoes a change in shape which, when transmitted to the nerve ending, produces a generator potential. In short, the sensory nerve ending is acting as a mechanoreceptor (Paintal, 1967, 1988). There is much to recommend this view since mechanoreception is widespread in sensory systems in nature, and essentially the hypothesis asserts that within the chemoreceptor bodies this general property is utilized for an unexpected purpose.

III. The Evidence

The present weight of evidence and opinion supports the first hypothesis. However, as with all of the others, there is no *direct* evidence in favor of it since the great majority of experiments to date have been upon the *whole* carotid body or aortic bodies.

These experiments are essentially of three types: first, recording from the afferent nerve in vivo or in vitro and examining the responses to different stimuli; second, superfusing the structure and collecting the effluent in order to see what substances are released under different conditions; third, operating on the structure or the nerve supply and examining the consequences of interference for the structure or the function. In all cases such experiments, in the nature of things, can only provide indirect evidence about transduction and the receptor since the whole carotid body complex is exposed to and may be affected by the intervention. It follows that the recorded variable may be modified by changes in cells other than the transducer because the whole carotid body is studied and not a discrete subset of the whole. Certainly these approaches do not lead to investigations uniquely confined to the presumed transducer cell. All conclusions about transduction are, accordingly, drawn by inference.

Perhaps the most compelling among these is the apparent congruence between the ability of agents to increase afferent activity in the sinus nerve and to increase the release of catecholamines (most notably, dopamine) from the structure (Fidone et al., 1982; Obeso et al., 1987). This suggests that any account of the transduction process should provide a mechanism to allow for this increased transmitter release, even though the role of dopamine remains embarrassingly confused (see below).

In recent years the refinement of the method of cell culture has been introduced. It has been shown that type I cells can be grown in groups and that they synthesize catecholamines (Pietruschka, 1974; Fishman and Schaffner, 1984; Nurse, 1986, 1987) and that these amines may be released by hypoxia (Fishman et al., 1985).

Still more recently, electrical recording methods have been applied to single cells either acutely dissociated or cultured for a short period. Such experiments have been carried out by the present authors and their co-workers; by Stea and Nurse (1988); by Hescheler et al. (1989) and Delpiano and Hescheler (1989); and by López-Barneo et al. (1988), Ureña et al. (1989), and López-López et al. (1989).

IV. Single-Cell Studies

We believe that major difficulties in identifying the transducer and solving the problem of transduction have arisen from the three-dimensional complexity of these bodies coupled with their small size, which has prevented *direct* analysis of the cellular mechanisms involved. We therefore set out to develop a preparation that would simplify the problems and in which we have control over the extracellular environment and access to the cell cytoplasm. To this end, we have employed a combination of enzymatic and mechanical dispersion procedures to obtain isolated cells from the adult rabbit carotid body. These cells have then been studied using structural meth-

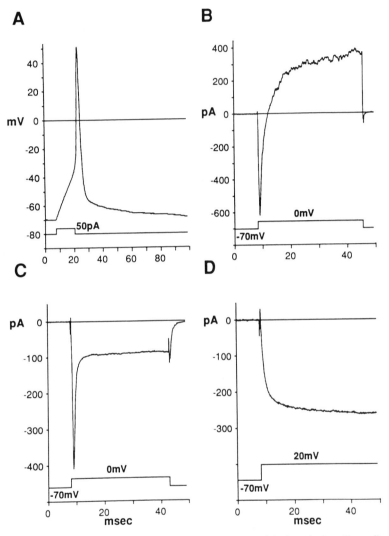

Figure 1 Responses of type I cells during recording with the whole cell configuration of the patch clamp technique. (A) Under current clamp showing the response to a depolarizing current step of 50 pA at the threshold for an action potential. The action potential overshot the membrane potential by 50 mV. (B) Under voltage clamp. The potential was clamped at -70 mV. A depolarizing step to 0 mV evoked a fast inactivating inward current (downward on the figure), followed by a delayed outward current that tended toward a plateau after about 20 msec. (C) Under voltage clamp. In this cell potassium currents were blocked by replacement of intracellular potassium with a combination of 105 mM cesium chloride and 25 mM tetraethylammonium chloride (TEA). A step depolarization from -70 mV to 0 mV evoked a fast inward current, which inactivated to a sustained inward current that remained fairly con-

ods, the whole cell configuration of the patch clamp technique (Hamill et al., 1981), and ion-sensitive fluorophores. Our results will now be discussed and compared with those of other workers. We will describe the recognition of cell types and their biophysical properties, leading to an account of their responses to cyanide, and of changes in intracellular calcium concentration caused by cyanide and by hypoxia.

We have used cyanide because it mimics the action of hypoxia, the natural stimulus to the carotid chemoreceptor, stimulating the afferent nerve output in vivo (von Euler and Liljestrand, 1937) and in vitro (Eyzaguirre and Koyano, 1965). Also, as with hypoxia, cyanide increases calcium-dependent catecholamine release from the whole carotid body (Obeso et al., 1987) and from type I cells grown in culture (Fishman et al., 1985). Cyanide was used first because it provides a maximum stimulus and is easier to control than PO_2.

Detailed accounts of these experiments have been published elsewhere (Biscoe et al., 1987 a,b, 1989, 1990; Duchen et al., 1988).

V. Intracellular Recordings from Type I Cells

On the basis of previous intracellular recordings of type I cells (Eyzaguirre and Zapata, 1984; Eyzaguirre et al., 1983), they were thought to have passive membrane properties. We studied the enzymatically dissociated rabbit carotid body and identified two populations of cells on the basis of appearance and size as seen with phase contrast and electron microscopy, and also according to the electrophysiological findings (Duchen et al., 1988). Under phase contrast one population consisted of spherical or ovoid cells 12-15 μm in diameter and another of spherical cells 8-10 μm in diameter. Formaldehyde fluorescence for catecholamines was demonstrated in the larger cells. Under the electron microscope two populations could again be distinguished. One population had cells 10-12 μm in diameter, found in clus-

---

stant throughout the 50-msec pulse. There is a small inward tail current on stepping back to -70 mV as the sustained current deactivates. These two currents are carried by sodium ions and calcium ions, respectively. (D) Under voltage clamp. This shows the characteristic form of the sustained inward current. The patch pipette contained 5 mM ATP to slow the decline of the current due to washout of essential phosphorylating compounds as the cytoplasm equilibrates with the electrode contents. The bathing solution contained 1 μM tetrodotoxin (TTX), which blocks the fast inactivating sodium current shown in B and C, and the pipette contained cesium with TEA to block potassium currents, as in C. The cell potential was clamped at -70 mV, and a 50-msec depolarizing step to +20 mV (the potential at which the current is maxiamally activated) was applied. This current is blocked by cobalt or cadmium and has the characteristics of the high-threshold calcium current (see text).

ters and possessing electron-dense cored vesicles characteristic of type I cells; another population was made up of smaller cells 5-7μm in diameter with peripherally condensed nuclear chromatin and fine cytoplasmic extensions typical of type II cells.

Two distinct populations were also distinguished in electrophysiological studies. The larger, type I, cells were found to be excitable, generating fast, sodium-dependent action currents or potentials (Fig. 1A) that were recorded in the cell-attached or whole cell recording configurations, respectively (Biscoe et al., 1987 a,b; Duchen et al., 1988). The smaller, type II, cells did not generate action potentials.

The currents underlying these events were investigated using voltage clamp techniques. These experiments showed that type I cells had an inactivating, tetrodotoxin-sensitive inward sodium current, a high-threshold sustained inward calcium current, and outward potassium currents (Fig. 1B-D). The last of these were blocked by tetraethylammonium and cesium. A component of the outward potassium current showed a dependence on voltage-gated calcium entry and was blocked by cobalt or cadmium (Fig. 2) (see Meech and Strumwasser, 1970; Lux et al., 1981; Barrett et al., 1982). A further component of the calcium-dependent potassium current was sensitive to apamin, suggesting that there may be two classes of calcium-dependent potassium channels expressed by the cells (Biscoe et al., 1987 a,b; Duchen et al., 1988). Type II cells showed only a high-threshold outward potassium current, with no evidence of calcium dependence.

Thus, the type I cell is essentially neuron like in its properties while the type II cell resembles glial elements elsewhere in the nervous system. These electrophysiological results have, in part, since been confirmed by Hescheler et al. (1989) and by López-Barneo et al. (1988) and Ureña et al. (1989) except that these authors did not attach significance to the calcium-activated potassium current.

A. The Action of Cyanide

We found that type I cells were hyperpolarized by 2 mM cyanide (Fig. 3) and the action potential was shortened (Biscoe et al., 1988c; Biscoe and Duchen, 1989a). This result was surprising at first because we know that cyanide increases transmitter output from the structure and excites the chemoreceptors.

The ionic basis for the hyperpolarization was determined by voltage clamp studies which showed that the principal effect of cyanide was to increase an outward current (Fig. 4). The current reversed at the potassium equilibrium potential and was thus an increased potassium current. This action of cyanide was shown to be calcium-dependent. The result suggests that intracellular calcium might be raised by cyanide, thus increasing the proba-

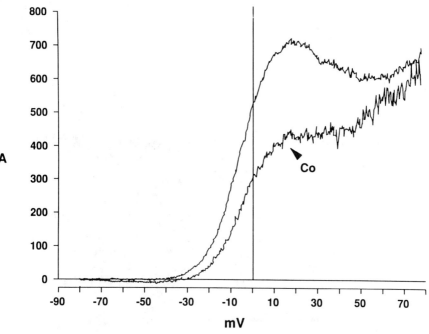

Figure 2 Recording from a type I cell under voltage-clamp control. A slow depolarizing ramp was applied, allowing us to sample the current that develops over a range of potentials from −80 to +80 mV. The current is then plotted against the applied voltage, illustrating the voltage dependence of the membrane currents. At about −30 mV, an outward current is activated that grows in amplitude with increasing depolarization. The current peaks at about +20 mV and then declines slightly toward +60 mV. If the response represented simply the activation of potassium channels, one would expect to see a monotonic increase in the current at potentials positive to those at which the channels are maximally activated, as the potential is moved progressively further away from the potassium equilibrium potential, thus increasing the driving force for the current. Application of 2 mM cobalt to the cell suppressed part of the outward current between about −20 to +60 mV, removing the hump on the i/v curve. Cobalt blocks voltage-dependent calcium entry into the cell and has no direct effect on potassium channels. This therefore shows that the hump on the i/v curve reflects the activation of calcium-dependent potassium channels following the entry of calcium into the cell through voltage-gated calcium channels.

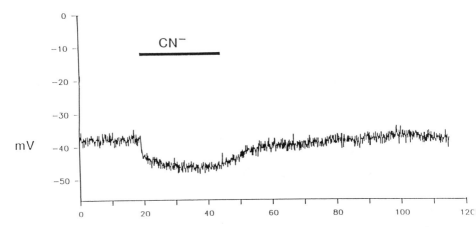

Figure 3 Recordings of membrane potential from a type I cell under current clamp control. Application of 2 mM sodium cyanide hyperpolarized the cell by a few mV at its resting potential of -58 mV, and by even more at -38 mV, as shown. As the cell was further depolarized by current injection, the response became larger.

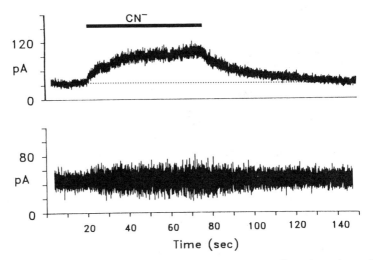

Figure 4 Recording of membrane current from a type I cell, under voltage-clamp control. The cell potential was clamped at $+40$ mV. Application of 2 mM sodium cyanide evoked an outward current (upper trace) that decayed slowly after the cyanide was removed. In the lower trace the same current recording is shown after being high-pass-filtered and amplified. The increase in membrane current fluctuations associated with the increased outward current is then easier to see.

bility of opening Ca^{2+}-dependent K^+ channels. The rise could be due to an increase in the calcium current described above (Fig. 1C) but this current was in fact found to be slightly attenuated by cyanide. This attenuation could incidentally be due to a calcium-dependent block of the calcium channel following the posited rise in intracellular calcium (Chad et al., 1984) and perhaps also to a decline in the levels of available ATP since the calcium channel requires rephosphorylation to preserve function (Chad and Eckert, 1986).

In any event, these results suggest that the increase in $gK_{(Ca)}$ caused by calcium must, most likely, arise from an alteration in cellular calcium homeostasis, and not from an increased entry of calcium through voltage-gated Ca^{2+} channels.

The increased outward current was associated with an increase in current fluctuations or noise (Fig. 4, lower trace), suggesting an increased probability of opening ion channels. Analysis of the noise showed that the K^+ channel opened by the action of cyanide had a very low conductance, only about 3 pS. A role for ATP-sensitive potassium channels (Noma, 1983) activated by a fall in intracellular ATP could be discounted since the addition of ATP to the patch pipette filling solutions did not modify the response to cyanide, and the conductance of the channels in type I cells (3 pS) (Biscoe and Duchen, 1989a) is much lower than that for ATP-sensitive channels (50-80 pS).

VI. Measurements of Intracellular Calcium

Since the modulation of currents by cyanide indicated that there must be an underlying rise in $[Ca^{2+}]_i$ (Biscoe and Duchen, 1989a), we have used the calcium-sensitive fluorophores fura-2 and indo-1 (Tsien, 1980; Grynkiewicz et al., 1985) to measure and study the regulation of $[Ca^{2+}]_i$ in single type I cells. In our earlier experiments we measured $[Ca^{2+}]_i$ using fura-2 (Biscoe et al., 1988a, 1989); more recently, we have used indo-1 (Biscoe and Duchen, 1990).

A. Response to Cyanide

Cyanide reversibly increased $[Ca^{2+}]_i$ from a mean value of about 90 nM to about 220 nM. The response was still seen when examined soon after removal of external calcium although it declined progressively until it was not recordable. This suggests that cyanide releases calcium from an intracellular site which can exchange fairly rapidly with the cytoplasm and so with the extracellular space.

Prolonged application of cyanide caused a sustained increase in $[Ca^{2+}]_i$, suggesting that cyanide impairs the removal or sequestration of calcium from the cytoplasm. The ability of the cell to handle a calcium load (e.g., in response to depolarization by raised extracellular K^+) was dramatically altered by metabolic blockade with cyanide. Further examination of calcium homeostasis showed that the clearance of a calcium load was slowed either by removal of $[Na^+]_o$ or by application of cyanide. Thus both Na-Ca exchange (Niedergerke, 1963; Baker et al., 1967; Requena and Mullins, 1979; Gill et al., 1981; Nicholls and Kerman, 1981) and a cyanide-sensitive process contribute to the regulation of $[Ca^{2+}]_i$.

One potential cause of a rise in $[Ca^{2+}]_i$ is an intracellular acidification, which might be expected to result from metabolic inhibition. Protons are known to displace calcium from cytoplasmic binding sites, and also to increase the release of calcium from intracellular organelles (e.g., see Mullins and Requena, 1987). We therefore used the pH-sensitive dye BCECF to measure pH_i. The viability of the technique was confirmed by demonstrating predictable changes in pH_i in response to ammonium chloride or CO_2 (Boron and de Weer, 1976). No consistent change in pH_i was seen in response to cyanide, confirming that the rise in $[Ca^{2+}]_i$ is not secondary to a change in pH_i.

Finally, carbachol releases calcium from intracellular stores in other cells (see Trautmann and Marty, 1984). In our experiments the drug hyperpolarized type I cells increasing a potassium conductance (Biscoe and Duchen, 1989a) and caused a small transient rise of $[Ca^{2+}]_i$ consistent with release of calcium from an exhaustible intracellular pool (Biscoe et al., 1989). The response to cyanide persisted in the presence of carbachol, suggesting that the two stimuli operate by distinct, though possibly overlapping, mechanisms.

B. The Effects of Hypoxia

Hypoxia is the physiological stimulus to the chemoreceptors. Therefore, we tested the effects of hypoxia on intracellular calcium concentration using the calcium fluorescent indicator indo-1 (Grynkiewicz et al., 1985) and compared the responses with those to cyanide. As was usual in our experiments, the superfusate for the cell preparations was normally equilibrated with room air, that is, at a PO_2 of about 150 mm Hg. The hypoxic stimulus was the same superfusate equilibrated with nitrogen (Biscoe and Duchen, 1989b, 1990). When precautions are taken to deliver the fluid to the bath through stainless steel tubes, it is possible to lower the perfusate oxygen partial pressure to around 15-20 mm Hg with the bath open to the air. A shroud over the chamber inflated with nitrogen makes very little difference. Lower partial pressures may be obtained by adding sodium dithionite to the superfusate, and at a concentration of 500-750 μM the pressure of oxygen is reduced to zero.

Higher partial pressures may easily be obtained by varying the intensity of the bubbling of the gas. The oxygen partial pressure in the bath achieved close to the cells was recorded continuously with an oxygen electrode.

Depolarization of cells by brief (5-20 s) exposure to solutions containing 30 mM potassium chloride (substituted for sodium chloride) caused a large, cobalt-sensitive rise in $[Ca^{2+}]_i$ of the same magnitude as in our earlier experiments with fura-2, having a fast onset and recovery. This response results from activation of the calcium current, and is used as a standard reference for other effects and as a test of the viability of the cells.

Both cyanide and exposure to zero oxygen partial pressure raised $[Ca^{2+}]_i$ reversibly and by a similar amount. In some cells at room temperature a small increase in $[Ca^{2+}]_i$ could also be seen at a P_{O_2} of 15-20 mm Hg. On warming the perfusate (temperature monitored with a thermocouple in the

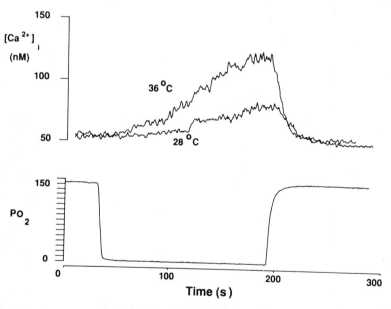

Figure 5 Recordings of fluorescence from type I cells loaded with indo-1. The fluorescence is excited at 365 nm and emission recorded at 400 and 500 nm. The ratio of fluorescence measured at 400/500 nm is shown expressed in terms of the intracellular calcium conentration. In response to a fall in P_{O_2}, monitored with an O_2 electrode in the bath and shown below, $[Ca^{2+}]_i$ rose at room temperature, which was 28 °C on an English summer day. Raising the temperature to 35 °C dramatically and reversibly increased the rate and amplitude of the response. The response to depolarization with 30 mM K^+ was largely unaffected by this change in temperature.

bath), from room temperature at around 25 °C to 35-37 °C, the responses to zero oxygen and to cyanide had a faster rise time, were quicker in offset, and were always larger (Fig. 5) at the higher temperature. Typically they were of a similar size to the response to potassium-induced depolarization. In addition, the preparation became more *sensitive* to hypoxia, thus responding more readily to changes in PO_2 above zero. This allows the construction of a stimulus-response curve (Fig. 6), in which intracellular Ca^{2+} is seen to increase monotonically with decreasing PO_2, resembling the frequency/PO_2 curves obtained from single units in the sinus nerve (Biscoe et al., 1970b). The response to potassium-induced depolarization was essentially unaffected by these temperature changes, confirming that the increase in the response with temperature is not simply the result of increased excitability of the cells.

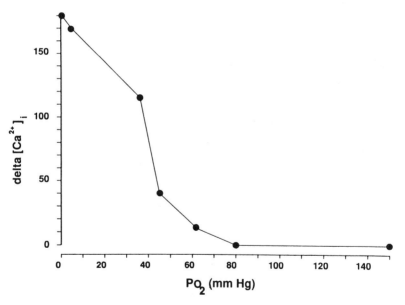

Figure 6   When recordings of indo-1 fluorescence were made at 35-37 °C, responses to a hypoxic stimulus were much increased, as shown in FIgure 5. By varying the intensity with which the perfusate was bubbled with N_2, and by mixing with air-equilibrated solutions, superfusates with a range of PO_2 values could be obtained. An increase in $[Ca^+]_i$ could now be detected following a fall in PO_2 from 150 to 58 mm Hg. The responses to solutions that were more severely hypoxic were clearly larger, following a stimulus/response relation similar to that seen with single chemoreceptor afferent fibers.

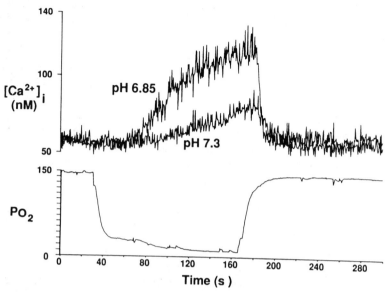

Figure 7   Recordings of indo-1 fluorescence from type I cells, as described in Figure 5. At 35 °C, $[Ca^{2+}]_i$ rose in response to exposure to a solution with a PO_2 of about 15 mm Hg. The control bathing solution had a pH of 7.3. Lowering the pH to 6.85, simply by altering the Hepes buffering of the superfusate, clearly increased the response to the same PO_2. The response to potassium-induced depolarization was slightly reduced at pH 6.85.

These results lend weight to the argument that the mechanism coupling the detection of oxygen partial pressure to the rise in $[Ca^{2+}]_i$ results from a metabolic mechanism (Smith et al., 1985), rather than from some alteration in membrane excitability. This is supported even further by the demonstration that an acid environment increases the response to hypoxia (Fig. 7), while *reducing* the response to potassium-induced depolarization, consistent with the well-established sensitizing effect of an acid environment on the responses of the carotid body in vivo.

The responses at 35-37 °C were then qualitatively much as described at room temperature, but were larger and faster. Thus, they were only slightly reduced by using a nominally calcium-free medium, that is, reducing $[Ca^{2+}]_o$ to μM levels, and declined slowly with time when EGTA was used to reduce $[Ca^{2+}]_o$ to submicromolar concentrations. Following restoration of $[Ca^{2+}]_o$ to normal, the $[Ca^{2+}]_i$ rose above control, to decline over several minutes back to control values. This change probably represents a process similar to that seen in cardiac myocytes, where it has been called the calcium paradox. This is due to an influx of Na^+ into the cell through calcium channels in the absence of external calcium. When extracellular calcium is restored, Na/ Ca exchange reverses, removing the acquired Na^+, but loading the cell with calcium. If an hypoxic test was given at this time during a period of stimulated calcium influx, and presumably when the intracellular stores had been loaded, the rise in $[Ca^{2+}]_i$ evoked by hypoxia was larger than before.

Tetrodotoxin (TTX) blocks the fast sodium inward current during the action potential in type I cells (Biscoe et al., 1987b; Duchen et al., 1988). We have found that TTX does not affect the hypoxia-induced rise in intracellular calcium, which therefore can occur quite independently of the occurrence of action potentials in the cell.

C. Conclusions from the Experiments with Hypoxia and Cyanide

Our interpretation of these data is that the actions of cyanide and of hypoxia are essentially on the processes that regulate $(Ca^{2+}]_i$ in type I cells. We would argue that the electrophysiological responses to these stimuli reflect, and do not initiate, the elevations of $[Ca^{2+}]_i$ that we have documented and that presumably underlie the increased transmitter release described by others. We still do not know precisely how $[Ca^{2+}]_i$ is mobilized in response to these stimuli, but we can offer speculations based on our data.

One suggestion is that the principal effect lies with the handling of calcium by mitochondria (as conjectured by Roumy and Leitner, 1977). Calcium is continuously cycled by most mitochondria studied (Crompton et al., 1978; Carafoli, 1979; Crompton, 1985). It is taken up into the organelles down a considerable electrochemical potential gradient largely resulting from the very negative mitochondrial membrane potential (Rottenberg and Scarpa, 1974). This in turn is maintained by the electron transport chain (Carafoli and Lehninger, 1971). Calcium is also continuously extruded from mitochondria by a Na/Ca exchange (in some cells), by proton/Ca exchange, and probably by other, less well-defined processes. If cyanide and hypoxia reduce the mitochondrial membrane potential sufficiently, the reduced rate of calcium uptake may be sufficient to raise $[Ca^{2+}]_i$ as demonstrated here. The increase in the response during periods of increased calcium influx following stimulation of Na/Ca exchange, the sensitivity to an acid pH, and the striking temperature sensitivity of the response are all consistent with this hypothesis.

Of course, it remains possible that some other, quite distinct mechanisms may be involved in addition to those operating in the mitochondria. The mitochondria may be less important as buffers for cytosolic calcium than was hitherto thought likely. Furthermore, their calcium store may be too small to account for the changes that we see in $[Ca^{2+}]_i$ and a more substantial reservoir may be sought. The obvious other candidate for a role as a cytosolic calcium buffer would be the endoplasmic reticulum and a calcium activated E_1-E_2 ATPase forming an integral part of a calcium pump (see review by Carafoli, 1987). Calcium release from nonmitochondrial stores is controlled at least in part by the phosphoinositides (Streb et al., 1983; Berridge and Irvine, 1984; Sekar and Hokin, 1986), which may play some as yet unspecified role in the responses described here.

Appropriate intervention with either of these intracellular pathways, mitochondrial and nonmitochondrial, could lead to a rise in cytosolic calcium which could follow a decline in the available ATP. Such a decline has been known to be possible since the work of Caldwell (1960) where he showed that in the squid axon 2mM NaCN could lead to a fall in ATP, and arginine phosphate, to very low levels.

We have not so far considered the effects of carbon dioxide on the carotid body. It is known to excite chemoreceptors and to enhance the actions of hypoxia. So far we have shown only a rise in intracellular hydrogen ion concentration on exposure to CO_2 (Biscoe et al., 1989) and have yet to examine the effects on intracellular calcium or the membrane currents. A change in intracellular pH could cause depolarization of the cell membrane leading to activation of the voltage-dependent calcium currents and a rise in $[Ca^{2+}]_i$, as suggested by Kaila et al. (1989) to account for the actions of carbonylcyanide-m-chlorophenyl-hydrazone (CCCP) in crayfish muscle. We have not so far seen changes in $[Ca^{2+}]_i$ on acidification of the perfusate with the Hepes buffer as described above, rather changes only in sensitivity to hypoxia.

Thus, the source of the rise in calcium is undecided though investigations promise to bear fruit.

If the type I cell is the receptor, then these observations suggest that changing regulation of $[Ca^{2+}]_i$ is the primary cellular mechanism underlying chemotransduction and imply that the excitability of the cells and their expression of voltage-gated Na^+ and Ca^{2+} channels (Biscoe et al., 1987 a,b; Duchen et al., 1988) provides a mechanism to vary the calcium loading of the cells. A fall in PO_2 would reduce the rate of removal of calcium from the cytoplasm, increasing transmitter release more if the cells are generating action potentials than if they are electrically silent.

VII. Other Recent Investigations

In their recent studies on carotid body cells cultured for short periods, López-Barneo et al. (1988) and Ureña et al. (1989) have found currents essentially similar to those which we have described in type I and type II cells except that they find only "a small percentage of calcium-activated (potassium) channels" in type I cells.

However, López-Barneo et al. (1988) and López-López et al. (1989) found that a potassium current was *reduced* by hypoxia. The effect was, they say, independent of the internal calcium and was graded in the range of PO_2 between 80 and 150 mm Hg. Between 10 and 70 mm Hg the relationship is curiously complex and, as the authors remark, is difficult to explain, especially since the same change in current would occur at three different values for the PO_2. This would render any detection system meaningless. Nevertheless, the authors suggested that the cells may possess a specific O_2-sensitive K^+ channel.

Before making their recordings they waited for currents to "stabilize." This may have included rundown of the calcium currents accounting for their failure to see more than a "small percentage" of $gK_{(Ca)}$ and leaving the dominant gK to be the voltage-gated conductance, gK_V. In a clonal pituitary cell line, the response to TRH that leads to an increase in $[Ca^{2+}]_i$ (Ritchie, 1987), and thus promotes $gK_{(Ca)}$ in some cells, also reduces gK_V in others, depending on which conductance is dominant in each individual cell (Dubinsky and Oxford, 1985). We found that in calcium-free media, cyanide reduced the remaining K^+ current. It is also uncertain how culture media might affect the expression of different classes of K^+ channels, perhaps reducing the density of calcium-dependent potassium channels in favor of voltage-gated channels. A further possibility is that the electrophysiological responses to cyanide and hypoxia may differ, although they do not with respect to changes in $[Ca^{2+}]_i$, as we have shown.

One other relevant point is that according to Whalen et al. (1973), the distribution of PO_2 values in the carotid body, as opposed to the arterial blood, lies between 0 and 109, with an average of 68 mm Hg in the outer 200 μm and 42 mm Hg deeper within the structure. Other estimates were: an average of 65 mm Hg (Whalen and Nair, 1975), a range of 10-104 mm Hg, mean 72 mm Hg (Whalen and Nair, 1976), and a mean level of 74.5 mm Hg (Nair et al., 1986). Acker et al. (1971) found a range of 25-90 mm Hg, with nearly 80% of the values below 40 mm Hg, apart from their low PO_2 outer zone not seen by Whalen and Nair. These measurements are all low on the curve of López-López et al. (1989) relating potassium current and PO_2, and within the zone that was difficult to explain.

At the 1988 IXth International Symposium on Arterial Chemoreceptors, Acker and his co-workers (Delpiano, Hescheler, and Pietruschka) showed a long-lasting Ca^{2+} current, a $gK_{(Ca)}$, gK_V, no gNa, and confirmed that cyanide reduces the Ca^{2+} current. These results have recently been published at length by Hescheler et al. (1989), and they describe what they conclude is a tonically active potassium current that was reduced by hypoxia. A fuller description of this current and of the channels involved is given by Delpiano and Hescheler (1989) in their most recent publication.

They show that reduction of the superfusate PO_2 caused a hyperpolarization during the first minute or so, followed by depolarization from a resting membrane potential of -40 mV to zero. This depolarization was followed by recovery to the resting level 1-2 min after return of the PO_2 to control. This result is similar to that of Fujiwara et al. (1987) in rat hippocampal neurons, who suggested that the delayed depolarization followed failure of the electrogenic Na/K pump. Delpiano and Hescheler (1989) go on to show that in the cell-attached mode with a symmetrical potassium concentration, they record the activity of a large conductance channel, which opens with decreased probability during the hypoxic stimulus. They propose that this channel is a K channel which is normally active at rest, and which closes during hypoxia, resulting in the observed depolarization. They do not directly demonstrate that the depolarization is accompanied by a decreased potassium conductance, and they do not appear to consider the initial hyperpolarization as significant. The precise details of the recording conditions are not clear from their publication.

In our experience using hypoxia as a stimulus we have so far been unable to replicate these results at room temperature even though we have lowered the ambient PO_2 to 20-25 mm Hg. This may be because we have not held the PO_2 low enough for long enough and because the temperature was too low in those experiments. We intend a further series of experiments on the effects of hypoxia on the biophysical properties of the cells.

It is also clear that the conditions in culture must be adjusted to ensure preservation of all the voltage-activated currents we have described.

To comprehend the behavior of the whole carotid body chemoreceptor complex it will be necessary to understand the direct actions of hypoxia, cyanide, pH, carbachol, and so forth on the nerve terminals as well as on type I and type II cells. At present the only possible approach to that problem is to study petrosal ganglion cells or possibly dorsal root ganglion cells. We have carried out some experiments on these lines, as indicated above and discussed further below.

VIII. The Centrifugal Pathway

Any model for the receptor must account for the evidence for a centrifugal, efferent pathway in the sinus nerve whose activation affects the carotid chemoreceptor (Biscoe and Sampson, 1968; Neil and O'Regan, 1971a,b; Sampson and Biscoe, 1970).

One explanation for this modification of chemoreceptor activity would be that the centrifugal pathway could alter the sensitivity of the system by changing the excitability of the type I cells. For example, by changing the rate of discharge of action potentials, the net calcium influx is altered. That suggestion requires centrifugal pathway activation to suppress type I cell excitability since the pathway suppresses chemoreceptor discharge and hence should suppress type I cell transmitter release (first hypothesis of transduction). Further, this view would require one of three alternatives: that there are two varieties of nerve ending on type I cells, one sensory and one motor; or that the sensory nerve ending lies elsewhere in the structure and that the nerve endings on the type I cell are efferent (Biscoe et al., 1970a; Biscoe, 1971); or that this mode of centrifugal action is wrong, that the type I cell is indeed the site of transduction, and that the efferent terminals are elsewhere producing their effects by action at a distance, perhaps by modifying blood flow. (Goodman, 1973; McCloskey, 1975; O'Regan and Majcherczyk, 1982; Belmonte and Eyzaguirre, 1974; and see Eyzaguirre and Zapata, 1984, for a full discussion of this problem).

If type I cells are the site of transduction but not of efferent effects, then the question remains as to where the efferents act. Equally, if type I cells are not transducers, they clearly have a crucial role (cf. Monti-Bloch et al., 1983) and the present analysis remains important because their behavior would be expected to modify the performance of the whole receptor complex.

IX. The Transmitters

Type I cells contain a variety of putative transmitters, especially varied according to a number of reports at the 1988 IXth International Symposium on Arterial Chemoreceptors, the most notable of which is dopamine (Eyzaguirre and Zapata, 1984; Fidone et al., 1982; Nurse, 1987; Fishman et al., 1985). These substances are released by hypoxia, cyanide, or depolarization, and the rise in $[Ca^{2+}]_i$ that we describe here would be expected to cause the release of transmitters.

If these events underlie transduction, the transmitter must excite the terminals of petrosal ganglion neurons making synaptic contact with type I cells. Much of the available evidence suggests that dopamine, applied to the whole structure in vivo, depresses chemoreceptor activity and does not excite, though the literature referring to the actions of dopamine and dopamine antagonists on the carotid body is often contradictory (Jacobs and Comroe, 1968; Sampson, 1972; Black et al., 1972; Sampson and Vidruk, 1977; Folgering et al., 1982; McQeen, 1983; Zapata and Llados, 1977; Mills et al., 1978; and see Eyzaguirre and Zapata, 1984, and Pallot, 1987).

Experiments will have to be performed to determine whether dopamine can excite nerve terminals, though this at first sight seems unlikely (see, for example, Marchetti et al., 1986). In the first instance iontophoretic application of dopamine to dissociated mouse dorsal root ganglion cells, the spinal homologs of the petrosal neurons, and examination of the responses under voltage clamp may reveal a population of cells that respond appropriately. It would not be expected that all sensory ganglion cells would respond in the same way, or even at all, and an increase in excitability might be seen in only a small proportion of cells.

Further studies of the pharmacology of petrosal neurons, of type I cells, and of transmitter uptake systems in type I and type II cells should help to reconcile these findings and demonstrate whether a mechanism exists whereby dopamine can act as an excitatory transmitter.

The resolution of this problem is crucial since the nature of the transmitter, if there is one, is a central question for the first hypothesis of transduction. It might be expected that spontaneous transmitter release would be apparent and presumably there would be events equivalent to the miniature end-plate potentials at the neuromuscular junction (Fatt and Katz, 1952) or there might be spontaneous postsynaptic potentials that can be recorded in central neurons (see, for example, Biscoe and Duchen, 1985). Such events might explain the continuous low-level chemoreceptor afferent discharge shown to be indistinguishable stochastically from a Poisson process (see Biscoe and Taylor, 1963).

X. Effects of Hypoxia on Neurons

In the central nervous system, transient hypoxic or hypoglycemic episodes lead to loss of consciousness, occasionally to seizures, and eventually to cell death. There is considerable evidence, including measurements of total calcium in affected brain, that calcium accumulates after arteriali occlusion (Rappaport et al., 1987) as a component of the pathophysiology. There is also evidence that resting transmitter release is promoted by hypoxia in the central nervous system (Benveniste et al., 1984) and at the neuromuscular junction (Nishimura, 1986), although it has also been found that evoked transmitter release from the central nervous system may be reduced (Bosley et al., 1983).

Cross and Silver (1962) showed that the resting level of PO_2 in rabbit forebrain gray matter was 13 mm Hg. Silver (1965) showed subsequently that the PO_2 varied between a high of 95-30 mm Hg over the capillaries and down to 2 mm Hg between capillaries. He also recorded fluctuations correlated with capillary size, with the respiratory cycle, and with large circulatory adjustments, such as a fall in blood pressure. These levels may be low enough to affect the intracellular calcium concentration. Thus, variations in $[Ca^{2+}]_i$ as a consequence of changes in *normal* arterial oxygen levels may well have as yet unknown consequences for neuronal function. Such results show that there is a high probability of many nerve cells being finely poised with respect to a hypoxia-induced rise in cytosolic calcium. The hyperpolarization seen as an early response of hippocampal neurons to hypoxia (Hansen, 1985; Fujiwara et al., 1987) may reflect activation of a calcium-dependent g (K) or of an ATP-inhibited g(K) (Noma, 1983; Fujiwara et al., 1987). Indeed, different populations of neurons may respond differently according to the types of ion channels expressed as suggested above to account for results on the carotid body.

In some cell preparations, but not all, the *N*-methyl-D-aspartate (NMDA) receptor antagonists, AP-5 or AP-7, protect cells from hypoxic injury (Wieloch, 1985; Simon et al., 1984; Mayer and Westbrook, 1987). NMDA action is associated with a rise in calcium ion permeability and a $[Ca^{2+}]_i$ increase (McDermott et al., 1986). Accordingly, release of L-glutamate, which activates the NMDA receptor following a rise in $[Ca^{2+}]_i$, from some other cause, would have a positive feedback effect on $[Ca^{2+}]_i$ and also presumably on transmitter release. The question, then, is could this transmitter release come about in the first place because of a rise in intracellular calcium following hypoxia? So it was of interest to make direct measurements and particularly relevant to the carotid body problem since we had shown the type I cells to be electrophysiologically very similar to neurons. It might well be that type I cells are merely excitable cells placed in an unusual position.

We have first studied the effects of cyanide and of glucose removal on isolated mouse dorsal root ganglion cells since the behavior of sensory nerve membrane may be especially relevant to understanding a sensory receptor. These cells are widely regarded as a model system for the study of the pharmacology of sensory terminals and of the central presynaptic membrane. All three major classes of Ca^{2+} currents (Nowycky et al., 1985) (T, N, and L) were attenuated, and a Ca^{2+}-dependent K^+ conductance was increased (Duchen and Somjen, 1988). Hansen et al., (1982) had earlier also found that anoxia increased a potassium conductance in hippocampal neurons. There were other effects, for example, on I_H and the GABA-evoked chloride current (Duchen, 1990). Measurements made using fura-2 in single cells showed both a rise in basal $[Ca^{2+}]_i$ and impaired clearance of a Ca^{2+} load in response to both stimuli (Biscoe et al., 1988b; Duchen et al., 1990). A major question is whether the alteration in $[Ca^{2+}]_i$ is responsible for the change in Ca^{2+} current (Eckert and Chad, 1984; Hess and Tsien, 1984; Chad and Eckert, 1986; Morad et al., 1988), or whether it is a separate reflection of the altered energy status of the cell. Manipulation of metabolites while monitoring both the Ca^{2+} current and $[Ca^{2+}]_i$ should answer this question.

The results of our experiments on the effects of hypoxia on neurons are qualitatively similar to our findings on the type I cell, though the Type I cell may be more responsive. In short, there is a rise in $[Ca^{2+}]_i$ that is enhanced by a rise in temperature and is still seen, albeit attenuated, by lowering extracellular calcium (5-7 μM). Such a rise in intracellular calcium might be expected to provoke transmitter release, as discussed above with respect to L-glutamate. Also, there are in neurons other powerful effects, not seen in type I cells and presumably reflecting a different expression of ion channels in the cell membrane.

Clarification of the conductance changes in different populations of cells and of the mechanisms by which they are produced in response to these metabolic stimuli should help identify the *basic* processes that lead to transient alterations in function of excitable cells. It is these *basic* processes which we believe are most likely to underlie chemotransduction in the carotid body, processes whose expression may differ among cell types, and whose differential expression must be understood if we are to understand chemoreception.

XI. Prospects for the Future

For the future it will be necessary to complete the study of isolated carotid body cells and sensory ganglion cells with respect to their biophysical properties and reactions to stimuli, in particular cyanide and hypoxia, and second, to contrive direct experiments on the type I cell-nerve cell combination. This will involve determining how the various components interact and how these interactions have affected the unitary properties. The reactions of the system to stimulation electrically, to direct application of chemicals, to flash release of caged compounds intracellularly, or to changes in the concentration of metabolites by intracellular perfusion (Soejima and Noma, 1984) will be necessary components of such an investigation.

Further, we have generally disregarded the contribution of the type II cell in this discussion, and it remains possible, indeed we would regard it as certain, that the type II cell plays a crucial role in the maintenance of an appropriate extracellular environment to optimize the sensitivity of the type I cell. This could involve regulation of extracellular potassium, hydrogen ion, and transmitter concentrations, and of tissue PO_2.

If it should turn out that the received view of the unitary structure is in error, then direct experimentation will make that at once apparent. That is to say, the nerve cell will be motor and excitation of the cell or fiber will cause transmitter release from the type I cell, a release heavily modulated by the aforementioned stimuli, hypoxia, cyanide, and so forth, as shown here. This would conform to the view that the unit structure forms part of an efferent pathway. In the latter case, then, the receptor would have to be sought elsewhere, perhaps in the small nerve fiber hypothesis.

Finally, if the view of Paintal (1967) that there is a mechanoreceptor process involved should be correct, then this would also have to be examined directly to establish the truth or falsehood of the idea.

There is now the possibility for direct tests of all these hypotheses rather than the indirect tests that have led to our present failure to resolve the problem. For the first time we are optimistic about finding a solution to many of the problems of chemotransduction.

References

Acker, H., and Eyzaguirre, C. (1987). Light absorbance changes in the mouse carotid body during hypoxia and cyanide poisoning. *Brain Res.* **409**:380-385.

Acker, H., Eyzaguirre, C., and Goldman, W. F. (1985). Redox changes in the mouse carotid body during hypoxia. *Brain Res.* **330**:158-163.

Acker, H., Lübbers, D. W., and Purves, M. J. (1971). Local oxygen tension field in the glomus caroticum of the cat and its change at changing arterial PO_2. *Pflügers Arch.* **329**:136-155.

Baker, P. F., Blaustein, M. P., Hodgkin, A. L., and Steinhardt, R. A. (1967). The effect of sodium concentration on calcium movements in gian axons of *Loligo forbesi*. *J. Physiol.* **192**:43-44P.

Barrett, J. N., Magleby, K. L., and Pallotta, B. S. (1982). Properties of single calcium-activated potassium channels in cultured rat muscle. *J. Physiol.* **331**:211-230.

Belmonte, C., and Eyzaguirre, C. (1974). Efferent influences on carotid body chemoreceptors. *J. Neurophysiol.* **37**:1131-1143

Benveniste, H., Drejer, J., Schousboe, A., and Diemer, N. H. (1984). Elevation of the extracellular concentrations of glutamate and aspartate in rat hippocampus during transient cerebral ischaemia monitored by intracerebral microdialysis. *J. Neurochem.* **43**:1369-1374.

Berridge, M. J., and Irvine, R. F. (1984). Inositol trisphosphate, a novel second messenger in cellular signal transduction. *Nature* **312**:315-321.

Biscoe, T. J. (1971). Carotid body; structure and function. *Physiol. Rev.* **51**:437-495.

Biscoe, T. J., Caddy, K. W. T., Duchen, M. R., Kirby, G. C., Patterson, D. L., and Ponte, J. (1987a). Preparation and identification of freshly dissociated cells from the rabbit carotid body for electrophysiological study. *J. Physiol.* **392**:11P.

Biscoe, T. J., and Duchen, M. R. (1985). The anion selectivity of GABA-mediated post-synaptic potentials in mouse hippocampal cells. *Q. J. Exp. Physiol.* **70**:305-312.

Biscoe, T. J., and Duchen, M. R. (1989a). Electrophysiological responses of dissociated type I cells of the rabbit carotid body to cyanide. *J. Physiol.* **413**:447-468.

Biscoe, T. J., and Duchen, M. R. (1989b). Hypoxia raises intracellular calcium in type I cells isolated from the rabbit carotid body. *J. Physiol.* **420**:128P.

Biscoe, T. J., and Duchen, M. R. (1990). Responses of type I cells dissociated from the rabbit carotid body to hypoxia. *J. Physiol.* **428**:39-59.

Biscoe, T. J., Duchen, M. R., Eisner, D. A., O'Neill, S. C., and Valdeol-millos, M. (1988a). Cyanide increases $[Ca^{2+}]_i$ in isolated type I cells from the rabbit carotid body. *J. Physiol.* **396**:99P.

Biscoe, T. J., Duchen, M. R., Eisner, D. A., O'Neill, S. C., and Valdeol-millos, M. (1988b). The effects of glucose removal and cyanide on intracellular Ca^{2+} in isolated, single mouse dorsal root ganglion cells. *J. Physiol.* **401**:60P.

Biscoe, T. J., Duchen, M. R., Eisner, D. A., O'Neill, S. C., and Valdeol-millos, M. (1989). Measurements of intracellular Ca^{2+} in dissociated type I cells of the rabbit carotid body. *J. Physiol.* **416**:421-434.

Biscoe, T. J., Duchen, M. R., Kirby, G. C., Patterson, D. L., and Ponte, J. (1987b). Electrophysiological characterization of freshly dissociated type I cells from the rabbit carotid body. *J. Physiol.* **392**:36P.

Biscoe, T. J., Duchen, M. R., Kirby, G. C., Patterson, D. L., and Ponte, J. (1988c). Voltage clamp study of the effects of cyanide on dissociated type I cells of the rabbit carotid body. *J. Physiol.* **396**:178P.

Biscoe, T. J., Lall, A., and Sampson, S. R. (1970a). Electron microscopic and electrophysiological studies on the carotid body following intra-cranial section of the glossopharyngeal nerve. *J. Physiol.* **208**:133-152.

Biscoe, T. J., Purves, M. J., and Sampson, S. R. (1970b). The frequency of nerve impulses in single carotid body chemoreceptor afferent fibres recorded in vivo with intact circulation. *J. Physiol.* **208**:121-131.

Biscoe, T. J., and Sampson, S. R. (1968). Rhythmical and nonrhythmical spontaneous activity recorded from the central cut end of the carotid sinus nerve. *J. Physiol.* **196**:327-338.

Biscoe, T. J., and Stehbens, W. E. (1966). The ultrastructure of the carotid body. *J. Cell Biol.* **30**:563-578.

Biscoe, T. J., and Taylor, A. (1963). The discharge pattern recorded in chemoreceptor afferent fibres from the cat carotid body with normal circulation and during perfusion. *J. Physiol.* **168**:332-344.

Black, A. M. S., Comroe, J. H. Jr., and Jacobs, L. (1972). Species difference in carotid body response of cat and dog to dopamine and serotonin. *Am. J. Physiol.* **223**:1097-1102.

Boron, W. R., and de Weer, P. (1976). Intracellular pH transients in squid giant axons caused by CO_2, NH_3 and metabolic inhibitors. *J. Gen. Physiol.* **67**:91-112.

Bosley, T. M., Woodhams, P. L., Gordon, R. D., and Balász, R. (1983). Effects of anoxia on the stimulated release of amino-acid neurotransmitters in the cerebellum in vitro. *J. Neurochem.* **40**:189-201.

Caldwell, P. C. (1960). The phosphorous metabolism of squid axons and its relationship to the active transport of sodium. *J. Physiol.* **152**:545-560.

Carafoli, E. (1979). The calcium cycle of mitochondria. *FEBS Lett.* **104**:1-5.

Carafoli, E. (1987). Intracellular calcium homeostasis. *Annu. Rev. Biochem.* **56**:395-433.

Carafoli, E., and Lehninger, A. L. (1971). A survey of the interaction of calcium ions with mitochondria from different tissues and species. *Biochem. J.* **122**:681-690.

Chad, J. E., and Eckert, R. (1986). An enzymatic mechanism for calcium current inactivation in dialysed *Helix* neurones. *J. Physiol.* **378**:31-51.

Chad, J. E., Eckert, R., and Ewald, D. (1984). Kinetics of calcium-dependent inactivation of calcium current in voltage-clamped neurones of *Aplysia californica*. *J. Physiol.* **347**:279-300.

Crompton, M. (1985). The regulation of mitochondrial calcium transport in heart. *Curr. Topics Membrane Transport* **25**:231-276.

Crompton, M., Moser, R., Lüdi, H., and Carafoli, E. (1978). The interactions between the transport of sodium and calcium in mitochondria of various mammalian tissues. *Eur. J. Biochem.* **82**:25-31.

Cross, B. A., and Silver, I. A. (1962). Some factors affecting oxygen tension in the brain and other organs. *Proc. Roy. Soc. B* **156**:483-499.

Davis, H. (1961). Some principles of sensory receptor action. *Physiol. Rev.* **41**:391-416.

De Castro, F. (1926). Sur la structure et l'innervation de la glande intercarotidienne (*Glomus caroticum*) de l'homme et des mammifères, et sur un nouveau système d'innervation autonome du nerf glossopharyngien. Études anatomiques et experimentales. *Trab. Lab. Invest. Biol. Univ. Madrid* **24**:365-432.

De Castro, F. (1928). Sur la structure et l'innervation du sinus carotidien de l'homme et des mammifères. Nouveaux faits sur l'innervation et la fonction du glomus caroticum. Études anatomiques et physiologiques. *Trab. Lab. Invest. Biol. Univ. Madrid* **25**:331-380.

Delpiano, M. A., and Hescheler, J. (1989). Evidence for a PO_2-sensitive K^+ channel in the type-I cell of the rabbit carotid body. *FEBS Lett.* **249**:195-198.

Dubinsky, J. M., and Oxford, G. S. (1985). Dual modulation of K channels by thyrotropin releasing hormone in clonal pituitary cells. *Proc. Natl. Acad. Sci. USA* **82**:4282-4286.

Duchen, M. R. (1989). Metabolic blockade attenuates $GABA_A$ evoked chloride currents in isolated mouse primary sensory neurones. *J. Physiol.* **415**: 48P.

Duchen, M. R. (1990). Effects of metabolic inhibition on the membrane properties of isolated mouse primary sensory neurons. *J. Physiol.* **424**: 387-409.

Duchen, M. R., Caddy, K. W. T., Kirby, G. C., Patterson, D. L., Ponte, J., and Biscoe, T. J. (1988). Biophysical studies of the cellular elements of the rabbit carotid body. *Neuroscience* **26**:291-313.

Duchen, M. R., and Somjen, G. G. (1988). Effects of cyanide and low glucose on the membrane currents of dissociated mouse primary sensory neurones. *J. Physiol.* **401**:61P.

Duchen, M. R., Valdeolmillos, M., O'Neill, S. C., and Eisner, D. A. (1990). Effects of metabolic blockade on the regulation of intracellular calcium in dissociated mouse sensory neurones. *J. Physiol.* **424**:411-426.

Eckert, R., and Chad, J. E. (1984). Inactivation of calcium channels. *Prog. Biophys. Mol. Biol.* **44**:215-267.

Euler, U. S. von and Liljestrand, G. (1937). The effects of cyanide on respiration. *Skand. Arch. Physiol.* **76**:27-36.

Eyzaguirre, C., and Koyano, H. (1965). Effects of some pharmacological agents on chemoreceptor discharges. *J. Physiol.* **178**:410-437.

Eyzaguirre, C., Monti-Bloch, L., Hayashida, Y., and Baron, M. (1983). Biophysics of the carotid body receptor complex. In *Physiology of Arterial Chemoreceptors*. Edited by H. Acker and R. G., O'Regan. N.Y. Elsevier, pp. 59-87.

Eyzaguirre, C., and Zapata, P. (1984). Perspectives in carotid body research. *J. Appl. Physiol.* **57**:931-957.

Fatt, P., and Katz, B. (1952). Spontaneous subthreshold activity at motor nerve endings. *J. Physiol.* **117**:109-128.

Fidone, S. J., Gonzalez, C., and Yoshizaki, K. (1982). Effects of low oxygen on the release of dopamine from the rabbit carotid body in vitro. *J. Physiol.* **333**:93-110.

Fishman, M. C., Greene, W. L., and Platika, D. (1985). Oxygen chemoreception by carotid body cells in culture. *Proc. Natl. Acad. Sci. USA* **82**:1448-1450.

Fishman, M. C., and Schaffner, A. E. (1984). Carotid body cell culture and selective growth of glomus cells. *Am. J. Physiol.* **246**:C106-C113.

Folgering, H., Ponte, J., and Sodie, T. (1982). Adrenergic mechanism and chemoreception in the carotid body of the cat and rabbit. *J. Physiol.* **325**:7-22.

Fujiwara, N., Higashi, H., Shimoji, K., and Yoshimura, M. (1987). Effects of hypoxia on rat hippocampal neurones in vitro. *J. Physiol.* **384**:131-151.

Gill, D. L., Grollman, E. F., and Kohn, L. D. (1981). Calcium transport mechanisms in membrane vesicles from guinea pig brain synaptosomes. *J. Biol. Chem.* **256**:184-192.

Goodman, N. W. (1973). Efferent control of arterial chemoreceptors mediated by glossopharyngeal fibres and artifacts introduced by stimulation techniques. *J. Physiol.* **230**:295-311.

Grynkiewicz, G., Poenie, M., and Tsien, R. Y. (1985). A new generation of Ca^{2+} indicators with greatly improved fluorescence properties. *J. Biol. Chem.* **260**:3440-3450.

Hamill, O. P., Marty, A., Neher, E., Sakmann, B., and Sigworth, F. J. (1981). Improved patch-clamp technique for high-resolution current recording from cells and cell-free membrane patches. *Pflügers Arch.* **391**:85-100.

Hansen, A. J. (1985). Effect of anoxia on ion distribution in the brain. *Physiol. Rev.* **65**:101-148.

Hansen, A. J., Hounsgaard, J., and Jahnsen, H. (1982). Anoxia increases potassium conductance in hippocampal nerve cells. *Acta Physiol. Scand.* **115**:301-310.

Hescheler, J., Delpiano, M. A., Acker, H., and Pietruschka, F. (1989). Ionic currents on type-I cells of the rabbit carotid body measured by voltage-clamp experiments and the effect of hypoxia. *Brain Res.* **486**:79-88.

Hess, P., and Tsien, R. W. (1984). Mechanisms of ion permeation through calcium channels. *Nature* **309**:453-456.

Jacobs, L., and Comroe, J. H., Jr. (1968). Stimulation of the carotid chemoreceptors of the dog by dopamine. *Proc. Natl. Acad. Sci. USA* **59**:1187-1193.

Kaila, K., Mattsson, K., and Voipio, J. (1989). Fall in intracellular pH and increase in resting tension induced by a mitochondrial uncoupling agent in crayfish muscle. *J. Physiol.* **408**:271-293.

Keilin, D., and Hartree, E. F. (1941). Absorption spectra of haemoglobin in solution and in red blood corpuscles. *Nature* **148**:75-77.

Lopez-Barneo, J., López-López, J. R., Ureña, J., and González, C. (1988). Chemotransduction in the carotid body: K^+ current modulated by PO_2 in type I chemoreceptor cells. *Science* **241**:580-582.

López-López, J., González, C., Ureña, J., and López-Barneo, J. (1989). Low PO_2 selectively inhibits K channel activity in chemoreceptor cells of the mammalian carotid body. *J. Gen. Physiol.* **93**:1001-1015.

Lux, H. D., Neher, E., and Marty, A. (1981). Single channel activity associated with the calcium dependent outward current in *Helix pomatia*. *Pflügers Arch.* **389**:293-295.

Marchetti, C., Carbone, E., and Lux, H. D. (1986). Effects of dopamine and noradrenaline on Ca channels of cultured sensory and sympathetic neurons of chick. *Pflügers Arch.* **406**:104-111.

Mayer, M. L., and Westbrook, G. L. (1987) Cellular mechanisms underlying excitotoxicity. *Trends Neurosci.* **10**:59-61.

McCloskey, D. I. (1975). Mechanisms of autonomic control of carotid chemoreceptor activity. *Respir. Physiol.* **25**:53-61.

McDermott, A. B., Mayer, M. L., Westbrook, G. L. Smith, S. J., and Barker, J. L. (1986). NMDA-receptor activation increases cytoplasmic calcium concentration in cultured spinal cord neurones. *Nature* **321**: 519-522.

McDonald, D. M. (1981). Peripheral chemoreceptors. Structure-function relationships of the carotid body. In *Regulation of Breathing.* Edited by T. F. Hornbein. New York, Marcel Dekker, pp. 105-319.

McQueen, D. S. (1983). Pharmacological aspects of putative transmitters in the carotid body. In *Physiology of the Peripheral Arterial Chemoreceptors.* Edited by H. Acker and R. G. O'Regan. New York, Elsevier, pp. 149-196.

Meech, R. W., and Strumwasser, F. (1970). Intracellular calcium injection activates potassium conductance in *Aplysia* nerve cells. *Fed. Proc.* **29**: 834.

Mills, E., and Jöbsis, F. F. (1970). Simultaneous measurement of cytochrome a_3 reduction and chemoreceptor afferent activity in the carotid body. *Nature* **225**:1147-1149.

Mills, E., and Jöbsis, F. F. (1972). Mitochondrial respiratory chain of carotid body and chemoreceptor response to changes in oxygen tension. *J. Neurophysiol.* **35**:405-428.

Mills, E., Smith, P. G., Slotkin, T. A., and Breese, G. (1978). "Commentary" role of carotid body catechol-amines in chemoreceptor function. *Neuroscience* **3**:1137-1146.

Monti-Bloch, L., Stensaas, L. J., and Eyzaguirre, C. (1983). Carotid body grafts induce chemosensitivity in muscle nerve fibres of the cat. *Brain Res.* **270**:77-92.

Morad, M., Davies, N. W., Kaplan, J. K., and Lux, H. D. (1988). Inactivation and block of calcium channels by photoreleased Ca^{2+} in dorsal root ganglion neurones. *Science* **241**:842-844.

Mullins, L. J., and Requena, J. (1987). Alterations in intracellular calcium produced by changes in intracellular sodium and intracellular pH. In *Cell Calcium and the Control of Membrane Transport.* Edited by L. J. Mandel and D. C. Eaton. New York, Rockefeller University Press, pp. 65-75.

Nair, P. K., Buerk, D. G., and Whalen, W. J. (1986). Cat carotid body O_2 metabolism and chemoreception described by a two cytochrome model. *Am. J. Physiol.* **250**:H202-H207.

Neil, E., and O'Regan, R. G. (1971a). The effects of electrical stimulation of the distal end of the cut sinus and aortic nerves on peripheral arterial chemoreceptor activity in the cat. *J. Physiol.* **215**:15-32.

Neil, E., and O'Regan, R. G. (1971b). Efferent and afferent impulse activity recorded from few fibre preparations of otherwise intact sinus and aortic nerves. *J. Physiol.* **215**:33-47.

Nicholls, D. G., and Kerman, K. E. O. (1981). Biochemical approaches to the study of cytosolic calcium regulation in nerve endings. *Phil. Trans. Roy. Soc. London B* **296**:115-122.

Niedergerke, R. (1963). Movements of Ca in frog heart ventricles at rest and during contractures. *J. Physiol.* **167**:515-550.

Nishimura, M. (1986). Factors influencing an increase in spontaneous transmitter release by hypoxia at the mouse neuromuscular junction. *J. Physiol.* **372**:303-314.

Noma, A. (1983). ATP-regulated K^+ channels in cardiac muscle. *Nature* **305**:147-148.

Nowycky, M. C., Fox, A. P., and Tsien, R. W. (1985). Three types of neuronal calcium channel with different calcium agonist sensitivity. *Nature* **316**:440-443.

Nurse, C. A. (1986). Dual function characteristics of glomus cells in cultures of dissociated rat carotid body. *Soc. Neurosci. Abstr.* **12**:627.

Nurse, C. A. (1987). Localization of acetylcholinesterase in dissociated cell cultures of the carotid body of the rat. *Cell Tissue Res.* **250**:21-27.

Obeso, A., Fidone, S., and Gonzalez, C. (1987). Pathways for calcium entry into type I cells: Significance for the secretory response. In *Chemoreceptors in Respiratory Control.* Edited by J. A. Ribeiro and D. J. Pallot. London, Croom Helm, pp. 91-97.

O'Regan, R. G., and Majcherczyk, S. (1982). Role of peripheral chemoreceptors and central chemosensitivity in the regulation of respiration and circulation. *J. Exp. Biol.* **100**:23-40.

Paintal, A. S. (1967). Mechanism of stimulation of aortic chemoreceptors by natural stimuli and chemical substances. *J. Physiol.* **189**:63-84.

Paintal, A. S. (1988). The responses of chemoreceptors with medullated and non-medullated fibres to chemical substances and the mechanical hypothesis. *Prog. Brain Res.* **74**:161-168.

Pallot, D. J. (1987). The mammalian carotid body. *Adv. Anat. Embryol. Cell Biol.* **102**:1-91.

Pietruschka, F. (1974). Cytochemical demonstration of catecholamines in cells of the carotid body in primary tissue culture. *Cell Tissue Res.* **151**:317-321.

Rappaport, Z. H., Young, W., and Flamm, E. S. (1987). Regional brain calcium changes in the rat middle cerebral artery occlusion model of ischemia. *Stroke* **18**:760-764.

Requena, J., and Mullins, L. J. (1979). Calcium movement in nerve fibres. *Q. Rev. Biophys.* **12**:371-460.

Rink, T. J., Tsien, R. Y., and Pozzan, T. (1982). Cytoplasmic pH and free Mg^{2+} in lymphocytes. *J. Cell Biol.* **95**:189-196.

Ritchie, A. K. (1987). Thyrotropin releasing hormone stimulates a calcium-activated potassium current in a rat anterior pituitary cell line. *J. Physiol.* **385**:611-626.

Rottenberg, H., and Scarpa, A. (1974). Calcium uptake and membrane potential in mitochondria. *Biochemistry* **13**:4811-4819.

Roumy, M., and Leitner, L. M. (1977). Role of calcium ions in the mechanism of arterial chemoreceptor excitation. In *Chemoreception in the Carotid Body*. Edited by H. Acker, S. Fidone, D. Pallot, C. Eyzaguirre, D. W. Lubbers, and R. W. Torrance. Berlin, Springer-Verlag, pp. 257-261.

Sampson, S. R. (1972). Mechanism of efferent inhibition of carotid body chemoreceptors in the cat. *Brain Res.* **45**:266-270.

Sampson, S. R., and Biscoe, T. J. (1970). Efferent control of the carotid body chemoreceptor. *Experientia* **26**:261-262.

Sampson, S. R., and Vidruk, E. H., (1977). Hyperpolarising effects of dopamine on chemoreceptor nerve endings from cat and rabbit carotid bodies in vitro. *J. Physiol.* **268**:211-221.

Sekar, C. M., and Hokin, L. E. (1986). The role of phosphoinositides in in signal transduction. *J. Membrane Biol.* **89**:193-210.

Silver, I. A. (1965). Some observations on the cerebral cortex with an ultra-micro, membrane-covered, oxygen electrode. *Med. Electron. Biol. Engineering* **3**:377-387.

Simon, R. P., Swan, J. H., Griffiths, T., and Meldrum, B. S. (1984), Blockade of N-methyl-D-aspartate receptors may protect against ischaemic damage in the brain. *Science* **226**:850-852.

Smith, J. B., Smith, L., and Higgins, B. L. (1985). Temperature and nucleotide dependence of calcium release by myo-inositol 1,4,5-trisphosphate in cultured vascular smooth muscle cells. *J. Biol. Chem.* **260**:14413-14416.

Soejima, M., and Noma, A. (1984). Mode of regulation of the ACh-sensitive K channel by the muscarinic receptor in rabbit atrial cells. *Pflügers Arch.* **400**:424-431.

Stea, A., and Nurse, C. A. (1988). Evidence for a large conductance chloride channel in cultured glomus cells of the rat carotid body. *Soc. Neurosci. Abstr.* **14**:643.

Streb, H., Irvine, R. F., Berridge, M. J., and Schulz, I. (1983). Release of Ca^{2+} from a nonmitochondrial intracellular store in pancreatic acinar cells by inositol-1,4,5-trisphosphate. *Nature* **306**:67-69.

Trautmann, A., and Marty, A. (1984). Activation of Ca-dependent K channels by carbamoylcholine in rat lacrimal glands. *Proc. Natl. Acad. Sci. USA* **81**:611-615.

Ureña, J., López-López, J., González, C., and López-Barneo, J. (1989). Ionic currents in dispersed chemoreceptor cells of the mammlian carotid body. *J. Gen. Physiol.* **93**:979-999.

Whalen, W. J., and Nair, P. (1975). Some factors affecting tissue PO_2 in the carotid body. *J. Appl. Physiol.* **39**:562-566.

Whalen, W. J., and Nair, P. (1976). PO_2 in the carotid body perfused and/or superfused with cell-free media. *J. Appl. Physiol.* **41**:180-184.

Whalen, W. J., Savoca, J., and Nair, P. (1973). Oxygen tension measurements in carotid body of the cat. *Am. J. Physiol.* **225**:986-991.

Wieloch, T. (1985). Hypoglycaemia induced neuronal damage prevented by an *N*-methyl-D-aspartate antagonist. *Science* **230**:681-683.

Zapata, P., and Llados, F. (1977). Blockade of carotid body chemoresensory inhibition. In *Chemoreception in the Carotid Body*. Edited by H. Acker, S. Fidone, D. Pallot, C. Eyzaguirre, D. W. Lübbers, and R. W. Torrance. Berlin, Springer-Verlag, pp. 152-159.

11

Discharge Properties of Carotid Bodies
Developmental Aspects

EILEEN M. MULLIGAN

Temple University School of Medicine
Philadelphia, Pennsylvania

I. Development of Carotid Body Chemoreceptors: Evidence from Ventilatory Studies

Studies of the chemoreflex suggest that the responses of the carotid body chemoreceptors develop postnatally. Many studies have been performed on both human infants and neonatal animals to test for the presence and the strength of the peripheral chemoreflex. The chemoreflex can be tested by measuring ventilatory changes during either a hypoxic exposure or a hyperoxic exposure. The initial increase in ventilation in response to hypoxia (Belenky et al., 1979; Brady and Ceruti, 1966; Bureau et al., 1985a, 1986) and the initial decrease in ventilation in response to hyperoxia (Bureau and Begin, 1982; Carroll and Bureau, 1987; Hertzberg and Lagercrantz, 1987) increase with postnatal age in newborns. Thus, these studies and others (Mayock et al., 1983; see Haddad and Mellins, 1984, and Hanson, 1986) have shown that the chemoreflex develops with time postnatally.

Newborns have a biphasic ventilatory response to hypoxia (see Haddad and Mellins, 1984). During hypoxia, ventilation increases for a period of time and then decreases toward, to, or even below the control normoxic level. Whereas the carotid body chemoreceptors are responsible for the initial increase in ventilation during hypoxia, the cause of the subsequent decline in ventilation is not fully understood. Some investigators have postulated that the carotid bodies may contribute to this decline in ventilation (Bureau et al., 1984, 1985b, 1986; Carroll and Bureau, 1987; Schramm and Grunstein, 1987). However, there are many links in the chain of events between the afferent activity of the carotid body chemoreceptors and the final product, ventilation. The only way to directly assess the activity of the carotid chemoreceptors and their potential effect on overall ventilation under various conditions is to record their afferent neural activity. A great deal is known about the responses of carotid chemoreceptors in adult animals (see Lahiri et al., 1983) whereas the responses in the newborn have not yet been studied in such great detail (see Hanson, 1986). What is known of these responses will be discussed in this chapter.

Studies have also been performed on animals to show that intact carotid sinus nerves are extremely important, if not essential for survival, in the neonatal period (Bureau et al., 1985a; Donnelly and Haddad, 1987; Haddad and Donnelly, 1988; Hofer, 1984). The relationship of this mortality to the development of the carotid body chemoreceptors is not fully understood. There have also been reports that the carotid bodies in infants who are victims of the sudden infant death syndrome are different from those of age-matched controls (Cole et al., 1979; Naeye et al., 1976; Perrin et al., 1984; Shannon and Kelly, 1982). However, the normal responses of the carotid chemoreceptors must be known before any abnormalities that may occur in carotid body development can be fully evaluated.

II. Carotid Chemoreceptor Recording: Methodology

A. Recording from Sinus Nerve Strands

The method that has been used most often to study the response characteristics of carotid body chemoreceptor afferents both in the newborn (Biscoe and Purves, 1967; Blanco et al., 1984a, 1984b) and in the adult (Heymans and Neil, 1958; Lahiri and DeLaney, 1975; Mulligan and Lahiri, 1981) is to completely section the central end of the carotid sinus nerve at or near its junction with the glossopharyngeal nerve. The nerve is then desheathed and mechanically separated into fine strands, which are tested for neural activity. The carotid sinus nerve contains both baroreceptor and chemoreceptor afferent neural activity. The nerve strands are separated further until single-fiber or few-fiber chemoreceptor activity free from baroreceptor activity is detected. Chemoreceptor fiber selection by this method is inevitably biased by fiber size. A chemoreceptor single-fiber afferent is verified by its stimulation by hypoxia and cyanide (Mulligan and Lahiri, 1981), by identity of the size and shape of the action potentials, and by the lack of action potential temporal overlap (Lahiri and DeLaney, 1975). This method of recording chemoreceptor afferent activity is technically difficult at best in adult animals and is even more difficult in the newborn given the delicate nature of the tissues. Nevertheless this method has provided a wealth of information on the responses of carotid chemoreceptors (see Lahiri et al., 1983). The fiber stranding method has the disadvantage that the carotid sinus nerve must be completely or at least partially cut in order to record from the chemoreceptor fibers. This removes the sinus nerve efferent input into the carotid body and eliminates any effect that these efferents may have on the afferent output of the carotid body. In the adult cat, stimulation of carotid sinus nerve efferent activity inhibits carotid chemoreceptor afferent activity (Neil and O'Regan, 1971). The role of efferent activity in chemoreceptor responsiveness in the newborn is not yet known.

Thus, while recording chemoreceptor afferent activity from cut strands of the carotid sinus nerve can provide valuable information on carotid chemoreceptor responses, it is more difficult in the newborn. Further, it requires surgery in the area of the carotid body and sectioning of all or part of the carotid sinus nerve, thereby interrupting efferent nerve traffic.

B. Recording from the Petrosal Ganglion

The petrosal ganglion is the sensory ganglion of the glossopharyngeal nerve. The cell bodies of the carotid body chemoreceptor afferent fibers are located in the petrosal ganglion. It is therefore possible to record from carotid body chemoreceptor afferents in the petrosal ganglion (Jordan et al., 1987; Mulligan et al., 1989). To do this, the ganglion is located, the outer sheath is removed, and glass or metal microelectrodes are inserted into the ganglion. Action potentials are recorded from the cell bodies of carotid sinus nerve afferents. The microelectrode is inserted and advanced into the ganglion in different places until a single-fiber chemoreceptor afferent is found. The carotid sinus nerve is identified and left intact. The trunk of the glossopharyngeal nerve distal to its junction with the carotid sinus nerve can be cut in order to leave intact only afferent neural traffic from the carotid sinus nerve. This greatly increases the probability of locating chemoreceptor afferents suitable for recording.

This method offers many advantages over the fiber stranding method. (1) The carotid sinus nerve is not cut and therefore both the efferent and afferent pathways to and from the carotid body remain intact. (2) No surgery is required in the area of the carotid body. It remains covered and in its "natural" environment with all innervation and blood supply intact. (3) If desired, this method can also be used to study carotid chemoreceptor afferents without intact carotid sinus nerve efferents by cutting the central connection of the petrosal ganglion. (4) Tissues in the neonate are more delicate and friable than in mature animals and obtaining strands of nerve fibers is especially difficult in neonates. (5) Many newborns are less stable than adults as experimental animals and so the potential time-saving element of this method is also crucial. (6) In most piglets (Mulligan et al., 1989) and in the kittens examined by this author (Mulligan, unpublished observations), the petrosal ganglion is usually located just outside the skull, making it readily accessible for microelectrode penetration and chemoreceptor recording via a lateral surgical approach. Petrosal ganglion recording has been used to study single-fiber carotid chemoreceptor responses in the neonatal piglet (Mulligan, 1988; Mulligan et al., 1989; Mulligan and Bhide, 1989). These results will be discussed later in this chapter.

III. Carotid Chemoreceptor Afferent Nerve Responses in the Fetus

The only carotid chemoreceptor afferent recording in a fetus has been performed using the fiber stranding technique on the cut carotid sinus nerve in the exteriorized lamb fetus (Biscoe et al., 1969; Blanco et al., 1984a). Early studies (Biscoe et al., 1969) suggested that carotid chemoreceptor activity could not always be detected in the fetus. In the near-term fetus, when chemoreceptor activity was detected, there was little response to increasing PO_2, to cyanide, or to nicotine in the cut carotid sinus nerve. However, when the umbilical cord was clamped, neural activity increased as the carotid artery PaO_2 decreased. Cervical sympathetic nerve stimulation increased the chemoreceptor neural discharge. Cord clamping also resulted in increased cervical sympathetic neural activity (Biscoe et al., 1969). In this study, no detailed O_2 or CO_2 response curves were reported. Also, chemoreceptor activity was not recorded at different levels of steady-state PaO_2 in the fetus. Another study that showed active and responsive chemoreceptor fibers was performed on carotid bodies in vitro that had been removed from fetal sheep (Jansen et al., 1980).

The responses of sheep fetal carotid chemoreceptors have been reexamined recently in more detail (Blanco et al., 1984a). These studies were performed on strands of the cut carotid sinus nerve in exteriorized sheep fetuses and showed carotid chemoreceptor activity in all fetuses examined. The chemoreceptor afferents were stimulated by CO_2-saturated saline and by sodium cyanide injected intraarterially close to the carotid body. They were also

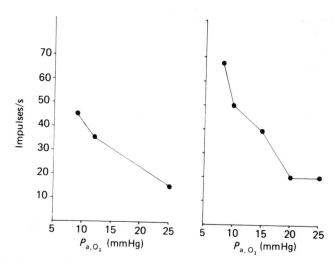

Figure 1 Steady-state hypoxic responses of few-fiber chemoreceptor preparations from two sheep fetuses aged 108 days (left) and 140 days (right). (Reprinted from Blanco et al., 1984a.)

stimulated by clamping the umbilical cord and by steady-state reductions in fetal arterial PaO_2 produced by ventilating the ewe with hypoxic gas mixtures. Figure 1 (Blanco et al., 1984a) shows two examples of steady-state carotid chemoreceptor responses versus PaO_2 in sheep fetuses. Chemoreceptor activity increased as PaO_2 decreased from 25 mm Hg to less than 10 mm Hg. The PaO_2 range over which the chemoreceptors responded was much lower than that of an adult animal (see Lahiri and DeLaney, 1975). The curves are left-shifted when compared to an adult. Some resetting of the carotid chemoreceptors must therefore occur between fetal and adult life. This will be discussed later in this chapter. In this fetal study, no data were presented on chemoreceptor activity under different steady-state levels of $PaCO_2$.

Thus it now appears that carotid chemoreceptors are active in the fetus and that their activity responds to PaO_2 levels that can be expected to occur in fetal life. The contribution of carotid chemoreceptors to physiological reflexes in the fetus is somewhat controversial and has been reviewed elsewhere (Hanson, 1986).

IV. Carotid Chemoreceptor Afferent Nerve Responses in the Neonate

A. Studies in the Lamb

Early studies performed on carotid chemoreceptor afferents from the cut carotid sinus nerve in newborn lambs showed that they increased their discharge rate in response to a transient (10 s) inhalation of a hypoxic gas mixture (10% oxygen in nitrogen) as well as to a transient inhalation (20 s) of a hypercapnic gas mixture (5% CO_2 in air) (Biscoe and Purves, 1967). Chemoreceptor activity decreased in response to inhalation of 100%O_2. No steady-state responses to different levels of PaO_2 and $PaCO_2$ or their potential interaction were reported. The authors stated that they saw no evidence of maturation of the chemoreceptor responses.

More recently, Hanson, Blanco and colleagues have studied the carotid chemoreceptor responses in the sheep fetus and newborn lamb in more detail (Blanco et al., 1984a, 1988; Hanson et al., 1987). When recording from carotid chemoreceptors in term fetuses during initiation of air breathing and clamping of the umbilical cord ("birth"), they found that carotid chemoreceptor activity was virtually silenced by the rise in arterial PO_2 associated with the transition from the "fetal" to the "neonatal" state (Blanco et al., 1984a). In these lambs, after air ventilation, chemoreceptor afferent activity could still be stimulated by injecting CO_2-saturated saline into the carotid artery. This group of investigators also showed that carotid chemoreceptor PaO_2 sensitivity in the lamb is reset over the first week or two after birth toward the adult range of responses (Blanco et al., 1984a, 1988; Hanson et al., 1987).

Figure 2 shows examples of single-fiber responses of carotid chemoreceptors recorded from strands of the cut carotid sinus nerves in a 3-day-old and a 20-day-old lamb. In both, carotid chemoreceptor activity increased as end-tidal PO_2 decreased. In the younger 3-day-old lamb, the chemoreceptor activity actually decreased at the most severe level of hypoxia. Increasing

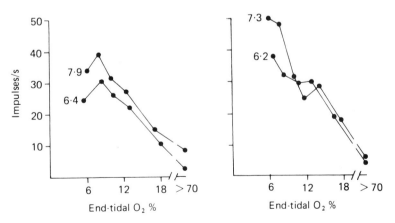

Figure 2 Steady-state isocapnic hypoxic responses of single carotid chemoreceptors from a 3-day-old lamb (left) and a 20-day-old lamb (right). End-tidal CO_2 percentage shown at the ends of the lines. (Reprinted from Blanco et al., 1984a.)

end-tidal P_{CO_2} had some stimulatory effect on the chemoreceptor activity, but there appears to be minimal multiplicative interaction between hypoxic and hypercapnic stimuli especially when compared to the responses of carotid chemoreceptors in the adult cat (Lahiri and DeLaney, 1975). This minimal enhancement of the CO_2 response by hypoxia is also apparent in the responses of the carotid chemoreceptors in newborn piglets (Mulligan, 1988; see below). Unfortunately, the data in Figure 2 on the lamb were reported with respect to end-tidal O_2 and CO_2 pressures instead of arterial O_2 and CO_2 partial pressures. However, it is obvious that these response curves are right-shifted when compared to the fetal response curves shown in Figure 1. Thus a resetting of carotid chemoreceptor P_{O_2} sensitivity does occur with the transition from fetal to postnatal life. Resetting or right shifting of the hypoxic response curves also occurs postnatally in chemoreceptors of newborn piglets (Mulligan, 1988; see below).

The mechanism of the postnatal resetting of the P_{O_2} sensitivity of the carotid chemoreceptors is not known. Blanco et al. (1988) presented evidence that the resetting was dependent at least in part on the increase in Pa_{O_2} that occurs after birth. They ventilated near-term lamb fetuses in utero with either hyperoxic gas or with an N_2/CO_2 mixture for varying periods of time. The lambs were then delivered and prepared for the recording of chemoreceptor afferent responses from strands of the cut carotid sinus nerve. They found that the response curves of carotid chemoreceptors from lambs that had been ventilated with hyperoxic gas for about 27 h were right-shifted when compared to both those ventilated with the same gas for a shorter period of time and those ventilated with an N_2/CO_2 mixture. Thus it appeared that chemoreceptor resetting had begun to occur in utero when the Pa_{O_2} of the fetus was raised in utero to about 180 mm Hg. These results suggest that the mechanism of chemoreceptor resetting is dependent at least in part on a rise in P_{O_2} and the duration of time the P_{O_2} is raised. More research is needed to investigate exactly how oxygen is involved in this mechanism and whether the resetting also involves a change in the cells of the carotid body, a change in blood flow (Acker et al., 1980), or perhaps a change in the neural afferents themselves (Katz, 1990).

B. Studies in the Kitten

Further evidence for the involvement of a rise in P_{O_2} in the postnatal resetting mechanism of the carotid chemoreceptors comes from studies performed on newborn kittens. Hanson et al. (1989) performed a study on carotid chemoreceptors in three groups of kittens that were born into and maintained in a normoxic, hypoxic, or hyperoxic environment. Figure 3 shows data from single- or few-fiber carotid chemoreceptor afferents recorded from strands

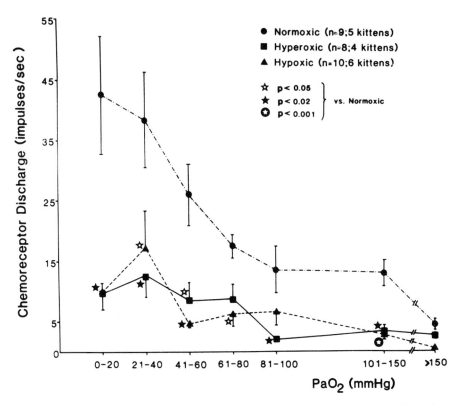

Figure 3 Single- or few-fiber carotid chemoreceptor responses to isocapnic hypoxia for the three groups of kittens on postnatal days 12-23. (●) Normoxic ($PaCO_2$ over the curve, 36.6 ± 1.4 mm Hg, mean $\pm$ SE); (▲) chronically hypoxic ($PaCO_2$, 40.2 ± 2.0 mm Hg); (■) chronically hyperoxic ($PaCO_2$, 36.9 ± 1.1 mm Hg). For statistical comparison by unpaired t-test, the results have been grouped into PaO_2 ranges. Over most ranges, the responses of the chemoreceptors from the chronically hypoxic and hyperoxic kittens were less than those from the normoxic kittens. (Reprinted from Hanson et al., 1989.)

of the cut carotid sinus nerve in kittens from the three aforementioned groups. The kittens were 12 to 23 days old when they were studied, and the control (normoxic) group showed a sustained ventilatory response to hypoxia (Hanson et al., 1989). Chemoreceptor activity (and ventilation) in the control group (Fig. 3) increased in response to isocapnic hypoxia, showing a characteristic hyperbolic-type response similar to that seen for the adult cat (Lahiri et al., 1983). In contrast, the carotid chemoreceptors (and ventilation) from the kittens born into a hypoxic environment ($FIO_2 = 0.13$-0.15) that were never exposed to an increase in PaO_2 showed very little response to hypoxia (Fig. 3). These results suggest that perhaps because the PO_2 was never raised in these chronically hypoxic kittens, the carotid chemoreceptors in these kittens never reset or were delayed in resetting from their fetal state (Hanson et al., 1989).

The carotid chemoreceptor responses (and ventilatory responses) to hypoxia in the chronically hyperoxic ($FIO_2 = 0.30$) group of kittens (Hanson et al., 1989) were also much less than those of the controls (Fig. 3). The explanation for this result is unclear. However, similar results have been found in adult cats that were exposed to hyperoxia and the effect in them was hypothesized to be due to oxygen toxicity (Lahiri et al., 1987).

The studies described above were performed on relatively "older" kittens, i.e., aged 12-23 days. A small amount of research has been done on chemoreceptors from younger kittens. Schwieler (1968) recorded from strands of the cut carotid sinus nerve in kittens from 1 to 19 days old while the kittens breathed air, 10% O_2 in N_2 or 100% O_2. In this study, it was reported that initially both ventilation and carotid chemoreceptor afferent activity increased when the PO_2 of the inspired gas was decreased and that the chemoreceptor activity appeared to be sustained even when ventilation had decreased as a result of the biphasic ventilatory response to hypoxia. Unfortunately no blood gases were taken and no chemoreceptor PO_2 response curves were reported. Thus the postnatal development of the chemoreceptor afferent responses was not described.

Blanco et al. (1984b) recorded from carotid chemoreceptor afferents in strands of the cut carotid sinus nerve in five kittens somewhere in the age range from 5 to 34 days. Chemoreceptor activity increased with decreasing PaO_2, and activity of these fibers was sustained even after ventilation had decreased during hypoxia. These authors did not show any PO_2 response curves and did not describe the postnatal development of the carotid chemoreceptors in the kitten. Also, none of the studies on kittens examined the effect of changes in $PaCO_2$ on carotid chemoreceptor activity or the multiplicative interaction of hypoxic and hypercapnic stimuli over the course of postnatal development. Admittedly, carotid chemoreceptor recording in the kitten is difficult. However, it would be of great value to have a more detailed description of carotid chemoreceptor responses and their development in the newborn kitten because the bulk of what is known about adult carotid chemoreceptor responses comes from research performed on adult cats (see Lahiri et al., 1983).

C. Studies in the Piglet

Recently, the piglet has become a popular model to study various aspects of postnatal development (Cohen et al., 1986; Gootman et al., 1989). The newborn piglet shows a biphasic response to hypoxia (Gootman et al., 1989; Lawson and Long, 1983) and offers the advantage that it is larger than a newborn kitten. As described in Section II. B, the technique of recording carotid chemoreceptor afferent activity from the petrosal ganglion using microelectrodes has been developed for the piglet by Mulligan and co-workers (1989). With this method, the carotid sinus nerve is not cut as it was in the previous studies of carotid chemoreceptors described for the lamb and kitten. Single-fiber carotid chemoreceptor responses to graded levels of PaO_2 and $PaCO_2$ were studied in piglets ranging from 1 to 21 days old (see below). These chemoreceptors were stimulated by hypoxia, by hypercapnia, and by cyanide and were only moderately stimulated by nicotine.

Figure 4 shows composite isocapnic hypoxic response curves from single carotid chemoreceptor afferents in piglets in two age groups: 1-5 days old and 12-20 days old. Chemoreceptor activity increases as PaO_2 decreases, and the curve is shifted to the right in the older piglets. This corresponds to the postnatal resetting of chemoreceptor responses described earlier for chemoreceptors in the lamb (Blanco et al., 1984a; Hanson et al., 1987). As for the lamb, it was much more difficult to find chemoreceptors in the piglet on the first day of postnatal life (Mulligan, unpublished observations). One noticeable difference between the lamb and the piglet is that the rate of firing of the chemoreceptor afferent fibers appears to be greater in the lamb than in the piglet (Figs. 2 and 4). This could be due to a species difference or to differences in the recording methods: The carotid sinus nerve and carotid sinus nerve efferents were intact in the piglet and were cut in the lambs.

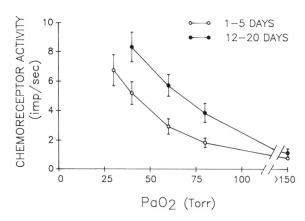

Figure 4   Single-cell carotid chemoreceptor afferent responses to isocapnic hypoxia in young (1-5 day old) and older (12-20 day old) piglets. (Mean ± SE, $n = 10$ for each age group.) Chemoreceptor activity was significantly greater in the older group at all PaO_2 levels except those associated with hyperoxia. (Data from E. Mulligan.)

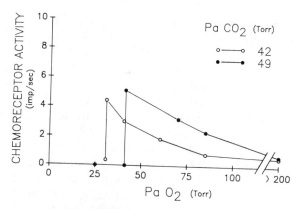

Figure 5 Single-cell carotid body chemoreceptor afferent responses in a 5-day-old piglet at different levels of steady-state PaO_2 and $PaCO_2$. See text for details. (Data from Mulligan et al., 1989.)

Figure 5 shows an example of hypoxic response curves from a single chemoreceptor afferent at two different $PaCO_2$ levels in a 5-day-old piglet (Mulligan et al., 1989). Chemoreceptor activity is somewhat greater at the higher level of $PaCO_2$ at any given level of PaO_2. At low PaO_2, however, chemoreceptor activity increased initially and was maintained at a relatively high level for a period of time and then decreased to a very low level of activity at the same PaO_2. This phenomenon, termed the nonsustained response to hypoxia, will be discussed further later in this chapter. Figure 6 shows an example of the responses of the same chemoreceptor as in Figure 5 plotted as $PaCO_2$ response curves at various levels of PaO_2. There is some increase

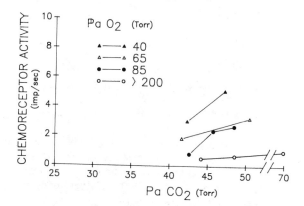

Figure 6 Single-cell carotid chemoreceptor afferent responses to CO_2 in a 5-day-old piglet at different levels of PaO_2. (Data from E. Mulligan.)

in the slope of the response to $PaCO_2$ as PaO_2 decreases, but this multiplicative interaction between these stimuli is much less than that seen, for example, in the responses of chemoreceptors from the adult cat (Lahiri and DeLaney, 1975). The slopes of the $PaCO_2$ response curves, i.e., the chemoreceptor CO_2 sensitivity at various levels of PaO_2 from piglet chemoreceptors, are contrasted with those of adult cat chemoreceptors (Lahiri et al., 1981) in Figure 7. No development of this response was found to occur in piglets up to 3 weeks of age, so data from all piglets were pooled (Mulligan, unpublished observations). The piglet chemoreceptors showed minimal multiplicative interaction between hypoxic and hypercapnic stimuli (Mulligan, 1988). The multiplicative interaction in the responses of the cat chemoreceptors are reflected in the ventilatory responses of the cat (Lahiri and DeLaney, 1975), which are qualitatively similar to those of adult humans (Neilsen and Smith, 1952). Whether the piglet chemoreceptor responses parallel ventilation in the piglet is not known. Also, whether more pronounced multiplicative interaction develops in the chemoreceptor responses of piglets older than 21 days is not known.

Although multiplicative interaction of hypoxic and hypercapnic stimuli in lamb chemoreceptors was not studied in detail, the data in Figure 2 suggest that as for the piglet, there is also minimal multiplicative interaction in

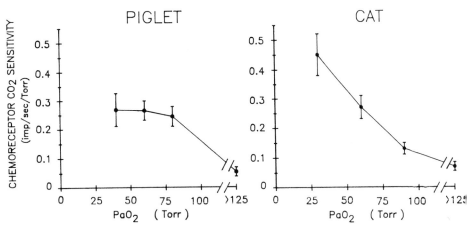

Figure 7 Effect of PaO_2 on the slopes of carotid body chemoreceptor afferent $PaCO_2$ response curves in piglets (1-21 days, mean $\pm$ SE, $n = 10$, left) and adult cats (mean $\pm$ SE, $n = 15$, right) (Slopes of $PaCO_2$ response curves are expressed as chemoreceptor sensitivity in impulses $\cdot$ sec^{-1} $\cdot$ torr PCO_2^{-1}.) (Piglet data from E. Mulligan; adult cat data from Lahiri et al., 1981.)

the responses of newborn lamb chemoreceptors. Future investigation of the differences in chemoreceptor O_2/CO_2 multiplicative interaction between species and during postnatal development may shed light on the mechanisms of O_2 and CO_2 sensing in the carotid body.

The phenomenon of the nonsustained response to hypoxia or "fade-out" of some piglet carotid chemoreceptors is intriguing (Fig. 5). All chemoreceptors studied did not show this type of response. When this response was observed, chemoreceptor activity increased with the onset of hypoxia and remained at the elevated level for a period of time, after which it decreased to a lower level (Fig. 5). If PaO_2 was raised to a hyperoxic level, chemoreceptor activity recovered and the response to hypoxia could be repeated (Mulligan and Bhide, 1989). Chemoreceptor afferent fibers that displayed this behavior were found more often in younger piglets than in older piglets (Mulligan, unpublished observations). The time course of the nonsustained chemoreceptor response to hypoxia is similar to that of the biphasic ventilatory response (Lawson and Long, 1983), but it is not yet known if these responses parallel each other or if there is any relationship between them. To evaluate this, the simultaneous recording of chemoreceptor activity and ventilation or phrenic nerve activity would be extremely useful.

Also, information is needed on how prevalent the chemoreceptor nonsustained response to hypoxia is both within and across species during postnatal development. More work needs to be done on single-fiber chemoreceptor responses as well as on the evaluation of the "whole nerve" or "sum total" of the afferent responses of the carotid body. In this way the total potential input from the carotid body into ventilation can be assessed. A recently developed electronic time-gating circuit to screen out baroreceptor activity in carotid sinus nerve recordings will prove useful in this regard (Philips and Mulligan, 1989).

The nonsustained hypoxic responses of some piglet chemoreceptors are different from the sustained hypoxic responses reported for the kitten and lamb chemoreceptors. However, the studies in the kitten and lamb were performed on chemoreceptor fibers in strands of the cut carotid sinus nerve (e.g., Blanco et al., 1984 a,b) whereas the studies in the piglet were performed with the carotid sinus nerves intact (Mulligan et al., 1989). Thus, in the piglet experiments, any efferent effects on carotid chemoreceptor afferent activity would be functional. Nothing is known of the effects of carotid sinus nerve efferents in the neonate. In the adult cat, however, carotid sinus nerve efferent input to the carotid body decreases carotid chemoreceptor afferent activity (Neil and O'Regan, 1971). It is possible, therefore, that carotid sinus nerve efferents could be involved in the "fadeout" of carotid chemoreceptor activity during hypoxia. The role of efferents in this and in all neonatal carotid body responses can be evaluated using the method described earlier in this chapter of recording carotid chemoreceptor afferent activity from the petrosal ganglion using microelectrodes (Jordan et al., 1987; Mulligan et al., 1989). The central connection of the ganglion either can be left intact leaving the efferents intact or can be cut to remove any efferent effects.

Although nonsustained responses to hypoxia of carotid chemoreceptors were not described in the work published on the kitten and lamb, the curves depicted in Figure 2 (Blanco et al., 1984a) for the carotid chemoreceptor responses in the 3-day-old lamb show that during severe hypoxia in this lamb, carotid chemoreceptor activity was decreased from its maximum level. Also, in ventilatory studies on the development of the chemoreflex in newborn lambs, Bureau and colleagues (1984, 1985b, 1986) and Carroll and Bureau (1987) have presented evidence that the ventilatory input from the carotid bodies diminishes with time during an exposure to hypoxia. They hypothesized that the carotid bodies contribute to the biphasic ventilatory response to hypoxia. The finding that not all chemoreceptor fibers in newborn piglets show sustained responses to hypoxia and that these types of fibers are more numerous in younger piglets also suggests that the carotid bodies may be involved in the biphasic response to hypoxia. However, the population response of chemoreceptor fibers needs to be better described before its role in the biphasic hypoxic response can be properly evaluated.

If the carotid body afferent activity in response to hypoxia in the newborn is largely sustained, it is still possible that carotid body afferent activity has less effect on ventilation as the exposure to hypoxia is maintained. For example, the central nervous system may process the same afferent signals differently due to a change in the threshold or properties of the postsynaptic neurons of the first synapse in the CNS (Lahiri et al., 1978; Schwieler, 1968; see also Chapter 20, this volume). There may be a decrease in the transmission of the chemoreceptor afferent signal due to a depletion of neurotransmitters in the immature afferent fibers themselves (Katz, 1990). It would therefore be of great interest in the newborn to study not only the carotid body chemoreceptor afferent activity, but also the transmission of this activity at the first synapse in the CNS under different PaO_2 and $PaCO_2$ conditions.

Recently, several laboratories have begun to study the intracellular and membrane properties of cells isolated from the carotid body (e.g., Delpiano and Hescheler, 1989; Duchen et al., 1988; Fishman and Schaffner, 1984; Lopez-Barneo et al., 1988; Nurse, 1987; Pietruschka, 1974). Perhaps studies such as these will reveal if there are differences between carotid body cells in the neonate and carotid body cells in the adult that can account for the differences in their afferent hypoxic and hypercapnic responses. This includes exploring a potential cellular origin of the hypoxic fadeout, $O_2/$

CO_2 multiplicative interaction, and the resetting of the chemoreceptor PO_2 response range from pre- to postnatal life.

Katz (1990) and Katz and Erb (1989) have begun to study the biochemical development of carotid body afferent neurons that contain the enzyme tyrosine hydroxylase. In the rat, at birth, there is a large increase in the tyrosine hydroxylase content of these neurons (Katz, 1990; Katz and Erb, 1989). Postnatally, both the tyrosine hydroxylase and catecholamine contents of these neurons increase (Katz, 1990). The biochemical immaturity of the carotid body afferent nerves at birth could contribute to their functional immaturity and might also contribute to the biphasic response to hypoxia in the neonate (Katz, 1990). Also, the postnatal increase in tyrosine hydroxylase is paralleled by an increase in the mRNA for tyrosine hydroxylase; this suggests that there may be ongoing regulation of the tyrosine hydroxylase gene postnatally (Katz, D. M., Moracic-Piperski, V., and Siegel, R. E., personal communication). More work such as this is needed in other neonatal animal models such as the lamb, kitten, and piglet.

V. Summary

This chapter has presented a discussion of the present state of knowledge regarding the discharge properties of carotid body chemoreceptors during development and has outlined many directions for future research in this area. To date we have some description of chemoreceptor afferent responses in the fetus and newborn. Carotid chemoreceptors in the neonate are stimulated by hypoxia and hypercapnia, and the stimulation of some neonatal chemoreceptors by hypoxia is not sustained as it is in the adult. A resetting, i.e., right shifting of the carotid chemoreceptor PaO_2 response curves, occurs postnatally. More detailed carotid chemoreceptor oxygen and carbon dioxide response curves and information on the interaction of hypoxia and hypercapnia of neonatal chemoreceptor responses are needed. At present, information on the afferent responses to CO_2 is especially lacking.

In addition to the issues discussed in this chapter, other areas of research need to be pursued with regard to the peripheral chemoreceptors in the neonate. To date, very little research has been performed on the responses and development of aortic body chemoreceptors in the fetus and newborn (Kumar and Hanson, 1989; Ponte and Purves, 1973). In the adult cat, differences between the carotid and aortic body chemoreceptor responses have been described (Lahiri et al., 1983). Whether this is true in the newborn is unclear. The pharmacology of the peripheral chemoreceptors in the neonate has not yet been directly examined. It will be interesting to investigate the effects on the carotid chemoreceptor responses of various drugs that affect ventilation in neonates and to evaluate how the chemoreceptors are involved in the overall response to various respiratory stimulants and depressants. Research is needed to investigate why there is such a high mortality rate in neonatal animals that have had their carotid sinus nerves sectioned (Bureau et al., 1985a; Donnelly and Haddad, 1987; Haddad and Donnelly, 1988; Hofer, 1984). Abnormalities have been reported in the carotid bodies of victims of the sudden infant death syndrome (SIDS) (Cole et al., 1979; Naeye et al., 1976; Perrin et al., 1984; Shannon and Kelly, 1982). It is as yet unknown whether there is any direct relationship between the development of the carotid body and SIDS.

Although an ever-increasing amount of information is becoming available on the development of the peripheral chemoreceptors, there is still an enormous amount of research that needs to be done in order to fully understand the responses and the roles of the carotid and aortic bodies in the chemoreflex control of the respiratory and circulatory systems in the fetus and neonate.

References

Acker, H., Lübbers, D. W., Purves, M. J., and Tan, E. D. (1980). Measurement of the partial pressure of oxygen in the carotid body of foetal sheep and newborn lambs. *J. Dev. Physiol.* 2:323-338.

Belenky, D. A., Standaert, T. A., and Woodrum, D. E. (1979). Maturation of hypoxic ventilatory response of the newborn lamb. *J. Appl. Physiol.* **47**:927-930.

Biscoe, T. J., and Purves, M. J. (1967). Carotid body chemoreceptor activity in the new-born lamb. *J. Physiol.* **190**:443-454.

Biscoe, T. J., Purves, M. J., and Sampson, S. R. (1969). Types of nervous activity which may be recorded from the carotid sinus nerve in the sheep foetus. *J. Physiol.* **202**:1-23.

Blanco, C. E. , Dawes, G. S., Hanson, M. A., and McCooke, H. B. (1984a). The response to hypoxia of arterial chemoreceptors in fetal sheep and new-born lambs. *J. Physiol.* **351**:25-37.

Blanco, C. E., Hanson, M. A., Johnson, P., and Rigatto, H. (1984b). Breathing pattern of kittens during hypoxia. *J. Appl. Physiol.* **56**:12-17.

Blanco, C. E., Hanson, M. A., and McCooke, H. B. (1988). Effects on carotid chemoreceptor resetting of pulmonary ventilation in the fetal lamb in utero. *J. Dev. Physiol.* **10**:167-174.

Brady, J. P., and Ceruti, E. (1966). Chemoreceptor reflexes in the new-born infant: Effects of varying degrees of hypoxia on heart rate and ventilation in a warm environment. *J. Physiol.* **184**:631-645.

Bureau, M. A., and Begin, R. (1982). Postnatal maturation of the respiratory response to O₂ in awake newborn lambs. *J. Appl. Physiol.* **52**: 428-433.

Bureau, M. A., Cote, A., Blanchard, P. W., Hobbs, S., Foulon, P., and Dalle, D. (1986). Exponential and diphasic ventilatory response to hypoxia in conscious lambs. *J. Appl. Physiol.* **61**:836-842.

Bureau, M. A., Lamarche, J., Foulon, P., and Dalle, D. (1985a). Postnatal maturation of respiration in intact and carotid body chemodenervated lambs. *J. Appl. Physiol.* **59**:869-874.

Bureau, M. A., Lamarche, J., Foulon, P., and Dalle, D. (1985b). The ventilatory response to hypoxia in the newborn lamb after carotid body denervation. *Respir. Physiol.* **60**:109-119.

Bureau, M. A., Zinman, R., Foulon, P., and Begin, R. (1984). Diphasic ventilatory response to hypoxia in newborn lambs. *J. Appl. Physiol.* **56**:84-90.

Carroll, J. L., and Bureau, M. A. (1987). Decline in peripheral chemoreceptor excitatory stimulation during acute hypoxia in the lamb. *J. Appl. Physiol.* **63**:795-802.

Cohen, H. L., Steele, A. M., Eberle, L. P., and Gootman, P. M. (1986). Development of central respiratory function in swine. In *Swine in Biomedical Research.* Vol. 2. Edited by M. E. Tumbleson. New York, Plenum Press, pp. 1289-1295.

Cole, S., Lindenberg, L. B., Galioto, F. M., Howe, P. E., DeGraff, A.C., Davis, J. M., Lubka, R., and Gross, E. M. (1979). Ultrastructural abnormalities of the carotid body in sudden infant death syndrome. *Pediatrics* **63**:13-17.

Delpiano, M. A., and Hescheler, J. (1989). Evidence for a PO_2-sensitive K^+ channel in the type-I cell of the rabbit carotid body. *FEBS Lett.* **249**:195-198.

Donnelly, D. F., and Haddad, G. G. (1987). Severe respiratory changes and neonatal death in chemodenervated young piglets. *Fed. Proc.* **46**:657.

Duchen, M. R., Caddy, K. W. T., Kirby, G. C., Patterson, D. L., Ponte, J., and Biscoe, T. J. (1988). Biophysical studies of the cellular elements of the rabbit carotid body. *Neuroscience* **26**:291-311.

Fishman, M. C., and Schaffner, A. E. (1984). Carotid body cell culture and selective growth of glomus cells. *Am. J. Physiol.* **246** *(Cell Physiol.* **15***)*: C106-C113.

Gootman, P. M., Sica, A. L., Steele, A. M., and Cohen, H. L. (1989). Responses of phrenic, recurrent laryngeal, and hypoglossal motoneurons to hypoxia in neonatal swine. In *Chemoreceptors and Reflexes in Breathing: Cellular and Molecular Aspects.* Edited by S. Lahiri, R. E. Forster II, R. O. Davies, and A. I. Pack. New York, Oxford University Press, pp. 351-360.

Haddad, G. G., and Donnelly, D. F. (1988). The interaction of chemoreceptors and baroreceptors with the central nervous system. *Ann. NY Acad. Sci.* **533**:221-227.

Haddad, G. G., and Mellins, R. B. (1984). Hypoxia and respiratory control in early life. *Annu. Rev. Physiol.* **46**:629-643.

Hanson, M. A. (1986). Peripheral chemoreceptor function before and after birth. In *Respiratory Control and Lung Development in the Fetus and Newborn.* Edited by B. M. Johnston and P. D. Gluckman. Ithaca, NY, Perinatology Press, pp. 311-330.

Hanson, M. A., Eden, G. J., Nijhuis, J. G., and Moore, P. J. (1989). Peripheral chemoreceptors and other oxygen sensors in the fetus and newborn. In *Chemoreceptors and Reflexes in Breathing: Cellular and Molecular Aspects.* Edited by S. Lahiri, R. E. Forster II, R. O. Davies, and A. I. Pack. New York, Oxford University Press, pp. 113-120.

Hanson, M. A., Kumar, P., and McCooke, H. B. (1987). Post-natal re-setting of carotid chemoreceptor sensitivity in the lamb. *J. Physiol.* **382**:57P.

Hertzberg, T., and Lagercrantz, H. (1987). Postnatal sensitivity of the peripheral chemoreceptors in newborn infants. *Arch. Dis. Child.* **62**:1238-1241.

Heymans, C., and Neil, E. (1958). *Reflexogenic Areas of the Cardiovascular System.* London, Churchill.

Hofer, M. A. (1984). Lethal respiratory disturbance in neonatal rats after arterial chemoreceptor denervation. *Life Sci.* **34**:489-496.

Jansen, A. H., Purves, M. J., and Tan, E. D. (1980). The role of sympathetic nerves in the activation of the carotid body chemoreceptors at birth in the sheep. *J. Dev. Physiol.* **2**:305-321.

Jordan, D., Donoghue, S., Felder, R. B., and Spyer, K. M. (1987). Central terminations of carotid body chemoreceptor afferents. In *Chemoreceptors in Respiratory Control.* Edited by J. A. Ribeiro and D. J. Pallot. London, Croom Helm, pp. 29-38.

Katz, D. M. (1990). Molecular mechanisms of carotid body afferent neuron development. In *Biology of Oxygen Adaptation: Organ to Organelle.* Edited by S. Lahiri, N. S. Cherniack, and R. S. Fitzgerald. New York, Oxford University Press (in press).

Katz, D. M., and Erb, M. J. (1990). Developmental regulation of tyrosine hydroxylase expression in primary sensory neurons of the rat. *Dev. Biol.* **137**(2):233-242.

Kumar, P., and Hanson, M. A. (1989). Resetting of the hypoxic sensitivity of aortic chemoreceptors in the newborn lamb. *J. Dev. Physiol.* **11** (4):199-206.

Lahiri, S., Brody, J. S., Motoyama, E. K., and Velasquez, T. M. (1978). Regulation of breathing in newborns at high altitude. *J. Appl. Physiol.* **44**:673-678.

Lahiri, S., and DeLaney, R. G. (1975). Stimulus interaction in the responses of carotid body chemoreceptor single afferent fibers. *Respir. Physiol.* **24**:249-266.

Lahiri, S., Mokashi, A., Mulligan, E., and Nishino, T. and (1981). Comparison of aortic and carotid chemoreceptor responses to hypercapnia and hypoxia. *J. Appl. Physiol.* **51**:55-61.

Lahiri, S., Mulligan, E., Andronikou, S., Shirahata, M., and Mokashi, A. (1987). Carotid body chemosensory function in prolonged normobaric hyperoxia in the cat. *J. Appl. Physiol.* **62**:1924-1931.

Lahiri, S., Smatresk, N. J., and Mulligan, E. (1983). Responses of peripheral chemoreceptors to natural stimuli. In *Physiology of the Peripheral Arterial Chemoreceptors*. Edited by H. Acker and R. G. O'Regan. Amsterdam, Elsevier, pp. 221-256.

Lawson, E. E., and Long, W. A. (1983). Central origin of biphasic breathing pattern during hypoxia in newborns. *J. Appl. Physiol.* **55**:483-488.

Lopez-Barneo, J., Lopez-Lopez, J. R., Urena, J., and Gonzalez, C. (1988). Chemotransduction in the carotid body: K^+ current modulated by Po_2 in type I chemoreceptor cells. *Science* **241**:580-582.

Mayock, D. E., Standaert, T. A., Guthrie, R. D., and Woodrum, D. E. (1983). Dopamine and carotid body function in the newborn lamb. *J. Appl. Physiol.* **54**:814-820.

Mulligan, E. (1988). Single fiber carotid body chemoreceptor responses in the piglet. *FASEB J.* **2**:A1293.

Mulligan, E., Alsberge, M., and Bhide, S. (1990). Carotid chemoreceptor recording in the newborn piglet. In *Arterial Chemoreception*. Edited by C. Eyzaguirre, S. J. Fidone, R. S. Fitzgerald, S. Lahiri, and D. McDonald. New York, Springer, pp. 285-289.

Mulligan, E., and Bhide, S. (1989). Non-sustained responses to hypoxia of carotid body chemoreceptor afferents in the piglet. *FASEB J.* **3**:A399.

Mulligan, E., and Lahiri, S. (1981). Dependence of carotid chemoreceptor stimulation by metabolic agents on Pao_2 and $Paco_2$. *J. Appl. Physiol.* **50**:884-891.

Naeye, R. L., Fisher, R., Ryser, M., and Whalen, P. (1976). Carotid body in sudden infant death syndrome. *Science* **191**:567-569.

Neil, E., and O'Regan, R. G. (1971). The effects of electrical stimulation of the distal end of the cut sinus and aortic nerves on peripheral arterial chemoreceptor activity in the cat. *J. Physiol. (London)* **215**:15-32.

Neilsen, M., and Smith, H. (1952). Studies on the regulation of respiration in acute hypoxia. *Acta. Physiol. Scand.* **24**:293-313.

Nurse, C. A. (1987). Localization of acetylcholinesterase in dissociated cell cultures of the carotid body of the rat. *Cell Tissue Res.* **250**:21-27.

Perrin, D. G., Cutz, E., Becker, L. E., Bryan, A. C., Madapallimatum, A., and Sole, M. J. (1984). Sudden infant death syndrome: Increased carotid body dopamine and noradrenaline content. *Lancet* **2**:535-537.

Philips, C. M., and Mulligan, E. M. (1990). An electronic gating circuit for separation of chemoreceptor from baroreceptor activity. In *Arter-*

ial Chemoreception. Edited by C. Eyzaguirre, S. J. Fidone, R. S. Fitzgerald, S. Lahiri, and D. McDonald. New York, Springer, pp. 243-246.

Pietruschka, F. (1974). Cytochemical demonstration of catecholamines in
cells of the carotid body in primary tissue culture. *Cell Tissue Res.* **151**:
317-321.

Ponte, J., and Purves, M. J. (1973). Types of afferent nervous activity which
may be measured in the vagus nerve of the sheep foetus. *J. Physiol.*
299:51-76.

Schramm, C. M., and Grunstein, M. M. (1987). Respiratory influence of
peripheral chemoreceptor stimulation in maturing rabbits. *J. Appl.
Physiol.* **63**:1671-1680.

Schwieler, G. H. (1968). Respiratory regulation during postnatal development in cats and rabbits and some of its morphological substrate. *Acta
Physiol. Scand.* **72** (Suppl. 304): 1-123.

Shannon, D. C., and Kelly, D. H. (1982). SIDS and Near-SIDS. *N. Engl. J.
Med.* **306**:959-965.

12

Central Respiratory Chemoreceptors
Cellular Mechanisms and Developmental Aspects

EUGENE E. NATTIE

Dartmouth Medical School
Hanover, New Hampshire

I. Introduction

The subject of central chemoreception in adult mammals remains a lively and controversial topic with excellent recent reviews (for example, Loeschcke, 1982; Bledsoe and Hornbein, 1981; Feldman, 1986; Millhorn and Eldridge, 1986; Bruce and Cherniack, 1987). In this chapter, within the context of a brief general review I shall focus on three areas in which recent data and new approaches provide possible new insights: (1) the location of central chemoreceptors, (2) the role of cells within the ventrolateral medulla (VLM) chemosensitive zones in more than just chemosensitivity, and (3) theories on the mechanism of central chemosensitivity. The available data on central chemoreception during development will then be examined. The hope is that new approaches now being applied to this problem in adults will be extended to the study of central chemoreception in development.

II. Evidence for the Existence of Central Chemoreception

The early history of this topic has been well reviewed by Bledsoe and Hornbein (1981). Leusen (1954) first unequivocally demonstrated central chemoreception by showing clear stimulation of ventilation by ventriculocisternal perfusion of acidic solutions in peripherally chemodenervated anesthetized dogs. This followed observations that anesthesia preferentially depressed central chemoreception in comparison to other respiratory reflexes (von Euler and Soderberg, 1952), which suggested chemoreception involved a separate process. This paper also emphasized the important effect of anesthesia in depressing central chemoreception, a theme that cannot be over-emphasized (see Pappenheimer et al., 1965; Cozine and Ngai, 1967; Nattie, 1983, for examples).

Many researchers have since shown that intracerebral infusion of acidic solutions can stimulate ventilation (for example, see Loeschcke et al., 1958; Loeschcke, 1982; Pappenheimer et al., 1965; Nattie, 1986a), with the effects being most dramatic in unanesthetized animals (Pappenheimer et al., 1965; Nattie, 1986a). The central chemoreceptor theme progressed to the extreme view of Fencl et al. (1966), who hypothesized that most, if not all, chemosensitivity could be attributed to central chemoreceptors sensing $[H^+]$ at a location estimated to be along the $[H^+]$ gradient between capillary and large cerebrospinal fluid cavities.

The existence of central chemoreceptors has also been shown by the continued presence of responses to increased CO_2 or intravenous infusions of acidic solutions in animals with peripheral chemodenervation. This evaluation of central chemoreceptors would seem to be a simple way to test the hypothesis of Fencl et al. (1966). However, the degree of the response in chemodenervated animals has been quite variable, with species, anesthesia, experimental duration, and stimulus intensity being possible explaining factors (see Nattie, 1983, for references). For example, in unanesthetized rabbits infused intravenously with acidic fluid, those with denervation of the carotid bodies (known in this species to represent the sole peripheral ventilatory chemoreceptor) had only 1/3 of the ventilatory response observed in intact rabbits (Nattie, 1983); this result does not support the concept of central chemoreceptor dominance. However, the study of Nattie (1983) was performed acutely while the preparation of Fencl et al. (1966) was more chronic. This issue of the relative roles and importance of central versus peripheral chemoreceptors is an important one that is difficult to study and could have special importance during development.

III. Location of Central Chemoreceptors

The evolution of what is now considered the traditional location of central chemoreceptors on the surface of the ventrolateral medulla began with the observation of the particular effectiveness of acid infusions at the lateral recesses of the fourth ventricle (Loeschcke et al., 1958). Mitchell et al. (1963a) then reported large effects with subarachnoid perfusions which bathed the entire VLM surface. The application of cotton pledgets soaked in an acidic mock cerebrospinal fluid to the VLM surface identified a rostral chemosensitive area (Mitchell et al., 1963b). (See Fig. 1.) A second, more caudal chemosensitive area was delineated by Loeschcke and colleagues (Loeschcke, 1982), and an area in between that was not chemosensitive but appeared important in the control of breathing was described (Schlaefke et al., 1970). The rostral area is referred to as Mitchell's area, the caudal as Loeschcke's area, and an area lying just rostral to Loeschcke's area as the intermediate area (IMA) or Schlaefke's area. (See Fig. 2.)

This VLM surface topography of central chemoreceptor location is a tidy picture, well entrenched in the educational literature, and has experimental support by others (see Bruce and Cherniack, 1987; Millhorn and Eldridge, 1986, for references), but not all have been able to duplicate these experiments. While Hori et al. (1970) observed responses to the surface application of acidic solutions in anesthetized rats, Malcolm et al. (1980) did not, and they, noting large effects of changes in surface temperature, wondered if the ventilatory effects of surface applications could be attributed to this temperature sensitivity. Lipscomb and Boyarski (1972) proposed that applications of substances on the VLM surface actually affected structures deep into the medulla, with substances being carried there by penetrating surface vessels. Cragg et al. (1977) emphasized the degree of vascular penetration from the ventral surface. Substances applied to the VLM surface could penetrate deeply via diffusion into the surface vessels or by following the spaces surrounding these vessels or the crypt-like spaces that have been demonstrated on the VLM surface (Dermietzel, 1976). Measurement of the depth of penetration of large substances placed on the VLM surface (Borison et al., 1980) and of smaller radiolabeled neuroactive substances (Keeler et al., 1984; Nattie et al., 1988a,b; Yamada et al., 1984) supports this concept.

In my laboratory we have found that bathing the entire VLM surface with acidic mock CSF augments phrenic activity in chloralose-urethane-anesthetized cats and that application of acidic cotton pledgets to the rostral (Mitchell's) area also augments phrenic activity. However, the responses to pledgets were small (5-12%), required low pH values (7.0), and were not reliable; i.e., second and third applications at the same site did not result in as great a response. This low sensitivity may be attributable in part to the depressive effect of anesthesia on central chemosensitivity (see, for example, Cozine and Ngai, 1967). The use of decerebrate cats to remove the effects of anesthesia did result in a small increase in the magnitude of the response but did not improve the reliability (Nattie et al., 1988). The low sensitivity to pledget applications could also be explained by central chemoreceptors being distributed at more than one site just below the VLM surface and/or

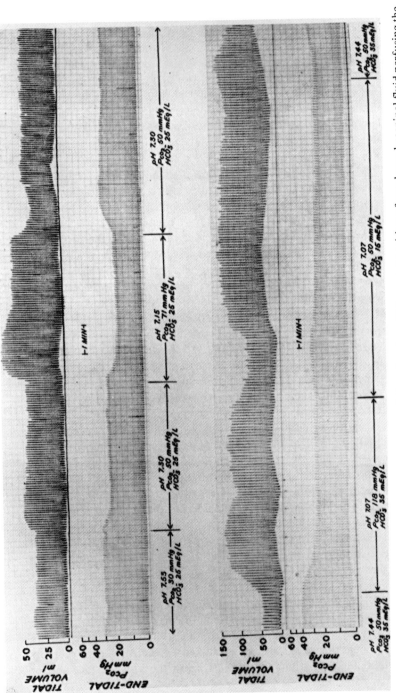

Figure 1   Effects on tidal volume and end-tidal P_{CO_2} of changing the acid-base composition of mock cerebrospinal fluid perfusing the subarachnoid space around the medulla in two anesthetized, vagotomized, glomectomized cats. In the top tracings HCO^-_3 was constant and pH and P_{CO_2} varied; in the bottom H^+ was constant and P_{CO_2} and HCO_3 varied. In each case perfusion with an acidic mock CSF had a rapid and dramatic stimulatory effect on ventilation. (From Mitchell et al., 1963a, with permission of the American Physiological Society.)

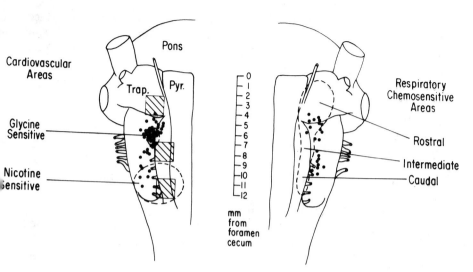

Figure 2 Diagramatic representation of the ventral surface of the medulla of the cat. On the right side the respiratory chemosensitive areas are outlined by dashed lines; on the left the pressor (glycine) and depressor (nicotine) cardiovascular areas are noted. The solid circles show the glutamate microinjection sites of McAllen (1986) just below the ventral surface that stimulated blood pressure (left side) and/or phrenic output (right side). Note the correspondence of these sites with the traditional chemosensitive areas. The hatched squares show the sites of pledget applications of kainic acid, DEPC, and muscarinic agonists and antagonists (Nattie, 1986b, 1988; Nattie et al., 1988a,b; 1989). (From Nattie et al., 1988b, with permission of the American Physiological Society.)

lying deep to the VLM surface (see below) so that the pledgets would stimulate only a small proportion of them.

In spite of some controversy, most investigators still consider the VLM surface to provide access to central chemoreceptors. This does not mean that access is entirely via the VLM surface or that this access site is most important in physiological terms. Access from blood is well demonstrated and can occur quickly (Borison et al., 1978; Nattie, 1983). Whether the VLM surface approach provides access to all chemoreceptors and whether all chemoreceptors need to be stimulated to obtain a maximum response are questions without clear answers.

Structures just below the VLM surface or in deeper regions have been examined in terms of responsiveness to CO_2 or acidic stimulation (see Loeschcke, 1982, for references). Lipscomb and Boyarski (1972) reported that none of 68 units recorded just below the rostral area responded to surface perfusion of acidic fluid, although examination of their results (Fig. 1 in their paper) suggests that four to eight units did increase firing, a proportion not too different from studies reviewed below. Pokorski (1976) found 9/44 responsive units just below the rostral chemosensitive area; von Euler and Soderberg (1952) found a smaller number of receptors deeper within the medulla at the level of the caudal chemosensitive area and IMA. Mitchell and Herbert (1974) observed no responses in 164 cells recorded intracellularly within the nucleus ambiguous and retroambiguous, i.e., the ventral respiratory group.

Arita and colleagues (1988a,b) infused CO_2-equilibrated saline via the verterbral artery and found 39/46 units that responded in phase with increased ventilation at depths essentially through the entire medulla below the IMA and caudal chemosensitive area. Of these 39 units, 18 were responsive to intravertebral CO_2 only ("C" units) and 21 were also responsive to carotid body stimulation ("CP" units). Other units were excited by nociceptive stimulation or combinations of stimuli. The C units clustered anatomically in three areas (see Fig. 3): (1) near the VLM surface lateral to the inferior olive, (2) in the rostral aspect of the ventral respiratory group ventral and dorsal to the nucleus ambiguous and the retrofacial nucleus, and (3) ventral to the nucleus tractus solitarious. The CP units were localized largely to the area of the nucleus paragigantocellularis lateralis. These workers did not examine more rostral sites but did evaluate brain tissue pH profiles (Arita et al., 1989) and showed that not all medullary regions had the same pH response to intravertebral CO_2 infusion and the areas with larger decreases in pH seemed to be located in proximity to the sites of the C units. Whether these C units are directly responsive to the stimulus or are neurons in the response pathway or are respiratory neurons is unknown.

Local H^+ microinjection in this same region (Arita, personal communication) showed that 16 of 164 units were pH sensitive. Of these 16, 10 were also stimulated by intravertebral artery CO_2 injection and were unreactive to nociceptive stimulation or to intracarotid CO_2 injection. These 10 units were located in two sites: (1) below the caudal VLM chemosensitive area near or within the lateral reticular nucleus dorsolateral to the inferior olive and (2) ventral or lateral to the nucleus tractus solitarius. In a similar study in approximately the same location, Marino and Lamb (1975) found 2 of 74 units responsive to CO_2 and local H^+ located in (1) the lateral tegmental field and (2) 1 mm ventral to the nucleus ambiguous. Whether these cells, sensitive to local pH changes, are respiratory or are involved in the chemoreceptor response is still uncertain.

Studies of brain slices are complementary to these in situ studies. Fukuda and Honda (1975, 1976) and Fukuda and Loeschcke (1977) found cells

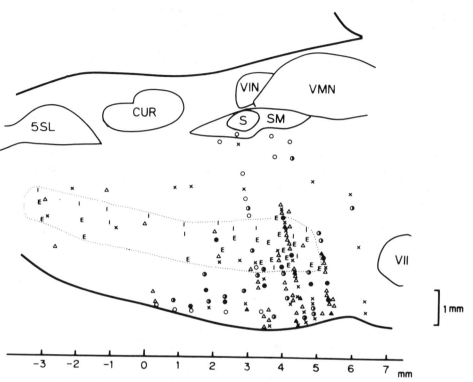

Figure 3 Schematic of a saggittal plane section of the medulla of the cat 3 mm lateral to the midline. The dotted line shows the ventral respiratory group. The symbols represent sites of extracellular recordings of units that are inspiratory (I), expiratory (E), responsive to nociceptive stimulation (△ ▲), to central chemoreceptor stimulation (○), to central and peripheral chemoreceptor stimulation (◑), to central and peripheral chemoreceptor and nociceptive stimulation (●), and nonresponsive units (x). Note the distribution of chemosensitive units both below the surface and deeper near the nucleus of the solitary tract. 5SL, Laminar spinal trigeminal nucleus; CUR, cuneate nucleus, rostral division; VIN, inferior vestibular nucleus; VMN, medial vestibular nucleus; S, solitary tract; SM, medial nucleus of the solitary tract; VII, facial nucleus. (From Arita et al., 1988b, with permission of Elsevier Science Publishers.)

responsive to acidic perfusates in slices from the VLM surface in the region of the IMA and caudal chemosensitive area in rat and cat as well as a smaller number of cells in a more dorsal slice near the nucleus tractus solitarius (Fukuda and Loeschcke, 1977). The use of a low Ca^{2+}/high Mg^{2+} medium to block synaptic transmission blocked the response in the ventral but not in the dorsal slice, which suggests that the ventral pH effect was synaptic. Miles (1983) found 8/29 responsive cells in VLM slices of rat as well as 13/46 responsive cells in control nonrespiratory nuclei and 15/23 responsive cells more dorsally in the medulla just ventral to the nucleus tractus solitarius. Dean et al. (1989) recorded intracellularly from cells in the nucleus ambiguous and the nucleus tractus solitarius in rat medullary slices with the ventral half removed to eliminate any possible influence from ventral chemoreceptors. Nucleus ambiguous cells were unresponsive to increased CO_2 but 48% of cells tested in the nucleus tractus solitarius were depolarized by increased CO_2 and exhibited an increase in membrane resistance; these effects were not blocked by low Ca^{2+}/high Mg^{2+} medium. Neubauer (1989, personal communication) found responsive neurons grown in culture from both dorsal and ventral medullary sites.

While controversy remains, the work cited above indicates that neurons both just below the VLM surface at the traditional rostral and caudal chemosensitive areas and deeper within the medulla near or within the nucleus tractus solitarius and other sites, but apparently not within the nucleus ambiguus, respond to acidic stimuli. These neurons are not necessarily chemoreceptor neurons. They may be interneurons in the chemoreception process or nonrespiratory neurons. Depolarization in response to local stimulation, known connectivity to respiratory neurons, and functionally significant effects resulting from stimulation of the cell in question would be the stringent requirements needed to prove absolutely the identity of a chemoreceptor neuron. If these neurons are distributed within the medulla both just below the surface and at deeper locations, it may be quite difficult to meet these requirements experimentally. It is interesting that the mechanism of pH sensitivity of cells near the VLM surface appears to involve synaptic events while that of pH-sensitive cells near the nucleus tractus solitarius appears to involve membrane events. These studies also suggest VLM neuronal involvement related to peripheral chemoreceptor function and to nociception.

IV. An Expanded Role for Neurons Within the VLM

Mitchell et al. (1963a,b), in their work examining the role of the VLM surface in respiratory chemosensitivity, noted that surface cooling or application of local anesthetics decreased blood pressure as well as ventilation.

Feldberg and Guertzenstein (1976) described VLM surface areas that responded to the application of neuroactive agents with changes in blood pressure. The caudal nicotine-sensitive depressor area (Guertzenstein and Silver, 1974) overlapped with the caudal respiratory chemosensitive zone and the rostral glycine sensitive pressor area overlapped with the IMA and perhaps the caudal aspect of the rostral respiratory chemosensitive zone (see Fig. 2). The study of the role of the VLM in the control of blood pressure has virtually exploded since then, producing a great deal of information useful to those interested primarily in the control of breathing.

Anatomical tracer studies (see Ross et al., 1984; Reis et al., 1987, for references) and neurophysiological approaches (Brown and Guyenet, 1984, 1985; Barman and Gebber, 1985; Morrison et al., 1988) have shown that cells below the rostral pressor area project to the intermediolateral area of the spinal cord and have connections to more dorsal medullary cardiovascular neuron groups. Immunohistochemical studies have demonstrated columns of cells just below the VLM surface extending through the regions below the chemosensitive areas that stain for 5-hydroxytryptamine (5-HT) (Ciriello et al., 1986, 1988), substance P (Marson and Loewy, 1985), dopamine-beta-hydroxylase (DBH), phenylethanolamine-*N*-methyl transferase (PNMT) (Ciriello et al., 1986), and choline acetyltransferase (Jones and Beaudet, 1987; Willenberg et al., 1985). (See Fig. 4.) A specific site (the C1 group) within the PNMT-staining cells located near or within the nucleus paragigantocellularis lateralis has been identified by stimulation, microinjection, and lesion studies as very important in the regulation of sympathetic outflow to the cardiovascular system (Guyenet et al., 1987; McAllen, 1986; Reis et al., 1987). Pacemaker cells located within this site have been described and hypothesized to represent the origin of tonic sympathetic outflow to the cardiovascular system (Sun et al., 1988). Others feel that the origin lies deeper within the medulla in the lateral tegmental field (Barman and Gebber, 1987). Recent work has shown that cells within the subretrofacial nucleus (McAllen, 1987), which may well be the same cells cited above as the C1 group, are anatomically organized with respect to their peripheral effector site (Dampney and McAllen, 1988). The C1 cells involved in tonic sympathetic tone are also involved in cardiovascular reflexes (Morrison and Reis, 1989; Granata et al., 1985), as a relay in the hypothalamically mediated defense reaction (Hilton and Smith, 1984), and in the link between respiratory output and sympathetic activity (McAllen, 1987). This brief overview demonstrates the importance in blood pressure regulation of cells accessible from the VLM surface respiratory chemosensitive zones that are clearly and unequivocally located just below the VLM surface and outlines approaches that have yet to be systematically applied to the study of the control of breathing in this region.

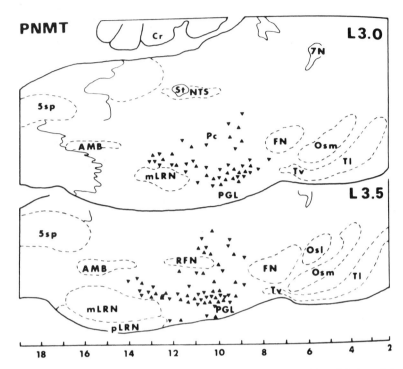

Figure 4 Drawings of saggital sections of the medulla of the cat 3.0 and 3.5 mm lateral to the midline. Each symbol represents a cell labeled with PNMT-like immunoreactivity. The horizontal axis shows numbers in mm from the frontal interaural line. AMB, Nucleus ambiguus; FN, facial nucleus; mLRN, magnocellular division of the lateral reticular nucleus; NTS, nucleus of the solitary tract; Osl, lateral superior olivary nucleus; Osm, medial superior olivary nucleus; Pc, nucleus parvicellularis; PGL, nucleus paragigantocellularis lateralis; pLRN, parvicellular division of the lateral reticular nucleus; RFN, retrofacial nucleus; St, solitary body; T1, lateral trapezoid nucleus; Tv, ventral trapezoid nucleus; 5sp, spinal trigeminal nucleus; 7N, facial nerve. (From Ciriello et al., 1986, with permission of A. R. Liss Inc.)

The early VLM surface cooling experiments using temperatures (20 °C) that block neurons but not axons of passage also produced apnea (Mitchell et al., 1963a,b) when cooling was localized to the IMA (Cherniack et al., 1979; Millhorn et al., 1982). In more recent work, a cooling probe extending rostrally just below the VLM surface from the IMA into the caudal border of the rostral chemosensitive zone also resulted in apnea (Budzinska et al., 1985). IMA cooling depressed phrenic output but not CO_2 sensitivity, while this more rostral probe also decreased CO_2 responses. What is the function of the cells being cooled in these experiments? Are they chemoreceptor cells? How are they related to the cardiovascular or pH-sensitive cells in the same region?

VLM surface application or microinjection of many neuroactive agents also results in dramatic effects on ventilation. The inhibitory neurotransmitter GABA decreases ventilation (Yamada et al., 1982; Gatti et al., 1987), as do barbiturates (Yamada et al., 1983), opiates (Hurle et al., 1982), and muscarinic receptor antagonists (Nattie et al., 1989), while the excitatory neurotransmitter glutamate stimulates ventilation (Gatti et al., 1986; Lawing et al., 1987; McAllen, 1986; Mitra et al., 1987), as do TRH (Holtman et al., 1986), acetylcholine (Dev and Loeschcke, 1979), carbachol (Nattie et al., 1989), substance P (Chen et al., 1988), and a growing list of other peptides. Figure 2 shows the sites at which glutamate microinjections just below the VLM surface augmented phrenic nerve activity (McAllen, 1986).

To determine whether cell bodies accessible from the VLM surface are involved in the chemoreceptor response to increased CO_2, we applied the excitatory amino acid neurotoxin kainic acid (KA) to the VLM surface in anesthetized, paralyzed, glomectomized, vagotomized, servoventilated cats (Nattie et al., 1988b). Rostral area application of KA, which excites neurons and then destroys them, produced dramatic augmentation followed by inhibition of phrenic output and of blood pressure leading to apnea in most animals. (See Fig. 5.) The response to CO_2 was also decreased or absent. Our dose was small, 4.7 mM, compared to prevous studies of surface kainic acid application (Gatti et al., 1985; McAllen et al., 1982) which found depression of blood pressure and apnea with 40-80 mM applied slightly more caudally.

KA microinjections were used to better localize the exact site of the effect (Nattie and Li, 1990a). After preliminary studies with 50-100 nl injections, unilateral 10-nl injections of kainic acid mixed with fluorescent microbeads (Quattrochi et al., 1989) identified a discrete region associated with decreased phrenic activity often to apnea and decreased CO_2 sensitivity but without effect on blood pressure. (See Fig. 6.)

The fluorescent microbead sites associated with these effects were within 400 μm of the surface just caudal to the superior olive, ventral to the facial nucleus, ventrolateral to the nucleus paragigantocellularis lateralis, and rostral to the retrofacial nucleus. This location coincided with the location of a group of cells called the retrotrapezoid nucleus (RTN) by Smith et al. (1989) and appears to be rostral and lateral to cells just below the ventral surface that contain 5-HT immunoreactivity called the nucleus ventralis suboolivaris (Gorcs and Millhorn, 1986). The kainic acid microinjection studies also identified a more rostral and a more caudal site that were associated with increased phrenic activity. The RTN connects anatomically to the dorsal and ventral respiratory groups, has cells with activity in phase with respiration,

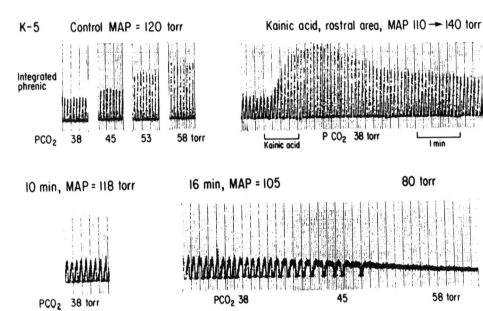

Figure 5 Tracings of integrated phrenic activity from an anesthetized, vagotomized, glomectomized, servoventilated cat demonstrating the effects of rostral area kainic acid application. Top left shows the control response to increased CO_2 and top right, the initial stimulation after application of the kainic acid to the medullary surface. The bottom tracings show the subsequent decrease in phrenic activity to apnea which lasted for 2 h. The numbers below the tracings show the end-tidal PCO_2 values; the numbers above show the mean arterial blood pressure values. (From Nattie et al., 1988b, with permission of the American Physiological Society.)

and lies at the rostral aspect of the paragigantocellularis lateralis/subretrofacial and retrofacial regions identified as the site of the blood pressure regulatory cells. It is interesting to note that cells within the more caudal retrofacial region are also respiratory in the nature of their firing pattern (Bianchi et al., 1989), that one of these cells labeled by intracellular injection has a long axon that reaches the VLM surface near the RTN (Grélot et al., 1988), and that large kainic acid injections in the RFN can also cause apnea and decreased CO_2 sensitivity (St. John et al., 1989).

These studies suggest the presence of groups of cells lying just below the VLM surface accessible from the traditional chemosensitive zones that are important in the control of breathing and are distinct from those involved in blood pressure regulation. Others have come to this conclusion (McAllen, 1986; Mitra et al., 1987). Are these cells chemosensitive? Or are they in some other way important in determining chemoreceptor responses? Could they

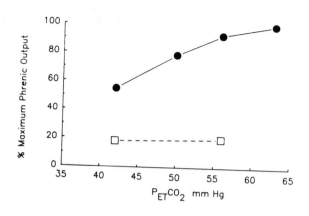

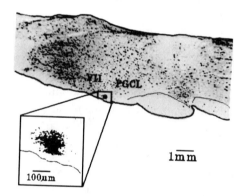

Figure 6 (Top) Solid symbols show the control response to increased CO_2 of integrated phrenic activity of an animal prepared as that described in Figure 5. Open symbols show the response after a single unilateral 10-nl microinjection of kainic acid (4.7 mM) just below the medullary surface in the caudal aspect of the rostral chemsensitive area. (Bottom) Location of the fluorescent microbeads (0.5 μm in diameter) mixed in with this microinjection. The injection site is just below the medullary surface ventromedial to the facial nucleus (VII) and ventrolateral to the nucleus paragigantocellularis lateralis (PGCL) in the region described as the retrotrapezoid nucleus (RTN) (Smith et al., 1989). This single small injection decreased baseline phrenic activity and abolished CO_2 sensitivity without an effect on blood pressure. (From Nattie and Li, unpublished observations).

provide tonic output to more dorsal respiratory neurons, e.g., the dorsal and ventral respiratory groups? Evidence suggests that some cells close to the VLM surface are involved in the respiratory responses to intravenous 2,4-dinitrophenol (Prabhaker et al., 1989) and the control of airway tone (Haxhiu et al., 1988), as well as the response to carotid sinus nerve stimulation (Millhorn et al., 1982; Nattie and Li, 1990a).

V. Mechanisms of Central Chemoreception

Without identification of chemoreceptor cells any discussion of the mechanism must be conjectural. The first question is whether CO_2 or H^+ or both are the actual stimuli. This has long been a controversial issue. Winterstein (1956) followed by Pappenheimer et al. (1965) and Berndt et al. (1972) favors the H^+ ion as the unique stimulus, while Nielsen (1936) and Borison et al. (1977) are among those who favor a unique role for CO_2. The effect of either could be on a specific receptor molecule or on some aspect of normal neuronal membrane or synaptic physiology. For example, in general, CO_2 hyperpolarizes mammalian cortical (Carpenter et al., 1974) and phrenic motoneurons (Gill and Kuno, 1963) and depresses monosynaptic reflexes (Carpenter et al., 1974). In contrast, in the abdominal ganglion of *Aplysia* CO_2 depolarizes neurons by an effect on membrane chloride or potassium conductance (Brown, 1974). Why the neuronal responses to CO_2 are so complex is uncertain, although it is clear that most neurons are sensitive to CO_2.

Recent experiments have provided evidence suggesting a role for both CO_2 and H^+ as independent stimuli. Exposure to solutions of the same pH but differing PCO_2 and HCO_3 values had different effects both in rat medullary slices (Fukuda, 1983) and in the isolated neonatal rat brainstem preparation (Harada et al., 1985).

Three different groups have measured VLM surface pH as representative of the pH sensed by the central chemoreceptors. Eldridge et al. (1985) measured integrated phrenic nerve activity, Teppema et al. (1988) assessed phrenic activity and ventilation, and Shams (1985) recorded ventilation; all these investigators found that when stimulated by increased CO_2, the response expressed per unit change in VLM surface pH was greater than with intravenous acid infusion. Eldridge et al. (1985) summarized the possible interpretations; (1) the chemoreceptor site is in extracellular fluid (ecf) but has different accessibility to CO_2 versus H^+, (2) intracellular pH is the stimulus, and (3) the chemoreceptor site is extracellular, with CO_2 and H^+ having separate effects. The observation that cells grown in culture from ventral and dorsal medulla respond to CO_2 but not H^+ (Neubauer, 1989; personal communication) supports the intracellular location if these cells are chemoreceptor cells.

These data are not easy to reconcile with the results of Pappenheimer et al. (1965), Fencl et al. (1966), and Berndt et al. (1972), who used ventriculocisternal perfusion in chronic acid-base disorders to estimate that ventilation was a unique function of brain ecf [H^+] at some point along a diffusion gradient between CSF and capillary blood. Such calculations are subject to error (Bledsoe et al., 1983), and recent direct measurements of medullary pH at different depths below the surface have not demonstrated large gradients (Cragg et al., 1977; Ichikawa et al., 1989). However, the gradient between CSF and capillary may not be easily measured by a relatively large electrode inserted into the tissue, and Arita et al., (1989) have observed regions within the medulla with quite different pH responses to CO_2 stimulation. It seems that the actual stimulus(i) for central chemoreceptors is still uncertain although the idea that the receptor process senses H^+ at a site more freely accessible to CO_2 than extracellular H^+, e.g., an intracellular location, has support.

In the following discussion H^+ will be arbitrarily considered the central chemoreceptor stimulus and three hypotheses will be presented for how H^+ could affect ventilation. Chemoreception is thought of in the broadest terms as any process sensitive to H^+ that subsequently alters ventilation.

The alpha-stat hypothesis of Reeves (1972, 1977) was formulated to explain how ectotherm blood pH in vitro and in vivo and homeotherm blood pH in vitro can vary so dramatically with temperature if pH is actually a physiologically important variable. Reeves (1972, 1977) noted that the pK ' of imidazole varied with temperature with approximately the same coefficient as did these blood pH values. Thus with changing temperature the fractional dissociation of imidazole (alpha-imidazole) remained constant even though pH changed. The closer the $\triangle pH/\triangle T$ and the $\triangle pK'/\triangle T$ values, the greater the constancy of alpha-imidazole (Burton, 1986).

Reeves further deduced that the structure and function of any protein determined by alpha-imidazole would be unaltered by changes in temperature but would be affected if pH changed isothermally. This concept has general biological significance (Burton, 1986; Hochachka and Guppy, 1987) and was applied to central ventilatory chemoreception by Hitzig (1982). He showed, by measuring ventilation in the unanesthetized fresh water turtle during different ambient temperatures and during ventriculocisternal perfusion, that ventilation was a unique function not of CSF or blood pH, but of calculated alpha-imidazole.

In a similar approach Kiley et al. (1984) decreased body temperature in anesthetized cats in normocapnic and hypocapnic conditions and used VLM surface pH and an assumed pK ' imidazole to calculate alpha-imidazole. Their calculated alpha-imidazole did not correlate with the phrenic activity and they concluded the alpha-stat hypothesis did not apply in these conditions. Burton (1986) emphasized that pK ' imidazole can vary widely depending on the site of the histidine within the protein and that calculations of alpha-imidazole based on assumed pK ' values can be misleading. If the data of Kiley et al. (1984) are recalculated using different assumed values of pK ' imidazole, the alpha-imidazole values do support the alphastat hypothesis (Burton, 1986).

Studies carried out in my laboratory have used an agent, diethyl pyrocarbonate (DEPC), thought to be relatively specific for binding to imidazole-histidine (see Nattie, 1986a, for references). DEPC binds to imidazole optimally at pH 7 and the carbethoxyhistidyl derivative has a half-life of 55 h at 25 °C, allowing time for ventilatory measurements. DEPC is attacked by any nucleophile including water, and its half-life in aqueous solutions is short, 10 min at 40 °C. Thus unbound DEPC is quickly converted to inactive amounts of ethanol and CO_2. This agent has been used by others to examine the possible role of histidine in ion transport (Grillo and Aronson, 1986). DEPC infused into the cisterna magna in conscious rabbits decreased ventilation, the ventilatory response to increased CO_2 in peripherally chemodenervated animals, and the response to intracisternal infusion of acidic mock CSF without affecting the ventilatory response to hypoxia (Nattie, 1986a). This effect of DEPC on central chemosensitivity was localized by VLM surface application to the rostral chemosensitive area in the anesthetized, paralyzed, glomectomized, servoventilated cat preparation (Nattie, 1986b). DEPC applied here, but not at the IMA or caudal chemosensitive areas, inhibited CO_2 sensitivity (Nattie, 1986b) and the response to intravenous acid infusion without affecting the VLM surface pH response to either stimulus or the phrenic nerve response to hypoxic stimulation (Nattie, 1988).

The secondary, long-lasting inhibition of ventilatory output after the brief initial stimulation observed with both the DEPC infusion via the cisterna magna in conscious rabbits and the direct application to the rostral VLM chemosensitive area in anesthetized cats suggests that the structures affected by DEPC may provide tonic input of a stimulatory nature. The KA inhibition results also suggest this. Whether the structures affected by DEPC and/or KA represent central chemoreceptors is unproven, but they certainly are involved in the central chemoreception process. These data also do not prove the alphastat hypothesis as other nucleophiles could possibly be involved in the DEPC effect. Still, it certainly is of interest that the DEPC effects are localized to the rostral chemosensitive area.

An older and still active hypothesis for central chemoreception is that pH changes ventilation via effects on cholinergic synaptic processes (Gesell and Hansen, 1945; Dev and Loeschcke, 1979). The application of acetylcholine to the VLM surface stimulated breathing and atropine applied bilaterally to both the rostral and caudal chemosensitive areas decreased ventilation and the responses to increased CO_2 (Dev and Loeschcke, 1979). This early study used high doses of atropine. In my laboratory using lower doses (4.4 mM) we localized the response to be predominantly at the rostral chemosensitive area and showed that pirenzepine, an antagonist of the M1 muscarinic receptor subtype, also inhibited the CO_2 responses when applied at the rostral area (Nattie et al., 1989). In our study the lower-dose atropine and the pirenzepine applications did not affect blood pressure or the phrenic response to carotid sinus nerve stimulation. Surface application of the M2 cardiac muscarinic receptor subtype antagonist AF-DX 116 had no effect on ventilation or blood pressure. We (Nattie et al., 1990b) and others (Sundaram et al., 1988) have reported blood pressure effects of microinjections of AF-DX 116 into the rostral VLM. These microinjection studies (Nattie et al., 1990b) also suggested that the M_3 muscarinic receptor subtype was primarily involved in the control of breathing. These preliminary studies indicate a role for muscarinic receptors in the rostral chemosensitive area in the ventilatory response to increased CO_2 but do not prove that muscarinic receptors or cholinergic synaptic events act as the chemoreception mechanism.

It is of interest with respect to the alphastat and muscarinic hypotheses that the histidines of the cloned muscarinic receptor molecule are located in the intracellular domain (Bonner et al., 1987; Peralta et al., 1987) and that following inhibition of CO_2 sensitivity by application of pirenzepine, the M1 receptor antagonist, at the rostral chemosensitive area DEPC application at the same location produces no further inhibition (Nattie et al., 1988a). This interaction raises the possibility that the DEPC effects on central chemosensitivity are via the histidines on the muscarinic receptor and this receptor is important in the chemoreception process.

A final hypothesis for the mechanism of central chemosensitivity involves the enzyme carbonic anhydrase (CA). CA has been localized to cells just below the surface of the VLM (Ridderstrale and Hanson, 1985) and is present in 25% of neurons cultured from the VLM as well as from more dorsal medulla and cortex and hypothalamus (Sterbenz et al., 1988; Neubauer, 1989). The location of the CA-positive cells within the medulla coincides approximately with the sites from which Arita et al. (1989) demonstrated large changes in local pH with intraarterial infusion of CO_2. The suggestion is that CA is somehow involved in central chemoreception. There is evidence to support this. CA inhibition by acetazolamide in peripherally chemodenervated animals does have a greater stimulatory effect on ventilation than in intact animals (Teppema et al., 1988). Further, Hanson et al. (1981) showed that the response to CO_2 injection in the vertebral artery is inhibited by intravenous administration of acetazolamide, which crosses the blood-brain barrier, but not of benzolamide, which does not. How CA is involved is unknown although in some in vitro conditions CA can be shown to affect ion transfer as well as CO_2 hydration (Diaz et al., 1982). It is interesting to note that CA seems to be present predominantly in sensory neurons (Neubauer, 1989).

VI. Central Chemoreception During Development

The work of Hohimer et al. (1983) clearly demonstrates the presence of central chemoreceptors in fetal lambs. Earlier work had suggested that fetal chemoreceptors are not functional (e.g., see Hodson et al., 1968; Woodrum et al., 1977), but these studies were in anesthetized preparations. Hohimer et al. (1983) showed in unanesthetized, chronically instrumented fetal lambs that ventriculocisternal perfusion of mock CSF with high, normal, or low $[HCO_3]$ had clear effects on fetal breathing movements (FBM). Perfusion with normal CSF had no significant effect, while the high-bicarbonate solution, which increased the CSF value to 30.7 mM, essentially eliminated the FBM. Perfusion of low bicarbonate CSF decreased the CSF value to 17.3 mM and increased the incidence of FBM from 22.1% to 45.7%. These perfusions had no effect on the amount of time the fetus was in the high- or low-voltage electrocorticogram (ECOG) state and the acidic perfusions increased FBM even when the fetus was in the usually unresponsive high-voltage state.

Increased CO_2 increases both the incidence and amplitude of FBM (Boddy et al., 1974; Bowes et al., 1981; Jansen et al., 1982) and increases the time spent in the low-voltage ECOG state. While the functional silence of the peripheral chemoreceptors in utero is well accepted, indicating that the FBM response to CO_2 is central in nature, the link of the FBM and the ECOG responses led to the idea that the FBM response was related to the state of brain arousal, not to central chemoreceptors (Jansen and Chernick, 1983). The specific response of FBM to ventriculocisternal perfusion of acidic mock CSF and the separation of the FBM response from the ECOG response to carbonic anhydrase inhibition (Hohimer et al., 1985) suggest that the FBM response is a specific central chemoreceptor function. And Rigatto et al. (1988), using a window technique that allows direct observation of the chronically instrumented fetus in utero, observed that CO_2 stimulation via the mother or directly to the fetus increased FBM without producing arousal by either ECOG or visual criteria.

The effect of brain arousal state on chemoreceptor responsiveness is very important. The fact that ventriculocisternal perfusion of acidic mock CSF clearly stimulated breathing only in the unanesthetized fetus emphasizes the strong depressant effect of anesthesia. In the unanesthetized state, as discussed by Jansen and Chernick (1983, 1988), powerful inhibitory mechanisms must exist during the high-voltage ECOG periods when FBM are absent and the responsiveness to CO_2 and other stimuli is absent or decreased. The source of this inhibition and its regulation are unknown. A recent study emphasizes this effect by showing the cardiovascular and metabolic responses to cooling in newborn lambs are crucially dependent on the arousal state. In the REM-like state the lambs were almost poikilothermic (Berger et al., 1989).

Studies of the site of central chemoreception have been performed in newborn animals. Perfusion of mock CSF with different pH values over the VLM surface has dramatic effects on breathing in anesthetized newborn rabbits and guinea pigs (Wennergren and Wennergren, 1980). Acidic fluid stimulated breathing and alkaline fluid inhibited it. Application of theophylline to the VLM surface stimulated breathing and application of meperiden inhibited it in the same preparations (Wennergren and Wennergren, 1983). (See Fig. 7.) In the isolated neonatal rat brainstem preparation, microinjections of 2-10 nl of high CO_2 mock CSF located an area that resulted in increased rate and depth of the integrated C2-C4 neural activity. The site was 0.50-0.75 mm lateral to the midline, 0.50-0.80 mm rostral to the obex, and 0.25-0.35 mm deep from the VLM surface in the preliminary report of

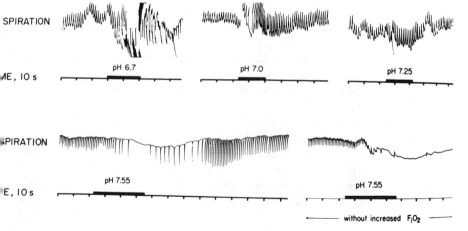

Figure 7 Recordings of respiration in an anesthetized newborn rabbit in response to superfusion of the medullary surface with mock cerebrospinal fluid of the pH values shown below the tracings. Note the marked pH sensitivity. (From Wennergren and Wennergren, 1980, with permission of *Acta Physiologica Scandinavica*.)

this work (Issa and Remmers, 1988). Whittaker et al. (1987) evaluated by topical application the caudal chemosensitive zone of the VLM surface in anesthetized neonatal and postnatal kittens. Nonphasic pH-sensitive units in this area responded to acidic CSF, increased CO_2, and acetylcholine in the postnatal and adult animals, but in the newborn increased CO_2 and acetylcholine caused only slight increases in neural activity. Cooling of the IMA on the VLM surface in newborn kittens has inhibitory effects similar to those reported in adults (Bruce, 1986).

These data indicate that in fetus and newborn, central chemoreceptors are present and functional although their effectiveness may be highly dependent on the arousal state (Jansen and Chernick, 1983, 1988). How effective are these central chemoreceptors? Bureau et al. (1979) infused HCl intravenously in unanesthetized newborn lambs and observed ventilation, blood gases, and CSF acid-base balance. CSF $[HCO_3]$ and pH decreased more than observed in adults but the ventilatory response did not seem to follow the increasing central and peripheral acidosis. Ventilation initially increased and arterial PCO_2 initially decreased, but as the acidosis worsened ventilation and arterial PCO_2 did not progressively change, as observed in adults. The authors concluded that central and peripheral chemoreceptors were functional but were not as effective as in the adult. Ventilatory responses to increased CO_2 are also less in anesthetized newborn animals than in adults (e.g., Nattie and Edwards, 1981; Guthrie et al., 1980), although in humans the response to CO_2 may not be decreased in comparison to adults (see Walker, 1984).

A number of substances appear to have important central effects on breathing in the fetus and neonate, including substance P, thyrotropin-releasing hormone, adenosine (see Lagercrantz, 1987), prostaglandins (Koos, 1985; Long, 1988), cholinergic agonists (Hanson et al., 1988), and endorphins (Moss, 1986). None of these effects have been localized to the VLM areas discussed above.

A peculiar set of findings in the fetus suggests the presence within the brainstem of hypoxic sensitivity. FBMs are abolished by hypoxia in the fetus (Boddy et al., 1974), an effect prevented by bilateral lesions in the pons just rostral to the trigeminal nuclei (Gluckman and Johnston, 1987). These data suggest that this pontine region senses the hypoxia and actively inhibits the fetal breathing movements. The neuronal mechanism involved is unknown, but at present, use of the term chemosensitivity to describe this process is not inappropriate. Jansen and Chernick (1975) reported that VLM surface application of cyanide in fetal sheep stimulated gasping respiration after a mean delay of 43 s.

VII. Conclusions

1. Central chemoreception exists in adults, the fetus, and the newborn.

2. The nature of this chemoreception process is unknown. CO_2 and/or H^+ could have effects on membrane, synaptic, or other cell or system events. Some current hypotheses include the alpha-stat hypothesis, muscarinic cholinergic physiology, and carbonic anhydrase involvement.

3. Neurons that are responsive to pH can be located below the VLM surface at the chemosensitive areas, but also in deeper areas, particularly near the nucleus tractus solitarius, but not the nucleus ambiguus or retroambiguus. It is unclear if these are respiratory chemoreceptor neurons or even respiratory neurons.

4. Neurons just below the surface of the ventrolateral medulla at approximately the level of the intermediate respiratory area are clearly very important in the regulation of blood pressure. These cells may be the origin of sympathetic cardiovascular tone and have been well studied by a number of techniques.

5. Unilateral kainic acid microinjection in a specific VLM region rostral to that involved in blood pressure regulation and within the retrotrapezoid nucleus, which contains cells firing in phase with respiration, dramatically inhibits respiration and CO_2 sensitivity.

6. Cells located more caudally in the VLM within the blood pressure responsive areas are also respiratory in nature.

7. The role of these VLM respiratory cells in central chemoreception or in general is unknown. They may provide tonic input to more dorsal respiratory neurons, e.g., the dorsal and ventral respiratory group.

8. Studies of these VLM cells involved in blood pressure and respiratory regulation and of central chemoreception in the fetus and newborn are few.

References

Arita, H., Ichikawa, K., Kuwana, S., and Kogo, N. (1989). Possible locations of pH-dependent central chemoreceptors: Intramedullary regions with acidic shift of extracellular fluid pH during hypercapnia. *Brain Res.* **485**:285-293.

Arita, H., Kogo, N., and Ichikawa, K. (1988a). Rapid and transient excitation of respiration mediated by central chemoreceptor. *J. Appl. Physiol.* **64**:1369-1375.

Arita, H., Kogo, N., and Ichikawa, K. (1988b). Locations of medullary neurons with non-phasic discharges excited by stimulation of central and/or peripheral chemoreceptors and by activation of nociceptors in cat. *Brain Res.* **442**:1-10.

Barman, S. M., and Gebber, G. L. (1985). Axonal projection patterns of ventrolateral medullospinal sympathoexcitatory neurons. *J. Neurophysiol.* **53**:1551-1566.

Barman, S. M., and Gebber, G. L. (1987). Lateral tegmental field neurons of cat medulla: A source of basal activity of ventrolateral medullospinal sympathoexcitatory neurons. *J. Neurophysiol.* **57**:1410-1424.

Berger, D. J., Horne, R. S. C., and Walker, A. M. (1989). Cardiorespiratory responses to cool ambient temperatures differ with sleep state in neonatal lambs. *J. Physiol.* **412**:351-363.

Berndt, J., Berger, W., Berger, K., and Schmidt, M. (1972). Untersuchungen zum zentralen chemosensible mechanismus der atmung. II. Die steuerung der atmung durch das extracellulare pH bewebe der medulla oblongata. *Pfluger's Arch.* **332**:146-170.

Bianchi, A. L., Grelot, L., Iscoe, S., and Remmers, J. R. (1988). Electrophysiological properties of rostral medullary respiratory neurons in the cat: an intracellular study. *J. Physiol.* **407**:293-310.

Bledsoe, S. W., Berger, A. J., and Hornbein, T. F. (1983). Limitations of ventriculocisternal perfusion for determining brain interstitial fluid composition. *J. Appl. Physiol.* **54**:1505-1510.

Bledsoe, S. W., and Hornbein, T. F. (1981). Central chemosensors and the regulation of their chemical environment. In *Regulation of Breathing.* Part I. Edited by T. F. Hornbein. New York, Marcel Dekker, pp. 347-428.

Boddy, K., Dawes, G. S., Fisher, R., Pinter, S., and Robinson, J. S. (1974). Fetal respiratory movements electrocortical and cardiovascular responses to hypoxia and hypercapnia in sheep. *J. Physiol.* **243**:599-618.

Bonner, T. I., Buckley, N. J., Young, A. C., and Brann, M. R. (1987). Identification of a family of muscarinic acetylcholine receptor genes. *Science* **237**:527-532.

Borison, H. L., Borison, R., and McCarthy, L. E. (1980). Brain stem penetration by horseradish peroxidase from the cerebrospinal fluid spaces in the cat. *Exp. Neurol.* **69**:271-289.

Borison, H. L., Gonsalves, S. F., Montgomery, S. P., and McCarthy, L. E. (1978). Dynamics of respiratory V_T responses to isocapnic pHa forcing in chemodenervated cats. *J. Appl. Physiol.* **45**:502-511.

Borison, H. L., Hurst, J. H., McCarthy, L. E., and Rosenstein, R. (1977). Arterial hydrogen ion versus CO_2 on depth and rate of breathing in decerebrate cats. *Respir. Physiol.* **30**:311-325.

Bowes, G., Wilkinson, M. H., Dowling, H., Ritchie, B. C., Brodecky, V., and Maloney, J. E. (1981). Hypercapnic stimulation of respiratory activity in unanesthetized fetal sheep in utero. *J. Appl. Physiol.* **50**: 701-708.

Brown, A. M. (1974). Carbon dioxide action on neuronal membranes. In *Carbon Dioxide and Metabolic Regulations.* Edited by G. Nahas and K. E. Schaefer. New York, Springer-Verlag, pp. 81-86.

Brown, D. L., and Guyenet, P. A. (1984). Cardiovascular neurons of the brainstem with projections to spinal cord. *Am. J. Physiol.* **247**:R1009-R1016.

Brown, D. L., and Guyenet, P. A. (1985). Electrophysiological study of cardiovascular neurons in the rostral ventrolateral medulla in rats. *Circ. Res.* **56**:359-369.

Bruce, E. N. (1986). Hypoglossal and phrenic nerve responses to chemical stimuli in kittens. In *Neurobiology of the Control of Breathing.* Edited by C. von Euler and H. Lagercrantz. New York, Raven Press, pp. 75-80.

Bruce, E. N., and Cherniack, N. S. (1987). Central chemoreceptors (brief review). *J. Appl. Physiol.* **62**:389-402.

Budzinska, K., von Euler, C., Kao, F. F., Pantaleo, T., and Yamamoto, Y. (1985). Effects of graded focal cold block in rostral areas of the medulla. *Acta Physiol. Scand.* **124**:329-340.

Bureau, M. A., Begin, R., and Berthiaume, Y. (1979). Central chemical regulation of respiration in term newborn. *J. Appl. Physiol.* **47**:1212-1217.

Burton, R. F. (1986). The role of imidazole ionization in the control of breathing. *Comp. Biochem. Physiol.* **83A**:333-336.

Carpenter, D. O., Hubbard, J. H., Humphrey, D. R., Thompson, H. K., and Marshall, W. H. (1974). Carbon dioxide effects on nerve cell function. In *Carbon Dioxide and Metabolic Regulations.* Edited by G. Nahas and K. E. Schaefer. New York, Springer-Verlag, pp. 49-62.

Chen, Z. B., Hedner, T., and Hedner, J. (1988). Hypoventilation and apnea induced by the substance P antagonist (D-Pro[2],D-Trp[7,9])-SP in the ventrolateral rat medulla. *Acta Physiol. Scand.* **134**:153-154.

Cherniack, N. S., von Euler, C., Homma, I., and Kao, F. F. (1979). Graded changes in central chemoreceptor input by local temperature changes on the ventral surface of the medulla. *J. Physiol.* **287**:191-211.

Ciriello, J., Caverson, M. M., Calaresu, F. R., and Kurkoff, T. L. (1988). Neuropeptide and serotonin immunoreactive neurons in the cat ventrolateral medulla. *Brain Res.* **440**:53-66.

Ciriello, J., Caverson, M. M., and Park, D. H. (1986). Immunohistochemical identification of noradrenaline and adrenaline synthesizing neurons in the cat ventrolateral medulla. *J. Comp. Neurol.* **253**:216-230.

Cozine, R. A., and Ngai, S. H. (1967). Medullary surface chemoreceptors and regulation of respiration in the cat. *J. Appl. Physiol.* **22**:117-121.

Cragg, P., Patterson, L., and Purves, M. J. (1977). The pH of brain extra-cellular fluid in the cat. *J. Physiol.* **272**:137-166.

Dampney, R. A. L., and McAllen, R. M. (1988). Differential control of sympathetic fibres supplying hindlimb skin and muscle by subretro-facial neurones in the cat. *J. Physiol.* **395**:41-56.

Dean, J. B., Lawing, W. L., and Millhorn, D. E. (1989). CO_2 decreases membrane conductance and depolarizes neurons in the nucleus tractus solitarii. *Exp. Brain Res.* **76**:656-661.

Dermietzel, R. (1976). Central chemosensitivity: Morphological studies. In *Acid-Base Homeostasis of the Brain Extracellular Fluid and the Respiratory Control System.* Edited by H. H. Loeschcke. Stuttgart, Thieme, pp. 52-65.

Dev, N. B., and Loeschcke, H. H. (1979). A cholinergic mechanism involved in the respiratory chemosensitivity of the medulla oblongata in the cat. *Pfluger's Arch.* **379**:29-36.

Diaz, E., Sandblom, J. P., and Wistrand, P. J. (1982). Selectivity properties of channels induced by a reconstituted membrane-bound carbonic anhy-drase. *Acta Physiol. Scand.* **116**:461-463.

Eldridge, F. L., Kiley, J. P., and Millhorn, D. E. (1985). Respiratory re-sponses to medullary hydrogen ion changes in cats: Different effects of respiratory and metabolic acidoses. *J. Physiol.* **358**:285-297.

Euler, C. von, and Soderberg, U. (1952). Medullary chemosensitive recep-tors. *J. Physiol.* **118**:545-554.

Feldberg, W., and Guertzenstein, P. G. (1976). Vasodepressor effects ob-tained by drugs acting on the ventral surface of the brainstem. *J. Physiol.* **258**:337-355.

Feldman, J. L. (1986). Neurophysiology of breathing in mammals. In *Hand-book of Physiology. The Nervous System. Intrinsic Regulatory Systems of the Brain.* Edited by V. B. Mountcastle, F. E. Bloom, and S. R. Geiger. Bethesda, MD, American Physiological Society, pp. 463-524.

Fencl, V., Miller, T. B., and Pappenheimer, J. R. (1966). Studies on the respiratory response to disturbances of acid-base balance, with deduc-tions concerning the ionic composition of cerebral interstitial fluid. *Am. J. Physiol.* **210**:459-472.

Fukuda, Y. (1983). Difference between actions of high PCO_2 and low [HCO_3] on neurons in the rat medullary chemosensitive areas in vitro. *Pfluger's Arch.* **398**:324-330.

Fukuda, Y., and Honda, Y. (1975). pH-sensitive cells at ventro-lateral sur-face of rat medulla oblongata. *Nature* **256**:317-318.

Fukuda, Y., and Honda, Y. (1976). pH sensitivity of cells located at the ventrolateral surface of the cat medulla oblongata in vitro. *Pfluger's Arch.* **364**:243-247.

Fukuda, Y., and Loeschcke, H. H. (1977). Effect of H^+ on spontaneous neuronal activity in the surface layer of the rat medulla oblongata in vitro. *Pfluger's Arch.* **371**:125-134.

Gatti, P. J., DaSilva, A. M. T., and Gillis, R. A. (1987). Cardiorespiratory effects produced by injecting drugs that affect GABA receptors into nuclei associated with the ventral surface of the medulla. *Neuropharmacology* **26**:423-431.

Gatti, P. J., DaSilva, A. M. T., Hamosch, P., and Gillis, R. (1985). Cardiorespiratory effects produced by application of L-glutamate and kainic acid to the ventral surface of the cat hindbrain. *Brain Res.* **330**:21-29.

Gatti, P. J., Norman, W. P., DaSilva, A. M. T., and Gillis, R. A. (1986). Cardiorespiratory effects produced by microinjecting l-glutamic acid into medullary nuclei associated with the ventral surface of the feline medulla. *Brain Res.* **381**:281-288.

Gesell, R., and Hansen, E. T. (1945). Anticholinesterase activity of acid as a biological instrument of nervous integration. *Am. J. Physiol.* **144**:126-163.

Gill, P. K., and Kuno, K. (1963). Properties of phrenic motoneurones. *J. Physiol.* **168**:258-273.

Gluckman, P. D., and Johnston, B. M. (1987). Lesions in the upper lateral pons abolish the hypoxic depression of breathing in unanesthetized fetal lambs in utero. *J. Physiol.* **382**:373-383.

Gorcs, T., and Millhorn, D. E. (1986). Serotonin immunoreactive neurons in the sub-olivary nucleus of the ventrolateral medulla in cat. *Fed. Proc.* **45**:5193.

Granata, A. R., Ruggiero, D. A., Park, D. H., Joh, T. H., and Reis, D. H. (1985). Brain stem area with C1 epinephrine neurons mediates baroreflex vasodepressor responses. *Am. J. Physiol.* **248**:H547-H567.

Grélot, L., Bianchi, A. L., Iscoe, S., and Remmers, J. R. (1988). Expiratory neurons of the rostral medulla: anatomical and functional correlates. *Neurosci. Lett.* **89**:140-145.

Grillo, F. G., and Aronson, P. A. (1986). Inactivation of the renal microvillus membrane Na^+-H^+ exchanger by histidine-specific reagents. *J. Biol. Chem.* **261**:1120-1125.

Guertzenstein, P. G., and Silver, A. (1974). Fall in blood pressure produced from discrete regions of the ventral surface of the medulla by glycine and lesions. *J. Physiol.* **242**:489-503.

Guthrie, R. D., Standaert, T. A., Hodson, W. A., and Woodrum, D. E. (1980). Sleep and maturation of eucapnic ventilation and CO_2 sensitivity in the premature primate. *J. Appl. Physiol.* **48**:347-354.

Guyenet, P. G., Filtz, T. M., and Donaldson, S. R. (1987). Role of excitatory amino acids in rat vagal and sympathetic baroreflexes. *Brain Res.* **407**:272-284.

Hanson, M. A., Moore, P. J., Nijhuis, J. G., and Parkes, M. J. (1988). Effects of pilocarpine on breathing movements in normal, chemodenervated and brainstem transected fetal sheep. *J. Physiol.* **400**:415-424.

Hanson, M. A., Nye, P. C. G., and Torrance, R. W. (1981). The location of carbonic anhydrase in relation to the blood-brain barrier at the medullary chemoreceptors of the cat. *J. Physiol.* **320**:113-125.

Harada, Y., Kuno, M., and Wang, Y. Z. (1985). Differential effects of carbon dioxide and pH on central chemoreceptors in the rat in vitro. *J. Physiol.* **368**:679-693.

Haxhiu, M. A., Deal Jr., E. C., Trivedi, R. D., van Lunteren, E., and Cherniack, N. S. (1988). Tracheal and phrenic responses to neurotensin applied to ventral medulla. *Am. J. Physiol.* **255**:R780-R786.

Hilton, S. M., and Smith, P. R. (1984). Ventral medullary neurons excited from the hypothalamic and mid-brain defence areas. *J. Auton. Nerv. Syst.* **11**:35-42.

Hitzig, B. M. (1982). Temperature-induced changes in turtle CSF pH and central chemical control of ventilation. *Respir. Physiol.* **49**:205-222.

Hochachka, P. W., and Guppy, M. (1987). *Metabolic Arrest and the Control of Biological Time.* Cambridge, MA, Harvard University Press, pp. 57-72.

Hodson, W. A., Fenner, A., Brumley, G., Chernick, V., and Avery, M. E. (1968). Cerebrospinal fluid and blood acid-base relationships in fetal and neonatal lambs and pregnant ewes. *Respir. Physiol.* **4**:322-332.

Hohimer, A. R., Bissonnette, J. M., Machida, C. M., and Horowitz, B. (1985). The effect of carbonic anhydrase inhibition on breathing movements and eclectrocortical activity in fetal sheep. *Respir. Physiol.* **61**:327-334.

Hohimer, A. R., Bissonnette, J. M., Richardson, B. S., and Machida, C. M. (1983). Central chemical regulation of breathing movements on fetal lambs. *Respir. Physiol.* **52**:99-111.

Holtman, J. R. Jr., Butler, A. L., Hamosh, P., and Gillis, R. A. (1986). Central respiratory stimulation produced by thyrotropin releasing hormone in the cat. *Peptides* **7**:207-212.

Hori, T., Roth, G. I., and Yamamoto, W. S. (1970). Respiratory sensitivity of rat brain stem surface to chemical stimuli. *J. Appl. Physiol.* **28**:721-724.

Hurle, M. A., Mediavilla, A., and Florez, J. (1982). Morphine, pentobarbital, and naloxone in the ventral medullary chemosensitive areas: Differential respiratory and cardiovascular effects. *J. Pharmacol. Exp. Ther.* **220**:642-647.

Ichikawa, K., Kuwana, S., and Arita, H. (1989). ECF pH dynamics within the ventrolateral medulla: A microelectrode study. *J. Appl. Physiol.* **67**:193-198.

Issa, F., and Remmers, J. R. (1988). Localization of a medullary area sensitive to changes in CO_2 in vitro brainstem-spinal cord preparation of newborn rats. *Neurosci. Abstr.* **14**:935.

Jansen, A. H., and Chernick, V. (1975). Site of central chemosensitivity in fetal sheep. *J. Appl. Physiol.* **39**:1-6.

Jansen, A. H., and Chernick, V. (1983). Development of respiratory control. *Physiol. Rev.* **63**:437-483.

Jansen, A. H., and Chernick, V. (1988). Onset of breathing and control of respiration. *Seminars in Perinatology.* **12**:104-112.

Jansen, A. H., Ioffe, S., Russell, S. J., and Chernick, V. (1982). Influence of sleep state on the response to hypercaobia in fetal lambs. *Respir. Physiol.* **48**:125-142.

Jones, B. E., and Beaudet, A. (1987). Distribution of acetylcholine and catecholamine neurons in the cat brainstem: a choline acetyltransferase and tyrosine hydroxylase immunohistochemical study. *J. Comp. Neurol.* **261**:15-32.

Keeler, J. R., Shults, C. W., Chase, T. N., and Helke, C. J. (1984). The ventral surface of the medulla in the rat: Pharmacological and autoradiographic localization of GABA-induced cardiovascular effects. *Brain Res.* **297**:217-224.

Kiley, J. P., Eldridge, F. L., and Millhorn, D. E. (1984). The effect of hypothermia on central neural control of respiration. *Respir. Physiol.* **58**:295-312.

Koos, B. J. (1985). Central stimulation of breathing movements in fetal lambs by prostaglandin synthetase inhibitors. *J. Physiol.* **362**:455-466.

Lagercrantz, H. (1987). Neuromodulators and respiratory control during development. *Trends Neurol. Sci.* **10**:368-372.

Lawing, W. L., Millhorn, D. E., Bayliss, D. A., Dean, J. B., and Trzebski, A. (1987). Excitatory and inhibitory effects on respiration of l-glutamate microinejected superficially into the ventral aspects of the medulla oblongata in cat. *Brain Res.* **435**:322-326.

Leusen, I. R. (1954). Chemosensitivity of the respiratory center. Influence of CO_2 in the cerebral ventricles on respiration. *Am. J. Physiol.* **176**:39-44.

Lipscomb, W. T., and Boyarski, L. L. (1972). Neurophysiological investigations of medullary chemosensitive areas of respiration. *Respir. Physiol.* **16**:362-376.

Loeschcke, H. H., (1982). Central chemosensitivity and the reaction theory. *J. Physiol.* **332**:1-24.

Loeschcke, H. H., Koepchen, H. P., and Gertz, K. H. (1958). Uber den einfluss von wasserstoffionenkonzentration und CO_2-druck im liquor cerebrospinalis auf die atmung. *Pfluger's Arch.* **266**:565-585.

Long, W. A. (1988). Prostaglandins and control of breathing in newborn piglets. *J. Appl. Physiol.* **64**:409-418.

Malcolm, J. L., Sarelius, I. H., and Sinclair, J. D. (1980). The respiratory role of the ventral surface of the medulla studied in the anesthetized rat. *J. Physiol.* **307**:503-515.

Marino, P. L., and Lamb, T. W. (1975). Effects of CO_2 and extracellular H^+ iontophoresis on single cell activity in the cat brainstem. *J. Appl. Physiol.* **38**:688-695.

Marson, L., and Loewy, A. D. (1985). Topographical organization of substance P and monoamine cells in the ventral medulla of the cat. *J. Auto. Nerv. Syst.* **14**:271-285.

McAllen, R. M. (1986). Location of neurones with cardiovascular and respiratory function at the ventral surface of the cat's medulla. *Neuroscience* **18**:43-49.

McAllen, R. M. (1987). Central respiratory modulation of subretrofacial bulbospinal neurons in the cat. *J. Physiol.* **388**:533-545.

McAllen, R. M., Neil, J. J., and Loewy, A. D. (1982). Effects of kainic acid applied to the ventral surface of the medulla oblongata on vasomotor tone, the baroreceptor reflex and hypothalamic autonomic responses. *Brain Res.* **238**:65-76.

Miles, R. (1983). Does low pH stimulate central chemoreceptors located near the ventral medullary surface? *Brain Res.* **271**:349-353.

Millhorn, D. E., and Eldridge, F. L. (1986). Role of ventrolateral medulla in regulation of respiratory and cardiovascular systems. *J. Appl. Physiol.* **61**:1249-1263.

Millhorn, D. E., Eldridge, F. L., and Waldrop, T. G. (1982). Effects of medullary area $I_{(S)}$ cooling on respiratory response to chemoreceptor inputs. *Respir. Physiol.* **49**:23-39.

Mitchell, R. A., and Herbert, D. A. (1974). The effect of carbon dioxide on the membrane potential of medullary respiratory neurons. *Brain Res.* **75**:345-349.

Mitchell, R. A., Loeschcke, H. H., Massion, W. H., and Severinghaus, J. W. (1963a). Respiratory responses mediated through superficial chemosensitive areas on the medulla. *J. Appl. Physiol.* **18**:523-533.

Mitchell, R. A., Loeschcke, H. H., Severinghaus, J. W., Richardson, B. W., and Massion, W. H. (1963b). Regions of respiratory chemosensitivity on the surface of the medulla. *Ann. NY Acad. Sci.* **109**:661-681.

Mitra, J., Prabhaker, N. R., Overholt, J. L., and Cherniack, N. S. (1987). Respiratory and vasomotor effects of excitatory amino acid on ventral medullary surface. *Brain Res. Bull.* **18**:681-684.

Morrison, S. F., Milner, T. A., and Reis, D. J. (1988). Reticulospinal vasomotor neurons of the rat rostral ventrolateral medulla: Relationship to sympathetic nerve activity and the C1 adrenergic cell group. *J. Neurosci.* **8**:1286-1301.

Morrison, S., and Reis, D. J. (1989). Reticulospinal vasomotor neurons in the RVL mediate the somatosympathetic reflex. *Am. J. Physiol.* **256**:R1084-R1097.

Moss, I. R. (1986). Endorphins in fetal and neonatal breathing. In *Neurobiology of the Control of Breathing.* Edited by C. von Euler and H. Lagercrantz. New York, Raven Press, pp. 119-124.

Nattie, E. E. (1983). Ventilation during acute HCl infusion in intact and chemodenervated conscious rabbits. *Respir. Physiol.* **54**:97-107.

Nattie, E. E. (1986a). Intracisternal diethylpyrocarbonate inhibits central chemosensitivity in conscious rabbits. *Respir. Physiol.* **64**:161-176.

Nattie, E. E. (1986b). Diethyl pyrocarbonate (an imidazole binding substance) inhibits rostral VLM CO_2 sensitivity. *J. Appl. Physiol.* **61**:843-850.

Nattie, E. E. (1988). Diethyl pyrocarbonate inhibits rostral ventrolateral medullary H^+ sensitivity. *J. Appl. Physiol.* **64**:1600-1609.

Nattie, E. E., and Edwards, W. H. (1981). CSF acid-base regulation and ventilation during acute hypercapnia in the newborn dog. *J. Appl. Physiol.* **50**:566-574.

Nattie, E. E., and Li, Aihua. (1990a). Fluorescence location of RVLM kainate microinjections that alter the control of breathing. *J. Appl. Physiol.* **68**:1157-1166.

Nattie, E. E., and Li, A. (1990b). Ventral medulla sites of muscarinic receptor subtypes involved in cardiorespiratory control. *J. Appl. Physiol.* **69**:33-41.

Nattie, E. E., Mills, J. W., and Ou. L. C. (1988a). Pirenzepine prevents diethylpyrocarbonate inhibition of central CO_2 sensitivity. *J. Appl. Physiol.* **65**:1962-1966.

Nattie, E. E., Mills, J. W., Ou, L. C., and St. John, W. M. (1988b). Kainic acid on the rostral ventrolateral medulla inhibits phrenic output and CO_2 sensitivity. *J. Appl. Physiol.* **65**:1525-1534.

Nattie, E. E., Wood, J., Mega, A., and Goritski, W. (1989). Rostral ventrolateral medulla muscarinic receptor involvement in central ventilatory chemosensitivity. *J. Appl. Physiol.* **66**:1462-1470.

Neubauer, J. A. (1990). Carbonic anhydrase and sensory function in the central nervous system. In *The Carbonic Anhydrases: Cellular Physiology and Molecular Genetics.* Edited by S. J. Dodgson. New York, plenum press (in press).

Nielsen, M. (1936). Untersuchungen uber die atmunsregulation beim menschen. *Skand. Arch. Physiol.* **74**(Suppl. 10):193-203.

Pappenheimer, J. R., Fencl, V., Heisey, S. R., and Held, D. (1965). Role of cerebral fluids in control of respiration as studied in unanesthetized goats. *Am. J. Physiol.* **208**:436-450.

Peralta, E. G., Winslow, J. W., Peterson, G. L., Smith, D. H., Ashkenazi, A., Ramachandran, J., Schimerlik, M. I., and Capon, D. J. (1987).

Primary structure and biochemical properties of an M_2 muscarinic receptor. *Science* **236**:600-605.

Pokorski, M. (1976). Neurophysiological studies on central chemosensor in medullary ventrolateral areas. *Am. J. Physiol.* **230**:1288-1295.

Prabhaker, N. R., Mitra, J., Adams, E. M., and Cherniack, N. S. (1989). Involvement of ventral medullary surface in respiratory responses induced by 2,4-dinitrophenol. *J. Appl. Physiol.* **66**:598-605.

Quattrochi, H. A., Mamelak, A. N., Madison, R., Macklis, J. D., and Hobson, J. A. (1989). Mapping neuronal inputs to REM sleep induction sites with carbachol-fluorescent microspheres. *Science* **245**:984-986.

Reeves, R. B. (1972). An imidazole alphastat hypothesis for vertebrate acid-base regulation: tissue carbon dioxide content and body temperature in bullfrogs. *Respir. Physiol.* **14**:219-236.

Reeves, R. B. (1977). The interaction of body temperature and acid-base balance in ectothermic vertebrates. *Annu. Rev. Physiol.* **39**:559-586.

Reis, D. J., Ross, C., Granata, A. R., and Ruggiero, D. A. (1987). Role of C_1 area of rostroventrolateral medulla in cardiovascular control. In *Brain Peptides and Catecholamines in Cardiovascular Regulation.* Edited by J. P. Buckley, and C. M. Ferrario. New York, Raven Press, pp. 1-14.

Ridderstrale, Y., and Hanson, M. (1985). Histochemical study of the distribution of carbonic anhydrase in the cat brain. *Acta Physiol. Scand.* **124**:557-564.

Rigatto, H., Lee, D., Davi, M., Moore, M., Rigatto, E., and Cates, D. (1988). Effect of increased arterial CO_2 on fetal breathing and behavior in sheep. *J. Appl. Physiol.* **64**:982-987.

Ross, C. A., Ruggiero, D. A., Joh, T. H., Park, D. H., and Reis, D. H. (1984). Rostral ventrolateral medulla: selective projections to the thoracic autonomic cell column from the region containing C1 adrenaline neurons. *J. Comp. Neurol.* **228**:168-185.

Schlaefke, M. E., See, W. R., and Loeschcke, H. H. (1970). Ventilatory response to alterations of H^+ ion concentration in small areas of the ventral medullary surface. *Respir. Physiol.* **10**:198-212.

Shams, H. (1985). Differential effects of CO_2 and H^+ as central stimuli of respiration in the cat. *J. Appl. Physiol.* **58**:357-364.

Smith, J. C., Morrison, D. E., Ellenberger, H. H., Otto, M. R., and Feldman, J. L. (1989). Brainstem projections to the major respiratory neuron populations in the medulla of the cat. *J. Comp. Neurol.* **281**:69-96.

Sterbenz, G. C., Neubauer, J. A., Geller, M. M., and Edelman, N. H. (1988). Carbonic anhydrase immunocytochemically localized in chemosensitive regions of medulla. *FASEB J.* **2**:A1295.

St. John, W. M., Hwang, Q., Nattie, E. E., and Zhou, D. (1989). Functions of the retrofacial nucleus in chemosensitivity and ventilatory neurogenesis. *Respir. Physiol.* **76**:159-172, 1989.

Sun, M. K., Hackett; J. T., and Guyenet, P. G. (1988). Sympathoexcitatory neurons of rostral ventrolateral medulla exhibit pacemaker properties in the presence of a glutamate-receptor antagonist. *Brain Res.* **438**:23-40.

Sundaram, K., Krieger, A. J., and Sapru, H. (1988). M_2 muscarinic receptors mediate pressor responses to cholinergic agonists in the ventrolateral medullary pressor area. *Brain Res.* **449**:141-149.

Teppema, L. J., Rochette, F., and Demedts, M. (1988). Ventilatory response to carbonic anhydrase inhibition in cats: Effects of acetazolamide in intact vs peripherally chemodenervated animals. *Respir. Physiol.* **74**:373-382.

Walker, D. W. (1984). Peripheral and central chemoreceptors in the fetus and newborn. *Annu. Rev. Physiol.* **46**:687-703.

Wennergren, G., and Wennergren, M. (1980). Respiratory effects elicited in newborn animals via the central chemoreceptors. *Acta Physiol. Scand.* **108**:309-311.

Wennergren, G., and Wennergren, M. (1983). Neonatal breathing control mediated via the central chemoreceptors. *Acta Physiol. Scand.* **119**:139-146.

Whittaker, J. A., Pan, Y., Barnard, D. G., Archer, P. W., Merrill, S., and Trouth, C. O. (1987). Age related differences in brainstem respiratory chemosensitive neuronal activity. *Neurosci. Abstr.* **13**:1640.

Willenberg, I. M., Dermietzl, R., Leibsten, A. G., and Effenberger, M. (1985). Mapping of cholinoceptive (nicotinoceptive) neurons in the lower brainstem: With special reference to the ventral surface of the medulla. *J. Auton. Nervs. Syst.* **14**:287-298.

Winterstein, H. (1956). Chemical control of pulmonary ventilation. III. The reaction theory of respiratory control. *N. Engl. J. Med.* **255**:331-337.

Woodrum, D. E., Standaert, T. A., Parks, C. R., Belenky, D., Murphy, J., and Hodson, W. A. (1977). Ventilatory response in the fetal lamb following peripheral chemadenervation.

Yamada, K. A., McAllen, R. M., and Loewy, A. D. (1984). GABA antagonists applied to the central surface of the medulla oblongata block the baroreceptor reflex. *Brain Res.* **297**:175-180.

Yamada, K. A., Moerschbaecher, J. M., Hamosh, P., and Gillis, R. A. (1983). Pentobarbital causes cardiorespiratory depression by interacting with a GABAergic system at the ventral surface of the medulla. *J. Pharmacol. Exp. Ther.* **226**:349-355.

Yamada, K. A., Norman, W. P., Hamosh, P., and Gillis, R. A. (1982). Medullary ventral surface GABA receptors affect respiratory and cardiovascular function. *Brain Res.* **248**:71-78.

Part Three

**RESPIRATORY MOTOR OUTPUT: CELLULAR
FUNCTION AND INTEGRATED BEHAVIOR**

13

Developmental Aspects of Diaphragm Muscle Cells
Structural and Functional Organization

GARY C. SIECK* and MARIO FOURNIER*

University of Southern California
Los Angeles, California

I. Introduction

During early postnatal development, the diaphragm, like other skeletal muscles, undergoes considerable change regarding its pattern of innervation and the morphological and metabolic properties of its muscle fibers. During this period, the diaphragm must continue to provide adequate contractile forces to ensure proper levels of ventilation. The means by which the diaphragm generates these forces changes considerably during postnatal development. In this review, we will focus on the basic functional organization of diaphragm neural control and examine its plasticity during postnatal development.

II. Motor Unit Physiological Properties

A. Classification of Motor Units

In skeletal muscles including the diaphragm, the final element of neural control is the motor unit, comprised of a motoneuron and the muscle fibers

---

*Present affiliation: Mayo Clinic and Mayo Medical School, Rochester, Minnesota
This research was supported by grants from the National Heart, Lung, and Blood Institute (HL34817 and HL34680) and the American Lung Association(State of California and Los Angeles County).

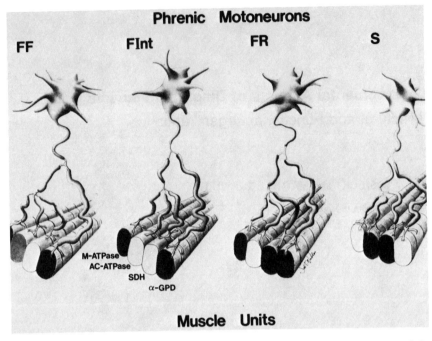

Figure 1 Diaphragm motor units are comprised of a phrenic motoneuron and the muscle fibers it innervates, i.e., the muscle unit. Muscle unit fibers have typical histochemical staining patterns for myofibrillar ATPase after alkaline (M-ATPase) and acid (AC-ATPase) preincubations, oxidative enzymes (SDH), and glycolytic enzymes (alpha-glycerophosphate dehydrogenase; α-GPD).

it innervates, i.e., the muscle unit (Liddell and Sherrington, 1925; Sherrington, 1929; Fig. 1). Burke et al. (1973) introduced standardized techniques for classifying four types of motor units in the adult based on differences in contractile and fatigue properties. In this scheme, fast-twitch units (type F) are distinguished from slow-twitch units (type S) by the presence of "sag" in unfused tetanic force responses. Generally, isometric twitch contraction time (CT, time to peak twitch tension) correlates with the presence of sag, although some overlap in the range of CT exists between type S and F units. Fast-twitch units are further subclassified as: fast-twitch fatigue resistant (FR), fast-twitch fatigue intermediate (FInt), and fast-twitch fatigable (FF). These different unit types vary considerably in the twitch and tetanic forces they generate, with FF and FInt units producing the greatest tensions and S units generating the lowest tensions. Thus, the tensions generated by the whole muscle will depend on the individual characteristics of those units which are recruited.

B. Motor Unit Tension

The tension generated by motor units depends on three factors: (1) the innervation ratio of the unit, i.e., the number of muscle fibers innervated by a motoneuron; (2) the cross-sectional area of muscle unit fibers; and (3) the specific tension of the unit, i.e., the force per unit cross-sectional area (Burke, 1981). With the introduction of the glycogen-depletion technique to identify muscle unit fibers (Edström and Kugelberg, 1968), it became possible to directly measure these variables. Yet, only a few studies have reported such direct measurements and only for motor units in hindlimb muscles (Table 1). These limited data do not provide a consistent underlying mechanism for the differences in tension produced by different unit types. For

Table 1 Summary of the Relative Differences in Maximum Tetanic Tension, Innervation Ratio, Cross-Sectional Area, and Specific Tension Among Different Types of Motor Units

Muscle	Unit type	Max. tetanic tension	Innervation ratio	Cross-sectional area	Specific tension
TA	S	1.0	1.0	1.0	1.0
	FR	2.5	2.1	1.0	1.2
	FF + FInt	5.5	2.9	1.3	1.4
MG	S	1.0	1.0	1.0	1.0
	FR	3.8	0.9	1.2	3.7
	FF + FInt	9.4	1.1	2.3	4.0
FDL	S	1.0	1.0	1.0	1.0
	FR	4.8	0.7	1.4	4.8
	FF + FInt	27.3	1.8	2.6	5.8
SOL	S	1.0	1.0	1.0	1.0
	FR	1.9	1.3	1.1	1.2
DIA	S	1.0	1.0	1.0	1.0
	FR	2.3	0.9	1.3	1.8
	FF + FInt	3.6	1.1	1.7	1.9

All values are referenced to those of type S units.
TA, Tibialis anterior muscle of cat (from Bodine et al., 1987); MG, medial gastrocnemius muscle of cat (from Burke, 1981); FDL, flexor digitorum longus muscle of cat (from Burke, 1981); SOL, soleus muscle of rat (from Chamberlain and Lewis, 1989); DIA, diaphragm muscle of cat (Sieck, unpublished observations).

example, in the medial gastrocnemius and flexor digitorum longus muscles of the cat, the larger tensions generated by type F units appear to be primarily related to greater specific tensions and, to a lesser extent, larger fiber cross-sectional areas (Burke, 1981). In contrast, in the tibialis anterior muscle of the cat (Bodine et al., 1987) and the soleus muscle of the rat (Chamberlain and Lewis, 1989), the larger tensions generated by type F units appear to be primarily related to a greater innervation ratio of these units and, to a lesser extent, larger fiber cross-sectional area.

C. Motor Unit Properties in the Adult Diaphragm

Recently, we described the mechanical properties of motor units in the adult cat diaphragm (Fournier and Sieck, 1988b; Sieck, 1988). We found very few FR units and a relatively high proportion of FInt units (Table 2). This was

Table 2 Summary of the Contractile and Morphometric Properties of Motor Units in the Cat Diaphragm

Motor unit type	S	FR	FInt	FF
Contractile properties				
Proportion (%)	30	4	25	41
Sag	—	+	+	+
CT (msec)	68	42	42	34
	± 18	± 5	± 7	± 3
P_t (g)	1.5	2.6	3.5	3.3
	± 0.4	± 2.0	± 2.3	± 1.9
P_o (g)	3.9	8.8	13.4	14.4
	± 1.2	± 6.0	± 7.2	± 8.0
Relative force (%)	11	3	31	55
Morphometric properties				
CSA (um²)	1324	1787	2086	2347
	± 170	—	± 510	± 793
% Total CSA	20	4	27	49
Innervation ratio	245	224	270	256
Specific tension (kg/cm²)	1.2	2.2	2.2	2.4

Values are means ± 1 standard deviation.
CT, Twitch contraction time; P_t, peak twitch tension; P_o, maximum tetanic tension; CSA, fiber cross-sectional area.

surprising since in appendicular muscles, very few FInt units are typically found (Burke, 1981; Burke et al., 1973). In addition, in a muscle such as the diaphragm, which is frequently activated with breathing, more FR units might have been expected. As in appendicular muscles, FF and FInt units in the diaphragm were found to produce the greatest twitch and tetanic tensions, followed in order by FR and S units (Tables 1 and 2). These differences in unit tensions in the diaphragm appeared to be related to a combination of a larger cross-sectional area of type F muscle unit fibers and a greater specific tension of these units (Table 1).

D. Diaphragm Innervation and Synapse Elimination

The diaphragm is innervated somatotopically by two to four segments of the cervical spinal cord. In the cat diaphragm, we found that the ventral portions of the costal and crural regions were primarily innervated by the C_5 ventral root and the dorsal portions of both regions were primarily innervated by the C_6 ventral root (Fournier and Sieck, 1987, 1988a). Within each of these larger innervated areas, the fibers comprising single motor units occupied a restricted territory. Motor axons situated more rostrally in the C_5 ventral root innervated muscle unit fibers in the sternocostal or ventral crural regions, whereas axons more caudally situated innervated muscle unit fibers in the midcostal or ventral crural regions. Thus, the territories occupied by diaphragm muscle unit fibers were also somatotopically organized (Fournier and Sieck, 1987).

Laskowski and Sanes (1987) demonstrated a similar somatotopic innervation of the costal region of the rat diaphragm, with C_4 innervating the more ventral portions and C_6 innervating the more dorsal portions. In the rat, C_5 innervated the midcostal region. Recently, we have observed somatotopy in the innervation of the hamster diaphragm as well, but in this species, the phrenic nerve is comprised of axons originating from C_3 to C_5 (Fournier and Sieck, unpublished observations).

In the adult, each muscle unit fiber is innervated by only one motoneuron (Fig. 1). In contrast, during early postnatal development, most muscle fibers are innervated by two or more motoneurons (Bennett and Pettigrew, 1974; Betz et al., 1979; Bixby and Van Essen, 1979; Brown et al., 1976; Dennis, 1981, Dennis et al., 1981; Gordon and Van Essen, 1985; O'Brien et al., 1978). Thus, muscle fibers can belong to more than a single motor unit. The time frame for synapse elimination varies from species to species. For example, in the cat diaphragm, synapse elimination is complete by the third postnatal week (Fournier and Sieck, unpublished observations). In the rat diaphragm, synapse elimination is completed by the second week (Bennett and Pettigrew, 1974; Rosenthal and Taraskevich, 1977), and in the rabbit by the first postnatal week (Bixby and Van Essen, 1979). Laskowski and High (1989) found that synapse elimination of the rat diaphragm was not random, but followed the underlying somatotopy of the adult muscle. Thus, the process of synapse elimination does not establish the somatotopic pattern of diaphragm innervation found in the adult, but instead sharpens it. The postnatal pattern of somatotopic axonal projection is not unique to the diaphragm, for it occurs in hindlimb muscles as well (Bennett et al., 1986; Brown and Booth, 1983).

The level of muscle activity during postnatal development influences the time frame for synapse elimination. For example, reducing muscle activity by deafferentation or spinal cord transection delays synapse elimination on fibers of the rat lumbrical muscle (Caldwell and Ridge, 1983). Similarly, complete muscle inactivation [using tetrodotoxin (TTX) to block axonal propagation of action potentials] has also been shown to delay the process of synapse elimination (Brown et al., 1982; Thompson et al., 1983). Conversely, increasing the level of muscle activation by chronic electrical stimulation advances the process of synapse elimination (O'Brien et al., 1978; Ridge and Betz, 1984; Thompson, 1983). One might expect that the dramatic increase in diaphragm activation at birth might accelerate the process of synapse elimination, especially if accompanied by increased mechanical loads to breathing.

E. Motor Unit Properties During Postnatal Development

In "mixed" muscles of the cat hindlimb, the contractile properties distinquishing fast- and slow-twitch motor units are not fully developed until 6-10 weeks of age (Bagust et al., 1974; Hammarberg and Kellerth, 1975; Westerman et al., 1979). Thus, motor unit types cannot be unambiguously classified during early postnatal development. It has been shown that during this period, both the average CT and half relaxation time (1/2 RT; time for twitch tension to relax to 1/2 peak value) of motor units decrease progressively. Hammarberg and Kellerth (1975) found that until 6 weeks of age, most units in cat hindlimb muscles were fatigue resistant. Thus, in these muscles, FF units appear only after the sixth postnatal week.

The mechanical properties of motor units in developing muscles are affected by synapse elimination (Balice-Gordon and Thompson, 1988; Betz et al., 1979; Brown et al., 1976; Fladby, 1987). As a result of polyneuronal innervation, the relative size of motor units, i.e., the amount of force (twitch or tetanic) generated by a unit relative to the total force produced by the whole muscle, is greater during early postnatal development than in the adult (Bagust et al., 1974; Betz et al., 1979; Brown et al., 1976; Fladby, 1987; Jones et al., 1987; Soileau et al., 1987). It should be kept in mind, however, that during postnatal development the overall contractile force of a muscle increases as a result of the increase in fiber cross-sectional area (see below). Yet the relative contribution of each unit to total muscle force is decreasing (due to the gradual elimination of polyneuronal innervation).

III. Neural Control of Muscle Tension

In the classic model of Sherrington (Liddell and Sherrington, 1925; Sherrington, 1929), it was suggested that the muscle forces generated during any motor behavior resulted from the recruitment of variable numbers of motor units, each of which contributed a quantal amount to the total force. The central nervous system produced the desired level of force by controlling the temporal and numerical combination of these functional quanta, i.e., motor units. Thus, force could be controlled by changing the number of activated motor units (recruitment coding) and/or by modifying the discharge frequency of recruited motor units (frequency coding). Subsequent research focused on the underlying mechanisms that control motor unit recruitment. In 1957, Henneman originally proposed the hypothesis that the susceptibility of neurons to discharge was a function of their size. Subsequently, this hypothesis was restated more specifically as the "size principle" relating motoneuron recruitment to their intrinsic size-related electrical properties (Henneman et al., 1965; Henneman and Mendell, 1981). Since that time, the size principle has become a major unifying theory of motor unit recruitment, and it has gained considerable experimental support in a variety of muscles, including the diaphragm (see below).

A. Intrinsic Properties of Motoneurons

As predicted by the size principle, there is an orderly progression of motor unit recruitment, in most muscles, that is related to the intrinsic size-related electrical properties of motoneurons (Henneman, 1957; Henneman et al., 1965; Henneman and Mendell, 1981). Smaller motoneurons with higher membrane input resistance, lower rheobase (current needed to induce firing), and slower axonal conduction velocities are recruited first (Burke, 1981; Fleshman et al., 1981; Henneman and Mendell, 1981; Sypert and Munson, 1981; Zajac and Faden, 1985; Zengel et al., 1985). These smaller motoneurons typically innervate type S muscle unit fibers, which produce lower tensions (Burke, 1981; Fleshman et al., 1981; Sypert and Munson, 1981; Zajac and Faden, 1985; Zengel et al., 1985). Larger motoneurons, having lower membrane input resistance, higher rheobase, and faster axonal conduction velocities, are recruited later and innervate type F muscle unit fibers, which generate greater tensions. Among the type F units, FR units have the lowest recruitment threshold, followed in order by FInt and FF units (Burke, 1981). Thus, the order of motor unit recruitment is directly related to the size (tetanic tension) and fatigue resistance of units.

B. Diaphragm Motor Unit Recruitment

Recently, Berger and colleagues (Dick et al., 1987a,b; Jodkowski et al., 1987) presented evidence that the order of motor unit recruitment in the diaphragm also depends on the intrinsic properties of phrenic motoneurons. Dick et al., (1987a,b) found that during spontaneous breathing, phrenic motoneurons in the cat were recruited in an order related to increasing axonal conduction velocity. In light of the demonstrated relationship between axonal conduction velocity and phrenic motoneuron membrane input resistance (Berger, 1979; Webber and Pleschka, 1976), these authors concluded that the size principle predicted the order of diaphragm motor unit recruitment. In the study by Jodkowski et al. (1987), it was reported that two subpopulations of phrenic motoneurons exist in the cat, segregated by differences in membrane input resistance and rheobase. These authors reported that most (i.e., 60%) of those phrenic motoneurons with higher input resistance and lower rheobase were recruited during spontaneous hypercapneic breathing. In contrast, none of the phrenic motoneurons with lower input resistance and higher rheobase were recruited during inspiration, even with end-tidal CO_2 at 7%. These authors suggested that phrenic motoneurons recruited with respiration innervated type S muscle unit fibers while those not recruited with breathing innervated type F muscle unit fibers. The proportion of all phrenic motoneurons that they found to be recruited with inspiration, i.e., 23%, approximated the proportion of type S units that we found in the cat diaphragm (Fournier and Sieck, 1988b; Sieck, 1988; Table 2).

An alternative model for diaphragm motor unit recruitment has been presented by Hilaire and co-workers (Hilaire et al., 1983, 1987; Monteau et al., 1985). Using techniques of cross correlation and spike-triggered averaging, these authors demonstrated that the medullary inspiratory neurons that discharge early during inspiration have an increased probability of monosynaptically driving phrenic motoneurons that also display early recruitment during inspiration. Conversely, the medullary inspiratory neurons that discharge late in inspiration monosynaptically drive phrenic motoneurons recruited later during inspiration. These results suggested that the recruitment order of diaphragm motor units might be dictated not only by the size and intrinsic properties of phrenic motoneurons, but also by the specific pattern of synaptic input from medullary inspiratory neurons.

In accomplishing the various ventilatory and nonventilatory behaviors of the diaphragm, the nervous system must select from the repertoire of motor unit types, which can vary in their contractile and fatigue properties (Tables 1 and 2). Some units produce greater tensions, but evince rapid fatigue when activated frequently. Recruitment of these FF and FInt units, therefore, may not be suitable for generating ventilatory forces. On the other hand, type S and FR units can be frequently activated with very little fatigue. The tradeoff, however, is that these units produce much lower tensions when recruited which may be insufficient for some ventilatory and nonventilatory behaviors. In a recent study (Sieck and Fournier, 1989), we measured the transdiaphragmatic pressures (Pdi) generated by the adult cat diaphragm during different ventilatory and nonventilatory behaviors. These Pdi's were normalized to the maximum Pdi generated by bilateral phrenic nerve stimulation, i.e., Pdi max. Figure 2 summarizes the Pdi's produced during different ventilatory and nonventilatory behaviors and presents a model of motor unit recruitment required to generate these pressures. We

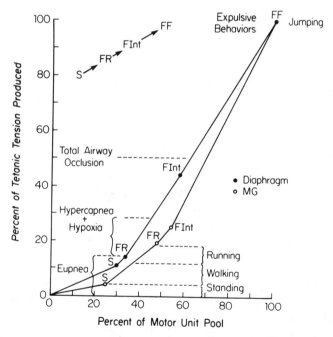

Figure 2 Models of motor unit recruitment in the diaphragm (data taken from Fournier and Sieck, 1988b) and medial gastrocnemius muscles (data taken from Walmsley et al., 1978) of the cat during different motor behaviors. These models assume: (1) a specific order of motor unit recruitment, i.e., S, FR, FInt, and FF; and (2) complete activation of one motor unit type before the next type is recruited.

found that during normal ventilation, the adult cat diaphragm generates only about 12% of its maximum force-generating capacity. This level of force could be generated by the recruitment of only fatigue-resistant motor units, i.e., type S and FR. When the drive for ventilation was increased by having the animals inspire a hypercapnic (5% CO_2), hypoxic (10% O_2) gas mixture, Pdi increased to 28% of Pdi max. This force level could be generated by the additional recruitment of some FInt units. Even during total airway occlusion, the forces generated by the diaphragm amounted to only 49% of Pdi max. Such a force level could be produced without the recruitment of FF units. Only during expulsive nonventilatory behaviors, e.g., gagging and sneezing, were maximum forces generated. From these results, we concluded that the normal ventilatory requirements of an animal could be met by the recruitment of only the fatigue-resistant motor units in the diaphragm. Furthermore, these results suggested that a large number of units in the adult diaphragm remain relatively inactive during normal ventilation. Figure 2 also compares the model of motor unit recruitment in the diaphragm to a similar model for motor unit recruitment in the cat medial gastrocnemius during different locomotor behaviors (Walmsley et al., 1978). As in the diaphragm, the recruitment model in the medial gastrocnemius muscle also predicts that a large fraction of the motor unit pool would be utilized only infrequently during behaviors of short duration requiring high levels of force, e.g., jumping. In contrast, all locomotor behaviors of the medial gastrocnemius muscle could be accomplished by the recruitment of fatigue-resistant motor units, i.e., S and FR.

C. Intrinsic Motoneuron Properties During Postnatal Development

During early postnatal development, motoneurons are much smaller than in the adult (Cameron et al., 1989; Conradi and Ronnevi, 1977; Conradi and Skoglung, 1969; Cullheim and Ulfhake, 1979; Huizar et al., 1975; Kellerth et al., 1971; Mellström and Skoglund, 1969; Remahl et al., 1985; Rose et al., 1984; Sato et al., 1977; Ziskind-Conhaim, 1988b). As motoneurons increase in size during postnatal development, input resistance progressively decreases and rheobase increases (Fulton and Walton, 1986; Shapovalov et al., 1981; Ziskind-Conhaim, 1988b). The importance of intrinsic motoneuron properties in determining the order of motor unit recruitment in the neonate is unclear. Sato et al. (1977) reported that a size difference between motoneurons innervating slow- and fast-twitch muscle unit fibers in the hindlimb did not develop until the tenth postnatal week in the kitten. This finding suggests that motor unit recruitment order in the neonate does not depend solely on the intrinsic membrane properties of motorneurons.

In recent studies on phrenic motoneurons in the kitten, Cameron and co-workers (personal communication) found that motoneuron input resistance could predict the order of unit recruitment. Phrenic motoneurons classified as "early," i.e., discharging within the initial 25% of the inspiratory period, had higher input resistance than motoneurons classified as "late," i.e., discharging in the last 75% of the inspiratory period. These authors also found that the input resistance of phrenic motoneurons decreased with age during postnatal development as mean somal surface area increased (Cameron et al., 1989). Rose et al. (1984) also reported a progressive increase in phrenic motoneuron size in the kitten that was still not complete by the eighth postnatal week. These authors concluded that maturation of the morphological properties of phrenic motoneurons did not correspond to maturation of motoneuron electrophysiological properties, reported to be complete by the fourth postnatal week (Duron and Marlot, 1980). However, in the recent studies by Cameron and co-workers (personal communication), adult values of membrane input resistance in phrenic motoneurons were not reached until the twelfth postnatal week in the kitten. These authors also found that the specific input resistance (i.e., motoneuron input resistance/somal surface area) of a subpopulation of phrenic motoneurons decreased between the second and fourth week of postnatal development in the kitten. They concluded that the increase in motoneuron size was the major factor contributing to the decrease in membrane input resistance. Clearly, these developmental changes in the intrinsic properties of phrenic motoneurons would influence motor unit recruitment.

D. Motor Unit Recruitment During Postnatal Development

In evaluating motor unit recruitment in the neonatal diaphragm, two important factors should be considered: (1) the relative forces required for normal ventilation, and (2) the influence of polyneuronal innervation. We speculate that during normal ventilation, the neonatal diaphragm produces a greater fraction of its maximum force-generating capacity than in the adult. This is likely for two reasons: First, diaphragm muscle fibers are much smaller during early postnatal development, and as a result, the maximum force-generating capacity of the muscle is less than it is in the adult. Second, the greater compliance of the chest wall during early postnatal development imposes a mechanical load on the breathing muscles (Mortola, 1987). Thus, it is likely that a greater Pdi is required for normal ventilation. During postnatal development, as muscle fiber size (and maximum tension) increases and chest wall compliance decreases, ventilatory forces would become a smaller fraction of the total force-generating capacity of the diaphragm.

Accordingly, the proportion of fibers recruited during normal ventilation would decrease. Cameron and co-workers (personal communication) have observed that in the 30-day-old kitten, about 50% of all phrenic motoneurons were recruited with inspiration when the chemical drive for breathing was increased. At a similar chemical drive, the inspiratory-related recruitment is approximately 23% of all phrenic motoneurons in the adult cat (Jodkowski et al., 1987). As in the adult cat (Jodkowski et al., 1989), the quiescent phrenic motoneurons in the kitten were recruited only during the aspiration reflex (Cameron et al., personal communication). Thus, even in the kitten, there is a functional reserve capacity within the diaphragm motor unit pool. However, the fraction of the diaphragm motor unit pool utilized for ventilatory behaviors appears to be greater in the kitten than in the adult.

The second important consideration in evaluating motor unit recruitment during the postnatal period is the influence of polyneuronal innervation. As mentioned above, because of polyneuronal innervation, the relative size of motor units is larger in the neonate than in the adult (Bagust et al., 1974; Betz et al., 1979; Brown et al., 1976; Fladby, 1987; Jones et al., 1987; Soileau et al., 1987). Thus, to generate a given level of relative force, the recruitment of fewer motor units would be required. However, this would hold true only during the period of polyneuronal innervation. After the completion of synapse elimination, i.e., third postnatal week in the kitten, the higher level of relative force generated by the diaphragm during ventilatory behaviors would require the recruitment of more motor units than in the adult (see above).

E. Frequency Coding of Motor Unit Tension

Because of their longer CT and 1/2 RT, the tension generated by type S units summates at lower frequencies of activation than type F units (Fig. 3). Maximum tetanic tensions are reached at lower rates of stimulation in type S units and the ratio of twitch tension to maximum tetanic tension is higher in type S units than in type F units. Thus, the range of frequency coding in type S units is smaller than in type F units.

In the adult cat, we recorded the spontaneous discharge of diaphragm motor units during different sleep and waking states (Sieck et al., 1984). We found that those diaphragm motor units recruited early during inspiration had modal discharge rates ranging from 8 to 12 pulses per sec (pps). Motor units recruited later during inspiration had modal discharge rates ranging from 13 to 16 pps. Occasionally, diaphragm motor units were recruited very late during inspiration with modal discharge rates of 20-25 pps. If it is assumed that the motor units consistently recruited during inspiration were type S, the forces generated by these rates of activation would be

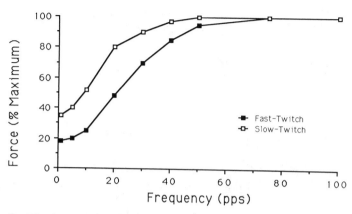

Figure 3 The forces generated by diaphragm motor units depend on the frequency of stimulation. The force/frequency response curve of slow-twitch (type S) motor units in the cat diaphragm is shifted to the left compared to that of fast-twitch (type F) motor units.

aproximately 50-75% of maximum tetanic tension. It is important to note that these discharge rates would be on the steep portion of the force/frequency curves for type S units (Fig. 3). If the "late"-recruited units, i.e., those with discharge rates of 20-25 pps, were type F units, the forces generated at these discharge rates would also be approximately 50-75% of maximum. Again, these discharge rates would be on the steep portion of the force/frequency curves of type F units (Fig. 3). Thus, the spontaneous discharge rates of diaphragm motor units allow for optimal frequency coding of tension.

F. Frequency Coding of Motor Units During Postnatal Development

In evaluating frequency coding of diaphragm motor units during postnatal development, two important factors should be considered: (1) The CT and 1/2 RT of motor units in the neonatal muscle are much longer than in the adult (Bagust et al., 1974; Hammarberg and Kellerth, 1975; Westerman et al., 1979). (2) The electrophysiological properties of motoneurons and motor nerve axons change during postnatal development (Conradi and Skoglund, 1969; Conradi and Ronnevi, 1977; Cullheim and Ulfhake, 1979; Fulton and Walton, 1986; Gallego et al., 1978; Glebovskii, 1961a; Huizar et al., 1975; Kellerth et al., 1971; Mellström and Skoglund, 1969; Redfern, 1970; Remahl et al., 1985; Sato et al., 1977; Walton and Fulton, 1986; Ziskind-Conhaim, 1988a,b), influencing their ability to follow higher rates of activation.

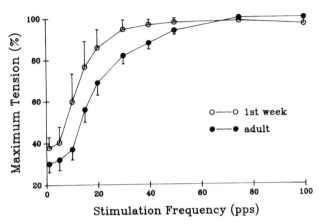

Figure 4 The forces generated by the diaphragm depend on the frequency of stimulation. The force/frequency response curve of the neonatal rat diaphragm is shifted to the left compared to the adult muscle.

The longer CT and 1/2 RT of motor units in the neonatal muscle allows for greater summation and earlier fusion of tetanic force responses than in the adult (Fig. 4). Thus, the force/frequency relationship of neonatal motor units is shifted to the left compared to the adult muscle. In the adult cat diaphragm, we demonstrated that slow-twitch motor units reach maximum tetanic tensions at lower rates of stimulation than fast-twitch motor units (Fournier and Sieck, 1988b; Sieck, 1988). As might be expected for a muscle predominantly composed of slower motor units, the force/frequency relationship of the neonatal rat diaphragm is shifted to the left compared to the adult muscle (Fig. 4).

In the adult, motor unit contractile properties are generally matched by the electrophysiological properties that limit motoneuron discharge rate, e.g., the duration of afterhyperpolarization (Burke, 1981). However, during postnatal development the duration of motoneuron afterhyperpolarization appears to be independent of muscle contractile properties (Hammarberg and Kellerth, 1975; Huizar et al., 1975). Yet the prolonged afterhyperpolarization of neonatal rat motoneurons undoubtedly limits their maximum discharge rates (Fulton and Walton, 1986; Walton and Fulton, 1986). This is consistent with the lower firing rate of motor units during postnatal development in the rat (Navarrete and Vrbova, 1983).

The smaller diameter of motor axons in the neonate would result in a greater longitudinal axonal resistance and a lower safety margin for propagation of action potentials along the axon, especially at higher rates of activation (Conradi and Skoglund, 1969; Conradi and Ronnevi, 1977; Ziskind-Conhaim, 1988a). In addition, because of polyneuronal innervation there is a greater extent of axonal branching, and thus axonal branch point failure (Krnjevic and Miledi, 1988; Smith, 1980) may contribute more to neurotransmission failure in the neonate (see below).

In the rat diaphragm, Kelly (1978) demonstrated that the quantal content of end-plate potentials is lower during postnatal development than it is at the adult neuromuscular junction. The mean amplitude of miniature end-plate potentials (mepp) decreased while the mean frequency of mepp's increased (Bennett and Pettigrew, 1974; Diamond and Miledi, 1962; Kelly, 1978; Wilson and Cardaman, 1984). The estimated safety factor for neuromuscular transmission increased with age, most rapidly during the first six weeks of life (Kelly, 1978). The lower safety factor for neuromuscular transmission in the neonate may be especially important during repetitive stimulation when the threshold for generating action potentials in muscle fibers increases (Krnjevic and Miledi, 1958).

IV. Factors Contributing to Diaphragm Muscle Fatigue

A. Central Versus Peripheral Fatigue

Muscle fatigue has been defined as an inability to maintain an expected or required level of force (Edwards, 1981). Fatigue may occur at any one of several potential sites which have been generally grouped as of either central or peripheral origin (Bigland-Ritchie, 1981). If motoneurons fail to generate action potentials, central fatigue of motor units results (Kernell and Monster, 1982b). This could be due to reduced synaptic drive or to an accommodation of motoneuron discharge (Kernell and Monster, 1982a,b). Peripheral fatigue results from failure in any of the following: axonal propagation, neuromuscular transmission, excitation-contraction coupling, sarcoplasmic reticulum release of calcium, muscle fiber contractile mechanisms, or energy production within muscle fibers. Central fatigue is bypassed by stimulating the motor axon. Discriminating between the potential sites of peripheral fatigue, however, is more difficult. Peripheral fatigue can be further categorized as a failure of either neuromuscular transmission (pre- or postsynaptic) or of muscular origin. In our studies, we have focused on presynaptic mechanisms of neurotransmission failure, i.e., excluding a failure of action potential propagation along the muscle fiber sarcolemma or transverse tubular system, and on the supply of energy within muscle fibers, i.e., oxidative enzyme activities, capillary distribution, and diffusion limitations.

B. Neurotransmission Failure and Diaphragm Fatigue

The contribution of neurotransmission failure to diaphragm fatigue can be assessed by comparing force loss during nerve stimulation to that occurring with direct muscle stimulation, i.e., bypassing the neuromuscular junction (Aldrich et al., 1986; Kelsen and Nochomovitz, 1982; Kuei et al., 1990; Pagala et al., 1984). In a study of the adult rat, we compared the forces generated by these two modes of stimulation and estimated the relative contribution of neurotransmission failure to diaphragm fatigue at various rates of stimulation (Kuei et al., 1990). At a stimulation rate of 20 pps, we estimated that neurotransmission failure contributed approximately 16% to diaphragm fatigue after 2 min of repetitive stimulation. At 40 pps, neurotransmission failure contributed approximately 35% and at 75 pps approximately 42%. Thus, in the adult rat diaphragm, most peripheral fatigue appears to be of muscular origin.

In recent studies of the neonatal rat diaphragm, we found that the forces generated by tetanic phrenic nerve stimulation (greater than 10 pps) were significantly less than those produced by direct muscle stimulation (Fig. 5; Fournier and Sieck, unpublished observations). The difference in forces generated by these two modes of stimulation increased markedly as stimulation frequency was increased. We estimated that in the neonate, neurotrans-

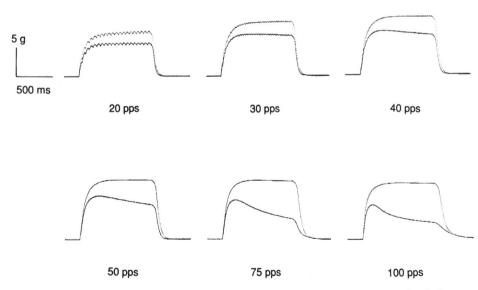

Figure 5 In the neonatal rat diaphragm, the forces generated by nerve stimulation were less than those generated by direct muscle stimulation. At higher rates of nerve stimulation, there was marked neurotransmission failure in the activation of the neonatal diaphragm.

mission failure would contribute 75-93% of the fatigue induced by 40 pps stimulation. By the third postnatal week, the contribution of neurotransmission failure to diaphragm fatigue approximated that found in the adult muscle. The greater susceptibility of the neonatal rat diaphragm to neurotransmission failure could be related to an inability of phrenic axons to follow higher rates of activation or to a failure at the neuromuscular junction (see above). Regardless, a greater susceptibility to neurotransmission failure would greatly curtail the efficacy of frequency coding in neonatal diaphragm force generation (see above).

Our study of the adult cat diaphragm demonstrated that motor units differed in their susceptibility to neurotransmission failure (Sieck and Fournier, 1990). Fatigue-resistant units, i.e., type S and FR, showed little change in evoked muscle unit action potentials (MUAP) during the fatigue test. In contrast, some FF and FInt units showed marked changes in evoked MUAP. These changes included a prolongation of MUAP duration, a decrease in peak MUAP amplitude, and a decrease in the integrated MUAP waveform. In most cases, these changes in MUAP did not correlate with motor unit force loss during the fatigue test. Some units showed very abrupt changes in MUAP waveform, a result that was also observed by Sandercock et al. (1985) in motor units of the cat medial gastrocnemius muscle. These authors reported that the abrupt changes in MUAP waveform correlated with a failure in the generation of action potentials in some muscle unit fibers, i.e., direct evidence for neurotransmission failure. In such cases, a concomitant decrease in motor unit force would also be expected. As for the muscle as a whole, the susceptibility of motor units to neurotransmission failure depends on the rate of stimulation (Clamann and Robinson, 1985; Sandercock et al., 1985). Thus, in these units, frequency coding of force, especially at higher rates of activation, would be less effective.

C. Motor Unit Fatigue Resistance and Oxidative Capacity

Using the technique of glycogen depletion to identify muscle unit fibers (Edström and Kugelberg, 1968), we examined the correlation between diaphragm motor unit contractile and fatigue properties and the metabolic properties of muscle unit fibers (Enad et al., 1989). In this study, we used a microphotometric procedure whereby the activity of the oxidative enzyme succinate dehydrogenase (SDH) could be quantified in individual muscle fibers (Blanco et al., 1988; Sieck et al., 1986, 1987). Using this procedure, we found a significant correlation between diaphragm motor unit fatigue resistance and the mean SDH activity of muscle unit fibers. In a recent study, Martin et al. (1988) used a similar quantitative histochemical procedure and reported comparable results for motor units in the cat tibialis anterior muscle. Using a microbiochemical procedure, oxidative enzyme activities have also been measured in dissected muscle unit fibers of the rat extensor digitorum longus (Nemeth et al., 1981) and cat tibialis posterior muscles (Hamm et al., 1988; Nemeth et al., 1986). These studies also reported that motor unit fatique resistance and oxidative capacity were correlated in these muscles. Several other studies have suggested a correlation between muscle unit fatique resistance and oxidative capacity based on a subjective assessment of oxidative enzyme staining intensities (Burke, 1981; Burke et al., 1973; Edström and Kugelberg, 1968). These subjective studies have suggested that adult muscle units are composed of fibers with homogeneous histochemical properties. Indeed, these studies have provided the functional basis justifying the histochemical classification of different muscle fiber types (see below).

D. SDH Variability Among Diaphragm Muscle Fibers

If neurogenic factors exert a predominant influence on the enzymatic activities of muscle fibers, it might be expected that muscle unit fibers would have homogeneous enzymatic properties. In the cat diaphragm, we found that the variability of SDH activity among diaphragm muscle unit fibers was significantly less than that noted for the muscle as a whole and similar to that observed along the length of muscle fibers (Fournier and Sieck, 1986; Sieck, 1988). This indicated that in the adult diaphragm, innervation exerts an important influence on the enzymatic properties of muscle fibers. The microbiochemical studies of Nemeth and co-workers (Hamm et al., 1988; Nemeth et al., 1981, 1986) also revealed that oxidative enzyme activities among muscle unit fibers were more homogeneous than in the muscle as a whole. However, in another quantitative histochemical study, Martin et al. (1988) reported that in the cat tibialis anterior muscle, the variability of SDH among muscle unit fibers was greater than that found along the length of muscle fibers. These authors concluded that the metabolic enzyme activities of muscle fibers are not under the complete control of neurogenic factors.

The uniformity of fiber oxidative enzyme activities in neonatal muscles (including the diaphragm) has been widely noted (Figs. 6-8; Cheung et al., 1988; Cooper et al., 1970; Dubowitz, 1965; Nyström, 1968; Sieck and Blanco, 1987a,b; Tomanek, 1975). For example, the coefficient of variation of SDH activities among all fibers in the neonatal cat diaphragm is approximately 20%, compared to 50% in the adult muscle. Likewise, in the neonatal rat diaphragm, the variance in SDH activities among fibers is about 15% compared to 40% in the adult. It is not until after 3 weeks of age in both the cat and the rat that variability of SDH activities among all diaphragm fibers increases to that found in the adult muscle. The increase in SDH variability during postnatal development is entirely due to the varia-

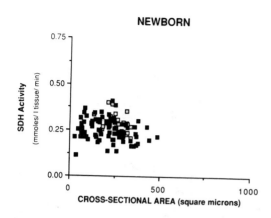

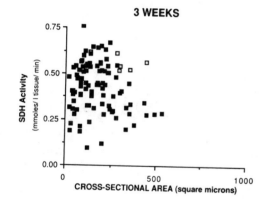

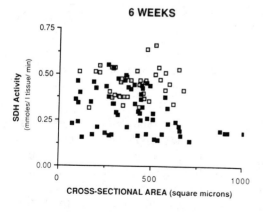

Figure 6 Scatter plots showing the relationships between muscle fiber cross-sectional area and SDH activity in the cat diaphragm at three different ages. Type I are represented by open squares and type II by closed squares.

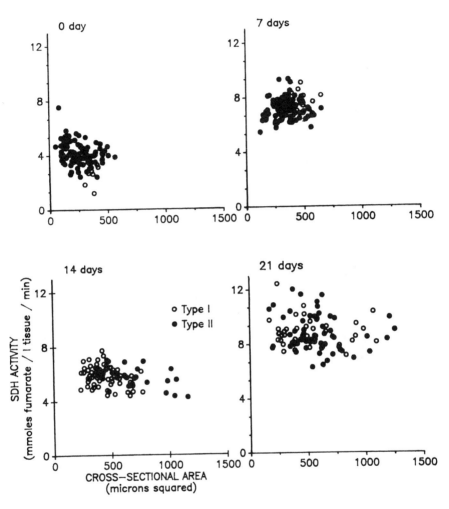

Figure 7 Scatter plots showing the relationships between muscle fiber cross-sectional area and SDH activity in the rat diaphragm at four different ages.

bility of type II muscle fibers (for definition of type I and II fibers, see section V) (Fig. 8). The coefficient of variation for SDH activity among type I fibers remains the same across all ages. During early postnatal development, the coefficient of variation for type II fibers is also low, being similar to that for type I fibers. Only after 3 weeks of age in the rat, and 6 weeks of age in the cat, does the marked variability of SDH activity among type II fibers appear.

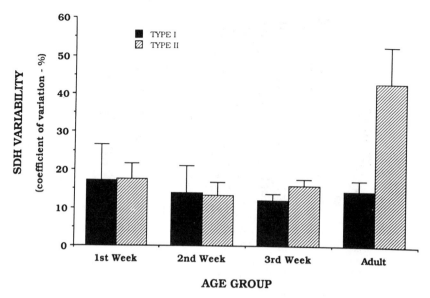

Figure 8 During postnatal development of the rat diaphragm, the variability of SDH activities among type I fibers showed little change. In contrast, the coefficient of variation of SDH activities among type II fibers increased dramatically.

If polyneuronal innervation contributes to the relative homogeneity of enzymatic activities among fibers, it might be expected that in hindlimb muscles, where synapse elimination occurs later, variability in fiber SDH activities might appear later as well. In preliminary studies on the development of the medial gastrocnemius muscle in the rabbit, we found that at 14 days of age, when polyneuronal innervation of the medial gastrocnemius persists but is absent in the diaphragm, the variability in SDH activities among diaphragm fibers is about 60% higher than among medial gastrocnemius fibers (Fig. 9). Just the opposite is true for enzymatic variability in adult diaphragm and medial gastrocnemius muscle fibers (Blanco and Sieck, 1987). We also found the SDH activity of neonatal rabbit diaphragm fibers to be higher than that of fibers in the medial gastronemius muscle (Fig. 9). The SDH activities of both diaphragm and medial gastrocnemius fibers increased with age. It is possible that the difference between diaphragm and medial gastrocnemius muscle fiber oxidative capacity reflects a higher level of activation of the diaphragm during early development. Yet, even muscle fibers in the neonatal medial gastrocnemius are affected by chronic disuse. Using TTX to block activation of medial gastrocnemius muscle fibers from birth to 7 days of age in the rabbit, we found that fiber SDH activities were ap-

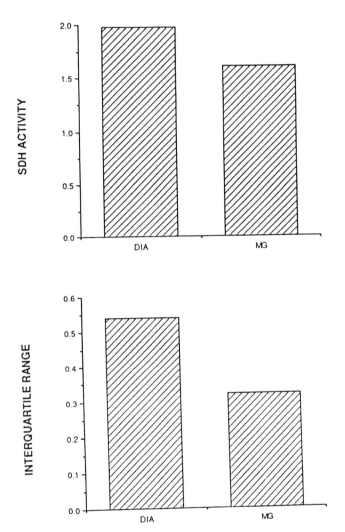

Figure 9 The mean SDH activity (in mmol fumarate/liter tissue/min) of muscle fibers in the 14-day-old rabbit diaphragm is compared to that of fibers in the medial gastrocnemius muscle (top). The variability of SDH activities in the diaphragm and medial gastrocnemius muscles of the 14-day-old rabbit were compared to determining the interquartile range, i.e., the median range of SDH activities over which 50% of all values lie.

proximately three times lower than in control muscles (Fig. 10). The range of SDH activities among medial gastrocnemius fibers in the 7-day-old rabbits was also markedly reduced following 1 week of inactivation (Fig. 10). In the adult rabbit and hamster diaphragms, we also found that inactivation (TTX block of nerve activity) decreased fiber SDH activities and the range of SDH activities among fibers (Sieck et al., 1989c; Zhan and Sieck, 1988; Figs. 11 and 12). It is possible that enzymatic heterogeneity among diaphragm muscle fibers is due to the combined influence of the elimination of multiple synapses, i.e., disappearance of polyneuronal innervation, and the emergence of differences in the activation history of diaphragm motor units.

One additional factor that may contribute to the homogeneity of metabolic enzyme activities of neonatal muscle fibers is the presence of electrical coupling between muscle fibers. In the rat soleus, lumbrical, and intercostal muscles, electrophysiological measurements have shown the presence of coupling between muscle fibers during the first postnatal week (Dennis, 1981; Dennis et al., 1981; Schmalbruch, 1982; Jones and Ridge, 1987; Ross et al., 1987). Morphological evidence for the presence of gap junctions between muscle fibers has also been shown (Kelly and Zacks, 1969, Ross, Duxson, and Harris, 1987, Schmalbruch, 1982). These gap junctions occur predominantly between primary and secondary myotubes such that the secondary myotubes are all joined via gap junctions to their host primary myotube (see Section V. C for brief discussion of myogenesis). The disappearance of these gap junctions parallels the formation of a basal lamina in secondary myotubes that is apposed to the host primary myofiber. This also coincides with the establishment of single innervation of these newly formed muscle fibers, which involves the transfer of nerve terminals from the primary myotube (Duxson et al., 1986). An important physiological consequence of the electrical coupling between muscle fibers in the neonate is that these fibers would be activated in synchrony with the discharge of the primary myofiber. The resulting common activation history may account for the homogeneity of enzymatic properties among neonatal muscle fibers.

Support for the homogenizing influence of polyneuronal innervation on muscle fiber metabolic properties comes from studies on the levator ani muscle of the rat. In this muscle, polyneuronal innervation of muscle fibers persists in the adult (Jordan et al., 1988). We have found that the variability of SDH activities among fibers of the levator ani muscle is very low, being comparable to that observed in neonatal muscle (Sieck et al., unpublished observations).

E. Fatigue Resistance of the Neonatal Diaphragm and Muscle Fiber Oxidative Capacity

Although several studies have clearly demonstrated that the adult diaphragm is susceptible to fatigue (Belman and Sieck, 1982; Roussos and Macklem,

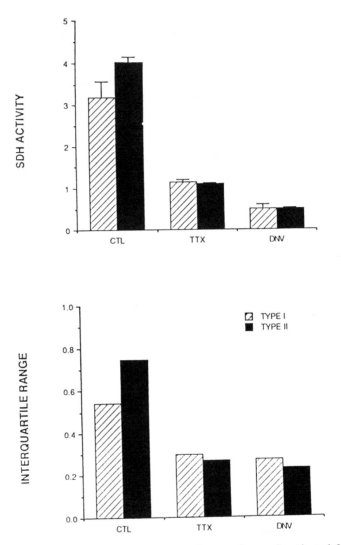

Figure 10   The neonatal medial gastrocnemius muscle was inactivated for 7 days either by superfusing TTX onto the sciatic nerve (via a chronic nerve cuff) or by denervation (DNV). Compared to controls (CTL), inactivity decreased the SDH activity (in mmol fumarate/liter tissue/min) of both type I and II fibers in the medial gastrocnemius muscle. Inactivity also decreased the variability (interquartile range; see legend for Fig. 9) of SDH activities among type I and II muscle fibers.

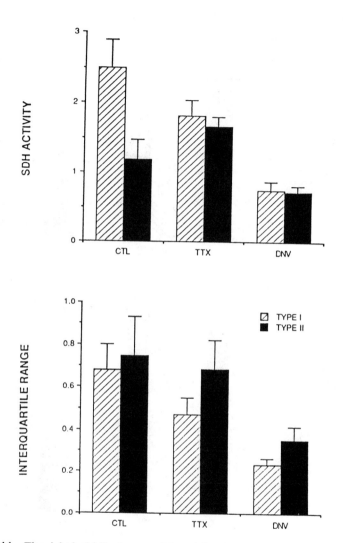

Figure 11 The right hemidiaphragm of the adult rabbit was inactivated for 30 days either by superfusing TTX onto the phrenic nerve (via a chronic nerve cuff) or by denervation (DNV). Inactivity decreased the SDH activity (in mmol fumarate/liter tissue/min) of type I fibers. The effects of TTX and DNV on the SDH activity of type II fibers differed, the SDH activity increasing after TTX-induced inactivity and decreasing after DNV. Inactivity reduced the variability (interquartile range; see legend for Fig. 9) of SDH activity among diaphragm fibers.

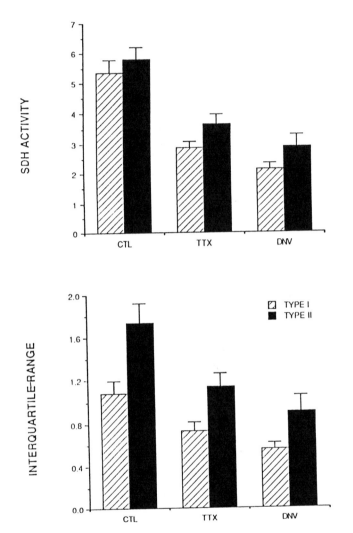

Figure 12 The right hemidiaphragm of the adult hamster was inactivated for 14 days either by superfusing TTX onto the phrenic nerve (via a chronic nerve cuff) or by denervation (DNV). Inactivity decreased the SDH activity (in mmol fumarate/ liter tissue/min) of both type I and II fibers and reduced the variability (interquartile range; see legend for Fig. 9) of SDH activity among diaphragm fibers.

1977), the susceptibility of the neonatal diaphragm remains controversial. Based on electromyographic evidence (i.e., a decline in the high frequency/ low frequency ratio), Muller et al. (1979) suggested that fatigue of the neonatal human diaphragm occurs in response to increased mechanical loads. Keens and colleagues (Keens et al., 1978; Keens and Ianuzzo, 1979) suggested that the neonatal human diaphragm might be more susceptible to fatigue than the adult muscle based on their observation of a relatively low proportion of type I fibers in the neonatal muscle. More recently, Le Souef et al. (1988) reported that the neonatal rabbit diaphragm was more susceptible to fatigue than the adult muscle. They also attributed the greater fatigability of the neonatal rabbit diaphragm to a lower proportion of high oxidative type I fibers. In contrast, Maxwell et al. (1983) found that the neonatal baboon diaphragm was much more fatigue resistant than the adult muscle. These authors attributed the greater fatigue resistance of the neonatal baboon diaphragm to a greater proportion of high oxidative type I and II fibers. Thus, in different ways, each of these previous studies attributed the fatigue properties of the neonatal diaphragm to its fiber type composition and oxidative capacity. However, muscle fiber oxidative capacity in the neonatal diaphragm was either inferred by the lower proportion of type I fibers (Keens et al., 1978; Keens and Ianuzzo, 1979; Le Souef et al., 1988) or assessed subjectively from the intensity of histochemical staining for oxidative enzymes (Maxwell et al., 1983).

In recent studies, we observed that the neonatal rat and cat diaphragms are more resistant to fatigue than the adult muscle (Fig. 13). During postnatal development, the diaphragm becomes progressively more susceptible to fatigue (Sieck and Blanco, 1987b). To determine whether these changes in diaphragm fatigue resistance are correlated with changes in muscle fiber oxidative capacity, we quantified the SDH activities of muscle fibers in the neonatal cat and rat diaphragms. We found that SDH activities were uniformly low at birth and then increased during postnatal development, reaching their highest levels by the third postnatal week (Figs. 6, 7, 14, and 15). Thereafter, fiber SDH activities became more variable, with the mean SDH activity of all fibers in the adult being approximately the same as that in the neonate. These developmental changes in fiber SDH activities did not correlate with the progressive decrease in muscle fatigue resistance that occurred during the same period (Fig. 13). We have also observed similar developmental changes in the SDH activities of fibers in the medial gastrocnemius muscle of the rabbit (Zhan and Sieck, unpublished observations). These results indicate that postnatal changes in diaphragm fatigue resistance do not correlate with fiber oxidative capacity.

Using our quantitative histochemical procedure, we have demonstrated that the SDH activities of type II muscle fibers in the adult cat and rat dia-

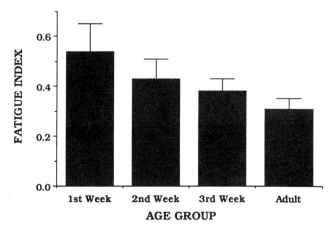

Figure 13 Fatigue resistance of the rat diaphragm was assessed during repetitive direct muscle stimulation at 40 pps delivered in 330-msec-duration trains repeated each second for a 2-min period. A fatigue index was calculated as the ratio of force generated after 2 min to the initial force. Note that fatigue resistance decreased progressively with age.

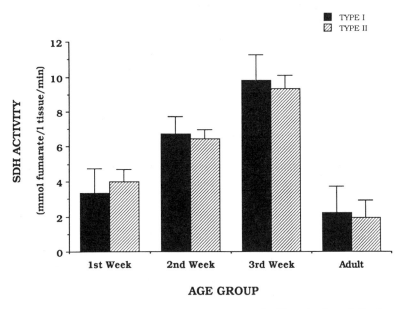

Figure 14 The SDH activities of type I and II muscle fibers in the rat diaphragm at various ages were quantified photometrically. During early postnatal development, SDH activities of both type I and II fibers increased before decreasing to adult values.

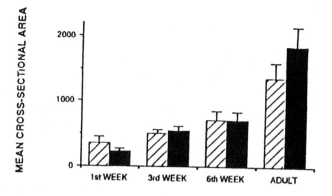

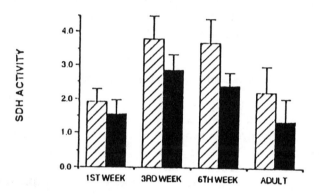

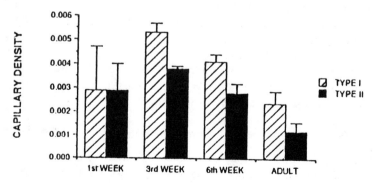

Figure 15 Postnatal changes in mean cross-sectional area (μm²), SDH activity (mmol fumarate/liter tissue/min), and capillary density (no. capillaries/fiber area) of type I and II fibers during postnatal development of the cat diaphragm.

phragms are unimodally distributed (Fig. 16; Blanco and Sieck, 1987; Sieck et al., 1986a, 1987, 1989b). Recently, we have also observed a similar distribution of SDH activities for type II muscle fibers in the cat and rat diaphragm during postnatal development (Fig. 16). Thus, at no age is there an objective basis for subclassifying type II fibers using differences in oxidative enzyme staining, e.g., as done by Maxwell et al. (1983).

In recent studies, we also found that the difference in mean SDH activity of type I and II fibers in the adult cat and rat diaphragm is not apparent during the early postnatal period (Figs. 6, 7, 14, and 15; Sieck and Fournier, 1988). Furthermore, the SDH activities of diaphragm fibers change considerably during postnatal development (Figs. 6, 7, 14, and 15; cf. Smith et al., 1988). Thus, it is inappropriate and misleading to infer information about the overall oxidative capacity of the neonatal diaphragm from the relative proportions of type I and II fibers (e.g., Keens et al., 1978; Keens and Ianuzzo, 1979; Le Souef et al., 1988).

F. Capillary Density and Diaphragm Fatigue Resistance

In a recent study (Enad et al., 1989), we examined the distribution of capillaries surrounding muscle unit fibers in the cat diaphragm. Muscle unit fibers were identified by glycogen depletion, and capillaries were identified in sections stained for ATPase after preincubation at pH 4.2. We found correlations between the fatigue resistance of diaphragm motor units and both the mean number of capillaries and the mean capillary density, i.e., number of capillaries per fiber cross-sectional area, of muscle unit fibers. In addition, we noted a correlation between capillary density and fiber SDH activity. These results suggested that in the adult cat diaphragm, motor unit fatigue resistance is related, at least in part, to the delivery and utilization of oxygen for energy production.

We have also examined the changes in capillary density of diaphragm fibers during postnatal development in the kitten. We found that changes in fiber capillary density paralleled the changes in fiber SDH activity (Fig. 15). However, during the postnatal development of the cat diaphragm, no correlation was observed between muscle fatigue resistance and either the capillary density or the SDH activity of diaphragm fibers (Figs. 13 and 15).

G. Fiber Cross-Sectional Area and Diaphragm Fatigue Resistance

In studies of the adult cat diaphragm, we measured the cross-sectional area of identified muscle unit fibers using an image-processing system calibrated for morphometry (Enad et al., 1989). We found that the more fatigable diaphragm motor units, i.e., FF and FInt, were composed of fibers with larger cross-sectional areas and lower SDH activities. In contrast, fatigue resis-

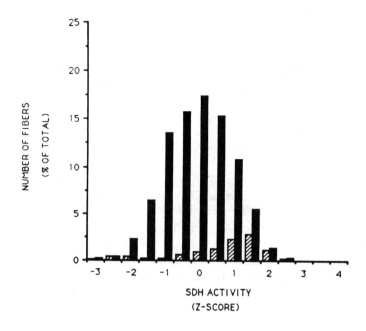

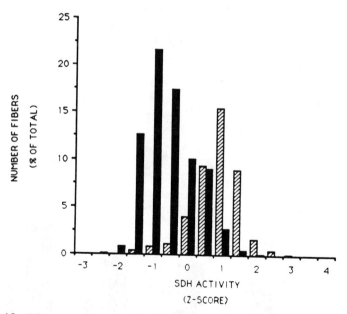

Figure 16 Histograms showing the distribution of SDH activities (normalized to Z-scores) for type I (cross-hatched bars) and II fibers (filled bars) in the neonatal (top histogram) and adult (bottom histogram) cat diaphragm. Note that at both ages, the distributions of SDH activities for type II fibers were unimodal.

tant units, i.e., S and FR, were composed of smaller fibers with higher SDH activities (Tables 1 and 2). The average fiber size of muscle units is correlated with both the maximum tetanic tensions produced by these units and their fatigability (Tables 1 and 2). These results suggest that motor unit fatigue resistance may be correlated to the distance for diffusion of oxygen and energy substrates into muscle fibers.

We have also observed that fiber cross-sectional areas in the cat and rat diaphragm increase progressively during postnatal development (Figs. 6, 7, 15, and 17). During early postnatal development, fiber sizes were relatively uniform, as were fiber SDH activities. Not until after the third postnatal week did the considerable range in fiber size, characteristic of the adult muscle, begin to emerge (Figs. 6 and 7). These data suggest that more fatigable fast-twitch units, i.e., FF and FInt, with larger-fiber cross-sectional areas, are the last to develop in the diaphragm.

The results on the postnatal changes in fiber SDH activity and cross-sectional area suggest that during the first 3 postnatal weeks, there is an increase in the number and/or size of mitochondria within diaphragm muscle fibers. After the third postnatal week, the size of some fibers continues to increase without a concomitant increase in mitochondrial volume. Thus, the oxidative capacity of these fibers does not match the increase in structural and contractile proteins. This process may be related to a differential activation history of muscle unit fibers. Muscle fibers belonging to units that are

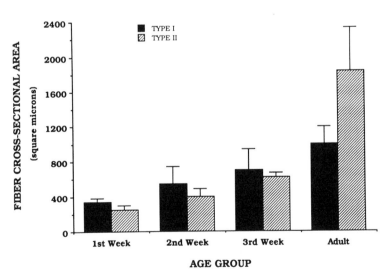

Figure 17   The mean cross-sectional area of type I and II fibers increased during postnatal development of the rat diaphragm.

infrequently activated, e.g., FF and FInt units, would grow at an accelerated rate with little change in mitochondrial volume, and their oxidative capacity would decrease as a function of size. Conversely, those muscle unit fibers more frequently activated would show slower growth, and their oxidative capacity (as a function of size) would decrease to a lesser extent. This speculation implies a correlation between activation history and the size and oxidative capacity of muscle fibers that is maintained throughout postnatal development. As mentioned above, when activity of the medial gastrocnemius muscle of the neonatal rabbit was blocked using TTX superfusion of the sciatic nerve, the postnatal changes in fiber size and SDH activity were prevented.

V. Diaphragm Fiber Type Proportions During Postnatal Development

A. Histochemical Classification of Muscle Fiber Types

In adult muscle, fibers can be categorized into two types based on differences in histochemical staining for myofibrillar ATPase after alkaline preincubation (Brooke and Kaiser, 1970; Brooke et al., 1971). Type I fibers stain lightly for ATPase whereas type II fibers stain darkly. In hindlimb muscles, it has been shown that type I fibers comprise slow-twitch motor units and type II fibers comprise fast-twitch motor units (Burke, 1981; Burke et al., 1973; Edström and Kugelberg, 1968). Recently, we demonstrated a similar fiber type composition of slow- and fast-twitch motor units in the cat diaphragm (Sieck et al., 1989a).

In the adult muscle, type II muscle fibers have been further classified as IIA nd IIB based on differences in ATPase staining after acid preincubation (Brooke and Kaiser, 1970). This classification scheme has also been shown to correspond to the classification of different fast-twitch motor unit types (Burke, 1981). Motor units classified as FR are comprised of IIA fibers, and FF units are comprised of IIB fibers. In a recent study (Sieck et al., 1989a), we found that FR and FF units in the adult cat diaphragm were also comprised of IIA and IIB fibers, respectively. However, the FInt units in the diaphragm had a mixed fiber type composition, containing both type IIA and IIB fibers. This suggests that motor units are not always homogeneous in their fiber type composition (at least as classified histochemically).

B. Histochemical Classification of Fiber Types in the Neonate

Based on the ATPase histochemical reaction, most neonatal muscle fibers would be classified as type II, e.g., 90% of all fibers in the neonatal cat and rat diaphragms (Sieck and Fournier, 1988; Fig. 18). During subsequent post-

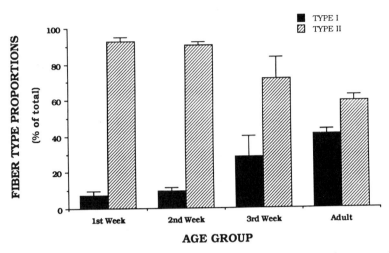

Figure 18 During postnatal development of the rat diaphragm, the proportion of type I fibers increased while the number of type II fibers decreased.

natal development, the proportion of type I fibers progressively increases, until the adult pattern is reached.

From the relatively high proportion of type II fibers, it might be expected that the neonatal diaphragm would have shorter CT and 1/2 RT and be more susceptible to fatigue than the adult muscle (e.g., Keens et al., 1978; Keens and Ianuzzo, 1979; Le Souef et al., 1988). Yet, several studies (Buller et al., 1960; Close, 1972; Glebovskii, 1961b; Gutmann et al., 1974; Jolesz and Sreter, 1981), including our own (Sieck and Blanco, 1987b), have shown that neonatal skeletal muscles have much longer CT and 1/2 RT than do adult muscles (Fig. 19). Similarly, just as the slower CT and 1/2 RT of the neonatal muscle do not match the predominance of type II fibers, the fatique resistance of the neonatal diaphragm (Maxwell et al., 1983; Sieck and Blanco, 1987b; Fig. 13) does not correlate with the high proportion of type II fibers or the lower SDH activity of these fibers. Paradoxical observations like these led Guth and Samaha (1972) to warn of possible erroneous interpretations concerning physiological properties of neonatal muscle based on histochemically determined fiber type proportions. Resolution of these paradoxical findings may relate to the type of contractile proteins found within developing muscle fibers. For example, Reiser et al. (1985, 1988) showed a correlation between the shortening velocity of single neonatal muscle fibers and the type of myosin heavy chain (MHC) found in these fibers. As the adult MHCs were expressed during postnatal development, the corresponding changes in muscle fiber physiological properties were observed. These

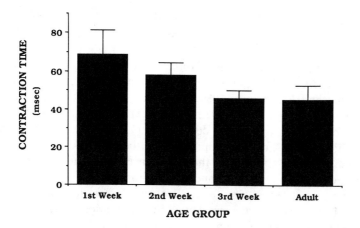

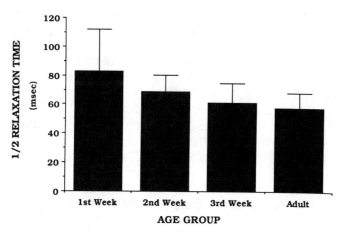

Figure 19 During postnatal development of the rat diaphragm, CT and 1/2 RT progressively decreased.

results suggest a dissociation between standard histochemical identification of fiber types in the neonate and the type of contractile proteins found within developing muscle fibers.

C. Immunohistochemical Classification of Muscle Fiber Types

Immunohistochemical procedures have been used to identify different forms of MHC within muscle fibers. In the adult muscle, it has been reported that three types of MHCs are expressed: "slow" type I, "fast" type IIA, and "fast" type IIB (Kelly, 1983; Mahdavi et al., 1986; Pette and Vrbova, 1985; Staron and Pette, 1987; Stockdale and Miller, 1987). Recently, an additional major isoform of adult "fast" MHC designated IIX has been described (Schiaffino et al., 1988b). Several studies have examined the postnatal differentiation of these MHCs (Dhoot, 1985; Gauthier et al., 1978; Hoh et al., 1988; Kelly and Rubinstein, 1980; Narusawa et al., 1987; Stockdale and Miller, 1987; Swynghedauw, 1986; Whalen et al., 1981). The etiology of fiber differentiation appears to depend on: (1) the genetic "program" of the specific muscle, i.e., the proportion of fiber types found in the adult; (2) the innervation pattern, e.g., the level of activation and neurotrophic influences, during development; and (3) hormonal influences, e.g., insulin, thyroid hormone, growth factors, corticosteroids. In adult muscles that are of "mixed" fiber type, it appears that all first-generation fibers, i.e., primary myotubes (multinucleated primary myotubes formed by the fusion of myoblasts), express the adult "slow" MHC. Some of these fibers also express an embryonic "fast" MHC. Secondary myotubes are formed by the fusion of myoblasts emanating from the basement membrane of the primary myotube. These secondary myotubes appear to express only the embryonic "fast" MHC isoform. Subsequently, in these secondary fibers, there is a sequential differentiation of a neonatal isoform and then an adult isoform of "fast" MHC, either IIA, IIB, or IIX. In those developing fibers that coexpress the embryonic "fast" and adult "slow" MHCs, the final expression of the adult "fast" MHC is dependent on the suppression of the "slow" MHC gene. If suppression of the "slow" MHC gene occurs, the fibers eventually express one form of the adult "fast" MHC, i.e., type IIA, IIB, or IIX. If suppression does not occur, the adult "slow" MHC is expressed (Kelly and Rubenstein, 1980).

D. Influence of Activity on Fiber Type Differentiation

As mentioned before, during postnatal development, some fibers express both the adult "slow" MHC and an embryonic or neonatal isoform of the "fast" MHC (Dhoot, 1985; Gauthier et al., 1978; Hoh et al., 1988; Kelly and Rubinstein, 1980; Kelly and Zacks, 1969; Narusawa et al., 1987; Stockdale and Miller, 1987; Swynghedauw, 1986; Whalen et al., 1981). The final expression or suppression of the adult "slow" or "fast" MHC in these fibers is affected by the level of neuromuscular activity during postnatal development (Gauthier et al., 1978; Kelly, 1983; Pette and Vrbova, 1985; Stockdale and Miller, 1987). If neural activation is abolished, e.g., using TTX or denervation, the proportion of fibers expressing "slow" MHC is decreased (Butler-Browne et al., 1982; Dhoot and Perry, 1983; Gauthier et al., 1978), and there is a delay in fiber differentiation, with continued coexpression of "slow" and neonatal MHCs in fibers (Butler-Browne et al., 1982; Dhoot and Perry, 1983). Recent studies have reported that after chronic inactivation (TTX- or denervation-induced) of the adult rat diaphragm (Carraro et al., 1985) or tibialis anterior (Schiaffino et al., 1988a) muscles, fibers begin to reexpress neonatal MHCs, which are coexpressed with "slow" MHC. In the adult rabbit and hamster diaphragm, as well as in the neonatal rabbit medial gastrocnemius muscle, we observed that chronic inactivation (TTX- or denervation-induced) of muscle fibers increased CT and 1/2 RT (Fig. 20). Whether these changes in diaphragm contractile properties with inactivity correlated to an altered expression of MHCs was not determined. Recently, Rosser et al. (1988) suggested that the earlier expression of adult "fast" MHC in the neonatal rat diaphragm, e.g., at 5 days of age versus 15 days of age in the extensor digitorum longus muscle, was due to the relatively greater activity of the diaphragm during early postnatal development.

E. Fiber Type Composition of Muscle Units During Postnatal Development

Whether the fibers comprising a motor unit in a developing muscle are of a uniform type remains controversial (Fladby and Jansen, 1988; Jones et al., 1987; Thompson et al., 1984). Thompson et al. (1984) suggested that motor units in the neonatal rat soleus muscle have a uniform fiber type composition even during the period of polyneuronal innervation. Their results, which were based on histochemical fiber typing, indicated that most, but not all, fibers, i.e., approximately 80%, comprising motor units of the soleus were either type I or II. More recently, Fladby and Jansen (1988) used immunohistochemical techniques to classify muscle unit fibers in the neonatal mouse soleus muscle and also concluded that motor units were comprised of a uniform fiber type, i.e., reacting for either the adult "slow" or "fast" MHC. Thus, these authors argued that motor unit organization occurs through a process of selective innervation of muscle fiber types, not through the process of selective synapse elimination. Jones et al. (1987) also used immunohistochemical procedures to type the fibers comprising units of the neonatal rat lumbrical muscle. These authors found that muscle units were composed of a mixture of "slow" and "fast", i.e., not reacting for "slow" MHC, fibers in proportions similar to that found in the muscle as a whole. Based on these results, they concluded that during polyneuronal innervation, there is no selective innervation of "fast" and "slow" fibers, and that the organization of adult motor units comes about through a process of selective synapse elimination. Note, however, that the results of these studies were actually very similar; i.e., motor units were found to be comprised of both "slow" and "fast" fibers, with one fiber type predominating. Their con-

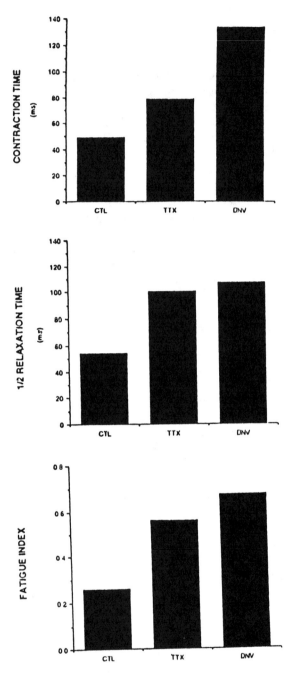

Figure 20 Inactivation of the right hemidiaphragm of the adult hamster for 14 days (by either TTX or denervation) increased twitch contraction time and half relaxation time.

flicting conclusions appear to be based on probabilistic interpretations of their data. For example, in the study of Fladby and Jansen (1988) they observed that motor units contained predominantly but not exclusively "slow" or "fast" fibers. They observed that the distribution of muscle unit fibers containing one MHC or the other differed from random, and thus, they concluded that "slow" and "fast" fibers were selectively innervated. Jones et al. (1987) also found that motor units were predominantly comprised of fibers containing one type of MHC. Again, based on a distributional analysis, these authors reached conclusions that were just opposite.

It should be emphasized that these studies focused only on how fast- and slow-twitch motor units in the adult become selectively organized during development. They did not address how different types of fast-twitch motor units, e.g., FR, FInt, and FF, become organized. Based on standard histochemistry, FR and FF units contain a specific fiber type, i.e., IIA and IIB, respectively (Burke, 1981; Burke et al., 1973; Sieck et al., 1989a). This suggests that these units are selectively organized, either by the initial pattern of innervation, e.g., by cell recognition through chemoaffinity or other cues, or by the pattern of synapse elimination, e.g., activity- or trophic-related. However, the mixed fiber type composition of FInt units (Sieck et al., 1989a) suggests that during development, innervation and synapse elimination may not be selective for fiber type. One important caveat, however, is that to date, no study has determined the MHC composition of different types of fast-twitch motor units using immunohistochemical procedures, i.e., antibodies specific for IIA, IIB, or IIX MHC. Therefore, it is possible that despite their histochemical staining pattern, the FInt units in the diaphragm were comprised of fibers containing only a single MHC type.

F. Correlation Between Diaphragm Contractile Properties and Fiber Type Proportions

We examined the isometric contractile and fatigue properties of the cat and rat diaphragms during postnatal development using an in vitro nerve-muscle strip preparation (Sieck and Blanco, 1987b). In some cases, we also characterized contractile properties in the cat diaphragm using an in situ preparation. As mentioned above, we found a progressive decrease in CT and 1/2 RT (Fig. 19). As CT and 1/2 RT of the diaphragm decreased during development, we found a progressive decrease in the proportion of type II fibers (putative "fast-twitch") and an increase in the number of type I fibers (putative "slow-twitch") (Fig. 18). These paradoxical findings illustrate that standard histochemical techniques for fiber typing do not always correlate with muscle physiology during early development (cf., Guth and Samaha, 1972). We found that almost all of the type II fibers in the newborn cat and rat diaphragms were classifiable as type IIC based on standard histochemistry

(Brooke et al., 1971). Note, however, that these type IIC fibers can later differentiate into either adult type I or type II fibers. We found that the differentiation of type IIC fibers into adult type II fibers did not correspond to the decrease in diaphragm CT during postnatal development. It is more likely that the changes in diaphragm contractile properties correspond with the development of other features of muscle fibers. For example, the decrease in CT and 1/2 RT most likely corresponds to the development of the sarcoplasmic reticulum and transverse tubular system (Luff and Atwood, 1971; Schiaffino and Margreth, 1969). In contrast, postnatal changes in the velocity of shortening of muscle fibers have been shown to correlate with the expression of neonatal and adult MHCs (Eddinger and Moss, 1987; Reiser et al., 1985, 1988; Schiaffino et al., 1988b).

G. Rate of Myofibrillar ATPase Activity in Diaphragm Muscle Fibers

It has been shown that the energy cost for contraction of type I fibers is approximately half that of type II fibers (Crow and Kushmerick, 1982). This difference in energy utilization between fiber types in adult muscle probably results from differences in cross-bridge turnover (velocity of shortening), and thus, the rate of myofibrillar ATPase activity (Barany, 1967). Biochemical studies have shown that the rate of myofibrillar ATPase activity in developing muscle is lower than in adult muscle (Baldwin et al., 1978; Belcastro, 1987). Therefore, it is entirely reasonable that neonatal fibers might vary from adult fibers in the energy costs for contraction. With lower energy costs, it might also be expected that the neonatal muscle would be more fatigue resistant. With the differentiation of adult fibers (especially type II), energy costs for contraction would increase and fatigue resistance would decrease. It might also be expected that as the energy costs for contraction increase, the metabolic support for energy production, e.g., oxidative capacity, within muscle fibers would also increase. The rapid growth and increased protein synthesis (structural and contractile proteins) within developing muscle fibers would also require additional metabolic support. Thus, it is not surprising that the SDH activities of diaphragm fibers increase during early stages of postnatal development (Figs. 6, 7, 14, and 15; cf. Smith et al., 1988).

We recently developed a photometric procedure to quantify the rate of myofibrillar ATPase activity in individual muscle fibers at physiological pH (Fig. 21). We found that not only is ATPase activity lower in type I than in type II fibers, but that the rate of ATPase activity in type IIB fibers was higher than that in IIA fibers. Preliminary studies have shown that the rate of ATPase activity in type IIC fibers of the neonatal cat diaphragm is lower than that of type IIA or IIB fibers of the adult muscle. These quantitative histochemical results are similar to results obtained from biochemical analy-

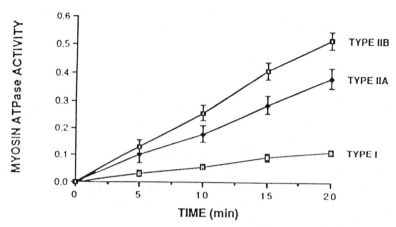

Figure 21 Myosin ATPase activity (expressed as optical density units) was quantified microphotometrically in different types of fibers in the adult cat diaphragm.

sis of ATPase activity of muscle homogenates (Baldwin et al., 1978; Belcastro, 1987). Therefore, it is possible that the fatigue resistance of the neonatal diaphragm is related to the low energy demands (reflected by ATPase activity) of neonatal contractile proteins. We speculate that as adult fast MHCs are expressed, the energy costs of contraction increase and SDH activity increases to match these higher energy costs. However, as fewer fibers are required to produce ventilatory forces, some fast-twitch (type II) fibers are no longer recruited during ventilation, and as a result, they decrease their oxidative capacities and become more susceptible to fatigue. We believe these fibers eventually comprise FF and FInt units in the adult diaphragm. Such a developmental sequence would account for the increasing heterogeneity of muscle fiber metabolic properties.

VI. Summary and Conclusions

It should be pointed out that in this chapter, we have focused only on some of the morphological, physiological, and metabolic changes that take place in the diaphragm during postnatal development. We have emphasized the development of the functional organization of diaphragm control at the level of motor units. By comparing the neonatal diaphragm to the fully differentiated adult muscle, we hoped to demonstrate the plasticity of diaphragm motor control that occurs during the postnatal period. Several major questions remain unresolved and were not addressed here. For example, what are the mechanisms responsible for determining the precise pattern of connections between phrenic nerve axon terminals and muscle fibers during both the initial period of innervation and the subsequent period of synapse elimination? Are there specialized molecules that "recognize" each other and, thus, allow the formation of synaptic connections? Is there competition among nerve terminals for the uptake of a trophic substance produced by muscle fibers, and does this regulate synapse elimination? Can activity- and trophic-related influences on synapse elimination be separated?

With regard to the differentiation of myoblasts into muscle fibers, other questions remain. What are the mechanisms that govern the transitions from embryonic to adult protein isoforms, e.g., the developmental sequence of embryonic, neonatal, and adult MHCs? How are the genes responsible for the expression of various protein isoforms influenced by functional demands, e.g., activity, and other epigenetic factors, e.g., hormones or trophic substances. Is there a preset developmental program that coordinates phenotypic expression? Future resolution of these and other questions will facilitate not only our understanding of what controls the continuous remodeling of diaphragm motor control during development, but also the factors that guide the dynamic changes observed in the adult muscle.

Acknowledgments

The authors acknowledge the important contributions of Drs. Wen-Zhi Zhan, Michael Lewis, and Joseph Kuei and the technical assistance of Cesar E. Blanco, Tracy Cheung, Jerome Enad, and Weizheng Wang in these studies.

References

Aldrich, T. K., Shander, A., Chaudhry, I., and Nagashima, H. (1986). Fatique of isolated rat diaphragm: Role of impaired neuromuscular transmission. *J. Appl. Physiol.* **61**:1077-1083.

Bagust, J., Lewis, D. M., and Westerman, R. A. (1974). The properties of motor units in a fast and slow twitch muscle during post-natal development in the kitten. *J. Physiol. (London)* **237**:75-90.

Baldwin, K. M., Hooker, A. M., Campbell, P. J., and Lewis, R. E. (1978). Enzyme changes in neonatal skeletal muscle: Effect of thyroid deficiency. *Am. J. Physiol.* **50**:197-218.

Balice-Gordon, R. J., and Thompson, W. J. (1988). Synaptic rearrangements and alterations in motor unit properties in neonatal rat extensor digitorum longus muscle. *J. Physiol. (London)* **398**:191-210.

Barany, M. (1967). ATPase activity of myosin correlated with speed of muscle shortening. *J. Gen. Physiol.* **235**:C97-C102.

Belcastro, A. N. (1987). Myofibril and sarcoplasmic reticulum changes during muscle development: Activity vs inactivity. *Int. J. Biochem.* **19**: 945-948.

Belman, M. J., and Sieck, G. C. (1982). The ventilatory muscles. Fatigue, endurance and training. *Chest* **82**:761-766.

Bennett, M., Ho, S., and Lavidis, N. (1986). Competition between segmental nerves at end-plates in rat gastrocnemius muscle during loss of polyneuronal innervation. *J. Physiol. (London)* **381**:351-376.

Bennett, M. R., and Pettigrew, A. G. (1974). The formation of synapses in striated muscle during development. *J. Physiol. (London)* **241**:515-545.

Berger, A. J. (1979). Phrenic motoneurons in the cat: Subpopulations and nature of respiratory drive potentials. *J. Neurophysiol.* **42**:76-90.

Betz, W. J., Caldwell, J. H., and Ribchester, R. R. (1979). The size of motor units during post-natal development of rat lumbrical muscle. *J. Physiol. (London)* **297**:463-478.

Bigland-Ritchie, B. (1981). EMG and fatigue of human voluntary and stimulated contractions. In *Human Muscle Fatigue: Physiological Mechanisms.* Ciba Foundation Symposium 82. Edited by R. Porter and J. Whelan. London, Pitman Medical, pp. 130-148.

Bixby, J. L., and Van Essen, D. C. (1979). Regional differences in the timing of synapse elimination in skeletal muscles of the neonatal rabbit. *Brain Res.* **169**:275-286.

Blanco, C. E., and Sieck, G. C. (1987) Comparison of succinic dehydrogenase activity between the diaphragm and medial gastrocnemius muscles of the rat. In *Respiratory Muscles and Their Neuromotor Control.* Edited by G. C. Sieck, S. C. Gandevia, and W. E. Cameron. New York, Alan R. Liss, pp. 281-289.

Blanco, C. E., Sieck, G. C., and Edgerton, V. R. (1988). Quantitative histochemical determination of succinic dehydrogenase activity in skeletal muscle fibres. *Histochem. J.* **20**:230-243.

Bodine, S. C., Roy, R. R., Eldred, E., and Edgerton, V. R. (1987). Maximal force as a function of anatomical features of motor units in the cat tibialis anterior. *J. Neurophysiol.* **57**:1730-1745.

Brooke, M. H., and Kaiser, K. K. (1970). Muscle fiber types: How many and what kind? *Arch. Neurol.* **23**:369-379.

Brooke, M. H., Williamson, E., and Kaiser, K. K. (1971). The behavior of four fiber types in developing and reinnervated muscle. *Arch. Neurol.* **25**:360-366.

Brown, M. C., and Booth, C. M. (1983). Postnatal development of the adult pattern of motor axon distribution in rat muscle. *Nature* **304**:741-742.

Brown, M. C., Hopkins, W. G., and Keynes, R. J. (1982). Short- and long-term effects of paralysis of the innervation of two different neonatal mouse muscles. *J. Physiol. (London)* **329**:439-450.

Brown, M. C., Jansen, J. K. S., and Van Essen, D. (1976). Polyneuronal innervation of skeletal muscle in newborn rats and its elimination during maturation. *J. Physiol. (London)* **261**:387-422.

Buller, A., Eccles, J., and Eccles, R. (1960). Differentiation of fast and slow muscles in the cat hind limb. *J. Physiol. (London)* **150**:399-416.

Burke, R. E. (1981). Motor units: Anatomy, physiology and functional organization. In *Handbook of Physiology*, Sec. 1, Vol. III, Part 1, *The Nervous System*. Edited by J. M. Brookhart and V. B. Mountcastle. Bethesda, MD, American Physiological Society, pp. 345-422.

Burke, R. E., Levine, D. N., Tsairis, P., and Zajac, F. E. (1973). Physiological types and histochemical profiles of motor units of cat gastrocnemius. *J. Physiol. (London)* **234**:723-748.

Butler-Browne, G. S., Bugaisky, L. B., Cuenoud, S., Schwartz, K., and Whalen, R. G. (1982). Denervation of newborn rat muscles does not block the appearance of adult fast myosin heavy chain. *Nature* **299**: 830-833.

Caldwell, J. H., and Ridge, R. M. A. (1983). The effects of deafferentation and spinal cord transection on synapse elimination in developing rat muscles. *J. Physiol. (London)* **339**:145-159.

Cameron, W. E., He, F., Brozanski, B. S., and Guthrie, R. D. (1989). The postnatal growth of motoneurons at three levels of the cat neuraxis. *Neurosci. Lett.* **104**:274-280.

Carraro, U., Morale, D., Mussini, I., Lucke, S., Cantini, M., Betto, R., Catani, C., Dalla Libera, L., Danieli Betto, D., and Noventa, D. (1985). Chronic denervation of rat hemidiaphragm: Maintenance of fiber heterogeneity with associated increasing uniformity of myosin isoforms. *J. Cell Biol.* **100**:161-174.

Chamberlain, S., and Lewis, D. M. (1989). Contractile characteristics and innervation ratio of rat soleus motor units. *J. Physiol. (London)* **412**: 1-21.

Cheung, T. S., Blanco, C. E., and Sieck, G. C. (1988). Changes in diaphragm muscle fiber capillarity and oxidative capacity during postnatal development. *FASEB J.* **2**:1278.

Clamann, H. P., and Robinson, A. J. (1985). A comparison of electromyographic and mechanical fatigue properties in motor units of the cat hindlimb. *Brain Res.* **327**:203-219.

Close, R. I. (1972). Dynamic properties of mammalian skeletal muscles. *Physiol. Rev.* **52**:129-197.

Conradi, S., and Ronnevi, L. O. (1977). Ultrastructural and synaptology of the initial axon segment of cat spinal motoneurons during early post-natal development. *J. Neurocytol.* **6**:195-210.

Conradi, S., and Skoglund, S. (1969). Observations on the ultrastructure of the initial motor axon segment and dorsal root boutons on the moto-neurons in the lumbosacral spinal cord of the cat during postnatal de-velopment. *Acta Physiol. Scand.* **333**:53-76.

Cooper, C. C., Cassens, R. G., Kastenschmidt, L. L., and Briskey, J. E. (1970). Histochemical characterization of muscle differentiation. *Dev. Biol.* **23**:169-184.

Crow, M. T., and Kushmerick, M. J. (1982). Chemical energies of slow- and fast-twitch muscles of the mouse. *J. Gen. Physiol.* **79**:147-162.

Cullheim, S., and Ulfhake, B. (1979). Relation between cell body size, axon diameter and axon conduction velocity of triceps surae alpha moto-neurons during postnatal development in the cat. *J. Comp. Neurol.* **188**:679-686.

Dennis, M. J. (1981). Development of the neuromuscular junction: Induc-tive interactions between cells. *Annu. Rev. Neurosci.* **4**:43-68.

Dennis, M. J., Ziskind-Conhaim, L., and Harris, A. J. (1981). Development ment of neuromuscular junctions in rat embryos. *Dev. Biol.* **82**:266-279.

Dhoot, G. K. (1985). Initiation of differentiation into skeletal muscle fiber types. *Muscle Nerve* **8**:307-316.

Dhoot, G. K., and Perry, S. V. (1983). Effect of denervation at birth on the development of skeletal muscle cell types in the rat. *Exp. Neurol.* **82**:131-142.

Diamond, J., and Miledi, R. (1962). A study of foetal and new-born rat muscle fibres. *J. Physiol. (London)* **162**:393-408.

Dick, T. E., Kong, F. J., and Berger, A. J. (1987a). Correlation of recruit-ment order with axonal conduction velocity for supraspinally driven motor units. *J. Neurophysiol.* **57**:245-259.

Dick, T. E., Kong, F. J., and Berger, A. J. (1987b). Recruitment order of diaphragmatic motor units obeys Henneman's size principle. In *Respiratory Muscles and Their Neuromotor Control.* Edited by G. C. Sieck, S. C. Gandevia, and W. E. Cameron. New York, Alan R. Liss, pp. 249-261.

Dubowitz, V. (1965). Enzyme histochemistry of skeletal muscle. I. Develop-ing animal muscle. *J. Neurol. Neurosurg. Psychiatry* **28**:516-519.

Duron, B., and Marlot, D. (1980). Nervous control of breathing during post-natal development in the kitten. *Sleep* **3**:323-330.

Duxson, M. J., Ross, J. J., and Harris, A. J. (1986). Transfer of differentiated synaptic terminals from primary myotubes to new-formed muscle cells during embryonic development in the rat. *Neurosci. Lett.* **71**: 147-152.

Eddinger, T. J., and Moss, R. L. (1987). Mechanical properties of skinned single fibers of identified types from rat diaphragm. *Am. J. Physiol.* **523**:C210-C218.

Edström, L., and Kugelberg, E. (1968). Histochemical composition, distribution of fibers and fatiguability of single motor units. *J. Neurol. Neurosurg. Psychiatry.* **31**:424-433.

Edwards, R. H. T. (1981). Human muscle function and fatigue. In *Human Muscle Fatigue: Physiological Mechanisms*. Ciba Foundation Symposium 82. Edited by R. Porter and J. Whelan. London, Pitman Medical, pp. 1-18.

Enad, J. G., Fournier, M., and Sieck, G. C. (1989). Oxidative capacity and capillary density of diaphragm motor units. *J. Appl. Physiol.* **67**:620-627.

Fladby, T. (1987). Postnatal loss of synaptic terminals in the normal mouse soleus muscle. *Acta Physiol. Scand.* **129**:229-238.

Fladby, T., and Jansen, J. K. S. (1988). Selective innervation of neonatal fast and slow muscle fibres before net loss of synaptic terminals in the mouse soleus muscle. *Acta Physiol. Scand.* **134**:561-562.

Fleshman, J. W., Munson, J. B., Sypert, G. W., and Friedman, W. A. (1981). Rheobase, input resistance, and motor unit type in medial gastrocnemius motoneurons in the cat. *J. Neurophysiol.* **46**:1326-1338.

Fournier, M., and Sieck, G. C. (1986). Variability in SDH activity among diaphragm muscle unit fibers. *Soc. Neurosci. Abstr.* **12**:1082.

Fournier, M., and Sieck, G. C. (1987) Topographical projections of phrenic motoneurons and motor unit territories in the cat diaphragm. In *Respiratory Muscles and Their Neuromotor Control*. Edited by G. C. Sieck, S. C. Gandevia, and W. E. Cameron. New York, Alan R. Liss, pp. 215-226.

Fournier, M., and Sieck, G. C. (1988a). Somatotopy in the segmental innervation of the cat diaphragm. *J. Appl. Physiol.* **64**:291-298.

Fournier, M., and Sieck, G. C. (1988b). Mechanical properties of muscle units in the cat diaphragm. *J. Neurophysiol.* **59**:1055-1066.

Fulton, B. P., and Walton, K. (1986). Electrophysiologic properties of neonatal rat motoneurones studied in vitro. *J. Physiol. (London)* **370**: 651-678.

Gallego, R., Huizar, P., Kudo, N., and Kuno, M. (1978). Disparity of motoneurone and muscle differentiation following spinal transection in the kitten. *J. Physiol. (London)* **281**:253-265.

Gauthier, G. F., Lowey, S., and Hobbs, A. W. (1978). Fast and slow myosin in developing muscle fibers. *Nature* **274**:25-29.

Glebovskii, V. D. (1961a). The physiological properties of motor fibers of the phrenic and intercostal nerves in adult and newborn animals. *Sechenov Physiol. J. USSR* **47**:41-48.

Glebovskii, V. D. (1961b). The contractile properties of the respiratory muscles in adult and new-born animals. *Sechenov Physiol. J. USSR* **47**: 470-480.

Gordon, H., and Van Essen, D. C. (1985). Specific innervation of muscle fiber types in a developmentally polyinnervated muscle. *Dev. Biol.* **111**: 42-50.

Guth, L., and Samaha, F. J. (1972). Erroneous interpretations which may result from application of the myofibrillar ATPase histochemical procedure to developing muscle. *Exp. Neurol.* **34**:465-475.

Gutmann, E., Melichna, J., and Syrovy, I. (1974). Developmental changes in contraction time, myosin properties and fibre pattern of fast and slow skeletal muscles. *Physiol. Bohemoslov.* **23**:19-27.

Hamm, T. M., Nemeth, P. M., Solanki, L., Gordon, D. A., Reinking, R. M., and Stuart, D. G. (1988). Association between biochemical and physiological properties of single motor units. *Muscle Nerve* **11**:245-254.

Hammarberg, C., and Kellerth, J. O. (1975). The postnatal development of some twitch and fatigue properties of single motor units in the ankle muscle of the kitten. *Acta Physiol. Scand.* **95**:243-257.

Henneman, E. (1957). Relation between size of neurons and their susceptibility to discharge. *Science* **126**:1345-1346.

Henneman, E., and Mendell, L. M. (1981). Functional organization of motoneuron pool and its inputs. In *Handbook of Physiology,* Sec. 1, Vol. III, Part 1, *The Nervous System.* Edited by J. M. Brookhart, and V. B. Mountcastle. Bethesda, MD, American Physiological Society, pp. 423-507.

Henneman, E., Somjen, G., and Carpenter, D. O. (1965). Functional significance of cell size in spinal motoneurons. *J. Neurophysiol.* **28**:560-580.

Hilaire, G., Gauthier, P., and Monteau, R. (1983). Central respiratory drive and recruitment order of phrenic and inspiratory laryngeal motoneurones. *Respir. Physiol.* **51**:341-359.

Hilaire, G., Monteau, R., and Khatib, M. (1987). Determination of recruitment order of phrenic motoneurons. In *Respiratory Muscles and Their Neuromotor Control.* Edited by G. C. Sieck, S. C. Gandevia, and W. E. Cameron. New York, Alan R. Liss, pp. 249-261.

Hoh, J. F., Hughes, S., Hale, P. T., and Fitzsimmons, R. B. (1988). Immunocytochemical and electrophoretic analyses of changes in myosin gene expression in cat limb fast and slow muscles during postnatal development. *J. Muscle Res. Cell Motil.* **9**:30-47.

Huizar, P., Kuno, M., and Miyata, Y. (1975). Differentiation of motoneurons and skeletal muscles in kittens. *J. Physiol. (London)* **252**:465-479.

Jodkowski, J., Guthrie, R. D., and Cameron, W. E. (1989). The activity pattern of phrenic motoneurons during the aspiration reflex: An intracellular study. *Brain Res.* **505**:187-194.

Jodkowski, J. S., Viana, F., Dick, T. E., and Berger, A. J. (1987). Electrical properties of phrenic motoneurons in the cat: Correlation with inspiratory drive. *J. Neurophysiol.* **58**:105-124.

Jolesz, F., and Sreter, F. A. (1981). Development, innervation, and activity pattern induced changes in skeletal muscle. *Annu. Rev. Physiol.* **43**: 531-552.

Jones, S. P., and Ridge, R. M. A. P. (1987). Motor units in a skeletal muscle of neonatal rat: Mechanical properties and weak neuromuscular transmission. *J. Physiol. (London)* **386**:355-375.

Jones, S. P., Ridge, R. M. A. P., and Rowlerson, A. (1987). The nonselective innervation of muscle fibres and mixed composition of motor units in a muscle of neonatal rat. *J. Physiol. (London)* **386**:377-394.

Jordan, C. L., Letinski, M. S., and Arnold, A. P. (1988). Synapse elimination occurs late in the hormone-sensitive levator ani muscle of the rat. *J. Neurobiol.* **19**:335-356.

Keens, T. G., Bryan, A. C., Levison, H., and Ianuzzo, C. D. (1978). Developmental pattern of muscle fiber types in human ventilatory muscles. *J. Appl. Physiol.* **44**:909-913.

Keens, T. G., and Ianuzzo, C. D. (1979). Development of fatigue resistant muscle fibers in human ventilatory muscles. *Am. Rev. Respir. Dis.* **119**: 139-141.

Kellerth, J. O., Mellström, A., and Skoglund, S. (1971). Postnatal excitability changes of kitten motoneurons. *Acta Physiol. Scand.* **83**:31-41.

Kelly, A. M. (1983). Emergence of muscle specialization. In *Handbook of Physiology,* Sec. 10. *Skeletal Muscle.* Edited by L. D. Peachey and R. H. Adrian. Bethesda, MD, American Physiological Society, pp. 507-537.

Kelly, A. M., and Rubinstein, N. A. (1980). Why are fetal muscles slow? *Nature* **288**:266-269.

Kelly, A. M., and Zacks, S. I. (1969). The histogenesis of rat intercostal muscle. *J. Cell Biol.* **42**:135-153.

Kelly, S. S. (1978). The effect of age on neuromuscular transmission. *J. Physiol. (London)* **274**:51-62.

Kelsen, S. G., and Nochomovitz, M. L. (1982). Fatigue of the mammalian diaphragm in vitro. *J. Appl. Physiol.* **53**:440-447.

Kernell, D., and Monster, A. W. (1982a). Time course and properties of late adaptation in spinal motoneurones of the cat. *Exp. Brain Res.* **46**: 191-196.

Kernell, D., and Monster, A. W. (1982b). Motoneurone properties and motor fatigue. An intracellular study of gastrocnemius motoneurones of the cat. *Exp. Brain Res.* **46**:197-204.

Krnjevic, K., and Miledi, R. (1958). Failure of neuromuscular propagation in rats. *J. Physiol. (London)* **140**:440-461.

Kuei, J. H., Shadmehr, R., and Sieck, G. C. (1990). Relative contribution of neurotransmission failure to diaphragm fatigue. *J. Appl. Physiol.* **68**:174-180.

Laskowski, M. B., and High, J. A. (1989). Expression of nerve-muscle topography during development. *J. Neurosci.* **9**:175-182.

Laskowski, M. B., and Sanes, J. R. (1987). Topographic mapping of motor pools onto skeletal muscles. *J. Neurosci.* **7**:252-260.

Le Souef, P. N., England, S. J., Stogryn, H. A. F., and Bryan, A. C. (1988). Comparison of diaphragmatic fatigue in newborn and older rabbits. *J. Appl. Physiol.* **65**:1040-1044.

Liddell, E. G. T., and Sherrington, C. S. (1925). Recruitment and some other factors of reflex inhibition. *Proc. Roy. Soc. Lond. (Biol.)* **97**: 488-518.

Luff, A. R., and Atwood, H. L. (1971). Changes in the sarcoplasmic reticulum and transverse tubular system of fast and slow skeletal muscles of the mouse during postnatal development. *J. Cell Biol.* **51**:369-383.

Mahdavi, V., Strehler, E. E., Periasamy, D., Wieizouk, S., Izumo, S., Gunds, S., Strehler, M. A., and Nadal-Ginard, B. (1986). Sarcomeric myosin heavy chain gene family: Organization and pattern of expression. In *Molecular Biology of Muscle Development*. Edited by C. Emerson, D. Fischman, B. Nadal-Ginard, and M. A. Q. Siddiqui. New York, Alan R. Liss, pp. 345-362.

Martin, T. P., Bodine-Fowler, S., Roy, R. R., Eldred, E., and Edgerton, V. R. (1988). Metabolic and fiber size properties of cat tibialis anterior motor units. *Am. J. Physiol.* **255**:C43-C50.

Maxwell, L. C., McCarter, J. M., Keuhl, T. J., and Robotham, J. L. (1983). Development of histochemical and functional properties of baboon respiratory muscles. *J. Appl. Physiol.* **54**:551-561.

Mellström, A., and Skoglund, S. (1969). Postnatal changes of neuron volume and nuclear volume in lamina IX of C7 and L7. *Acta Physiol. Scand.* **33** (Suppl.):28-49.

Monteau, R., Khatib, M., and Hilaire, G. (1985). Central determination of recruitment order: intracellular study of phrenic motoneurons. *Neurosci. Lett.* **56**:341-346.

Mortola, J. P. (1987). Dynamics of breathing in newborn mammals. *Physiol. Rev.* **67**:187-243.

Muller, N., Gulston, G., Cade, D., Whitton, J., Froese, A. B., Bryan, M. H., and Bryan, A. C. (1979). Diaphragmatic muscle fatigue in the newborn. *J. Appl. Physiol.* **46**:688-695.

Narusawa, M., Fitzsimmons, R. B., Izumo, S., Nadal-Ginard, B., Rubinstein, N. A., and Kelly, A. M. (1987). Slow myosin in developing rat skeletal muscle. *J. Cell Biol.* **104**:447-459.

Navarrete, R., and Vrbova, G. (1983). Changes of activity patterns in slow and fast muscles during postnatal development. *Dev. Brain Res.* **8**: 11-19.

Nemeth, P., Pette, D., and Vrbova, G. (1981). Comparison of enzyme activities among single muscle fibers within defined motor units. *J. Physiol. (London)* **311**:489-495.

Nemeth, P. M., Solanki, L., Gordon, D. A., Hamm, T. M., Reinking, R. M., and Stuart, D. G. (1986). Uniformity of metabolic enzymes within individual motor units. *J. Neurosci.* **6**:892-898.

Nyström, B. (1968). Histochemistry of developing cat muscles. *Acta Neurol. Scand.* **44**:405-439.

O'Brien, R. A. D., Ostberg, A. J. C., and Vrbova, G. (1978). Observations on the elimination of polyneuronal innervation in developing mammalian skeletal muscle. *J. Physiol. (London)* **282**:571-582.

Pagala, M. K. D., Namba, T., and Grob, D. (1984). Failure of neuromuscular transmission and contractility during muscle fatigue. *Muscle Nerve* **7**:454-464.

Pette, D., and Vrbova, G. (1985). Invited review: Neural control of phenotypic expression in mammalian muscle fibers. *Muscle Nerve* **8**:676-689.

Redfern, P. A. (1970). Neuromuscular transmission in newborn rats. *J. Physiol. (London)* **209**:701-709.

Reiser, P. J., Kasper, C. E., Greaser, M. L., and Moss, R. L. (1988). Functional significance of myosin transitions in single fibers of developing soleus muscle. *Am. J. Physiol.* **254**:C605-C613.

Reiser, P. J., Moss, R. L., Giulian, G. G., and Greaser, M. L. (1985). Shortening velocity and myosin heavy chains of developing rabbit muscle fibers. *J. Biol. Chem.* **260**:14403-14405.

Remahl, S., Cullheim, S., and Ulfhake, B. (1985). Dimensions and branching patterns of triceps surae alpha-motor axons and their recurrent collaterals in the spinal cord during the postnatal development of the cat. *Dev. Brain Res.* **23**:193-200.

Ridge, R. M. A. P., and Betz, W. J. (1984). The effect of selective, chronic stimulation on motor unit size in developing rat muscle. *J. Neurosci.* **4**:2614-2620.

Rose, D., Larnicol, N., and Duron, B. (1984). An HRP study of the cat's spinal respiratory motoneurones during postnatal development. *Exp. Brain Res.* **56**:458-467.

Rosenthal, J. L., and Taraskevich, P. S. (1977). Reduction of multiaxonal innervation at the neuromuscular junction of the rat during development. *J. Physiol. (London)* **270**:299-310.

Ross, J. J., Duxson, M. J., and Harris A. J. (1987). Formation of primary and secondary myotubes in rat lumbrical muscles. *Development* **100**: 383-394.

Rosser, B. W. C., Choksi, R. M., Kelly, A. M., and Nemeth, P. M. (1988). Early metabolic differentiation of the rat diaphragm. *Soc. Neurosci. Abstr.* **14**:1147.

Roussos, C. S., and Macklem, P. T. (1977). Diaphragmatic fatigue in man. *J. Appl. Physiol.* **43**:189-197.

Sandercock, T. G., Faulkner, J. A., Albers, J. W., and Abbrecht, P. H. (1985). Single motor unit and fiber action potentials during fatigue. *J. Appl. Physiol.* **58**:1073-1079.

Sato, M., Mizuno, N., and Konishi, A. (1977). Postnatal differentiation of cell body volumes of spinal motoneurons innervating slow-twitch and fast-twitch muscles. *J. Comp. Neurol.* **175**:27-36.

Schiaffino, S., Ausoni, S., Gorza, L., Saggin, L., Gundersen, K., and Lømo, T. (1988a). Myosin heavy chain isoforms and velocity of shortening of type 2 skeletal muscle fibers. *Acta Physiol. Scand.* **134**:575-576.

Schiaffino, S., Gorza, L., Pitton, G., Saggin, L., Ausoni, S., Sartore, S., and Lømo, T. (1988b). Embryonic and neonatal myosin heavy chain in denervated and paralyzed rat skeletal muscle. *Dev. Biol.* **127**:1-11.

Schiaffino, S., and Margreth, A. (1969). Coordinated development of the sarcoplasmic reticulum and T-system during postnatal differentiation of rat muscle. *J. Cell Biol.* **41**:855-875.

Schmalbruch, H. (1982). Skeletal muscle fibers of newborn rats are coupled by gap junctions. *Dev. Biol.* **91**:485-490.

Shapovalov, A. I., Shiriaev, B. I., and Tamarova, Z. A. (1981). A study of neuronal activity of mammalian superfused or intraarterially perfused CNS preparations. In *Electrophysiology of Isolated Mammalian CNS Preparations*. Edited by G. A. Kerkut and H. V. Wheal. London, Academic Press, pp. 367-394.

Sherrington, C. S. (1929). Ferrier lecture: Some functional problems attaching to convergence. *Proc. R. Soc. Lond. (Biol.)* **105**:332-362.

Sieck, G. C. (1988). Diaphragm muscle: Structural and functional organization. In *Respiratory Muscle: Function in Health and Disease*. Clinics in Chest Medicine, Vol. 9. Edited by M. J. Belman. Philadelphia, W. B. Saunders, pp. 195-210.

Sieck, G. C., and Blanco, C. E. (1987a). Change in diaphragm muscle fiber size and oxidative capacity during postnatal development. *Am. Rev. Respir. Dis.* **135**:A332.

Sieck, G. C., and Blanco, C. E. (1987b). Fatigue resistance and oxidative capacity of the diaphragm during postnatal development. *Soc. Neurosci. Abstr.* **13**:1506.

Sieck, G. C., and Fournier, M. (1988). Metabolic profile of muscle fibers in the fetal cat diaphragm. In *Sudden Infant Death Syndrome. Risk Factors and Basic Mechanisms.* Edited by R. M. Harper and H. J. Hoffman. New York, PMA Publishing Corp. pp. 361-378.

Sieck, G. C., and Fournier, M. (1989). Diaphragm motor unit recruitment during ventilatory and nonventilatory behaviors. *J. Appl. Physiol.* **66**:2539-2545.

Sieck, G. C., and Fournier, M. (1990). Changes in diaphragm motor unit EMG during fatigue. *J. Appl. Physiol.* **68**:2539-2545.

Sieck, G. C., Fournier, M., and Enad, J. G. (1989a). Fiber type composition of muscle units in the cat diaphragm. *Neurosci. Lett.* **97**:29-34.

Sieck, G. C., Lewis, M. I., and Blanco, C. E. (1989b). Effects of undernutrition on diaphragm fiber size, SDH activity, and fatigue. *J. Appl. Physiol.* **66**:2196-2205.

Sieck, G. C., Sacks, R. D., and Blanco, C. E. (1987). Absence of regional differences in the size and oxidative capacity of diaphragm muscle fibers. *J. Appl. Physiol.* **63**:1076-1082.

Sieck. G. C., Sacks, R. D., Blanco, C. E., and Edgerton, V. R. (1986). SDH activity and cross-sectional area of muscle fibers in cat diaphragm. *J. Appl. Physiol.* **60**:1284-1292.

Sieck, G. C., Trelease, R. B., and Harper, R. M. (1984). Sleep influences on diaphragmatic motor unit discharge. *Exp. Neurol.* **85**:316-335.

Sieck, G. C., Zhan, W-Z., and Fournier, M. (1989c). Adaptations of diaphragm muscle to altered use. *Nature J. (China)* **12**:656-663.

Smith, D. O. (1980). Mechanisms of action potential propagation failure at sites of axon branching. *J. Physiol. (London)* **301**:243-259.

Smith, D., Green, H., Thomson, J., and Sharratt, M. (1988). Oxidative potential in developing rat diaphragm, EDL, and soleus muscle fibers. *Am. J. Physiol.* **254**:C661-C668.

Soileau, L. C., Silberstein, L., Blau, H. M., and Thompson, W. J. (1987). Reinnervation of muscle fiber types in the newborn rat soleus. *J. Neurosci.* **7**:4176-4194.

Staron, R. S., and Pette, D. (1987). The multiplicity of combination of myosin light chains and heavy chains in histochemically typed single fibres. *Biochem. J.* **243**:695-699.

Stockdale, F. E., and Miller, J. B. (1987). The cellular basis of myosin heavy chain isoform expression during development of avian skeletal muscle. *Dev. Biol.* **123**:1-9.

Swynghedauw, B. (1986). Developmental and functional adaptation of contractile proteins in cardiac and skeletal muscles. *Physiol. Rev.* **66**:710-771.

Sypert, G. W. and Munson, J. B. (1981). Basis of segmental motor control: Motoneuron size or motor unit type? *Neurosurgery* **8**:608-621.

Thompson, W. (1983). Synapse elimination in neonatal rat muscle is sensitive to pattern of muscle use. *Nature* **303**:614-616.

Thompson, W. J., Kuffler, D. P., and Jansen, J. K. S. (1983). The effect of prolonged reversible block of nerve impulses on the elimination of polyneuronal innervation of skeletal muscles in neonatal rats. *J. Physiol. (London)* **335**:343-352.

Thompson, W. J., Sutton, L. A., and Riley, D. A. (1984). Fibre type composition of single motor units during synapse elimination in neonatal rat soleus muscle. *Nature* **309**:709-711.

Tomanek, R. J. (1975). A histochemical study of postnatal differentiation of skeletal muscle with reference to functional overload. *Dev. Biol.* **42**:305-314.

Walmsley, B., Hodgson, J. A., and Burke, R. E. (1978). Forces produced by medial gastrocnemius and soleus muscles during locomotion in freely moving cats. *J. Neurophysiol.* **41**:1203-1216.

Walton, K., and Fulton, B. P. (1986). Ionic mechanisms underlying the firing properties of rat neonatal motoneurons studied in vitro. *Neuroscience* **19**:669-683.

Webber, C. L., Jr., and Pleschka, K. (1976). Structural and functional characteristics of individual phrenic motoneurons. *Pflugers Arch.* **364**:113-121.

Westerman, R. A., Chan, H. S., Ziccone, S. P., Spiratana, D., Dennett, X., and Tate, K. A. (1979). Plasticity of motor reinnervation in the kitten. In *Neural Growth and Differentiation*. Edited by E. Meisami and M. A. B. Brazier. New York, Raven Press, pp. 397-432.

Whalen, R. G., Sell, S. M., Butler-Browne, G. S., Schwartz, K., Bouveret, P., and Pinset-Härström, I. (1981). Three myosin heavy chain isozymes appear sequentially in rat muscle development. *Nature* **292**:805-809.

Wilson, D. F., and Cardaman, R. C. (1984). Age-associated changes in neuromuscular transmission in the rat. *Am. J. Physiol.* **247**:C288-C292.

Zajac, F. E., and Faden, J. S. (1985). Relationship among recruitment order, axonal conduction velocity, and muscle-unit properties of type-identified motor units in cat plantaris muscle. *J. Neurophysiol.* **53**:1303-1322.

Zengel, J. E., Reid, S. A., Sypert, G. W., and Munson, J. B. (1985). Membrane electrical properties and prediction of motor-unit type of medial gastrocnemius motoneurons in the cat. *J. Neurophysiol.* **53**:1323-1344.

Zhan, W. Z., and Sieck, G. C. (1988). Diaphragm fiber size and oxidative capacity following phrenic denervation or inactivation. *Soc. Neurosci. Abstr.* **14**:1147.

Ziskind-Conhaim, L. (1988a). Physiological and morphological changes in developing peripheral nerves of rat embryos. *Dev. Brain Res.* **42**:15-28.

Ziskind-Conhaim, L. (1988b). Electrical properties of motoneurons in the spinal cord of rat embryos. *Dev. Biol.* **128**:21-29.

14

Functional Aspect of Human Muscles and Their Central Nervous System Interactions
Implications for Muscle Failure

BRENDA BIGLAND-RITCHIE

John B. Pierce Laboratory, Inc.
New Haven, Connecticut

I. Introduction

Failure of respiratory function is a major clinical problem, particularly in the newborn. It is often unclear to what extent it may arise from neuromuscular fatigue and, if so, whether this results from impaired function of the muscles themselves or from failure of the central nervous system (CNS) to drive them. The purpose of this chapter is to summarize our view of the various possible causes of neuromuscular fatigue, and to raise questions that may indicate useful lines of research for future studies.

A. Definition

Neuromuscular fatigue is often defined as an inability to sustain or repeat the required force or power output (Edwards, 1981). This definition has two drawbacks. First, it implies that fatigue is delayed in onset and ignores contributing physiological changes which clearly start from the onset of exercise. Second, the timing and extent of the changes at various sites depend directly on the type and intensity of exercise being performed. For our purposes, we define neuromuscular fatigue as any reduction in a subject's force-generating capacity in response to maximum voluntary effort, regardless of the task from which fatigue develops. We measure it from changes in the magnitude of a brief maximal voluntary contraction (MVC) executed periodically throughout the exercise. Measured this way, fatigue clearly starts from the onset of exercise and is an integral part of most neuromuscular activity.

B. Sites of Fatigue and Methods for Investigation

A voluntary contraction is initiated by processes within higher brainstem centers that activate lower motoneurons in the spinal cord to discharge impulses to the muscle or group of muscles involved. Within the spinal cord excitatory influences from descending pathways are reinforced, or compete with, excitatory and inhibitory spinal reflexes arising from other sources. Thus the force that can be exerted depends critically on being able to maintain an adequate central motor drive. The degree to which a muscle is activated also depends on effective impulse transmission across the neuromuscular junction, over the muscle surface membrane, and down each fiber's t-tubular system. Consequently, charge movements at the "foot" junctions between the t-tubules and terminal cisternae initiate Ca^{2+} release from the sarcoplasmic reticulum (SR) which, interacting with troponin, allow cross-bridge formation between actin and myosin. But continued cross-bridge cycling is only possible in the presence of sufficient energy from ATP, and may be inhibited by the accumulation of metabolites such as lactic acid (LA) or inorganic phosphate (Pi). This series of events and the means whereby they can be investigated in human experiments are illustrated in Figure 1.

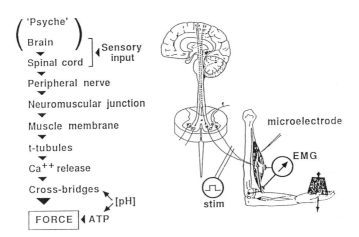

Figure 1   Potential sites of fatigue and methods for testing (see text).

In human experiments it is difficult to investigate functional changes within the CNS during prolonged activity. But the force from a maximum voluntary contraction can be compared with that from supramaximal stimulation of the motor nerve. If the MVC force declines more quickly than that from supramaximal nerve stimulation, the force loss must be due, in part, to failure of CNS motor drive. However, if both decline in parallel, fatigue must be due to processes distal to the point of stimulation. The effectiveness of impulse transmission across the neuromuscular junction and over the muscle surface membrane can be monitored by measuring the amplitude and area of compound muscle action potentials (M-waves) elicited by periodically applying single maximal shocks to the motor nerve during and following a voluntary contraction. These shocks also elicit contractions from which muscle contractile properties can be measured. Changes in motor drive are measured from the muscle surface integrated electromyogram (EMG). Spike trains recorded with intramuscular microelectrodes from individual muscle fibers monitor spinal motoneuron firing rates, provided impulse propagation is not impaired. Comparing these with changes in integrated EMG indicates any changes in motor unit recruitment or rate coding as activity progresses. Serial biopsies or nuclear magnetic resonance (NMR) measurements can detect changes in either whole muscle biochemical composition or that of individual muscle fiber types (biopsies only).

II. Problems in Fatigue Studies

A. Multiple Causes

Despite its enormous practical implications, there is no general agreement among different investigators as to the mechanisms responsible for neuromuscular fatigue. Much of the confusion comes from the difficulties of interpreting data obtained under different experimental conditions; and from the common practice of referring to fatigue as a single phenomenon, to which a single unique solution can be found. However, since a voluntary or reflex contraction depends on a chain of events involving numerous processes, functional changes occurring at many sites simultaneously may all contribute to reduced performance during prolonged activity. It therefore seems likely that fatigue results from multiple factors and that different types of exercise place varying amounts of stress on different sites. Thus, the rate-limiting factor(s) in any particular situation may depend both on the muscle groups involved and, more particularly, on the type of exercise from which fatigue develops. Evidence supporting this view will be presented. Other problems to be discussed include those of comparing fatigue processes seen when muscles are stimulated with those seen during contractions elicited naturally, and possible differences between the contractile properties of human and animal muscles.

B. Stimulated Versus Voluntary Contractions

Because of the difficulties of investigating fatigue processes in intact humans or freely moving animals, many of the more detailed studies must be performed on isolated or anesthetized animal preparations where muscles are stimulated electrically. A major problem with this approach concerns the frequencies of stimulation selected. This factor can be vital since stimulus frequency plays a major role in determining the rate at which functional changes develop at many peripheral sites. Unfortunately, it is not possible to stimulate a nerve at the discharge rates employed by the central nervous system, where individual motoneuron firing rates vary widely in relation to the unique contractile properties of each motor unit supplied (Bigland-Ritchie and Woods, 1984). Thus, one must be cautious when extrapolating data obtained from isolated or stimulated preparations to the interpretation of fatigue processes in intact humans.

Other factors may also complicate the clinical relevance of data obtained from stimulated isolated diaphragm preparations since phrenic motoneurons are innervated by two separate pathways, one from respiratory centers, the other directly from the corticospinal tract (Gandevia and Rothwell, 1987). Phrenic nerve stimulation can excite all diaphragm motor units to respond with maximum tetanic force but maximal diaphragm activation has not been reported in vivo in response to CO_2 loading or exercise. Moreover, based on animal studies, Sieck and Fournier (chapter 13, this volume) conclude that only part of the total strength of the diaphragm is used for respiratory purposes, even under maximal stress. In contrast, the human diaphragm can generate forces similar to those elicited by maximum phrenic nerve stimulation (Bellemare and Bigland-Ritchie, 1984), but this has only been demonstrated in contractions, such as expulsive maneuvers, that also involve voluntary effort. Thus, presumably phrenic motoneurons must be excited by both pathways before the diaphragm is activated maximally. Hence procedures that change muscle strength in response to nerve stimulation may not necessarily indicate a corresponding change in respiratory capacity when muscles are used to breath naturally.

C. Interspecies Variations

When considering the clinical relevance of data obtained from isolated or anesthetized animal preparations, it is also reasonable to question what differences there may be between the contractile and fatigue properties of human and animal muscles. In Chapter 13, Sieck and Fournier describe important differences between properties of motor units from different animals. Similar or perhaps greater differences might be found if these experiments could be done on human muscle. Unfortunately, little is known about the contractile properties of human motor units, mainly because few methods are available that are appropriate for use in humans. Thus, human motor units can generally only be classified from differences in their histochemical characteristics, assuming the same relationship between their histochemical and contractile properties as that established in most animal studies.

In animal studies a good correlation has generally been found between the contractile properties of motor units and their histochemical characteristics. Units with the highest oxidative capacity (type I) generate small forces, have long twitch contraction times, and are highly fatigue resistant (S units), while those that produce the most force are fast but fatigue rapidly (FF units). These units have low oxidative, but high glycolytic capacity (type IIb). A third group (FR units), which are also fast but fatigue resistant, contain high concentrations of both oxidative and glycolytic enzymes (type IIa). Thus, a cat or rat motor unit can generally be confidently identified as belonging to one of these three distinct types by examining *either* its contractile *or* histochemical properties, and whole muscle properties generally reflect those of its fiber type composition identified histochemically. But the contractile properties of whole human muscles cannot always be predicted so clearly. For example, both the human adductor pollicis and tibialis anterior muscles are composed of about the same mixture of fiber types as the slow soleus muscle (about 80% type I, slow oxidative fibers), but their twitch contraction times and fatigue resistance (in sustained isometric maximal voluntary contractions) are similar to those of biceps brachii, first dorsal interosseous (FDI), or quadriceps, all composed of much higher proportions of type II fibers (Bellemare et al., 1986; Thomas et al., 1989). These findings prompted us to compare the contractile and fatigue properties of human motor units with those of cats; we then tried to classify human muscle fibers into distinct types using the same criteria as those applied in animal studies.

III. Contractile Properties of Human Motor Units

Twitch properties of human motor units have been measured by spike-triggered averaging during ongoing voluntary contractions (Stein et al., 1972; Thomas et al., 1987). This method provides no information about tetanic responses or how motor unit properties change with fatigue. The units sampled are often biased in favor of those recruited at low forces. Moreover, because motor units can seldom be made to fire voluntarily at rates below 8-10 Hz, twitch amplitudes and their contraction times may be distorted by substantial fusion between sequential twitches (Calancie and Bawa, 1986; Thomas et al., 1990b). Limited data have also been obtained from intramuscular stimulation of fine nerve terminals (Taylor and Stephens, 1976). With this method it is difficult to isolate the responses of only single units and these cannot be held long enough to examine their changes during fatigue. Thus, alternative methods are needed to determine the extent to which the results from animal studies can safely be used to interpret whole human muscle behavior.

Recently we examined the contractile properties of individual motor units of human thenar muscles when single motor axons of the median nerve were stimulated above the elbow (Westling et al., 1990). All-or-none force and EMG responses were obtained from 45 different motor units. These responses remained stable over extended periods and did not change when the stimulus current was varied over a wide margin. Motor axons were stimulated at frequencies from 1 to 100 Hz, including the protocol used to measure Burke fatigue indices (Burke et al., 1973). This allowed accurate measurement of twitch and tetanic forces and contractile speeds both before and following fatigue. Individual motor axon conduction velocities were also measured. These varied between 36.1 and 61.9 m/s, distributed relatively normally. Since this range agreed well with those recorded from other studies (e.g., Johansson and Vallbo, 1983), it seems likely that the motor units sampled were not biased toward either high- or low-threshold units. Typical records are shown in Figure 2.

Initial twitch amplitudes ranged from 2.9 to 34.0 mN, but 75% of the units were small, generating less than 15.6 mN (Thomas et al., 1990a). Twitch contraction and half-relaxation times ranged from 35.0 to 80.0 msec and from 25.0 to 107.5 msec, respectively, each distributed unimodally. Following tetanic stimulation the twitch amplitudes of most smaller units potentiated markedly, some by up to threefold. But, both before or after twitch potentiation, contraction and half-relaxation times, together with other indices of contractile speed, were all distribution unimodally. Nor did units show "sag" when stimulated at low rates (see 15 Hz record in Fig. 2). Thus, these units could not be divided into distinct groups of either fast or slow unit types based on differences in their contractile speeds.

Classification of animal motor units is also determined from differences of fatigue resistance, with most units having fatigue indices either <0.25 (FF) or >70.75 (FR and S). However, most human thenar motor units were highly fatigue resistant, with Burke fatigue indices ranging unimodally from 0.34 to 1.37. None behaved like FF units recorded from other mammalian muscles. Another difference was the large number of units (20%) with fatigue indices between 0.25 and 0.75, or fatigue-intermediate (Fint). Thus, these units could not be classified into distinct types based on differences in either their contractile speed or fatigue resis-

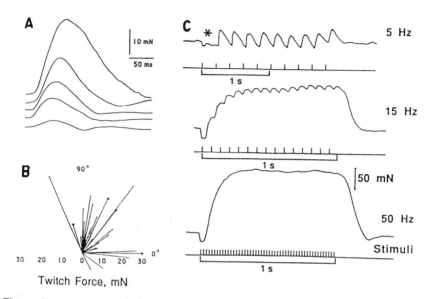

Figure 2 Responses of single human thenar motor units. (A) Representative twitches from different units. (B) Angles at which twitch forces were generated. (C) Responses to repetitive stimulation. Note lack of "sag" at 15 Hz.

tance properties. However, more experiments must be done, including measurements from muscle where more detailed histochemistry is known, before generalized conclusions can be drawn as to the extent to which fundamental differences exist between human motor unit properties and those of other species.

In animal muscles the smallest motor units are usually slow and the large ones fast. But the twitch amplitudes of human thenar motor units were not significantly correlated with their contraction or half-relaxation times either before or after twitch potentiation (Fig. 3). This species difference is less surprising when one considers that twitch force depends not only on the number of fibers per motor unit, but also on the cross-sectional area of each unit's constituent muscle fibers. In most animal muscles the diameter of FF fibers is much larger than those of S units; there are no systematic differences between fiber diameters of type I and type II fibers in humans.

Another surprising outcome from these experiments was finding that different units exert force in widely different directions within the muscle (Fig. 2). This was revealed by recording force independently in the directions of both flexion and abduction. Thus, the total force output of many units may be markedly underestimated, without influencing contractile speed measurements, by the usual practice of recording force in only one direction.

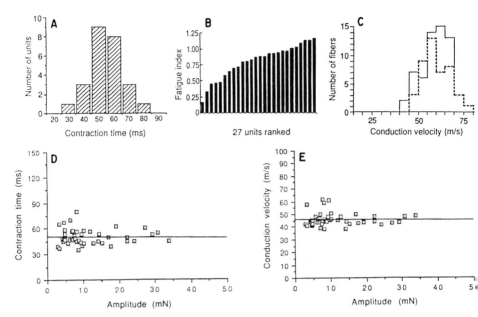

Figure 3 Distribution of values recorded from single human thenar motor units and the relationship between them. (A) Twitch contraction times. (B) Fatigue indices ranked in order. (C) Motor axon conduction velocities (dotted line) compared with those measured from a much larger sample by Johansson and Vallbo (1983) (solid line). (D and E.) Relation between twitch amplitudes and contraction times (D) and axon conduction velocity (E), respectively.

Moreover, if units exert force in different directions, this implies that each does different amounts of work in response to a given stimulus. There may be even greater differences between the work done by different fibers within a unit, since the territory of motor unit fibers is relatively widely distributed within the muscle. For example, some fibers may be stretched while others shorten. These different use patterns could also influence their biochemical composition. Thus the question as to whether the biochemical properties of all fibers of a given unit are necessarily homogeneous may depend on the geometry of motor unit fiber distribution within the particular muscle being examined.

IV. Potential Causes of Fatigue

A. Influence of Stimulus Frequency

Our early experiments (Jones et al., 1979: Bigland-Ritchie et al., 1979) showed that the force reduction during a sustained maximal voluntary contraction could not be imitated by stimulating a motor nerve at any constant rate (Fig. 4). Initially the MVC force could only be matched by supramaximal stimulation at 50-100 Hz, rates that elicit maximum tetanic force from all motor units. However, when these rates were applied continuously, the force declined more rapidly than in a MVC, but force was well maintained when lower rates were used. We also found that the excessive force loss during continuous high-frequency stimulation could be almost instantly restored when the stimulus rate was reduced from 50 to 20 Hz; i.e., fatigued muscle generates more force when stimulated at low, rather than high, excitation rates. In fact, the MVC force loss was well matched by stimulating initially at high rates, then progressively reducing the rate during the first 20 s. This provided our first clue that the same strategy may be employed by the central nervous system (see below).

The reasons for the rapid force loss during continuous high-frequency stimulation become apparent from examining the muscle compound action potentials (M-waves) recorded simultaneously. These decline rapidly during high-frequency stimulation, but are restored once these rates are reduced. This type of transmission failure (referred to as "high-frequency fatigue") is often assumed to occur at the neuromuscular junction, but the same action

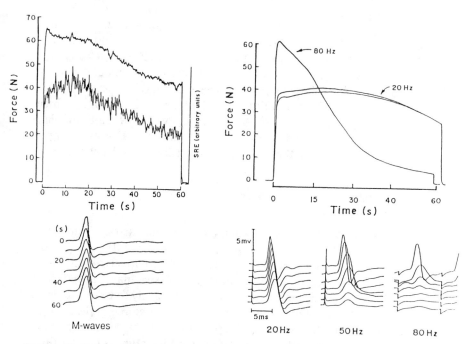

Figure 4 Comparison of force and M-waves recorded during maximal voluntary contractions (left) compared with those during stimulation at different rates (right).

potential responses are seen when isolated, fully curarized muscle preparations are stimulated in the same way. In this case any transmission at the neuromuscular junction has been eliminated. In other situations it is often difficult to demonstrate at which of these sites propagation has become impaired. If impulses are blocked at the neuromuscular junction, some muscle fibers are not excited and can therefore no longer exert any force. But at the muscle membrane there is a large safety factor for effective muscle activation. Action potential amplitudes may decline or increase without necessarily influencing force produced.

Ample evidence shows that the rate at which defects of action potential propagation develop at both the neuromuscular junction and muscle surface membrane is exquisitely sensitive to excitation rates. This may be largely due to extracellular K^+ accumulation, since each action potential is generated by an equal and opposite flux of K^+ and Na^+ across the membrane. If the Na^+/K^+ pumps cannot keep up with the fluxes, the membrane potential must fall and become progressively less excitable. Indeed, changes in electrolyte composition of the bathing medium cause changes in action potential amplitude and area similar to those recorded during high-frequency stimulation (Jones and Bigland-Ritchie, 1986). Large K^+ effluxes have also been recorded from fatiguing muscles during voluntary exercise where $[K^+]$ may increase to more than 8 mM in both arterial and veinous blood (Sjøgaard, 1986, 1988; Vøllestad et al., 1989a, b). Membrane depolarization from K^+ accumulation has been directly demonstrated (Juel 1988) and is probably largely responsible for the slowing of muscle fiber conduction velocity. Within the t-tubules the spaces for outward diffusion of accumulated K^+ are extremely small. Hence excitability changes must develop far more readily at this site. Indeed, Adrian and Peachy (1973) calculated that, in the absence of Na^+/K^+ pump activity, t-tubular membranes would become completely depolarized after only a relatively few nerve impulses. However, impulse propagation failure within the t-tubules is not easily detectable when recording from the muscle surface.

B. Excitation/Contraction Coupling and Metabolic Changes

When action potentials fail to propagate throughout the t-tubular system, this must reduce force output because Ca^{2+} release from the sarcoplasmic reticulum of some myofibrils is not initiated. Alternatively, even when t-tubular propagation is intact, other fatigue processes may still reduce the amount of Ca^{2+} released per impulse. This is more likely to be due to defects of Ca^{2+} channel activation than to depletion of Ca^{2+} stored in the SR (or to its diminished reuptake) because the application of caffeine, which opens SR Ca^{2+} channels, generally restores full force in fatigued muscle fibers. Thus failure of excitation/contraction coupling may result from failure of either t-tubular propagation or Ca^{2+} channel activation in the SR. When t-tubular propagation is impaired, the superficial myofibrils may generate normal force, while those lying deep within the fiber are not activated at all. Alternatively, if t-tubular propagation remains intact but less Ca^{2+} is released from the SR, the force from all myofibrils will decline.

Recent experiments by Allen and co-workers (1989a,b) have now separated these two factors. Using the Ca^{2+}-sensitive dyes Aquorin and Fura II on stimulated single frog muscle fibers, they first showed a direct relationship between intracellular Ca^{2+} levels and muscle force before fatigue, and also that these Ca^{2+} levels declined in parallel with force loss as fatigue developed. However, when fatigue was induced by continuous stimulation at high frequency, the Ca^{2+} signals arising from deep within the fiber declined more rapidly than those from superficial sites, indicating defects of t-tubular action potential propagation. When the stimulus frequency was suddenly reduced, as in our whole human and curarized mouse muscle preparations, both force and Ca^{2+} signals showed substantial recovery. Under these conditions the intensity of Ca^{2+} signals became evenly distributed throughout the muscle. Thus, reducing the stimulus frequency restored t-tubular function, but the Ca^{2+} release per impulse from the SR remained somewhat depressed. The application of caffeine restored muscle force to control levels, with Ca^{2+} signals now much higher than those resulting from any type of stimulation. This confirmed that fatigue had not significantly depleted the total Ca^{2+} stores in the SR. These experiments provided an example of force reduction which could not have been due to insufficient energy supplies or accumulation of metabolites since cross-bridge activity returned to control levels after caffeine was applied. But metabolic changes probably influence muscle t-tubular membrane excitability, since Na^+/K^+ pump activity is ATP dependent and also sensitive to the pH changes induced by fatigue.

V. Fatigue Processes in Human Voluntary Contractions

The above experiments show clearly that fatigue in isolated muscle preparations stimulated artificially can result from failure of excitation/contraction (e/c) coupling in the absence of metabolic changes. Can this also occur in fatigue from voluntary contractions elicited under normal exercise conditions?

A. Role of Excitation/Contraction Coupling and Metabolic Changes

We examined the reduction in force-generating capacity caused by fatigue from repeated low-intensity (30% MVC) static contractions of the quadriceps muscle (Vøllestad et al., 1988) where the contraction intensity and duty cycle were such that the metabolic demands could apparently be met entirely from aerobic sources replenished between contractions (Vøllestad et al., 1989). The contraction protocol is illustrated in Figure 5, together with the fatigue tests executed periodically throughout the exercise. During the first 30 min of exercise the force of brief test MVCs declined by about 50%. But during this period sequential biopsies showed no significant changes in the metabolic composition of the muscle. Since both the MVC force and that in response to brief bursts of tetanic stimulation declined in parallel, there was no indication that MVC force loss was due to insufficient muscle activation by the CNS or to impaired impulse propagation. Thus the fall in MVC force must have resulted from defective e/c coupling.

This conclusion has since been supported by results from several other studies in which muscles were fatigued by similar contraction protocols, including those in which metabolic events were monitored by NMR analysis. The results of Miller et al. (1987) for contractions of adductor pollicis were similar to ours. For contractions of the first dorsal interosseous muscle (FDI),

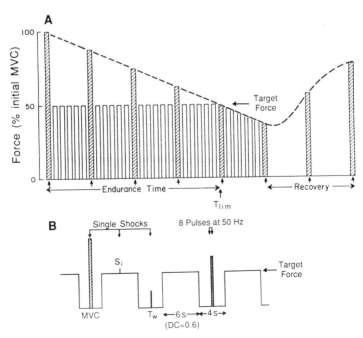

Figure 5   (A) Contraction protocol to measure fatigue from intermittent submaximal contractions. (B) Fatigue test carried out periodically at times shown. See text.

Newham et al. (personal communication) found some initial metabolic changes, but these were confined to only the first 1-2 min of exercise. No further changes were seen during the remaining exercise period, despite continued progressive MVC force decline. When these experiments were repeated at higher force intensities, they were accompanied by greater metabolic changes, but these were still confined to only the initial exercise period, and final steady-state metabolic levels were not well related to the corresponding force reductions. Further, Hultman and Spriet (1986) caused fatigue of the human quadriceps by stimulating with trains at 20 Hz. Sequential muscle biopsies demonstrated that the substantial metabolic changes, seen during the initial contraction period, later partially recovered during a time when the force from stimulation continued to decline. All these experiments implicate impaired e/c coupling, rather than only metabolic changes, as a major factor contributing to force reduction.

Fatigue from sustained high-force isometric contractions in which only anaerobic energy sources are available results in rapid and profound metabolic changes. Similar metabolic changes can often be seen during dynamic exercise where metabolic demands are high. In these cases, the depletion of energy supplies and lactic acid accumulation appear sufficient to account for the observed force reduction. Indeed, numerous studies on unfatigued isolated muscle or skinned fiber preparations have shown a decline in tension when they are exposed to either sufficiently low ATP or creatine phosphate, or high lactate concentrations. However, in voluntary contractions, no unique quantitative relation has been demonstrated between these metabolic changes and force reduction when fatigue is induced by different forms of exercise (static, dynamic, sustained, intermittent, etc). This raises the possibility that e/c coupling failure may still be the rate-limiting factor, the concomitant metabolic changes being only coincidental. If indeed this is the case, it would explain why voluntary exercise, no matter how severe, never depletes muscle energy supplies to the point of inducing rigor. Fortunately, the process of fatigue is always reversible. Clearly, some factor limits the metabolic changes that can be safely tolerated. A logical candidate for this role would be failure of Ca^{2+} release.

B. Role of Impaired Action Potential Transmission

Force reduction during fatigue may also result from failure of action potential propagation at the neuromuscular junction, mainly at presynaptic nerve terminal branch points, or from reduced excitability of postjunctional receptors. Propagation failure can also occur at the muscle surface membrane itself (Krnjevic and Miledi, 1958). But again, the rate at which it develops depends on the frequency at which nerve impulses are delivered, and the safety factor at these sites appears to be high, compared to that in the t-tubules. While propagation failure can easily be demonstrated when muscles are stimulated electrically, we have never found convincing evidence that it occurs during fatigue induced by human voluntary contractions (Bellemare and Bigland-Ritchie, 1987; Bigland-Ritchie et al., 1979, 1982, 1983a, 1986a,b; Bigland-Ritchie and Woods, 1984; Thomas et al., 1989), although others disagree. The reduced likelihood of peripheral propagation failure during voluntary contractions can probably be attributed to the lower motor discharge rates delivered by the CNS, compared to those used during most types of nerve or muscle stimulation.

In all our experiments, wherever possible, the integrity of neuromuscular transmission is routinely checked by monitoring the amplitude, duration, and area of muscle compound action potentials (M-waves) evoked by supramaximal single shocks applied to the motor nerves (Bigland-Ritchie et al., 1982). Unfortunately large stimulus artifacts prevent reliable M-wave measurement when intramuscular nerve terminals are stimulated percutaneously. In our hands, provided adequate precautions are taken to maintain stimulus maximality (Bigland-Ritchie, 1987), M-wave amplitudes usually remain relatively stable, even during sustained maximal contractions. But in the early part of the exercise, M-wave durations invariably increase owing to slowing of muscle conduction velocity. During this period we have sometimes found that the M-wave amplitudes of some muscles do decline, but with no reduction in their total area. We attribute this to increased dispersion between impulse arrival times in different muscle fibers [e.g., first dorsal interosseous (FDI) in Fig. 6]. Only if the total M-wave area is reduced do we interpret this as indicating transmission failure.

If transmission failure occurs in fatigue of voluntary contractions, it is most likely to be detected in sustained maximal contractions where high-threshold units are recruited and motoneuron discharge rates are high. In previous experiments we found no decrement of M-wave parameters recorded during MVC contractions of adductor pollicis or FDI sustained for 60 s (Bigland-Ritchie et al., 1979, 1982). But the responses to periodic bursts of tetanic stimulation demonstrated a progressive reduction in the transmission safety factor (Bigland-Ritchie et al., 1979). Thus, actual failure might have been apparent had the contractions been maintained longer. Recently we tested this suggestion, and also the possibility that responses of different muscles may vary, by comparing M-waves recorded from both FDI and tibialis anterior (TA) when contractions were maintained as high as possible for up to 5 min (Thomas et al., 1989).

The force from both muscles declined continuously at similar rates (despite the differences in their respective fiber type compositions; see above). For TA M-wave amplitudes remained constant throughout the contractions

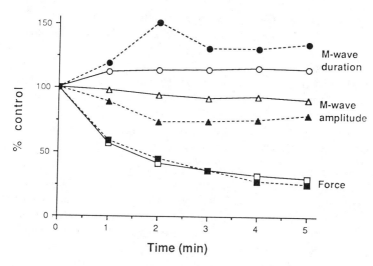

Figure 6 Changes in force and action potential amplitude and duration during maximal voluntary contractions of the human first dorsal interosseous (FDI) (closed symbols) and tibialis anterior (TA) (open symbols) muscles sustained for 5 min.

with only minor increases in duration. For FDI the amplitudes of M-waves declined by about 20%, but only during the first 2 min. This was accompanied by a 50% increase in M-wave durations, with little or no change in total area. Thereafter all parameters remained constant, while the force continued to fall progressively. It is therefore unlikely that force loss was due to transmission failure at either the neuromuscular junction or the muscle surface membrane.

Many studies conclude that, if a reduction in EMG signals accompanies force reduction, this proves failure of neuromuscular transmission. This conclusion is not necessarily justified for there is no unique relation between muscle force and EMG. Many examples can be found where the EMG signals decline while the force remains constant. Conversely, force decline is common under conditions where the EMG is stable. Examples illustrating both situations can be found in whole muscle studies (e.g., Cooper et al., 1988; Gibson et al., 1988) and in single cat motor units (e.g., Clamann and Robinson, 1985). Moreover, both the amplitude and area of muscle potentials may show marked potentiation during periods of force reduction (Hicks et al., 1998; Gibson et al., 1988). Thus there is no unique relationship between EMG and force during fatigue. Rather, there appears to be a large safety factor between the extent to which muscle membranes can become depolarized and any consequent impairment of muscle activation. Interestingly, Kernell et al.

(1987) finds that the efficacy of peripheral impulse transmission, like most other factors, can be improved by training. Normally when cat single motor units are stimulated for 2 min with the Burke fatigue protocol (13 pulses at 40 Hz delivered 1/s; Burke et al., 1973), there is a decrement in both their force and EMG responses. However, after weeks of chronic stimulation the same rate of force decline may still be present, but the EMG responses remain unaltered.

C. Role of CNS Motor Drive

Under normal exercise conditions it is likely that reduced performance is mainly due to an inability, or unwillingness, to maintain an adequate motor drive. However, under laboratory conditions we find that highly motivated subjects can generally fully activate most muscles by voluntary effort, even when fatigue has reduced their MVC force by up to 50% (Bigland-Ritchie et al., 1978, 1986a; Vøllestad et al., 1988; Woods et al., 1987; Thomas et al., 1989). This is demonstrated by the absence of additional force when electrical stimuli are delivered to the muscle or motor nerve during the contraction. When the diaphragm and soleus muscles were fatigued by the same intermittent submaximal contractions protocol as that used for quadriceps (Fig. 4), additional force could still be elicited (Bellemare and Bigland-Ritchie, 1987; Bigland-Ritchie et al., 1986a). Thus, the degree to which reduced CNS motor drive contributes to fatigue may vary depending on the muscle groups employed. But when subjects do not have the constant visual force-feedback and loud verbal encouragement provided in laboratory experiments, it is likely that an inability to maintain adequate motor drive plays an important role in fatigue generated under all conditions.

VI. Relation Between CNS Motor Drive and Muscle Contractile Properties

A. Motor Unit Firing Rates

The firing rates of individual motor units can be recorded with tungsten microelectrodes, even when the voluntary contraction is maximal. Mean MVC firing rates vary between muscles in relation to their intrinsic contractile speed (Bigland-Ritchie et al., 1983a,b). During the first 40 s of a sustained MVC, mean firing rate of adductor pollicis declined from about 27 Hz to 15 Hz. During this time the muscle contractile speed also slowed so that these lower firing rates still achieved maximum tetanic muscle activation. Similar results have been obtained from other muscles (Bigland-Ritchie et al., 1986b; Woods et al., 1987), suggesting that some mechanism exists within the CNS that regulates motoneuron firing rates to match changes of contractile speed during fatigue.

B. Fatigue-Induced Reflex Inhibition

The reduction of motoneuron firing rates that can be achieved by voluntary effort during fatigue may result solely from processes confined within the CNS. Kernell (1965a,b), Kernell and Monster (1982a,b) and others have demonstrated reduced motoneuron excitability and increased afterhyperpolarization during prolonged activity. Alternatively, motoneuron firing rates may be regulated by some peripheral feedback originating from the fatigued muscle.

These possibilities were tested for maximal contractions of biceps brachii (Bigland-Ritchie et al., 1986b) and quadriceps (Woods et al., 1987) using the protocol shown in Figure 7. The rationale for this protocol relied on various well-established facts. First, in high-force contractions, the muscle blood supply is occluded by the high intramuscular pressure generated by the contraction. Thus, the behavior of the muscle is unaffected by whether or not the blood supply to the limb is occluded by a cuff. Second, following a contraction a muscle can be maintained in a state of fatigue by keeping it ischemic. Neither force nor contractile speed recovers until the blood supply is restored. In the first 40 s of maximal contractions (MVC 1) the force declined by 40%. The muscle then remained relaxed for 3 min of ischemic rest, allowing any excitability changes in the CNS to recover. MVC 2, executed just before cuff release, showed little or no recovery of force or contractile speed, but both were fully recovered in the final maximal contraction, 3 min after the blood supply was restored.

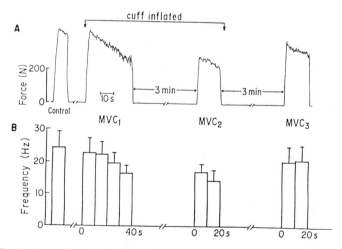

Figure 7 Evidence for reflex inhibition of motoneuron firing rates. (A) Force records demonstrating contraction protocol. (B) Motoneuron firing rates. (See text for explanation.)

During MVC 1 motoneuron firing rates declined by 40%. They did not recover in MVC 2, following 3 min of ischemic rest, but were fully recovered 3 min after cuff release once the fatigued state of the muscle had been relieved. Since 3 min was clearly enough time for full recovery of any central excitability changes, the reduced firing rates in MVC 2 must have been due to some feedback to the CNS from the fatigued muscle. Similar results were found when these experiments were repeated using MVC contractions of quadriceps, which also demonstrated that the reduced firing rates in MVC 2 were not due to either impaired neuromuscular transmission or an inability of the subject to fully activate the muscles (Woods et al., 1987).

The initial hypothesis behind these experiments was that reflex inhibition of motoneuron firing rates is initiated during fatigue by some direct response to a corresponding slowing of muscle contractile speed. This now seems less likely because we have since found that motoneuron firing rates are unaffected by contractile slowing induced by either cooling or changes in muscle length in the absence of fatigue (Bigland-Ritchie et al., 1986c, 1989). Presumably some other fatigue-induced change within the muscle must be responsible—perhaps some metabolic factor that itself changes in relation to contractile slowing. We are tempted to suggest that the reflex may respond to K^+ accumulation within the muscle, since this may well be the most important factor contributing to impaired e/c coupling and is obviously reduced if motoneuron firing rates are minimized. This, however, remains to be established.

It is not known which sensory endings and afferent pathways are involved in reflex inhibition of motoneuron firing rate (Garland and Mc-Comas, 1989). The small-diameter group III and IV muscle afferents seem obvious candidates since they have been shown to respond to many of the metabolic changes that accompany fatigue (Kniffki et al., 1978) and are known to have input to inhibitory interneurons within the spinal cord. However, microneurographic recordings by Hagbarth et al. (1986) suggest a role for changing input from Ia afferents which may reduce the tonic excitatory drive from muscle spindles. This is supported to some extent by more recent results from Gandevia et al. (1989), who recorded the usual decline in muscle spike train frequencies during fatigue but spike frequencies were not reduced after total nerve block distal to the recording site. Their evidence suggested changes in both excitatory and inhibitory input from muscles during fatigue.

VII. Fatigue of Respiratory Muscles

Most of these results and conclusions are drawn from experiments on hand or limb muscles. We also performed experiments in which the diaphragm was fatigued by the same 50% and 30% MVC contraction patterns shown in Figure 5. For these, the abdomen was strapped to minimize shortening, so that contractions were largely isometric. Diaphragmatic force was measured using Pdi and EMG electrodes. Throughout the exercise, single supramaximal shocks were periodically applied bilaterally to both phrenic nerves, to measure M-wave amplitudes and areas; changes in the amplitude of Pdi twitches were measured when these were superimposed on ongoing diaphragm contractions. Unfortunately, in these experiments, supramaximal tetanic stimulation of both phrenic nerves proved unreliable, but changes in diaphragmatic muscle activation by the CNS were assessed by the twitch occlusion method.

Exercise involving 50% MVC target force contractions could be continued for about 10 min before a maximal effort was required. Initially the diaphragm behaved like limb muscles, in that both the twitches elicited between contractions and those superimposed on target force voluntary contractions declined as expected. However, after the first 20% of the endurance time, the amplitudes of both types of twitches stabilized, and large superimposed twitches were still evident even at the limit of endurance when maximal efforts failed to reach the target Pdi force. Thus, maximal activation of the fatigued diaphragm seems impossible following this type of exercise, although complete twitch occlusion could always be achieved in the rested state.

From these experiments we concluded that about half the reduction in force that could be generated by the diaphragm was due to impaired diaphragm muscle contractility and the rest was due to a mandatory reduction of motor drive. This conclusion was also supported by the behavior of the integrated EMG, which at first increased as expected, indicating recruitment of additional motor units as those active previously became fatigued. However, when the twitch amplitudes stabilized, the EMG did not yet match that in control maximal contractions. From this point on the EMG either stabilized or started to decline. These results clearly differed from those found when the same subjects fatigued their quadriceps muscles by the same contraction protocol. In those experiments complete twitch occlusion could be achieved throughout the experiment even when the MVC force had declined by 50% or more. Thus there appear to be differences in the processes responsible for fatigue of respiratory muscles compared with most limb muscles, at least under these exercise conditions.

At first we found this difference surprising, since all other properties of the human diaphragm so closely resembled those recorded from other skeletal muscles, e.g., twitch contraction and half-relaxation times, twitch/tetanus ratios, and so forth. However, it can be explained if one postulates that, during fatigue, the same inhibitory reflex operates for the human diaphragm as for other skeletal muscles, but with increased gain. In this case, we suggest the reflex is sufficiently powerful to reduce motor unit firing rates to the point where the muscle can now be only partly activated. Such a mechanism might have important functional consequences in protecting the diaphragm from the extreme peripheral contractile failure that can easily be experienced by limb muscles, for it is vital that respiratory muscles be able to continue functioning adequately, often under substantially elevated metabolic loads, at times when limb muscles may be exhausted.

VIII. Implications for Neonatal Respiratory Muscle Failure

The fatigue resistance of the neonatal diaphragm cannot be predicted from its motor unit properties. Their uniformly slow contractile speed suggests high forces can be generated even if neural firing rates are low, but their low oxidative capacity implies poor fatigue resistance. Although the fatigue indices of neonatal rat diaphragms, in response to repeated 40-Hz pulse trains, were high and declined with age (see Chapter 13), this may not reflect their responses to naturally occurring neural discharge patterns. In addition, perhaps not all motor units can be effectively excited by the immature neural control networks of the newborn. Even in adults, comparison between responses to human voluntary effort and those of anesthetized animals to CO_2 loading suggests that maximal diaphragm activation cannot be achieved by excitatory drive emanating from respiratory centers alone, but may require additional input from corticospinal pathways. If, as for limb muscles, corticospinal connections are not yet operational, the neonatal diaphragm may be functionally weaker than indicated when stimulated directly, and the responses of the diaphragm in vivo may depend more on deficits of central motor drive than on the intrinsic fatigue properties of its constituent motor units. This may be a particular problem for the developing neonate since endurance times, before performance fails, are critically dependent on the relationship between muscle strength and the load of the task required (Rohmert, 1960; Bellemare and Grassino, 1982). In neonates the relative respiratory load is increased by its more compliant rib cage, compared to that of adults, a problem compounded during the first few weeks of development when the rib cage may grow faster than the diaphragm. Indeed, it has been suggested that this may explain why respiratory failure in human neonates is seen most often several weeks after birth (England et al., 1987).

IX. Summary and Conclusions

Information is slowly becoming available about important factors which may influence neonatal muscle properties, such as the rates of fiber type differentiation, rejection of multiple fiber innervation, changing safety factors for neuromuscular transmission, muscle and motoneuron membrane properties and so forth. But these are usually studied in isolated or whole animal preparations stimulated at rates not employed by the CNS in normal life. It is therefore important to know more about the neural discharge rates that occur naturally in both brainstem and spinal motoneurons of neonates, including the maximum rates elicited by stress, for it is these rates which most influence changes in muscle and motoneuron properties. Human studies also indicate that this is a reciprocal relationship, with changes in muscle properties also influencing neural discharge rates. Unfortunately, no methods have yet been developed suitable for studying respiratory neural discharge rates in intact humans. In the absence of contrary evidence, it seems reasonable to assume that changes in neural control mechanisms found during fatigue of limb muscles may also operate during fatigue of respiratory muscles. While detailed studies of individual events can probably only be studied in isolated preparations, for clinical purposes more attention should be given to how changes in each process influence those at other sites. Also, until more data are available, we should bear in mind that the properties of human muscles and how they are influenced by fatigue may not necessarily always parallel closely those seen in animal models.

References

Adrian, R. H., and Peachy, L. D. (1973). Reconstruction of the action potential of frog sartorius muscle. *J. Physiol.* **235**:103-131.

Allen, D. G., Lee, J. A., and Westerblad, H. (1989a). Intracellular calcium and tension during fatigue in isolated single muscle fibres from *Xenopus laevis*. *J. Physiol.* **415**:433-458.

Allen, D. G., Lee, J. A., and Westerblad, H. (1989b). The effects of fatigue on intracellular calcium measured with fura-2 in isolated single muscle fibres from *Xenopus*. *J. Physiol.* **414**:49P.

Bellemare, F., and Bigland-Ritchie, B. (1984). Assessment of human diaphragm strength and activation using phrenic nerve stimulation. *Respir. Physiol.* **58**:263-277.

Bellemare, F., and Bigland-Ritchie, B. (1987). Central components of diaphragmatic fatigue assessed from bilateral phrenic nerve stimulation. *J. Appl. Physiol.* **62**:1307-1316.

Bellemare, F., Bigland-Ritchie, B., and Woods, J. J. (1986). Contractile properties of the human diaphragm in vivo. *J. Appl. Physiol.* **61**:1153-1161.

Bellemare, F., and Grassino, A. (1982). Effect of pressure and timing on contraction on human diaphragm fatigue. *J. Appl. Physiol.* **53**:1190-1195.

Bigland-Ritchie, B. (1987). Respiratory muscle fatigue: Posters, methods and more. In *Respiratory Muscles and Their Neuromotor Control*. Edited by G. C. Sieck, S. C. Gandevia, and W. E. Cameron. New York, Alan R. Liss, pp. 379-389.

Bigland-Ritchie, B., Dawson, N. J., Johansson, R. S., and Lippold, O. C. J. (1986b). Reflex origin for the slowing of motoneurone firing rates in fatigue of human voluntary contractions. *J. Physiol.* **379**:451-459.

Bigland-Ritchie, B., Furbush, F., Gandevia, S. C., and Thomas, C. K. (1989). Firing rates of motor units of human tibialis anterior at different muscle lengths. *J. Physiol.* **409**:33P.

Bigland-Ritchie, B., Furbush, F., and Woods, J. J. (1986a). Fatigue of intermittent, submaximal voluntary contractions: Central and peripheral factors in different muscles. *J. Appl. Physiol.* **61**:421-429.

Bigland-Ritchie, B., Johansson, R., Lippold, O. C. J., Smith, S., and Woods, J. J. (1983b). Changes in motoneurone firing rates during sustained maximal voluntary contractions. *J. Physiol.* **340**:335-346.

Bigland-Ritchie, B., Johansson, R., Lippold, O. C. J., and Woods, J. J. (1983a). Contractile speed and EMG changes during fatigue of sustained maximal voluntary contractions. *J. Neurophysiol.* **50**:313-324.

Bigland-Ritchie, B., Jones, D. A., Hosking, G. P., and Edwards, R. H. T. (1978). Central and peripheral fatigue in sustained maximum voluntary contractions of human quadriceps muscle. *Clin. Sci. Mol. Med.* **54**: 609-614.

Bigland-Ritchie, B., Jones, D. A., and Woods, J. J. (1979). Excitation frequency and muscle fatigue: Electrical responses during human voluntary and stimulated contractions. *Exp. Neurol.* **64**:414-427.

Bigland-Ritchie, B., Kukulka, C. G., Lippold, O. C. J., and Woods, J. J. (1982). The absence of neuromuscular transmission failure in sustained maximal voluntary contractions. *J. Physiol.* **330**:265-278.

Bigland-Ritchie, B., and Woods, J. J. (1984). Changes in muscle contractile properties and neural control during human muscular fatigue. Invited review. *Muscle Nerve* **7**:691-699.

Bigland-Ritchie, B., Woods, J. J., Furbush, F., and Karrmann, D. (1986c). Effect of muscle temperature on motor neuron MVC firing rates. *Neurosci. Abstr.* **12**:680.

Burke, R. E., Levine, D. N., Tsairis, P., and Zajak, F. E. (1973). Physiological types and histochemical profiles in motor units of cat gastrocnemius. *J. Physiol.* **234**:723-748.

Calancie, B., and Bawa, P. (1986). Limitations of the spike-triggered averaging technique. *Muscle Nerve* **9**:78-83.

Clamann, P. H., and Robinson, A. J. (1985). A comparison of electromyographic and mechanical fatigue properties in motor units of the cat hindlimb. *Brain Res.* **327**:203-219.

Cooper, R. G., Edwards, R. H. T., Gibson, H., and Stokes, M. J. (1988). Human muscle fatigue: Frequency dependance of excitation and force generation. *J. Physiol.* **397**:585-599.

Edwards, R. H. T. (1981). Human muscle function and fatigue. In *Human Muscle Fatigue: Physiological Mechanisms.* Ciba Foundation Symposium. Edited by R. Porter and J. Whelan. New York, Pitman Press, pp. 1-18.

England, S. J., Guslits, B. G., and Bryan, A. C. (1987). Diaphragmatic function in newborn infants. In *Respiratory Muscles and Their Neuromotor Control.* Edited by G. C. Sieck, S. C. Gandevia, and W. E. Cameron. New York, Alan R. Liss, pp. 415-422.

Gandevia, S. C., Macefield, G., Burke, D., and McKenzie, D. K. (1990). Voluntary activation of human motor axons in the absence of muscle afferent feedback and the control of the deafferented hand. *Brain* **113** (in press).

Gandevia, S. C., and Rothwell, J. C. (1987). Activation of the human diaphragm from the motor cortex. *J. Physiol.* **384**:109-118.

Garland, S. J., and McComas, A. J. (1988). Role of small diameter afferents in reflex inhibition during fatigue of human soleus muscles. *Neurosci. Abstr.* **14**:62.

Gibson, H., Cooper, R. G., Stokes, M. J., and Edwards, R. H. T. (1988). Mechanisms resisting fatigue in isometric contracting human skeletal muscle. *Q. J. Exp. Physiol.* **73**:903-914.

Hagbarth, K., Kunesch, E. J., Nordin, M., Schmidt, R., and Wallin, E. U. (1986). Gamma loop contributing to maximal voluntary contractions in man. *J. Physiol.* **380**:575-591.

Hicks, J., Fenton, J., Garner, S., and McComas, A. J. (1988). M-wave potentiation during and after activity. *J. Appl. Physiol.* **66**:2606-2610.

Hultman, E., and Spriet, L. (1986). Skeletal muscle metabolism, contraction force and glycogen utilization during prolonged electrical stimulation in humans. *J. Physiol.* **374**:493-501.

Johansson, R. S., and Vallbo, Å (1983). Tactile sensory coding in the glabrous skin of the human hand. *Trends Neurosci.* **6**:27-32.

Jones, D. A., and Bigland-Ritchie, B. (1986). Electrical and contractile changes in muscle fatigue. *Biochem. Exercise* **14**:377-392.

Jones, D. A., Bigland-Ritchie, B., and Edwards, R. H. T. (1979). Excitation frequency and muscle fatigue: Mechanical response during voluntary and stimulated contractions. *Exp. Neurol.* **64**:401-413.

Juel, C. (1988). Muscle action potential propagation velocity changes during activity. *Muscle Nerve* **11**:714-719.

Kernell, D. (1965a). The adaptation and the relation between discharge frequencies and current strength of cat lumbosacral motoneurons stimulated by long lasting injected current. *Acta Physiol. Scand.* **65**:65-73.

Kernell, D. (1965b). The limits of firing frequency in cat lumbrosacral moto-
    neurones possessing different time course of after hyperpolarization.
    Acta Physiol. Scand. **65**:87-100.
Kernell, D., Danselaar, Y., and Eerbeck, O. (1987). Effects of physiologi-
    cal amounts of high- and low- rate chronic stimulation on fast-twitch
    muscle of cat hind limb. II. Endurance-related property. *J. Neurophy-
    siol.* **58**:614-627.
Kernell, D., and Monster, A. W. (1982a). Time course and properties of
    late adaptation in spinal motoneurones of the cat. *Exp. Brain Res.* **46**:
    191-196.
Kernell, D., and Monster, A. W. (1982b). Motoneurone properties and mo-
    tor fatigue. An intracellular study of gastrocnemius motoneurones of
    the cat. *Exp. Brain Res.* **46**:197-204.
Kniffki, K. D., Mense, S., and Schmidt, R. F. (1978). Responses of group
    IV afferent units from skeletal muscle to stretch, contraction and chem-
    ical stimulation. *Exp. Brain Res.* **31**:511-522.
Krnjevic, K., and Miledi, R. (1958). Failure of neuromuscular propagation
    in rats. *J. Physiol.* **140**:440-461.
Merton, P. A. (1954). Voluntary strength and fatigue. *J. Physiol.* **123**:553-
    564.
Miller, R. G., Giannini, D., Milner-Brown, H. S., Layzer, R. B., Koretsky,
    A. P., Hooper, D., and Weiner, M. W. (1987). Effects of fatiguing
    exercise on high-energy phosphates, force, and EMG: Evidence for three
    phases of recovery. *Muscle Nerve* **10**:810-821.
Rohmert, W. (1960). Ermittlung von Erholungspausen für statische Ar-
    beit des Menschen. *Int. Z. Angew. Physiol. Einschl. Arbeitphysiol.* **18**:
    123-164.
Sjøgaard, G. (1986). Water and electrolyte fluxes during exercise and their
    relation to fatigue. *Acta Physiol. Scand.* **128**(Suppl. 556):129-136.
Sjøgaard, G. (1988). Muscle energy metabolism and electrolyte shifts dur-
    ing low-level prolonged static contractions in man. *Acta Physiol. Scand.*
    134:181-187.
Stein, R. B., French, A. S., Mannard, D., and Yemm, R. (1972). New meth-
    ods for analysing motor function in man and animals. *Brain Res.* **40**:
    187-192.
Taylor, A., and Stephens, J. A. (1976). Study of human motor unit contrac-
    tions by controlled intramuscular microstimulation. *Brain Res.* **117**:
    331-335.
Thomas, C. K., Johansson, R. S., Westling, G., and Bigland-Ritchie, B.
    (1990a). Twitch properties of human thenar motor units measured in
    response to intraneural motor-axon stimulation. *J. Neurophysiol.* **64**:
    119-123.

Thomas, C. K., Ross, B. H., and Calancie, B. (1987). Human motor unit recruitment during isometric contractions and repeated dynamic movements. *J. Neurophysiol.* **57**:311-324.

Thomas, C. K., Bigland-Ritchie, B., Westling, G., and Johansson, R. G. (1990b). A comparison of human thenar motor-unit properties studied by intraneural motor-axon stimulation and spike-triggered averaging. *J. Neurophysiol.* **64**:124-127.

Thomas, C., Woods, J. J., and Bigland-Ritchie, B. (1989). Impulse propagation and muscle activation in long maximal voluntary contractions. *J. Appl. Physiol.* (in press).

Vøllestad, N. K., and Sejersted, O. M. (1989a). Plasma K$^+$ during exercise of various intensity in normal humans. *Biochem. Exercise* (in press).

Vøllestad, N. K., and Sejersted, O. M. (1989b). Plasma K$^+$ shifts in muscle and blood during and after exercise. *Int. J. Sports Med.* (in press).

Vøllestad, N. K., Sejersted, O. M., Bahr, R., Woods, J. J., and Bigland-Ritchie, B. (1988). Motor drive and metabolic responses during repeated submaximal voluntary contractions in man. *J. Appl. Physiol.* **64**:1421-1427.

Vøllestad, N. K., Wesche, J., and Sejersted, O. M. (1990). Gradual increase in leg oxygen uptake during repeated submaximal contractions in humans. *J. Appl. Physiol.* **68**:1150-1156.

Westling, G., Johansson, R. S., Thomas, C. K., and Bigland-Ritchie, B. (1990). Measurement of contractile and electrical properties of single human thenar motor axons in response to intraneural motor-axon stimulation. *J. Neurophysiol.* **64**:115-118.

Woods, J. J., Furbush, F., and Bigland-Ritchie, B. (1987). Evidence for a fatigue-induced reflex inhibition of motoneuron firing rates. *J. Neurophysiol.* **58**:125-136.

15

Integrated Response of the Respiratory Muscles to Load

ALIA R. BAZZY and JONATHAN D. FELDMAN

Yale University School of Medicine
New Haven, Connecticut

I. Introduction

Changes in lung volume with consequent gas exchange cannot occur without properly functioning respiratory muscles. These muscles constitute the effector organ, which receives processed signals from the central nervous system, following which they contract and act as the pump that drives the lungs to expand. The realization that these muscles may fail under certain conditions is fairly recent and has been demonstrated now by many investigators in different models in which the work of breathing has been increased. However, the mechanisms leading to respiratory muscle failure and the sites of failure remain unclear. Metabolic, neural, and mechanical factors have all been implicated in this problem, with evidence supporting each of these processes. In this chapter, we will review the various components involved in the response of the respiratory muscles to acute loads, and how these components can be modulated by chronic exercise such as endurance training.

This research was supported by grants from the National Heart, Lung and Blood Institute (HL 34467 and HL 01518) and The Meyer Foundation.

We will also review what is known about the response of the neonatal respiratory muscles to loading and the changes that occur with development.

II. Response to Acute Loads

A. Ventilatory Response

The ventilatory response to acute respiratory loads has been examined extensively. However, most of the available studies have focused on the initial response of the respiratory system to these loads, i.e., what has been termed "load compensation." Several mechanisms are involved in the initial compensation to loads. These mechanisms include: (1) those which increase force generation by the respiratory muscles, with or without an increase in the activity of respiratory motoneurons, and (2) mechanisms which lead to changes in respiratory timing (Cherniack and Milic-Emili, 1985). The first group of mechanisms depend on the intrinsic properties and geometry of the respiratory muscles, thereby allowing changes in length or velocity of contraction to lead to greater force generation. These mechanisms also depend on the response of the respiratory muscles to sensory input from mechanoreceptors. Changes in respiratory timing are mainly dependent on changes in chemical drive and on sensory input from mechanoreceptors (Cherniack and Milic-Emili, 1985).

One cannot necessarily predict from the load compensatory mechanisms what the response to a sustained ventilatory load might be. With sustained loads, chemical drive may change if there are accompanying decreases in arterial PO_2 and increases in arterial PCO_2. In fact, in anesthetized animals these changes may occur rapidly, over the first few breaths (Callahan and Read, 1974). Such an increase in drive leads to a progressive increase in respiratory neural efferent activity. Maintaining constant blood gas composition, as with breathing 100% oxygen or with cross perfusion, minimizes or abolishes the increase in respiratory activity (Orthner and Yamamoto, 1974).

Nonchemical factors play a role in the response to sustained ventilatory loads. For instance, Kelsen et al., have shown that during tracheal occlusions in anesthetized dogs, the increase in respiratory muscle activity was disproportionate to the increase in PCO_2 (Kelsen et al., 1976). Such changes may be due to spinal reflexes originating in the chest wall or from the J-receptors in the lung (Cherniack and Altose, 1981).

The level of consciousness of the subject affects these responses as well. Anesthesia diminishes the response to hypoxia and hypercapnia, probably secondary to a decrease in chemosensitivity (Cherniack and Altose, 1981). With constant chemical drive, inspiratory flow resistive loads do not affect the average rate of diaphragmatic electrical activity in anesthetized animals, whereas in conscious goats respiratory neural efferent activity increases independently of chemical drive (Isaza et al., 1976).

Because of our interest in the role of the respiratory muscles in maintaining ventilation during sustained loaded breathing, we have examined the ventilatory response to inspiratory flow resistive (IFR) loads in unanesthetized, chronically instrumented sheep (Sadoul et al., 1985). To avoid the hypoxia associated with these loads, inspired gas was supplemented with oxygen. In these experiments we found that whether the IFR load was moderate or severe, the initial response in the first 30 s consisted of a prolongation of inspiratory time (Ti). Subsequently, Ti returned toward, but did not reach, the values observed before loading. In contrast, expiratory time (Te) shortened throughout, with the greatest decrease seen in the first 30 s. As expected, tidal volume decreased, with the lowest values seen in the first 30 s, followed by a gradual slight recovery as loaded breathing was sustained. These changes led to a reduction in minute ventilation which was associated with hypercapnia, the degree of which depended on the severity of the load. With IFR loads greater than 150 cm H_2O/liter/s, a terminal increase in $PaCO_2$ (reaching up to 80-85 torr) was noted at the time when the animal had evidence of diaphragmatic fatigue (see below). Interestingly, this terminal hypercapnia was not associated with a further reduction in minute ventilation, but actually preceded it (Fig. 1). The reason for this discrepancy may be due to a switch to glycolytic metabolism leading to an increase in CO_2 production. On the other hand, recruitment of respiratory muscles other than the diaphragm may explain the increase in CO_2 production and the rise in $PaCO_2$.

B. Diaphragmatic Function

Although it has been well recognized that the respiratory muscles constitute the ventilatory pump (Roussos and Macklem, 1982), the fact that these muscles may fail in human subjects was not recognized until the late 1970s. At that time Roussos and Macklem reported that diaphragmatic fatigue could be induced in healthy adult humans in response to loaded breathing (Roussos and Macklem, 1977). Since then the study of respiratory muscle function has grown at both clinical and basic science levels. It is now clear that respiratory muscle fatigue, and in particular diaphragmatic fatigue (the diaphragm being the major inspiratory muscle), contributes to the development of respiratory failure in adult and newborn humans (Cohen et al., 1982; Lopes et al., 1981; Moxham et al., 1981; Muller et al., 1979a), as well as in experimental animals (Aldrich and Appell, 1989; Aubier et al., 1981b; Bazzy and Haddad, 1984; Robertson et al., 1977; Rochester and Bettini, 1976).

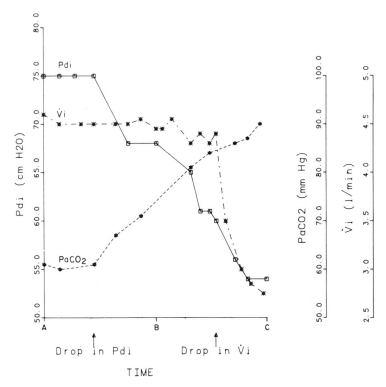

Figure 1 Time course of Pdi, minute ventilation (Vi), and Pa_{CO_2} during the last 20 min of severe IFR loaded breathing in one sheep. Note that Pa_{CO_2} increases when Pdi decreases, and that Vi decreases several minutes later. A, B, and C refer to the last 20, 10, and 1 min, respectively, before removal of the IFR load. (From Sadoul et al., 1985.)

As with skeletal muscles, diaphragmatic fatigue has been defined in a variety of ways. Currently, the two most utilized indices of diaphragmatic function are transdiaphragmatic pressure (Pdi), an index of force generation, and the electromyogram (EMG), an index of muscle activation. These indices were chosen based on data from limb skeletal muscle where fatigue has been defined as a deterioration in force-generating capability (Edwards, 1978) or as changes in the frequency components of the EMG (Lindstrom et al., 1977). Hence a decrease in Pdi after being maintained at a predetermined level for a period of time can be interpreted as the inability of the diaphragm to maintain force generation. However, it has also been recognized that changes in Pdi may be secondary to factors not related to fatigue.

For instance, a change in diaphragmatic muscle fiber length secondary to a change in thoracoabdominal configuration may lead to a decrease in Pdi. The effect of muscle fiber length on tension generation can be avoided by electrically stimulating the diaphragm isometrically via the phrenic nerve and measuring the Pdi generated in response to single twitches or various frequencies of stimulation, thus obtaining a better measure of diaphragmatic force-generating capacity. Although this method is widely used experimentally (Aubier et al., 1981a; Moxham et al., 1981; Watchko et al., 1986, 1987), it requires restriction of changes in thoracoabdominal configuration by application of binders or casts in addition to occlusion of the airway and therefore is not practical. In addition, our experience has been that awake animals may become agitated and try to spontaneously breathe during stimulation, which makes the interpretation of Pdi more difficult. Since, to our knowledge, this method has not been attempted in human infants, we do not know how valid the technique is in the neonate.

Analysis of the diaphragm EMG has been another means of assessing diaphragm function and has been performed in adult and neonatal humans (Cohen et al., 1982; Gross et al., 1979; Muller et al., 1979b; Schweitzer et al., 1979;), as well as in experimental animals during loaded breathing (Aubier et al., 1981b; Bazzy and Haddad, 1984; Watchko et al., 1986). This technique derives from studies of skeletal muscle which have shown that an increase in the low-frequency components of the EMG precedes the decrease in force generation (Lindstrom et al., 1977). Hence a leftward shift in the EMG power spectrum has been considered as an index of impending skeletal muscle fatigue. Such a shift has been observed in adults and infants who have failed weaning from mechanical ventilation (Cohen et al., 1982; Muller et al., 1979b), in human volunteers subjected to loaded breathing (Gross et al., 1979), and in experimental animal models (Bazzy and Haddad, 1984) (Fig. 2). The meaning of this change as well as the mechanism(s) that leads to it remains elusive at present. In addition, changes in the EMG power spectrum can be influenced by problems unrelated to fatigue. We and others have shown that a change in muscle fiber length can lead to an increase in the low-frequency components of the EMG (Bazzy et al., 1985). Electrode position and interelectrode distance also influence the EMG power spectrum. These issues may take on special significance when surface electrodes are used, as is the case in many human studies.

In our own studies of diaphragmatic function, we have subjected unanesthetized, chronically instrumented sheep to IFR loads, measured Pdi, analyzed diaphragm EMG, and monitored arterial blood gas tensions. We have observed that when loads were severe, ventilatory failure ensued and was associated with a decrease in Pdi (Bazzy and Haddad, 1984; Sadoul et al., 1985). We also found that the EMG low-frequency components in-

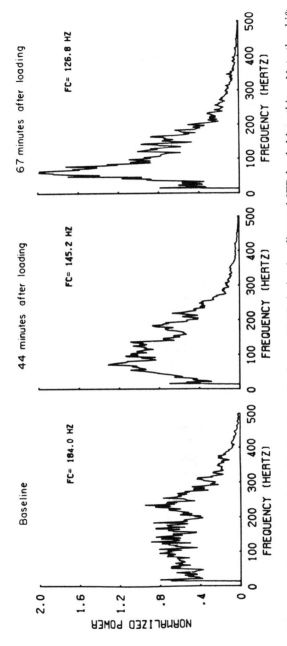

Figure 2 Power spectral density of a sheep diaphragm EMG during baseline and IFR loaded breathing. Note the shift to the left and decrease in centroid frequency (FC) at 44 and 67 min of loading. (From Bazzy and Haddad, 1984.)

creased prior to the drop in Pdi. Hence we have used a conglomeration of indices of diaphragmatic function to define fatigue, rather than a single index. These indices include: (1) a drop in Pdi after being maintained at high levels for prolonged periods of time, (2) a decrease in integrated EMG activity which precedes the drop in Pdi, (3) a leftward shift to lower frequencies in the EMG power spectrum, (4) a rightward shift or decrease in Pdi generated at various frequencies during phrenic nerve stimulation, and finally (5) a terminal rise in PaCO$_2$ indicating ventilatory failure secondary to pump failure.

C. Metabolic Basis of Diaphragmatic Fatigue

It has been hypothesized that when the energy demand of the working diaphragm exceeds its energy supply, fatigue will ensue (Roussos and Macklem, 1982). To test this hypothesis several investigators have examined various mechanisms involved in the delivery and utilization of energy-producing fuels by the diaphragm. These mechanisms primarily involve changes in blood flow, with resulting changes in the delivery of nutrients to the muscle and elimination of waste products, and the utilization of blood-borne and tissue energy substrates by the muscle itself.

Blood Flow

Diaphragmatic circulation has recently been shown to consist of a complex network of arteries and veins with elaborate anastomoses between them (Comtois et al., 1987). This circular pattern, which may be compared to the circle of Willis in the brain, may help to ensure delivery of blood to all areas of the diaphragm even at high work loads, when localized vascular occlusion may result from muscular contraction (Bark et al., 1987; Bellemare et al., 1983). The magnitude of this effect has been shown to be a function of the duration of the tension generated during the contraction, i.e., the duty cycle. Since the amount of energy expended is also directly proportional to the force generated, Bellemare and Grassino have studied an index known as the tension time index, which consists of the product of the duty cycle and Pdi/Pdi max. They have found that endurance time of the diaphragm is inversely related to this index (Bellemare and Grassino, 1982). In subsequent work Bellemare and co-workers have shown that the decrement in blood flow resulting from compression during diaphragmatic contraction is almost linearly related to the tension time index, raising the possibility that fatigability of the diaphragm may be due to compromised blood flow at high levels of contraction (Bellemare et al., 1983). However, in vivo measurements of blood flow during loaded breathing (Robertson et al., 1977; Rochester and Bettini, 1976), including work from our own laboratory (Pang et al., 1987), shows that blood flow increases as the intensity of the load increases, and that, at least in our animal model, blood flow to the diaphragm and other respiratory muscles remains elevated when Pdi decreases. These findings, coupled with the anatomical data, suggest that the diaphragm is protected from a reduction in energy supply when the work of breathing is increased.

The question remains as to what is the important signal which regulates blood flow to the diaphragm. Whether it is blood-borne substrates, oxygen supply, or other factors remains unclear from available data. In recent experiments, we explored the possibility that oxygen availability may limit the performance of the diaphragm during loaded breathing (Bazzy et al., 1989). Interestingly, we found that phrenic venous PO_2 increased in these animals before and during the time when Pdi was decreasing (Fig. 3). This is in contrast to reports from skeletal muscle showing that PO_2 in the venous effluent decreases to very low levels during exercise (Thomson et al., 1982; Tibes et al., 1977). In addition, we found no increase in arterial or phrenic venous lactate. These results indicated to us that the diaphragm does not become hypoxic during IFR loaded breathing and that the blood supply of O_2 is at least sufficient for, if not in excess of, the demands of this diaphragm. An excessive delivery of oxygen may be necessary to deliver other energy substrates such as glucose or free fatty acids. To our knowledge the utilization of these substrates by the working diaphragm has not yet been explored.

Tissue Substrates

In skeletal muscle, glycogen has been shown to be an important source of energy during exercise (Bergstrom and Hultman, 1967; Felig and Wahren, 1975). Prolonged and exhaustive exercise has been associated with glycogen depletion (Felig and Wahren, 1975). Since the diaphragm is basically a skeletal muscle, the utilization of glycogen by the diaphragm has been the subject of many studies (Bazzy et al., 1988; Fregosi and Dempsey, 1986; Green et al., 1987; Ianuzzo et al., 1987; Moore and Gollnick, 1982). Most of the models, however, used whole body exercise as a means of increasing the work of breathing, which may explain the lack of consensus in the results, i.e., showing either a decrease or no change in the glycogen content of the diaphragm. In experiments where we used IFR loaded breathing, thereby targeting the exercise specifically at the diaphragm, we found that glycogen content did in fact decrease in the diaphragm as well as in the intercostal muscles (Bazzy et al., 1988) (Fig. 4). However, this decrease was not to the level of near-total depletion, as has been observed in fatiguing skeletal muscle. Glycogen content remained at about 40% of control values, which is higher than has been reported for skeletal muscle (Gollnick et al., 1973). To determine whether glycogen depletion was selective to a specific fiber

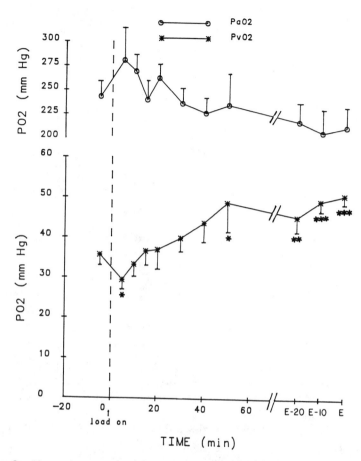

Figure 3 Change in mean arterial PO_2 (PaO_2) and phrenic venous PO_2 (PvO_2) during IFR loaded breathing in 11 sheep. Note that PvO_2 begins to increase after about 40 min of loaded breathing and continues to rise, significantly exceeding baseline levels. x-axis: E – 20, E – 10, and E refer to the last 20, 10 and 1 min, respectively, before removal of the load, *** $P < 0.01$. Bar = 1 SD. (From Bazzy et al., 1989.)

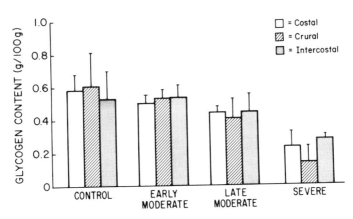

Figure 4 Mean glycogen content in the diaphragm (costal and crural) and inter-costal muscles (internal and external) in control sheep (n = 6), sheep subjected to moderate IFR loads of different durations (early: short duration, 10-20 min, n = 4; late: long duration, 75 min, n = 4) and severe IFR loads (n = 4). Note that glycogen content decreases in all muscles as the duration and severity of loading increase. The content in all muscles from animals with severe loads is statistically significantly different from control ($P < 0.05$) Bar = 1 SD. (From Bazzy et al., 1988.)

type, we examined glycogen content in type I and II fibers, identified by the histochemical reaction for myosin ATPase at pH 9.4. In general, type I fibers correspond to slow-twitch fibers, whereas type II fibers are fast-twitch fibers. We found that most of the glycogen depletion was in type II fibers in the diaphragm, with a considerable number of type I fibers remaining full of glycogen. The latter finding is not consistent with most studies from skeletal muscle, which show that glycogen depletion is primarily in type I fibers or in both type I and II fibers, depending on the type and intensity of the exercise (Gollnick et al., 1972a, 1973, 1974a, b). Possible reasons for this discrepancy could be either that the recruitment pattern of fibers is different in the diaphragm, with type II fibers being recruited early during loaded breathing, or that glycogen is not an important energy substrate for type I fibers during IFR loaded breathing. This is supported by the finding that the oxidative capacity in the diaphragm is higher than in the other skeletal muscle (Farkas and Roussos, 1984; Keens et al., 1978a) and therefore may rely on other energy sources, such as circulating lipids or glucose. Additionally, it has been shown that in skeletal muscle, prolonged electrical stimulation of fibers with high oxidative capacity does not deplete them of glycogen, whereas low oxidative capacity fibers are easily depleted (Kugelberg and Edstrom, 1968).

D. Neural Output and Neuromuscular Transmission in Response to Load

Analysis of the electrical activity in the phrenic nerve has been used as an index of change in the neural drive to the diaphragm (Evanich et al., 1976). By correlating changes in the phrenic neurogram with changes in the diaphragm electromyogram during loaded breathing, one may be able to deduce whether a decrease in muscle force generation is related to a decrease in central neural drive, neuromuscular transmission failure, or a peripheral failure in the muscle itself (e.g., excitation contraction coupling or contractile failure). Sites of muscle fatigue may also be examined by comparing the tension generated by the muscle when stimulated directly to the tension generated when the muscle is stimulated via its supplying nerve. Although central drive cannot be examined using this technique, neuromuscular transmission failure can be differentiated from contractile failure. Both these methods have been used in vivo and in vitro in an attempt to understand the sites where respiratory muscle fatigue may occur. For instance, it has been shown that in the rat phrenic nerve hemidiaphragm preparation, neuromuscular transmission failure can be elicited following indirect muscle stimulation via the phrenic nerve (Aldrich et al., 1986; Kelsen and Nochomovitz, 1982; Kuei et al., in press). However, the various studies do not agree on the relative contribution of neural transmission failure (i.e., failure up to and including the neuromuscular junction) versus contractile failure (beyond the junction). Estimates of the contribution of the former vary between 16% and 82%, depending on the study conditions and the stimulation frequency.

The findings from in vitro studies contrast with available in vivo data. Whereas Aldrich has shown in recent work that neuromuscular transmission failure contributes to the development of diaphragmatic fatigue in response to IFR loaded breathing (Aldrich, 1987), other studies do not support Aldrich's findings (Aubier et al., 1981b; Watchko et al., 1986, 1987). In a model of cardiogenic shock, Aubier et al. found that both the phrenic neurogram and the diaphragm EMG continued to increase while Pdi decreased, attributing diaphragmatic fatigue to contractile failure in the muscle itself (Aubier et al., 1981b). Watchko and colleagues found no evidence of neuromuscular transmission failure in piglets. They based their conclusions on data from anesthetized piglets subjected to hypoxia and hypercapnia without added respiratory loads. Lack of transmission failure was demonstrated by electrical stimulation of the diaphragm (Watchko et al., 1986, 1987). Clearly, the major differences in the models used may explain the differences in the results, especially with the presence of confounding factors such as a low perfusion state and anesthesia.

In our animal model of diaphragmatic fatigue in awake sheep, we have been intrigued by the observation that the integrated activity of the diaphragm EMG decreased before Pdi decreased. This suggested that either a decrease in central neural output or neuromuscular transmission failure may contribute to diaphragmatic fatigue. To explore these possibilities, we studied phrenic neural activity during IFR loaded breathing in awake sheep using chronically implanted nerve cuff electrodes. We found that while the integrated EMG was decreasing, phrenic nerve integrated activity continued to increase. This increase persisted when Pdi was decreasing. We interpreted these data to suggest that in the unanesthetized sheep subjected to IFR loaded breathing, neuromuscular transmission failure did in fact play a role in the development of diaphragmatic fatigue, and that this failure preceded muscular contractile failure (Bazzy et al., in press). This is now an area of active investigation in our laboratory.

III. How Respiratory Muscle Training Modulates the Response of Respiratory Muscles to Acute Loads

Since a possible consequence of respiratory muscle fatigue is respiratory failure with accompanying high morbidity and mortality, an important question has been whether the development of respiratory muscle fatigue can be prevented by improving muscle strength or endurance. A wealth of information indicates that skeletal muscles in general display tremendous plasticity, and that their endurance and strength can be improved by training, i.e., by exposure to a chronic load. These studies also show that an increase in the oxidative capacity of the muscle is associated with the functional improvement in endurance. Based on this knowledge, the potential for training the respiratory muscles began to be explored in both human subjects and experimental animal models (Pardy et al., 1988). In the following sections we will review how respiratory muscle training affects function, metabolic profile, and neuromuscular transmission in the respiratory muscles.

A. Functional Adaptation of the Respiratory Muscles to Chronic Loads

The functional response to chronic loading varies with the type, intensity, and duration of the stimulus used to train the muscle. In this respect, it is important to differentiate whether the stimulus leads to an improvement in the strength of the muscle, i.e., its ability to generate greater force for a given excitation, or to better endurance, i.e., its ability to maintain a given level of force generation for longer periods of time. Clinical studies have used both types of training and have demonstrated that respiratory muscle strength can be increased in normal subjects and in patients with obstructive lung disease with a stimulus that is specifically targeted at the respiratory muscles (Pardy et al., 1988). Endurance training has been attempted using both specific and nonspecific programs. The nonspecific programs exercise the respiratory muscles by achieving an increase in ventilation using whole body exercise such as running or swimming (Belman and Kendergan, 1981; Keens et al., 1977; Orenstein et al., 1981). The improvement in respiratory muscle function as a consequence of this has been variable, probably because the intensity of the stimulus was not sufficient in all cases to induce a load effect on the muscles. In addition, ventilatory limitations in patients with lung disease may preclude the increases in ventilation necessary to induce a training effect. With specific programs such as voluntary isocapnic hyperpnea (Belman and Mittman, 1980; Keens et al., 1977; Leith and Bradley, 1976), IFR loading (Asher et al., 1982; Belman et al., 1986; Pardy et al., 1981), or inspiratory threshold loading (Clanton et al., 1985), more consistent improvement in endurance of the respiratory muscles has been noted. This improvement, however, has usually been measured as changes in tests such as the maximal sustained voluntary contraction (MSVC) (which is the maximal level of ventilation that can be sustained under isocapnic conditions for 15 min) or a change in the level of resistance that can be tolerated (Pardy et al., 1988). These indices are not direct measures of respiratory muscle function, and they may be affected by factors such as breathing pattern (Belman et al., 1986), severity of disease, or the presence or absence of hypoxia or hypercapnia. In addition, these studies do not shed any light on the mechanisms involved in the response to training in the respiratory muscles.

Because of the limitations of the human studies, animal models have been developed to directly examine: (1) changes in diaphragmatic function by a chronic load and (2) the biochemical and structural correlates of this training process. The three main models that have been used are whole body exercise (Ianuzzo et al., 1982; Metzger and Fitts, 1986; Moore and Gollnick, 1982), tracheal banding (Keens et al., 1978b), and emphysema (Farkas and Roussos, 1982). These studies again show variable results with respect to improvement in diaphragmatic function. With whole body exercise no improvements in contractile properties and endurance were noted, probably because the training stimulus to the diaphragm was insufficient. Chronic loading targeted to the respiratory muscles, as with tracheal banding and emphysema, did lead to changes in contractile properties of the muscle. However, these respiratory loads were associated with high morbidity and mortality, which may have been secondary to accompanying hypoxemia and acidosis. These confounding factors can have major effects on muscle function and metabolism (Fregosi and Dempsey, 1986; Schnader et al., 1979), making interpretation of the results difficult. We have recently applied IFR loads intermittently over a 3-week period to adult sheep, thereby targeting the load specifically to the respiratory muscles (Akabas et al., 1989). We also adjusted these loads so that the animals did not become hypoxemic or acidemic during the training sessions. To determine whether diaphragmatic performance improved as a result of this regimen, we subjected the sheep to the same severe IFR load prior to and 2 days following the 3-week period. Interestingly, we found that for the same load, each animal was able to generate higher Pdi (20-40% higher) following training, and that this high Pdi could be sustained for 1/2-2 h longer than pretraining before finally decreasing. In addition, $PaCO_2$ levels were 10-40% lower in the posttraining study than in the pretraining study, therefore showing not only an improvement in diaphragmatic endurance (and possibly strength), but also an improvement in ventilation.

B. Metabolic Adaptations to Chronic Loads

In skeletal muscle, the specific metabolic adaptation to chronic loading depends on the pattern of recruitment during the activity, as well as on the energy requirements of that activity. In strength training, such as weight lifting or sprinting, short, intensive bursts of energy are required. Since muscle strength is directly related to cross-sectional area, an increase in myofibrillar protein leading to muscle hypertrophy is observed. The energy source for such activities is primarily high-energy phosphagens stored within the muscle, the concentrations of which do not appear to be significantly altered with exercise (Karlsson et al., 1972). Enzymes of glycolysis may be increased, as in sprinting for instance (Costill et al., 1976). When an increase in endurance of the muscle is the goal, the level of activity is lower than in strength training but the duration is longer. The ability to perform these kinds of activities requires a supply of energy for prolonged periods of time without accumulating toxic metabolites. This entails an increase in mitochondrial enzymes involved in oxidative metabolism, such as citrate synthase, 3-hydroxyacyl CoA dehydrogenase (HAD), succinate dehydrogenase, and cytochrome oxidase (Bylund-Fellenius et al., 1977; Costill et al., 1976; Gollnick et al., 1972a). The increase in enzymatic activity is associated with an increase in mitochondrial surface area (Gollnick and King, 1969). In addition, capillary density increases, probably to ensure better delivery of energy substrates to individual fibers (Andersen and Henriksson, 1977; Klausen et al., 1981).

The physiological consequences of such specific metabolic adaptations can be appreciated. Following strength training, the increase in muscle fiber cross-sectional area and hence muscle strength allows the subject to generate greater force more easily, without requiring excessive energy expenditure.

Following endurance training, the capacity to utilize lipids while sparing glycogen is enhanced. The importance of this lies in the observation that in tasks requiring prolonged endurance, glycogen depletion in the muscle has been correlated with muscle fatigue (Bergstrom and Hultman, 1967). The ability to spare glycogen and utilize lipids has been found to be associated with prolonged endurance (Hermansen et al., 1967).

The few studies that have examined the metabolic profile of the diaphragm suggest that the oxidative capacity of the diaphragm is higher than that of limb muscle of similar fiber type composition (Farkas and Roussos, 1982; Keens et al., 1978a). The implication of this finding is that the diaphragm may rely on oxidation of free fatty acids for energy, sparing intramuscular glycogen, thus behaving as an endurance-trained skeletal muscle. This raises the question of whether additional increases in the activities of the enzymes involved in oxidative metabolism are possible. Although there are only a few reports describing the effects of training on respiratory muscle metabolism, models that have targeted the exercise to the respiratory muscles demonstrate an increase in the activity of oxidative enzymes (Farkas and Roussos, 1982; Keens et al., 1978a). Interestingly, if the load used for training is very severe, glycolysis can also be enhanced (Ianuzzo et al., 1982). These models, however, were complicated by many variables, such as lung disease, weight loss, high morbidity and mortality rates of the animals, and probable hypoxemia and acidosis. Therefore, whether these enzymatic changes were a direct result of the training stimulus or a consequence of the complications of the models is unclear. We performed similar analyses in our sheep model of respiratory muscle training with IFR loads and found that, in fact, the improvement in diaphragmatic function was associated with significant increases in the activities of enzymes of oxidative metabolism (citrate synthase, HAD, and cytochrome oxidase). We found no increase in phosphofructokinase, the key enzyme in glycolysis (Akabas et al., 1989) (Fig. 5). We also found that these changes were induced in both type I and II fibers, the main histochemical fiber types in the sheep diaphragm (Bazzy and Kim, submitted). This conclusion was reached by analyzing the histochemical reaction for cytochrome oxidase activity in individual fibers using computerized image-processing techniques. Hence, we believe that although the oxidative capacity of the diaphragm may be high, it can be further increased by the appropriate stimulus, and that this is associated with improvement in function.

C. Neuromuscular Transmission and Chronic Loads

While much is known about the effects of endurance training on the motoneuron and the muscle fiber, very little is known about its effect on the neuromuscular junction. Numerous studies of skeletal muscles under various

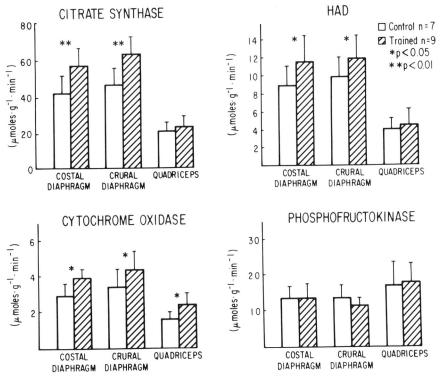

Figure 5 Mean enzymatic activities in the costal and crural diaphragm and quadriceps of control (n = 7) and trained (n = 9) sheep. Note that citrate synthase, HAD, and cytochrome oxidase were higher in the diaphragm of trained animals, whereas phosphofructokinase did not increase. Bar = 1 SD, *P <0.05, ** P <0.01. (From Akabas et al., 1989.)

conditions have suggested that differences in neuromuscular junction biochemistry (Lomo and Slater, 1980; Snyder et al., 1973; Washio et al., 1987), morphology (Fahim and Robbins, 1982; Robbins and Fahim, 1985; Rosenheimer and Smith, 1985; Tuffery, 1971; Wernig et al., 1980; Wigston, 1989), transmitter release (Gertler and Robbins, 1978; Robbins and Kelly, 1983; Vyskocil and Gutmann, 1972;), and neonatal development (Benoit and Changeux, 1975; O'Brien et al., 1978; Riley, 1978) are due to differences in muscle activity. For instance, Crockett and co-workers have examined the effect of endurance training on the metabolism of acetylcholine, acetylcholinesterase (AChE), and choline acetyltransferase activities in the neuromuscular junctions of slow-twitch oxidative or type I (soleus), fast-twitch oxidative, glycolytic or type II A (red vastus lateralis) muscles in the rat (Crockett et al., 1976). The only muscle in which a significant change was induced by the training was the fast-twitch glycolytic muscle, as evidenced by an increase in end-plate AChE. This contrasts with studies of muscle fiber oxidative metabolism which show endurance training to have the least absolute effects on this fiber type (Baldwin et al., 1972; Holloszy and Booth, 1976). Gardiner et al. studied the effect of endurance training on the recovery of end-plate AChE activity after its reduction by prior hindlimb immobilization (Gardiner et al., 1982). Four weeks after remobilization, end-plate AChE activity in the exercise group had returned to control values while that in the sedentary group was only 84% of control.

In recent years, a number of groups have begun to investigate the relationship between endurance training and neuromuscular junction morphology. Stebbins et al. trained rats by treadmill running at a level estimated to require ≤80% VO_2 max (Stebbins et al., 1985). Compared to control, the exercised soleus had a significantly higher percentage of junctions showing growth configuration, defined as terminal sprouts (i.e., unmyelinated outgrowths from a node of Ranvier) terminating in the parent end-plate. The percentage of junctions showing growth configuration was unchanged in the gastrocnemius. This was an interesting finding as the soleus (a slow-twitch tonic muscle) would be expected to be highly recruited at this level of training while the gastrocnemius (a mixed phasic muscle) would not. These data apparently confirmed previous suggestions that growth configurations are stimulated by increased muscular work (Cardasis, 1983; Cardasis and Padykula, 1981; Tuffery, 1971).

Chronic stress has been shown to be associated with decreased numbers of nerve terminal branches per neuromuscular junction in rat soleus (slow-twitch), extensor digitorum longus (EDL) (fast-twitch), and diaphragm (fast-twitch) (Rosenheimer, 1985). Treadmill running, superimposed on chronic stress, prevented the decrease in terminal branching for hindlimb muscles but accentuated it in the diaphragm. Andonian and Fahim studied the effect of treadmill running on the junction morphology of soleus and extensor digitorum longus muscle in mice from 12 to 24 months of age (Andonian and Fahim, 1987). In 12-month-old animals, exercise resulted in larger-than-normal neuromuscular junctions in both muscles, while in 18- and 24-month-old animals, exercise resulted in no change of soleus junctions and a reduction in the size of EDL junctions. In a separate study of rats (Andonian and Fahim, 1988), these investigators found that endurance training caused an increase in nerve terminal area in both soleus and EDL muscle and increases in length, terminal branches per junction, and percentage of junctions with growth configurations in the EDL only. These findings were associated with no changes in muscle fiber diameter. Thus, it is apparent that many variables, including muscle fiber type, the muscle's physiological function, species, and age, impact on the effect of training on a specific population of neuromuscular junctions.

In a unique study, Appell simulated endurance training of the diaphragm by subjecting Japanese waltzer mice to 1-2 weeks of hypobaric hypoxia (Appell, 1984). He then evaluated neuromuscular junctions from the sternal portion of the diaphragm and found their mean area to be increased from control of 218 ± 87 μm to 414 ± 174 μm at 7 days and 313 ± 104 μm at 14 days. The fascinating finding was that the mean number of junctions increased from 612 ± 177 in control animals and 628 ± 140 at 7 days to 862 ± 128 at 14 days. This remains the only study to demonstrate an increase in neuromuscular junction number with endurance training in the diaphragm.

In studies of nonmammalian systems, there is evidence of dramatic, yet reversible, changes in the amounts of vesicular and axolemmal membrane, vesicle number, and terminal sprouting after relatively short periods of increased muscle activity (Palacios-Pru et al., 1983; Palacios-Pru and Cola-sante, 1988; Reinecke and Walther, 1981). In a mammalian system (rat), Pockett and Pygott studied the effect of 6-13 days of chronic sciatic nerve stimulation on the quantal content of synaptic vesicles in EDL junctions. They found a slight increase in the mean quantal content in the stimulated nerve terminals (Pockett and Pygott, 1982).

Thus, there is good evidence that endurance training has effects on neuromuscular junction enzyme activity and morphology. There is some evidence that endurance training has effects on the number of junctions as well as on neurotransmitter release properties, but these have not been well studied. In addition, whether similar changes would be seen in the diaphragm following endurance training remains to be determined.

IV. The Developing Respiratory Muscles

The response of the respiratory muscles to loaded breathing in the neonatal period may differ significantly from the response of the adult muscle. However, it is not clear at this time what the difference may be. Data related to the properties and fatigability of the respiratory muscles in the newborn have been controversial. Whereas some biochemical and histochemical studies suggest that fatigable fibers predominate in early life (Keens et al., 1978a; Mayock et al., 1987a), other studies suggest that fatigue resistance decreases with increasing age (Maxwell et al., 1983; Sieck and Blanco, 1987). There are other properties of the respiratory system in the neonate which differ from the adult and may affect the response to load. For instance, the rib cage in the newborn is notorious for its high compliance, therefore providing little support of the chest wall for diaphragmatic action (Fisher and Mortola, 1980; Gerhardt and Bancalari, 1980). In addition, in contrast to the adult, horizontal arrangement of the ribs does not allow them to be raised in order to increase lung volume. The newborn infant spends most of the time supine, a position that has been shown in adults to lead to up to a 15% decrease in functional residual capacity as compared to the upright position (Agostoni et al., 1965). A further confounding issue is the fact that newborn infants spend a significant proportion of the day in REM sleep. During this behavioral state, intercostal muscle activity appears to be inhibited, leading to an increase in the work done by the diaphragm per minute ventilation (Muller et al., 1979a), as well as to a further decrease in lung volume (Henderson-Smart and Read, 1979).

The response to acute loads has been examined in anesthetized piglets (Mayock et al., 1987a). In these studies, newborn piglets subjected to inspiratory flow resistive loads were found to have a decrease in Pdi/EMG ratio after 1 h of loaded breathing, indicating that the muscle contractile mechanisms may be failing. The cause of the decrease in Pdi could have been the accompanying hypercapnia, and this was addressed in a second set of experiments in very young and older piglets (Watchko et al., 1987). The findings from these experiments suggested that older piglets tolerate loads more than the younger ones. In contrast, when the effect of hypoxemia was examined in these piglets, diaphragmatic function was affected more in the older than younger animals (Watchko et al., 1986). Whether similar effects of blood gas perturbations are seen in sick human infants is not known.

In summary, the respiratory muscles play a major role in the response of the ventilatory system to loaded breathing. Although information is available on some of the factors that determine how these muscles will respond to stress, several questions remain unanswered, especially with regard to the developing muscle, its susceptibility to failure, and the sites where it may fail. It is of prime importance that we explore these questions to be able to understand the mechanisms involved in the adaptation of respiratory muscles to stress. Without such an understanding, we will not be able to treat and prevent respiratory muscle dysfunction in the neonatal period, an age when respiratory problems remain a major cause of morbidity and mortality.

References

Agostoni, E., Mognoni, P., Torri, G., and Saracino, F. (1965). Relation between changes in rib cage circumference and lung volume. *J. Appl. Physiol.* **20**:1179-1186.

Akabas, S. R., Bazzy, A. R., DiMauro, S., and Haddad, G. G. (1989). Metabolic and functional adaptation of the diaphragm to training with resistive loads. *J. Appl. Physiol.* **66**:529-535.

Aldrich, T. K. (1987). Transmission fatigue of the rabbit diaphragm. *Respir. Physiol.* **69**:307-319.

Aldrich, T. K., and Appel, D. (1989). Diaphragm fatigue induced by inspiratory resistive loading in spontaneously breathing rabbits. *J. Appl. Physiol.* **59**:1527-1532.

Aldrich, T. K., Shander, A., Chaudhry, I., and Nagashima, H. (1986). Fatigue of the isolated diaphragm: Role of impaired neuromuscular transmission. *J. Appl. Physiol.* **61**:1077-1083.

Andersen, P. J., and Henriksson, J. (1977). Capillary supply of the quadriceps femoris muscle of man: Adaptive response to exercise. *J. Physiol. (London)* **270**:677-690.

Andonian, M. H., and Fahim, M. A. (1987). Effects of endurance exercise on the morphology of mouse neuromuscular junctions during aging. *J. Neurocytol.* **16**:589-599.

Andonian, M. H., and Fahim, M. A. (1988). Endurance exercise alters the morphology of fast- and slow-twitch rat neuromuscular junctions. *Int. J. Sports Med.* **9**:218-233.

Appell, H. J. (1984). Proliferation of motor end-plates induced by increased muscular activity. *Int. J. Sports Med.* **5**:125-129.

Aubier, M., Farkas, G. DeTroyer, A., Mozes, R., and Roussos, C. (1981a). Detection of diaphragmatic fatigue in man by phrenic stimulation. *J. Appl. Physiol.* **50**:538-544.

Aubier, M., Trippenbach, T., and Roussos, C. (1981b). Respiratory muscle fatigue during cardiogenic shock. *J. Appl. Physiol.* **51**:499-508.

Baldwin, K. M., Klinkerfuss, G. H., Terjung, R. L., Mole, P. A., and Holloszy, J. O. (1972). Respiratory capacity of white, red, and intermediate muscle: Adaptive response to exercise. *Am. J. Physiol.* **222**:373-378.

Bark, H., Supinski, G., LaManna, J. C., and Kelsen, S. G. (1987). Relationship of changes in diaphragmatic muscle blood flow to muscle contractile activity. *J. Appl. Physiol.* **62**:291-299.

Bazzy, A. R., Akabas, S. R., Hays, A. P., and Haddad, G. G. (1988). Respiratory muscle response to load and glycogen content in type I and II fibers. *Exp. Neurol.* **101**:17-28.

Bazzy, A. R., and Haddad, G. G. (1984). Diaphragmatic fatigue in unanesthetized adult sheep. *J. Appl. Physiol.* **57**:182-190, 1984.

Bazzy, A. R., and Kim, Y. J. Effect of chronic respiratory loads an oxidative capacity of the diaphragm. *J. Appl. Physiol.* (submitted).

Bazzy, A. R., Kim, Y. J., Donnelly, D. F., and Haddad, G. G. Neuromuscular transmission failure can be a mechanism for diaphragmatic fatigue in vivo. *Am. Rev. Respir. Dis.* (in press).

Bazzy, A. R., Pang, L. M., Akabas, S. R., and Haddad, G. G. (1989). O_2 metabolism of the sheep diaphragm during flow resistive loaded breathing. *J. Appl. Physiol.* **66**:2305-2311.

Bellemare, F., and Grassino, A. (1982). Effect of pressure and timing of contraction on human diaphragm fatigue. *J. Appl. Physiol.* **53**:1190-1193.

Bellemare, G., Wight, D., Lavigne, C. M., and Grassino, A. (1983). Effect of the tension and timing of contraction on the blood flow of the diaphragm. *J. Appl. Physiol.* **54**:1597-1606.

Belman, M. J., and Kendergan B. A. (1981). Exercise training fails to increase skeletal muscle enzymes in patients with chronic obstructive pulmonary disease. *Am. Rev. Respir. Div.* **123**:256-261.

Belman, M. J., and Mittman, M. J. (1980). Ventilatory muscle training improves exercise capacity in chronic obstructive pulmonary disease patients. *Am. Rev. Respir. Dis.* **121**:273-280.

Belman, M. J., Thomas, S. F., and Lewis, M. I. (1986). Resistive breathing training in patients with chronic obstructive pulmonary disease. *Chest* **90**:663-670.

Benoit, P., and Changeux, J-P. (1975). Consequences of tenotomy on the evolution of multiinervation in developing rat soleus muscle. *Brain Res.* **99**:354-358.

Bergstrom, J., and Hultman, E. (1967). A study of glycogen metabolism during exercise in man. *Scand. J. Clin. Lab. Invest.* **19**:218-228.

Bryan, A. C., and Gaultier, C. (1985). Chest wall mechanics in the newborn. In *The Thorax*. Lung Biology in Health and Disease, Vol. 29. Edited by C. Roussos and P. T. Macklem. New York, Marcel Dekker, pp. 871-888.

Bylund-Fellenius, A. C., Bjuro, T., Cederblad, G., Holm, J., Lundholm, K., Sjostrom, M., Angqvist, K. A., and Schersten, T. (1977). Physical training in man. Skeletal muscle metabolism in relation to muscle morphology and running ability. *Eur. J. Appl. Physiol. Occup. Physiol.* **36**:151-169.

Callahan, D., and Read, D. J. C. (1974). The role of arterial chemoreceptors in the breath-by-breath augmentation of inspiratory effort in rabbits during airway occlusion or elastic loading. *J. Physiol. (London)* **241**:33-44.

Cardasis, C. A. (1983). Ultrastructural evidence of continued reorganization at the aging (11-26 mos) rat soleus neuromuscular junction. *Anat. Rec.* **207**:399-415.

Cardasis, C. A., and Padykula, H. A. (1981). Ultrastructural evidence indicating reorganization of the neuromuscular junction in the normal rat soleus muscle. *Anat. Rec.* **200**:41-59.

Cherniack, N. S., and Altose, M. D. (1981). Respiratory responses to ventilatory loading. In *Regulation of Breathing,* Part II. Lung Biology in Health and Disease, Vol. 17. Edited by T. F. Hornbein. New York, Marcel Dekker, pp. 905-964.

Cherniack, N. S., and Milic-Emili, J. (1985). Mechanical aspects of loaded breathing. In *The Thorax.* Lung Biology in Health and Disease, Vol. 29. Edited by C. Roussos and P. T. Macklem. New York, Marcel Dekker, pp. 751-786.

Clanton, T. L., Dixon, G., Drake, J., and Gadek, J. E. (1985). Inspiratory muscle conditioning using a threshold loading device. *Chest* **87**: 62-66.

Cohen, C. A., Zagelbaum, G., Gross, D., Roussos, C., and Macklem, P. T. (1982). Clinical manifestations of inspiratory muscle fatigue. *Am. J. Med.* **73**:308-316.

Comtois, A., Gorczyca, W., and Grassino, A. (1987). Anatomy of diaphragmatic circulation. *J. Appl. Physiol.* **62**:238-244.

Costill, D. L., Daniels, J., Evans, W., Fink, W., Krahenbuhl, G., and Saltin, B. (1976). Skeletal muscle enzymes and fiber composition in male and female track athletes. *J. Appl. Physiol.* **40**:149-154.

Crockett, J. L., Edgerton, V. R., Max, S. R., and Barnard, R. J. (1976). The neuromuscular junction in response to endurance training. *Exp. Neurol.* **51**:207-215.

Edwards, R. H. T. (1978). Physiologic analysis of skeletal muscle weakness and fatigue. *Clin. Sci. Mol. Med.* **54**:463-470.

Evanich, M. J., Lopata, M., and Lorenco, R. V. (1976). Analytical methods for the study of electrical activity in respiratory nerves and muscles. *Chest* 70 (Suppl.):158-162.

Fahim, M. A., and Robbins, N. (1982). Ultrastructural studies of young and old mouse neuromuscular junction. *J. Neurocytol.* **11**:641-656.

Farkas, G. A., and Roussos, C. (1982). Adaptability of the hamster diaphragm to exercise and/or emphysema. *J. Appl. Physiol.* **53**:1263-1272.

Farkas, G. A., and Roussos, C. (1984). Histochemical and biochemical correlates of ventilatory muscle fatigue in emphysematous hamsters. *J. Clin. Invest.* **74**:1214-1220.

Felig, P., and Wahren, J. (1975). Fuel homeostasis in exercise. *N. Engl. J. Med.* **293**:1078-1084.

Fisher, J. T., and Mortola, J. P. (1980). Statics of the respiratory system in newborn animals. *Respir. Physiol.* **41**:155-172.

Fregosi, R. F., and Dempsey, J. A. (1986). Effects of exercise in normoxia and acute hypoxia on respiratory muscle metabolites. *J. Appl. Physiol.* **60**:1274-1283.

Gardiner, P. F., LaPointe, M., and Gravel, D. (1982). Exercise effects on recovery of muscle acetylcholinesterase from reduced neuromuscular activity. *Muscle Nerve* **5**:363-368.

Gerhardt, T., and Bancalari, E. (1980). Chest wall compliance in full term and preterm infants. *Acta. Pediatr. Scand.* **69**:359-364.

Gertler, R. A., and Robbins, N. (1978). Differences in neuromuscular transmission in red and white muscles. *Brain Res.* **142**:160-164.

Gollnick, P. D., Armstrong, R. B., Saubert, C. W. IV, Piehl, K., and Saltin, B. (1972a). Enzyme activity and fiber composition in skeletal muscle of untrained and trained men. *J. Appl. Physiol.* **33**:312-319.

Gollnick, P. D., Armstrong, R. B., Saubert, IV., C. W., Sembrowich, W. L., Shepherd, R. E., and Saltin, B. (1973). Glycogen depletion patterns in human skeletal muscle fibers during prolonged work. *Pflugers Arch.* **344**:1-12.

Gollnick, P. D., Karlsson, J., Piehl, K., and Saltin, B. (1974a). Selective glycogen depletion in skeletal muscle fibers of man following sustained contractions. *J. Physiol.* **241**:59-67.

Gollnick, P. D., and King, D. W. (1969). Effect of exercise training on mitochondria of rat skeletal muscle. *Am. J. Physiol.* **216**:1502-1509.

Gollnick, P. D., Piehl, K., and Saltin, B. (1974b). Selective glycogen depletion pattern in human muscle fibers after exercise of varying intensity and at varying pedalling rates. *J. Physiol.* **241**:45-57.

Gollnick, P. D., Piehl, K., Saubert, IV, C. W., Armstrong, R. B., and Saltin, B. (1972b). Diet, exercise and glycogen changes in human muscle fibers. *J. Appl. Physiol.* **33**:421-425.

Green, H., Ball-Burnett, M. E., Morrissey, M. A., Spalding, M. J., Hughson, R. L., and Fraser, I. G. (1987). Fiber type specific glycogen utilization in rat diaphragm during treadmill exercise. *J. Appl. Physiol.* **63**:75-83.

Gross, D., Grassino, A., Ross, W. R. D., and Macklem, P. T. (1979). Electromyogram pattern of diaphragmatic fatigue. *J. Appl. Physiol.* **46**:1-7.

Henderson-Smart, D. J., and Read, J. C. (1979). Reduced lung volume during behavioral active sleep in the newborn. *J. Appl. Physiol.* **46**:1081-1085.

Hermansen, L., Hultman, E., and Saltin, B. (1967). Muscle glycogen depletion during prolonged severe exercise. *Acta. Physiol. Scand.* **71**:129-139.

Holloszy, J. O., and Booth, F. W. (1976). Biochemical adaptations of endurance exercise in muscle. *Annu. Rev. Physiol.* **38**:273-291.

Ianuzzo, C. D., Noble, E. G., Hamilton, N., and Dabrowski, B. (1982). Effects of streptozotocin diabetes, insulin treatment, and training on the diaphragm. *J. Appl. Physiol.* **52**:1471-1475.

Ianuzzo, C. D., Spalding, M. J., and Williams, H. (1987). Exercise-induced glycogen utilization by the respiratory muscles. *J. Appl. Physiol.* **62**: 1405-1409.

Isaza, G. D., Posner, J. D., Altose, M. D., Kelsen, S. G., and Cherniack, N. S. (1976). Airway occlusion pressure in awake and anesthetized goats. *Respir. Physiol.* **27**:87-98.

Karlsson, J., Nordesjo, L. O., Jorfedlt, L., and Saltin, B. (1972). Muscle lactate, ATP and CP levels during exercise after physical training in man. *J. Appl. Physiol.* **33**:199-203.

Keens, T. G., Bryan, A. C., Levison, H., and Ianuzzo, C. D. (1978a). Developmental pattern of muscle fiber types in human ventilatory muscles. *J. Appl. Physiol.* **44**:909-913.

Keens, T. G., Chen, V., Patel, P., O'Brien, P., Levison, H., and Ianuzzo, C. D. (1978b). Cellular adaptations of the ventilatory muscles to a chronic increased respiratory load. *J. Appl. Physiol.* **44**:905-908.

Keens, T. G., Krastins, I. R. B., Wannamaker, E. M., Levison, H., Crozier, D. N., and Bryan, A. C. (1977). Ventilatory muscle endurance training in normal subjects and patients with cystic fibrosis. *Am. Rev. Respir. Dis.* **116**:853-860.

Kelsen, S. G., Altose, M. D., Stanley, N. N., Levison, R. S., Cherniack, N. S., and Fishman, A. P. (1976). The electromyographic response of the respiratory muscles during elastic loading. *Am. J. Physiol.* **230**: 675-683.

Kelsen, S. G., and Nochomovitz, M. L. (1982). Fatigue of the mammalian diaphragm in vitro. *J. Appl. Physiol.* **53**:440-447.

Klausen, K., Andersen, L. B., and Pelle, I. (1981). Adaptive changes in work capacity, skeletal muscle capillarization and enzyme levels during training and detraining. *Acta Physiol. Scand.* **113**:9-16.

Kuei, J. H., Shadmehr, R., and Sieck, G. C. Relative contribution of neurotransmission failure to diaphragm fatigue. *J. Appl. Physiol.* (in press).

Kugelberg, E., and Edstrom, L. (1968). Differential histochemical effects of muscle contractions on phosphorylase and glycogen in various types of fibers: Relation to fatigue. *J. Neurol. Neurosurg. Psychiatry* **31**: 415-423.

Leith, D. E., and Bradley, M. (1976). Ventilatory muscle strength and endurance training. *J. Appl. Physiol.* **41**:508-516.

Lindstrom, L., Kadefors, B., and Petersen, I. (1977). An electromyographic index for localized muscle fatigue. *J. Appl. Physiol.* **43**:750-754.

Lomo, T., and Slater, C. R. (1980). Control of junctional acetylcholinesterase by neural and muscular influences in the rat. *J. Physiol. (London)* **303**:191-202.

Lopes, J. M., Muller, N. L., Bryan, M. H., and Bryan, A. C. (1981). Synergistic behavior of inspiratory muscles after diaphragmatic fatigue in the newborn. *J. Appl. Physiol.* **51**:547-551.

Maxwell, L. C., McCarter, R. J. M., Kuehl, T. J., and Robotham, J. L. (1983). Development of histochemical and functional properties of baboon respiratory muscles. *J. Appl. Physiol.* **54**:551-561.

Mayock, D. E., Badura, R. J., Watchko, J. F., Standant, T. A., and Woodrum, D. E. (1987a). Response to resistive loading in the newborn piglet. *Pediatr. Res.* **21**:121-125.

Mayock, D. E., Hall, J., Watchko, J. F., Standaert, T. A. and Woodrum, D. E. (1987b). Diaphragmatic muscle fiber type development in swine. *Pediatri. Res.* **22**:449-454.

Metzger, J. M., and Fitts, R. H. (1986). Contractile and biochemical properties of diaphragm: Effects of exercise training and fatigue. *J. Appl. Physiol.* **60**:1752-1758.

Moore, R. L., and Gollnick, P. D. (1982). Response of the ventilatory muscles of the rat to endurance training. *Pflugers Arch.* **392**:268-271.

Moxham, J., Morris, A. J. R., Spiro, S. G., Edwards, R. H. T., and Green, M. (1981). Contractile properties and fatigue of the diaphragm in man. *Thorax* **36**:164-168.

Muller, N., Gulston, G., Cade, D., Whitton, J., Froese, A. B., Bryan, M. H., and Bryan, A. C. (1979a). Diaphragmatic muscle fatigue in the newborn. *J. Appl. Physiol.* **46**:688-695.

Muller, N., Volgyes, G., Bryan, M. H., and Bryan, A. C. (1979b). The consequences of diaphragmatic fatigue in the newborn infant. *J. Pediatr.* **95**:793-797.

O'Brien, R. A. D., Ostberg, A. J. C., and Vrbova, G. (1978). Observations on the elimination of polyneuronal innervation developing mammalian skeletal muscle. *J. Physiol. (London)* **282**:571-582.

Orenstein, D. M., Franklin, B. A., Doerschuk, C. F., Hellerstein, H. K., German, K. J., Horowitz, J. G., and Stern, R. C. (1981). Exercise conditioning and cardiopulmonary fitness in cystic fibrosis. *Chest* **80**:392-397.

Orthner, F. H., and Yamamoto, W. S. (1974). Transient respiratory response to mechanical loads at fixed blood gas levels in cats. *J. Appl. Physiol.* **36**:280-287.

Palacios-Pru, E., and Colasante, C. (1988). Ultrastructural reversible changes in fish neuromuscular junctions after chronic exercise. *J. Neurosci. Res.* **19**:245-251.

Palacios-Pru, E., Mendoza, R. V., Palacios, L., and Colasante, C. (1983). Morphological changes in neuromuscular junctions during exercise. *J. Neurosci. Res.* **9**:371-380.

Pang, L. M., Bazzy, A. R., Akabas, S. R., and Haddad, G. G. (1987). Blood flow to respiratory muscles and major organs in conscious sheep subjected to inspiratory flow resistive (IFR) loads. *Am. Rev. Respir. Dis.* **135**:A296 (Abstract).

Pardy, R. L., Reid, W. D., and Belman, M. J. (1988). Respiratory muscle training. In *Respiratory Muscles: Function in Health and Disease.* Clinics in Chest Medicine, Vol. 9. Edited by M. J. Belman. Philadelphia, W. B. Saunders, pp. 287-296.

Pardy, R. L., Rivington, R. N., Despos, P. J., and Macklem, P. T. (1981). The effect of inspiratory muscle training on exercise performance in chronic airflow limitation. *Am. Rev. Respir. Dis.* **123**:426-434.

Pockett, S., and Pygott, V. (1982). Chronic nerve stimulation increases quantal output in rat muscles. *Exp. Neurol.* **78**:477-480.

Reinecke, M., and Walther, C. (1981). Ultrastructural changes with high activity and subsequent recovery at locust motor nerve terminals. A stereological analysis. *Neuroscience* **6**:489-503.

Riley, D. A. (1978). Tenotomy delays the postnatal development of the motor innervation of the rat soleus. *Brain Res.* **143**:162-167.

Robbins, N., and Fahim, M. A. (1985). Progression of age changes in mature mouse motor nerve terminals and its relation to locomotor activity. *J. Neurocytol.* **14**:1019-1036.

Robbins, N., and Kelly, S. S. (1983). Progression of age changes in synaptic transmission of mouse neuromuscular junctions. *Abstr. Soc. Neurosci.* **9**:929.

Robertson, Jr., C. H., Foster, G. H., and Johnson, Jr., R. L. (1977). The relationship of respiratory failure to the oxygen consumption of, lactate production by, and distribution of blood flow among respiratory muscles during increasing inspiratory resistance. *J. Clin. Invest.* **59**:31-42.

Rochester, D. F., and Bettini, G. (1976). Diaphragmatic blood flow and energy expenditure in the dog. *J. Clin. Invest.* **57**:661-672.

Rosenheimer, J. L. (1985). Effects of chronic stress and exercise on age-related changes in end-plate architecture. *J. Neurophysiol.* **53**:1582-1589.

Rosenheimer, J. L., and Smith, D. O. (1985). Differential changes in the end-plate architecture of functionally diverse muscles during aging. *J. Neurophysiol.* **53**:1567-1581.

Roussos, C., and Macklem, P. T. (1977). Diaphragmatic fatigue in man. *J. Appl. Physiol.* **43**:189-197.

Roussos, C., and Macklem, P. T. (1982). The respiratory muscles. *N. Engl. J. Med.* **307**:786-797.

Sadoul, N., Bazzy, A. R., Akabas, S. R., and Haddad, G. G. (1985). Ventilatory response to fatiguing and non-fatiguing inspiratory flow-resistive loads in unanesthetized sheep. *J. Appl. Physiol.* **59**:969-978.

Schnader, J. Y., Juan, G., Howell, S., Fitzgerald, R., and Roussos, C. (1985). Arterial CO2 partial pressure affects diaphragmatic function. *J. Appl. Physiol.* **58**:823-829.

Schweitzer, T. W., Fitzgerald, J. W., Bowden, J. A., and Lynne-Davies, P. (1979). Spectral analysis of human inspiratory diaphragmatic electromyograms. *J. Appl. Physiol.* **46**:152-165.

Stebbins, C. L. Schultz, E., Smith, R. T., and Smith, E. L. (1985). Effects of chronic exercise during aging on muscle and end-plate morphology in rats. *J. Appl. Physiol.* **58**:45-51.

Synder, D. H., Rifenberick, D. H., and Max, S. R. (1973). Effects of neuromuscular activity on choline acetyltransferase and acetylcholinesterase. *Exp. Neurol.* **40**:36-42.

Thomson, J. M., Stone, J. A., Ginsburg, A. D., and Hamilton, P. (1982). O_2 transport during exercise following blood reinfusion. *J. Appl. Physiol.* **53**:1213-1219.

Tibes, U., Hemmer, B., and Boning, D. (1977). Heart rate and ventilation in relation to venous $[K^+]$, osmolality, pH, P_{CO_2}, P_{O_2}, [orthophosphate], and [lactate] at transition from rest to exercise in athletes and nonathletes. *Eur. J. Appl. Physiol. Occup. Physiol.* **36**:127-140.

Tuffery, A. R. (1971). Growth and regeneration of motor end-plates in normal cat hindlimb muscles. *J. Anat.* **110**:221-247.

Vyskocil, F., and Gutmann, E. (1972). Spontaneous transmitter release from nerve endings and contractile properties in the soleus and diaphragm muscles of senile rats. *Experientia* **28**:280-281.

Washio, H., Imazato-Tanaka, C., Kanda, K., and Nomoto, S. (1987). Choline acetyltransferase and acetylcholinesterase activities in muscles of aged mice. *Brain Res.* **416**:69-74.

Watchko, J. F., LaFramboise, W. A., Standaert, T. A., and Woodrum, D. E. (1986). Diaphragmatic function during hypoxemia: neonatal and developmental aspects. *J. Appl. Physiol.* **60**:1599-1604.

Watchko, J. F., Standaert, T. A., and Woodrum, D. E. (1987). Diaphragmatic function during hypercapnia: Neonatal and developmental aspects. *J. Appl. Physiol.* **62**:768-775.

Wernig, A., Pecot-Dechavassine, M., and Stover, H. (1980). Sprouting and regression of the nerve at the frog neuromuscular junction in normal conditions and after prolonged paralysis with curare. *J. Neurocytol.* **9**:277-303.

Wigston, D. J. (1989). Remodeling of neuromuscular junctions in adult mouse soleus. *J. Neurosci.* **9**:639-647.

16

Neural Control of Airway Smooth Muscle in the Newborn

MARK A. WALDRON and JOHN T. FISHER

Queen's University
Kingston, Ontario, Canada

I. Introduction

The efferent control of airway smooth muscle in the adult mammal is the subject of many excellent reviews documenting the advances and rich history of this field (Einthoven, 1975; Russell, 1980a; Otis, 1983; Widdicombe, 1963, 1966b, 1985; Richardson, 1979; Nadel and Barnes, 1984; Kirkpatrick, 1984; Barnes, 1986; Andersson and Grundstrom, 1987; Stephens and Hoppin, 1986). The number of studies of control mechanisms in the newborn, however, pales in comparison to the adult literature (Reid, 1985; Nadel, 1980; Barnes, 1986; Diamond and Altiere, 1988). Although the presence of pulmonary diseases such as bronchopulmonary dysplasia in premature infants has sparked interest in the control of airway caliber and reactivity (Hodgman, 1981; Motoyama et al., 1987), very few studies have examined directly

Work from our laboratory is supported by the Medical Research Council of Canada, the Hospital for Sick Children Foundation, and the Ontario Thoracic Society. John Fisher was the recipient of an MRC Development Award and Mark Waldron was supported by an MRC predoctoral studentship award and the Ontario Thoracic Society.

the functional status of airway innervation in the newborn. Indeed, Reid (1985) recently stated that the "competence of muscular function in the airways represents a key area for research in the newborn."

In this chapter we have concentrated on literature specifically related to the neural control of airway smooth muscle in the fetus or newborn. As a result we have excluded discusion of inflammatory mediators such as histamine, prostaglandins, leukotrienes, and platelet-activating factor. Although histamine has been examined to some extent (Hayashi and Toda, 1980; Sauder et al., 1986; Prendiville et al., 1987a,c; LeSoeuf et al., 1989; Sparrow and Mitchell, 1990; Stecenko et al., 1989), studies of other agonists are either not available or severely limited in number (Duncan and Douglas, 1985b). When reference is made to studies in the adult, we have cited review articles as a gateway to the original literature.

II. Species Differences

Airway innervation generally displays a remarkable degree of variability between species. Not only does the presence of specific neural elements depend on location within the airway (i.e., trachea versus bronchus), but an entire type of innervation (e.g., sympathetic or nonadrenergic inhibitory innervation) may be absent in the airway smooth muscle of a particular species (Nadel, 1980; Barnes, 1986; Diamond and Altiere, 1988). This situation is further complicated in the newborn by the fact that marked differences exist between species in their relative maturity at birth. Species that are precocial at birth (e.g., sheep, pig, horse, and guinea pig) are capable of sophisticated locomotor behavior immediately after delivery, and morphological studies indicate they possess a more developed lung (Alcorn et al., 1981) than other species (Burri, 1974). In contrast, altricial species (e.g., rat, rabbit, cat, and dog) undergo considerable postnatal lung development and may exhibit striking morphological changes in alveolar and vascular compartments of the lung (Burri et al., 1974; Burri, 1974, 1984). Marsupial species typify the altricial end of the maturational spectrum since much of their postnatal maturation resembles that occurring during intrauterine life in placental mammals. In the opossum, 50-60 postnatal days are required to reach a developmental status similar to that of a newborn dog or cat (McCrady, 1938; Reynolds, 1952; Krous et al., 1985; see Chapter 8, this volume).

In general, innervation and maturational differences between species necessitate caution in extrapolating findings from a single mammalian species to the human neonate, for whom the specific goal may be to improve or develop treatment. On the other hand, studies that employ several species provide a unique overview of the spectrum of biological design, which may be useful in delineating general principles of the control of airway smooth muscle.

III. Autonomic Airway Innervation in the Adult: An Overview

A. Peripheral Organization

Vagal autonomic innervation of airway smooth muscle consists of preganglionic fibers whose cell bodies are located within the medulla, intrachondral and extrachondal nerve plexuses containing airway ganglia, and postganglionic efferent fibers. Postganglionic fibers possess axonal varicosities containing vesicles releasing neurotransmitters which act on post-"junctional" smooth muscle receptors (Gabella, 1987; Burnstock, 1988). The large distances between nerve endings and muscle cells (100 nm-1μm) preclude the use of the term synapse in describing this neuroeffector junction (Gabella, 1987; Burnstock, 1988). Differences between species exist with respect to the number of axons per muscle cell (Gabella, 1987). In some species, this ratio also varies regionally from the trachea to bronchi (Daniel et al., 1986; Gabella, 1987). Although there do not appear to be any comparable morphological data for fetal or neonatal airway smooth muscle, morphological studies describe a rich network of nerve fibers around newborn mammalian airways (Larsell and Dow, 1933; Pessacq, 1971; Taylor and Smith, 1971), and biochemical estimates of the number of muscarinic receptors are available (see section IVA).

The most ubiquitous feature of vagal efferent innervation, apparent in all mammalian species studied, is an excitatory cholinergic innervation. Stimulation of the vagus nerves, or the application of cholinergic agonists, causes smooth muscle contraction and mucus secretion via muscarinic receptors (Olsen et al., 1965; Cabezas et al., 1971; Russell, 1978; Woolcock et al., 1969; Nadel, 1980). A second vagal excitatory component has been identified in several species; however, this appears to be due to the release of tachykinins from unmyelinated afferent endings that can be antidromically activated by electrical stimulation (Lundberg and Saria, 1987).

In a variety of species, including human (Richardson and Beland, 1976), baboon (Middendorf and Russell, 1980), cat (Diamond and O'Donnell, 1980; Irvin et al., 1980), and guinea pig (Coburn and Tomita, 1973), a neurally mediated relaxation of airway smooth muscle can be elicited that is neither adrenergic nor cholinergic in origin (Diamond and Altiere, 1988). The nonadrenergic noncholinergic (NANC) inhibitory system can be activated by stimulation of the vagus nerves, and some evidence suggests that the neurotransmitter involved is coreleased from the same nerve terminals as acetyl-choline (Diamond and Altiere, 1988). The identity of the NANC inhibitory neurotransmitter is the subject of some controversy. Although it was origi-nally thought to be purinergic, the weight of evidence now suggests this is unlikely (Irvin et al., 1982), and vasoactive intestinal peptide (VIP) has emerged as a strong candidate as the NANC neurotransmitter (Diamond and Altiere, 1988). Confirmation of VIP in this role awaits development of a specific VIP antagonist.

Morphological studies in the adult cat reveal that 80-90% of the affer-ent and efferent components of vagal bronchial branches are unmyelinated fibers (Jammes et al., 1982). The efferent component of vagal bronchial branches has been estimated to comprise 38% of all fibers (Jammes et al., 1982). Assuming a similar efferent: afferent ratio for the cervical vagus, tracheobronchial fibers should represent approximately 27% of the entire vagus nerve in the cat (10% efferent and 17% afferent; Jammes et al., 1982). In the newborn, the cervical vagus nerve and its bronchial branches have considerably fewer myelinated fibers, ranging from 0 to 20% of the adult value (Schwieler, 1968; Marlot and Duron, 1979; DeNeef et al., 1982; Krous et al., 1985). Unfortunately, there are no data regarding the number of un-myelinated fibers present at birth, or the ratio of efferent to afferent fibers. Studies comparing axon counts from intact bronchial branches with those after supranodose vagotomy, to eliminate efferent axons, would significantly enhance the interpretation of functional studies as well as generate new testable hypotheses regarding efferent airway innervation.

Sympathetic innervation to the airways varies markedly among species. Preganglionic fibers exit from the thoracic spinal cord (upper six segments) to synapse within prevertebral ganglia. Postganglionic fibers innervate air-way smooth muscle in some species (Olsen et al., 1965; Cabezas et al., 1971; Russell, 1980b), but this does not appear to be true in humans (Richardson and Ferguson, 1980; Barnes, 1986). Nevertheless, β-adrenergic receptors are present on smooth muscle, despite the lack of direct innervation. Re-laxation of airway smooth muscle in response to β agonists suggests that circulating catecholamines play a functional role in modulating airway tone. Sympathetic fibers innervate airway ganglia and may influence vagal trans-mission at this point (Baker, 1986; Coburn, 1987; Skoogh, 1986). Few stu-dies of sympathetic innervation have been performed in the newborn (Sch-wieler et al., 1970; Pandya, 1977; see Section V.), and much of what is known about adrenergic control of the airways comes from the clinical use of bron-chodilators in premature infants.

The importance of airway ganglia as a control point and site of integra-tion of neural signals has only recently been appreciated (Baker, 1986; Co-burn, 1987; Skoogh, 1988; Coburn, 1988). Indeed, Coburn noted that the first electrophysiological studies of airway ganglia appeared only 5 years prior to his review (Coburn, 1987). Physiological studies have not been conducted on airway ganglia in the newborn. Suffice it to say, this is extremely fertile ground for future research since airway ganglia may be the site of considerable neural integration of cholinergic, sympathetic, and nonadrenergic efferent inputs, as well as possibly receiving afferent input.

B. Central Organization

Histological and electrophysiological studies have demonstrated that pulmonary preganglionic vagal neurons have cell bodies located predominantly within the nucleus ambiguous (NA) and the dorsal motor nucleus of the vagus (DmnX); a few neurons are also located within the reticular formation between these nuclei (Kalia and Mesulam, 1980; McAllen and Spyer, 1978; Jordan et al., 1986). Medullary preganglionic cells are labeled by the retrograde transport of horseradish peroxidase (HRP) applied to peripheral branches of the vagus presumed to innervate airway smooth muscle or applied to the airway directly. The former method provides more control of the areas exposed to HRP and the medullary areas labeled (Kalia and Mesulam, 1980). The medullary distribution of these cell bodies has also been mapped electrophysiologically by recording from cells antidromically activated by stimulation of a peripheral efferent nerve branch (McAllen and Spyer, 1978; Jordan et al., 1986). In both types of studies, bronchial branches of the vagus are identified as efferent to the airways or heart on the basis of functional or anatomical criteria. Functional criteria consist of a rise in pulmonary resistance (or bradycardia for cardiac branches) during stimulation of the nerve; anatomical criteria consist of visual confirmation that the branch has a pulmonary or cardiac destination. There are no data in the newborn or fetus with respect to the location or electrophysiology of medullary cell bodies with axons projecting to airway smooth muscle (Jansen and Chernick, 1983; see Chapter 9, this volume).

IV. Vagal Control of Airway Smooth Muscle

A. Cholinergic Excitatory Innervation

Histochemical and Pharmacological Evidence

Histochemical evidence suggestive of cholinergic innervation to airway smooth muscle in the newborn is available for several species. In lung tissue from human fetuses, acetylcholinesterase (suggestive of cholinergic innervation) has been detected at 10-12 weeks of gestation (Sheppard et al., 1983). Cholinesterase staining of fetal nerves and ganglia suggests bronchial innervation is well established at the 100-mm stage (12-14 weeks). Indeed, nerves are present in adventitial tissue of 35-mm fetuses (8-9 weeks) although they have not yet penetrated tracheal tissue (Taylor and Smith, 1971). In newborn infants, peribronchial plexuses, from which terminal nerve fibers travel to innervate smooth muscle, have been described by light microscopy studies (Pessacq, 1971). Evidence from other species also suggests considerable intrauterine development. Acetylcholinesterase-positive neuroblasts are present in fetal rat lungs as early as days 12-13 (term = 22 days) of embryonic development (Morikawa et al., 1978a,b). Furthermore, airway ganglia are apparent at days 15-16, and by day 17, fibers projecting from these ganglia to smooth muscle tissue can be detected (Morikawa et al., 1978a,b).

Pharmacological estimates of muscarinic receptor density, based on the number of binding sites for the tritiated muscarinic antagonist quinuclidinyl benzilate (^{3}HQNB), differ among species with respect to when peak values occur and the influence of growth. Data for the rat indicate that peak densities occur in fetal lungs on gestational day 17, followed by a marked decrease until term (days 21-22; Whitsett and Hollinger, 1984). The number of binding sites subsequently remains constant from term through adulthood (Whitsett and Hollinger, 1984). In bovine tissue, however, there is a postnatal decrease in the density of cholinergic muscarinic receptors (from animals 2 weeks, 3-5 months, and >5 years old) that is accompanied by a decrease in the specific activity of acetylcholinesterase (Rothberg et al., 1987). Data for fetal bovine tissue are not available. It would be of interest to determine whether the peak receptor density in the rat is accompanied by a greater functional response to cholinergic agonists, as has been reported for the cow (Wills and Douglas, 1988). In the cow membrane potential does not appear to change postnatally (Souhrada et al., 1988); however, the contribution of the Na^+/K^+ electrogenic pump to resting membrane potential of tracheal smooth muscle decreases from 38% at 2 weeks to 31% at 5 months to 21% in mature animals. The neural/hormonal mechanisms controlling these changes and their functional significance remain to be elucidated.

In Vitro Studies of Cholinergic Agonists

In vitro studies have clearly demonstrated functional cholinergic muscarinic receptors in the newborn, although disagreement exists about whether the magnitude of the response to cholinergic agonists is greater in the newborn than in the adult. Perhaps the most significant problem in comparing responsivity of airway smooth muscle to agonists in the newborn and adult is the choice of a normalization factor(s) independent of body and tissue size. Comparisons of the raw value of force generated by tissue in vitro simply reflect the amount of tissue. Normalization methods typically express force per unit mass of tissue, or per unit cross-sectional area (stress). However, these values may be biased by disproportionate differences in noncontractile components, such as cartilage and connective tissue, especially when comparing animals of different ages (Bradley et al., 1974; Brink et al., 1980). Normalization methods based on morphological estimates of the amount of smooth muscle in the tissue sample, or on biochemical measurements of the amount of contractile protein (i.e., actin or myosin), are two promising approaches. Indeed, morphological normalization has been used in the adult (Gunst and Stropp, 1988), and myosin content has recently been used as a normalization factor in a study of airway smooth muscle maturation (Sparrow and Mitchell, 1990).

In porcine airway tissue, the amount of myosin per unit weight increases, and the proportion of the two myosin heavy chains changes with age (Mohammad and Sparrow, 1988; Sparrow and Mitchell, 1990). In fetal and young pigs, normalization of maximal force by cross-sectional area of myosin influenced conclusions regarding which group had the greatest force-generating potential in vitro and was suggested to provide a more meaningful basis for comparison. Studies of vascular smooth muscle indicate that maximal force-generating capabilities increase with growth in that tissue (Cohen and Berkowitz, 1974; Seidel and Murphy, 1979; Seidel et al., 1987). Furthermore, increases in force-generating capabilities are accompanied by increases in the total amount of actomyosin (Seidel and Murphy, 1979); in aortas from 3-7-week-old rats there is an increase in maximal in vitro force that is correlated with an increase in content of actomyosin (expressed as a percentage of the tissue weight or total protein). Although this is a promising correlation, developmental changes in maximal force are not necessarily correlated with increases in the relative amount of actin or myosin, depending on the age range and species studied. For example, in vascular smooth muscle the ratio of actomyosin to total protein does not seem to explain the increase in smooth muscle contractile responses in rats between 1 and 3 weeks of age (Seidel and Allen, 1979). A similar situation was reported for comparisons between newborn and adult dog (Seidel et al., 1987). Whether this is true of airway smooth muscle remains to be seen.

Hayashi and Toda (1980) reported that tracheal muscle strips from rabbits, ranging in age from 3 to 360 days, all responded to acetylcholine, a finding confirmed in several other species (see below). In a series of papers examining the response of airway smooth muscle to several excitatory agonists, Duncan and Douglas, and their co-workers, consistently reported an enhanced airway reactivity in immature or young guinea pigs (<1 week old or ≈ 100 g) compared to adults (>12 weeks) (Brink et al., 1980; Duncan and Douglas, 1985a,b; Bayol et al., 1985; see also Wills and Douglas, 1988, for bovine tissue). The enhanced reactivity was generally characterized by an increased smooth muscle contractility in response to agonists (i.e., maximal stress) and increased sensitivity of the tissue to the agonist (i.e., agonist potency; Brink et al., 1980; Duncan and Douglas, 1985a,b; Bayol et al., 1985; Wills and Douglas, 1988).

Contractility of airway smooth muscle depends on species, age, and location within the tracheobronchial tree. In the guinea pig, the enhanced contractility (i.e., force/mm² tissue) of immature airways exposed to muscarinic agonists is linked to location (Fig. 1); larger maximal carbachol-induced contractions occur in the bronchi, but not the trachea, of immature versus mature guniea pigs (Duncan and Douglas, 1985a). In bovine trachea, maximal stress in response to several muscarinic agonists is greater in infant (<2 weeks) than mature animals (Wills and Douglas, 1988), and in the pig, airways from the trachea to bronchioles are more responsive to carbachol in suckling (4 weeks old) than in fetal (95 days; 83% term), young (20-26 weeks), or mature (>6 months) animals, independent of the type or normalization factor (Sparrow and Mitchell, 1990). Enhanced cholinomimetic responsiveness has also been reported for tracheal smooth muscle of 2-week-old compared to 10-week-old pigs, although responses to potassium chloride are similar (Murphy et al., 1989). Changes in acetylcholinesterase function appear to contribute to increased reactivity of younger animals since inhibition of the enzyme enhanced the response to acetylcholine of tracheal smooth muscle from 10-week-old animals but had no effect on the response of tissue from 2-week-old animals (Murphy et al., 1989).

Although specific comparative studies of airway reactivity among neonates of different mammalian species are not available, species differences do exist. Airways of fetal pigs are generally less responsive than those of mature animals when force is normalized per unit cross-sectional area

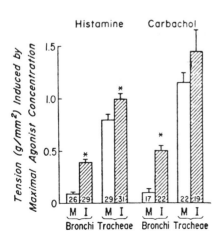

Figure 1 Maximal stress produced by histamine or carbachol in airway tissues from mature (M) or immature (I) guinea pigs. Values are means ± SEM. Asterisk indicates significantly greater stress in immature tissues compared to mature tissues contracted with the same agonist ($P<0.05$ using Student's t-test for unpaired values). The number of experiments is indicated in each bar. (Adapted from Duncan and Douglas, 1985a.)

(stress). This finding is reversed, however, when normalization is based on the amount of myosin (force per mm^2 myosin; Sparrow and Mitchell, 1990). As mentioned above, 4-week-old animals are most responsive. In contrast, sheep (Fig. 2) tracheal smooth muscle is reported to generate significantly lower stress in preterm(102-116 days, 70-81% term) than in adult animals (Panitch et al., 1989). On this basis, it was suggested the preterm trachea is more susceptible to deformation than the adult trachea and, therefore, more susceptible to damage during mechanical ventilation (Panitch et al., 1989).

The second factor typically associated with increased airway reactivity, enhanced agonist potency, is assessed by agonist concentration-effect curves. "Concentration" refers to the molar concentration of agonist in the organ bath, while "effect" is the magnitude of contraction at each agonist concentration, expressed as a percentage of the maximal tension induced. The concentration of agonist required to produce 50% of the maximal response (EC_{50}) represents potency (often expressed as the negative of the log of this concentration, or the pD_2). On this basis, tracheas from immature guinea pigs (Duncan and Douglas, 1985a) and cows (Wills and Douglas, 1988) are more sensitive than those of mature animals. Bronchi from guinea pigs of different ages do not differ in their sensitivities (Duncan and Douglas, 1985a), while no data are available for bronchial tissue in the cow. Whether an increased potency of cholinergic agonists in the trachea of the newborn is a typical mammalian feature is not clear. For example, preterm sheep have a higher EC_{50} than adult animals (Panitch et al., 1989). Furthermore, no clear trend is apparent for the pig (Sparrow and Mitchell, 1990), although data

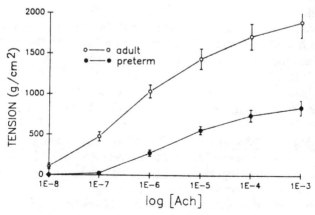

Figure 2 Effect of acetylcholine (Ach) concentration on stress produced by trachealis muscle from adult and preterm sheep. Tissue from preterm animals developed significantly less stress than tissue from adults. Values are mean ± SEM (n = 5 adults, 7 preterms). (Reprinted from Panitch et al., 1989.)

from newborn animals are not available for this species. Data from altricial species (e.g., rabbit, cat, or dog) would be useful in assessing the influence of maturation on airway reactivity, as would studies of human neonatal tissue. In the latter case, data from the literature for airway smooth muscle of adult humans are available for comparison (Davis et al., 1982).

It is significant that all of the studies described above used isometric measurements of airway smooth muscle contraction. In vitro studies of the dynamic (isotonic) properties (i.e., force-velocity, velocity-length-time relationships) of airway smooth muscle are not available in the newborn, although they have been conducted in the adult (Stephens and Kroeger, 1980; Stephens and Hoppin, 1986). Such studies would provide new information regarding cellular mechanisms underlying smooth muscle contraction in the newborn, such as cross-bridge cycling rates and maximal shortening ability (Seow and Stephens, 1988; Stephens, 1988).

In Vivo Studies of Cholinergic Innervation

Functional cholinergic innervation to nonrespiratory viscera (e.g., heart and gut) is present in the newborn (Downing, 1960; Schwieler et al., 1970; Gershon and Thompson, 1973; Geiss et al., 1975; Levin et al., 1982; Maggi et al., 1984). Of special relevance are studies of the development of innervation to the gastrointestinal tract, due to the common embryological origin of the lung and gut. The components of autonomic innervation to the gastrointestinal tract develop at different rates; cholinergic excitatory innervation and nonadrenergic inhibitory innervation are the first to be detected functionally, followed by sympathetic innervation (Gershon and Thompson, 1973; Levin et al., 1982).

In vivo studies indicate functional vagal cholinergic *innervation* to airway smooth muscle is present in several species at birth (Schwieler et al., 1970; Isabel et al., 1972; Fisher and Sant'Ambrogio, 1982b; Fisher et al., 1983, 1987; Rossi and Mortola, 1987; Waldron and Fisher, 1988; Fisher et al., 1990), and muscarinic agonists elicit changes consistent with bronchoconstriction (Fig. 3; Sauder et al., 1986; Tepper, 1987; Fisher et al., unpublished).

Schwieler and co-workers (1970) recorded pressure from a balloon placed in an in situ isolated rabbit tracheal segment, as an index of smooth muscle activity. They reported that pressure increased during stimulation of the vagus nerves of newborn rabbits, as young as 24 h. Maximal contractions occurred at stimulus frequencies of 20-30 impulses/s, but they were not maintained at higher frequencies. In contrast, in adult rabbits (>4 weeks), maximal contractions occurred at approximately 50 impulses/s and were maintained at stimulation frequencies as high as 150-200 impulses/s (Sch-

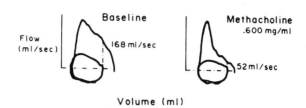

Volume (ml)

Figure 3 Evidence for functional muscarinic airway receptors in infants. Methacholine aerosol significantly reduced maximal expiratory flow at functional residual capacity during a partial flow-volume curve in a 3.5-month-old infant. (Adapted from Tepper, 1987.)

wieler et al., 1970). They suggested the smaller size of axons in immature nerves, and metabolic insufficiencies, may have caused conduction failure at high frequencies (Schwieler et al., 1970). Fisher and co-workers (1990) provided direct evidence of vagal efferent innervation to tracheobronchial airways in the newborn dog and cat (Fig. 4, top). They stimulated the peripheral vagus nerves at frequencies ranging from 2 to 20/s and measured the associated increases in lung resistance and decreases in dynamic compliance

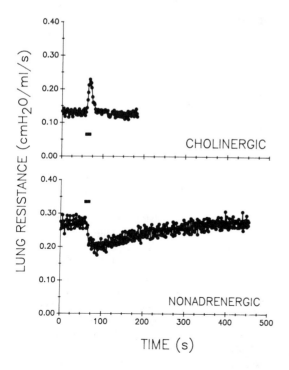

Figure 4 Functional cholinergic excitatory and nonadrenergic inhibitory innervation to the airways in an anesthetized, ventilated 4-day-old cat. Each panel presents breath-by-breath plot of lung resistance. Note the slower onset and longer duration of the stimulation-induced effects of NANC innervation. Ten-second periods of maximal vagal stimulation at 10 impulses/s are indicated by the horizontal bars. To illustrate a nonadrenergic inhibitory response (lower panel), atropine (2 mg/kg) and propranolol (2 mg/kg) were administered and airway tone was increased by an infusion of serotonin (90 mg/kg/min).

(Fisher et al., 1990). The changes in lung mechanics were due solely to cholinergic innervation since they were eliminated by muscarinic antagonists (Duthie et al., 1989; Fisher et al., 1990). The increase in total pulmonary resistance was dependent on stimulus frequency, reaching a plateau at 15/s, with a maximal increase of approximately 300, (Fig. 5). The decrease in dynamic compliance was independent of stimulus frequency and much smaller in magnitude (20-30%) than the increase in resistance, a finding consistent with the predominantly central distribution of cholinergic innervation to the lung (Olsen et al., 1965; Cabezas et al., 1971; Hahn et al., 1976). The maximal increase in resistance occurred at stimulus frequencies similar to peak discharge frequencies recorded from efferent fibers in the adult (Widdicombe, 1961; Widdicombe, 1966a). Although there are no reports of firing patterns of vagal efferent axons in the newborn, they probably discharge at frequencies lower than those in the adult since both phrenic efferents (Marlot and Duron, 1981) and pulmonary afferents (Fisher and Sant'Ambrogio, 1982a) have lower discharge frequencies in the newborn.

It is difficult to draw quantitative conclusions about the functional capabilities of airway innervation on the basis of stimulation-induced changes in lung mechanics in the newborn and adult. Although qualitatively similar, the newborn appears to be somewhat less responsive to vagal stimulation (i.e., smaller % increase in pulmonary resistance) than the adult (Olsen et al., 1965; Fisher et al., 1990). Several factors complicate this interpretation. Pulmonary resistance is an indirect index of airway smooth muscle contrac-

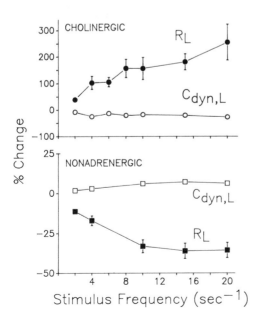

Figure 5 Plot of the mean changes in lung resistance (R_L) and dynamic compliance ($C_{dyn,L}$) caused by stimulation of the cholinergic excitatory (upper panel, n = 14), and nonadrenergic inhibitory (lower panel, n = 10) systems at increasing frequencies in newborn cats (1-14 days). Values are means ± SEM. See legend of Figure 4 for details of preparation. (Adapted from Waldron et al., 1989, and Fisher et al., 1990.)

tion, and it is inversely related to the fourth power of airway radius. Therefore, a similar activation of airway smooth muscle may induce different changes in resistance if the mechanical properties of noncontractile airway elements differ among age groups. Indeed, airways in the newborn are highly compliant structures (Bhutani et al., 1981a) that are easily deformed (Bhutani et al., 1981b; 1986). Furthermore, owing to the smaller airway radius in the newborn, Laplace's law ($P = 2\,T/r$, where P is the pressure required to maintain the airway patent, T is the tension, and r the airway radius) would predict that larger collapsing pressures would be exerted for a given smooth muscle tension in the newborn compared to the adult, leading to greater narrowing. This situation may be complicated even further if the amount of narrowing of the internal surfaces of the airways is different from the external surfaces (i.e., the interior diameter changes more than the outer diameter). Consequently, comparisons of airway contractility between the newborn and adult are probably better made using in vitro methods. However, in vivo studies provide important information on the range and magnitude of the changes produced by vagal pulmonary innervation in the newborn and provide a basis from which physiological and pathological studies may be interpreted.

In the adult, a reflex increase or decrease in bronchomotor tone is possible due to the presence of a "resting" level of airway smooth muscle contraction produced by tonic cholinergic activity (for review see Nadel, 1980; Coleridge et al., 1989). Several studies have addressed whether the newborn has "resting" cholinergic bronchomotor tone, and a consensus is lacking (Clement et al., 1986; Rossi and Mortola, 1987; Waldron and Fisher, 1988). These studies have compared respiratory system (Clement et al., 1986) and lung mechanics (Rossi and Mortola, 1987; Waldron and Fisher, 1988) before and after vagotomy or atropine administration in anesthetized animals. In spontaneously breathing kittens, lung resistance decreases 12% and chest wall compliance increases 35% following vagotomy (Rossi and Mortola, 1987). The former change has been attributed to a withdrawal of bronchomotor tone, while the latter is thought to reflect the larger tidal volumes seen postvagotomy. The arterial blood gas status of these animals, a factor that can influence bronchomotor tone (Waldron and Fisher, 1988), was not assessed. In contrast, we found vagotomy or atropine administration did not alter lung mechanics in newborn dogs that were mechanically ventilated to maintain volume history and CO_2 constant (Waldron and Fisher, 1988). In ventilated premature infants, resting bronchomotor tone is present, as indicated by the reduction in respiratory system resistance following inhalation of the muscarinic antagonist ipratropium bromide (Fig. 6; Wilkie and Bryan, 1987; Brundage et al., 1990). However, Hiatt et al. (1988) reported that a control group of infants (ranging in age from 2 to 18 months), which were studied for comparison with infants with cystic fibrosis (CF), showed no

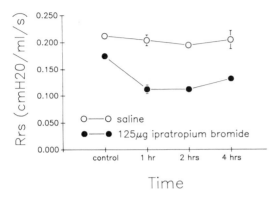

Figure 6 Effect of nebulized ipratropium bromide on total respiratory system resistance in a premature, ventilated infant (GA = 28.5 weeks, age = 34 days) with bronchopulmonary dysplasia. The positive response to ipratropium indicates the presence of cholinergic tone. Values are the mean ± SEM (Adapted from Brundage et al., 1990).

change in the maximal flow at functional residual capacity (FRC) following inhalation of a β agonist, while the CF infants did. Clearly, studies that employ measurements of pulmonary mechanics before and after inhalation of muscarinic antagonists in unanesthetized newborn animals or term infants would be helpful in settling this issue.

In adult mammals, chemoreceptor stimulation due to systemic hypoxia or hypercapnia causes a reflex increase in bronchomotor tone via vagal cholinergic mechanisms (Nadel and Widdicombe, 1962, 1963; Green and Widdicombe, 1966; for review see Coleridge et al., 1989). In the anesthetized newborn dog, the majority of animals respond to hypercapnia with an increase in total pulmonary resistance (Fig. 7, bottom) that is remarkably similar to values reported for the adult (Waldron and Fisher, 1988). Since the average increase in resistance in the newborn during hypercapnia is only about 20% of the maximal values recorded during vagal stimulation, this reflex response may be adaptive rather than detrimental, in that it would reduce dead space ventilation and/or reduce airway deformation during increased ventilation. The latter mechanism would be useful only if dynamic flow limitation was present during exposure to CO_2 in the newborn.

It is important to recall that in the unanesthetized adult dog, the ventilatory response to chemoreceptor stimulation eliminates the bronchoconstrictor effect (Sorkness and Vidruk, 1986). This is thought to be due to an increase in airway slowly adapting receptor activity as ventilation rises. This increased activity, in turn, causes a reflex reduction in bronchomotor tone. In contrast, hypoxia appears to be rather ineffective in causing an increase in bronchomotor tone in the anesthetized newborn (Fig. 7, middle) as compared to the adult (Fisher et al., 1987; Waldron and Fisher, 1988). This phenomenon is difficult to explain. Immature carotid body afferents are not likely responsible since recordings of afferent activity during hypoxia in

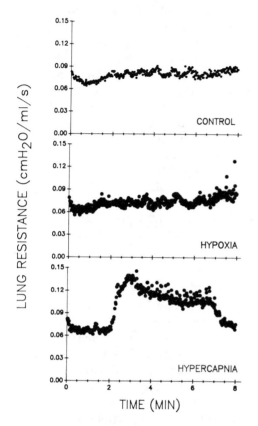

Figure 7 Effect of systemic hypoxia and hypercapnia on lung resistance in an anesthetized, ventilated 9-day-old dog. Each point represents a measurement for a single breath. Hypoxia and hypercapnia were induced by ventilation with hypoxic (10% O_2, 90% N_2) or hypercapnic (5% CO_2, 40% O_2, 65% N_2) gas between the first and sixth minutes of the recording periods shown. Control = (40% O_2, 60% N_2). Hypoxia caused an increase in resistance while hypoxia did not. (Adapted from Waldron and Fisher, 1988.)

kittens and lambs show sustained increases in activity (Blanco et al., 1984a,b; Walker, 1984). Dawes and co-workers (1983) have suggested that an active central inhibitory mechanism may suppress the response to hypoxia in utero. Whether this mechanism remains active in the immediate postnatal period is not known. In the adult, stimuli such as hypoxia cause an increase in drive to respiratory muscles as well as airway smooth muscle. Indeed, Mitchell and co-workers (Richardson et al., 1984; Mitchell et al., 1985) initially suggested that a common central inspiratory drive activated both respiratory skeletal and smooth muscle in parallel. Subsequent studies have reported several exceptions that make this suggestion unlikely as a general principle (for review, see Coleridge et al., 1989). However, Mitchell's hypothesis may be helpful in interpreting the lack of airway response to acute hypoxia in the newborn. The general respiratory response to acute hypoxia in the newborn consists of a drop in metabolism accompanied by no change, a decrease, or an increase in ventilation (Haddad and Mellins, 1984; Rigatto, 1984; Mortola et al., 1986; Saetta and Mortola, 1987; Piazza et al., 1988). Therefore, the reduced or constant central drive to respiratory muscles, which is appropriate for a drop in metabolism, may also underlie the lack of a bronchomotor response. Interestingly, the presence of chronic lung diseae may alter the response. Teague et al. (1988) found that infants with chronic lung disease increased their lung resistance during acute hypoxia, but did not change ventilation. On the other hand, infants free of lung disease did not change lung resistance but had a vigorous increase in ventilation. The latter response is reminiscent of that reported by Sorkness and Vidruk (1986) for unanesthetized adult dogs, where the reflex bronchodilation due to increased airway receptor activity overrides chemoreceptor-induced bronchoconstriction (see above). Studies of unanesthetized newborn animals, employing techniques similar to those used in the adult (Sorkness and Vidruk, 1986; Coleridge et al., 1989), are required to clarify the influence of chemoreflex control on airway caliber in the newborn.

B. Noncholinergic Excitatory Innervation

The ability of neuropeptides to cause bronchoconstriction is well documented in the adult (Lundberg and Saria, 1987; Andersson and Grundstrom, 1987), and the release of substance P and other members of the tachykinin family from airway afferents has been suggested to exert excitatory control via an axon reflex (for review, see Barnes, 1986; Lundberg and Saria, 1987; Andersson and Grundstrom, 1987; Coleridge et al., 1989). Two components of the airway response to tachykinins have been described: an increase in airway resistance and neurogenic inflammation involving an increase in airway vascular permeability (Lundberg and Saria, 1987). However, major species differences are present. Indeed, the guinea pig is the only mammalian species in which stimulation of the peripheral vagus causes a vigorous increase in resistance after cholinergic and adrenergic blockade (Lundberg and Saria, 1987; Andersson and Grundstrom, 1987). Other species have weak (e.g., rat) or absent (e.g., cat or dog) responses. Nevertheless, histochemical studies have revealed C-fibers containing peptides in the mucosa of several species, including humans (Lundberg et al., 1984; Lundberg and Saria, 1987), and it seems unlikely that such a rich innervation would not be "employed" physiologically.

    Grunstein et al. (1984) found a decrease in airway conductance during the infusion of substance P in rabbit pups ranging in age from 2 to 29 days. They also reported an age-dependent decrease in sensitivity to substance P. Atropine caused a decrease in the apparent potency of substance P, which the authors suggested reflects an "accelerated" spontaneous release of acetylcholine from vagal efferents due to substance P (Grunstein et al., 1984). The bronchoconstrictor response to substance P was unaffected by bilateral cervical vagotomy, indicating that vagal reflexes were not involved.

To our knowledge, data on the in vitro response of airway smooth muscle to substance P or other tachykinins are not available for the newborn. However, in the gastrointestinal tract, atropine-resistant contractions during electrical field stimulation are present in fetal and neonatal rats (stomach strips from embryonic day 15 to 7 days postnatal; Ito et al., 1988), and the magnitude of atropine-resistant contractions in the rabbit bladder increases in the first 2 weeks after birth (Maggi et al., 1984). In fetal rabbits, however, atropine completely abolishes the contractile response of intestinal tissue to electrical stimulation (Gershon and Thompson, 1973). On the basis of these studies, changes in in vitro responsivity (i.e., contractility and potency) may be expected for fetal and neonatal airway smooth muscle.

Studies of reflex changes in bronchomotor tone due to afferent C-fiber stimulation in the newborn are not available in the literature (Coleridge et al., 1989; see also Chapter 7, this volume). Reflex changes in ventilation due to chemical stimulation of C-fiber afferents suggest reduced reflex effects (Kalia, 1976; Trippenbach, 1988). In preliminary studies, we found that capsaicin (a stimulant of C-fiber afferents) caused a reflex bronchoconstriction in the newborn dog via increased efferent cholinergic activity, since atropine abolished the response (Fig. 8). It is not clear whether the newborn is more or less sensitive to C-fiber-induced changes in bronchomotor tone.

It is unfortunate that so little is known of the possible influences of C-fiber activity on bronchomotor tone in the newborn. These afferents outnumber myelinated axons 10 to 1 in the adult and play an extremely important role in controlling bronchomotor tone (Coleridge et al., 1989).

C. Nonadrenergic Noncholinergic Inhibitory Innervation

VIP-like immunoreactivity has been reported in the airways of human fetuses at 20 weeks' gestation (Sheppard et al., 1983). While the presence of this putative NANC neurotransmitter suggests early development of this system, no functional studies are available for human fetuses or neonates. The gastrointestinal literature has preceded evidence for NANC innervation to airway smooth muscle in the newborn. Gastrointestinal NANC inhibitory innervation is functional in the fetal rat (Ito et al., 1988) and rabbit (Gershon and Thompson, 1973), and the appearance of NANC control in the fetal rabbit is coincident with that of cholinergic innervation (Gershon and Thompson, 1973). In the newborn cat, vagal stimulation causes bronchodilation of precontracted airways (Fig. 4, bottom) after atropine and propranolol administration (Waldron et al., 1989). Vagal postganglionic fibers mediate

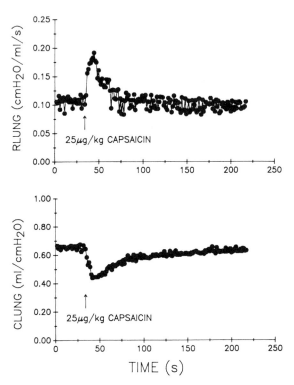

Figure 8 Effect of right heart capsaicin injection on lung resistance and dynamic compliance in a 4-day-old dog. Each point represents a measurement for a single breath. Capsaicin causes a reflex increase in bronchomotor tone (Fisher et al., unpublished).

the response since hexamethonium abolished the bronchodilation. The frequencies at which maximal relaxation is obtained (15-20/s; Fig. 5, bottom) are similar to those observed for cholinergic bronchoconstriction responses (Fisher et al., 1990). Studies of NANC inhibition in the adult cat report maximal relaxation at much higher frequencies (Diamond and O'Donnell, 1980). Although the physiological role of this innervation is not entirely clear, it has been suggested that the NANC mediator may act to reduce the excitatory effects of acetylcholine and that both transmitters are coreleased from the same endings (Diamond and Altiere, 1988; Lundberg and Hokfelt, 1986). A defect in NANC innervation has been suggested to play a role in airway hyperreactivity (Richardson and Beland, 1976; Barnes, 1986), and in support of this hypothesis, VIP immunoreactivity appears to be absent from airways of asthmatics (Ollerenshaw et al., 1989). These findings raise the possibility that the NANC system plays a role in the elevated airway resistance of infants with bronchopulmonary dysplasia (Hodgman, 1981; Motoyama et al., 1987).

Reflex activation of this system in the adult occurs during laryngeal irritation (Szarek et al., 1986) and stimulation of C-fiber afferents (Ichinose et al., 1987) in the lung. This system can also be reflexly recruited in adult humans (Michoud et al., 1987; Ichinose et al., 1988). Whether the NANC system is reflexly recruited in the newborn is not known. Clerici et al. (1989) have described a reflex activation of NANC innervation in young guinea pigs (13 days old) triggered by serotonin-induced bronchoconstriction. Adrenergic and cholinergic influences were excluded by the use of appropriate antagonists; the involvement of vagal postganglionic efferent fibers is implicated since the ganglionic blocker hexamethonium eliminated the response.

Clearly, very little is known of this innervation in the newborn, and until it is better understood, it will continue to be invoked as a possible mechanism underlying or contributing to airway disease.

D. Summary of Vagal Airway Innervation

Functional cholinergic innervation to airway smooth muscle in the newborn is consistently found among the mammalian species studied. This innervation may be reflexly activated by afferent input from chemoreceptors or pulmonary C-fibers. However, much remains to be learned regarding its reflex control. Studies that examine and compare reflexes affecting airway caliber in unanesthetized and anesthetized animal models will be important in understanding the physiological and pathological factors that influence bronchomotor tone. In vitro studies have been extremely useful in delineating the functional capabilities of vagal cholinergic innervation to airway smooth muscle. However, studies that employ innovative and appropriate normalization techniques, in precocial as well as altricial species, are required to answer the question of whether the newborn is more or less "reactive" than the adult to cholinergic and other agonists.

Our knowledge of NANC vagal innervation is minimal and many questions remain regarding its function. The relative effectiveness of NANC innervation or VIP (a putative NANC mediator) in the newborn compared to the adult, or to other inhibitory systems, is not available, and our knowledge of the reflex control of this system is rudimentary. Answers to these questions and others will increase our understanding of this enigmatic innervation in the newborn and provide an appreciation of the array of bronchomotor responses available to the neonate.

V. Adrenergic/Sympathetic Control of Airway Smooth Muscle

Biochemical evidence from several species suggests significant prenatal development of sympathetic innervation to the lungs. Dopamine-β-hydroxylase (an enzyme involved in neuronal catecholamine synthesis) has been detected in human fetal lungs at 10 weeks' gestation (Sheppard et al., 1983). In addition, a sharp increase in the content of norepinephrine and cyclic nucleotides occurs in rat lungs during the perinatal period (Tordet et al., 1981). Consistent with these findings, biochemical evidence of adrenergic innervation is present late in the gestation period for rabbit intestine (Gershon and Thompson, 1973). Studies specifically addressing the morphology of sympathetic airway innervation during development are not yet available.

A. β-Adrenergic Function

Histochemical and Pharmacological Evidence

High-affinity β-adrenergic receptors have been found in crude membrane fractions of lung tissue from fetal rats (Whitsett et al., 1982b), rabbits (Whitsett et al., 1981), and newborn guinea pigs (Gatto et al., 1984). In the rat and rabbit, the density of these receptors increases with gestational age and continues to increase until adulthood (Whitsett et al., 1981, 1982b). A similar increase in overall pulmonary β-receptor density during postnatal maturation occurs in the guinea pig, as well as a differential distribution of receptor density in pulmonary tissues (parenchyma > bronchi > trachea; Duncan et al., 1982). None of these studies found a significant change in the affinity of β receptors during development.

In an effort to more clearly delineate the tissue types in which β-receptor development occurs, Gatto et al. (1984) performed binding studies on slices of whole guinea pig lungs. Similar to the results of previous reports, they found an increase in overall β-receptor density with age. However, most of this increase could be accounted for by a proliferation of receptor sites in peripheral tissues such as alveolar membranes and type II pneumocytes. Beta receptors on airway smooth muscle did not appear to increase significantly with age and, in fact, decreased in a few cases (Gatto et al., 1984). A preferential increase in peripheral β receptors during late gestation and postnatal development is consistent with the observation that the basic structure of the tracheobronchial tree is complete in the newborn, while alveoli and parenchymal structures continue to grow and multiply long after birth (Thurlbeck, 1975; Langston et al., 1984).

In vitro studies

In vitro studies of airway smooth muscle from newborns of several species indicate that β receptors are functional at birth. Precontracted airway smooth muscle from newborn rabbit (Hayashi and Toda, 1980), guinea pig (Brink et al., 1980), and dog (Pandya, 1977) relaxes in response to the specific β agonist isoproterenol. Differences between the newborn and adult, with respect to the potency of β-receptor agonists mediating airway smooth muscle relaxation, may be species dependent. Pandya observed no age-related difference in the relaxation produced by a single dose of isoproterenol in acetylcholine-contracted dog tracheal muscle (Pandya, 1977). In contrast, newborn rabbit tracheal muscle, precontracted with acetylcholine, is relaxed more potently by isoproterenol than adult tissue (Hayashi and Toda, 1980). However, in this study, the degree of precontraction produced (a critical factor in evaluating relaxation responses) was not closely controlled. Guinea pig bronchial muscle (but not tracheal) precontracted with histamine is more sensitive to isoproterenol in young (118 g) than in old (757 g) guinea pigs (Brink et al., 1980). However, the relaxant effect of β agonists also depends on the agent used to create smooth muscle tone, or to precontract the airways (Duncan and Douglas, 1985a). If maximal contractions are induced with histamine, bronchial tissues appear to be more sensitive to salbutamol or isoproterenol in the newborn (< 1 week old) than in the adult guinea pig. However, when carbachol is used to create tone, the sensitivity of immature tissues to β agonists decreases as the precontraction is increased from 30 to 90% of maximum. Interaction between carbachol and the β agonists is also present in adult guinea pig tissues, but to a much smaller degree (Duncan and Douglas, 1985a). Other investigators have described a similar adrenergic/cholinergic interaction in airway smooth muscle from adult dogs (Russell, 1984) and cows (Andersson et al., 1975), in which the potency of isoproterenol as a tracheal smooth muscle relaxant is reduced if precontraction is produced with a cholinergic agonist rather than histamine or serotonin. Although data from other species are not available for comparison, the findings of Duncan and Douglas (1985a) suggest that a cholinergic inhibition of β-receptor effects may be more pronounced in the newborn and decrease during maturation. The mechanism responsible for the cholinergic interference with β-receptor-induced relaxation is not clear, although effects on receptor coupling and second-messenger systems may play a role (Russell, 1984). Further studies of the age-related interaction of adrenergic and cholinergic agents in different species are required.

In vivo functional studies

In vitro studies of the effects of β-adrenergic agonists on isolated human newborn or fetal airway tissue are not available; thus, our knowledge of the functional status of β receptors in the human newborn comes from observations of the effects of these agonists on lung mechanics in the intact infant.

Beta-adrenergic agonists are widely used in the treatment of infant respiratory ailments, variously described as "acute airway obstruction" or "wheeze" (Silverman, 1984). Surprisingly, in several clinical trials no decrease in airway resistance was observed following β-adrenergic agonist inhalation (Rutter et al., 1975; Radford, 1975; Lenney and Milner, 1978; Stokes et al., 1983). Nevertheless, the weight of more recent evidence (Fig. 9) indicates the presence of functional β-adrenergic receptors in airway smooth muscle in the human newborn. In infants with chronic airway obstruction, associated with bronchopulmonary dysplasia, adrenergic bronchodilators significantly decrease airway resistance (Kao et al., 1984; Gomez-Del Rio et al., 1986; Wilkie and Bryan, 1987). In other studies, baseline resistance of "wheezy" infants was not improved by salbutamol; however, this agent did protect against further airway constriction in response to administration of histamine (Prendiville et al., 1987a) or nebulized water (O'Callaghan et al., 1988). A variety of explanations have been proposed to account for the inability of β agonists to cause bronchodilation in the newborn in some clinical trials. The pathology of acute airway obstruction in the newborn may be such that, even if smooth muscle were relaxed by β agonists, the benefit of this would be overshadowed by the obstructing effects of edema, inflammation, and mucus accumulation (Silverman and Prendiville, 1987; Silverman, 1984). In addition, the methods used to measure lung mechanics and effectiveness of aerosol delivery may contribute to differences between studies.

Recently, a technique for measuring forced expiratory flows in infants, by means of thoracic compression at end inspiration, has gained favor as an indicator of changes in thoracic airway tone (Taussig et al., 1982). Although this method holds promise, results for nonintubated infants should be interpreted cautiously since "squeeze-related" upper airway reflexes

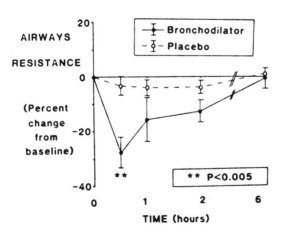

Figure 9 Evidence for functional β-adrenergic airway receptors in premature infants. Inhalation of 0.1% isoproterenol aerosol caused a significant reduction in airway resistance in infants with bronchopulmonary dysplasia. Values are mean ± SEM of measurements from 10 infants. Asterisk indicates significant difference from placebo. (Reprinted from Kao et al., 1984.)

(such as laryngeal adduction) could also influence maximal flows. Indeed, in any technique in which it is not bypassed by tracheal intubation, the upper airway may influence the measured results. Using the "squeeze" technique, Prendiville et al. (1987b) reported a paradoxical decrease in maximal flow at functional residual capacity ($V_{max}FRC$) in wheezy infants following nebulized salbutamol. They interpreted this as evidence for functional β-adrenergic receptors, which caused a relaxation of airway smooth muscle that would otherwise serve to "splint" the airways during forced expiration to maintain $V_{max}FRC$. No formal studies of the effects of β-adrenergic agonists on lung mechanics in an intact newborn animal model are available. However, we have observed a drop in pulmonary resistance following intravenous administration of salbutamol to anesthetized, ventilated kittens in which airway tone had been increased by an infusion of serotonin (see Waldron et al., 1989).

We are aware of only one study in which the response of airway smooth muscle to direct stimulation of sympathetic innervation has been examined in the newborn. Schwieler et al. (1970) found that sympathetic stimulation did not cause a relaxation of tracheal muscle in rabbits less than 1 week old, although an increase in heart rate was observed. They cautioned that sympathetic responses are very sensitive to the level of anesthesia, and the lack of response in newborn rabbits may simply reflect increased sensitivity to anesthesia at this age.

In their study of the innervation of fetal rabbit intestine, Gershon and Thompson (1973) found that functional development of adrenergic inhibitory innervation lagged far behind that of the cholinergic excitatory and nonadrenergic inhibitory innervations. Relaxation of fetal intestinal strips by electrical field stimulation was not affected by adrenergic antagonists until 30 days postnatally. However, the possibility remains that the relaxation produced by the nonadrenergic inhibitory system (for which antagonists are not available) was maximal, and hence, antagonism of any simultaneously activated adrenergic innervation would not have been detected. Development of adrenergic innervation after cholinergic innervation would be consistent with the order of their appearance in the phylogenetic evolution of visceral innervation (Burnstock, 1969). Whether a similar scenario is true for airway smooth muscle is not known.

B. Development of α-Adrenergic Receptors

Alpha receptors are relatively sparse in the normal lung and their relevance to autonomic control of airways has been questioned. The contractile effects of α-adrenergic stimulation on smooth muscle tone can be demonstrated only under special conditions (Barnes, 1986). Alpha receptors on airway ganglia, when activated by norepinephrine, inhibit the firing of ganglionic cells (Baker et al., 1983; Skoogh, 1986). In this way, sympathetic nervous activity may inhibit vagal bronchoconstriction despite the lack of direct innervation of airway smooth muscle β receptors.

In contrast to the persistent increase in numbers of β receptors, pulmonary α-adrenergic receptor density appears to peak in the early postnatal period (day 15 in the rat) and thereafter decrease until adult receptor levels are reached on postnatal day 28 (Whitsett et al., 1982b). Whitsett et al. (1982a) suggested that the postnatal decrease in α-receptor density may be a result of down-regulation as increased catecholamine is released from maturing neurons.

The pattern of increased numbers of α receptors in the neonatal period is consistent with the in vitro studies of Pandya (1977), who reported that mixed α- and β-adrenergic agonists such as epinephrine and norepinephrine elicited contractions in tracheal muscle from very young dogs (2-10 days) unless an α-adrenergic blocker was present. In contrast, tracheal muscle from older dogs was relaxed by these agonists unless a β-adrenergic antagonist was used. Pandya suggests that, in the dog, α receptors predominate in the newborn and β receptors become more prevalent as the animal ages (Pandya, 1977). The differential development of adrenergic receptor subpopulations is an intriguing possibility which may have clinical ramifications, and further research in this area is needed.

C. Summary

Biochemical evidence of sympathetic innervation has been found prenatally in several species. Pulmonary α- and β-adrenergic receptors develop before birth and are present in newborn mammals. Alpha receptors continue to increase in number for a short time after birth and then decline, while β receptors continue to proliferate until adulthood, although this increase is mostly in lung parenchymal tissues. Newborn mammalian airway smooth muscle relaxes in response to β-adrenergic agonists and may be more susceptible to the excitatory influences of α-receptor activation than the adult, although more information is needed. The single existing study of the influence of sympathetic nerve stimulation on airway smooth muscle in the newborn suggests that this innervation is functional, although results for the very early postnatal period (1 week) are inconclusive. Clearly, insufficient data exist to speculate on the relative role played by adrenergic versus nonadrenergic innervation in the control of airway smooth muscle in the neonatal period.

VI. Conclusion

Attention to the topic of neural control airway smooth muscle in the newborn is overdue. While many of the findings we have described for the newborn are the result of very recent work, analogous information for the adult is described in reviews of the literature a decade old (Nadel, 1980; Boushey et al., 1980). Research on the control of airway smooth muscle in the newborn will move rapidly given the wealth of information available from studies in the adult; however, many questions are unique to the fetus and newborn. In vivo studies on topics such as the central and peripheral control of airway tone, and the role of neuropeptides, will be important in delineating the limits to, and mechanisms of, normal and pathological airway function. The use of traditional techniques in studies of airway function, coupled with neurobiological methods, such as patch-clamping of airway cells, or molecular biological techniques, such as those which led to the identification of four or five distinct muscarinic receptors (Mitchelson, 1988), should cause rapid growth in our understanding of airway smooth muscle control in the newborn.

Acknowledgments

We thank K. Brundage for critically reading the manuscript and Ms. Sandra Vincent for preparing the figures.

References

Alcorn, D. G., Adamson, T. M., Maloney, J. E., and Robinson, P. M. (1981). A morphologic and morphometric analysis of fetal lung development in the sheep. *Anat. Rec.* **201**:655-667.

Andersson, R. G. G., and Grundstrom, N. (1987). Innervation of airway smooth muscle. Efferent mechanisms. *Pharmacol. Ther.* **32**:107-130.

Andersson, R., Nilsson, K., Wikberg, K., Johansson, S., Lundholm, E., and Lundholm, L. (1975). Cyclic nucleotides and the contraction of smooth muscle. *Adv. Cyclic Nucleotide Res.* **5**:491-518.

Baker, D. G. (1986). Parasympathetic motor pathways to the trachea: Recent morphologic and electrophysiologic studies. *Clin. Chest Med.* **7**: 223-229.

Baker, D. G., Basbaum, C. B., Herbert, D. A., and Mitchell, R. A. (1983). Transmission in airway ganglia of ferrets: Inhibition by norepinephrine. *Neurosci. Lett.* **41**:139-143.

Barnes, P. J. (1986). Neural control of human airways in health and disease. *Am. Rev. Respir. Dis.* **134**:1289-1314.

Bayol, A., Benveniste, J., Brink, C., Cerrina, J., Gateau, O., Labat, C., and Raffestin, B. (1985). Response and sensitivity of guinea-pig airway muscle preparations to 5-hydroxytryptamine during ontogenesis. *Br. J. Pharmacol.* **85**:569-574.

Bhutani, V. K., Koslo, R. J., and Shaffer, T. H. (1986). The effects of tracheal smooth muscle tone on neonatal airway collapsibility. *Pediatr. Res.* **20**:492-495.

Bhutani, V. K., Rubenstein, S. D., and Shaffer, T. H. (1981a). Pressure-volume relationships of,tracheae in fetal, newborn, and adult rabbits. *Respir. Physiol.* **43**:221-231.

Bhutani, V. K., Rubenstein, S. D., and Shaffer, T. H. (1981b). Pressure induced deformation in immature airways. *Pediatr. Res.* **15**:829-832.

Blanco, C. E., Dawes, G. S., Hanson, M. A., and McCooke, H. B. (1984b). The response to hypoxia of arterial chemoreceptors in fetal sheep and new-born lambs. *J. Physiol. (London)* **351**:25-37.

Blanco, C. E., Hanson, M. A., Johnson, P., and Rigatto, H. (1984a). Breathing pattern of kittens during hypoxia. *J. Appl. Physiol.* **56**:12-17.

Boushey, H. A., Holtzman, M. J., Sheller, J. R., and Nadel, J. A. (1980). Bronchial hyperreactivity. *Am. Rev. Respir. Dis.* **121**:389-413.

Bradley, K. H., McConnell, S. D., and Crystal, R. G. (1974). Lung collagen composition and biosynthesis: characterization and changes with age. *J. Biol. Chem.* **249**:2674-2683.

Brink, C., Duncan, P. G., Midzenski, M., and Douglas, J. S. (1980). Response and sensitivity of female guinea pig respiratory tissues to agonists during ontogenesis. *J. Pharmacol. Exp. Ther.* **215**:426-433.

Brundage, K. L., Mohsini, K. G., Froese, A. B., and Fisher, J. T. (1990). Bronchodilator response to ipratropium bromide in infants with bronchopulmonary dysplasia. *Am. Rev. Respir. Dis.* **142**(in press).

Burnstock, G. (1969). Evolution of the autonomic innervation of visceral and cardiovascular systems in vertebrates. *Pharmacol. Rev.* **21**:247-324.

Burnstock, G. (1988). Autonomic neural control mechanisms: with special reference to the airways. In *The Airways. Neural Control in Health and Disease.* Edited by M. A. Kaliner and P. J. Barnes. New York, Marcel Dekker, pp. 1-22.

Burri, P. H. (1974). The postnatal growth of the rat lung III. Morphology. *Anat. Rec.* **180**:77-98.

Burri, P. H. (1984). Fetal and postnatal development of the lung. *Annu. Rev. Physiol.* **46**:617-628.

Burri, P. H., Dbaly, J., and Weibel, E. R. (1974). The postnatal growth of the rat lung. I. Morphometry. *Anat. Rec.* **178**:711-730.

Cabezas, G. A., Graf, P. D., and Nadel, J. A. (1971). Sympathetic versus parasympathetic nervous regulation of airways in dogs. *J. Appl. Physiol.* **31**:651-655.

Clement, M. G., Mortola, J. P., Albertini, M., and Aguggini, G. (1986). Effects of vagotomy on respiratory mechanics in newborn and adult pigs. *J. Appl. Physiol.* **60**:1992-1999.

Clerici, C., MacQuin-Mavier, I., and Harf, A. (1989). Nonadrenergic bronchodilation in newborn guinea pigs. *J. Appl. Physiol.* **67**:1764-1769.

Coburn, R. F. (1987). Peripheral airway ganglia. *Annu. Rev. Physiol.* **49**: 573-582.

Coburn, R. F. (1988). Cholinergic neuroeffector mechanisms in airway smooth muscle. In *The Airways: Neural Control in Health and Disease.* Edited by M. A. Kaliner, and P. J. Barnes. New York, Marcel Dekker, pp. 159-186.

Coburn, R. F., and Tomita, T. (1973). Evidence for nonadrenergic inhibitory nerves in guinea pig trachealis muscle. *Am. J. Physiol.* **224**:1072-1080.

Cohen, M., and Berkowitz, B. A. (1974). Age-related changes in vascular responsiveness to cyclic nucleotides and contractile agonists. *J. Pharmacol. Exp. Ther.* **191**:147-155.

Coleridge, H. M., Coleridge, J. C. G., and Schultz, H. D. (1989). Afferent pathways involved in reflex regulation of airway smooth muscle. *Pharmacol. Ther.* **42**:1-63.

Daniel, E. E., Kannan, M., Davis, C., and Posey-Daniel, V. (1986). Ultrastructural studies on the neuromuscular control of human tracheal and bronchial muscle. *Respir. Physiol.* **63**:109-128.

Davis, C., Kannan, M. S., Jones, T. R., and Daniel, E. E. (1982). Control of human airway smooth muscle: In vitro studies. *J. Appl. Physiol.* **53**:1080-1087.

Dawes, G. S., Gardner, W. N., Johnston, B. M., and Walker, D. W. (1983). Breathing in fetal lambs: The effect of brain stem sectioning. *J. Physiol. (London)* **335**:535-553.

DeNeef, K. J., Jansen, J. R. C., and Versprille, A. (1982). Developmental morphometry and physiology of the rabbit vagus nerve. *Dev. Brain Res.* **4**:265-274.

Diamond, L., and Altiere, R. J. (1988). Airway nonadrenergic noncholinergic inhibitory nervous system. In *The Airways: Neural Control in Health and Disease.* Edited by M. A. Kaliner and P. J. Barnes. New York, Marcel Dekker, pp. 343-394.

Diamond, L., and O'Donnell, M. (1980). A non-adrenergic vagal inhibitory pathway to feline airways. *Science* **208**:185-188.

Downing, S. E. (1960). Baroreceptor reflexes in new-born rabbits. *J. Physiol. (London)* **150**:201-213.

Duncan, P. G., Brink, C., and Douglas, J. S. (1982). Beta-receptors during aging in respiratory tissues. *Eur. J. Pharmacol.* **78**:45-52.

Duncan, P. G., and Douglas, J. S. (1985a). Age-related changes in guinea pig respiratory tissues: Considerations for assessment of bronchodilators. *Eur. J. Pharmacol.* **108**:39-48.

Duncan, P. G., and Douglas, J. S. (1985b). Influences of gender and maturation on responses of guinea-pig airway tissues to LTD4. *Eur. J. Pharmacol.* **112**:423-427.

Duthie, K. L., Watson, L. A., and Fisher, J. T. (1989). Muscarinic blockade of vagally mediated bradycardia and bronchoconstriction in the newborn dog. *FASEB. J.* **3**:A276.

Einthoven, W. (1975). On the action of the bronchial muscles, studied with a new method and on neurogenic asthma. In *Translations in Respiratory Physiology*. Edited by J. B. West. Stroudsburg, PA, Dowden, Hutchinson and Ross, pp. 319-377.

Fisher, J. T., Brundage, K. L., Waldron, M. A., and Connelly, B. J. (1990). Vagal cholinergic innervation of the airways in newborn cat and dog. *J. Appl. Physiol.* **69**(in press).

Fisher, J. T., and Sant'Ambrogio, G. (1982a). Location and discharge properties of respiratory vagal afferents in the newborn dog. *Respir. Physiol.* **50**:209-220.

Fisher, J. T., and Sant'Ambrogio, G. (1982b). Effects of inhaled CO_2 on airway stretch receptors in the newborn dog. *J. Appl. Physiol.* **53**: 1461-1465.

Fisher, J. T., Sant'Ambrogio, F. B., and Sant'Ambrogio, G. (1983). Stimulation of tracheal slowly adapting stretch receptors by hypercapnia and hypoxia. *Respir. Physiol.* **53**:325-339.

Fisher, J. T., Waldron, M. A., and Armstrong, C. J. (1987). Effects of hypoxia on lung mechanics in the newborn cat. *Can. J. Physiol. Pharmacol.* **65**:1234-1238.

Gabella, G. (1987). Innervation of airway smooth muscle: Fine structure. *Annu. Rev. Physiol.* **49**:583-594.

Gatto, C., Johnson, M. G., Seybold, V., Kulik, T. J., Lock, J. E., and Johnson, D. E. (1984). Distribution and quantitative developmental changes in guinea pig pulmonary beta-receptors. *J. Appl. Physiol.* **57**: 1901-1907.

Geiss, W. P., Tatooles, G. J., Priola, D. V., and Friedman, W. F. (1975). Factors influencing neurohumoral control of the heart in the newborn dog. *Am. J. Physiol.* **228**:1685-1689.

Gershon, M. D., and Thompson, E. B. (1973). The maturation of neuromuscular function in a multiply innervated structure: Development of the longitudinal smooth muscle of the foetal mammalian gut and its cholinergic excitatory, adrenergic inhibitory, and non-adrenergic inhibitory innervation. *J. Physiol. (London)* **234**:257-277.

Gomez-Del Rio, M., Gerhardt, T., Hehre, D., Feller, R., and Bancalari, E. (1986). Effect of a beta-agonist nebulization on lung function in neonates with increased pulmonary resistance. *Pediatr. Pulmonol.* **2**:287-291.

Green, M., and Widdicombe, J. G. (1966). The effect of ventilation of dogs with different gas mixtures on airway calibre and lung mechanics. *J. Physiol. (London)* **186**:363-381.

Grunstein, M. M., Tanaka, D. T., and Grunstein, J. S. (1984). Mechanisms of substance P-induced bronchoconstriction in maturing rabbits. *J. Appl. Physiol.* **57**:(4):1238-1246.

Gunst, S. J., and Stropp, J. Q. (1988). Pressure-volume and length-stress relationships in canine bronchi in vitro. *J. Appl. Physiol.* **64**:2522-2531.

Haddad, G. G., and Mellins, R. B. (1984). Hypoxia and respiratory control in early life. *Annu. Rev. Physiol.* **46**:629-43.

Hahn, H. L., Graf, P. D., and Nadel, J. A. (1976). Effect of vagal tone on airway diameters and on lung volume in anaesthetized dogs. *J. Appl. Physiol.* **41**:581-589.

Hayashi, S., and Toda, N. (1980). Age-related alterations in the response of rabbit tracheal smooth muscle to agents. *J. Pharmacol. Exp. Ther.* **214**:675-681.

Hiatt, P., Eigen, H., Yu, P., and Tepper, R. S. (1988). Bronchodilator responsiveness in infants and young children with cystic fibrosis. *Am. Rev. Respir. Dis.* **137**:119-122.

Hodgman, J. E. (1981). Chronic lung disorders. In *Neonatology: Pathophysiology and Management of the Newborn*. Edited by G. B. Avery. Philadelphia, J. B. Lippincott pp. 398-411.

Ichinose, M., Inoue, H., Miura, M., and Takashima, T. (1988). Nonadrenergic bronchodilation in normal subjects. *Am. Rev. Respir. Dis.* **138**:31-34.

Ichinose, M., Inoue, H., Miura, M., Yafuso, N., Nogami, H., and Takishima, T. (1987). Possible sensory receptor of nonadrenergic inhibitory nervous system. *J. Appl. Physiol.* **63**:923-929.

Irvin, C. G., Boileau, R., Tremblay, J., Martin, R. R., and Macklem, P. T. (1980). Bronchodilatation: Noncholinergic, nonadrenergic mediation demonstrated in vivo in the cat. *Science* **207**:791-792.

Irvin, C. G., Martin, R. R., and Macklem, P. T. (1982). Nonpurinergic nature and efficacy of nonadrenergic bronchodilation. *J. Appl. Physiol.* **52**:562-569.

Isabel, J. B., Towers, B., Adams, F. H., and Gyepes, M. T. (1972). The effects of ganglionic blockade on tracheobronchial muscle in fetal and newborn lambs. *Respir. Physiol.* **15**:255-267.

Ito, S., Kimura, A., and Ohga, A. (1988). Development of non-cholinegic, non-adrenergic excitatory and inhibitory responses to intramural nerve stimulation in rat stomach. *Br. J. Pharmacol.* **93**:684-692.

Jammes, Y., Fornaris, E., Mei, N., and Barrat, E. (1982). Afferent and efferent components of the bronchial vagal branches in cats. *J. Auton. Nerv. Syst.* **5**:165-176.

Jansen, A. H., and Chernick, V. (1983). Development of respiratory control. *Physiol. Rev.* **63**:437-483.

Jordan, D., Spyer, K. M., Withington-Wray, D. J., and Wood, L. M. (1986). Histochemical and electrophysiological identification of cardiac and pulmonary vagal preganglionic neurones in the cat. *J. Physiol. (London)* **372**:87.

Kalia, M. (1976). Visceral and somatic reflexes produced by J pulmonary receptors in newborn kittens. *J. Appl. Physiol.* **41**:1-6.

Kalia, M., and Mesulam, M. M. (1980). Brain stem projections of sensory and motor components of the vagus complex in the cat. II. Laryngeal, tracheobronchial, pulmonary, cardiac and gastrointestinal branches. *J. Comp. Neurol.* **193**:467-508.

Kao, L. C., Warburton, D., Platzker, A. C. G., and Keens, T. G. (1984). Effect of isoproterenol inhalation on airway resistance in chronic bronchopulmonary dysplasia. *Pediatrics* **73**:509-514.

Kirkpatrick, C. T. (1984). Nervous control of airway muscle tone. *Bull. Eur. Physiopathol. Respir.* **20**:389-394.

Krous, H. F., Jordan, J., Wen, J., and Farber, J. P. (1985). Developmental morphology of the vagus nerve in the opossum. *Dev. Brain Res.* **20**:155-159.

Langston, C., Kida, K., Reed, M., and Thurlbeck, W. M. (1984). Human lung growth in late gestation and in the neonate. *Am. Rev. Respir. Dis.* **129**:607-613.

Larsell, O., and Dow, R. S. (1933). The innervation of the human lung. *Am. J. Anat.* **52**:125-135.

Lenney, W., and Milner, A. D. (1978). At what age do bronchodilators work?. *Arch. Dis. Child.* **53**:532-535.

LeSoeuf, P. N., Geelhoed, G. C., Turner, D. J., Morgan, S. E. G., and Landau, L. I. (1989). Response of normal infants to inhaled histamine. *Am. Rev. Respir. Dis.* **139**:62-66.

Levin, R. M., Malkowick, D., Jacobowitz, D., and Wein, A. J. (1982). The ontogeny of the autonomic innervation and contractile response of the rabbit urinary bladder. *J. Pharmacol. Exp. Ther.* **219**:250-257.

Lundberg, J. M., and Hokfelt, T. (1986). Multiple co-existence of peptides and classical transmitters in peripheral autonomic and sensory neurons—Functional and pharmacological implications. In *Coexistence of Neuronal Messengers: A New Principal in Chemical Transmission.* (Prog. in Brain Res, Vol. 68. Edited by T. Hokfelt K. Fuxe, and B. Pernow. New York, Elsevier Science, pp. 241-262.

Lundberg, J. M., Hokfelt, T., Martling, C-R., Saria, A., and Cuello, C. (1984). Substance P-immunoreactive sensory nerves in the lower respiratory tract of various mammals including man. *Cell Tissue Res.* 235:251-261.

Lundberg, J. M., and Saria, A. (1987). Polypeptide-containing neurons in airway smooth muscle. *Annu. Rev. Physiol.* 49:557-572.

Maggi, C. A., Santicioli, P. and Meli, A. (1984). Postnatal development of myogenic contractile activity and excitatory innervation of rat urinary bladder. *Am. J. Physiol.* 247:R972-R978.

Marlot, D., and Duron, B. (1979). Postnatal development of vagal control of breathing in the kitten. *J. Physiol. (Paris)* 75:891-900.

Marlot, D., and Duron, B. (1981). Postnatal development of the discharge pattern of phrenic motor unit in the kitten. *Respir. Physiol.* 46:125-136.

McAllen, R. M., and Spyer, K. M. (1978). Two types of vagal preganglionic motoneurons projecting to the heart and lungs. *J. Physiol. (London)* 282:353-364.

McCrady, E., Jr. (1938). Embryology of the opossum. *Am. Anat. Mem.* 16:1-233.

Michoud, M-C., Amyot, R., Jeanneret-Grosjean, A., and Couture, J. (1987). Reflex decrease of histamine-induced bronchoconstriction after laryngeal stimulation in humans. *Am. Rev. Respir. Dis.* 136:618-622.

Middendorf, W. F., and Russell, J. A. (1980). Innervation of airway smooth muscle in the baboon: Evidence for a nonadrenergic inhibitory system. *J. Appl. Physiol.* 48:947-956.

Mitchell, R. A., Herbert, D. A., and Baker, D. G. (1985). Inspiratory rhythm in airway smooth muscle tone. *J. Appl. Physiol.* 58:911-920.

Mitchelson, F. (1988). Muscarinic receptor differentiation. *Pharmacol. Ther.* 37:357-423.

Mohammad, M. A., and Sparrow, M. P. (1988). Changes in myosin heavy chain stoichiometry in pig tracheal smooth muscle during development. *F.E.B.S.* 228:109-112.

Morikawa, Y., Donahoe, P. K., and Hendren, W. H. (1978a). Cholinergic nerve development of fetal lung in vitro. *J. Pediatr. Surg.* 13:653-661.

Morikawa, Y., Donahoe, P. K., and Hendren, W. H. (1978b). Cholinergic nerve development in fetal lung. *Dev. Biol.* 65:541-546.

Mortola, J. P., Morgan, C. A., and Virgona, V. (1986). Respiratory adaptation to chronic hypoxia in newborn rats. *J. Appl. Physiol.* 61:1329-1336.

Motoyama, E. K., Fort, M. D., Klesh, K. W., Mutich, R. L., and Gutherie, R. D. (1987). Early onset of airway reactivity in premature infants with bronchopulmonary dysplasia. *Am. Rev. Respir. Dis.* 136:50-57.

Murphy, T. M., Mitchell, R. W., Blake, J. S., Mack, M. M., Kelly, E. A., Munoz, N. M., and Leff, A. R. (1989). Expression of airway contractile properties and acetylcholinesterase activity in swine. *J. Appl. Physiol.* 67:174-180.

Nadel, J. A. (1980). Autonomic regulation of smooth muscle. In *Physiology and Pharmacology of the Airways*. Edited by J. A. Nadel. New York, Marcel Dekker, pp. 217-257.

Nadel, J. A., and Barnes, P. J. (1984). Autonomic regulation of the airways. *Annu. Rev. Med.* **35**:451-467.

Nadel, J. A., and Widdicombe, J. G. (1962). Effect of changes in blood gas tensions and carotid sinus pressure on tracheal volume and total lung resistance to airflow. *J. Physiol.* **163**:13-33.

Nadel, J. A., and Widdicombe, J. G. (1963). Reflex control of airway size. *Ann. NY. Acad. Sci.* **109**:712-723.

O'Callaghan, C., Milner, A. D., and Swarbrick, A. (1988). Nebulised salbutamol does have a protective effect on airways in children under 1 year old. *Arch. Dis. Child.* **63**:479-483.

Ollerenshaw, S., Jarvis, D., Woolcock, A., Sullivan, C., and Scheibner, T. (1989). Absence of immunoreactive vasoactive intestinal polypeptide in tissue from the lungs of patients with asthma. *N. Engl. J. Med.* **320**: 1244-1248.

Olsen, C. R., Colebatch, H. J. H., Mebel, P. E., Nadel, J. A., and Staub, N. C. (1965). Motor control of pulmonary airways studied by nerve stimulation. *J. Appl. Physiol.* **20**(2):202-208.

Otis, A. B. (1983). A perspective of respiratory mechanics. *J. Appl. Physiol.* **54**(5):1183-1187.

Pandya, K. H. (1977). Postnatal developmental changes in adrenergic receptor responses of the dog tracheal muscle. *Arch. Int. Pharmacodyn. Ther.* **230**:53-64.

Panitch, H. B., Allen, J. L., Ryan, J. P., Wolfson, M. R., and Shaffer, T. H. (1989). A comparison of preterm and adult airway smooth muscle mechanics. *J. Appl. Physiol.* **66**:1760-1765.

Pessacq, T. P. (1971). The innervation of the lung of newborn children. *Acta Anat.* **79**:93-101.

Piazza, T., Lauzon, A. M., and Mortola, J. P. (1988). Time course of adaptation of hypoxia in newborn rats. *Can. J. Physiol. Pharmacol.* **66**: 152-158.

Prendiville, A., Green, S., and Silverman, M. (1987a). Airway responsiveness in wheezy infants: evidence for functional beta adrenergic receptors. *Thorax* **42**:100-104.

Prendiville, A., Green, S., and Silverman, M. (1987b). Paradoxical response to nebulised salbutamol in wheezy infants, assessed by partial expiratory flow-volume curves. *Thorax* **42**:86-91.

Prendiville, A., Green, S., and Silverman, M. (1987c). Bronchial responsiveness to histamine in wheezy infants. *Thorax* **42**:92-99.

Radford, M. (1975). Effect of salbutamol in infants with wheezy bronchitis. *Arch. Dis. Child.* **50**:535-538.

Reid, L. (1985). Developmental anatomy, embryology of the foregut. *Am. Rev. Respir. Dis.* **131**:S60-S61.

Reynolds, H. C. (1952). Studies on reproduction in the opossum (*Didelphis virginiana virginiana*). *Univ. Calif. Berkeley Publ. Zool.* **52**:223-283.

Richardson, C. A., Herbert, D. A., and Mitchell, R. A. (1984). Modulation of pulmonary stretch receptors and airway resistance by parasympathetic efferents. *J. Appl. Physiol.* **57**(6):1842-1849.

Richardson, J. B. (1979). Nerve supply to the lungs. *Am. Rev. Respir. Dis.* **119**:785-803.

Richardson, J., and Beland, J. (1976). Non-adrenergic inhibitory nervous system in human airways. *J. Appl. Physiol.* **41**:764-771.

Richardson, J. B., and Ferguson, C. C. (1980). Morphology of the airways. In *Physiology and Pharamacology of the Airways*. Edited by J. A. Nadel. New York, Marcel Dekker, pp. 1-30.

Rigatto, H. (1984). Control of ventilation in the newborn. *Annu. Rev. Physiol.* **46**:661-674.

Rossi, A., and Mortola, J. P. (1987). Vagal influence on respiratory mechanics in newborn kittens. *Bull. Eur. Physiopathol. Respir.* **23**:61-66.

Rothberg, K. G., Morris, P. L., and Douglas, J. S. (1987). Characterization of cholinergic muscarinic receptors in cow tracheal muscle membranes. *Biochem. Pharmacol.* **36**:1687-1695.

Russell, J. A. (1978). Responses of isolated canine airways to electric stimulation and acetylcholine. *J. Appl. Physiol.* **45**:690-698.

Russell, J. A. (1980a). Innervation of airway smooth muscle in the dog. *Bull. Eur. Physiopathol. Respir.* **16**:671-692.

Russell, J. A. (1980b). Noradrenergic inhibitory innervation of canine airway. *J. Appl. Physiol.* **48**(1):16-22.

Russell, J. A. (1984). Differential inhibitory effect of isoproterenol on contractions of canine airways. *J. Appl. Physiol.* **57**(3):801-807.

Rutter, N., Milner, A. D., and Hiller, E. J. (1975). Effect of bronchodilators on respiratory resistance in infants and young children with bronchiolitis and wheezy bronchitis. *Arch. Dis. Child.* **50**:719-722.

Saetta, M., and Mortola, J. P. (1987). Interaction of hypoxic and hypercapnic stimuli on breathing pattern in the newborn rat. *J. Appl. Physiol.* **62**:506-512.

Sauder, R. A., McNicol, K. J., and Stecenko, A. A. (1986). Effect of age on lung mechanics and airway reactivity in lambs. *J. Appl. Physiol.* **61**:2074-2080.

Schwieler, G. H. (1966). The effect of vagal stimulation on the bronchial tree of the cat. *Acta Physiol. Scand.* **66**:495-500.

Schwieler, G. H. (1968). Respiratory regulation during postnatal development in cats and rabbits and some of its morphological substrates. *Acta Physiol. Scand.* (Suppl. 304):1-123.

Schwieler, G. H., Douglas, J. S., and Bouhuys, A. (1970). Postnatal development of autonomic efferent innervation in the rabbit. *Am. J. Physiol.* **219**(2):391-397.

Seidel, C. L., and Allen, J. C. (1979). Pharmacological characteristics and actomyosin content of aorta from neonatal rats. *Am. J. Physiol.* **237**: C81-C86.

Seidel, C. L., and Murphy, R. A. (1979). Changes in rat aortic actomyosin content with maturation. *Blood Vessels* **16**:98-108.

Seidel, C. L., Ross, B., Michael, L., Freedman, J., Burdick, B., and Miller, T. (1987). Maturational chyanges in the pharmacological characteristics and actomyosin content of canine arterial and venous tissue. *Pediatr. Res.* **21**:152-158.

Seow, C. Y., and Stephens, N. L. (1988). Velocity-length-time relations in canine tracheal smooth muscle. *J. Appl. Physiol.* **64**:2053-2057.

Sheppard, M. N., Marangos, P. J., Bloom, S. R., and Polak, J. M. (1983). Neuron specific enolase: A marker for the early development of nerves and endocrine cells in the human lung. *Life Sci.* **34**:265-271.

Silverman, M. (1984). Bronchodilators for wheezy infants? *Arch. Dis. Child.* **59**:84-87.

Silverman, M., and Prendiville, A. (1987). Airway responsiveness in infancy. *Am. Rev. Respir. Dis.* **4**:S71-S73.

Silverman, M., Prendiville, A., and Green, S. (1986). Partial expiratory flow-volume curves in infancy: Technical aspects. *Bull. Eur. Physiopathol. Respir.* **22**:257-262.

Skoogh, B-E. (1986). Parasympathetic ganglia in the airways. *Bull. Eur. Physiopathol. Respir.* (Suppl. 7):143-147.

Skoogh, B-E. (1988). Airway parasympathetic ganglia. In *The Airways: Neural Control in Health and Disease.* Edited by M. A. Kaliner and P. J. Barnes. New York, Marcel Dekker, pp. 217-240.

Sorkness, R. L., and Vidruk, E. H. (1986). Ventilatory responses to hypoxia nullify hypoxic tracheal constriction in awake dogs. *Respir. Physiol.* **66**:41-52.

Souhrada, M., Rothberg, K. G., and Douglas, J. S. (1988). Membrane properties of bovine smooth muscle cells: Effects of maturation. *Pulmon. Pharmacol.* **1**:47-52.

Sparrow, M. P., and Mitchell, H. W. (1990). Contraction of smooth muscle of pig airway tissues from before birth to maturity. *J. Appl. Physiol.* **68**:468-477.

Stecenko, A., McNicol, K., and Polk, S. (1989). Evaluation of the mechanism of decreased airway responsiveness in lambs. *J. Appl. Physiol.* **66**: 727-731.

Stephens, N. L. (1988). Structure of airway smooth muscle. In *Asthma: Basic Mechanisms and Clinical Management.* Edited by P. J. Barnes, I. W. Rodger, and N. C. Thomson. London, Academic Press, pp. 11-34.

Stephens, N. L., and Hoppin, F. G., Jr. (1986). Mechanical properties of airway smooth muscle. In *Handbook of Physiology,* Section 3: *The Respiratory System,* Vol. III. Mechanics of breathing, Part 1. Edited by A. P. Fishman, P. T. Macklem, J. Mead, and S. R. Geiger. Bethesda, MD, American Physiological Society, pp. 263-276.

Stephens, N. L., and Kroeger, E. A. (1980). Ultrastructure, biophysics and biochemistry of airway smooth muscle. In *Lung Biology in Health and Disease. Physiology and Pharmacology of the Airways.* Edited by J. A. Nadel. New York, Marcel Dekker.

Stokes, G. M., Milner, A. D., Hodges, I. G. C., and Henry, R. L. (1983). Nebulised ipratropium bromide in wheezy infants and young children. *Eur. J. Respir. Dis.* (Suppl. 128):494-498.

Szarek, J. L., Gillespie, M. N., Altiere, R. J., and Diamond, L. (1986). Reflex activation of the non-adrenergic non-cholinergic inhibitory nervous system in feline airways. *Am. Rev. Respir. Dis.* **133**:1159-1162.

Taussig, L. M., Laundau, L. I., Godfrey, S., and Arad, I. (1982). Determinants of forced expiratory flow in newborn infants. *J. Appl. Physiol.* **53**:1220-1227.

Taylor, I. M., and Smith, R. B. (1971). Intrinsic innervation of the human feotal lung between 35 and 140 mm crown-rump length stages. *Biol. Neonate* **18**:193-198.

Teague, W. G., Pian, M. S., Heldt, G. P., and Tooley, W. H. (1988). An acute reduction in the fraction of inspired oxygen increases airway constriction in infants with chronic lung disease. *Am. Rev. Respir. Dis.* **137**:861-865.

Tepper, R. S. (1987). Airway reactivity in infants: A positive response to methacholine and metaproterenol. *J. Appl. Physiol.* **62**:1155-1159.

Thurlbeck, W. M. (1975). Postnatal growth and development of the lung. *Am. Rev. Respir. Dis.* **111**:803-844.

Tordet, C., Bertin, R., and Gardey, C. (1981). Lung catecholamines and cyclic nucleotides during perinatal development in the rat. Possible relationships with biochemical and morphological differentiation. *Pediatr. Res.* **15**:787-793.

Trippenbach, T. (1988). The pulmonary chemoreflex in newborn rabbits. *Physiologist* **31**:A6.

Vidruk, E. H., and Sorkness, R. L. (1985). Histamine-induced reflex tracheal constriction is attenuated by hyperoxia and exaggerated by hypoxia. *Am. Rev. Respir. Dis.* **132**:287-291.

Waldron, M. A., Connelly, B. J., and Fisher, J. T. (1989). Nonadrenergic inhibitory innervation to the airways of the newborn cat. *J. Appl. Physiol.* **66**:1995-2000.

Waldron, M. A., and Fisher, J. T. (1988). Differential effects of CO_2 and hypoxia on bronchomotor tone in the newborn dog. *Respir. Physiol.* **72**:271-282.

Walker, D. W. (1984). Peripheral and central chemoreceptors in the fetus and newborn. *Annu. Rev. Physiol.* **46**:687-703.

Whitsett, J. A., and Hollinger, B. (1984). Muscarinic cholinergic receptors in developing rat lung. *Pediatr. Res.* **18**:1136-1140.

Whitsett, J. A., Machulskis, A., Noguchi, A., and Burdsall, J. A. (1982b). Ontogeny of alpha 1- and beta-adrenergic receptors in rat lung. *Life Sci.* **30**:139-145.

Whitsett, J. A., Manton, M. A., Darovec-Beckerman, C., Adams, K. G., and Moore, J. J. (1981). Beta-adrenergic receptors in the developing rabbit lung. *Am. J. Physiol.* **240**:E351-E357.

Whitsett, J. A., Noguchi, A., and Moore, J. J. (1982a). Developmental aspects of alpha- and beta-adrenergic receptors. *Semin. Perinatol.* **6**: 125-141.

Widdicombe, J. G. (1961). Action potentials in vagal efferent nerve fibres to the lungs of the cat. *Arch. Exp. Pathol. Pharmakol.* **241**:415-432.

Widdicombe, J. G. (1963). Regulation of tracheobronchial smooth muscle. *Physiol. Rev.* **43**:1-37.

Widdicombe, J. G. (1966a). Action potentials in parasympathetic and sympathetic fibres to the trachea and lungs of dogs and cats. *J. Physiol. (London)* **186**:56-88.

Widdicombe, J. G. (1966b). The regulation of bronchial calibre. In *Advances in Respiratory Physiology.* Edited by C. G. Caro. Baltimore, Williams & Wilkins, pp. 48-82.

Widdicombe, J. G. (1985). Control of airway caliber. *Am. Rev. Respir. Dis.* **131**:S33-S35.

Wilkie, R. A., and Bryan, M. H. (1987). Effect of bronchodilators on airway resistance in ventilator-dependant neonates with chronic lung disease. *J. Pediatr.* **111**:278-282.

Wills, M., and Douglas, J. S. (1988). Aging and cholinergic responses in bovine trachealis muscle. *Br. J. Pharmacol.* **93**:918-924.

Woolcock, A. J., Macklem, P. T., Hogg, J. C., Wilson, N. J., and Nadel, J. A., Frank, N. R., and Brain, J. (1969). Effect of vagal stimulation on central and peripheral airways in dogs. *J. Appl. Physiol.* **26**:806-813.

Part Four

**NEUROPHYSIOLOGICAL BASIS FOR
STATE-DEPENDENT RESPIRATORY BEHAVIOR**

17

State-Dependent Electrophysiological Changes in Central Nervous System Activity

RONALD M. HARPER

UCLA School of Medicine
Los Angeles, California

I. Introduction

A classic assumption behind our understanding of control of respiration is that breath-to-breath variation and responses to metabolic demands are largely mediated by structures in the brainstem, particularly by medullary structures. This assumption is based on the understanding that cyclic respiratory patterns can be maintained in a preparation with suprapontine transection (Hoff and Breckenridge, 1952), and manipulations of CO_2 and O_2 presentation to such preparations alter respiratory patterning. However, states of consciousness and behaviors manifested in intact, drug-free preparations can influence respiratory patterning far out of proportion to immediate metabolic demands, and these influences frequently involve descending forebrain contributions to brainstem control mechanisms. Obvious examples for such higher brain influences include voluntary breathing or vocalization, but an equally important case can be made for sleep states, which, combined with certain pathological conditions, can totally suppress normal respiratory patterning in some or all of the ventilatory musculature in sleep apnea (Severinghaus and Mitchell, 1962; Harper and Sauerland, 1978; Remmers et al., 1978).

Even in the absence of pathology, sleep states exert powerful influences on respiratory patterning. Sleep states are associated with dramatic changes in timing and amplitude of ventilatory muscle action (Remmers et al., 1976; Orem et al., 1977), as well as large changes in responsiveness to afferent stimuli (Phillipson, 1978; Phillipson and Sullivan, 1978; Bowes et al., 1980; Marks and Harper, 1989). Moreover, sleep states differentially affect upper airway, diaphragmatic, and body wall musculature cycle timing and degree of activation (Tusiewicz et al., 1976; Duron and Marlot, 1980; Sherrey and Megirian, 1980). Activity in the genioglossal muscles of the tongue, the major oral airway dilators, is greatly reduced or ceases completely during rapid eye movement (REM) sleep (Sauerland and Harper, 1976; Harper and Sauerland, 1978), whereas the diaphragm retains much of its activity, with selective functional loss of some fibers (Sieck et al., 1984).

Since unique characteristics of respiratory patterning occur in different sleep and waking states, and since rostral structures govern many of the characteristics of at least quiet sleep, it is important to extend our examination of brain structures modifying state-related respiratory patterns to forebrain as well as brainstem regions. The examination of rostral brain contributions to respiratory patterning is particularly important for the study of respiratory control during development, since the potential for suprapontine changes during early life is substantial. Moreover, dramatic changes in the temporal distribution and total amount of particular sleep states occur early in life; REM sleep, for example, dominates life in the newborn and only gradually is replaced by increasing amounts of waking and quiet sleep.

Electrophysiologists have traditionally defined sleep and waking states using constellations of behavioral and electrical signs. Thus, quiet sleep is described as periods in which the healthy subject exhibits regular, slow respiration with relatively long inspiratory times, paucity of both whole body and eye movements, and high-amplitude, low-voltage slow waves in the electroencephalogram (EEG), punctuated by 1-3 s bursts of 12-14-Hz "sleep spindles," whereas REM sleep is characterized by low-voltage, high-frequency EEG activity, a widespread paralysis of muscles, eye movements, phasic twitches of the peripheral musculature, and highly variable heart rate and respiration, with large-amplitude spikes observed in the pons and lateral geniculate of animals [pontogeniculooccipital (PGO) waves]. Phasic spike-like activity is observed in other sensory and motor systems as well, in particular the auditory relay nuclei, and manifestations of concomitant activity can be recorded in muscles that regulate tympanic membrane tension (Baust et al., 1964). The occurrence of PGO spikes is a major feature of active periods of REM sleep. These spikes may have a particular relationship to ongoing respiratory activity; Orem (1980a) showed that the more active the medullary respiratory neurons, the more positively correlated was the discharge with PGO activity, and also demonstrated temporal relationships between diaphragmatic EMG suppression and PGO waves (Orem, 1980b).

These physiological characteristics are pervasive across a number of species, and some aspects are so characteristic as to form criteria for state-classification schemes, both by manual means and by automatic computer procedures (Harper et al., 1987). However, the gross characteristics of each sleep state are not uniform over development; 12-14-Hz spindles require some time to mature in the infant (Sterman et al., 1977), and the development of other EEG frequencies, such as low-frequency waves during quiet sleep, is nonlinear over the first 6 months of life (Harper et al., 1981). Similarly, the amplitude of slow waves begins to diminish in the second decade of life in the human, and the pronounced diminution of such low-frequency EEG activity is characteristic of the elderly (Feinberg et al., 1977). Thus, gross central nervous system (CNS) electrical changes accompany development. Similar maturational changes may occur in brain structures modifying respiration and may be observed within the context of state-related CNS changes.

Our examination of neural mechanisms underlying state modulation of respiration will consider "normal" states; however, one must realize that the physiological "signs" of states, although usually present, may be disturbed under circumstances of illness, drug action, or CNS lesion. Although by no means an exhaustive list, a few examples might be mentioned. Monod et al. (1967) describe high muscle tonus during REM sleep in infants with particular neurological syndromes; undoubtedly this high tonus will alter state-related respiratory patterning. Both animals and humans will manifest high-voltage, slow-wave EEG activity with 10-14-Hz "spindles" during waking following administration of atropine, a condition frequently accompanied by extreme alertness. In animals, the large-amplitude slow waves of atropine during extreme alertness are also accompanied by burst-pause neuronal discharge in thalamic structures characteristic of quiet sleep as well. Benzodiazepine administration will cause an increased incidence of 12-14 Hz spindle activity that may burst into EEG traces in which their presence is normally absent or rare. Thus, lesion or pharmacological action can interact with states or alter gross characteristics of that state and may interact with respiratory characteristics as well.

II. Electrophysiologic Aspects of REM Sleep

Any examination of respiratory patterning during sleep must consider the profound paralysis of musculature associated with REM sleep. This substantial atonia affecting skeletal musculature also acts on respiratory musculature; intercostal and abdominal muscles lose input from motoneurons, to the extent that infants with pliable thoracic cages exhibit paradoxical rib cage movement during REM sleep, with abdominal movements out of phase with thoracic cage motions. Upper airway muscles, particularly those of the facial musculature (e.g., dilator nares) and the genioglossal muscles of the tongue, which dilate the airway (Sauerland and Harper, 1976), also undergo profound loss of tone during REM sleep. This suppression of upper airway tone predisposes to upper airway obstruction from diaphragmatic movements, which creates increased negative pressure that can collapse a flaccid upper airway. The diaphragmatic muscles undergo some loss of tone as well, with some muscle fibers coming into play and others leaving during that state (Sieck et al., 1984). However, the CNS is extremely adaptable in adjusting to loss of diaphragmatic action in REM sleep. Bilateral phrenicotomy does not kill rats through asphyxia during REM sleep; instead, they recruit skeletal muscles other than the diaphragm to assist respiration (Sherrey and Megirian, 1989).

In addition to loss of tonic activity during REM sleep, the respiratory muscles are affected by phasic activation that characterizes that state. Thus, genioglossal fibers will phasically regain tone for several breaths (Harper and Sauerland, 1978), and respiratory rates will momentarily increase. Thus, REM sleep is characterized by both a tonic loss of EMG activity in accessory respiratory musculature and phasic excitation. It appears that atonia of respiratory (and skeletal) muscles during REM sleep is an *active* suppression. Behavioral evidence of this statement comes from the brainstem lesion literature indicating that the atonia of REM sleep can be partially or extensively (depending on lesion size and placement) abolished in cats with bilateral lesion of a region ventral and caudal to the locus coeruleus (Sastre and Jouvet, 1979; Hendricks et al., 1982). Cats with extensive lesions actively move about, pounce at imaginary prey, and generally give the appearance of acting out dream activity. A "REM-disorder syndrome" in humans with particular brainstem lesions shows similar behavioral signs, sometimes accompanied by exaggerated movements and aggression directed to bed partners (Mahowald and Schenck, 1989); this syndrome also is presumed to be a release of the normal atonia of REM sleep. Initially, this ventral-coeruleus region was assumed to be the locus for generation of atonia, especially since a number of neurons in that region appeared to fire selectively during atonia (Sakai, 1980). However, the lesion effects may partially result from damage to fibers descending from rostral structures; more selective lesions are associated not only with REM without atonia, but also with release of specific behaviors such as affective behavior (Hendricks et al., 1982), which was also manifested in waking. Thus, the release of behaviors may not be confined to REM sleep, but may be a general release of both REM sleep and waking behaviors. A second region in which lesions cause loss of atonia normally associated with REM sleep has recently been found in the medial caudal medulla (Schenkel and Siegel, 1989).

A. Control of Motoneurons at the Membrane Level During Sleep

Additional evidence that motoneurons supplying skeletal musculature are actively suppressed comes from recordings of membrane and synaptic potentials in the drug-free cat. At this time, the evidence is principally derived from lumbar spinal motoneurons (Morales and Chase, 1981), medullary reticular formation (Chase et al., 1981), and neurons of the trigeminal motor nucleus (Chandler et al., 1980; Chase et al., 1980), rather than from "respiratory" motoneurons. However, at least a subset of trigeminal motoneurons discharge phasically with the respiratory cycle and assist airway dilation in the human (Hairston and Sauerland, 1981); thus, findings in these cells may be similar at least for neurons controlling upper airway respiratory musculature.

Relative to quiet sleep, lumbar and trigeminal motoneurons are hyperpolarized during REM sleep, with the extent of hyperpolarization varying from 2 to 10 mV (Chase et al., 1980) (Fig. 1). Membrane potentials become even more hyperpolarized during phasic twitches of REM sleep, except for short epochs of depolarization that may be associated with generation of spikes. Following the transition to waking, membrane potentials of motoneurons depolarize to a level similar to or below that observed during quiet sleep. Hyperpolarization of the motoneuron soma leads to a decreased probability of spike generation, and thus the target muscle fibers do not contract. After the transition from quiet to active sleep, the interval between two portions of spikes observed on the soma, the "initial segment" and "soma-dendritic spikes," is lengthened, suggesting that the motoneuron soma is inhibited during REM sleep. During phasic events, the interval between initial segment and soma-dendritic spikes is further increased (Morales et al., 1987), suggesting increases in conductance through the cell membrane, a mechanism for postsynaptic inhibition (Llinas and Terzuolo, 1964). A dramatic decrease in membrane resistivity occurs in REM sleep over quiet sleep (43%), with fluctuations during phasic events (Morales and Chase, 1981). Finally, there is a substantial increase in the number of spontaneous inhibitory postsynaptic potentials (IPSPs) recorded from motoneurons during REM sleep relative to quiet sleep or waking, and the occurrence of very large-amplitude IPSPs that appear to occur only during REM sleep (together with smaller-amplitude IPSPs) (Morales et al., 1987). The collective evidence suggests that motoneurons have a pronounced decrease in excitability during REM sleep, and that postsynaptic inhibition is involved.

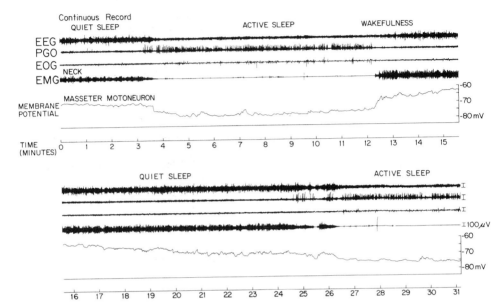

Figure 1 Traces describing EEG, PGO, EOG, and neck EMG together with membrane potentials recorded by an intracellular electrode from a masseter motoneuron in a cat. The membrane potential increased abruptly at 3.5 min, concomitant with a decrease in neck muscle tone and a transition from quiet to REM sleep. At 12.5 min, the membrane depolarized and the animal awakened. After the animal passed into quiet sleep again, a brief, aborted episode of active sleep occurred at 25.5 min that was accompanied by a phasic period of hyperpolarization. A minute later, the animal reentered active sleep and the membrane potential increased. (Reprinted with permission from Chase et al., 1980.)

REM sleep is characterized not only by atonia, however, but also by phasic twitches, abrupt movements of the limbs, accelerations of respiratory effort, and bursts of eye movements that are superimposed on periods of paralysis. These phasic periods appear to result from the simultaneous impingement on the motoneuron of excitatory influences as well as the pronounced inhibitory influences described earlier. These depolarizing influences result in generation of spikes; depolarizing potentials on the membrane remain, even after chloride injection or application of strychnine (Chase et al., 1989). Thus, both inhibitory and excitatory influences simultaneously barrage motoneurons during REM sleep, which results in generalized atonia mixed with erratic twitches, not movements of a coordinated nature. The influences on respiratory musculature reflect this uncoordinated excitation, with irregular patterning during these excitatory bursts.

Application of strychnine on the cell membrane by microiontophoresis while simultaneously recording membrane potentials blocks most IPSPs, suggesting that glycine or glycine analogs are most likely released from inhibitory interneurons of the cord that synapse on motoneurons. Picrotoxin or bicuculline (γ-aminobutyric acid antagonists), applied similarly, has no effect (Chase et al., 1989). Similar blockade of IPSPs by strychnine occurs following stimulation of the gigantocellular field of the medullary reticular formation (Soja et al., 1987). Although local inhibitory circuits are probably involved in spinal motoneuron inhibition, Renshaw cells are unlikely candidates for this inhibitory role (Morales et al., 1988).

Identification of a transmitter responsible for the excitatory postsynaptic potentials (EPSPs) that cause action potentials during REM sleep (and thus cause phasic increases in respiratory rate and momentary restoration of tone to particular respiratory and other skeletal musculature) is not as certain. It appears that the substance is not an N-methyl-D-aspartate (NMDA) transmitter, since the excitatory amino acid antagonist kynurenic acid suppresses naturally occurring phasic depolarizing membrane shifts and action potentials during REM sleep, which NMDA antagonists fail to diminish; however, NMDA antagonists can block EPSPs evoked by sciatic nerve stimulation. Kynurenic acid does not block IPSPs elicited by stimulation (Soja et al., 1988).

The source of the inhibitory influences that hyperpolarize motoneurons during REM sleep, and the excitatory drives that act to depolarize already-inhibited membranes to generate phasic activity during that state, are not well defined. It is clear from transection experiments that the structures responsible for the motoneuron inhibition found in REM sleep lie in a brain region located between the rostral end of the mesencephalon and the caudal medulla (Siegel et al., 1984). Particular regions have been implicated, in both the pons and the medulla; in the pons, these sites include the nucleus pontis oralis. Electrical stimulation of the nucleus pontis oralis elicits state-specific motor reflex modifications that reverse from an excitatory to an inhibitory response during the transition from quiet to REM sleep (Chase and Babb, 1973). The "peri-locus coeruleus" borders the nucleus pontis oralis and has been suggested as a site responsible for atonia (Sakai et al., 1981). Profound atonia mimicking many of the characteristics of REM sleep can be generated by application of carbachol in the dorsolateral pontine tegmentum or in the nucleus pontis oralis; these regions may induce activation of neurons in the nucleus reticularis gigantocellularis, which in turn promotes atonia (Lydic and Baghdoyan, 1989). The carbachol response appears to affect upper airway muscles in a fashion very similar to that observed during REM sleep (Lydic et al., 1989). However, a region of the medial caudal medulla also, on lesion, releases the atonia of REM sleep (Schenkel and Siegel, 1989).

Thus, the brain structures responsible for the suppression of muscle tone in respiratory musculature during REM sleep lie in the brainstem; the precise structures involved remain to be defined, and carbachol studies appear to indicate that structures capable of playing a role in antonia are cholinoceptive (but possibly not cholinergic). Lesions in relatively precise regions will abolish the atonia of REM sleep, possibly by blocking fibers that disturb the equilibrium of excitatory and inhibitory influences. The atonia of REM sleep appears to be mediated by postsynaptic inhibition through glycine or glycine analogs as neurotransmitters.

It should be noted that we have been considering structures that mediate the *atonia* of REM sleep, as opposed to structures that underlie the generation of that state. The literature on neural mechanisms underlying the initiation of the REM state, rather than one of its characteristics (atonia), is vast and far beyond the scope of this chapter. The characteristics of REM sleep most important for respiratory control, however, relate to tonic suppression of action and phasic activation.

III. Descending "Arousal" Systems: What Do Descending Forebrain Systems Do for Respiratory Control?

A major component necessary for maintenance of tone to upper airway musculature during different states appears to be "wakefulness." Although respiratory-related neurons, in particular medullary sites (possibly those of motor neurons to the laryngeal musculature), show minimal changes in neuronal burst characteristics as the animal passes through different sleep states, other neurons manifest profound alteration or loss of phasic discharge during such state transitions from wakefulness (Orem, 1980a). The reduction of this breath-to-breath dependency during sleep is suggestive of a withdrawal of influences normally present during the waking state. Under this concept, a patient undergoing an episode of obstructive sleep apnea must "arouse" to restore tone to the upper airway and initiate airflow, and an infant must "arouse" to initiate oral breathing through blocked nostrils.

A. A Descending "Arousal" System to Respiratory Musculature

To determine the nature of descending "wakefulness" influences, we first must determine what rostral structures associated with "waking" functions project to respiratory brainstem nuclei. One prominent limbic descending system with substantial projections to the midbrain and medullary structures involved in respiratory control is a system with cell bodies in the central nucleus of the amygdala (ACE) and lateral portion of the bed nucleus of the stria terminalis (BNST), which project caudally to respiratory regions such as the lateral division of the nucleus of the solitary tract (NTS) and the parabrachial pons (but *not* to the primary motor neurons of the nucleus ambiguus). The ACE and BNST also receive afferents from the same brainstem regions (Hopkins and Holstege, 1978; Holstege et al., 1985). This system appears to serve a visceral "arousal" role; i.e., it participates in the affective aspect of motor activity in response to a dangerous situation. Since the respiratory musculature assists in preparation for flight, these pathways may influence breathing pattern, both by enhancing respiratory effort and by providing tone to diaphragmatic and abdominal musculature to maintain thoracic pressure, and thus assist in "readiness" to move (Fig. 2).

There is considerable evidence that this limbic system contributes to modifying breathing patterns. The ACE has a high proportion of neurons that discharge on a breath-by-breath basis with the respiratory and cardiac cycles (Pascoe and Kapp, 1985; Zhang et al., 1986a; Frysinger et al., 1988) (Fig. 3), and tonic discharge of these neurons is correlated (not highly, but significantly) with respiratory rate (Fig. 4) and blood pressure (Frysinger et al., 1988). Repetitive single-pulse stimulation of the ACE entrains respiration, a phenomenon that is greatly reduced by quiet sleep (Harper et al., 1984). Cold blockade of the ACE abolishes an aversively conditioned respiratory and blood pressure response, but *not* a heart rate response (Zhang

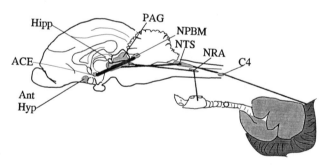

Figure 2   Schematic outline of a selected set of forebrain and brainstem structures that have the potential to modify respiratory patterning. For simplicity, some projections, e.g., anterior hypothalamic (Ant Hyp), hippocampal (Hipp), and nucleus parabrachialis medialis (NPBM), are not included. The central nucleus of the amygdala (ACE) projects to the periaqueductal gray region (PAG), the NPBM, and the nucleus of the solitary tract (NTS); the NPBM has interactive connections with the NTS. The PAG also projects to nucleus retroambiguus (NRA), which contains motoneurons for the laryngeal airway musculature. The NTS projects to the spinal C4 phrenic pool motoneurons innervating the diaphragm. (Reprinted with permission from Harper, 1989.)

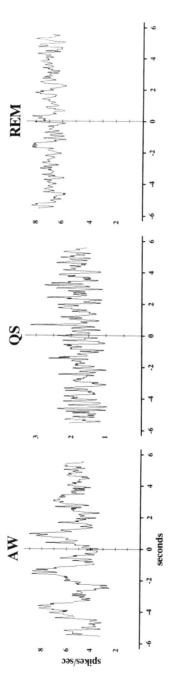

Figure 3 Cross-correlation histograms of discharge of a single ACE neuron with onset of inspiration as measured by integrated diaphragmatic EMG. The histogram shows a strong dependency between the two traces during waking (AW), but the relationship is abolished in quiet sleep (QS) and REM sleep (REM). (Reprinted with permission from Frysinger et al., 1988.)

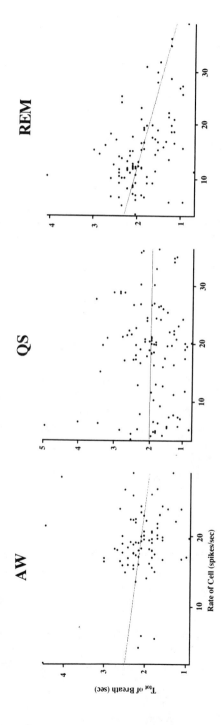

Figure 4 Scatterplots of tonic cell discharge rate of an ACE cell with breath-by-breath values of respiratory period. Correlation values are low and insignificant in waking (AW) and quiet sleep (QS), but significant in REM sleep (REM). Pearson's $r = -0.45$, $df = 96$, $p < 0.01$. (Reprinted with permission from Frysinger et al., 1988.)

et al., 1986b). Thus, the ACE appears to play both tonic and phasic roles in respiratory patterning and blood pressure control, but this influence may operate under certain behavioral conditions, especially those mediated by stressful aversive conditioning, and the influence is state specific.

There are several implications for breathing in the developing infant in these findings of an "affective" descending system. We might speculate that severe breath-holding spells in young children may be mediated by such affective structures (Thach, personal communication). Such breath holding is often initiated by extreme negative emotion and is associated with apneusis, rather than apnea. The ACE, on train stimulation or kindling-induced epilepsy, will produce apneusis (Harper and Frysinger, 1988); the phasic discharge accompanied by kindling-induced discharge results in a one-spike-for-one-inspiration coupling (Harper, 1986). A second implication, both for the ACE and for the structures described below, is that this region, which serves what could be called a "limbic arousal" role, receives input from a vast array of cortical and subcortical regions (Graybiel, 1975; Saper and Loewy, 1980; Price and Amaral, 1981; Russchen, 1982); projections from these regions require time to develop and may exert diminished influence early in life. The developmental course of influence for these rostral regions is relatively unknown.

A more rostral structure (located near the anterior commissure), the lateral division of the BNST, has brainstem projections that virtually overlap projections of medial portions of the ACE, i.e., significant projections to the nucleus parabrachialis medialis (NPBM), lateral division of the NTS, dorsal vagus, and nucleus ambiguus, with reciprocal connections between many of these structures (Holstege et al., 1985). The BNST contains neurons that discharge strongly in phase with aspects of the respiratory cycle. The lateral BNST receives afferents from the medial portion of the ACE, the basolateral and basomedial amygdaloid nuclei (Krettek and Price, 1978), the NPBM (Saper and Loewy, 1980), and the NTS (Ricardo and Koh, 1978). Projections from the locus coeruleus to the BNST have also been reported (Weller and Smith, 1982; Jones and Yang, 1985). Neurons in the ACE and BNST may contribute to a state-dependent behavioral modification of a variety of somatic and visceral responses.

Major recipients of projecting fibers from the ACE and BNST are particular regions of the lateral periaqueductal gray region (PAG) (Hopkins and Holstege, 1978; Holstege, 1987). The PAG serves a role in analgesia (Liebeskind et al., 1973) and integrates defense reactions, including cardiac and respiratory responses, even in the absence of the hypothalamus (Bandler and Carrive, 1988). Viscerotopically organized regions have been identified for hypotensive and hypertensive responses, and these areas project to sub-retrofacial areas of the ventrolateral medulla implicated in blood pressure regulation (Carrive et al., 1988). Of particular importance to respiratory physiologists, however, is the identification of a "vocalization" region in the PAG that projects directly to the nucleus retroambiguus, a premotor nucleus to the facial, tongue, laryngeal, and abdominal musculature; all these muscles are also involved in respiration. Electrical or excitatory amino acid stimulation of this area results in coordinated patterns of action in these muscles and vocalization (Holstege, 1989). Moreover, particular regions of the PAG are extremely sensitive to application of opiates and to naloxone; the respiratory depressant effect of opioids, particularly in interaction with sleep states, is well known. A subset of neurons in the lateral PAG discharge on a breath-by-breath basis with the respiratory cycle; other neurons discharge tonically (i.e., with an overall cell discharge rate relationship) with respiratory rate (Ni et al., 1989) (Fig. 5). These dependencies are state related; i.e., a discharge dependency is present during some states but absent in others.

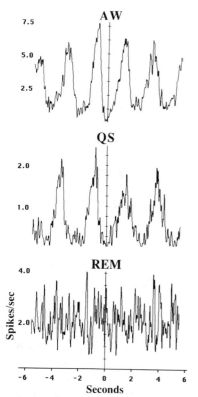

Figure 5 Cross-correlation histograms calculated between onset of inspiration and discharge of a periaqueductal gray (PAG) neuron during three states. A strong correlation is present during waking (AW) and quiet sleep (QS), but is virtually absent during REM sleep (REM). (Reprinted with permission from Ni et al., 1989.)

Descending input from the ACE to the PAG may provide affective "drive" that can provide tonic or phasic influences on neurons of the upper airway; because the nucleus retroambiguus projects to abdominal musculature as well as to upper airway dilators in the larynx and tongue (Holstege, 1989), that nucleus has the potential for modifying expiratory activity during different states.

The so-called "pneumotaxic area" of the parabrachial pons, consisting of the NPBM and part of the Kölliker-Fuse nucleus, receives heavy projections from the ACE and BNST. This region appears to be involved in aspects of respiratory phase switching; stimulation of this area elicits premature transition from inspiration to expiration and can pace respiratory rate (Bassal and Bianchi, 1982). NPBM lesions in vagotomized preparations result in prolonged inspiration (Bertrand and Hugelin, 1971; St. John et al., 1972; Gautier and Bertrand, 1975; von Euler et al., 1976), whereas NPBM lesions in the intact cat result in apneas during quiet sleep that are accentuated in REM sleep (Baker et al., 1981). Neurons in the NPBM discharge phasically with the respiratory cycle, and this relationship is state dependent (Lydic and Orem, 1979; Sieck and Harper, 1980). Neurons in the NPBM often change phase with the respiratory cycle in the transition from one sleep state to another. The implication of these findings is not clear. It may be the case that the NPBM contributes excitation to either phase of the respiratory cycle, depending on respiratory demands. Interactions between locally recorded pairs of NPBM neurons also exhibit a variety of state-dependent relationships (Harper and Sieck, 1980) that appear to be strongly mediated by input from external sources (Frostig et al., 1983). Sleep state modulation of respiration could partly be mediated through the NPBM, although after NPBM lesion, the respiratory cycle is still present (Baker et al., 1981), suggesting that this structure may mediate only part of the respiratory oscillation. The Kölliker-Fuse and NPBM nuclei may receive tonic activation from ACE and BNST projections, and these influences may be modified by sleep states.

B. Respiratory Patterns from Descending Temperature Influences

The activity of temperature-sensitive neurons is of particular interest to respiratory physiologists, because core temperature has profound influences on respiratory patterning. Small increases in core temperature result in increased respiratory rates, whereas further temperature increases result in panting. Such respiratory responses provide a primary mechanism for reducing core temperature in some species. The panting response in the cat involves extreme recruitment of upper airway dilators and accessory abdominal respiratory musculature, thus providing a convenient model for examination of activation to such muscles. Lowered temperatures can slow respiratory rate, and this effect is easily observed in the infant; a cold infant readily stops breathing. The possibility exists that temperature influences on breathing pattern are the *major* influences operating on respiratory patterning in the newborn infant, overriding a weakly developed chemoreceptor response (Johnson, 1988), for at least the first few days of life.

Temperature influences on respiratory musculature are, of course, neurally mediated. Although it is possible that temperature-sensitive neurons are situated in several areas of the brain, one population of such neurons is located in the anterior hypothalamic-preoptic region (Parmeggiani et al., 1983). These neurons discharge in a pattern related to core temperature, and their responsiveness to changes in brain temperature is preserved in quiet sleep and waking, but not in REM sleep. The transition from quiet sleep to REM sleep appears to "dissociate" changes in brain temperature induced by cold/warm probes in the forebrain from discharge dependencies of anterior hypothalamic temperature-sensitive neurons. These studies suggest that REM sleep renders the animal poikilothermic. At least some respiratory patterning is modified in this fashion; respiratory responses to hyperthermia are suppressed when the animal enters REM sleep (Parmeggiani et al., 1976).

If animals do indeed become poikilothermic during REM sleep, the implications for respiratory patterning are profound, particularly in the infant. One would have to believe that the respiratory system is losing a major source of "drive" during REM sleep. Since an infant remains in REM sleep during a substantial portion of its sleep time, the suggestion is that factors other than temperature mediate respiratory pattern changes during that state, and thus during a substantial portion of the infant's life. A subset of neurons in the anterior hypothalamus, located near temperature-sensitive neurons, discharge on a breath-by-breath basis with the respiratory cycle in the anesthetized cat preparation (Kastella et al., 1974). We do not yet know how neurons in this area discharge to respiration in normal sleep and waking states.

IV. Ascending Projection Systems

We have thus far considered influences that cause prolonged suppression of muscle tone or phasic activation either from descending affective stimuli or from excitatory mechanisms of REM sleep. Quiet sleep, however, is accompanied by slowing of respiratory rate, sustained inspiratory periods, and extreme regularity at the same time that slow waves are generated in the EEG. Generation of slow waves is presumably caused by active processes, and any examination of respiratory control effects exerted by quiet sleep must consider these processes.

Several diffusely projecting ascending systems impinge on cortical regions. Recent attention has been focused on three projection systems: the raphe serotonergic system, the adrenergic system of the locus coeruleus, and the ventral cholinergic system.

Neurons in the dorsal raphe of the cat, a midline brainstem structure containing a substantial number of neurons primarily using serotonin as a neurotransmitter, have widespread rostral projections to the cerebral cortex. The dorsal raphe has the distinction of being one of the few areas in which cells diminish their discharge rate during REM sleep. These cells become virtually silent during the phasic aspects of active sleep; indeed, neurons become silent before the onset of phasic events. The slowing of discharge in presumably serotonergic neurons, coincident with acceleration in discharge of reticular gigantocellularis neurons, has led to a proposal of a "reciprocal interaction" model of REM sleep control (Hobson et al., 1975), in which these two systems, mediated by different transmitters, alternate states.

Spontaneous discharge of dorsal raphe neurons during waking and quiet sleep is also unique relative to neurons in cortical and thalamic nuclei: firing rate is slow (waking < 2.4/s, quiet sleep < 1.3/s) and somewhat regular (Fig. 6), showing none of the short, high-frequency bursts characteristic of thalamic or cortical neurons during quiet sleep. At the time of the first description of raphe activity (McGinty and Harper, 1976), no other descriptions of similar neuronal patterns existed; since then, however, neurons in a number of limbic areas, particularly in the locus coeruleus (McGinty et al., 1974; Hobson et al., 1975) and amygdala (Zhang et al., 1986a), have been found to exhibit similar slow, regular rates. Although the discharge of dorsal raphe neurons is closely related to phasic motor and PGO events during REM sleep, few studies describe interactions with cardiovascular or respiratory patterning. Neurons in the raphe groups exhibit substantial interactions with sympathetic activity in the anesthetized, paralyzed preparation (Morrison and Gebber, 1982).

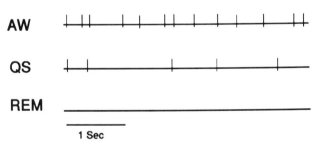

Figure 6   Traces of discharge of a single dorsal raphe neuron (presumably serotonergic) during waking (AW), quiet sleep (QS), and REM sleep (REM). Note the overall slow rates in AW and QS and the virtual absence of discharge during REM. Burst-pause activity characteristic of cells in specific thalamic nuclei or sensory cortex during QS is also absent in these cells. (Modified with permission from McGinty and Harper, 1976.)

The relationship of raphe cells to general cortical activation or to slow waves is still unclear. Although once proposed as a major system underlying the generation of quiet sleep (Jouvet, 1969), the serotonin system appears to mediate other functions, possibly related to modulation of sensory systems and perhaps motor reflexes (Holtman et al., 1986).

Examination of the mechanisms behind the generation of slow waves during quiet sleep and desynchronization during arousal has been a major focus of many electrophysiologists, since these electrical signs are such major characteristics of those states. The long-standing view of cortical activation was that ascending projections from the brainstem reticular formation act on nonspecific thalamic neurons, and these thalamic neurons in turn discharge through widespread projections to desynchronize the cortex for arousal; thalamic neurons, according to this scheme, also act to synchronize cortical potentials upon withdrawal of these ascending excitatory influences on thalamic structures (Moruzzi and Magoun, 1949). However, it now appears from a variety of evidence that the thalamus is not essential for all aspects of electroencephalographic arousal and that cortical "arousal" is heavily dependent on ascending cholinergic fibers coursing in a ventral pathway through the basal forebrain. Cortical desynchronization and behavioral arousal remain, for example, following total bilateral lesion of the thalamus by chemical means (Vanderwolf and Stewart, 1988). Moreover, cortical desynchronization appears to be related to release of acetylcholine in cortical regions (Szerb, 1967) even if the desynchronization occurs in REM sleep (Jasper and Tessier, 1971). This desynchronization would be dependent on the presence of cholinergic neurons in the thalamus; however, the thalamus contains no cholinergic neurons. Indeed, atropine-sensitive desynchronized activity in the cortex is abolished by destruction of cholinergic cell bodies in the basal forebrain (Stewart et al., 1984). Destruction of the basal forebrain leads to profound disturbance of state organization (Shouse et al., 1984), while either low- or high-frequency stimulation will lead to induction of sleep (Sterman and Clemente, 1962a,b).

A subset of neurons, located in the basal forebrain in the vicinity of the diagonal band and medial to the ACE, discharge at higher rates at the onset of quiet sleep (Sterman and Clemente, 1974; Szymusiak and McGinty, 1986) (Fig. 7). A subset of neurons located in the globus pallidus and substantia innominata fire preferentially during generation of slow waves (Detari and Vanderwolf, 1987). Collectively, this evidence suggests that regions within the basal forebrain serve a critical role in both cortical arousal and initiation of quiet sleep. Neurons in this region also serve a number of other physiological functions, including temperature regulation, and interactions of neuronal groups in this area undoubtedly modify respiratory patterning.

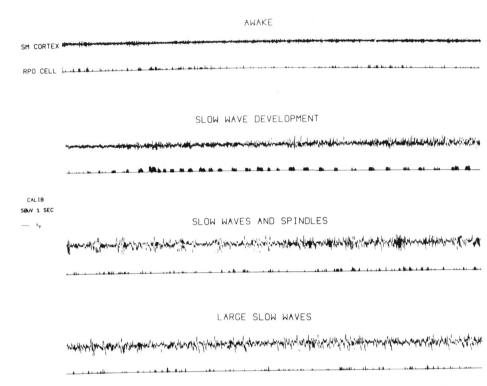

Figure 7 Discharge of a preoptic neuron from a cat during waking, transition into slow-wave development, and slow waves. Note the distinctive increase in firing rate and bursting discharge rate during the transition to slow wave sleep. (Reprinted with permission from Sterman and Clemente, 1974.)

V. Relationship of Slow-Wave Activity to Respiratory Patterning

Normally, rapid respiratory rates are accompanied by desynchronized EEG activity of waking or REM sleep, whereas large-amplitude, low-frequency activity of quiet sleep is accompanied by slower, deeper respiratory patterns. Thus, tonic EEG characteristics bear general relationships to ongoing respiratory patterning. However, the question arises whether breath-by-breath changes in respiratory musculature may be reflected in a phasic manner in slow-wave EEG potentials. At least on a gross level, the answer appears to be yes. Under operant conditioning, cats can be trained to produce bursts of 12-14-Hz "spindle" EEG activity from the sensorimotor cortex in the waking state. Both onset and offset of these EEG bursts occur in synchrony with change in phase of the respiratory effort (Chase and Harper, 1971), and neurons in thalamic relay nuclei change their discharge pattern to that which is normally observed during sleep (Harper, 1973). It is not clear whether the respiratory pattern is being driven by these EEG patterns, whether muscle afferents feed back to alter neural activity that would be reflected in EEG patterns, or whether this correlation is an epiphenomenon created by some by-product of generation of the rhythm.

It appears also that respiratory-event-by-EEG-event coupling appears under other circumstances that lie at the limits of respiratory patterning. Periodic breathing, a waxing and waning of the respiratory signal that occurs during quiet sleep [and typically disappears during REM sleep (Berssenbrugge et al., 1983)], is characterized by a close correlation on an event-by-event basis with characteristics of the EEG signal (Pack, 1989). Periodic breathing is not a normal feature of adult breathing, but is commonly observed in patients with congestive heart failure, breathing at altitude, and a variety of other situations. The pattern is frequently observed in infants, but, as yet, correlations with EEG characteristics have not been described during development.

Selected subcortical structures demonstrate marked moment-by-moment changes in slow-wave electrical characteristics with respiratory efforts. During REM sleep, hippocampal structures give rise to prolonged trains of 4-9-Hz slow-wave activity in the cat, as well as in many other small mammals. This activity, frequently termed "theta activity" or "rhythmic slow activity" (RSA), dominates slow-wave patterns of the hippocampus in small animals. This rhythmical activity has been associated with specific types of motor behavior in the waking animal, with lower frequencies (4-7-Hz in the rat) associated with immobility or "automatic" movements such as grooming, and higher frequencies (7-9-Hz) related to active or voluntary movements such as head movements or walking (Vanderwolf, 1969). The respiratory musculature plays a substantial role in ordinary locomotion and other body movements. Sustained tension of the diaphragm and adduction of the laryngeal muscles are used, for example, to maintain thoracic pressure in changing body position (to prevent unilateral collapse of the body wall), in landing from a leap, or in cessation of a step. Thus, control of respiratory musculature and skeletal locomotor musculature *must* be closely integrated, and CNS electrical activity must be expected to reflect respiratory patterning.

The amplitude of hippocampal RA is profoundly modified by the respiratory cycle in rats (Harper et al., 1989). Moreover, this amplitude modulation is phase dependent, with peak amplitude of hippocampal RA occurring on the inspiratory or expiratory phase of the respiratory cycle, depending on respiratory rate. Higher frequencies of RA accompanied more rapid respiratory rates, such as those associated with phasic twitches of REM sleep. A relationship between hippocampal slow wave activity and respiration has been previously promoted (Komisaruk, 1970) and disclaimed (Vanderwolf et al., 1975); however, the close interaction of respiration with somatic motor activity suggests closely interrelated mechanisms.

The encoding of respiratory phase and rate at such a rostral brain structure as the hippocampus is remarkable and immediately brings to question whether this electrical activity is merely an epiphenomenon, resulting possibly from parallel activation of brainstem motor systems. The medial septum, a major source for theta generation, receives strong ascending projections from the nucleus of Gudden, a pontine structure related to motor control (Leichnetz et al., 1989). It is possible that this ascending system integrates some function of the hippocampus, perhaps related to motor patterning, and incorporates respiratory patterning in that integration.

VI. Summary

Control of respiratory pattern during sleep and wakefulness states is subject to the interaction of an extremely complex set of brain networks. During REM sleep, motoneurons to the respiratory musculature may be hyperpolarized in the same fashion as the skeletal musculature (although motoneurons of the phrenic pool and motoneurons to the laryngeal musculature have yet to be examined).

If motoneurons to the respiratory musculature are subjected to the same influences as lumbar motoneurons, one would expect to find a combination of pronounced excitation as well as postsynaptic inhibition during REM sleep. The sources for initiating this excitation or inhibition lie in the brainstem, but the precise stimuli that initiate and maintain this inhibition and excitation are unclear, although the complex balancing of systems can be unmasked through lesions that can give rise to an abolition of the paralysis during REM sleep.

Brainstem structures that mediate final control to medullary and spinal motoneurons to respiratory muscles appear to receive strong projections from rostral structures, particularly from the ACE and the BNST. These projections appear to modulate excitation that could be broadly defined as affect, and this drive is highly dependent on state.

Respiratory patterning and, in particular, release of the suppression of respiratory "drive" during sleep states may be dependent on activation of very rostral structures, including cortical regions. Ascending projections from brainstem regions may activate rostral regions, initiating a chain of events that may modify respiratory patterning. These systems include projections from the dorsal raphe and ascending cholinergic systems. In particular, the ventral cholinergic system through the basal forebrain appears to be particularly important in mediating aspects of quiet sleep.

These ascending and descending "arousal" systems may have different temporal sequences in establishing control during development. Since all these systems may mediate aspects of respiratory control, the time course of maturation of each of these systems may have considerable influence on respiratory patterning.

Acknowledgments

We thank Ms. Karen Kluge and Dr. Chris Richard for their assistance. This research was supported by HL-22418.

References

Baker, T. L., Netick, A., and Dement, W. C. (1981). Sleep-related apneic and apneustic breathing following pneumotaxic lesion and vagotomy. *Respir. Physiol.* **46**:271-294.

Bandler, R., and Carrive, P. (1988). Integrated defence reaction elicited by excitatory amino acid microinjection in the midbrain periaqueductal grey region of the unrestrained cat. *Brain Res.* **439**:95-106.

Bassal, M., and Bianchi, A. L. (1982). Inspiratory onset or termination induced by electrical stimulation of the brain. *Respir. Physiol.* **50**:23-40.

Baust, W., Berlucchi, G., and Moruzzi, G. (1964). Changes in the auditory input in wakefulness and during the synchronized and desynchronized stages of sleep. *Arch. Ital. Biol.* **102**:657-674.

Berssenbrugge, A., Dempsey, J., Iber, C., Skatrud, J., and Wilson, P. (1983). Mechanisms of hypoxia-induced periodic breathing during sleep in humans. *J. Physiol. (London)* **343**:507-524.

Bertrand, F., and Hugelin, A. (1971). Respiratory synchronizing function of nucleus parabrachialis medialis: pneumotaxic mechanisms. *J. Neurophysiol.* **34**:189-207.

Bowes, G., Woolf, G. M., Sullivan, C. E., and Phillipson, E. A. (1980). Effect of sleep fragmentation on ventilatory and arousal responses of sleeping dogs to respiratory stimuli. *Am. Rev. Respir. Dis.* **122**:899-908.

Carrive, P., Bandler, R., and Dampney, R. A. L. (1988). Anatomical evidence that hypertension associated with the defence reaction in the cat is mediated by a direct projection from a restricted portion of the midbrain periaqueductal grey to the subretrofacial nucleus of the medulla. *Brain Res.* **460**:339-345.

Chandler, S. H., Chase, M. H., and Nakamura, Y. (1980). Intracellular analysis of synaptic mechanisms controlling trigeminal motoneuron activity during sleep and wakefulness. *J. Neurophysiol.* **44**:359-371.

Chase, M. H., and Babb, M. (1973). Masseteric reflex response to reticular stimulation reverses during active sleep compared with wakefulness or quiet sleep. *Brain Res.* **59**:421-426.

Chase, M. H., Chandler, S. H., and Nakamura, Y. (1980). Intracellular determination of membrane potential of trigeminal motoneurons during sleep and wakefulness. *J. Neurophysiol.* **44**:349-358.

Chase, M. H., Enomoto, S., Murakami, T., Nakamura, Y., and Taira, M. (1981). Intracellular potential of medullary reticular neurons during sleep and wakefulness. *Exp. Neurol.* **71**:226-233.

Chase, M. H., and Harper, R. M. (1971). Somatomotor and visceromotor correlates of operantly conditioned 12-14 c/sec sensorimotor cortical activity. *Electroencephalogr. Clin. Neurophysiol.* **31**:85-92.

Chase, M. H., Soja, P. J., and Morales, F. R. (1989). Evidence that glycine mediates the postsynaptic potentials that inhibit lumbar motoneurons during the atonia of active sleep. *J. Neurosci.* **9**:743-751.

Detari, L., and Vanderwolf, C. H. (1987). Activity of identified cortically projecting and other basal forebrain neurones during large slow waves and cortical activation in anesthetized rats. *Brain Res.* **437**:1-8.

Duron, B., and Marlot, D. (1980). Intercostal and diaphragmatic electrical activity during wakefulness and sleep in normal unrestrained cats. *Sleep* **3**:269-280.

Euler, C. von, Marttila, I., Remmers, J. E., and Trippenbach, T. (1976). Effects of lesions in the parabrachial nucleus on the mechanisms for central and reflex termination of inspiration in the cat. *Acta Physiol. Scand.* **96**:324-337.

Feinberg, I., Hibi, S., and Carlson, V. R. (1977). Changes in EEG amplitude during sleep with age. In *The Aging Brain and Senile Dementia*. Edited by K. Nandy and I. Sherwin. New York, Plenum Press, pp. 85-98.

Frostig, R. D., Frostig, Z., Sieck, G. C., and Harper, R. M. (1983). Extrinsic and intrinsic interactions among simultaneously recorded neurons in the nucleus parabrachialis medialis during sleep-waking states. *Soc. Neurosci. Abstr.* **9**:539.

Frysinger, R. C., Zhang, J., and Harper, R. M. (1988). Cardiovascular and respiratory relationships with neuronal discharge in the central nucleus of the amygdala during sleep-waking states. *Sleep* **11**:317-332.

Gautier, H., and Bertrand, F. (1975). Respiratory effects of pneumotaxic center lesions and subsequent vagotomy in chronic cats. *Respir. Physiol.* **23**:71-85.

Graybiel, A. M. (1975). Wallerian degeneration and anterograde tracer methods. In *The Use of Axonal Transport for Studies of Neuronal Connectivity.* Edited by W. M. Cowan and M. Cuénod. Amsterdam, Elsevier, pp. 173-216.

Hairston, L. E., and Sauerland, E. K. (1981). Electromyography of the human palate: Discharge patterns of the levator and tensor veli palatini. *Electromyogr. Clin. Neurophysiol.* **21**:287-297.

Harper, R. M. (1973). Relationship of neuronal activity to EEG waves during sleep and wakefulness. In *Brain Unit Activity During Behavior.* Edited by M. I. Phillips. Springfield, IL, Charles C. Thomas, pp. 130-154.

Harper, R. M. (1986). State-related physiological changes and risk for the sudden infant death syndrome. *Aust. Paediatr. J.* **22** (Suppl 1): 55-58.

Harper, R. M. (1989). Forebrain mechanisms related to respiratory patterning during sleep-waking states. In *Modeling and Parameter Estimation in Respiratory Control.* Edited by M. C. K. Khoo. New York, Plenum Press (in press).

Harper, R. M., and Frysinger, R. C. (1988). Suprapontine mechanisms underlying cardiorespiratory regulation: implications for the sudden infant death syndrome. In *Sudden Infant Death Syndrome: Risk Factors and Basic Mechanisms.* Edited by R. M. Harper and H. J. Hoffman. New York, PMA Publishing, pp. 399-414.

Harper, R. M., Frysinger, R. C., Trelease, R. B., and Marks, J. D. (1984). State-dependent alteration of respiratory cycle timing by stimulation of the central nucleus of the amygdala. *Brain Res.* **306**:1-8.

Harper, R. M., Leake, B., Miyahara, L., Hoppenbrouwers, T., Sterman, M. B., and Hodgman, J. (1981). Development of ultradian periodicity and coalescence at 1 cycle/hour in electroencephalographic activity. *Exp. Neurol.* **73**:127-143.

Harper, R. M., Richard, C. A., Frysinger, R. C., Terreberry, R. R., Garfinkel, A., and Harper, R. K. (1989). Slow wave electrical activity in the hippocampus: relation to respiratory patterning. *Soc. Neurosci. Abstr.* **15**:1194.

Harper, R. M., and Sauerland, E. K. (1978). The role of the tongue in sleep apnea. In *Sleep Apnea Syndromes.* Edited by C. Guilleminault and W. C. Dement. New York, Alan R. Liss, pp. 219-234.

Harper, R. M., Schechtman, V. L., and Kluge, K. A. (1987). Machine classification of infant sleep state using cardiorespiratory measures. *Electroencephalogr. Clin. Neurophysiol.* **67**:379-387.

Harper, R. M., and Sieck, G. C. (1980). Discharge correlations between neurons in the nucleus parabrachialis medialis during sleep-waking states. *Brain Res.* **199**:343-358.

Hendricks, J. C., Morrison, A. R., and Mann, G. L. (1982). Different behaviors during paradoxical sleep without atonia depend on pontine lesion site. *Brain Res.* **239**:81-105.

Hobson, J. A., McCarley, R. W., and Wyzinski, P. W. (1975). Sleep cycle oscillation: reciprocal discharge by two brainstem neuronal groups. *Science* **189**:55-58.

Hoff, H. E., and Breckenridge, C. G. (1952). Levels of integration of respiratory patterns. *J. Neurophysiol.* **15**:47-56.

Holstege, G. (1987). Some anatomical observations on the projections from the hypothalamus to brainstem and spinal cord: an HRP and autoradiographic tracing study in the cat. *J. Comp. Neurol.* **260**:98-126.

Holstege, G. (1989). Anatomical study of the final common pathway for vocalization in the cat. *J. Comp. Neurol.* **284**:242-252.

Holstege, G., Meiners, L., and Tan, K. (1985). Projections of the bed nucleus of the stria terminalis to the mesencephalon, pons, and medulla oblongata in the cat. *Exp. Brain Res.* **58**:379-391.

Holtman, J. R., Jr., Dick, T. E., and Berger, A. J. (1986). Involvement of serotonin in the excitation of phrenic motoneurons evoked by stimulation of the raphe obscurus. *J. Neurosci.* **6**:1185-1193.

Hopkins, D. A., and Holstege, G. (1978). Amygdaloid projections to the mesencephalon, pons and medulla oblongata in the cat. *Exp. Brain Res.* **32**:529-547.

Jasper, H. H., and Tessier, J. (1971). Acetylcholine liberation from cerebral cortex during paradoxical (REM) sleep. *Science* **172**:601-602.

Johnson, P. (1988). Environmental temperature and the development of breathing. In *Sudden Infant Death Syndrome: Risk Factors and Basic Mechanisms*. Edited by R. M. Harper and H. J. Hoffman. New York, PMA Publishing, pp. 233-248.

Jones, B. E., and Yang, T-Z. (1985). The efferent projections from the reticular formation and the locus coeruleus studied by anterograde and retrograde axonal transport in the rat. *J. Comp. Neurol.* **242**:56-92.

Jouvet, M. (1969). Biogenic amines and the states of sleep. *Science* **163**:32-41.

Kastella, K. G., Spurgeon, H. A., and Weiss, G. K. (1974). Respiratory-related neurons in anterior hypothalamus of the cat. *Am. J. Physiol.* **227**:710-713.

Komisaruk, B. R. (1970). Synchrony between limbic system theta activity and rhythmical behavior in rats. *J. Comp. Physiol. Psychol.* **70**:482-492.

Krettek, J. E., and Price, J. L. (1978). Amygdaloid projections to subcortical structures within the basal forebrain and brainstem in the rat and cat. *J. Comp. Neurol.* **178**:225-254.

Leichnetz, G. R., Carlton, S. M., Katayama, Y., Gonzalo-Ruiz, A., Holstege, G., DeSalles, A. A. F., and Hayes, R. L. (1989). Afferent and efferent connections of the cholinoceptive medial pontine reticular formation (region of the ventral tegmental nucleus) in the cat. *Brain Res. Bull.* **22**:665-688.

Liebeskind, J. C., Guilbaud, G., Besson, J-M., and Oliveras, J-L. (1973). Analgesia from electrical stimulation of the periaqueductal gray matter in the cat: behavioral observations and inhibitory effects on spinal cord interneurons. *Brain Res.* **50**:441-446.

Llinas, R., and Terzuolo, C. A. (1964). Mechanisms of supraspinal actions upon spinal cord activities. Reticular inhibitory mechanisms on alpha-extensor motoneurons. *J. Neurophysiol.* **27**:579-591.

Lydic, R., and Baghdoyan, H. A. (1989). Cholinoceptive pontine reticular mechanisms cause state-dependent respiratory changes in the cat. *Neurosci. Lett.* **102**:211-216.

Lydic, R., Baghdoyan, H. A., and Zwillich, C. W. (1989). State-dependent hypotonia in posterior cricoarytenoid muscles of the larynx caused by cholinoceptive reticular mechanisms. *FASEB J.* **3**:1625-1631.

Lydic, R., and Orem, J. (1979). Respiratory neurons of the pneumotaxic center during sleep and wakefulness. *Neurosci. Lett.* **15**:187-192.

Mahowald, M. W., and Schenck, C. H. (1989). REM sleep behavior disorder. In *Principles and Practice of Sleep Medicine.* Edited by M. H. Kryger, T. Roth, and W. C. Dement. Philadelphia, W. B. Saunders, pp. 389-401.

Marks, J. D., and Harper, R. M. (1989). State-dependent diaphragmatic and laryngeal abductor responses to lactic acidemia (submitted for publication).

McGinty, D. J., and Harper, R. M. (1976). Dorsal raphe neurons: depression of firing during sleep in cats. *Brain Res.* **101**:569-575.

McGinty, D. J., Harper, R. M., and Fairbanks, M. K. (1974). Neuronal unit activity and the control of sleep states. In *Advances in Sleep Research,* Vol. 1. Edited by E. D. Weitzman. New York, Spectrum Publications, pp. 173-216.

Monod, N., Eliet-Flescher, J., and Dreyfus-Brisac, C. (1967). Le sommeil du nouveau-né et du prémature. III. Les troubles de l'organisation du sommeil chez le nouveau-né pathologique: analyse des études polygraphiques. *Biol. Neonate* **11**:216-247.

Morales, F. R., Boxer, P., and Chase, M. H. (1987). Behavioral state-specific inhibitory postsynaptic potentials impinge on cat lumbar motoneurons during active sleep. *Exp. Neurol.* **98**:418-435.

Morales, F., and Chase, M. H. (1981). Postsynaptic control of lumbar motoneuron excitability during active sleep in the chronic cat. *Brain Res.* **225**:279-295.

6

46 *Harper*

Morales, F. R., Engelhardt, J. K., Pereda, A. E., Yamuy, J., and Chase, M. H. (1988). Renshaw cells are inactive during motor inhibition elicited by the pontine microinjection of carbachol. *Neurosci. Lett.* **86**: 289-295.

Morrison, S. F., and Gebber, G. L. (1982). Classification of raphe neurons with cardiac-related activity. *Am. J. Physiol.* **243**:R49-R59.

Moruzzi, G., and Magoun, H. W. (1949). Brain stem reticular formation and activation of the EEG. *Electroencephalogr. Clin. Neurophysiol.* **1**:455-473.

Ni, H., Zhang, J., and Harper, R. M. (1989). Respiratory-related discharge of periaqueductal gray neurons during sleep-waking states. *Brain Res.* (in press).

Orem, J. (1980a). Medullary respiratory neuron activity: relationship to tonic and phasic REM sleep. *J. Appl. Physiol.* **48**:54-65.

Orem, J. (1980b). Neuronal mechanisms of respiration in REM sleep. *Sleep* **3**:251-267.

Orem, J., Netick, A., and Dement, W. C. (1977). Breathing during sleep and wakefulness in the cat. *Respir. Physiol.* **30**:265-289.

Pack, A. I. (1989). State and periodic ventilation. In *Modeling and Parameter Estimation in Respiratory Control*. Edited by M. C. K. Khoo. New York, Plenum Press (in press).

Parmeggiani, P. L., Azzaroni, A., Cevolani, D., and Ferrari, G. (1983). Responses of anterior hypothalamic-preoptic neurons to direct thermal stimulation during wakefulness and sleep. *Brain Res.* **269**:382-385.

Parmeggiani, P. L., Franzini, C., and Lenzi, P. (1976). Respiratory frequency as a function of preoptic temperature during sleep. *Brain Res.* **111**:253-260.

Pascoe, J. P., and Kapp, B. S. (1985). Electrophysiological characteristics of amygdaloid central nucleus neurons during Pavlovian fear conditioning in the rabbit. *Behav. Brain Res.* **16**:117-133.

Phillipson, E. A. (1978). Respiratory adaptations in sleep. *Annu. Rev. Physiol.* **40**:133-156.

Phillipson, E. A., and Sullivan, C. E. (1978). Arousal: the forgotten response to respiratory stimuli. *Am. Rev. Respir. Dis.* **118**:807-809.

Price, J. L., and Amaral, D. G. (1981). An autoradiographic study of the projections of the central nucleus of the monkey amygdala. *J. Neurosci.* **1**:1242-1259.

Remmers, J. E., Bartlett, D., Jr., and Putnam, M. D. (1976). Changes in the respiratory cycle associated with sleep. *Respir. Physiol.* **28**:227-238.

Remmers, J. E., deGroot, W. J., Sauerland, E. K., and Anch, A. M. (1978). Pathogenesis of upper airway occlusion during sleep. *J. Appl. Physiol.* **44**:931-938.

Ricardo, J. A., and Koh, E. T. (1978). Anatomical evidence of direct projections from the nucleus of the solitary tract to the hypothalamus, amygdala, and other forebrain structures in the rat. *Brain Res.* **153**:1-26.

Russchen, F. T. (1982). Amygdalopetal projections in the cat. I. Cortical afferent connections. A study with retrograde and anterograde tracing techniques. *J. Comp. Neurol.* **206**:159-179.

Sakai, K. (1980). Some anatomical and physiological properties of ponto-mesencephalic tegmental neurons with special reference to the PGO waves and postural atonia during paradoxical sleep in the cat. In *The Reticular Formation Revisited: Specifying Function for a Nonspecific System.* Edited by J. A. Hobson and M. A. B. Brazier. New York, Raven Press, pp. 427-447.

Sakai, K., Sastre, J-P., Kanamori, N., and Jouvet, M. (1981). State-specific neurons in the ponto-medullary reticular formation with special reference to the postural atonia during paradoxical sleep in the cat. In *Brain Mechanisms of Perceptual Awareness and Purposeful Behavior.* Edited by O. Pompeiano and C. A. Marsan. New York, Raven Press, pp. 405-429.

Saper, C. B., and Loewy, A. D. (1980). Efferent connections of the parabrachial nucleus in the rat. *Brain Res.* **197**:291-317.

Sastre, J-P., and Jouvet, M. (1979). Le comportement onirique du chat. *Physiol. Behav.* **22**:979-989.

Sauerland, E. K., and Harper, R. M. (1976). The human tongue during sleep: electromyographic activity of the genioglossus muscle. *Exp. Neurol.* **51**:160-170.

Schenkel, E., and Siegel, J. M. (1989). REM sleep without atonia after lesions of the medial medulla. *Neurosci. Lett.* **98**:159-165.

Severinghaus, J. W., and Mitchell, R. A. (1962). Ondine's curve - failure of respiratory center automaticity while awake. *Clin. Res.* **10**:122.

Sherrey, J. H., and Megirian, D. (1980). Respiratory EMG activity of the posterior cricoarytenoid, cricothyroid and diaphragm muscles during sleep. *Respir. Physiol.* **39**:355-365.

Sherrey, J. H., and Megirian, D. (1989). After phrenicotomy the rat alters respiratory muscle output, not its sleep-waking pattern (submitted for publication).

Shouse, M. N., Sterman, M. B., Hauri, P. J., and Belsito, O. (1984). Sleep disruption with basal forebrain lesions decreases latency to amygdala kindling in cats. *Electroencephalogr. Clin. Neurophysiol.* **58**:369-377.

Sieck, G. C., and Harper, R. M. (1980). Pneumotaxic area neuronal discharge during sleep-waking states in the cat. *Exp. Neurol.* **67**:79-102.

Sieck, G. C., Trelease, R. B., and Harper, R. M. (1984). Sleep influences on diaphragmatic motor unit discharge. *Exp. Neurol.* **85**:316-335.

Siegel, J. M., Nienhuis, R., and Tomaszewski, K. S. (1984). REM sleep signs rostral to chronic transections at the pontomedullary junction. *Neurosci. Lett.* **45**:241-246.

Soja, P. J., López, F., Morales, F. R., and Chase, M. H. (1988). Depolarizing synaptic events influencing cat lumbar motoneurons during rapid eye movement episodes of active sleep are blocked by kynurenic acid. *Soc. Neurosci. Abstr.* **14**:941.

Soja, P. J., Morales, F. R., Baranyi, A., and Chase, M. H. (1987). Effect of inhibitory amino acid antagonists on IPSPs induced in lumbar motoneurons upon stimulation of the nucleus reticularis gigantocellularis during active sleep. *Brain Res.* **423**:353-358.

St. John, W. M., Glasser, R. L., and King, R. A. (1972). Rhythmic respiration in awake vagotomized cats with chronic pneumotaxic area lesions. *Respir. Physiol.* **15**:233-244.

Sterman, M. B., and Clemente, C. D. (1962a). Forebrain inhibitory mechanisms: Cortical synchronization induced by basal forebrain stimulation. *Exp. Neurol.* **6**:91-102.

Sterman, M. B., and Clemente, C. D. (1962b). Forebrain inhibitory mechanisms: Sleep patterns induced by basal forebrain stimulation in the behaving cat. *Exp. Neurol.* **6**:103-117.

Sterman, M. B., and Clemente, C. D. (1974). Forebrain mechanisms for the onset of sleep. In *Basic Sleep Mechanisms*. Edited by O. Petre-Quadens and J. D. Schlag. New York, Academic Press, pp. 83-97.

Sterman, M. B., Harper, R. M., Havens, B., Hoppenbrouwers, T., McGinty, D. J., and Hodgman, J. E. (1977). Quantitative analysis of infant EEG development during quiet sleep. *Electroencephalogr. Clin. Neurophysiol.* **43**:371-385.

Stewart, D. J., Macfabe, D. F., and Vanderwolf, C. H. (1984). Cholinergic activation of the electrocorticogram: Role of the substantia innominata and effects of atropine and quinuclidinyl benzilate. *Brain Res.* **322**:219-232.

Szerb, J. C. (1967). Cortical acetylcholine release and electroencephalographic arousal. *J. Physiol. (London)* **192**:329-343.

Szymusiak, R., and McGinty, D. (1986). Sleep-related neuronal discharge in the basal forebrain of cats. *Brain Res.* **370**:82-92.

Tusiewicz, K., Moldofsky, H., and Bryan, A. C. (1976). Mechanics of the rib cage and diaphragm during sleep. *Fed. Proc.* **35**:396.

Vanderwolf, C. H. (1969). Hippocampal electrical activity and voluntary movement in the rat. *Electroencephalogr. Clin. Neurophysiol.* **26**:407-418.

Vanderwolf, C. H., Kramis, R., Gillespie, L. A., and Bland, B. H. (1975). Hippocampal rhythmical slow activity and neocortical low-voltage fast activity: Relations to behavior. In *The Hippocampus,* Vol. 2: *Neurophysiology and Behavior.* Edited by R. L. Isaacson and K. H. Pribram. New York, Plenum Press, pp. 101-128.

Vanderwolf, C. H., and Stewart, D. J. (1988). Thalamic control of neocortical activation: a critical re-evaluation. *Brain Res. Bull.* **20**:529-538.

Weller, K. L., and Smith, D. A. (1982). Afferent connections to the bed nucleus of the stria terminalis. *Brain Res.* **232**:255-270.

Zhang, J. X., Harper, R. M., and Frysinger, R. C. (1986a). Respiratory modulation of neuronal discharge in the central nucleus of the amygdala during sleep and waking states. *Exp. Neurol.* **91**:193-207.

Zhang, J. X., Harper, R. M., and Ni, H. (1986b). Cryogenic blockade of the central nucleus of the amygdala attenuates aversively conditioned blood pressure and respiratory responses. *Brain Res.* **386**:136-145.

18

Respiratory Motor Output
Effect of State and Maturation in Early Life

ERIC C. EICHENWALD and ANN R. STARK

Harvard Medical School
The Children's Hospital
Boston, Massachusetts

I. Introduction

In breathing, as in many other functions, the infant differs greatly from the adult. The process of maturation profoundly affects respiratory regulation both during gestation and in early postnatal life. At birth, the most crucial change occurs when the fetus must switch from its irregular, episodic, non-respiratory "breathing" movements to the sustained respiratory rhythm essential to life.

The effects of behavioral state on respiratory motor output in the fetus and newborn have been recognized only recently. Sleep state influences both the mechanical relationships of the respiratory muscles and the control of ventilation. Although the structural and cellular mechanisms for ventilatory homeostasis are in place at birth, neurophysiological immaturity and mechanical disadvantages of the immature chest wall present special challenges to the control of breathing during sleep in early life. Furthermore, since most studies on respiration in the newborn are performed while the subject

Research reported here was supported by NIH SCOR HL34616.

sleeps, understanding the effects of behavioral state on the developing respiratory system is especially important.

This chapter will review the effects of sleep state on the respiratory pump muscles, the upper airway, and the generation of respiratory rhythm. Although certain aspects of the control of breathing during sleep will be discussed, a comprehensive examination is beyond the scope of this chapter. Several excellent recent reviews of respiratory control during sleep are available (Bryan et al., 1986; Phillipson and Rowes, 1986).

II. Determination of State

The term *state* is used "to describe constellations of certain functional patterns and physiological variables which may be relatively stable and which seem to repeat themselves" (Prechtl et al., 1968). In the adult, stable and well-defined states of wakefulness, non-rapid eye movement (NREM) sleep, and rapid eye movement (REM) sleep are readily defined using neurophysiological measurements (Rechtschaffen and Kales, 1968). These measurements include the electroencephalogram (EEG), electrooculogram (EOG), and postural muscle electromyogram (EMG). NREM sleep is characterized by a synchronized high voltage-low-frequency EEG pattern, low levels of tonic postural muscle tone, and slow or no eye movements on EOG. Deeper levels of NREM sleep, or "slow-wave sleep," can be recognized by the presence of "sleep spindles" on the EEG. REM sleep is characterized by desynchronized low-voltage-high-frequency EEG patterns similar to the awake state, lack of postural muscle tone, and rapid conjugate eye movements (REMs). Despite similar EEGs, REM sleep differs from wakefulness in the profound inhibition of sensorimotor responses to external stimuli as well as the obvious loss of consciousness. NREM and REM sleep in the adult represent distinct states of neurophysiological organization, although the neuronal mechanisms involved are unclear.

Sleep states in the perinatal period are more poorly defined. Differences in the neurophysiological characteristics of sleep states in the newborn period make it impossible to use strict adult standards for state definition. In addition, the level of maturity at birth and the pace of maturation after birth differ considerably among species, making comparisons between newborn animals and human infants hazardous (Dreyfus-Brisac, 1975).

A. State Definition in the Infant

Neurophysiological criteria can be useful in the determination of sleep state in the human infant born after full-term gestation (Table 1). After 36-37 weeks' gestational age (term 40 weeks), cyclic organization of NREM and

Table 1 Sleep State Determination in the Newborn

Criteria	NREM/quiet sleep	REM/active sleep
Neurophysiological		
EEG	High voltage-low frequency	Low voltage-high frequency
EMG	Low-level tonic activity	No tonic activity
EOG	Slow/no eye movements	Rapid eye movements
Behavioral		
Body movement	Little or none	Intermittent twitches
Eye movement	No	Yes
Respiratory	Regular rate and depth	Irregular rate and depth
	Rib cage/abdomen in phase	Rib cage/abdomen paradox

EEG, Electroencephalogram; EMG, electromyogram; EOG, electrooculogram; NREM, non-rapid eye movement; REM, rapid eye movement.

REM sleep, as defined by EEG, EOG, and EMG criteria, is present (Dreyfus-Brisac, 1970, 1975). Full maturation of NREM sleep into true slow-wave sleep does not occur, however, until after several weeks of age (Dreyfus-Brisac, 1964). Prior to 36 weeks' gestational age, NREM sleep is poorly organized, with a discontinuous EEG pattern. In contrast, continuous, organized REM sleep EEG patterns are first recognized in the human as early as 32 weeks' gestation and become fully mature at 34-35 weeks (Dreyfus-Brisac, 1975). Other neurophysiological measurements, such as the EOG and postural muscle EMG, are not useful to discriminate state until after 32-33 weeks' gestation (Parmalee and Stern, 1972).

Because of difficulties in the use of neurophysiological criteria in sleep state determination throughout development in the human infant, use of behavioral criteria has been advocated (Prechtl, 1974). Two distinct states, active and quiet sleep, corresponding to REM and NREM sleep, respectively, have been defined using changes in body and eye movements and respiratory patterns (Table 1). Respiratory patterns are especially useful since use of respiratory cycle time variability has been shown to be almost as accurate as neurophysiological criteria in the determination of sleep state in the term infant (Haddad et al., 1987). Using behavioral criteria, the first differentiation between active and quiet sleep can be made as early as 30 weeks' gestational age (Dreyfus-Brisac, 1975). In the premature newborn, however, marked individual differences in neurological reflex responses can occur within states (Prechtl, 1974). With behaviorally defined states, the preterm infant spends 50-60% of the time in active sleep, the term infant slightly less. A significant proportion of sleep time in newborns, especially the preterm, may be indeterminate using standard criteria.

B. State Definition in the Fetus

Differentiation in discrete neurophysiological states is difficult in the developing fetus. First, the fetus is relatively inaccessible to direct measurement. Second, this inaccessibility is superimposed on a template of rapidly changing neurological development.

Measurements have been made of the electrocorticogram (ECoG) in the chronically instrumented lamb fetus. These have revealed a developmental pattern of neurophysiological states (Ruckebusch et al., 1977). The ECoG early in gestation is an undifferentiated (*trace alternans*) pattern. At approximately 120-125 days (0.8 gestation, term 146 days) a more differentiated pattern is observed. Discrete periods of low-voltage-high-frequency ECoG activity are detected; these are associated with lateral eye movements characteristic of REM sleep. High-voltage-low-frequency ECoG characteristics of NREM sleep also develop (Clewlow et al., 1983). By term, the lamb fetus spends approximately half its time in each discrete state, although a small proportion of the low-voltage electrocortical state may be equivalent to wakefulness (Ruckebusch, 1972). A "REM-like" state in the absence of a mature ECoG pattern can be recognized as early as 75 days, by measurement of lateral eye movements and nuchal muscle tone (Ioffe et al., 1987).

Because the lamb's neurophysiological development is precocious, care must be exercised in extrapolating data from the lamb to other species, including the human. In the human fetus, for example, although discrete periods of behavioral activity can be identified by ultrasound as early as the late second trimester (Nijhuis et al., 1983), neurophysiological differentiation into discrete REM and NREM states measured by electroencephalogram after birth does not occur before 36 weeks' gestation (see Section I.A) (Dreyfus-Brisac, 1975).

III. Fetal Breathing Movements

Fetal respiratory muscle activity was first described in the latter part of the 19th century. It was "rediscovered" in the 1970s with the advent of experimental animal preparations instrumented in a way that made possible chronic fetal monitoring. Cyclical fetal respiratory activity was observed, first through measurement of tracheal pressure swings (Dawes et al., 1972), and later correlated with diaphragmatic and laryngeal muscle EMG activity (Maloney et al., 1975; Harding, 1980; Harding et al., 1980). The physiology of fetal breathing movements has been most extensively studied in the fetal lamb, but the presence of fetal breathing has been confirmed in other species, including the human (Boddy and Robinson, 1971). In fact, fetal respiratory activity is essential for normal lung growth and differentiation during gestation. Developmental abnormalities of the lungs have been observed after bilateral phrenectomy in the lamb (Alcorn et al., 1980). In the human fetus, congenital agenesis of the phrenic nerve has been associated with severe pulmonary architectural abnormalities (Goldstein and Reid, 1980).

In the fetal lamb, respiratory muscle activity is strongly associated with neurophysiological state. Breathing is limited to periods of low-voltage-high-frequency ECoG activity with associated eye movements; respiratory movements cease with the onset of high-voltage-low-frequency NREM state (Dawes et al., 1972). Prior to the development of electrophysiological state differentiation in the fetus, high-frequency diaphragm EMG activity is almost continuous (Bowes et al., 1981a). Coincident with the differentiation of the NREM state, the amount and length of apneic periods increase with development. The increase in time spent in apnea is linearly correlated with increasing gestational age (Maloney et al., 1980). Although increasing periods of apnea result in a decrease in the average number of breaths per day, during periods of breathing the respiratory rate is similar throughout the last approximate one-third of gestation (Bowes et al., 1981a) (Fig. 1). This suggests that there is no developmental change in central pattern generation through this period.

The ontogeny of the respiratory inhibition during the NREM state remains somewhat obscure. Central inhibition appears to be mediated at the level of the upper pons. Decortication does not affect the amount or intrinsic cycling of fetal respiratory activity, although this results in dissociation of the ECoG and breathing bursts (Ioffe et al., 1984). In fetuses with mid-collicular transections, however, the rapid irregular respiratory activity characteristic of the REM state becomes continuous (Dawes et al., 1983). Although this evidence points to suprabulbar inhibition of respiratory activity, it is possible that apnea during high-voltage ECoG activity may be due to the absence of significant afferent input to sustain the respiratory rhythm. When afferent stimuli are increased by direct cooling of the fetus *in utero,* for example, breathing is continuous during the high-voltage ECoG state (Gluckman et al., 1983). State-dependent apnea in the fetus may depend on an interplay of central inhibition and decreased intrinsic afferent input during the NREM state.

Little is known about the neurochemical changes within the brainstem that accompany electrocortical maturation and associated changes in fetal breathing. In the fetal lamb, infusions of 5-hydroxytryptophan, a biochem-

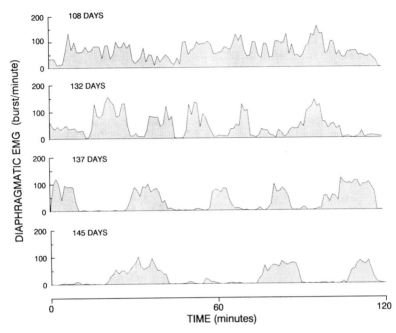

Figure 1   Analysis of diaphragm EMG activity from a lamb fetus at four different gestational ages. Respiratory rate is shown on ordinate for each minute of a 2-h recording period. Gestational age at time of study is displayed. Note that respiratory rate is similar throughout gestation and duration of apneic periods increased with development. (Adapted from Bowes et al., 1981a.)

ical precursor of serotonin known to influence sleep state in adults, increases the time spent in the high-voltage state; conversely, however, the amount and amplitude of fetal breathing increase (Quilligan et al., 1981). Prostaglandin E_2 (PGE$_2$), found in high concentrations in the fetus at term, may be an important mediator in fetal breathing control. Blockade of prostaglandin synthesis with indomethacin or meclofenamate results in state-independent fetal breathing (Kitterman et al., 1979). Infusion of PGE$_2$ after blockade results in a dose-dependent increase in fetal breathing and restoration of physiological cycling with electrophysiological state (Wallen et al., 1986). It is unclear whether PGE$_2$ effects are direct or are mediated through local alterations in cerebral blood flow. Further research is required to clarify what role the dramatic fall in prostaglandin levels after birth may play in the onset of continuous respiration (Long, 1988).

A. Metabolic Control

Hypoxia

Fetal breathing differs from respiration after birth, in that decreased arterial oxygen content in the fetus virtually eliminates rather than stimulates breathing activity (Koos et al., 1987). Hypoxemia hastens the onset and duration of the high-voltage NREM state; the decrease in fetal breathing, however, is out of proportion to the increased time spent in the NREM state (Boddy et al., 1974). Peripheral chemoreceptors, although clearly active in the fetus, do not appear to be important in the fetal respiratory response to hypoxemia (Murai et al., 1985).

Ablation experiments have been used to localize regions in the brain responsible for the response to hypoxia. Hypoxia-induced apnea appears to be centrally mediated in the upper lateral pons. Breathing is augmented during hypoxemia in fetuses with caudal midbrain transections (Dawes et al., 1983). Lesions placed with stereotactic techniques in the upper lateral pons abolish the depressant effect of hypoxemia, but do not alter state-related fetal breathing in the unstimulated setting (Gluckman and Johnston, 1987). This suggests that the neurons responsible for depression of fetal breathing during hypoxemia are discrete from those responsible for apnea during the NREM state. Further studies are required to determine whether the effect of decreased O_2 content is a direct effect or mediated through associated cardiovascular and regional blood flow alterations.

The fetal response to hypoxemia is adaptive to its environment in that it serves to decrease O_2 consumption at times of stress. For example, decreased uterine blood flow during uterine contractions results in a marked diminution of both fetal breathing movements and skeletal muscle activity (Nathanielsz et al., 1980). This response may be important in the ability of the fetus to withstand the decreased oxygen delivery associated with labor.

Hypercarbia

Hypercarbia, induced via the pregnant ewe or directly in the fetus, results in increased time spent in the low-voltage REM state and a concomitant rise in fetal breathing movements. In the mature lamb fetus, the increase in respiratory activity during hypercarbia is largely due to a decrease in fetal apnea associated with the decrease in the NREM state (Boddy et al., 1974). Stimulatory effects of hypercarbia, however, occur as early as 75 days' gestation, well before ECoG differentiation into discrete states (Ioffe et al., 1987). Once state differentiation develops in the chronically instrumented lamb fetus, if hypercarbia is induced while the fetus is in the NREM state, there is no respiratory stimulation or sleep state effect until the P_{CO_2} is above 90 torr (Rigatto et al., 1988). The stimulatory effects of hypercarbia may result from an overall increase in central nervous system activity, rather than specific "respiratory" stimulation.

B. Upper Airway Activity

Special attention to fetal upper airway activity is warranted given the important role it plays at birth in the establishment of lung volume. During fetal breathing in the lamb, the laryngeal abductors (posterior cricoarytenoids, PCA) (Harding, 1980) and alae nasi (Johnston et al., 1986) are phasically active with the diaphragm, while the adductors are inhibited. With the onset of the NREM state, PCA activity ceases, and low-intensity tonic activity of the laryngeal adductors develops, with associated partial adduction of the larynx. The genioglossus does not behave as a respiratory muscle under resting conditions *in utero* in that it is rarely phasically active during the REM state. Phasic activity, however, can be induced with surface cooling (Johnston et al., 1988) or significant hypercarbia (Johnston et al., 1986).

Behavior of the upper airway muscles in the fetus can be understood through their functional role. The fetal lungs are filled with a fluid distinctly different from the amniotic fluid. Maintenance of a positive pressure gradient between the pulmonary and amniotic space is essential to normal lung development (Fewell et al., 1983). The larynx and upper airway interact to prevent a net efflux of pulmonary fluid or influx of amniotic fluid. During the NREM state, laryngeal and supralaryngeal resistance to fluid efflux is activity maintained; this is at least partially due to tonic activity of the adductor muscles (thyroarytenoids, lateral cricoarytenoids, cricothyroids, and thyrohyoids) (Harding et al., 1986a). During fetal breathing, phasic PCA activity causes laryngeal resistance to efflux to decrease. A substantial resistance to an influx of amniotic fluid remains, however, mediated by mechanical properties of the upper airway that are independent of muscle activity (Fewell and Johnson, 1982).

IV. Newborn Period

A. Respiratory Muscles

The function of the entire motor system is dramatically altered by sleep, especially REM sleep. During REM sleep, tonic postural muscle activity ceases (Jouvet, 1967), and monosynaptic reflexes are depressed (Kubota et al., 1965). This modulation of motor activity during REM sleep is the result of postsynaptic hyperpolarization of both alpha and gamma motoneurons (Glenn et al., 1978). Motor inhibition also involves phasic inhibition of 1a spindle afferents. Despite the significant motor inhibition that prevails with REM sleep, unorganized phasic motor discharges do occur, concomitant with other brainstem-mediated phasic activity such as REMs and pontogeniculooccipital (PGO) activity (Orem, 1980). This "phasic REM" period corresponds to the behavioral array seen in the newborn infant during active sleep. In NREM sleep, in contrast, tonic postural muscle activity persists, although at a lower level than in the awake state. In this respect, NREM sleep more closely resembles quiet wakefulness than it does REM sleep.

Diaphragm and Intercostal Muscles

Alteration in skeletal muscle activity and tone with state changes exerts a profound influence on the respiratory muscles and respiratory mechanics. In REM sleep, tonic intercostal muscle activity is essentially abolished, and phasic activity is markedly depressed (Fig. 2). This appears to be true across species and throughout development, having been demonstrated in human adults (Tusiewicz et al., 1977), cats (Duron and Marlot, 1980), neonatal lambs (Henderson-Smart and Read, 1976), and newborn infants (Prechtl et al., 1977; Lopes et al., 1981). The mechanical effect on the rib cage caused by loss of intercostal activity is exaggerated in newborn infants because of

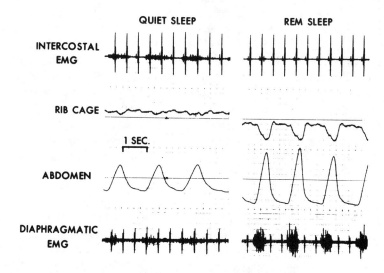

Figure 2 Diaphragm and intercostal muscle EMGs and rib cage and abdominal motion measured with magnetometers in a preterm infant in quiet and REM sleep. Note the paradoxical rib cage and abdominal movement and increased abdominal excursion during REM sleep. In REM sleep, phasic intercostal EMG activity markedly decreases and diaphragm EMG activity increases compared to quiet sleep. (From Bryan et al., 1986.)

their increased chest wall compliance (Agostoni, 1959). As the diaphragm shortens during REM sleep in the newborn, the rib cage moves inward and abdominal displacement increases, leading to the characteristic "paradoxical breathing" of REM sleep. With maturation, rib cage distortion decreases, improving the mechanical stability of the respiratory system (Heldt, 1988). In NREM sleep, tonic and phasic activity of the intercostals helps to stabilize and "stiffen" the rib cage (Prechtl et al., 1977; Thach et al., 1980). With development, rib cage contribution to tidal volume is increased (Honma et al., 1984; Hershenson et al., 1990). Increased activity of the intercostal muscles may not be enough to offset the increased flexibility of the ribs in the premature infant, in that premature infants may continue to demonstrate paradoxical chest movement during neurophysiologically determined NREM sleep (Davi et al., 1979).

Phasic diaphragmatic activity is not subject to the generalized inhibition of the skeletal muscles during REM sleep. Indeed, evidence in both human adults (Tabachnik et al., 1981) and newborns (Muller et al., 1979b) suggests that phasic diaphragm EMG activity may be increased during REM sleep, representing a compensatory response to the loss of intercostal muscle contribution to ventilation (Fig. 2). In newborns, however, a significant proportion of diaphragmatic work is expended on distortion of the chest wall. This may be more problematic in preterm infants, in whom diaphragmatic volume displacement during paradoxical breathing can be up to twice the volume change of the lung, leading to inefficient diaphragmatic work (Heldt and McIlroy, 1987). It has been suggested that increased diaphragmatic work in the newborn during REM sleep may result in muscle fatigue, demonstrated by a change in the power spectrum of the diaphragm EMG (Muller et al., 1979b) and generation of decreased occlusion pressure (Le-Souef et al., 1988). Although likely to be of little significance in the healthy newborn, increased diaphragmatic work and resultant fatigue during paradoxical breathing may contribute to respiratory failure in infants with lung diseases characterized by decreased pulmonary compliance (Muller et al., 1979c).

Although phasic activity of the diaphragm is similar across sleep states in the newborn, there appear to be differences in both its tonic and post-inspiratory activity. The basal level of diaphragm tone, measured between inspiratory bursts from surface EMGs, is lower in REM compared to NREM sleep in both human adults (Muller et al., 1979a) and newborns (Prechtl et al., 1977) (Fig. 3). In contrast, cats demonstrate little diaphragmatic tone in either NREM or REM sleep (Duron and Marlot, 1980). This discrepancy may represent species or maturational differences or may be secondary to measurement technique. With thoracic surface electrodes, measured diaphragmatic activity may be contaminated with signals generated from pos-

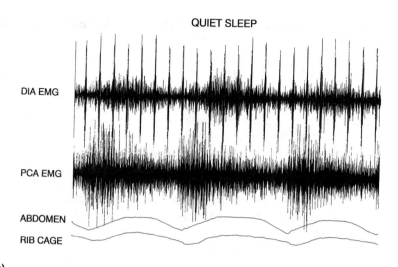

Figure 3 EMG activity of the diaphragm (DIA) and posterior cricoarytenoid muscles (PCA) and motion of the rib cage and abdomen (magnetometer recording) in a premature infant in (a) quiet and (b) active sleep. In quiet sleep, there is increased DIA and PCA EMG tone and increased PCA EMG amplitude. Note paradoxical rib cage and abdominal motion in active sleep.

tural muscles, including the intercostals, which are tonically active during NREM sleep.

It is more evident that postinspiratory inspiratory activity (PIIA) of the diaphragm is considerably more variable during REM sleep than NREM. This alteration in PIIA is evident in early life, demonstrated in infants using surface electrodes (Lopes et al., 1981; Kosch and Stark, 1984; Stark et al., 1987) and in puppies by direct recordings of the diaphragm (England et al., 1985). It is unclear whether the alteration in PIIA behavior is a centrally mediated state-related phenomenon or secondary to changes in peripheral afferent input. Neural inspiratory time and postinspiratory activity measured from the diaphragm EMG are shortened with increased chest wall distortion associated with REM sleep in the newborn (Hagan et al., 1977). The inspiratory inhibition is likely mediated by an intercostal-to-phrenic inhibitory reflex (Remmers, 1970). Diaphragmatic PIIA can be elicited during REM sleep, however, if rib cage stability is increased with continuous positive airway pressure (Hagan et al., 1977). These results suggest that changes in diaphragmatic activity that occur with state may be mediated indirectly by mechanical feedback, such as differences in chest distortion or lung volume, rather than directly by the neurophysiological changes themselves.

Upper Airway Muscles

The upper airway, consisting of the larynx, pharynx, and nose, significantly contributes to total pulmonary resistance to airflow (van Lunteren and Strohl, 1986). Upper airway muscle activity results in dynamic changes in the resistance to airflow during breathing. Similar to other respiratory muscles, the muscles of the upper airway are subject to state-related modulation of their activity. Alterations in upper airway muscle activity with sleep have been most extensively studied in adults, but recently there has been increased recognition of the role state-related changes in upper airway muscle activity play in breathing strategies in early life.

Larynx

The larynx is the most important variable resistor to respiratory airflow (Bartlett et al., 1973). The five intrinsic muscles of the larynx (cricothyroids, lateral cricoarytenoids, transverse arytenoid, thyroarytenoids, and posterior cricoarytenoids) dynamically vary laryngeal resistance during breathing through their effects on vocal cord movement. The posterior cricoarytenoids (PCA), the sole abductors of the vocal cords, and the thyroarytenoids (TA), the principal adductors, have been the most extensively studied of the intrinsic laryngeal muscles.

Muscular activity of the PCA measured by EMG in animals occurs in phase with increasing glottic aperture (Farber, 1978; Harding, 1984; Harding et al., 1986b) and decreased laryngeal resistance (Bartlett et al., 1973). Phasic PCA EMG activity precedes inspiratory flow in adult and newborn animals (England et al., 1985; Harding et al., 1986b) and has recently been demonstrated in human adults (Brancatisano et al., 1984) and newborn infants (Carlo et al., 1987; Kosch et al., 1988; Howell et al., 1988) using an esophageal surface electrode. The vocal cords are partially adducted and laryngeal resistance to airflow is increased during expiration (England et al., 1982). The interplay of decreased PCA and, under some circumstances, increased TA activity helps retard expiratory airflow. Considerable species and developmental differences exist, however, in laryngeal muscle behavior during sleep.

State-related changes in PCA activity demonstrated to date are inconsistent, making comparisons across species and stages of development difficult. A decrease in phasic as well as tonic PCA EMG activity during sleep has been observed in the adult rat (Megirian and Sherrey, 1980; Sherry and Megirian, 1980) and cat (Orem and Lydic, 1978). The relative decrease in activity was greater during REM than NREM sleep. Modifications of PCA activity with sleep parallel measured changes in total upper airway resistance in the cat (Orem et al., 1977), with lowest measured resistance in wakefulness and highest in REM sleep. Decreased PCA activity in NREM sleep compared to the awake state has also been observed in the adult dog (Harding et al., 1986b). In REM sleep, however, PCA EMG activity was greater than measured levels during wakefulness. Newborn studies reveal similar inconsistencies. Increased PCA activity was observed during both NREM and REM sleep compared to wakefulness in puppies, although the relative increase was less during REM sleep (England et al., 1985). Lambs, in contrast, appear to have increased PCA activity during REM compared to NREM sleep (Harding, 1980; Harding et al., 1980). Recent observations in preterm infants suggest that phasic and tonic PCA activity are decreased during active sleep (Fig. 3), similar to sleep state effects on PCA activity observed in puppies.

Inconsistencies in the observations of relative changes in PCA activity with sleep state may be partly due to changes in PCA EMG activity related to head and neck posture (Bonora et al., 1985), which have not been controlled in most studies. It is likely that total upper airway resistance is increased during REM compared to NREM sleep (Orem et al., 1977; Lopes et al., 1983), but the contribution of changes in laryngeal aperture to this increase is unclear.

Considerably more evidence exists to indicate that state-related changes that occur in expiratory laryngeal muscle activity are affected by development. Expiratory airflow in eupneic breathing in the adult is modulated predominantly by expiratory PCA tone. In adult humans (Kuna et al., 1988a) and most adult animals (Harding, 1980; Harding et al., 1986b), phasic expiratory TA activity appears to be limited to wakefulness. In newborn animals, however, active expiratory laryngeal adduction by phasic TA activity plays a more prominent role, although important species differences exist. Phasic expiratory TA activity was observed in quiet wakefulness and NREM sleep in puppies, but disappeared upon onset of REM sleep (Fig. 4) (England et al., 1985). Active laryngeal adduction causing increased laryngeal resistance was also observed in lambs during NREM sleep, was absent in REM sleep, and was variably present during wakefulness (Harding, 1980; Harding et al., 1980). Maturation affected the level of state-related TA activity in both the puppy and lamb. With maturation in the puppy, TA activity progressively declined during both wakefulness and NREM sleep. In contrast, phasic TA activity during NREM sleep increased over the first few weeks of life in the lamb, but was absent in adult ewes. The reasons for the discrepancy in maturational patterns of TA activity in the lamb and puppy are unclear, but the discrepancy may represent a true species difference.

Demonstration of active laryngeal adduction in the human infant is difficult because of the inaccessibility of the TA to surface measurements of its activity. Indirect evidence suggests that active laryngeal adduction may contribute to the establishment of lung volume in the human infant during the immediate postnatal period (Fisher et al., 1982; Mortola et al., 1982). Profound reductions in expiratory flow, often approaching zero, were observed in newborns studied in the first hour after birth. Expiratory flow interruption was most likely accomplished via glottic adduction. A similar expiratory pattern is observed in the newborn opossum (Farber, 1978). Active adduction of the larynx also occurs during the expiratory "grunting" observed in infants with surfactant deficiency. Beyond the immediate postnatal period, however, the activity of the laryngeal adductors and any effect of state in the human infant remains poorly defined.

Pharynx and Nose

Phasic respiratory activity of the pharyngeal airway has been observed in newborn infants with fluoroscopy (Bosma and Lind, 1962). Cineroentgen films demonstrated pharyngeal expansion with inspiration and expiratory constriction. Pharyngeal dilatation with inspiration suggests an active muscular process. Negative pressure generated with diaphragm contraction would promote pharyngeal collapse unless actively opposed. Postmortem studies of human infants confirm the important role of pharyngeal muscle activity during breathing. After death, the newborn upper airway is vulnerable to collapse at negative pressures less than those typically generated during tidal breathing (Wilson et al., 1980). The pharyngeal airway is the most likely site of collapse (Reed et al., 1985).

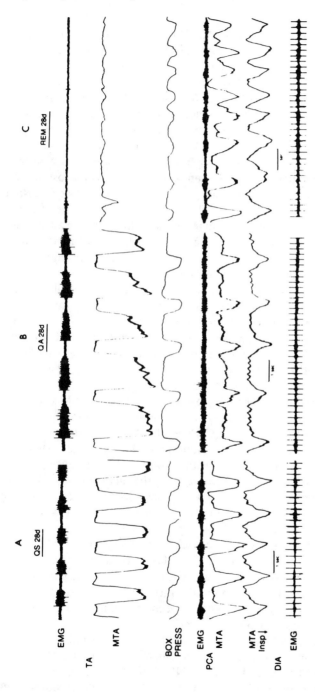

Figure 4 EMG activities of the thyroarytenoid (TA), posterior cricoarytenoid (PCA), and diaphragm (DIA) with associated moving time averages (MTA) in a 28-day-old puppy. Pressure deflection (Box Press) from a plethysmograph is also shown. Measurements made in NREM sleep (A), quiet wakefulness (B), and REM sleep (C). Note prominent TA EMG activity during NREM sleep and quiet wakefulness and absence of TA EMG activity in REM sleep. In REM sleep, PCA EMG activity is also slightly decreased. (From England et al., 1985.)

Generalized sleep-induced muscular inhibition, most prominent during REM sleep, appears to affect the pharyngeal muscles as well. Inhibition of both pharyngeal constrictors and dilators (tensor palatini, stylopharyngeus, styloglossus, geniohyoid) has been observed in the human adult during REM sleep (Mathew and Remmers, 1984). Loss of phasic and tonic activity of the pharyngeal dilators may make upper airway collapse more likely during REM sleep (Hudgel and Hendricks, 1988). Indeed, in one study of preterm infants, increased negative pressure generated during inspiratory efforts against an added elastic load resulted in a higher frequency of upper airway obstruction during active compared to quiet sleep (Knill et al., 1976). Most infants, however, are able to maintain airway patency during active sleep despite increased negative pressures generated during occluded efforts (Roberts et al., 1985; Cohen and Henderson-Smart, 1986). Although airway obstruction appears more likely during REM sleep when considered on mechanical grounds alone, experimental data suggest that reflexes which promote upper airway patency continue to operate during sleep.

The genioglossus, the major external muscle of the tongue, also contributes to the maintenance of upper airway patency. Phasic respiratory activity of the genioglossus, measured with either surface (Gauda et al., 1987; Cohen and Henderson-Smart, 1989) or needle (Roberts et al., 1986) electrodes, is uncommon during eupneic breathing in infants. Data in adult humans (Sauerland and Harper, 1976) and animals (Goh et al., 1986; Issa et al., 1988) suggest that phasic and tonic genioglossus activity during quiet breathing is reduced during sleep, but to a greater degree in REM sleep. The majority of studies of genioglossus activity in infants have examined its response to external airway occlusion (Carlo et al., 1985; Roberts et al., 1986; Gauda et al., 1987; Cohen and Henderson-Smart, 1989). No consistent differences in genioglossus recruitment with airway occlusion between active and quiet sleep have been observed, suggesting that genioglossus reflex activity is preserved during sleep. Similar results have been reported in sleeping human adults, despite the diminution of baseline genioglossus activity in REM sleep (Kuna and Smickley, 1988). Species differences may be important, however; adult dogs demonstrate decreased genioglossus response to airway occlusion in REM compared to NREM sleep (Issa et al., 1988). Other developmental factors may also alter the reflex responses of the genioglossus during sleep. For example, preterm infants with apnea were less likely to recruit genioglossus activity during airway occlusion compared to preterm infants without apnea (Gauda et al., 1987).

Activity of the nasal alae decreases nasal resistance to airflow (Carlo et al., 1983a). Phasic alae nasi activity, measured by surface electrodes, increased during REM sleep compared to NREM in preterm infants (Carlo et al., 1983b). Sleep state differences were obscured when breathing was stimulated by CO_2. Increased nasal dilatation may be a compensatory response to the overall increase in total pulmonary resistance during REM sleep.

V. Respiratory Muscle Output

A. Lung Volume

State-related alterations in respiratory muscle behavior and their mechanical consequences affect lung volume. In newborn infants, in contrast to adults, end-expiratory lung volume is dynamically maintained above the relaxation volume determined mechanically by opposing recoils of the lung and chest wall (Griffiths et al., 1983; Mortola et al., 1985; Kosch and Stark, 1984). End-expiratory lung volume is determined by an interplay of the expiratory time constant, expiratory flow rate, and expiratory duration (Mortola, 1983; Mortola et al., 1984). Lung volume is decreased in REM sleep compared to NREM, most likely as a result of the loss of early retardation of expiratory flow (braking) from diminished PIIA, decreased expiratory laryngeal resistance, and loss of chest wall stability. Changes in expiratory duration may contribute to the volume change as well (Stark et al., 1987).

Using occlusion plethysmography, Henderson-Smart and Read (1979a) demonstrated a 30% decrease in lung volume during behaviorally determined active sleep compared to quiet sleep in term infants. Also using behaviorally identified sleep states, smaller differences in lung volumes between quiet and active sleep have been shown in preterm infants (Lopes et al., 1981; Stark et al., 1987) using relative decreases in lung volumes during apneas in the two sleep states (Fig. 5). These results are in contrast to studies using helium dilution techniques to measure FRC in preterm (Moriette et al., 1983) and term infants (Beardsmore et al., 1989). Using neurophysiologically determined sleep states, the latter studies could not confirm a significant volume change upon transition from NREM to REM sleep. The discrepancy between these results may be partly explained by the fact that behavioral criteria for sleep state rely to some extent on mechanical and ventilatory changes in the respiratory system, such as paradoxical movement of the rib cage and abdomen and altered cycle timing. These effects, in turn, may influence lung volume. Neurophysiological determination of state, on the other hand, may include periods when these changes in breathing are not as extreme.

The decrease in lung volume associated with active sleep in the newborn may have important implications for oxygenation and gas exchange. Newborn infants have small oxygen stores relative to their metabolic rate and thus might be more vulnerable to hypoxemia when lung volumes are decreased in active sleep (Henderson-Smart, 1980; Findley et al., 1983). Trans-

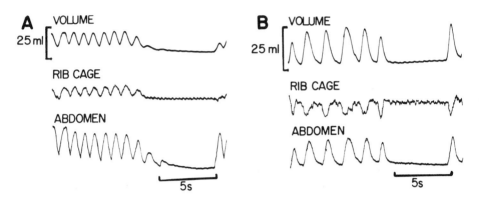

Figure 5 Recordings of volume, rib cage, and abdominal motion during breathing and unobstructed apnea in a preterm infant. (A) Quiet sleep; (B) active sleep. End-expiratory lung volume during apnea falls from that during breathing in quiet sleep and remains unchanged in active sleep. (From Stark et al., 1987.)

cutaneous oxygen tensions are lower and more unstable during active compared to quiet sleep (Martin et al., 1979). Newborns also demonstrate a more rapid decline in oxygen saturation than adults during the brief periods of apnea and hypoventilation characteristic of REM sleep (Henderson-Smart, 1980). In addition to decreased oxygen stores, regional atelectasis and alterations in ventilation caused by chest distortion during active sleep may contribute to the decreased PO_2 in this sleep state (LeSouef et al., 1983).

B. Ventilation

Changes in sleep state have significant effects on ventilation and gas exchange. Although little information exists concerning ventilatory changes upon transition from wakefulness to sleep in the newborn, available evidence suggests that the effects are similar to those observed in adults. With the onset of sleep from quiet wakefulness in adult humans and animals, there is a fall in alveolar ventilation associated with a modest increase in $PaCO_2$ of 2-4 mm Hg (Tabachnik et al., 1981; Colrain et al., 1987). In the premature newborn monkey, a fall in ventilation and a 3-torr increase in $PaCO_2$ has also been observed on transition to sleep (Guthrie et al., 1980). The relative differences in ventilation in the newborn monkey between wakefulness and sleep increase with maturation.

Decreased ventilation with sleep has been attributed to a loss of tonic excitation of medullary respiratory neurons by the brainstem reticular system (Fink et al., 1963). With sleep, withdrawal of the "wakefulness" stimulus to breathing with reticular deactivation or inhibition results in a primary reduction in medullary respiratory neuronal activity (Phillipson, 1978) and, as a result, decreased ventilation. The withdrawal of the "wakefulness" stimulus to breathe during sleep is of little significance under steady-state conditions, but may become important in primary hypoventilation disorders associated with insensitivity of peripheral chemoreceptors. In infants with congenital failure of automatic ventilation (Ondine's curse), for example, ventilation under waking conditions may be normal, but profound hypoventilation occurs with sleep onset (Shannon et al., 1976).

Several investigators have examined differences in ventilation between sleep states in the newborn. Several studies in both term and preterm infants have reported a 14-20% increase in minute ventilation in active compared to quiet sleep (Bolton and Herman, 1974; Hathorn, 1974; Finer et al., 1976; Davi et al., 1979), comparable to similar adult data. The increase in ventilation in REM sleep is primarily caused by a higher respiratory rate, with little change in tidal volume. Others have argued that maturation does affect sleep state differences in ventilation (Haddad et al., 1979; Guthrie et al., 1980). In a study of term newborn infants examined monthly until 4 months of age, minute ventilation was similar in REM and NREM sleep until 3 months of age (Haddad et al., 1979). Although increased respiratory rate was observed during REM sleep, it was accompanied by a corresponding decrease in tidal volume until 3 months. The discrepancy between these results may represent true developmental changes, but could be related to methodology. Because of the irregular respiratory pattern characteristic of REM sleep, true "steady-state" ventilation may not exist. The lack of homogeneity of ventilation during REM sleep makes comparisons to breathing during NREM sleep complicated.

C. Pattern of Breathing

Alterations in breathing pattern associated with sleep state have been described in both adults (Aserinsky, 1965) and infants (Bolton and Herman, 1974; Hoppenbrouwers et al., 1977; Haddad et al., 1979, 1982a, 1987). Maturation, however, affects the rate, depth, and regularity of breathing during both REM and NREM sleep in early life.

During NREM sleep in term infants, the rate and depth of breathing are generally regular. The regular pattern is interrupted by occasional sighs or body movements, often followed by a brief period of repeated short apneas (< 5 s), or "periodic breathing" (Barrington et al., 1987). Superimposed on this pattern are periodic oscillations of both tidal volume and respiratory rate, which occur out of phase with one another such that minute ventilation remains stable (Hathorn, 1978). The short apneas observed during NREM breathing (Fenner et al., 1973) are related to the minimum phase of oscillatory breathing patterns (Waggener et al., 1982). The periodic cycle duration and the frequency of short apneas decrease with maturation over the first 4 months of life (Barrington et al., 1987; Lee et al., 1987). Respiratory rate also decreases over the first few months of life, due to a proportional increase in both inspiratory and expiratory duration (Haddad et al., 1979).

In contrast to NREM sleep, REM sleep in newborns is characterized by marked irregularity in the rate and depth of breathing (Hathorn, 1974; Stevenson and McGinty, 1978). Respiratory rate is higher compared to NREM sleep, secondary to a proportional decrease in inspiratory and expiratory duration (Haddad et al., 1979; Curzi-Dascalova et al., 1983). Periodic breathing and longer apneas are more frequent in REM sleep (Hoppenbrouwers et al., 1977), and the oscillatory pattern of the rate and depth of breathing is increased (Hathorn, 1978; Waggener et al., 1982). Cycle duration of the oscillatory patterns, however, is similar to that observed in NREM sleep (Barrington et al., 1987). In contrast to NREM sleep, the oscillatory patterns of tidal volume and rate frequently are in phase during REM sleep, accounting for the variability in ventilation observed (Hathorn, 1978). Changes in respiratory patterns with maturation are similar in REM and NREM sleep. Over the first few months of life, respiratory rate decreases (Haddad et al., 1979), and shortening of the frequency and duration of apneas (Hoppenbrouwers et al., 1977; Lee et al., 1987) as well as periodic breathing cycle duration, is observed (Barrington et al., 1987).

The marked differences in breathing patterns observed in REM and NREM sleep suggest that sleep state strongly influences central generation of respiratory rhythm. In animals, the slower, more regular breathing during NREM sleep continues despite removal of a variety of afferent stimuli, including vagal and spinal reflexes (Foutz et al., 1979) and pneumotaxic center ablation (Baker et al., 1981). Irregular phasic respiratory patterns of REM sleep occur in synchrony with other brainstem phasic activity, such as REMs and pontogeniculooccipital activity, suggesting a common brainstem source (Orem, 1980; Haddad et al., 1982a). In newborns, the maturation of respiratory patterns with age may represent either an alteration in the central generation of respiratory rhythm or maturation of peripheral moderating influences, such as metabolic control mechanisms (Fagenholz et al., 1976; Barrington et al., 1987) and mechanical relationships and feedback (Bodegard, 1976; Knill et al., 1976).

Respiratory motor output and rhythm generation during sleep deserve special attention in the premature infant. Over 50% of premature infants of less than 32 weeks' gestational age will have clinically significant apnea, defined as pauses longer than 20 s with associated bradycardia (Henderson-Smart, 1981). Periodic breathing is more pronounced in the premature infant in both quiet and active sleep compared to the term infant, as are relative differences between the sleep states (Lee et al., 1987). The majority of prolonged apneas, however, occur during active sleep (Gabriel et al., 1976; Gould et al., 1977). Apneas tend to occur at times of maximal spinal inhibition (Schulte et al., 1977), corresponding to periods of increased phasic brainstem activity.

The ontogeny of increased apnea during active sleep in premature infants is multifactorial. Reduced lung volume and the lower, more variable O_2 levels during active sleep can lead to rapid development of hypoxemia, even during brief respiratory pauses (Martin et al., 1979; Henderson-Smart, 1980). In addition, inhibition of muscle activity during REM sleep also affects the upper airway muscles, which may make airway obstruction more likely to occur (see Section IV.B). Upper airway obstruction has been implicated as an important contributory factor in prolonging central apneic events (Thach and Stark, 1979). In some respects, however, apnea in the preterm infant is an exaggeration of the normally phasic inhibitory-excitatory central mechanisms occurring during REM sleep.

Immaturity may influence both respiratory pattern generation and other brainstem activity. Preterm infants with apnea, for example, have prolonged latency responses of brainstem auditory-evoked potentials compared to infants of similar gestational age without apnea (Henderson-Smart et al., 1983). This finding suggests that differences in overall brainstem maturity may explain abnormalities in respiratory rhythm generation in preterm infants with apnea.

VI. Metabolic Control: Effect of State

A. Hypoxia

Newborns of most mammalian species studied exhibit a "biphasic" response to steady-state hypoxia. Initial ventilatory stimulation is followed by a return to eupneic levels despite continued hypoxia (Henderson-Smart and Read, 1979b; Bureau et al., 1984; Bonora et al., 1984; Saetta and Mortola, 1987). The newborn response to hypoxia is in contrast to the sustained ventilatory stimulation with hypoxia observed in the adult. Human infants and newborn primates differ from other species in that ventilation during prolonged hypoxia is depressed to levels below control values in room air (Cross and Warner, 1951; Miller and Smull, 1955; Rigatto et al., 1975; Woodrum et al., 1981).

The origin of the biphasic response to hypoxia in newborns is unclear. It has been attributed to diminished peripheral chemoreceptor drive (Bureau and Begin, 1982), depression of central respiratory output (Lawson and Long, 1983), and changes in pulmonary mechanics with hypoxia (LaFramboise et al., 1983). Regardless of the mechanism, as the newborn matures the biphasic response to hypoxia is replaced by a sustained increase in ventilation similar to the adult response (Belenky et al., 1979; Bureau and Begin, 1982; Haddad et al., 1982b). In human infants born at term, a sustained increase in ventilation with induced hypoxia is observed after 1 week of life (Brady and Cervti, 1966); in premature infants the switch to a more mature response occurs later in development (Rigatto et al., 1975).

Few studies to examine the effect of state on the ventilatory response to hypoxia have been performed either in early life or in adults. The ventilatory response to progressive hypoxia in the human adult and dog appears to be slightly less in REM compared to NREM sleep (Phillipson, 1978; Douglas et al., 1982). Both lambs (Henderson-Smart and Read, 1979b) and calves (Jeffrey and Read, 1980) demonstrate a progressive increase in ventilation in NREM sleep during increasing hypoxia, but the response in REM sleep is inconsistent. In contrast, in puppies a progressive increase in ventilation with rising hypoxia is observed in both NREM and REM sleep (Henderson-Smart and Read, 1979b). Species differences in the ventilatory response to progressive hypoxia have been attributed to changes in relative chest wall stability across sleep states. Both lambs and calves have a compliant rib cage and exhibit paradoxical breathing during REM sleep, similar to the human infant. Paradoxical breathing is exaggerated at times of respiratory stress, which can uncouple the translation of increased drive into increased ventilation. The puppy, however, has a more stable rib cage and does not breathe paradoxically during REM sleep. Decreased rib cage compliance may provide additional mechanical stability to meet the increased ventilatory demands of progressive hypoxia.

The newborn response to steady-state levels of hypoxia may also be influenced by sleep state. In puppies, the ventilatory response to exposure to 15% O_2 is different in REM compared to NREM sleep (Haddad et al., 1982b). The investigators examined the hypoxic response after the initial increase in ventilation characteristic of the newborn period. In REM sleep, the puppies had a sustained increase in instantaneous minute ventilation at the earliest study age (14 days) and this response did not change with maturation. In NREM sleep, the youngest subjects decreased minute ventilation in response to hypoxia. The decrease in ventilation during NREM sleep was accompanied by a proportional decrease in metabolic rate. Sleep state differences in the hypoxic response were eliminated with maturation, with the NREM ventilatory response becoming similar to that observed in REM sleep. A maturational effect on sleep state-mediated responses to hypoxia is suggested by similar studies performed in kittens (Bonora et al., 1984). Severe respiratory depression and cardiopulmonary failure in kittens caused by prolonged exposure to a hypoxic environment are observed during the NREM state, but rapidly reverse upon onset of REM sleep (Baker and McGinty, 1977, 1979).

Comparable studies examining sleep state effects on the hypoxic response have not been performed in human infants. In one study of preterm infants, the early hyperventilation observed on exposure to 15% O_2 was prolonged in quiet sleep; the classic biphasic response with late hypoventilation was observed only in active sleep (Rigatto et al., 1982). Comparison with studies in newborn animals cannot be made, however, since only short periods of exposure to the hypoxic environment could be performed in human infants. In a study of the progressive breath-by-breath response to airway occlusion at FRC, which provides primarily a hypoxic stimulus, preterm infants had a decreased response to occlusion during active sleep compared to term infants as measured by occlusion pressure (Frantz et al., 1976a). Occlusion pressure became greater with increasing postnatal and gestational age, suggesting the response to hypoxia may be influenced by maturational changes.

Sleep state differences in arousal threshold during hypoxia may have clinical relevance. In adult animals, an arousal response to hypoxia occurs at a lower oxygen saturation during REM sleep compared to NREM (Phillipson et al., 1978; Bowes et al., 1980); arousal is dependent on integrity of peripheral chemoreceptors (Bowes et al., 1981b). Newborns also awaken at lower arterial saturations during REM sleep compared to NREM (Henderson-Smart and Read, 1979b; Jeffrey and Read, 1980). NREM sleep arousal occurs at similar saturations in both newborns and adults, but data suggest that REM sleep arousal thresholds may be at lower saturations in the newborn.

A maturational change in arousal threshold through gestation and postnatally is suggested by studies in human infants (Miller and Smull, 1955). Although not controlled for sleep state, term infants tended to awaken sooner than preterm infants when placed in a 12% O_2 environment. When retested several weeks later, most infants awoke significantly sooner with the hypoxic challenge. Such experimental data may have important clinical correlates. Diminished arousal response to hypoxia in REM sleep has been implicated in the pathogenesis of prolonged apnea in the preterm newborn and hypothesized to contribute to the sudden infant death syndrome (Hunt et al., 1981).

B. CO₂ Response

The slope of the CO_2 response curve is similar in term newborns and adults (Avery et al., 1963; Krauss et al., 1975). Preterm infants demonstrate decreased responsiveness to CO_2 compared to term infants (Rigatto et al., 1975; Frantz et al., 1976b; Krauss et al., 1976). The cause of the depressed response to hypercarbia or hypercapnia in the preterm infant is unclear, but evidence suggests the response is a product of central insensitivity as well as mechanical disadvantages (Moriette et al., 1985; Rigatto, 1984).

Sleep state affects the ventilatory response to CO_2 in adult humans and animals (Phillipson, 1978). The response to CO_2 in REM sleep is slightly decreased overall compared to NREM. Relative differences between REM and NREM sleep are more pronounced during phasic REM compared to tonic REM sleep, suggesting that behavioral influences in phasic REM may influence metabolic control of breathing, much as in the awake state (Sullivan et al., 1979).

State-related alterations in CO_2 response are less clear in the newborn. When responses to steady-state levels of CO_2 are measured, no differences between NREM and REM sleep are observed in either human infants (Fagenholz et al., 1976; Davi et al., 1979; Haddad et al., 1980; Anderson et al., 1983) or newborn monkeys (Guthrie et al., 1980, 1981). In the young primate, state-related differences in CO_2 response appear with maturation (Guthrie et al., 1980, 1981). Similar maturational changes were not observed in human infants studied to 4 months of age (Haddad et al., 1980). When rebreathing techniques are used, however, exposing the subject to progressively higher inhaled CO_2, term newborns have a decreased ventilatory response in active compared to quiet sleep (Moriette et al., 1985; Honma et al., 1984). The decreased response observed in active sleep may be secondary to an inability to recruit greater rib cage contributions to increase tidal volume during active sleep (Honma et al., 1984; Hershenson et al., 1989). The lack of a sleep state-related difference when steady-state levels of CO_2 are breathed may be somewhat misleading, since steady-state ventilation does not exist during the irregular breathing characteristic of the REM state. Further studies will be required to conclude whether there are true maturational differences in the effect of state on ventilatory response to CO_2 between the newborn and adult.

VII. Conclusions

Behavioral state profoundly influences respiratory motor output in the developing fetus and newborn. As a result, studies of respiration in early life must be controlled for sleep state to be adequately interpreted. The relative immaturity and instability of the developing nervous system, however, can make it difficult to determine sleep state accurately and reliably. Differences in the level of maturity at birth may also complicate comparisons across species. Although these problems seem daunting, careful study of breathing during sleep will provide significant insight into the developing organism.

Perhaps the most significant sleep-related change in respiratory motor output has been termed the "paradox of REM sleep" (Bryan and England, 1984). Through effects on the respiratory pump and upper airway, REM sleep disables the primary mechanisms used by the newborn to defend lung volume. In the healthy infant, loss of active retardation of expiratory air-flow may be partially offset by the irregular respiratory rhythm and rapid respiratory rate inherent to the REM state. The alterations in respiratory muscle activity and lung volume control during REM sleep may be harmful to the newborn only when other respiratory control mechanisms are disabled as well, however, such as with prematurity or perhaps sudden infant death. Further research on how sleep and maturation affect the developing respiratory system will enhance our understanding of its disorders in early life.

Acknowledgments

We gratefully acknowledge the assistance of Rosemarie Ungarelli in the preparation of the figures and the secretarial assistance of Julie Reid.

References

Agostoni, E. (1959). Volume-pressure relationships of the thorax and lung in the newborn. *J. Appl. Physiol.* **14**:909-913.

Alcorn, D., Adamson, T. M., Malone, J. E., and Robinson, P. M. (1980). Morphologic effects of chronic bilateral phrenectomy or vagotomy in fetal lamb lung. *J. Anat.* **130**:683-695.

Anderson, J. V., Jr., Martin, R. J., Abboud, E. F., Dyme, I. Z., and Bruce, E. N. (1983). Transient ventilatory response to CO_2 as a function of sleep state in full term infants. *J. Appl. Physiol.* **54**:1482-1488.

Aserinsky, E. (1965). Periodic respiratory pattern occurring in conjunction with eye movements during sleep. *Science* **150**:763-766.

Avery, M. E., Chernick, V., Dutton, R. E., and Permutt, S. (1963). Ventilatory response to inspired carbon dioxide in infants and adults. *J. Appl. Physiol.* **18**:895-903.

Baker, T. L., and McGinty, D. J. (1977). Reversal of cardiopulmonary failure during active sleep in hypoxic kittens: Implications for sudden infant death. *Science* **199**:419-421.

Baker, T. L., and McGinty, D. J. (1979). Sleep-waking patterns in hypoxic kittens. *Dev. Psychobiol.* **12**:561-575.

Baker, T. L., Netick, A., and Dement, W. C. (1981). Sleep related apneic and apneustic breathing following pneumotaxic center ablation and vagotomy. *Respir. Physiol.* **46**:271-294.

Barrington, K. J., Finer, N. N., and Wilkinson, M. H. (1987). Progressive shortening of the periodic breathing cycle duration in normal infants. *Pediatr. Res.* **21**:247-251.

Bartlett, D., Remmers, J. E., and Gautier, H. (1973). Laryngeal regulation of respiratory airflow. *Respir. Physiol.* **18**:194-204.

Beardsmore, C. S., MacFadyen, U. M., Moosavi, S. S. H., Wimpress, S. P., Thompson, J., and Simpson, H. (1989). Measurement of lung volumes during active and quiet sleep in infants. *Pediatr. Pulmonol.* **7**:71-77.

Belenky, D. A., Standaert, T. A., and Woodrum, D. E. (1979). Maturation of hypoxic ventilatory response of the newborn lamb. *J. Appl. Physiol.* **47**:927-930.

Boddy, K., Dawes, G. S., Fischer, R. L., Pinter, S., and Robinson, J. S. (1974). Fetal respiratory movements, electrocortical and cardiovascular responses to hypoxia and hypercapnea in sheep. *J. Physiol. (London)* **243**:599-618.

Boddy, K., and Robinson, J. S. (1971). External method for detection of fetal breathing in utero. *Lancet* **2**:1123-1233.

Bodegard, G. (1976). Control of respiration in newborn babies. *Acta. Paediatr. Scand.* **65**:257-266.

Bolton, D. P. G., and Herman, S. (1974). Ventilation and sleep state in the new-born. *J. Physiol.* **240**:67-77.

Bonora, M., Bartlett, D., and Knuth, S. L. (1985). Changes in upper airway muscle activity related to head position in adult cats. *Respir. Physiol.* **60**:181-192.

Bonora, M., Marlot, D., Gautier, H., and Duron, B. (1984). Effects of hypoxia on ventilation during postnatal development in conscious kittens. *J. Appl. Physiol.* **56**:1464-1471.

Bosma, J., and Lind, J. (1962). Upper respiratory mechanisms of newborn infants. A clinical essay. *Acta Paediatr. Suppl.* **135**:32-44.

Bowes, G., Adamson, T. M., Ritchie, B. C., Dowling, M., Wilkinson, M. H., and Maloney, J. E. (1981a). Development of patterns of respiratory activity in unanesthetized fetal sheep in utero. *J. Appl. Physiol.* **50**:693-700.

Bowes, G., Townsend, E. R., Kozar, L. F., Bromley, S. M., and Phillipson, E. A. (1981b). Effect of carotid body denervation on arousal response to hypoxia in sleeping dogs. *J. Appl. Physiol.* **51**:40-45.

Bowes, G., Woolf, G. M., Sullivan, C. E., and Phillipson, E. A. (1980). Effect of sleep fragmentation on ventilatory and arousal responses of sleeping dogs to respiratory stimuli. *Am. Rev. Respir. Dis.* **122**:899-908.

Brady, J. P., and Cervti, E. (1966). Chemoreceptor reflexes in the newborn infants: Effect of varying degrees of hypoxia on heart rate and ventilation in a warm environment. *J. Physiol. (London)* **184**:631-645.

Brancatisano, T. P., Dodd, D. S., and Engel, L. A. (1984). Respiratory activity of the posterior cricoarytenoid muscle and vocal cords in humans. *J. Appl. Physiol.* **57**:1143-1149.

Bryan, A. C., Bowes, G., and Maloney, J. E. (1986). Control of breathing in the fetus and newborn. In *Handbook of Physiology.* Section 3. The Respiratory System, Vol. II. Control of Breathing, Part 2. Washington, DC, American Physiological Society.

Bryan, A. C., and England, S. J. (1984). Maintenance of an elevated FRC in the newborn. *Am. Rev. Respir. Dis.* **129**:209-210.

Bureau, M. A., and Begin, R. (1982). Postnatal maturation of the respiratory response to O_2 in awake newborn lambs. *J. Appl. Physiol.* **52**: 428-433.

Bureau, M. A., Zinman, R., Foulon, P., and Begin, R. (1984). Diphasic ventilatory response to hypoxia in newborn lambs. *J. Appl. Physiol.* **56**:84-90.

Carlo, W. A., Kosch, P. C., Bruce, E. N., Strohl, K. P., and Martin, R. J. (1987). Control of laryngeal muscle activity in preterm infants. *Pediatr. Res.* **22**:87-91.

Carlo, W. A., Martin, R. J., Bruce, E. N., Strohl, K. P., and Fanaroff, A. A. (1983a). Alae nasi activation (nasal flaring) decreases nasal resistance in preterm infants. *Pediatrics* **72**:338-343.

Carlo, W. A., Martin, R. J., Abboud, E. F., Bruce, E. N., and Strohl, K. P. (1983b). Effect of sleep state and hypercapnea on alae nasi and diaphragm EMGs in preterm infants. *J. Appl. Physiol.* **54**:1590-1596.

Carlo, W. A., Miller, M. J., and Martin, R. J. (1985). Differential response of respiratory muscles to airway occlusion in infants. *J. Appl. Physiol.* **59**:847-852.

Clewlow, F., Dawes, G. S., Johnston, B. M., and Walker, D. W. (1983). Changes in breathing, electrocortical and muscle activity in unanesthetized fetal lambs with age. *J. Physiol. (London)* **341**:463-476.

Cohen, G., and Henderson-Smart, D. J. (1986). Upper airway stability and apnea during nasal occlusion in newborn infants. *J. Appl. Physiol.* **60**:1511-1517.

Cohen, G., and Henderson-Smart, D. J. (1989). Upper airway muscle activity during nasal occlusion in newborn babies. *J. Appl. Physiol.* **66**:1328-1335.

Colrain, I. M., Trinder, J., Fraser, G., and Wilson, G. V. (1987). Ventilation during sleep onset. *J. Appl. Physiol.* **63**:2067-2074.

Cross, K. W., and Warner, P. (1951). The effects of inhalation of high and low concentrations of oxygen on the respiration of the newborn infant. *J. Physiol. (London)* **114**:283-295.

Curzi-Dascalova, L., Lebrun, F., and Korn, G. (1983). Respiratory frequency according to sleep states and age in normal premature infants: A comparison with full term infants. *Pediatr. Res.* **17**:152-156.

Davi, M., Sankaran, K., MacCullum, M., Cates, D., and Rigatto, H. (1979). Effect of sleep state on chest distortion and on the ventilatory response to CO_2 in neonates. *Pediatr. Res.* **13**:982-986.

Dawes, G. S., Fox, H. E., Leduc, B. M., Liggins, G. C., and Richards, R. T. (1972). Respiratory movements and rapid eye movement sleep in the foetal lamb. *J. Physiol. (London)* **220**:119-143.

Dawes, G. S., Gardner, W. N., Johnson, B. M., and Walker, D. W. (1983). Breathing in fetal lambs; the effect of brainstem section. *J. Physiol.* **335**:535-553.

Douglas, N. J., White, D. P., Weil, J. V., Pickett, C. K., Martin, R. J., Hudgel, D.W., and Zwillich, C.W. (1982). Hypoxic ventilatory response decreases during sleep in normal men. *Am. Rev. Respir. Dis.* **125**:286-289.

Dreyfus-Brisac, C. (1964). The electroencephalogram of the premature infant and full-term newborn. In *Neurologic and Electroencephalographic Correlative Studies in Infancy*. Edited by P. Kellaway and I. Peterson. New York, Grune & Stratton.

Dreyfus-Brisac, C. (1970). Ontogenesis of sleep in human prematures after 32 weeks of conceptional age. *Dev. Psychobiol.* **3**:91-121.

Dreyfus-Brisac, C. (1975). Neurophysiological studies in human premature and full-term newborns. *Biol. Psychiatry* **10**:485-496.

Duron, B., and Marlot, D. (1980). Intercostal and diaphragmatic electrical activity during wakefulness and sleep in normal unrestrained adult cats. *Sleep* **3**:269-280.

England, S. J., Bartlett, D., and Daubenspeck, J. A. (1982). Influence of human vocal cord movements on airflow and resistance during eupnea. *J. Appl. Physiol.* **52**:773-777.

England, S. J., Kent, G., and Stogryn, H. A. F. (1985). Laryngeal muscle and diaphragmatic activities in conscious dog pups. *Respir. Physiol.* **60**:95-108.

Fagenholz, S. A., O'Connell, K., and Shannon, D. C. (1976). Chemoreceptor function and sleep state in apnea. *Pediatrics* **58**:31-36.

Farber, J. P. (1978). Laryngeal effects and respiration in the suckling opossum. *Respir. Physiol.* **35**:189-201.

Fenner, A., Schalk, U., Hoenicke, H., Wendenburg, A., and Roehling, T. (1973). Periodic breathing in premature and neonatal babies: incidence, breathing pattern, respiratory gas tensions, response to changes in the composition of ambient air. *Pediatr. Res.* **7**:174-183.

Fewell, J. E., Hislop, A. A., Kitterman, J. A., and Johnson, P. (1983). Effect of tracheostomy on lung development in fetal lambs. *J. Appl. Physiol.* **55**:1103-1108.

Fewell, J. E., and Johnson, P. (1982). Upper airway dynamics during fetal breathing and during apnea in fetal lambs. *J. Physiol.* **339**:495-504.

Findley, L. J., Ries, A. L., Tisi, G. M., and Wagner, P. D. (1983). Hypoxemia during apnea in normal subjects: mechanisms and impact of lung volume. *J. Appl. Physiol.* **55**:1777-1783.

Finer, N. N., Abrams, I. F., and Taeusch, H. W. (1976). Ventilation and sleep states in newborn infants. *J. Pediatr.* **89**:100-108.

Fink, B. R., Hanks, E. C., Ngai, S. H., and Papper, E. M. (1963). Central regulation of respiration during anesthesia and wakefulness. *Ann. NY Acad. Sci.* **109**:892-899.

Fisher, J. T., Mortola, J. P., Smith, J. B., Fox, G. S., and Weeks, S. (1982). Respiration in newborns. Development of the control of breathing. *Am. Rev. Respir. Dis.* **125**:650-657.

Foutz, A. S., Netick, A., and Dement, N. C. (1979). Sleep state effects on breathing after spinal cord transection and vagotomy in the cat. *Respir. Physiol.* **37**:89-100.

Frantz, I. D., III, Adler, S. M., Abroms, I. F., and Thach, B. T. (1976a). Respiratory response to airway occlusion in infants: Sleep state and maturation. *J. Appl. Physiol.* **41**:634-638.

Frantz, I. D., III, Adler, S. M., Thach, B. T., and Taeusch, H. W., Jr. (1976b). Maturational effects and respiratory responses to carbon dioxide in premature infants. *J. Appl. Physiol.* **41**:41-45.

Gabriel, M., Albani, M., and Schulte, F. J. (1976). Apneic spells and sleep states in preterm infants. *Pediatrics* **57**:142-147.

Gauda, E. B., Miller, M. J., Carlo, W. A., Difiore, J. M., Johnsen, D. C., and Martin, R. J. (1987). Genioglossus response to airway occlusion in apneic versus nonapneic infants. *Pediatr. Res.* **22**:683-687.

Glenn, L. L., Foutz, A. S., and Dement, W. C. (1978). Membrane potential of spinal motorneurons during natural sleep in cats. *Sleep* **1**:199-204.

Gluckman, P. D., Gunn, T. R., and Johnston, B. M. (1983). The effect of cooling on breathing and shivering in unanaesthetized fetal lambs in utero. *J. Physiol.* **343**:495-506.

Gluckman, P. D., and Johnston, B. M. (1987). Lesions in the upper lateral pons abolish the hypoxic depression of breathing in unanaesthetized fetal lambs in utero. *J. Physiol.* **382**:373-383.

Goh, A. S. F., Issa, F. G., and Sillivan, C. E. (1986). Upper airway dilating forces during wakefulness and sleep. *J. Appl. Physiol.* **61**:2148-2155.

Goldstein, J. D., and Reid, L. M. (1980). Pulmonary hypoplasia resulting from phrenic nerve agenesis and diaphragmatic amyoplasia. *J. Pediatr.* **97**:282-287.

Gould, J. B., Lee, A. F. S., James, O., Sander, L., Teager, H., and Fineberg, N. (1977). The sleep state characteristics of apnea during infancy. *Pediatrics* **59**:182-194.

Griffiths, G. B., Noworaj, A., and Mortola, J. P. (1983). End-expiratory level and breathing pattern in the newborn. *J. Appl. Physiol.* **55**:243-249.

Guthrie, R. D., Standaert, T. A., Hodson, W. A., and Woodrum, D. E. (1980). Sleep and maturation of eucapnic ventilation and CO_2 sensitivity in the premature primate. *J. Appl. Physiol.* **48**:347-354.

Guthrie, R. D., Standaert, T. A., Hodson, W. A., and Woodrum, D. E. (1981). Development of CO_2 sensitivity: Effects of gestational age, postnatal age, and sleep state. *J. Appl. Physiol.* **50**:956-961.

Haddad, G. G., Epstein, R. A., Epstein, M. A. F., Leistner, H. L., Marino, P. A., and Mellins, R. B. (1979). Maturation of ventilation and ventilatory pattern in normal sleeping infants. *J. Appl. Physiol.* **46**:998-1002.

Haddad, G. G., Lai, T. L., and Mellins, R. B. (1982a). Determination of ventilatory pattern in REM sleep in normal infants. *J. Appl. Physiol.* **53**:52-56.

Haddad, G. G., Ghandi, M. R., and Mellins, R. B. (1982b). Maturation of ventilatory response to hypoxemia in puppies during sleep. *J. Appl. Physiol.* **52**:309-314.

Haddad, G. G., Jeng, H. J., Lai, T. L., and Mellins, R. B. (1987). Determination of sleep state in infants using respiratory variability. *Pediatr. Res.* **21**:556-562.

Haddad, G. G., Leistner, H. L., Epstein, R. A., Epstein, M. A. F., Grodin, W. K., and Mellins, R. B. (1980). CO_2-induced changes in ventilation and ventilatory pattern in normal sleeping infants. *J. Appl. Physiol.* **48**:684-688.

Hagan, R., Bryan, A. C., Bryan, M. H., and Gulston, G. (1977). Neonatal chest wall afferents and regulation of respiration. *J. Appl. Physiol.* **42**: 362-367.

Harding, R. (1980). State-related and developmental changes in laryngeal function. *Sleep* **3**:307-322.

Harding, R. (1984). Function of the larynx in the fetus and newborn. *Annu. Rev. Physiol.* **46**:645-659.

Harding, R., Bocking, A. D., and Sigger, J. N. (1986a). Upper airway resistance in fetal sheep: The influence of breathing activity. *J. Appl. Physiol.* **60**:160-165.

Harding, R., England, S. J., Straddling, J. R., Kozar, L. F., and Phillipson, E. A. (1986b). Respiratory activity of laryngeal muscles in awake and sleeping dogs. *Respir. Physiol.* **66**:315-326.

Harding, R., Johnson, P., and McClelland, M. E. (1980). Respiratory function of the larynx in developing sheep and the influence of sleep state. *Respir. Physiol.* **40**:165-179.

Hathorn, M. K. S. (1974). The rate and depth of breathing in newborn infants in different sleep states. *J. Physiol.* **243**:101-113.

Hathorn, M. K. S. (1978). Analysis of periodic changes in ventilation in newborn infants. *J. Physiol.* **285**:85-99.

Heldt, D. P., and McIlroy, M. B. (1987). Distortion of chest wall and work of diaphragm in preterm infants. *J. Appl. Physiol.* **62**:164-169.

Heldt, G. P. (1988). Development of stability of the respiratory system in preterm infants. *J. Appl. Physiol.* **65**:441-444.

Henderson-Smart, D. J. (1980). Vulnerability to hypoxaemia during sleep in the newborn. *Sleep* 1:195-208.

Henderson-Smart, D. J. (1981). The effect of gestational age on the incidence and duration of recurrent apnoea in newborn babies. *Aust. Paediatr. J.* 17:273-276.

Henderson-Smart, D. J., Pettigrew, A. G., and Campbell, D. J. (1983). Clinical apnea and brain stem neural function in preterm infants. *N. Engl. J. Med.* 308:353-357.

Henderson-Smart, D. J., and Read, D. J. C. (1976). Depression of respiratory muscles and defective responses to nasal obstruction during active sleep in the newborn. *Aust. Paediatr. J.* 12:261-266.

Henderson-Smart, D. J., and Read, D. J. C. (1979a). Reduced lung volume during behavioral active sleep in the newborn. *J. Appl. Physiol.* 46:1081-1085.

Henderson-Smart, D. J., and Read, D. J. C. (1979b). Ventilatory responses to hypoxaemia during sleep in the newborn. *J. Dev. Physiol.* 1:195-208.

Hershenson, M. B., Colin, A. A., Wohl, M. E. B., and Stark, A. R. (1990). Changes in the contribution of the rib cage to tidal breathing during infancy. *Am. Rev. Respir. Dis.* 141:922-925.

Hershenson, M. B., Stark, A. R., and Mead, J. (1989). Action of the inspiratory muscles of the rib cage during breathing in newborns. *Am. Rev. Respir. Dis.* 139:1207-1212.

Honma, Y., Wilkes, D., Bryan, M. H., and Bryan, A. C. (1984). Rib cage and abdominal contributions to ventilatory response to CO_2 in infants. *J. Appl. Physiol.* 56:1211-1216.

Hoppenbrouwers, T. Hodgman, J., Harper, R. M., Hofmann, E., and Sterman, M. B., and McGinty, D. J. (1977). Polygraphic studies of normal infants during the first six months of life. III. Incidence of apnea and periodic breathing. *Pediatrics* 60:418-425.

Howell, R. G., Leszczynski, L., Kosch, P. C., and Stark, A. R. (1988). Uncoupling of laryngeal abduction and inspiratory effort in premature infants. *Pediatr. Res.* 23:564A.

Hudgel, D. W., and Hendricks, C. (1988). Palate and hypopharynx—Sites of inspiratory narrowing of the upper airway during sleep. *Am. Rev. Respir. Dis.* 138:1542-1547.

Hunt, J. R., McCulloh, K., and Brouillette, R. T. (1981). Diminished hypoxic ventilatory responses in near-miss sudden infant death syndrome. *J. Appl. Physiol.* 50:1313-1317.

Ioffe, S., Jansen, A. H., and Chernick, V. (1984). ECoG and breathing activity in fetal lambs after undercut of cerebral cortex. *J. Appl. Physiol.* 57:1195-1201.

Ioffe, S., Jansen, A. H., and Chernick, V. (1987). Maturation of spontaneous fetal diaphragmatic activity and fetal response to hypercapnea and hypoxemia. *J. Appl. Physiol.* 62:609-622.

Issa, F. G., Edwards, P., Szeto, E., Lawff, D., and Sullivan, C. (1988). Genioglossus and breathing responses to airway occlusion: Effect of sleep and route of occlusion. *J. Appl. Physiol.* **64**:543-549.

Jeffrey, H. E., and Read, D. J. C. (1980). Ventilatory responses of newborn calves to progressive hypoxia in quiet and active sleep. *J. Appl. Physiol.* **48**:892-895.

Johnston, B. M., Gunn, T. R., and Gluckman, P. D. (1986). Genioglossus and alae nasi activity in fetal sheep. *J. Dev. Physiol.* **8**:323-331.

Johnston, B. M., Gunn, T. R., and Gluckman, P. D. (1988). Surface cooling rapidly induces coordinated activity in the upper and lower airway muscles of the fetal lamb in utero. *Pediatr. Res.* **23**:257-261.

Jouvet, M. (1967). Neurophysiology of the states of sleep. *Physiol. Rev.* **47**:117-177.

Kitterman, J. A., Liggins, G. C., Clements, A., and Tooley, W. H. (1979). Stimulation of breathing movements in fetal sheep by inhibitors of prostaglandin synthesis. *J. Dev. Physiol.* **1**:453-466.

Knill, R., Andrews, W., Bryan, A. C., and Bryan, M. H. (1976). Respiratory load compensation in infants. *J. Appl. Physiol.* **40**:357-361.

Koos, B. J., Sameshima, H., and Power, G. G. (1987). Fetal breathing, sleep state and cardiovascular responses to graded hypoxia in sheep. *J. Appl. Physiol.* **62**:1033-1039.

Kosch, P. C., Hutchison, A. A., Wozniak, J. A., Carlo, W. A., and Stark, A. R. (1988). Posterior cricoarytenoid and diaphragm activities during tidal breathing in neonates. *J. Appl. Physiol.* **64**:1968-1978.

Kosch, P. C., and Stark, A. R. (1984). Dynamic maintenance of end-expiratory lung volume in full-term infants. *J. Appl. Physiol.* **57**:1126-1133.

Krauss, A. N., Klain, D. B., Waldman, S., and Auld, P. A. M. (1975). Ventilatory response to carbon dioxide in newborn infants. *Pediatr. Res.* **9**:46-50.

Krauss, A. N., Waldman, S., and Auld, P. A. M. (1976). Diminished response to carbon dioxide in premature infants. *Biol. Neonate* **30**:216-223.

Kubota, K., Iwamura, Y., and Niimi, Y. (1965). Monosynaptic reflex and natural sleep in the cat. *J. Neurophysiol.* **28**:125-138.

Kuna, S. T., Insalaco, G., and Woodson, G. E. (1988). Thyroarytenoid muscle activity during wakefulness and sleep in normal adults. *J. Appl. Physiol.* **65**:1332-1339.

Kuna, S. T., and Smickley, J. (1988). Response of genioglossus muscle activity to nasal airway occlusion in normal sleeping adults. *J. Appl. Physiol.* **64**:347-353.

LaFramboise, W. A., Guthrie, R. D., Standaert, T. A., and Woodrum, D. E. (1983). Pulmonary mechanics during the ventilatory response to hypoxemia in the newborn monkey. *J. Appl. Physiol.* **55**:1008-1014.

Lawson, E. E., and Long, W. A. (1983). Central origin of biphasic breathing pattern during hypoxia in newborns. *J. Appl. Physiol.* **55**:483-488.

Lee, D., Caces, R., Kwiatkowski, K., Cates, D., and Rigatto, H. (1987). A developmental study on types and frequency distribution of short apneas (3 to 15 seconds) in term and preterm infants. *Pediatr. Res.* **22**: 344-349.

LeSouef, P. N., England, S. J., Stogryn, H. A. F., and Bryan, A. C. (1988). Comparison of diaphragmatic fatigue in newborn and older rabbits. *J. Appl. Physiol.* **65**:1040-1044.

LeSouef, P. N., Lopes, J. M., England, S. J., Bryan, M. H., and Bryan, A. C. (1983). Influence of chest wall distortion on esophageal pressure. *J. Appl. Physiol.* **55**:353-358.

Long, W. A. (1988). Prostaglandins and control of breathing in newborn piglets. *J. Appl. Physiol.* **64**:409-418.

Lopes, J. M., Muller, N. L., Bryan, M. H., and Bryan, A. C. (1981). Importance of inspiratory muscle tone in maintenance of FRC in the newborn. *J. Appl. Physiol.* **51**:830-834.

Lopes, J. M., Tabachnik, E., Muller, N. L., Levison, H., and Bryan, A. C. (1983). Total airway resistance and respiratory muscle activity during sleep. *J. Appl. Physiol.* **54**:773-777.

Maloney, J. E., Adamson, T. M., Brodecky, V., Cranage, S., Lambert, T. F., and Ritchie, B. C. (1975). Diaphragmatic activity and lung liquid flow in the unanesthetized foetal sheep. *J. Appl. Physiol.* **39**:552-558.

Maloney, J. E., Bowes, G., and Wilkinson, M. (1980). Fetal breathing and development of patterns of respiration before birth. *Sleep* **3**:299-306.

Martin, R. J., Okken, A., and Rubin, D. (1979). Arterial oxygen tension during active and quiet sleep in the normal neonate. *J. Pediatr.* **94**:271-274.

Mathew, O. P., and Remmers, J. E. (1984). Respiratory function of the upper airway. In *Sleep and Breathing*. Edited by N. A. Saunders and C. E. Sullivan. New York, Marcel Dekker.

Megirian, D., and Sherrey, J. H. (1980). Respiratory functions of the laryngeal muscles during sleep. *Sleep* **3**:289-298.

Miller, H. C., and Smull, N. W. (1955). Further studies on the effects of hypoxia on the respiration of newborn infants. *Pediatrics* **6**:93-102.

Moriette, G., Chaussain, M., Radvanyi-Bouvet, M. F., Walti, H., Pajot, N., and Relier, J. P. (1983). Functional residual capacity and sleep states in the premature newborn. *Biol. Neonate* **43**:125-133.

Moriette, G., Van Reempts, P., Moore, M., Cates, D., and Rigatto, H. (1985). The effect of rebreathing CO_2 on ventilation and diaphragmatic electromyography in newborn infants. *Respir. Physiol.* **62**:387-397.

Mortola, J. P. (1983). Comparative aspects of the dynamics of breathing in newborn mammals. *J. Appl. Physiol.* **54**:1229-1235.

Mortola, J. P., Fisher, J. T., Smith, J. B., Fox, G. S., Weeks, S., and Willis, D. (1982). Onset of respiration in infants delivered by cesarean section. *J. Appl. Physiol.* **52**:716-724.

Mortola, J. P., Magnante, D., and Saetta, M. (1985). Expiratory pattern of newborn mammals. *J. Appl. Physiol.* **58**:528-533.

Mortola, J. P., Milic-Emili, J., Noworaj, A., Smith, B., Fox, G. S., and Weeks, S. (1984). Muscle pressure and flow during expiration in infants. *Am. Rev. Respir. Dis.* **129**:49-53.

Muller, N., Volgyesi, G., Becker, L., Bryan, M. H., and Bryan, A. C. (1979a). Diaphragmatic muscle tone. *J. Appl. Physiol.* **47**:279-284.

Muller, N., Gulston, G., Cade, D., Whitton, J., Froese, A. B., Bryan, M. H., and Bryan, A. C. (1979b). Diaphragmatic muscle fatigue in the newborn. *J. Appl. Physiol.* **46**:688-695.

Muller, N., Volgyesi, G., Bryan, M. H., and Bryan, A. C. (1979c). The consequences of diaphragmatic muscle fatigue in the newborn infant. *J. Pediatr.* **95**:793-797.

Murai, D. T., Lee, C. H., Wallen, L. D., and Kitterman, J. A. (1985). Denervation of peripheral chemoreceptors decreases breathing movements in fetal sheep. *J. Appl. Physiol.* **59**:575-579.

Nathanielsz, P. W., Bailey, A., Poore, E. R., Thorburn, G. D., and Harding, R. (1980). The relationship between myometrial activity and sleep state in fetal sheep throughout the last third of gestation. *Am. J. Obstet. Gynecol.* **138**:653-659.

Nijhuis, J. G., Martin, C. B., Gommers, J., Bouws, P., Bots, R. S., and Jongsma, H. W. (1983). The rhythmicity of fetal breathing varies with behavioral state in the human fetus. *Early Hum. Dev.* **9**:1-7.

Orem, J. (1980). Medullary respiratory neuron activity: relationship to tonic and phasic REM sleep. *J. Appl. Physiol.* **48**:54-65.

Orem, J., and Lydic, R. (1978). Upper airway function during sleep and wakefulness: Experimental studies on normal and anesthetized cats. *Sleep* **11**:49-68.

Orem, J., Netick, A., and Dement, W. C. (1977). Increased upper airway resistance to breathing during sleep in the cat. *Electroencephalogr. Clin. Neurophysiol.* **43**:14-22.

Parmalee, A. H., Jr., and Stern, E. (1972). Development of states in infants. In *Sleep and the Maturing Nervous System*. Edited by C. D. Clement, D. P. Purpura, and F. E. Mayer. New York, Academic Press.

Phillipson, E. A. (1978). Respiratory adaptions in sleep. *Annu. Rev. Physiol.* **40**:133-56.

Phillipson, E. A., and Bowes, G. (1986). Control of breathing during sleep. In *Handbook of Physiology*. Section 3. The Respiratory System. Vol. II. Control of Breathing, part 2. Washington, DC, American Physiological Society.

Phillipson, E. A., Sullivan, C. E., Read, D. J. C., Murphy, E., and Kozar, L. F. (1978). Ventilatory and waking responses to hypoxia in sleeping dogs. *J. Appl. Physiol.* **44**:512-520.

Prechtl, H. F. R. (1974). The behavioral states of the newborn infant (a review). *Brain Res.* **76**:185-212.

Prechtl, H. F. R., Akiyama, Y., Zinkin, P., and Grant, D. K. (1968). Polygraphic studies of the full term newborn. I. Technical aspects and quantitative analysis. In *Studies in Infancy*. Edited by M. Bax and R. C. MacKeith. Clinics in Developmental Medicine No. 27. Spastics Society. London, Heinemann.

Prechtl, H. F. R., Van Eykern, L. A., and O'Brien, M. J. (1977). Respiratory muscle EMG in newborns: A non-intrusive method. *Early Hum. Dev.* **1**:265-283.

Quilligan, E. J., Clewlow, F., Johnston, B. M., and Walker, D. (1981). Effect of 5-hydroxytryptophan on electrocortical activity and breathing movements of fetal sheep. *Am. J. Obstet. Gynecol.* **41**:271-275.

Rechtschaffen, A., and Kales, A. (1968). *Manual of Standardized Terminology. Techniques and Scoring System for Sleep Stages in Human Subjects*. Washington, DC, National Institutes of Health, publ. No. 204.

Reed, W. R., Roberts, J. L., and Thach, B. T. (1985). Factors influencing regional patency and configuration of the human infant upper airway. *J. Appl. Physiol.* **58**:635-644.

Remmers, J. E. (1970). Inhibition of inspiratory activity by intercostal muscle afferents. *Respir. Physiol.* **10**:358-383.

Rigatto, H. (1984). Control of ventilation in the newborn. *Annu. Rev. Physiol.* **46**:661-674.

Rigatto, H., Brady, J. P., and de la TorreVerduzco, R. (1975). Chemoreceptor reflexes in preterm infants: the effect of gestational and postnatal age on the ventilatory response to inhalation of 100% and 15% oxygen. *Pediatrics* **55**:604-613.

Rigatto, H., Kalapesi, Z., Leahy, F. N., Durand, M., MacCollum, M., and Cates, D. (1982). Ventilatory response to 100% and 15% O_2 during wakefulness and sleep in preterm infants. *Early Hum. Dev.* **7**:1-10.

Rigatto, H., Lee, D., Davi, M., Moore, M., Rigatto, E., and Cates, D. (1988). Effect of increased arterial CO_2 on fetal breathing and behavior in sheep. *J. Appl. Physiol.* **64**:982-987.

Roberts, J. L., Reed, W. R., Mathew, O. P., Menon, A. A., and Thach, B. T. (1985). Assessment of pharyngeal airway stability in normal and micrognathic infants. *J. Appl. Physiol.* **58**:290-299.

Roberts, J. L., Reed, W. R., Mathew, O. P., and Thach, B. T. (1986). Control of respiratory activity of the genioglossus muscle in micrognathic infants. *J. Appl. Physiol.* **61**:1523-1533.

Ruckebusch, Y. (1972). Development of sleep and wakefulness in the foetal lamb. *Electroencephalogr. Clin. Neurophysiol.* **32**:119-128.

Ruckebusch, Y., Goujoux, M., and Eghboli, B. (1977). Sleep cycles and kinesis in the fetal lamb. *Electroencephalogr. Clin. Neurophysiol.* **42**: 226-237.

Saetta, M., and Mortola, J. P. (1987). Interaction of hypoxic and hypercapnic stimuli on breathing pattern in the newborn rat. *J. Appl. Physiol.* **62**:506-512.

Sauerland, E. K., and Harper, R. M. (1976). The human tongue during sleep: Electromyographic activity of the genioglossus muscle. *Exp. Neurol.* **51**:160-170.

Schulte, F. J., Busse, C., and Eichhorn, W. (1977). Rapid eye movement sleep, motoneurone inhibition, and apneic spells in preterm infants. *Pediatr. Res.* **11**:709-713.

Shannon, D. C., Marsland, D. W., Gould, J. B., Callahan, B., Todres, I. D., and Dennis, J. (1976). Central hypoventilation during quiet sleep in two infants. *Pediatrics* **55**:589-594.

Sherry, J. H., and Megirian, D. (1980). Respiratory EMG activity of the posterior cricoarytenoid and diaphragm muscles during sleep. *Respir. Physiol.* **39**:355-365.

Stark, A. R., Cohlan, B. A., Waggener, T. B., Frantz, I. D., III, and Kosch, P. C. (1987). Regulation of end-expiratory lung volume during sleep in premature infants. *J. Appl. Physiol.* **62**:1117-1123.

Stevenson, M., and McGinty, D. (1978). Polygraphic studies of kitten development: Respiratory rate and variability during sleep-waking states. *Dev. Psychobiol.* **11**:393-403.

Sullivan, C. E., Murphy, E., Kozar, L. F., and Phillipson, E. A. (1979). Ventilatory responses to CO_2 and lung inflation in tonic versus phasic REM sleep. *J. Appl. Physiol.* **47**:1304-1310.

Tabachnik, E., Muller, N. L., Bryan, A. C., and Levison, H. (1981). Changes in ventilation and chest wall mechanics during sleep in normal adolescents. *J. Appl. Physiol.* **51**:557-564.

Thach, B. T., Abroms, I. F., Frantz, I. D., III, Sotrel, A., Bruce, E. N., and Goldman, M. D. (1980). Intercostal muscle reflexes and sleep breathing patterns in the human infant. *J. Appl. Physiol.* **48**:139-146.

Thach, B. T., and Stark, A. R. (1979). Spontaneous neck flexion and airway obstruction during apneic spells in preterm babies. *J. Pediatr.* **94**:275-281.

Tusiewicz, K., Moldofsky, H., Bryan, A. C., and Bryan, M. H. (1977). Mechanics of the rib cage and diaphragm during sleep. *J. Appl. Physiol.* **48**: 139-146.

van Lunteren, E., and Strohl, K. P. (1986). The muscles of the upper airways. *Clin. Chest Med.* **7**:171-188.

Waggener, T. B., Frantz, I. D., III, Stark, A. R., and Kronauer, R. E. (1982). Oscillatory breathing patterns leading to apneic spells in infants. *J. Appl. Physiol.* **52**:1288-1295.

Wallen, L. D., Murai, D. T., Clyman, R. I., Lee, C. H., Mauray, F. E., and Kitterman, J. A. (1986). Regulation of breathing movements in fetal sheep by prostaglandin E_2. *J. Appl. Physiol.* **60**:526-531.

Wilson, S. L., Thach, B. T., Brouillette, R. T., and Abu-osba, Y. K. (1980). Upper airway patency in the human infant: Influence of airway pressure and posture. *J. Appl. Physiol.* **48**:500-504.

Woodrum, D. E., Standaert, T. A., Mayock, D. E., and Guthrie, R. D. (1981). Hypoxic ventilatory response in the newborn monkey. *Pediatr. Res.* **15**:367-370.

Part Five

HYPOXIA AND ASPHYXIA: CELLULAR, INTEGRATED, AND PATHOPHYSIOLOGICAL ASPECTS

19

Ventilatory Response to Hypoxia
Integrated, Cellular, and Molecular Aspects

GABRIEL G. HADDAD and CAROL L. ROSEN

Yale University School of Medicine
New Haven, Connecticut

I. Introduction

With the present state of knowledge about hypoxia and its effect on the body, it would be naive to think that the ventilatory response to hypoxia can be viewed simply through the scope of a respiratory physiologist focusing on the airways, lungs, and respiratory muscles. In addition, we are starting to appreciate the effect of hypoxia on cellular function, whether at the level of the respiratory muscles, brain, or heart. From a practical viewpoint, however, we will not be able to address a number of issues in this chapter, not only because of space limitation, but also because it may not be thematically appropriate in this volume. Therefore, we will briefly review in the first part of this chapter some of the integrated ventilatory and metabolic responses to hypoxia and focus in the second part on the changes in brain function at the cellular and molecular level during O_2 deprivation. We have specifically

This work was supported by: NIH Grants HL 39924 and HD 15736, a grant from the Meyer Foundation and an Established Investigatorship Award from the American Heart Association to G.G.H.

chosen to review neuronal function since brain hypoxia and its effects on brainstem neurons would impact in a major way on the recruitment of respiratory and upper airway muscles and therefore influence airway patency and magnitude of ventilation.

II. Integrated Response

The response to hypoxia is an interaction of a variety of respiratory, circulatory, and metabolic adjustments. The purpose of these adaptations is to minimize the effect of decreased O_2 supply at the cellular level on vital organs. The overall strategy includes increasing ventilation to replenish O_2 supplies, increasing cardiac output to improve O_2 delivery, and possibly decreasing metabolism (O_2 consumption) to minimize demand. Although the circulatory response is a major part of the overall response, no attempt is made here to comprehensively address cardiovascular issues.

A number of factors modify the character of the responses to hypoxia: (1) maturation, (2) severity of the hypoxia, (3) duration (acute versus chronic), (4) sleep state, and (5) associated conditions such as hypo- or hypercarbia. Hypoxia is defined as any condition in which the supply of O_2 to a cell, tissue, organ, or the whole body is insufficient to meet the O_2 demand (Cain, 1987). A variety of clinical situations can result in different types of hypoxia (Fidone and Gonzalez, 1986). *Hypoxic hypoxia*, characterized by a decreased O_2 supply (low PaO_2), occurs in lung disease, ventilation/perfusion mismatch, and high-altitude environments. Decreased O_2 transport to the tissues (*stagnant hypoxia*) occurs in shock or myocardial dysfunction. Decreased O_2 carrying capacity (*anemic hypoxia*) is seen in sickle cell disease, other anemias, and carbon monoxide poisoning. Finally, decreased O_2 utilization (*histocytoxic hypoxia*) develops in cyanide poisoning. Although there are similarities in the responses to each type of hypoxia, most of the remainder of this discussion will focus on the responses to hypoxia secondary to a decrease in PaO_2.

A. Ventilatory Response

The ventilatory response to hypoxia is a balance of excitatory and inhibitory inputs. In the intact animal, the overall effect of hypoxia is excitatory, mostly related to a reflex increase in ventilation via the carotid body chemoreceptor. These sensors respond to changes in PaO_2, rather than changes in oxygen saturation or O_2 content (Fidone and Gonzalez, 1986). Although the peripheral chemoreceptor activity is the major stimulus for the ventilatory response to hypoxia, it is also influenced by higher centers. For example, decortication markedly potentiates the ventilatory response to hypoxia (Tenney and Ou, 1977). Since ventilation is a composite of tidal volume and respiratory rate, either variable can affect the response to hypoxia. Previous studies have suggested that the respiratory rate is the main variable altered; the tidal volume remains nearly constant (Haddad et al., 1982; Blanco et al., 1984; Bonora et al., 1984; Bureau and Begin, 1982).

Maturation has a profound influence on the ventilatory response to acute hypoxia. The response ranges from inhibition of breathing movements in the intact fetus (Boddy et al., 1974; Towell and Salvador, 1974), to transient stimulation followed by a decrease in ventilation in the neonate (Blanco et al., 1984; Bonora et al., 1984; Bureau et al., 1984; Lawson and Long, 1983; Sankaran et al., 1979; Woodrum et al., 1981), to a sustained increase in ventilation in the adult. The direction of the response depends on the balance of inhibitory central mechanisms versus excitatory peripheral chemoreceptor input.

The ventilatory response to acute hypoxia is generally "biphasic" in the newborn as well as in the adult. The response is characterized by an initial brisk increase followed by a gradual decline (Haddad and Mellins, 1984). In the newborn, the increase in ventilation is followed within a few minutes by a decrease to baseline or below. This decline in ventilation is less marked in species with greater maturity at birth (Mortola et al., 1987). The magnitude of this decline in the second phase also depends on maturation. In the adult, a decrease follows an initial increase, but ventilation remains above baseline. Possible mechanisms for this difference between newborn and adults in their ability to maintain sustained hyperventilation include maturational differences in (1) the associated fall in metabolic rate (Haddad et al., 1981; Rosen et al., 1989), (2) peripheral chemoreceptor activity (Bureau and Begin, 1982; Bureau et al., 1986), (3) integration of the peripheral responses at the central nervous system (CNS) level (Lawson and Long, 1983), (4) response to the central depressant effect of hypoxia, and (5) respiratory mechanics (LaFramboise et al., 1983; Cote et al., 1988).

The biphasic response is a transient response to an acute stimulus. Longer exposure to low PaO_2 also modifies the ventilation response. Hypoxia for hours to days is associated with a gradual increase in ventilation in adult humans (Dempsey et al., 1974). Even the newborn is capable of sustained hyperventilation in response to a more prolonged stimulus. Using prolonged hypoxic stimuli (>30 min), sustained hyperventilation has been observed in newborns from several species (Rosen et al., 1989; Moss et al., 1987). Newborn rats, well known to have ventilatory depression following an acute stimulus, hyperventilated when studied after 5-7 days of chronic hypoxia (Mortola et al., 1986). The ventilatory response to long-term hypoxia (days to years) is best characterized in people who live at high altitude. Unlike newly acclimatized travelers to the same altitude, highlanders appear to have lost their ventilatory response to acutely induced hypoxia and ventilate less at rest, during exercise, and when breathing CO_2. This insensitivity to hypoxia appears to be acquired as newborns at high altitude have normal responses to hypoxia during infancy, but lose them during childhood (Lahiri et al., 1978).

The ventilatory response to hypoxia is also greatly modified by associated conditions such as hypo- or hypercapnia. For example, acute hypoxia produces a hyperbolic increase in ventilation at a threshold of ~ 60 torr under hypocapnic conditions, while in hypercapnic hypoxia, no threshold is seen and the peripheral stimulation is augmented (Cunningham et al., 1986). Some of the conflicting results of studies of ventilatory responses to acute hypoxia can actually be explained by differences in experimental conditions related to CO_2 level.

Sleep state modifies ventilatory response to hypoxia. Studies from our laboratory have shown that in beagle puppies exposed to moderate hypoxia, instantaneous ventilation increased at all ages during REM. In contrast, during quiet (NREM) sleep, ventilation was depressed in the youngest group and progressively increased as puppied matured (Haddad et al., 1982). In adult animals, it appears that the ventilatory response to acute hypoxia is reduced in NREM sleep compared to wakefulness (Phillipson and Bowes, 1986). This response to hypoxia in NREM sleep is dependent on the integrity of the carotid chemoreceptors. Denervation or removal of the carotid body results in attenuation or abolition of the hypoxic ventilatory response and exaggerated respiratory depression (Bowes et al., 1981). The mechanisms by which NREM reduces the ventilatory response to hypoxia are uncertain, but decreased sympathetic tone, reduced peripheral chemoreceptor activity, and removal of the tonic stimulatory influences of wakefulness may be contributing factors (Phillipson and Bowes, 1986).

With regard to hypoxic response during REM sleep in adults, the data seem to indicate that there is generally a smaller increase in ventilation in this than in other states (Phillipson and Bowes, 1986). That fluctuations in PaO_2 during REM are greater after sinoaortic denervation suggests that chemoreceptor activity plays a role in stabilizing ventilation in REM (Guazzi and Freis, 1969). Another explanation for a smaller increase in ventilation during REM sleep may relate to the effect of sleep state on the rib cage's contribution to ventilation (Bryan et al., 1986; Henderson-Smart and Read, 1979). In NREM sleep, the ventilatory response of newborn lambs is quite brisk. In REM sleep, during which paradoxical motion of the rib cage occurs, the response to hypoxia is much flatter and irregular. Finally, sleep state modifies the ventilatory response not only to acute hypoxia, but also to chronic hypoxia. Healthy adults sleeping in high altitude develop periodic breathing during NREM, but these changes disappear during REM (Reite et al., 1975). Similar observations have been made in sleeping kittens (Baker and McGinty, 1979).

B. Metabolic Response

When oxygen supplies are limited, decreasing metabolic rate (VO_2) is another mechanism to reduce O_2 demand. Newborn and adults of many species have been reported to drop VO_2 within minutes after exposure to hypoxia (Hill, 1959; Mortola et al., 1987; Moss et al., 1987; Rosen et al., 1989). This decrease in metabolism is modified by maturation, environmental temperature, and severity of hypoxia. Approximately 80% of the basal cellular oxygen utilization takes place in the mitochondria. The other 20% of metabolism is extramitochrondrial, involving multiple functions such as enzymatic reactions (oxygenases, oxidases) and growth (Robin, 1980). During short-term mild or moderate hypoxia, it is possible for the body to conserve energy without necessarily resorting to anaerobiosis by decreasing VO_2, but long-term use of this strategy may result in impairment of growth (Baker, 1969; Mortola et al., 1986; Pepelko, 1970; Sidi et al., 1983b). During more severe hypoxia, anaerobic metabolism develops, energy-inefficient glycolysis is stimulated, and lactate production is increased. Lactate accumulation, an index of tissue hypoxia, indicates that tissue energy demand exceeds production. Under these conditions, maturation differences in oxygen consumption and lactate accumulation (Sidi et al., 1983a; Moss et al., 1987; Fahey and Lister, 1989; Rosen et al., 1989) suggests greater tolerance to hypoxic stress in newborn animals.

III. Cellular and Molecular Response

There has been an enormous interest recently in examining the cellular and molecular effects of O_2 deprivation on a number of tissues. Because of their vital importance, the heart, the kidneys, and the brain have probably been the most studied. We will limit the following discussion, therefore, to these tissues but draw especially on data obtained from studies on the CNS because of the clear association between breathing and the state of neuronal function. Clearly, the importance of studying brainstem neurons each as vagal motoneurons, premotor neurons in the ventral to the tractus solitarius, and hypoglossal motoneurons stems from the fact that these are some of the neurons that are critical for the cardiorespiratory response to hypoxia, a response that can alleviate the effect of hypoxia on the rest of the CNS.

With the increasing recent use of cellular and subcellular techniques such as nuclear magnetic resonance (NMR) spectroscopy, intracellular and extracellular ion selective microelectrodes, and optical and biophysical methods, evidence has accumulated in favor of three important concepts. First, it is clear now that there is a cellular and a molecular basis for the adaptation or the responsiveness to hypoxia that occurs at a systemic or global level, such as change in blood flow distribution or increase in cardiac output and ventilation. More important perhaps is the fact that the bases for adaptation are tractable and testable. Second, although a number of cellular mechanisms are activated during hypoxia, it is possible that a major single pathway leads to irreversible cellular disintegration. Third, it is becoming clear that there are hypoxia-tolerant and hypoxia-susceptible tissues and animals. Within mammals, age seems to be important in early life or even also in adulthood. These concepts will emerge through the following discussion in which we will present supporting data from in vivo and reduced in vitro preparations.

A. Hypoxia and CNS Function

Adult mammals do not tolerate O_2 deprivation well, whether hypoxia is induced by low inhaled O_2 or by low blood flow. Within seconds, humans lose consciousness if the hypoxia or ischemia is severe (Rossen et al., 1943). The electroencephalogram shows slowing and decreased amplitude within 30 s post-hypoxia and silence (close to an isoelectric line) in 60-90 s (Hansen, 1985; Rossen et al., 1943; Sanocka et al., 1988; Siemkowicz and Hansen, 1981). Evoked potentials in in vitro hippocampal brain slice decrease after the first minute into hypoxia and become very small by 3-3.5 min (Hansen, 1985). These changes in the overall function of the CNS have been relatively well characterized in a number of mammals and lower vertebrates (Hansen, 1985). In spite of this, major questions remain unanswered about the sequence of events at the cellular and molecular levels. For example, what is the role of cellular properties in the electrophysiological response to hypoxia? How are the repetitive firing properties changed with anoxia? Are the young more resistant to hypoxia, and if so, what are the mechanisms for this resistance? What is the role of Ca^{2+} ions in the susceptibility to anoxia?

Regional blood flow differences in response to hypoxia have been appreciated and could play a role in the response of neuronal groups to hypoxia (Balestrino et al., 1989; Cote and Haddad, 1990; Hansen, 1985; Santiago and Edelman, 1986; unpublished observations from our laboratory). We are now starting to recognize that the inherent properties of neurons may also be critical in determining the response to hypoxia (Haddad and Donnelly, 1990; Hansen, 1985). Similarly, cellular differences may be important in determining the response to hypoxia and anoxia (Haddad and Donnelly, 1990). For example, there is some evidence now that brainstem neurons respond to hypoxia differently from hippocampal CA1 cells (Haddad and Donnelly, 1990; LeBlond and Krnjevic, 1989).

B. Changes in Membrane Potential During Hypoxia

The most studied neurons with respect to O_2 deprivation are the hippocampal CA1 cells. Several groups have now used intracellular techniques to determine the response to hypoxia and the mechanisms responsible for the neuronal functional alterations in the hippocampus (Fujiwara et al., 1987; Krnjevic and LeBlond, 1989; LeBlond and Krnjevic, 1989; Donnelly et al., 1990). More recently, we have started to examine the nature of the response to hypoxia in brainstem neurons in vitro (Haddad and Donnelly, 1990). We have so far studied in detail hypoglossal (HYP) and vagal (DMNX) neurons. With hypoxia, HYP neurons always depolarized and no hyperpolarization was seen. Adult HYP neurons began to slowly depolarize by 30 s and more rapidly between 1.5 and 3.5 min after the start of hypoxia (Fig. 1A). Indeed, most of the depolarization (80-90%) takes place by 4 min. By 5 min, adult HYP neurons had an average maximum depolarization of over 30 mV. Typically, as neurons depolarized gradually, they started to fire spontaneously (Fig. 1A). A depolarization block was also observed toward the end of the 5-min experimental run in about half of the neurons, as in Figure 1. This block was apparent as neurons ceased to fire and were unresponsive to increasing injected currents. Also of interest, all adult HYP neurons had an increase in synaptic noise that started 1-2 min into hypoxia. This manifested as an increase in amplitude of postsynaptic potentials. Toward the end of the experimental run (>4.5 min), however, there was a decline in the synaptic noise especially in the neurons that depolarized most. DMNX neurons also showed major depolarization within a few minutes into hypoxia and a majority experienced blockade. Overall changes in membrane potential in DMNX neurons were similar to those in HYP neurons.

Our data on vagal and HYP motoneurons (Haddad and Donnelly, 1990) show, therefore, that these adult brainstem neurons depolarize in response to hypoxia. This is in clear contrast with cortical hippocampal CA1 neurons (Fujiwara et al., 1987; LeBlond and Krnjevic, 1989). Fujiwara et al. (1987) and LeBlond and Krnjevic (1989) as well as ourselves (Donnelly et al., 1990) have shown that CA1 neurons depolarize immediately for 5-10 s after hypoxia. Following this short period of depolarization, a major hyperpolarization ensues. If hypoxic exposure is severe and long (15-20 min) enough, a second depolarization occurs, more like that of spreading depression (Leblond and Krnjevic, 1989). Our studies show that HYP and DMNX neurons demonstrate only a depolarization that is quickly reversible upon reoxygenation. These data demonstrate that mammalian CNS neurons behave differently in response to hypoxia and that brainstem neurons show solely a depolarization.

C. Membrane Ionic Fluxes: Channels and Pumps

There are at least four major determinants of cellular neuronal function during graded hypoxia: membrane potential or voltage, neurotransmitters released,

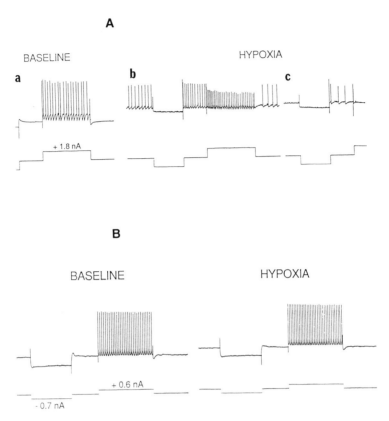

Figure 1 (A) HYP adult brain during baseline (a) and during hypoxia (b,c); 1-2 min into hypoxia (b), this HYP neuron depolarized and excitability increased; 4-5 min into hypoxia, this neuron went into depolarization block as evidenced by the lack of generation of action potential during a step depolarization (c). (B) Neonatal HYP neuron. There was little depolarization for 5 min into hypoxia. Time base for current pulses = 500-1000 msec; membrane potential changes from -78 mV (a) to -54 mV (b) and -47 mV (c) in the adult. The change in membrane potential in the newborn was 7 mV during 5 min of hypoxia. Hyperpolarizing pulses are used to determine input resistance.

ion fluxes, and intracellular levels of ATP, all of which are likely to be inter-dependent. For example, a membrane potential that is rendered less negative will have a major impact on the flux of ions across cellular membranes. Similarly, a decreasing level of ATP or PCr may lead to marked changes in ion pumping.

Several studies have now shown that K^+ in the extracellular space, K^+_o, increases to high levels in adult animals in response to hypoxia or ischemia (Hansen, 1978, 1985; Vyskocil et al., 1972), and this increase in K^+_o is dependent on the severity of tissue hypoxia (Hansen, 1985). The K^+_o profile is fairly consistent during hypoxia. In our studies, after a latency of about 30 s, ECF K^+ concentration starts to increase (Haddad and Donnelly, 1990; see also Hansen, 1985). This increase is most dramatic between 1 and 3 min into moderate hypoxia and generally increases by about 3 mM after 4 min at which time K^+ levels level off (Fig. 2) (Haddad and Donnelly, 1990). In more severe hypoxia (95% N_2-5% CO_2), K^+_o levels can reach 20 mM or higher in adult brains.

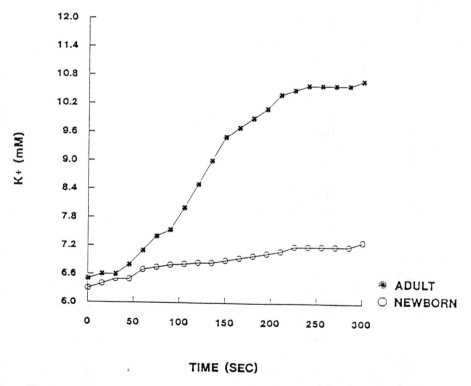

Figure 2 Effect of hypoxia on extracellular fluid K^+ in the HYP area. Time 0 = start of hypoxia. Note marked increase in K^+ concentration in the adult but lack thereof in the newborn HYP area.

The source of the increasing ECF K^+ and the mechanism(s) underlying this increase remain unknown. It has been suggested that increased outward flux of K^+ from neurons may account for the increased ECF K^+ during increased neuronal activity (Baylor and Nicholls, 1969; Orkand, 1980; Connors et al., 1982) as in our adult neurons during hypoxia. Evidence that ECF K^+ derives from neurons stems from the fact that ECF K^+ increases in the area of stimulated neurons (Baylor and Nicholls, 1969; Orkand, 1980). However, an increase in K^+ efflux from neurons does not exclude the possibility that glial cells, well known to take up K^+, can be a source of K^+ during hypoxic exposure. Furthermore, the extracellular space which shrinks during hypoxia may contribute to the increase in K^+_o (Hansen and Olsen, 1980; Hansen, 1985).

Ca^{2+}, H^+, Na^+, and Cl^- have also been studied to various degrees in adult brain tissue (Bourke et al., 1978; Hansen and Zeuthen, 1981; Mutch and Hansen, 1984; Nicholson et al., 1977; Van Harreveld and Schade, 1959) (see also Chapter 21, this volume). Although Ca^{2+}, Na^+, and Cl^- in the extracellular space do not seem to change in a major way in the first 30-60 s after hypoxia, Ca^{2+} decreases to very low levels at a later stage and Na^+, Cl^-, and H_2O enter brain cells. Hydrogen ion concentration, especially during anoxia, increases in the intracellular compartment due to accumulation of lactic acid or CO_2 (the latter reaching levels of over 100 torr in ischemic hypoxia) (Ljunggren et al., 1974).

It is not difficult to imagine that, with these changes in intracellular and extracellular ionic movements, ionic pumps can be activated or inactivated in response to hypoxia or as a consequence of ionic fluxes themselves. Similarly, with depolarization or hyperpolarization and with release of neurotransmitters during hypoxia, ion channels can be activated, open or closed, and hence can contribute in a major way to the overall ionic homeostasis across neuronal membranes and to neuronal excitability.

Recently, LeBlond and Krnjevic (1989) and Fujiwara et al. (1987) have dissected out the effect of hypoxia on hippocampal neurons in the CA1 region. An increase in K^+ conductance and outward K^+ current were responsible for the hyperpolarization seen during hypoxia (Fujiwara et al., 1987; Hansen et al., 1982; Krnjevic and LeBlond, 1989). These K^+ currents were not fully eliminated by blockers such as tetraethylammonium, cesium, 4-aminopyridine, tolbutamide, and apamin, indicating that these K^+ currents did not depend totally on ATP, Ca^{2+}, or an inward rectifier current. Other outward currents, such as the A-current (a fast inactivating K^+ current), were reduced with hypoxia. Inward currents, such as calcium currents of the N type, were resistant to hypoxia but those of the L type were almost totally abolished (Krnjevic and LeBlond, 1989).

Neurotransmitters may have an effect in the slice preparation during hypoxia, especially through their action on the ligand-gated ion channels. One way to determine whether neurotransmitters in the slice are important in the observed changes is to study single isolated cells. We have started to examine the effect of hypoxia or anoxia on membrane currents in acutely dissociated single neurons of the CA1 region using patch clamp technique and have obtained interesting results (Cummins et al., 1990). Both holding potential and an inward current decrease during hypoxia (Cummins et al., 1990). These results suggest that isolated neurons are responding to hypoxia in a similar way as in the slice.

Ion pumps are also likely to play a major role in the ionic homeostasis during hypoxia or anoxia. For example, Kass and Lipton (1989) have suggested in their studies that K^+ leakage during hypoxia is due, in part, to a decrease in intracellular ATP and inhibition of ATPase. The increase in intracellular H^+ during hypoxia can activate $Na^+:H^+$ exchange. Similarly $Na^+:Ca^{2+}$, a regulator of intracellular Ca^{2+} (Blaustein, 1988; Carafoli, 1987), may also increase extrusion of Ca^{2+} when Ca^{2+} increases intracellularly. A number of questions could be posed regarding the activity of intracellular Ca^{2+} pumps during hypoxia. Little, if anything, is known, however, about such issues.

D. Intracellular Metabolism

In Chapter 20 in this volume Nioka and Chance present in detail the metabolic changes that occur with hypoxia. These are clearly crucial issues since they impact on energy requirements and availability at a time when O_2 is deprived. Mammalian cells, in particular neurons, cannot survive O_2 lack and generally resort to one of two strategies: (1) shift from oxidative to glycolytic metabolism to try to maintain an ATP level commensurate with some "essential" cellular function or (2) decrease in energy requirements; by so doing, neurons decrease the need to generate large amounts of ATP through glycolysis (which may not be possible) and decrease intracellular lactic acid and H^+ pollution. As will be discussed below, these capabilities differ with phylogeny and ontogeny.

Neuronal metabolism during hypoxia can be linked to neuronal membrane excitability in at least two ways. First, various membrane ionic pumps are energy- or ATP-driven and these may cease to function when ATP production is jeopardized. For example, $NA^+:K^+$ ATPase, if inhibited, can induce an increase in K^+_o and this, in turn, depolarizes neurons. Depending on the level of K^+_o, cells may reach a stage of increased or decreased excitability. Second, metabolism can be linked to membrane excitability throught membrane channels. For example, ATP-dependent K^+ channels have been shown to exist in the heart and other organs (pancreas) and more recently in neurons also. At rest, when ATP stores are normal, these K^+ channels are closed. They open and a major outward current can flow via these specific K^+ channels (Sturgess et al., 1985; Trube et al., 1986; Weiss and Lamp, 1987; Ashcroft, 1988) with a decrease in intracellular ATP levels. It is interesting to note that in cardiac myocytes, glycolysis was more effective than oxidative phosphorylation in preventing these channels from opening (Weiss and Lamp, 1987). It is argued that it is the proximity of the channels to the ATP-generating site that would account for the preference between oxidative and glycolytic pathways.

A number of studies have shown that neurons (as well as other types of cells) can shut down metabolic processes to save on energy requirements (Carter and Muller, 1989; Hochachka, 1986). For example, synthesis of proteins, espeically those that are not essential for survival, and membrane stability may be decreased. It would be interesting to find out how selective this metabolic arrest is and what protein syntheses are considered "temporarily expendable" by neurons.

E. Reversibility of Neuronal Hypoxic Changes

Reversibility of hypoxic changes with reinstitution of oxygenation depends on a number of factors, not the least of which are duration and severity of hypoxia, region involved, and stage of maturation or development of the brain. After a period of minutes of hypoxia or anoxia, as in our studies on brain slices, there is generally a "hysteresis" in the pattern of reversibility of hypoxic changes. The sequence of alterations during reoxygenation is opposite to the sequence of events that occurs during hypoxia. Within seconds after reinstitution of O_2, K^+_o and membrane potential start to go back to normal. In HYP neurons for instance, membrane potential and K^+_o are close to baseline within 2-3 min. On the other hand, membrane excitability lags behind and in vivo studies have shown that the EEG also recovers after K^+_o has gone back to baseline (Hansen, 1985). This may indicate that the reactivation of Na^+ channels may lag behind the mechanisms responsible for the reversibility of K^+_o and membrane potential.

With longer exposure to anoxia, not only does function return to baseline more slowly, but it may never return to full capacity. Recent studies from this institution by Stys and colleagues on the optic nerve of rats have indicated that function, as assessed by compound action potential, returns very slowly and incompletely to baseline after 60 min of anoxia (Stys et al., submitted). Interestingly, this return to baseline is very dependent on the extracellular concentration of Ca^{2+} (Stys et al., 1990), reversibility being quicker with low Ca^{2+} and very slow or nonexistent with high Ca^{2+} levels.

With a very severe, prolonged anoxic insult, neurons do not recover. Although the sequence of events is still not well understood, it is possible that the following reactions serve as a common final denominator. It seems that Ca^{2+} ions may play a major role when they enter neurons in an uncontrollable flux (Farber et al., 1981; Hansen, 1985; Mattson et al., 1989; Stys et al., 1990). One of the most toxic effects is the activation of phospholipases by Ca^{2+}, which leads to increasing free fatty acids and disruption of mitochrondrial and cellular membranes (Hochachka, 1986; Siesjo and Bengtsson, 1989). Activation of phospholipase C also leads to increased levels of diacylglycerol (DAG) and inositol triphosphate which signal the release of additional Ca^{2+} from intracellular pools, thus further potentiating the stimulus. In addition, DAG is thought to increase $Na^+:H^+$ exchange, which can decrease $Na^+:Ca^{2+}$ exchange and lead to further accumulation of Ca^{2+}. In spite of the seeming importance of Ca^{2+} in hypoxic injury, direct measurement of intracellular calcium is lacking.

F. Comparative Hypoxia Studies: Does Ontogeny Follow Phylogeny?

In Chapter 21, Trippenbach presents data on the newborn mammal and draws a parallel with the adult in vivo. In the following presentation, we will detail some of our work (Haddad and Donnelly, 1990) and that of others (Cherubini et al., 1989; Krnjevic et al., 1989) on newborn and adult rat brain slices in vitro and compare these with some relevant data from our work and those of others on turtles (Haddad and Donnelly, 1989; Sick et al., 1982a,b).

Newborn Rat Brainstem In Vitro Studies

We have focused mostly on HYP neurons. Neonatal HYP neurons in the first 2 weeks of life behaved very differently from adult HYP cells (Haddad and Donnelly, 1990). Young neurons depolarized much less than adult neurons and the mean maximal change in membrane potential between 3 and 16 days of age was about one-third that of adult neurons (Fig. 1B). Only 3 of 47 neonatal neurons (up to 16 days) had a maximum depolarization above 20 mV, which is the smallest maximum depolarization value for adult neurons.

Although neonatal neurons also exhibited an increase in synaptic potentials, none of them became spontaneously active. With positive current injections, there was an increase, albeit small, in both peak and steady-state frequencies and none showed a decline in spike frequency or depolarization block. Depolarizations of magnitudes close to those in adult HYP neurons could be induced in neonatal neurons by exposure to hypoxia for much longer periods of time. When hypoxia was kept for 15-30 min (three to six times longer), neonatal cells depolarized by 23-27 mV (about 80% of the adult value) and continued to fire when evoked.

The depolarization observed at 21 and 28 days was much larger than that observed in neurons of younger rats. The average maximal change in membrane potential at 21 days of age was significantly different from that of younger rats. It was also significantly different from neurons of 28-day old and adult animals. No significant difference existed in the mean maximal neuronal depolarization between 28-day-old and adult animals.

In contrast to the results in the adult brainstem slice, K^+_o in the neonate increases minimally and averages about 10-20% of the increase in the adult (Fig. 2). Even with severe anoxia, K^+_o in the neonatal rat does not increase beyond 30% of the adult level reached during moderate hypoxia. Interestingly, both in the adult and in the newborn, K^+_o levels in the HYP plateau after a few minutes of exposure, indicating that a saturation level is reached, and this has been described in in vivo experiments as well (Haddad and Donnelly, 1990; Hansen, 1977, 1985).

We have little information about pH and calcium levels in the extra- and intracellular space of newborn neurons. These changes will most likely also prove to be different from those in the adult since they are tied to ion channel function, pumps, and energy metabolism. It is important to realize that although differences between the newborn and adult brain regarding blood or blood flow exist, there are other major differences that are independent from the circulation and blood. Such differences between neonatal and adult neurons include generic differences inherent in the neurons themselves. For example, the intracellular energy pathways, their requirements, and how these cells respond to hypoxia may be different. Is the neonatal neuron more glycolytic than the adult? Are the channel proteins all developed, and if not, how do these impact on the response to hypoxia? Are the various ionic pumps mature? Can the newborn decrease its energy requirements during O_2 limitations more than the adult? Work in this area is still much needed although answers to some of these questions are now available. For example, ionic homeostasis seems to be better maintained across neuronal membranes in the newborn than in the adult (Haddad and Donnelly, 1989, 1990; Kass and Lipton, 1989). This may reflect the ability of the newborn to either decrease its metabolic rate substantially, shift metabolism to glycolysis and maintain an adequate ATP level, or both. Recent data (Kass and Lipton, 1989) describing the effect of anoxia on ATP levels in slices taken from young and adult rats shed some light on the mechanisms underlying the ability of neurons to maintain ionic gradients across neuronal membranes. Young rats in this study showed a lower rate of drop in ATP levels than adult rats. This difference in ATP levels during hypoxia between the young and the older animals (Kass and Lipton, 1989) could serve as an explanation for the difference in the degree of maintenance of ionic gradients across neuronal membranes. It would be tempting to speculate that the $Na^+ - K^+$ pump fails in the adult neuron but is much less affected in the newborn.

Turtle Brain During Hypoxia

Turtles, such as the *Pseudemys scripta elegans*, are well known to tolerate anoxia. Turtles can dive for minutes to hours without major neuronal dysfunction. When exposed to 95% N_2, 5% CO_2, turtles' EEG does not get attenuated before 30-45 min whereas it is completely isoelectric in adult mammals in less than 1 min (Feng et al., 1988; Sick et al., 1982b; Sanocka et al., 1988). This does not seem to be due to a more beneficial redistribution of blood flow or to preserved cellular O_2 stores in the turtle since cytochrome aa_3 get reduced, albeit to a smaller level than in the rat, during hypoxia (Sick et al., 1982a,b, 1985). In addition, our studies on turtle slices have indicated that brainstem neurons maintain membrane potential and repetitive firing properties when exposed to complete anoxia for about 2 h (Haddad and Donnelly, 1989) (Fig. 3). Turtles have been shown actually to remain alive for more than 6 months when submerged at 3 °C in spite of severe anoxia and acidosis (Ultsch and Jackson, 1982; Jackson and Heisler, 1983)! A recent study of turtle hatchlings demonstrated that they can survive a supercooling temperature of $-3°$ to $-4°C$ with more than 50% of their body as ice and that recovery appeared complete 20 h after thawing (Storey et al., 1988)! Freezing at $-11°C$ produced icing of more than 2/3 of their body and was lethal (Storey et al., 1988).

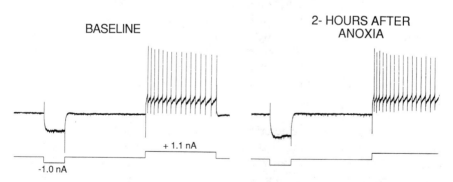

BASELINE

2- HOURS AFTER ANOXIA

+ 1.1 nA

-1.0 nA

Figure 3 Adult brainstem neuron in the turtle. Note that after 2 h of *complete* O_2 deprivation, turtle neuron was still firing with little change from baseline. Time base of pulses is 200 msec (hyperpolarizing pulses) and 600 msec (depolarizing pulses). Hyperpolarizing pulses are used to calculate input resistance.

Since maximal glycolytic rate does not generate more than 28% of total ATP needs, a major defense mechanism in the turtle is a drop in metabolic rate. When anoxic, turtles indeed decrease their metabolic needs and hence the demand on ATP production (Jackson, 1968; Sick and Kreisman, 1981; Sick et al., 1982a,b, 1985; Chih et al., 1989; Robin, 1979). Glycolysis is a main survival mechanism since the turtle has a comparatively much lower capacity for oxidative metabolism as assessed by citrate synthase (Suarez et al., 1989) or cytochrome C oxidase (Haddad et al., 1990). How metabolic needs are decreased or how the need to lower metabolic rate is transduced is not well defined.

Although Hochachka and others have suggested that ion channel density in turtle neuronal membranes may be lower than in the rat and may account for lower energy requirement for ion pumping (Hochachka, 1986), there have been no data on this issue until recently. Data appeared recently showing that turtle neuronal membranes had a lower maximum binding to brevetoxin than rat membranes, indicating that turtle membranes may have a lower density of Na channels (Edwards et al., 1989). However, turtle membranes did not have a lower density of calcium channels (Suarez et al., 1989). How the lower density of Na channels helps maintain ionic homeostasis across neuronal membranes in turtles is not clear at present.

G. Protective Mechanisms: Clinical Implications

It is clear from the above data that the cardiorespiratory response to hypoxia depends on how central neurons alter their function during O_2 deprivation. It is also obvious that neuronal function is better preserved if certain mechanisms are activated. Undoubtedly, these fundamental alterations will be exploited for clinical use. We now know, for instance, that cooling or pentobarbital preserves neuronal function during O_2 deprivation. Accumulating evidence is also suggesting that certain glutamate antagonists (Ikonomidou et al., 1989a,b; Mosinger and Olney, 1989; Steinberg et al., 1989; Westerberg et al., 1989), Ca^{2+} inhibitors, and galanin (Ari and Lazdunski, 1989) may alleviate neuronal damage. Much more work needs to be performed in this area.

In summary, we have described how the respiratory system responds to hypoxia at the integrated level and also attempted to detail some of the more recent cellular and molecular work that can be of use when we try to understand how respiratory neurons in the central nervous system behave when deprived of O_2 and the impact of their altered function on the respiratory system as a whole.

References

Ari, Y. B., and Lazdunski, M. (1989). Galanin protects hippocampal neurons from the functional effects of anoxia. *Eur. J. Pharm.* **165**:331-332.

Ashcroft, F. M. (1988). Adenosine 5-triphosphate-sensitive potassium channels. *Annu. Rev. Neurosci.* **11**:97-118.

Baker, P. T. (1969). Human adaptation to high altitude. *Science* **163**:1149-1156.

Baker, T. L., and McGinty, D. J. (1979). Sleep-waking patterns in hypoxic kittens. *Dev. Psychobiol.* **12**:561-575.

Balestrino, M., Aitkin, P. G., and Somjen, G. G. (1989). Spreading depression-like hypoxic depolarization in CA1 and fascia dentata of hippocampal slices: Relationship to selective vulnerability. *Brain Res* **497**: 102-107.

Baylor, D. A., and Nicholls, J. G. (1969). Changes in extracellular potassium concentration produced by neuronal activity in the central nervous sytem of the leech. *J. Physiol.* **203**:555-569.

Blanco, C. E., Hanson, M. A., Johnson, P., and Rigatto, H. (1984). Breathing pattern of kittens during hypoxia. *J. Appl. Physiol.* **56**:12-17.

Blaustein, M. P. (1988). Calcium transport and buffering in neurons. *Trends in Neurosc.* **11**:438-443.

Boddy, K., Dawes, G. S., and Fisher, R., Pinter, S., and Robinson, J. S. (1974). Fetal respiratory movements, electrocortical, and cardiovascular responses to hypoxemia and hypercapnea in sheep. *J. Physiol.* **243**:599-618.

Bonora, M., Marlot, D., Gautier, H., and Duron, B. (1984). Effects of hypoxia on ventilation during postnatal development in conscious kittens. J. Appl. Physiol. **56**:1464-1471.

Bourke, R. S., Kimelberg, H. K., and Daze, M. A. (1978). Effects of inhibitors and adenosine on $(HCO_{-3} - /CO_2)$-stimulated swelling and Cl^- uptake in brain slices and cultured astrocytes. *Brain Res.* **154**:196-202.

Bowes, G., Townsend, E. R., Kozar, L. F., Bromley, S. M., and Phillipson, E. A. (1981). Effect of carotid body denervation on arousal response to hypoxia in sleeping dogs. *J. Appl. Physiol.* **51**:40-45.

Bryan, A. C., Bowes, G., and Maloney, J. E. (1986). Control of breathing in the fetus and the newborn. In *Handbook of Physiology.* Section 3. Vol. II. Edited by N. S. Cherniak, and J. G. Widdicombe. Washington, DC, American Physiological Society, Chapter 18, pp. 621-647.

Bureau, M. A., and Begin, R. (1982). Postnatal maturation of the respiratory response to O_2 in awake newborn lambs. *J. Appl. Physiol.* **52**:428-433.

Bureau, M. A., Côtè, A., Blanchard, P. W., Hobbs, S., Foulon, P., and Dalle, D. (1986). Exponential and diphasic ventilatory response to hypoxia in conscious lambs. *J. Appl. Physiol.* **61**:836-842.

Bureau, M. A., Zinman, R., Foulon, P., and Begin, R. (1984). Diphasic ventilatory response to hypoxia in newborn lamb. *J. Appl. Physiol.* **56**:84-90.

Cain, S. M. (1987). Gas exchange in hypoxia, apnea, and hyperoxia. In *Handbook of Physiology.* Section 3. Vol. IV. Edited by L. E. Farhi, and S. M. Tenney. Washington, DC, American Physiological Socieity, Chapter 19, pp. 403-420.

Carafoli, E. (1987). Intracellular calcium homeostasis. *Annu. Rev. Biochem.* **56**:395-433.

Carter, A. J., and Muller, R. E. (1989). Inhibition of NMDA receptor activation in hippocampal slices facilitates the recovery of protein synthesis from damage caused by anoxia. *Soc. Neurosci. Abstr.* **15**:45.

Cherubini, E., Ben-Air, Y., and Krnjevic, K. (1989). Anoxia produces smaller changes in synaptic transmission, membrane potential, and input resistance in immature rat hippocampus. *J. Neurophysiol.* **62**:882-895.

Chih, C. P., Feng, Z. C., Rosenthal, M., Lutz, P. L., and Sick, T. J. (1989). Energy metabolism, ion homeostasis, and evoked potentials in anoxic turtle brain. *Am. J. Physiol.* **257**:R854-R860.

Connors, B. W., Ransom, B. R., Kunis, D. M., and Gutnick, M. J. (1982). Activity-dependent K^+ accumulation in the developing rat optic nerve. *Science* **216**:1341-1343.

Cote, A., and Haddad, G. G. (1990). Effect of sleep on regional blood flow distribution in piglets. *Pediatr. Res.* **28**:218-222.

Cote, A., Yunis, K., Blanchard, P. W., Mortola, J. P., and Bureau, M. A. (1988). Dynamics of breathing in the hypoxic awake lamb. *J. Appl. Physiol.* **64**:354-359.

Cummins, T. R., Donnelly, D. F., and Haddad, G. G. (1990). Effect of cynanide (CN) on hippocampal membrane currents studied by whole cell patch: Developmental aspects. *Fed. Proc.* **4**:A405.

Cunningham, D. J. C., Robbins, P. A., and Wolff, C. B. (1986). Integration of respiratory responses to changes in alveolar partial pressures of CO_2 and O_2 and in arterial pH. In *Handbook of Physiology.* Section 3. Vol. II. Edited by N. S. Cherniak and J. G. Widdicombe. Washington, DC, American Physiological Society, Chapter 15, pp. 475-528.

Dempsey, J. A., Forster, H. V., and DoPico, G. A. (1974). Ventilatory acclimatization to moderate hypoxemia in man. *J. Clin. Invest.* **53**:1091-1100.

Donnelly, D. F., Cummins, T. R., and Haddad, G. G. (1990). Brainstem neurons respond differently to O_2 deprivation than cortical neurons. *Fed. Proc.* (in press).

Edwards, R. A., Lutz, P. L., and Baden, D. G. (1989). Relationship between energy expenditure and ion channel density in the turtle and rat brain. *Am. J. Physiol.* **257**:R1354-R1358.

Fahey, J. T., and Lister, G. (1989). Response to low cardiac output: Developmental differences in metabolism during oxygen deficit and recovery in lambs. *Pediatr. Res.* **26**:180-187.

Farber, J. L., Chien, K. R., and Mittnacht, S. (1981). The pathogenesis of irreversible cell injury in ischemia. *Am. J. Pathol.* **102**:271-281.

Feng, Z. C., Rosenthal, M., and Sick, T. J. (1988). Suppression of evoked potentials with continued ion transport during anoxia in turtle brain. *Am. J. Pathol.* **255**:R478-R484.

Fidone, S. J., and Gonzalez, C. (1986). Initiation and control of chemoreceptor activity in the carotid body. In *Handbook of Physiology.* Section 3. Vol. II. Edited by N. S. Cherniak and J. G. Widdicombe. Washington DC, American Physiological Society, Chapter 9, pp. 247-312.

Fujiwara, N., Higashi, H., Shimoji, K., and Yoshimura, M. (1987). Effects of hypoxia on rat hippocampal neurons in vitro. *J. Physiol.* **384**:131-151.

Guazzi, M., and Freis, E. D. (1969). Sino-aortic reflexes and arterial pH, pO_2, and PCO_2 in wakefulness and sleep. *Am. J. Physiol.* **217**:1623-1627.

Haddad, G. G., and Donnelly, D. F. (1989). Tolerance to anoxia in newborn mammals and reptiles: Intracellular neuronal studies. *Soc. Neurosci. Abstr.* **15**:1194.

Haddad, G. G., and Donnelly, D. F. (1990). O_2 deprivation induces a major depolarization in brainstem neurons in the adult but not in the neonate. *J. Physiol.* (*London*) (in revision).

Haddad, G. G., Gandhi, M. R., and Mellins, R. B. (1981). O_2 consumption during hypoxia in sleeping puppies. *Am. Rev. Respir. Dis.* **123**:183A.

Haddad, G. A., Gandhi, M. R., and Mellins, R. B. (1982). Maturation of ventilatory response to hypoxia in puppies during sleep. *J. Appl. Physiol.* **52**:309-314.

Haddad, G. G., and Mellins, R. B. (1984). Hypoxia and respiratory control in early life. *Annu. Rev. Physiol.* **46**:629-643.

Haddad, G. G., Xia, Y., and Bang, R. (1990). Cytochrome oxidase activity is much higher in rat than in turtle central nervous system (CNS). *Fed. Proc.* (in press).

Hansen, A. J. (1977). Extracellular potassium concentration in juvenille and adult rat brain cortex during anoxia. *Acta Physiol. Scand.* **99**:412-420.

Hansen, A. J. (1978). The extracellular potassium concentration in brain cortex following ischemia in hypo- and hyper-glycemic rats. *Acta Physiol. Scand.* **102**:324-329.

Hansen, A. J. (1985). Effect of anoxia on ion distribution in the brain. *Physiol. Rev.* **65**:101-148.

Hansen, A. J., Hounsgaard, J., and Jahnsen, H. (1982). Anoxia increases potassium conductance in hippocampal nerve cells. *Acta Physiol. Scand.* **115**:301-310.

Hansen, A. J., and Olsen, C. E. (1980). Brain extracellular space during spreading depression and ischemia. *Acta Physiol. Scand.* **108**:355-365.

Hansen, A. J., and Zeuthen, T. (1981). Extracellular ion concentrations during spreading depression and ischemia in the rat brain cortex. *Acta Physiol. Scand.* **113**:437-445.

Henderson-Smart, D. J., and Read, D. J. C. (1979). Ventilatory response to hypoxia during active sleep in the newborn. *J. Dev. Physiol.* **1**:195-208.

Hill, J. R. (1959). The oxygen consumption of the new-born and adult mammals. Its dependence on the oxygen tension in the inspired air and on the environmental temperature. *J. Physiol.* **149**:346-373.

Hochachka, P. W. (1986). Defense strategies against hypoxia and hypothermia. *Science* **231**:234-241.

Ikonomidou, C., Price, M. T., Mosinger, J. L., Frierdich, G., Labruyere, J., Shahid Salles, K., and Olney, J. W. (1989a). Hypobaric-ischemic conditions produce glutamate-like cytopathology in infant rat brain. *J. Neurosci.* **9**:1693-1700.

Ikonomidou, C., Mosinger, J. L., Shahid Salles, K., Labruyere, J., and Olney, J. W. (1989b). Sensitivity of the developing rat brain to hypobaric/ischemic damage parallels sensitivity to *N*-methyl-aspartate neurotoxicity. *J. Neurosci.* **9**:2809-2818.

Jackson, D. C. (1968). Metabolic depression and oxygen depletion in the diving turtle. *J. Appl. Physiol.* **24**:503-509.

Jackson, D. C., and Heisler, N. (1983). Intracellular and extracellular acid-base and electrolyte status of submerged anoxic turtles at 3 °C. *Resp. Physiol.* **53**:187-201.

Kass, I. S., and Lipton, P. (1989). Protection of hippocampal slices from young rats against anoxic transmission damage is due to better maintenance of ATP. *J. Physiol.* **413**:1-11.

Krnjevic, K., Cherubini, E., and Ben-Air, Y. (1989). Anoxia on slow inward currents of immature hippocampal neurons. *J. Neurophysiol.* **62**:896-906.

Krnjevic, K., and LeBlond, J. (1989). Changes in membrane currents of hippocampal neurons evoked by brief anoxia. *J. Neurophysiol.* **62**:15-30.

LaFramboise, W. A., Guthrie, R. D., Standaert, T. A., and Woodrum, D. E. (1983). Pulmonary mechanics during the ventilatory response to hypoxemia in the newborn monkey. *J. Appl. Physiol.* **55**:1008-1014.

Lahiri, S., Brody, J. S., Motoyama, E. K., and Velasquez, T. M. (1978). Regulation of breathing of newborns at high altitude. *J. Appl. Physiol.* **44**:673-678.

Lawson, E. E., and Long, W. A. (1983). Central origin of biphasic breathing pattern during hypoxia in newborns. *J. Appl. Physiol.* **55**:483-488.

LeBlond, J., and Krnjevic, K. (1989). Hypoxic changes in hippocampal neurons. *J. Neurophysiol.* **62**:1-14.

Ljunggren, B., Norberg, K., and Siesjo, B. K. (1974). Influence of tissue acidosis upon restitution of brain energy metabolism following total ischemia. *Brain Res.* **77**:173-186.

Mattson, M. P., Guthrie, P. B., and Kater, S. B. (1989). A role for Na^+-dependent Ca^{2+} extrusion in protection against neuronal excitotoxicity. *FASEB J.* **3**:2519-2526.

Mortola, J. P., Morgan, C. A., and Virgona, V. (1986). Respiratory adaptation to chronic hypoxia. *J. Appl. Physiol.* **61**:1329-1336.

Mortola, J. P., Okubo, S., and Carroll, J. L. (1987). *Respiratory Adaptation to Hypoxia in Newborn Mammals.* Hypoxia Symposium. Edited by J. R. Sutton, C. S. Houston, and N. L. Jones. New York, Alan R. Liss.

Mosinger, J. L., and Olney, J. W. (1989). Combined treatment with MK-801 and CNQX prevents ischemic neuronal degeneration in the in-vivo rat retina. *Soc. Neurosci. Abstr.* **15**:45.

Moss, M., Moreau, G., and Lister, G. (1987). Oxygen transport and metabolism in the conscious lamb: the effects of hypoxemia. *Pediatr. Res.* **22**:177-183.

Mutch, W. A. C., and Hansen, A. J. (1984). Extracellular pH changes during spreading depression and cerebral ischemia: Mechanisms of brain pH regulation. *J. Cereb. Blood. Flow Metab.* **4**:17-27.

Nicholson, C., Ten Bruggencate, G., Steinberg, R., and Stockle, H. (1977). Calcium modulation in brain extracellular microenvironment demonstrated with ion-selective micropipette. *Proc. Natl. Acad. Sci. USA* **74**:1287-1290.

Orkand, P. K. (1980). Extracellular potassium accumulation in the nervous system. *Fed. Proc.* **39**:1515-1518.

Pepelko, W. E. (1970). Effects of hypoxia and hpercapnia, single and combined on growing rats. *J. Appl. Physiol.* **28**:646-651.

Phillipson, E. A., and G. Bowes. (1986). Control of breathing during sleep. In *Handbook of Physiology.* Section 3. Vol. II. Edited by N. S. Cherniak, and J. G. Widdicombe. Washington, DC, American Physiological Society, Chapter 19, pp. 649-689.

Reite, M., Jackson, D., Cahoon, R. L., and Weil, J. V. (1975). Sleep physiology at high altitude. *Electroencephalogr. Clin. Neurophysiol.* **38**:463-471.

Robin, E. D. (1979). Bioenergetic pattern of turtle brain and resistance to profound loss of mitochondrial. *Proc. Natl. Acad. Sci. USA* **76**:3922-3926.

Robin, E. D. (1980). Of men and mitochondria: Coping with hypoxic dysoxia. *Am. Rev. Respir. Dis.* **122**:517-531.

Rosen, C. L., Schecter, W. S., Mellins, R. B., and Haddad, G. G. (1989). Respiratory and metabolic responses to hypoxia in awake piglets. *Pediatr. Res.* **25**:1927A.

Rossen, R., Kabat, H., and Anderson, J. P. (1943). Acute arrest of cerebral circulation in man. *Arch. Neurol. Psychiatry* **50**:510-528.

Sankaran, K., Wiebe, H., Seshia, M. M. K., Boychuk, R. B., Cates, D., and Rigatto, H. (1979). Immediate and late ventilatory response to high and low O_2 in preterm infants and adult subjects. *Pediatr. Res.* **13**:875-878.

Sanocka, U. M., Donnelly, D. F., and Haddad, G. G. (1988). Cardiovascular and neurophysiologic changes during graded duration of apnea in piglets. *Pediatr. Res.* **23**:402-407.

Santiago, T. V., and Edelman, N. H. (1986). Brain blood flow and control of breathing. In *Handbook of Physiology*. Section 3. Vol. II. Edited by N. S. Cherniak and J. G. Widdicombe. Washington, DC, American Physiological Society, Chapter 6, pp. 163-179.

Sick, T. J., Chasnoff, E. P., and Rosenthal, M. (1985). Potassium ion homeostasis and mitochondrial redox status of turtle brain during and after ischemia. *Am. J. Physiol.* **248**:R531-R540.

Sick, T. J., and Kreisman, N. R. (1981). Potassium ion homeostatis in amphibian brain: Contribution of active transport and oxidative metabolism. *J. Neurophysiol.* **45**:998-1012.

Sick, T. J., Lutz, P. L., LaManna, J. C., and Rosenthal, M. (1982a). Comparative brain oxygenation and mitochondrial redox activity in turtles and rats. *J. Appl. Physiol.* **53**:1354-1359.

Sick, T. J., Rosenthal, M., LaManna, J. C., and Lutz, P. L. (1982b). Brain potassium ion homeostasis, anoxia, and metabolic inhibition in turtles and rats. *Am. J. Physiol.* **243**:R281-R288.

Sidi, D., Kuipers, J. R. G., Heyman, M. A., and Rudolph, A. M. (1983a). Effects of ambient temperature on O_2 consumption and the circulation in newborn lambs at rest and during hypoxemia. *Pediatr. Res.* **17**:254-258.

Sidi, D., Kuipers, J. R. G., Teitel, D., Heymann, M. A., and Rudolph, A. M. (1983b). Developmental changes in oxygenation and circulatory responses to hypoxemia in lambs. *Am. J. Physiol.* **245**:H674-H682.

Siemkowicz, E., and Hansen, A. J. (1981). Brain extracellular ion composition and EEG activity following 10 minutes ischemia in normo- and hyperglycemic rats. *Stroke* **12**:236-240.

Siesjo, B. O. K., and Bengtsson, F. (1989). Calcium fluxes, calcium antagonists, and calcium-related pathology in brain ischemia, hypoglycemia, and spreading depression: a unifying hypothesis. *J. Cereb. Blood Flow Metab.* **9**:127-140.

Steinberg, G. K., Saleh, J., Kunis, D., and Delapaz, R. (1989). Protection against cerebral ischemia by the NMDA antagonist dextrorphan is dose dependent and correlated with plasma and brain levels. *Soc. Neurosci. Abstr.* **15**:45.

Storey, K. B., Story, J. M., Brooks, S. P. J., Churchill, T. A., and Brooks, R. J. (1988). Hatchling turtles survive freezing during winter hibernation. *Proc. Natl. Acad. Sci. USA* **85**:8350-8354.

Sturgess, N. C., Ashford, M. L. J., Cook, D. L., and Hales, C. N. (1985). The sulphonylurea receptor may be an ATP-sensitive potassium channel. *Lancet* **8453**:474-475.

Suarez, R. K., Doll, C. J., Buie, A. E., West, T. G., Funk, G. D., and Hochachka, P. W. (1989). Turtles and rats: a biochemical comparison of anoxia-tolerant and anoxia-sensitive brains. *Am. J. Physiol.* **257**:R1083-R1088.

Stys, P. K., Ransom, B. R., Waxman, S. G., and Davis, P. (1990). The role of extracellular calcium in anoxic injury of mammalian central white matter. *Proc. Natl. Acad. Sci. (USA)* **87**:4212-4216.

Tenney, S. M., and Ou, L. C. (1977). Ventilatory response of decorticate and decerebrate cats to hypoxia and CO_2. *Respir. Physiol.* **29**:81-92.

Towell, M. E., and Salvador, H. S. (1974). Intrauterine asphyxia and respiratory movement in the fetal goat. *Am. J. Obstet. Gynecol.* **118**:1124-31.

Trube, G., Rosman, P., and Ohno-Shosaku, T. (1986). Opposite effects of tolbutamide and diazoxide on the ATP-dependent K^+ channel in mouse pancreatic β-cells. Pflugers Arch. **407**:493-499.

Ultsch, G. R., and Jackson, D. C. (1982). Long-term submergence at 3 °C of the turtle, chrysemys picta bellii, in normoxic and severely hypoxic water. I. Survival, gas exchange and acid-base status. *J. Exp. Biol.* **96**:11-28.

Van Harreveld, A., and Schade, J. P. (1959). Chloride movements in cerebral cortex after circulatory arrest and during spreading depression. *J. Cell. Comp. Physiol.* **54**:65-84.

Vyskocil, F., Kriz, N., and Bures, J. (1972). Potassium-selective microelectrodes used for measuring the extracellular brain potassium during spreading depression and anoxic depolarization in rats. *Brain Res.* **39**:255-259.

Weiss, J. N., and Lamp, S. T. (1987). Glycolysis preferentially inhibits ATP-sensitive K^+ channels in isolated guinea pig cardiac myocytes. *Science* **238**:67-69.

Westerberg, E., Monaghan, D. T., Kalimo, H., Cotman, C. W., and Wieloch, T. W. (1989). Dynamic changes of excitatory amino acid receptors in the rat hippocampus following transient cerebral ischemia. *J. Neurosci.* **9**:798-805.

Woodrum, D. E., Standaert, T. A., Mayock, D. E., and Guthrie, R. D. (1981). Hypoxic ventilatory response in the newborn monkey. *Pediatr. Res.* **15**:367-370.

20

Energy Metabolism and Ionic Homeostasis During Hypoxic Stress in the Developing Mammalian Brain

SHOKO NIOKA and BRITTON CHANCE

University of Pennsylvania
Philadelphia, Pennsylvania

I. Introduction

Numerous processes are involved in mammalian brain development. Neurogenesis, migration of neurons, axonal growth, dendritic elaboration, synaptogenesis, and myelination contribute to the progressive development of the brain (Sidman and Rakic, 1973). Some regressive events are also observed during neuronal development (Cowan et al., 1984; Purves and Lichtman, 1980). For example, more than half the neuroblasts originally generated die during development (Janowsky and Finley, 1986). Development of the nervous system begins early in embryogenesis and continues well into the postnatal period. At birth, neurons are in various stages of maturation. Among nonprecocious mammals, such as humans, dogs, and cats, neuronal migration is still observed postnatally.

The biochemistry and physiology of the brain change during the development of motor and sensory functions. Energy requirements increase with age. Between birth and 5 months of age, the amount of lipids, amino acids, RNA, and DNA present in humans doubles, as do electroencephalographic (EEG) and evoked potential activity in the visual and auditory nerves (Dobbing and

Sands, 1973). On the other hand, the anoxic resistance of the nervous system, present early in development, is significantly diminished by adulthood.

The degree of brain damage due to hypoxic stress during delivery or early postnatal life may depend on the neural plasticity of the neonate as well as the neonate's metabolic and physiological capabilities. In the adult, brain tissue that is destroyed is not regenerated or replaced. Such brain damage usually results in predictable patterns of functional deficit. By contrast, tissue destruction in the developing brain results in deficits of lesser severity but the patterns of functional loss are perhaps less predictable (Janowsky and Finley, 1986). At the anatomical and cellular level, the developing brain exhibits considerable plasticity. This plasticity is believed to be responsible for the neonate's apparent resistance to hypoxic insult. However, the significance of the specific type of plasticity observed in experimental studies, e.g., alteration in cell death rate, anomalous projections, and ectopic structures, has yet to be fully understood.

This chapter focuses on the mechanisms of tolerance for hypoxic stress in young mammals. We will elucidate some of the mechanisms that are activated with oxygen deprivation at various stages of development. In order to be able to examine such mechanisms, it will be important and relevant to provide the reader with the appropriate background on energy balance of the developing brain. Therefore, the beginning of this chapter will concentrate on mechanisms of steady-state energy metabolism in the mitochondria, energy utilization, ionic homeostasis, and the development of brain circulation.

II. Energy Metabolism in Early Life

A. Principles of Energy Metabolism During Development

Energy requirements vary in different areas of the brain. These requirements change as the brain matures. A critical aspect of neuronal function and maturation is ATP synthesis, which is associated with glycolytic and mitochondrial capacity. The mitochondrial oxidative capacity determines the risk of a hypoxic catastrophe.

Considering the fundamental equation for the maintenance of life, as defined by C. Bernard (1878), both energy conservation from oxidative metabolism and energy utilization for cell function must be considered. In the steady state, the production rate and the utilization rate of ATP must be equal.

The mitochondrial and glycolytic capacities can be quantitated by measuring the concentrations and turnover rates of particular enzymes and substrates. The level of ATP depends on the steady-state rates of ATP production and utilization. ATP synthesis can be represented in two simplified equations:

(1) for electron transfer and (2) for phosphorylation. These two equations can then be written as an overall equation of (3).

$$NADH^+ + H^+ + \frac{1}{2} O_2 \text{-------} > NAD + H_2O \tag{1}$$

$$3ADP + 3Pi + 3H^+ \text{---------} > 3ATP + 3H_2O \tag{2}$$

$$NADH + 3H^+ + \frac{1}{2} O_2 + 3ADP + 3Pi \text{---} > NAD + 3ATP + 4H_2O \tag{3}$$

The expression relating the concentrations of reactants and the relative velocity of ATP production may be formulated as a Michaelis-Menten equation:

$$\frac{V}{V_{max}} = \cfrac{1}{1 + \cfrac{Km_1}{[ADP]} + \cfrac{Km_2}{[Pi]} + \cfrac{Km_3}{[NADH]} + \cfrac{Km_4}{[O_2]}} \tag{4}$$

where Km is the concentration of substrates that gives half the velocity of ATP production.

The reserved mitochondrial capacity can be calculated from V/V_{max}, which is a ratio of the ATP synthesis rate (oxygen consumption) to the maximum rate. The great advantage of using the ratio of V to V_{max} is that only the substrates [ADP] and [Pi] need to be measured and this can be done by ^{31}P nuclear magnetic resonance (NMR) noninvasively. This does not require any further biochemical analysis, cerebral blood flow (CBF) measurement, or blood sampling (Chance et al., 1986).

Phosphorylation Potential

The phosphorylation potential ([ATP]/[Pi] x [ADP]) is a measure of free energy in thermodynamic terms, relating cellular energy levels, oxygen tension, substrate concentrations, and cellular work (energy utilization). This relationship was first evaluated by Klingenberg (1969) and was further developed by others (Erecinska and Wilson, 1982). Recently, ^{31}P NMR has been used to evaluate cellular energy levels and the ratio of PCr (phosphocreatine) to Pi (inorganic phosphate). PCr/Pi is found to be correlated to phosphorylation potential in isolated mitochondria (Gyulai et al., 1985). In the in vivo dog, this correlation is also found in carbonic acid-loaded brains (Nioka et al., 1988) and in hypoxic brains (Nioka et al., 1990). In neonatal dogs, the PCr/Pi ratio is about 1.2 (Mayevsky et al., 1988), while adults have a much higher PCr/Pi ratio of 2-3.5 (Nioka et al., 1988). This indicates the high energy capacity of the anesthetized adult brain as compared to the puppy brain.

Creatine Kinase Equilibrium System

Phosphocreatine is a high-energy molecule that supplies [ATP] in the mitochondria to the acceptors, such as Na-K ATPase, located in the cytosolic membrane.

$$[PCr] + [ADP] + [H^+] <\text{---------}> [ATP] + [Cr] \qquad (5)$$

Creatine kinase (CK) acts as a facilitative transporter of ATP through the outer mitochondrial membrane and the cytosolic space. This enzyme is located in the cytosolic space (CK1, CK3) and in the outer membrane next to ATP translocase (mitochondrial, CK4). Although total CK increases six-fold during development, the isozyme CK1 is the predominant enzyme in both the neonate (86%) and the adult (64%) (Norwood et al., 1983). While there is no noticeable CK4 in the neonate, the adult contains 15% of CK4 in its mitochondria. The amount of CK3 present increases with age. Also, the turnover of creatine kinase in the adult brain is much faster than ATP synthesis (Shoubridge et al., 1982). The transport rate of ATP (PCr being the carrier) from the mitochondria to the ATPase location should be sufficient in the in vivo adult brain. The role of the Cr-PCr creatine kinase system is to maintain ATP levels at any rate of ATP synthesis. The only time we see disequilibrium of the system is when ATP synthesis is not equal to ATP utilization. Disequilibrium is observed in neonatal rats during ischemia (Norwood et al., 1983), but is not apparent in ischemic neonatal dogs (Nioka et al., 1986).

B. Mitochondrial Alterations in the Developing Brain

Previous microscopic studies using rats have revealed that the cell number per gram of tissue decreases with age (4.6 million at birth, 2.4 million in adults) (Gregson and Williams, 1969). Detailed morphological studies show that mitochondria in neurons increase not only in number, but also in size (Table 1). Note that as the distance between cells increases in the cortex from $2\mu M$ at birth to 10-30 μm at 2 weeks, mitochondrial number and size also increase. On the other hand, the extracellular space decreases (Table 2).

The overall mitochondrial capacity that determines V_{max} of oxidative phosphorylation is related to the volume, size, and number of cristae. From these histological studies we can conclude that cells increase in size with age by an order of magnitude if we include the space of the axon, dendrites, and cell body (calculated from the distance between cells and extracellular

Table 1 Age-Dependent Mitochondrial Morphology in Rat

	Age						
	1/day	7/days	14/days	20/days	25/days	35/days	Adult
Mitochondria /total volume(%)	1.5	2.5	4.3	6.2	7.5	8.3	8.8
Number of mitochondria/cell	12	19	36	49	53	57	65
Size of mitochondria	60	65	60	63	70	72	68

Source: Pysh, 1969.

Table 2 ECS Space; % Volume of Rat Brain

	Age					
	1/day	4/days	7/days	10/days	14/days	Adult
Inferior colliculus	14.6	16.0	15.7	11.8	11.0	8.0
Cortex				40.5	31.8	21.7

Source: Pysh, 1969.

space). Also, the number of mitochondria increases but mitochondrial density per cell does not change. It is possible that neuronal mitochondria may transfer electrons on the mitochondrial membranes at similar rates and consume similar amounts of oxygen regardless of age.

C. Biochemical Changes in the Mitochondria

A number of studies of isolated brain mitochondria have demonstrated mitochondrial alterations with age. Table 3 shows mitochondrial activity with phosphorylation efficiency (ADP/O), the ratio of state 3 to state 4 activity (RCR), and oxygen consumption (Vo_2). State 3 is defined as a state where ATP synthesis is at maximum velocity. State 4 is a resting state. These investigations agree that the efficiency of ADP/O increases with age. However, RCR does not seem to depend on age and is probably an index of the quality of the preparation.

The concentration of electron transport enzymes determines the maximum activity and velocity of ATP production. The concentration of these

Table 3 Age-Dependent Mitochondrial
Function in Rat Brain

Age	ADP/O	RCR	VO$_2$
Holzman and Wood, 1973			
0-2/weeks	1.5	5.8	
2-4/weeks	2.1	10	
4-7/weeks	2.8	4.4	
Milstein et al., 1968			
1/day	1.67	2.32	14.6
5/days	1.86	2.97	26.2
9/days	2.21	3.22	47.5
14/days	2.31	3.82	85.5
20/days	2.40	4.7	110

enzymes increases with age, and this is associated with an increase in ATP demand. In conclusion, as neuronal size increases during development, mitochondrial number and size increase almost proportionally. Higher concentrations of oxidative enzymes (Table 4) and maximum activity are also found and these are necessary due to a higher energy demand.

D. Development of Carbohydrate, Fat, and Amino Acid Metabolism

There are parallel increases in glycolytic activity and oxidative phosphorylation during the postnatal period (Table 5). Neonatally, the hexose monophosphate pathway is intact (Winick, 1970). In contrast to adult brain slices (Itoh and Quastel, 1970), the rate of $^{14}CO_2$ evolution in the respiring infant brain cortex from [1-^{14}C] glucose exceeds considerably that from [6-^{14}C] glucose. Significant differences are also observed with glycogen concentrations. Glycogen is two times higher in the fetus than in the neonate or adult (Vannucci and Duffy, 1976). Glycolytic activity is monitored by measuring the initial decay of major high-energy metabolites following decapitation (Lowry et al., 1964). The decay of total high-energy phosphate is two times faster in the adult than in the 10-day-old mouse. Neonatal glycogen and lactate concentrations are more than three times greater than in the adult. The neonate has a much lower ratio of glucose consumption to oxygen consumption. Neonatal rats have only 10% of the glucose utilization of adults

Table 4 Concentration of Mitochondrial Enzymes in Rat Brain

Age	SDH	ATPase	CYTa/b/c	Protein
		Gregson and Williams, 1969		
			upi/g/h	
Neonate		1.2	0.2/0.15/0.4	0.3
Adult		1.2	0.2/0.16/0.4	0.1

	Wilson, 1972
	units/g wet wt (cyta)
0 days	14
4 days	16.4
6.5 days	18.9
9 days	24.8
11 days	29.6
20 days	50.7

	Potter et al., 1944	
	units/mg	units/mg
5 days	3	1
0 days	3	1.1
5 days	3	1.3
10 days	3.5	2.5
15 days	5	3
20 days	7	5
30 days	10	7

(Moore et al., 1971), but glucose utilization suddenly increases just before weaning. During anoxia, glycogenolysis is slower in the fetus than in the neonate (Vannucci and Duffy, 1976). The mechanism underlying this rate of glycogenolysis is probably an inhibition by acidosis. Brain tricarboxylic acid intermediates and amino acids have also been measured at different ages. A progressive increase in α-ketoglutarate is found in the freeze-blown brains (Miller and Shamban, 1977). The concentrations of malate (Miller and Shamban, 1977; Ozand et al., 1975), glutarate, and aspartate increase

Table 5 Activities of Selected Enzymes in the Developing Rat Brain[a]

Postnatal age (days)[b]	Hexokinase	Pyruvate kinase	Lactate dehydrogenase	Adenylate kinase	Isocitrate dehydrogenase	Cytochrome oxidase
0	3.10 ± 0.08	29.2 ± 1.2	30.4 ± 1.2	71.1 ± 4.3	4.07 ± 0.09	14.0 ± 1.2
2.5	3.23 ± 0.16	29.9 ± 0.8	27.6 ± 0.6	70.4 ± 2.4	3.89 ± 0.10	—[c]
4	3.88 ± 0.14	28.7 ± 0.6	28.4 ± 0.8	83.1 ± 2.8	3.99 ± 0.13	16.4 ± 1.1
6.5	4.39 ± 0.09	29.4 ± 0.3	32.7 ± 0.6	85.3 ± 3.3	3.99 ± 0.11	18.9 ± 0.5
9	5.17 ± 0.10	33.0 ± 0.8	40.9 ± 1.3	109 ± 2	4.05 ± 0.03	24.8 ± 0.9
11	5.84 ± 0.12	37.9 ± 0.7	50.2 ± 1.5	132 ± 3	4.07 ± 0.25	29.6 ± 1.3
13	6.85 ± 0.19	35.1 ± 0.3	60.0 ± 1.1	131 ± 3	3.95 ± 0.11	34.6 ± 1.8
15	8.74 ± 0.21	45.4 ± 1.1	75.6 ± 1.0	155 ± 2	3.61 ± 0.08	47.7 ± 1.1
17	8.61 ± 0.18	56.2 ± 1.5	74.8 ± 1.7	168 ± 2	3.36 ± 0.11	48.1 ± 2.7
19	8.70 ± 0.08	58.8 ± 1.2	81.6 ± 2.3	164 ± 3	2.65 ± 0.37	46.7 ± 0.6
20	9.21 ± 0.15	64.6 ± 1.5	86.8 ± 0.7	181 ± 2	2.82 ± 0.07	50.7 ± 1.9

[a]The values given are average activities (units per g fresh wt.), ± average deviation for five animals at each age, except the 17-day-old values obtained with four animals.

Source: Wilson, 1972.

with age (Agrawal et al., 1966; Miller and Shamban, 1977) while alanine decreases. Miller and Shamban report that the glutamine levels increase with age, but others do not report a change (Bayer and McMurry, 1967).

Fatty acid metabolism was investigated with a brain cortex slice (Itoh and Quastel, 1970). Acetoacetate and β-hydroxybutyrate increase the rate of oxygen consumption, but the effect does not appear to be age-related. Acetoacetate and β-hydroxybutyrate do not inhibit glucose metabolism, but lactate accumulates because of an increase in NADH in the cytosol.

E. Ionic Homeostasis in Developing Brains

Ionic gradients between vascular space and extracellular space (ECS) and between ECS and inside of cells are essential for cell function. The membrane potential generated by ion gradients is necessary for transport and neuronal function. For example, cytosolic Ca^{2+} regulates many enzymatic and hormonal steps for protein synthesis. A high potassium concentration in cells is required for protein biosynthesis (Pestka, 1971). In the mitochondria, $NADH^+/NAD$ high redox energy is stored, while in the inner membrane, a proton can be translocated to transform the redox energy to chemical energy.

Diffusion, active transport, and carrier-mediated transport comprise cellular membrane transport. In order to keep the concentrations of substrates necessary for chemical reactions high, the water volume must be restricted. Protein concentration in mitochondria is much higher than in the cytosol.

The ion gradients of K^+, Na^+, Ca^{2+}, and Mg^{2+} rely on the ATP demand system. Ion turnover is also found to be [ATP]-dependent (Hodgkin and Keynes, 1955). In transient ischemia, however, the extracellular K^+ increases seconds before [ATP] starts to decrease (order of minutes). Extracellular calcium decreases as K^+ increases (see Fig. 1). These changes occur much earlier than one would anticipate based on the absence of high-energy phosphates observed at this time. Therefore, Na-K ATPase activity during oxygen deprivation and development is important and will be described later. These membrane ion transport processes possibly control mitochondrial respiration (Van Rossum, 1972). The glycolytic and gluconeogenesis activities are regulated by K^+ ion (Soling and Kleineke, 1976).

F. Development of Na-K ATPase

Free energy, stored in the sodium gradient, is used to drive the uphill transport of amino acids, glucose, vital ions, and substrates into the cell. Another critical function of the Na^+ gradient is electrical transmission through the

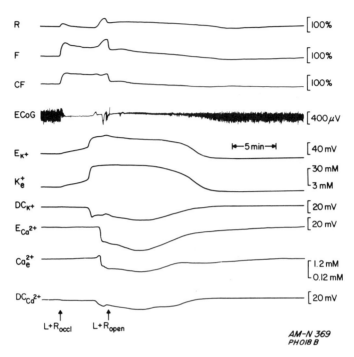

Figure 1 Responses of the Mongolian gerbil brain to bilateral carotid artery occlusion. The multiprobe assembly (NADH light guide, ECoG, K$^+$, Ca$^+$ electrodes) is located on the pia mater of the cortex. E_{K^+}, $E_{Ca^{2+}}$, Uncorrected potential measured by two surface electrodes; K_e^+, Ca_e^{2+}, corrected extracellular potassium and calcium (mM); DC_{K^+}, $DC_{Ca^{2+}}$, steady potential measured around the two electrodes used; ECoG, electrocorticogram; R, reflectance; F, NADH fluorescence, CF, corrected NADH fluorescence. When left and right carotid arteries are occluded, NADH and ECoG respond first, followed by potassium effusion and calcium infusion. During recovery, following 5 min of ischemia, NADH recovers first. ECoG and resting ion concentrations recover much later (10-15 min). (Permission to use granted by Dr. Mayevsky.)

neuron. Between 30% and 90% of ATP is utilized by Na-K ATPase. Stoicheometrically, one phosphate yields 3 Na$^+$ ions transported in the nerves. Half-saturation constant values are 20 mM for internal Na ($[Na]_i$) and 2 mM for external K ($[K]_o$). Typical physiological values of those ions are close to or less than these constants. There are no reports showing that ion gradients are age-dependent, but their pumping activities are very different. The Na-K ATPase activity increases with age and this rate increases linearly with respect to potential activity recorded in the cortex (Huttenlocker and Rawson, 1968). Na-K ATPase's low activity is due to low V_{max} and does not seem to be due to different Km's (saturation constants) (Samson and Quinn, 1967). Na-K ATPase in the neonate is less sensitive to temperature and inhibition by ouabain and calcium.

G. The Development of Circulation-Energy Metabolism Coupling

The circulatory system develops proportionally as the mass of brain increases from fetal to neonatal and postnatal periods. This increase in tissue volume triggers vasculogenesis. The fetal cerebral vascular system is comprised of the inner connected polygonal vessels. The third bifurcation initially consists of capillaries. At first, flow is passively distributed by resistances imposed mainly by the geometry of the vascular bed and transmural pressure gradients. These shunting systems help to distribute flow equally in the immature brain. Unlike that of adults, the capillary endothelium in the very young is highly vulnerable and is associated with the highly developed metabolic machinery commensurate with the capillaries' special tasks. These include sprouting, synthesis of new endothelia, lysis and attachment at junctions, synthesis of basement membrane, and development of some selectivity in the blood-brain barrier. Specific nutrient deficiencies result in microvascular stunting. On the other hand, high oxgen partial pressure causes hypervascularization typically seen in the retina of premature infants, causing blindness. Generally, blood flow is linearly related to the activity of microvascularization and the neonate's blood flow is 70% higher per weight than that of the adult. At birth, vaginally delivered neonates experience a 50% decrease in cerebral blood flow and microvasculogenesis is impaired for 24 h (Baker, 1978) (see Table 6). At the same time, the high-energy phosphate concentration decreases (Vannucci and Duffy, 1974).

Generally, energy substrate delivery couples energy metabolism. As shown in Table 7, the relationship between oxygen consumption and blood

Table 6 Relationship Between Flow Rate and Microvascularization

Age	Blood flow	Microvasculogenesis increment
Rat		
fetus	930 μl/min	3.3 μg $\times$ dry wt/min
	100%	100%
Rat,	400-767	
0 day	75%	40%

Source: Baker, 1978.

Table 7   Cerebral Blood Flow (CBF) and Oxygen Consumption ($CMRO_2$)

Age	$CMRO_2$ ml/kg bw/min)	CBF (ml/kg bw/min)	$CMRO_2$/CBF	Ref.[a]
In utero				
Whole brain	4.04	121 ± 8.7	0.033	1
Sheep	2.61	130	0.020	2
Newborn				
1-7 days, dog	1.13 ± 0.3	23 ± 8	0.049	3
1-5 days, dog	2.15	48	0.0448	4
Adult				
Human	3.2	53	0.061	5
Human	3.02	95	0.0318	6
Rat	10.3	114	0.0903	7

[a]References: 1. Makowski et al., 1972. 2. Mann et al., 1978. 3. Hernandez et al., 1978. 4. Gregoire et al., 1978. 5. Kety and Schmidt, 1948. 6. Cohen et al., 1967. 7. Johannson and Siesjo, 1974.

flow seems to change with age. In the fetus, there is higher blood flow compared to brain O_2 demand than in the neonate or adult. The oxygen consumption is expressed by $CMRO_2$ values, and their ratio to CBF ($CMRO_2$/CBF) is a measure of extraction efficiency. An O_2 extraction of 2-3% occurs in the fetus whereas there is 4-5% extraction in the neonate and adult. Values vary between 3% and 9% for several species. These values show that in the immature brain, where vascularization is ongoing, susceptibility to hypoxia increases due to high metabolism for endothelial generation and high blood flow rate to stimulate brain vascularization.

III. Hypoxic Tolerance in the Developing Brain

Developing and mature animals demonstrate remarkable differences in their ability to tolerate hypoxic stress. Young animals tolerate long periods of anoxia and hypoxia (Duffy et al., 1974; Glass et al., 1944) (see Table 8), while adults are extremely sensitive to hypoxia and ischemia (Siesjo, 1978).

The hypoxia studies described below will detail aspects of mitochondrial capacity when oxygen is lacking, as well as the mechanisms regulating energy metabolism. We will also describe each possible step of the regulatory mechanism, as well as the critical parameter levels that can be helpful to clinicians.

Table 8 Duration of Tolerance to Anoxia Versus Age

	Survival time (min)			
Age	1[a] (rabbit)	1[a] (guinea pig)	1[a] (dog)	2[a] (rat)
Fetus				
− 2 days	40			45
1 day	27	4.5	30	25
5 days	10	3.5	17	
7 days				8.8
10 days	4		14	
15 days	4		8	
Adult	6	3	3	

[a]References: 1. Glass et al., 1944. 2. Duffy et al., 1974.

The differences in hypoxic tolerance due to age involve hemoglobin's oxygen affinity, oxygen consumption, oxygen availability, and the capability of mitochondria. We will discuss the known mechanisms individually to understand tolerance to oxygen deprivation.

A. Fetal Hemoglobin During Brain Tissue Hypoxia

The necessity for high oxygen affinity in fetal hemoglobin is attributed to low available oxygen tension in fetal and early postnatal periods. The fetus has low oxygen partial pressure in the arteries with PaO_2 values of 10-20 mm Hg. These values increase with age (Mela et al., 1976). The oxygen affinity change of fetal hemoglobin at Pa_{50} (PaO_2 at which oxygen saturation is 50%) of 15-22 mm Hg enables full oxygen saturation in the arteries at low PaO_2, while dissociating enough oxygen to brain tissue with a minimum drop of oxygen partial pressure. The definition of hypoxia could not be made by absolute oxygen partial pressure. For example, a PaO_2 of 20 mm Hg is critically low in adults, but physiologically normal in the fetus. The critical oxygen level in arteries, veins, and brain tissue vasculature has been investigated by monitoring PaO_2 and oxygen saturation of hemoglobin circulating in the brain tissue of dogs by double-beam spectroscopy ($S_{tv}O_2$). "Critical hypoxia" can be defined as the failure to maintain the cellular energy level generated from oxidative phosphorylation in the mitochondria. A ratio of phosphocreatine (PCr) to inorganic phosphate (Pi) can be used to monitor the cellular energy levels since PCr/Pi is linearly related to phosphorylation potential (Gyulai et al., 1985). PCr/Pi can be measured by ^{31}P NMR spectroscopy, which can be acquired noninvasively and repeatedly.

Relationships between PaO_2, $S_{tv}O_2$, and PCr/Pi were investigated in our laboratory. The results, which follow, revealed many age-dependent features of oxygen distribution in arteries, tissues, and cells (Nioka et al., 1990b).

The relationship between $S_{tv}O_2$ and PaO_2 during oxygen deprivation is shown in Table 9. The age-dependent relationship between $S_{tv}O_2$ and PaO_2 indicates that Pa_{50} is left shifted in the 0-6-day-old group compared to older groups. Under more hypoxic conditions, for example at $S_{tv}O_2 = 25\%$, PaO_2's in three age groups are 12, 22, and 24 mm Hg, showing an age-dependent relationship. In dogs less than 1 week old, the critical PaO_2 is 9 mm Hg. At 1-3 weeks, critical PaO_2 is 17 mm Hg, while in adults critical PaO_2 is 23 mm Hg. At these PaO_2 levels, PCr/Pi drops 50% and hemoglobin saturation in brain tissue is 8% in puppies and 16% in adults. In fetal sheep, a PaO_2 of 4.5 mm Hg is recorded when their EEG becomes isoelectric (Mann et al., 1978). These very low critical PaO_2's in fetal and postnatal animals are attributed to the high oxygen affinity of fetal hemoglobin.

B. Developing Oxygen Gradients in Vasculatures and in Brain

In principle, the oxygen gradient is a driving force for oxygen diffusion in the blood, medium, and cells. Our laboratory results, described as follows, illustrate this. In dogs given FiO_2 0.33, PaO_2's and SaO_2's in three age groups (0-6 days old, 1-3 weeks, and adults) were similar, with PaO_2 at 120-150 mm Hg and $SaO_2 = 100\%$. On the other hand, in brain vasculature, the mean oxygen pressure and oxygen saturation of hemoglobin ($P_{tv}O_2$, $S_{tv}O_2$) are age-

Table 9 Mean Oxygen Saturation of Hemoglobin Circulating in the Brain ($S_{tv}O_2$) at Different PaO_2

PaO_2 (mm Hg)	$S_{tv}O_2$ (%) (0-6 days)	$S_{tv}O_2$ (%) (1-3 weeks old)	$S_{tv}O_2$ (%) (adult)
60	70	58	55
40	61	42	45
20	40	12	12
10	12	7	-
The lowest possible PaO_2 at which Pcr/Pi is 50% of initial:			
$S_{tv}O_2$ %	8	8	16
(PaO_2) mm Hg	(9)	(17)	(23)

dependent. A significant difference is observed in brain $S_{tv}O_2$'s among three age groups with values of 86% in dogs less than 1 week old, 80% in dogs 1-3 week olds, and 77% in adults. $P_{tv}O_2$'s were 27, 32, and 39 mm Hg, respectively. This implies that early in development, low oxygen consumption keeps oxygen saturation of hemoglobin high in the brain, 86% in the neonate compared to 77% in the adult.

During oxygen deprivation, oxygen gradients between arteries and tissue vasculature in the three age groups proved also to be age-dependent (Table 10). At $FiO_2 = 33\%$, however, the oxygen gradient between arteries and the brain vasculature is not age-dependent. During severe oxygen deprivation a slight gradient is seen in neonates, while a steeper gradient is observed with increasing mitochondrial activity and age.

The oxygen gradient between the inside and outside of brain cells can be measured. Since the Km of mitochondrial oxidative phosphorylation enzymes for oxygen is less than 0.1 mm Hg, oxygen gradients through the cell can be as large as the value of oxygen pressure in the capillaries. Thus, when mitochondrial oxidative phosphorylation enzymes produce half of maximum ATP synthesis or half of phosphorylation potential (PCr/Pi half value), under oxygen-limiting conditions, $[O_2]$ is determined to be near Km. We can therefore calculate this oxygen gradient when cells respond to hypoxia, changing the redox state of cytochrome oxidase or the phosphorylation potential. Brain cellular oxygen gradients located between cytosol and the capillary during hypoxia are age-dependent (7.2 mm Hg at 0-6 days, 8.8 mm Hg at 1-3 weeks, and 11.7 mm Hg in adults) (Nioka et al., 1990b). Previous studies show that the ratio of oxygen consumption among these three age groups is approximately 1:2:3 (Gregoire et al., 1978; Hernandez et al., 1978; Mann et al., 1978). There is a much higher oxygen saturation of hemoglobin (16%) in adult brain tissue than in neonates (8%) during critical hypoxia. The relatively low oxygen affinity (high Pa_{50}) of hemoglobin raises tissue vascular oxygen tension, establishing a cell environment rich in oxygen

Table 10 Oxygen Gradient Between Arterial O_2 (PaO_2) and Mean Vascular O_2 in Brain ($P_{tv}O_2$) (mm Hg)

Oxygen status	0-6 days old	1-3 weeks old	Adult
Normal ($FiO_2 = 0.33$)	101	100	98
Mild hypoxia ($S_{tv}O_2 = 50\%$)	5.2	11.5	12.8
Critical hypoxia (PCr/Pi halved)	1.3	7.9	11

$P_{tv}O_2$ is estimated from $S_{tv}O_2$.

which provides the driving force for mitochondrial diffusion. These age-dependent oxygen gradients function in accordance with the amount of oxygen consumption.

C. Mitochondrial Responses During Oxygen Deprivation

While hypoxia implies low oxygen tension with normal or high blood flow, the definition of ischemic hypoxia is "low flow" causing metabolic catastrophe. The mitochondrial response to oxygen deprivation is entirely dependent on cellular oxygen concentration. The critical intracellular oxygen concentration that causes a response is near the Km of the cytochrome oxidase reaction for O_2 (Nioka et al., 1990a). There are two opposing directions for response: stimulation and inhibition.

Typically, stimulation and inhibition can be observed in hypoxia, while only inhibition seems to take place in ischemia. Another difference between these two types of oxygen deprivation is that hypoxia shows homogeneous tissue metabolism, while ischemic hypoxia shows heterogeneous metabolic function. Heterogenous cell death occurs first. One should exercise caution when interpreting ischemic metabolic data. For these reasons, hypoxia and ischemia will be discussed separately.

D. Hypoxia Response in the In Vivo Brain and in the Isolated Mitochondria

During mild hypoxia (PaO_2 values ranging from 60 mm Hg to 40 mm Hg), physiological compensation mechanisms such as carotid body stimulation, vasodilation in arterioles, and high blood pressure are activated. As a result, there is a 50-100% increase in cerebral blood flow (CBF) which compensates for the lack of oxygen delivery to the tissue. Although there is no significant O_2 delivery loss at this level of hypoxia, impairment of acetylcholine synthesis has been reported even at this early, mild level of hypoxia (Gibson and Duffy, 1981). In chronic hypoxic cases, such as in children with cyanotic congenital heart disease, high hemoglobin content compensates for reduced PaO_2. However, there are reports indicating that in these children the ratio of cytochrome oxidase to protein increases (Park et al., 1973), but that total concentration of cytochrome decreases (Mela et al., 1976, 1977). The biochemical compensation in these patients is due to a high turnover in the electron transport system during chronic hypoxia.

In more severe hypoxia (PaO_2 lower than 40 mm Hg), intracellular PO_2 is between 1 and 0.1 mm Hg (Km of oxidative enzymes), and the redox state of the mitochondrial electron transport system may be reduced. In spite of the physiological compensation of a high blood flow, oxygen delivery is reduced by 30%. Since cytochrome oxidase (aa_3) is a substrate for the oxygen reaction, the reduced form of cytochrome aa_3 can be increased in order to maintain a constant rate of oxidation (Chance, 1965; Potter et al., 1944). This change increases the ratio of cytochrome $aa_3/[O_2]$, thereby affecting all the ratios of the redox states of the carriers in the mitochondrial electron chain: cytochrome c, cytochrome b, and NADH. As shown in Eq. (3), all the substrates, [Pi], [ADP], and [NADH], increase to compensate for the reduction of $Km_{O_2}/[O_2]$. Anaerobic glycolysis is a non-steady-state phenomenon and may not help to restore steady-state metabolism. Inefficient ATP production by glycolysis may lead to a disequilibrium between ATP synthesis and ATP breakdown. This results in an accumulation of lactate ions and protons. Acid elimination is rate-limited in the blood-brain barrier, and consequently, the accumulation of acid causes phosphocreatine to decrease significantly (Nioka et al., 1987). In general, suppression of excessive glycolysis and activation of mitochondrial oxidative phosphorylation may maintain steady-state metabolism. Indeed, pHi (intracellular pH) decreases from 7.10 in the normal dog brain to 6.95 in carefully controlled hypoxia (Nioka et al., 1990a). In a study of isolated mitochondria, an increase in the turnover of cytochrome aa_3 and a high RCR were observed during hypoxia, along with an increase in succinate dehydrogenase (Mela et al., 1976, 1977; Purshottam and Ghosh, 1975; Goodwin et al., 1978). Also, [Pi] was doubled and [ADP] increased 10%. Nevertheless, the calculated V_{max} at this time decreased as a result of lower oxygen consumption. Possibly, some parts of the mitochondrial electron transport system experienced acceleration due to low pH (Chance and Conrad, 1959) and increased concentrations of substrates, while in other parts of the electron transport system, inhibition occurred. This inhibition could result from insufficient delivery of substrates (i.e., O_2 and Ca^{2+}) (Aw et al., 1987a), inhibited ATPase, or inhibited substrate transport through the mitochondrial membrane (Aw et al., 1987b). In isolated mitochondria, hypoxia caused both high and low RCR (Nakazawa and Nunokawa, 1977; Meta et al., 1975, 1976, 1977) and high and low Ca^{2+} transport (Mela et al., 1976, 1977; Aw et al., 1987a). The possibility of mitochondrial damage due to tissue disruption during the process of isolation might have obscured the validity of this information.

E. Age-Dependent Mitochondrial Function During Hypoxia

When the degree of hypoxia is evaluated by the critical oxygen levels (i.e., oxygen levels causing unstable levels of ATP synthesis), age-dependent mitochondrial function can be assessed during an hypoxic insult. As mentioned earlier in the section on hemoglobin contribution, oxygen partial pressure in the vasculature does not normalize the degree of hypoxia. With in vivo ^{31}P NMR, changes in phosphocreatine (PCr) and inorganic phosphate (Pi)

or in the ratio of the two (PCr/Pi) can be monitored in time, and the point
at which energy metabolism becomes unstable is easily detected. Three groups
of puppies and adults are compared in Table 11. In the control, differences
of V/V_{max} [calculated from Eq. (4)] and V_{max} (calculated from VO_2) are
presented. These are the values obtained in the in vivo dog brain and calculated with assumptions described in the principles of energy metabolism.
These values indicate greater differences due to age than those observed in
the isolated mitochondrial and histological studies. During critical hypoxia
(i.e., when PCr/Pi decreases to 40% of the initial value), creatine kinase
equilibrium begins to break down and ATP begins to decrease. In critical
hypoxia, when all of the critical oxygen parameters shown in Table 11 are
decreased, V/V_{max} shows an increase. V/V_{max} increases only 3% in the 0-2-
day-old neonate, while older puppies adjust more with a 10% increase in
V/V_{max}. A decrease in V_{max} is observed in all of the age groups (Nioka et
al., 1991). This observation provides evidence that newborns and adults
are more capable of maintaining mitochondrial function under conditions
of minimal to moderate hypoxia than 3-21 day old neonates. Some other
mechanisms, occurring during hypoxia, are also responsible for the neonatal ability to decrease oxygen demand (Duffy et al., 1974).
al., 1974).

Table 11   Mitochondrial Activity In Vivo

	Age			
	0-2 days	3-10 days	11-21 days	Adult
CONTROL				
V_{max}	2.7	6.0	9.1	19.8
% V/V_{max}	30	31	32	27.5
VO_2 (Himwich and Fasekas, 1941)	27	27	29	52
Critical hypoxia at Pcr/Pi = 40% of control				
V_{max}	2.6	4.8	7.4	14.0
%V/V_{max}	33	38	41	33.2
%FiO_2	3.21	5.9	5.7	7.12
PaO_2	11.4	18.3	19.3	21.3
Temperature	33.0	35.5	33.0	35.8
Na-K ATPase[a]	1.0	1.8	3.0	—

[a]Permission to cite NA-K ATPase data granted by Dr. Mayevsky.

F. Ischemic Responses and Energy Metabolism in the Developing Brain

Without a timely recovery, transient ischemia may lead to an irreversible phenomenon that results in cell death. Therefore, it is important to understand the process of reversibility and the order of events. Two important events are involved: (1) plasma membrane damage and (2) irreversible mitochondrial damage. The second depends heavily on the first. For example, damaged plasma membranes allow an unusual influx of Ca^{2+} from the extracellular environment.

Mitochondral alterations begin with the loss of substrates, namely oxygen. Compensatory mechanisms described in the hypoxia section last only a few minutes before cells lose their creatine kinase equilibrium, leading to a sudden increase of anaerobic glycolysis. Consequently, when PCr is depleted, ATP depletion follows. For greater detail, refer to Siesjo (1978). In short, the anaerobic glycolysis provides ATP for a short time to delay the onset of irreversible cell damage until glycogen is depleted and inhibition of glycolysis occurs. Anaerobic glycolysis is eventually inhibited by negative feedback due to the increased concentration of protons produced by glycolysis. When this happens, mitochondria completely lose control of respiration. NADH CoQ reductase decreases markedly only 20 min after ischemic onset (Rouslin, 1983). Succinate utilization is still intact, but ketoglutarate cannot be consumed. State 3 respiration decreases and state 4 respiration increases, resulting in low RCR and low ADP/O ratio (Mergner et al., 1979; Ozawa et al., 1967; Schutz et al., 1973). Mitochondrial proteins and cytochrome contents decrease as well (Mela et al., 1975). Uncoupler-sensitive ATPase and other ATPase activities are lost. Adenine nucleotide translocase activity drops (Bitter et al., 1976). Other enzyme systems, such as the malate-asparatate shuttle, cannot maintain membrane transport of ions (Yakushev et al., 1974). Consequently, mitochondrial ions and substrate concentrations are altered (Aw et al., 1987a) and redox potential is increased. Ion changes involve decreases in Mg^{2+}, Ca^{2+}, and K^+ and an increase in Na^+ (Aw et al., 1987a; Baue et al., 1974), resulting in low membrane potential. Finally, membrane permeability is disrupted and massive Ca^{2+} entry causes irreversible mitochondrial damage.

The phenomena of irreversible cell damage can be correlated with morphological studies (Mergner et al., 1979). As cells pass the point of irreversible damage, inner membrane spheres disappear, allowing the ATPase complex (F_1) to be seen by electron microscopy. In the gerbil, edema occurs much more rapidly than the decrease in mitochondrial Ca^{2+} and state 3 respiration decreases (Ginsberg et al., 1977). Cellular energy levels in the dog after 2 h of mild global ischemia were correlated with cell death 3 h later using histological and in vivo NMR techniques (Garcia et al., 1988). It seems also that reversibility can be established with complete ischemia better than with incomplete ischemia. No clear argument has been given so far to explain why mitochondria that experience incomplete ischemia build up very high Ca^{2+} during recovery (Farber et al., 1981). No single "Archilles heel" of mitochondrial function can be identified, nor can reversible and irreversible effects be readily distinguished. Finally, preparative damage cannot be separated easily from in situ damage. Thus, in vitro assays must be guided by in vivo in situ studies of redox state, Ca^{2+} and K^+ levels, and finally V/V_{max} and determinations by ^{31}P NMR spectroscopy. When V approaches V_{max}, loss of steady-state conditions occurs and the tortuous pathway to mitochondrial damage and cell death is inevitable.

G. Age-Dependent Response to Ischemia

Low oxygen consumption in the neuron and an ability of the heart to function for a longer period of time under anoxia result in a remarkably high tolerance against ischemia in neonates (Glass et al., 1944). The twofold-greater glycogen level in fetuses shows higher energy storage than that in adults. This high capacity of energy storage prolongs the second phase of potassium leakage and thus delays the onset of irreversible cell derangement. Acute ischemia studies have shown that the neonatal brain is more sensitive to acidosis, which inhibits cellular activities, than the mature brain (Vannucci and Duffy, 1974). This phenomenon is demonstrated in a 2-week-old dog model whose EEG activity and [ADP] drop substantially with hypercapnia (Yoshioka et al., 1988). This inhibition does not occur in adult dogs (Nioka et al., 1987), suggesting that the neonate has greater capability than the adult to decrease energy demand during oxygen deprivation. During asphyxia, hypercapnia in the fetus depresses metabolism (Vannucci and Duffy, 1976). This protective mechanism, lowering the energy demand for survival, can be seen in non-glucose-treated rat neonates during seizures (Wasterlain and Duffy, 1976).

The recovery of neuronal function after an ischemic insult is more closely related to the stage of development than to the extent of cell damage. The fact that more neurons are formed than are necessary, and that these die naturally if unused, seems to support our hypothesis that cells originally "destined" to die can be recruited after an injury causes cell death. As discussed in the section on circulation metabolism coupling, 99.99% of vaginal deliveries in the rat involve a 50% decrease in blood flow at birth, producing an ATP breakdown of 40% (Duffy et al., 1974; Vannucci and Duffy, 1974). This striking phenomenon is physiologically normal, but no doubt ischemic enough to cause anaerobic glycolysis. There is no reported study on post-delivery cell death or regression of brain function. However, plasticity exists during this critical period.

In the chronic ischemia model, high vulnerability of bilateral carotid artery occlusion (BCAO) was observed in developing dogs. BCAO-treated puppies at 2 weeks of age exhibited the lack of visual perception and slow response of motoneurons (Zaman et al., 1988). The white matter width, determined histologically, decreased in these puppies, as did the corpus callosum (Yoshioka et al., in press). Similar results were observed in cats (Nioka, personal communication). On the other hand, in neonatal gerbils BCAO had less impact that in adults (Matsuyama et al., 1983).

IV. Summary

This chapter describes the possible mechanisms responsible for the maintenance of energy metabolism and ionic homeostasis in the developing mammalian brain. These mechanisms can be elucidated by experimentally and clinically evaluating the brain during hypoxia and ischemia. Mitochondrial function, neuronal function, and neuronal activity develop with age. In the perinatal period, severe hypoxia and ischemia (50% CBF), cause a decrease in high-energy phosphates, leading to physiological neuronal cell death during normal vaginal delivery. The neonatal tolerance may be explained by high oxygen affinity of hemoglobin, low oxygen consumption, and high glycogen storage, but not high mitochondrial capability.

References

Agrawal, H. C., Davis, J. M., and Himwich, W. A. (1966). Postnatal changes in the free amino acid pool of rat brain. *J. Neurochem.* **13**:607-615.

Aw, T. Y., Anderson, B. S., and Jones, D. P. (1987a). Mitochondrial transmembrane ion distribution during anoxia. *Am. J. Physiol.* **252**:c356-c361.

Aw, T. Y., Anderson, B. S., and Jones, D. P. (1987b). Suppression of mitochondria respiratory function after short term anoxia. *Am. J. Physiol.* **252**:c362-c368.

Baker, J. N. (1978). Regulation and stunting of cerebral microvasculogenesis by blood flow and endothelial nutrition during the fetal/premature period and birth. In *Fetal and Newborn Cardiovascular Physiology.* Vol. 2. Edited by L. D. Longo and D. D. Reneau. New York, Garland Press.

Baue, A. E., Caudry, I. H., Wurth, M.A., and Sayeed, M. M. (1974). Cellular alterations with shock and ischemia. *Angiology* **25**:31-41.

Bayer, S. M., and McMurry, W. C. (1967). The metabolism of amino acids in developing rat brain. *J. Neurochem.* **14**:695-706.

Bernard, C. (1878). *Les Phenomenes de la Vie.* Vols. 1, 2. Paris, J. B. Balliere.

Bitter, N., Shug, A. L., Koke, J. R., Folts, J. D., and Shrago, E. S. (1976). Inhibited adenine nucleotide translocation in mitochondria isolated from ischemic myocardium. *Recent. Adv. Strokes Cardiac Struct. Metab.* **7**:137-143.

Chance, B. (1965). Reaction of oxygen with the respiratory chain in cells and tissues. *J. Gen. Physiol.* **49**:163-188.

Chance, B., and Conrad, H. (1959). I. Acid-linked functions of intermediates in oxidative phosphorylation. II. Experimental studies of the effect of pH upon respiratory, phosphorylative, and transfer activities of liver and heart mitochondria. *J. Biol. Chem.* **234**:1568-1570.

Chance, B., Leigh, J. S., Jr., Kent, J., McCully, K., Nioka, S., Clark, B. J., Maris, J. M., and Graham, T. (1986). Multiple controls of oxidative metabolism in living tissues as studied by phosphorus magnetic resonance. *Proc. Natl. Acad. Sci. USA* **83**:9458-9462.

Chaudry, I. H. (1984). Cellular mechanisms in shock and ischemia and their correction. *Am. J. Physiol.* **245**:r117-r134.

Cohen, P. J., Alexander, S. C., Smith, T. C., Reivich, M., and Wollman, H. (1967). Effects of hypoxia and normocarbia on cerebral blood blow and metabolism in conscious man. *J. Appl. Physiol.* **23**:183-189.

Cowan, W. M., Fawcett, J. W., and O'Leary, D. M., and Stanfield, B. B. (1984). Regressive events in neurogenesis. *Science* **225**:1258-1265.

Dobbing, J., and Sands, J. (1973). Quantitative growth and development of the human brain. *Arch. Dis. Child.* **48**:757-767.

Duffy, T. E., Kohle, S. J., and Vannucci, R. C., (1974). Carbohydrate and energy metabolism in perinatal rat brain: Relation to survival in anoxia. *J. Neurochem.* **24**:271-276.

Erecinska, M., and Wilson, D. F. (1982). Regulation of cellular energy metabolism. *J. Memb. Biol.* **70**:1-14.

Farber, J. L., Chien, K., and Mittnacht, S. (1981). The pathogenesis of irreversible cell injury in ischemia. *Am. J. Pathol.* **102**:271-281.

Garcia, J. H., Smith, D. S., Nioka, S., Halsey, J. H., Anderson, M. L., and Chance, B. (1988). Brain swelling in experimental brain ischemia. In *Cortico Steroids in Neurological and Neurosurgical Diseases.* Edited by A. Harmann and S. Sierz.

Gibson, G., and Duffy, T. E. (1981). Impaired synthesis of ACH by mild hypoxic hypoxia or N_2O. *J. Neurochem.* **36**:28-33.

Ginsberg, M. D., Mela, L., Wrobel-Kuhl, K., and Reivich, M. (1977). Mitochondrial metabolism following bilateral cerebral ischemia in the gerbil. *Ann. Neurol.* **1**:519-527.

Glass, H. G., Snyder, F. F., and Webster, E. (1944). The rate of decline in resistance to anoxia of rabbits, dogs, and guinea pigs from the onset of viability to adult life. *Am. J. Physiol.* **140**:609-615.

Goodwin, C. W., Mela, L., and Miller, L. D. (1978). Adaptive activation of succinic dehydrogenase in chronic systemic hypoxia. *Fed. Proc.* **37**:414.

Gregoire, N. M., Giedde, A., Blum, F., and Duffy, T. G. (1978). Cerebral blood flow and cerebral metabolic rates for oxygen, glucose and ketone bodies in newborn dogs. *J. Neurochem.* **30**:63-69.

Gregson, N. A., and Williams, D. L. (1969). A comparative study of brain and liver mitochondria from new-born and adult rats. *J. Neurochem.* **16**:617-626.

Gyulai, L., Roth, Z., Leigh, J. S., and Chance, B. (1985). Bioenergetic studies of mitochondrial oxidative phosphorylation using ^{31}P NMR. *J. Biol. Chem.* **260**:3947-3954.

Hernandez, M. J., Brennan, R. W., Vanucci, R. C., and Bowman, G. S. (1978). Cerebral blood flow and oxygen consumption in the newborn dog. *Am. J. Physiol.* **234**:R209-R215.

Himwich, H. E., and Fazekas, J. F. (1941). Comparative studies of the metabolism of the brain of infant and adult dogs. *Am. J. Physiol.* **132**:454-459.

Hodgkin, A. L., and Keynes, R. D. (1955). Active transport of cations in giant axons from sepia and loligo. *J. Physiol.* **128**:28-60.

Holtzman, D., and Moore, C. L. (1973). Oxidative phosphorylation in immature rat brain mitochondria. *Biol. Neonat.* **22**:230-242.

Horowitz, J. H. (1981). The extracellular fluid space: Measurement and transport function in hemorrhagic shock. In *Proceedings of the Workshop on Albumin*. Bethsda, MD, National Institutes of Health, pp. 168-181.

Hossman, K. A. (1971). Cortical steady potential, impedence and excitability changes during and after tital ischemia of cat brain. *Exp. Neurol.* **32**:163-175.

Huttenlocker, P. R., and Rawson, M. D. (1968). Neuronal activity and adenosine triphosphatase in immature cerebral cortex. *Exp. Neurol.* **22**:118-129.

Itoh, T., and Quastel, J. H. (1970). Acetoacetate metabolism in infant and adult rat brain in vitro. *Biochem. J.* **116**:641-655.

Janowsky, J. S., and Finley, B. L. (1986). The outcome of perinatal brain damage: The role of normal neuron loss and axon retraction. *Dev. Med. Child Neurol.* **28**:375-389.

Johannson, H., and Siesjo, B. K. (1974). Blood flow and oxygen consumption of the rat brain in profound hypoxia. *Acta Physiol. Scand.* **90**:281-282.

Kety, S., and Schmidt, C. F. (1948). The effects of arterial tensions of carbon dioxide and oxygen on cerebral blood flow and cerebral oxygen consumption of normal young man. *J. Clin. Invest.* **27**:484-492.

Kjellmer, I., Karlsson, K., Olsson, T., and Rosen, K. G. (1974). Cerebral reactions during intrauterine asphyxia in sheep. I. Circulation and oxygen consumption in the fetal brain. *Pediatr. Res.* **3**:50-57.

Klingenberg, M. (1969). Respiratory control as a function of the phosphorylation potential. In *The Energy Level and Metabolic Control in Mitochondria.* Edited by S. Papa, J. M. Tager, E. Quagliariello, and E. C. Slater. Adriatic Editrice, pp. 189-193.

Lowry, O. H., Passonneau, J. V., Hasselberger, F. X., and Schultz, D. W. (1964). Effect of ischemia on known substrates and cofactors of the glycolytic pathway in brain. *J. Biol. Chem.* **239**:18-30.

Mackler, B., Grace, R., and Duncan, H. M. (1971). Studies of mitochondrial development during embryogenesis in the rat. *A. Biochem. Biophys.* **144**:603-610.

Makowski, E. L., Schneider, J. M., Tsous, N. G., Colwill, J. R., Battaglia, F. C., and Meschia, G. (1972). Cerebral blood flow, oxygen consumption, and glucose utilization of fetal lambs in utero. *Am. J. Obstet. Gynecol.* **114**:292-301.

Mann, L. I., Peress, N., Bhakthavathsalan, A., Szeto, H., and Liu, M. (1978). Fetal brain function, metabolism and neuropathology following acute hypoxia. In *Fetal and Newborn Cardiovascular Physiology.* Vol. II. Edited by L. D. Longo and D. D. Renau. New York, Garland Press.

Matsuyama, T., Matsumoto, M., Fujisawa, A., Handa, N., Tanaka, K., Yoneda, S., Kimura, K., and Abe, H. (1983). Why are infant gerbils more resistant than adults to cerebral infraction after carotid ligation? *J. Cereb. Blood Flow Metab.* **3**:381-385.

Mayevsky, A., Nioka, S., Subramanian, V. H., and Chance, B. (1988). Brain oxidative metabolism of the newborn dog: Correlation between ^{31}P NMR spectroscopy and pyridine nucleotide redox state. *J. Cereb. Blood Flow Metab.* **8**:201-207.

Mayevsky, A., and Zipi, B. (in press). The mongolian gerbil as a model for cerebral ischemia. In *Cerebral Ischemia and Resuscitation.* Edited by A. Schurr and B. M. Rigor. Orlando, FL, CRC Press.

Mela, L., Delivoria-Papadopolos, M., and Miller, L. D. (1978). Fetal and neonatal mitochondrial electron transfer chain. In *Fetal and Newborn Cardiovascular Physiology.* Vol. II, Edited by L. D. Longo and D. D. Reneau. New York, Garland Press.

Mela, L., Goodwin, C. W., and Miller, L. D. (1975). Correlation of mitochondrial cytochrome concentration and activity to oxygen availability in the newborn. *Biochem. Biophys. Res. Commun.* **64**:384-390.

Mela, L., Goodwin, C. W., and Miller, L. D. (1976). In vivo control of mitochondrial enzyme concentration and activity by oxygen. *Am. J. Physiol.* **231**:1811-1816.

Mela, L., Goodwin, C. W., and Miller, L. D. (1977). In vivo adaptation of O_2 utilization to O_2 availability. Comparison of adult and newborn mitochondria. In *Oxygen and Physiological Function*. Edited by F. F. Jobsis. Dallas, Professional Information Library, pp. 285-291.

Mergner, W. L., Chang, S., Marzella, L., Kahng, M. W., and Trump, B. F. (1979). Studies on the pathogenesis of ischemic cell injury. *Lab. Invest.* **40**:686-94.

Miller, A. L., and Shamban, A. (1977). A comparison of methods for stopping intermediary metabolism of developing rat brain. *J. Neurochem.* **28**:1327-1334.

Milstein, J. M., White, J. G., and Swarman, K. F. (1968). Oxidative phosphorylation in mitochondria of developing rat brain. *J. Neurochem.* **15**:411-415.

Moore, T. J., Lione, A. D., Regen, D. M., Tarpley, H. L., and Raines, P. L. (1971). Brain glucose metabolism in the newborn rat. *Am. J. Physiol.* **221**:1746-1753.

Myers, R. Z. (1979). A unitary theory of causation of anoxic and hypoxic brain pathology. *Adv. Neurol.* **26**:195-217.

Nakazawa, T., and Nunokawa, T. (1977). Energy transduction and adenine nucleotides in mitochondria from rat liver after hypoxic perfusion. *J. Biochem.* **82**:1575-1583.

Nioka, S., Chance, B., Hilberman, M., Subramanian, H. V., Leigh, J. S., Veech, R. L., and Forster, R. E. (1987). Relationship between intracellular pH and energy metabolism in dog brain as measured by ^{31}P NMR. *J. Appl. Physiol.* **62**:2094-2102.

Nioka, S., Mayevsky, A., Chance, B., Subramanian, H. V., Alter, C., Gilbert, D., Sinwell, T., Ghosh, M., Donlon, E., and Butler, S. (1986). Age dependent metabolic control parameters in the neonate puppy brain from birth to 21 days age. *Soc. Nucl. Magnetic Res. Med.* **5**:674-675.

Nioka, S., Smith, D., Butler, S., and Closter, J. (1988). Studies of mitochondrial metabolic uncoupling in vivo using ^{31}P NMR. *Soc. Nucl. Magnetic Med.* **7**:294.

Nioka, S., Smith, D. S., Chance, B., Subramanian, H. V., Butler, S., and Katzenberg, M. (1990a). Oxidative phosphorylation system during steady-state hypoxia in the dog brain. *J. Appl. Physiol.* **68**:2527-2535.

Nioka, S., Chance, B., Smitn, D. S., Mayevsky, A., Reilly, M. P., Alter, C., and Asakura. T. (1990b). Cerebral energy metabolism and oxygen state during hypoxia in neonate and adult dogs. *Pediatr. Res.* **28**:54-62.

Nioka, S., Smith, D. S., Mayevsky, A., Dobson, G., Veech, R. L., Subramanian, H. V., Chance, B. (1991). Age dependence of steady state mitochondrial oxidative metabolism during brain hypoxia in dogs. *Neurological Research.* in press.

Norwood, W. I., Ingwall, J. S., Norwood, C. R., and Fossel, E. T. (1983). Developmental changes of creatine kinase metabolism in rat brain. *Am. J. Physiol.* **244**:c205-c210.

Ozand, P. T., Stevenson, J. H., Tildon, J. T., and Cornblath, M. (1975). The effects of hyperketonemia on glutamate and gluamine metabolism in developing brain. *J. Neurochem.* **25**:67-71.

Ozawa, K., Seta, K., Araki, H., and Honda, H. (1967). The effect of ischemia on mitochondrial metabolism. *J. Biochem.* **61**:512-514.

Park, C. D., Mela, L., Wharton, R., Reiley, J., Fishbein, P., and Aberdeen, E. (1973). Cardiac mitochondrial activity in acute and chronic cyanosis. *J. Surg. Res.* **14**:139-146.

Pestka, S. (1971). Protein biosynthesis: Mechanism, requirements and potassium dependency. In *Membranes and Ion Transport.* Vol. III. Edited by E. E. Bittar. London, Wiley-Interscience, pp. 279-296.

Potter, V. R., Schneider, W. C., and Lieble, G. J. (1944). Enzyme changes during growth and differentiation in the tissues of the newborn rat. *Cancer Res.* **5**:21-24.

Purshottam, T., and Ghosh, N. C. (1975). Effect of acute hypoxia on the enzymes involved in the metabolic and nervous functioning of rat brain. *Env. Physiol. Biochem.* **5**:73-77.

Purves, D., and Lichtman, J. W. (1980). Elimination of synapses in the developing nervous system. *Science* **210**:153-157.

Pysh, J. J. (1969). The development of the extracellular space in neonatal rat inferior colliculus: An electron microscopic study. *Am. J. Anat.* **124**:411-430.

Pysh, J. J. (1970). Mitochondrial changes in rat inferior colliculus during postnatal development: An electron microscopic study. *Brain Res.* **18**: 325-342.

Rodier, P. M. (1980). Chronology of neuron development: animal studies and their implications. *Dev. Med. Child. Neurol.* **22**:525-545.

Rouslin, W. (1983). Mitochondrial complexes 1,2,3,4 and 5 in myocardial ischemia and autolysis. *Am. J. Physiol.* **244**:H743-H748.

Samson, F. E., and Quinn, D. J. (1967). Na-K activated ATPase in rat brain development. *J. Neurochem.* **14**:421-427.

Schutz, H., Silverstein, P. R., Vapalahti, M., Bruce, D. A., Mela, L., and Langfitt, T. W. (1973). Brain mitochondrial function after ischemia and hypoxia. *Arch. Neurol.* **29**:408-416.

Shoubridge, E. A., Briggs, R. W., and Radda, G. K. (1982). ^{31}P NMR saturation transfer measurements of the steady-state rates of creatine kinase and ATP synthetase in the rat brain. *FEBS. Lett.* **104**:288-292.

Sidman, R. L., and Rakic, P. (1973). Neuronal migration with special reference to developing human brain: A review. *Brain Res.* **62**:1-35.

Siesjo, B. K. (1978). *Brain Energy Metabolism.* New York, Wiley.

Soling, H. D., and Kleineke, J. (1976). Species-dependent regulation of hepatic gluconeogenesis in higher animals. In *Gluconeogenesis: Its Regulation in Mammalian Species.* Edited by R. W. Hanson and M. A. Mehlman. New York, Wiley, pp. 369-462.

Vannucci, R. C., and Duffy, T. E. (1974). Influence of birth on carbohydrate and energy metabolism in rat brain. *Am. J. Physiol.* **226**:933-940.

Vannucci, R. C., and Duffy, T. E. (1976). Carbohydrate metabolism in fetal and neonatal rat brain during anoxia and recovery. *Am. J. Physiol.* **230**:1269-1275.

Van Rossum, G. D. V. (1972). The relation of sodium and potassium ion transport to the respiration and adenine nucleotide content of liver slices treated with inhibitors of respiration. *Biochem. J.* **129**:427-438.

Wasterlain and Duffy, T. E. (1976). Status epileptics in immature rats. *Arch. Nerol.* **33**:821-827.

Whittam, R. (1970). Enzymatic aspects of the sodium pump. In *Essays in Cell Metabolism.* Edited by W. Bradley, H. L. Kornberg, and R. Ouayley. New York, Wiley-interscience, pp. 235-254.

Wilson, J. E. (1972). The relationship between glycolytic and mitochondrial enzymes in the developing rat brain. *J. Neurochem.* **19**:223-227.

Winick, M. (1970). Nutrition and nerve cell growth. *Fed. Proc.* **29**:1510-1515.

Yakushev, V. S., Solobodin, R. L., Liftshits, R. L., Kemilof, F. K., Bochina, T. V., Valdman, B. M., and Efmenko, G.P. (1974). Change in the activity of malate dehydrogenase and of its iosenzymes in early periods of experimental burn. *Chem. Abstr.* **80**:294.

Yoshioka, H., Goma, H., Ochi, M., Sawada, T., Nioka, S., Zaman, A., Miyake, H., Masumura, M., and Chance, B. (in press). Bilateral carotid artery occlusion interferes with neuronal development in the puppy. The First International Stroke Congress. Tokyo.

Yoshioka, H., Nioka, S., Miyake, H., Zaman, A., and Masumura, M. (1988). The effects of pHi on energy metabolism in *in vivo* puppy brain. *Soc. Nucl. Magnetic Res. Med.* **7**:474.

Zaman, A., Nioka, S., Yoshioka, H., Masumura, M., Miyake, H., Shiffman, D., Williams, S., and Schor, N. (1988). ^{31}P-NMR study of cerebral metabolism in developing puppy brain and its relationship to neurological development. *Soc. Nucl. Magnetic Res. Med.* **7**:473.

21

Ionic Environment in the Central Nervous System and Effects of Hypoxia in Early Life

TERESA TRIPPENBACH

McGill University
Montreal, Quebec, Canada

I. Introduction

Extracellular ion activities in the brain rely on water distribution in the interstitial space composed of 20-50-nm-wide clefts between the neurons and glial cells. Owing to the enormous surface area of brain cells, the extracellular space (ECS) represents about 20% of brain volume (Van Harreveld, 1972). Because local changes in ion levels in the ECS are transmitted to the surrounding cells, the ECS serves as a communication channel and is important for the integrative functions of the central nervous system (CNS) (Nicholson, 1980). Furthermore, ion equilibrium in the ECS defines characteristics of the neuronal membrane and is responsible for the functional performance of the CNS. Ion equilibrium is maintained by active ion transport, which in turn is an energy-dependent process. The relationships among glucose utili-

Work presented here was carried out with the collaboration of Drs. Helmut Acker and Diethelm W. Richter at the Max-Planck-Institut fur Systemphysiologie, Dortmund, F.R.G., and was supported by the MRC of Canada and the Deutsche Forschungsgemeinschaft and the Max-Planck-Gesellschaft.

zation (Sokoloff et al., 1977), oxygen consumption (Cartheuser, 1987), and functional activity of the brain, as well as between extracellular potassium activity and O_2 consumption (Lewis and Schuette, 1976), have been evaluated in the cortex of adult rats and cats. Because brain tissue has almost no oxygen storage and only small stores of glucose, even a brief interruption of cerebral blood flow or anoxia can cause severe disturbances in ionic homeostasis. Under conditions of substrate deficiency, K^+ leaks from the cells and Na^+ and Ca^{2+} enter them. Accumulation of K^+ in the ECS (K_e^+) may result in depolarization of presynaptic membranes (Llinas, 1968) and reduction of the amount of a transmitter available for release by nerve impulses (Hubbard and Willis, 1962). It follows that changes in the ionic composition of the ECS may have a large effect on central synaptic transmission. Therefore, the glial cells, which play an important role in the regulation of ion levels in the ECS (Kuffler et al., 1966), may be considered a component of the overall integratory mechanisms of the CNS.

Ionic homeostasis is important for the functional performance of the central control of all systems. In this chapter, extracellular ion activities in the ECS of the respiration-related neurons are described during normoxic as well as hypoxic conditions in early life.

II. Anatomical Organization of Respiration-Related Neurons

Most investigations concerned with central control of ventilation have been performed in anesthetized, paralyzed, and artificially ventilated adult cats. On the basis of these studies, neurons involved in the control of the basic respiratory pattern have been localized in the medulla oblongata in two bilateral neuronal aggregates. According to their anatomical locations, they have been called the dorsal (DRG) and the ventral (VRG) respiratory groups (Baumgarten and von Kanzow, 1958; Merrill, 1981). The DRG, located in the site of the nucleus tractus solitarius (NTS), contains mainly neurons with inspiration-related activity. Inspiratory bulbospinal premotor neurons, R-alpha neurons (Baumgarten and von Kanzow, 1958), project to the phrenic and the inspiratory intercostal motor neurons on the contralateral side (Berger, 1977). They represent about half the total cell population. The second large group of inspiration-related neurons of the DRG, R-beta neurons (Baumgarten and von Kanzow, 1958), are located in the ventrolateral subnucleus of the NTS (von Euler et al., 1973). They receive monosynaptic input from the vagal slowly adapting pulmonary mechanoreceptors (Berger and Dick, 1987). In the same location, a small population of "P" neurons was also described (Berger et al., 1984). These neurons are activated by vagal volume-related information during both spontaneous respiration and mechanical ventilation. Without this information they remain silent. Input from the vagal pulmonary stretch receptor afferents is not restricted to the ventrolateral side of the NTS since their fibers also terminate in the region of the intermediate nucleus of the NTS (Kalia and Richter, 1985).

The VRG forms a large complex of both inspiration- and expiration-related neurons. Inspiratory cells are located in the intermediate part of this complex (the nucleus para-ambigualis; Kalia, 1982) and in the most rostral part of the VRG (the region of nucleus retrofacialis; Bianchi and Barillot, 1982). These are inspiratory premotor neurons with projection to the phrenic and the inspiratory intercostal motor neuron pools and propriobulbar interneurons with different types of inspiratory activity (Merrill, 1981). Recently, inspiration-related neurons have also been found in the most caudal part of the VRG, at the C1 and C2 levels in cats. These neurons project to the phrenic motor neurons of the ipsilateral side (Aoki et al., 1980).

Expiration-related neurons are found in the site of the nucleus para-ambigualis and in the region of the nucleus retrofacialis. A large conglomerate of expiration-related cells is also found in the caudal part of the VRG (nucleus retroambigualis). Nucleus retroambigualis accommodates bulbo-spinal expiratory premotor neurons projecting to the ipsi- and the contra-lateral spinal motor neuron pools for the expiratory intercostal and abdominal muscles (Merrill, 1981).

Anatomical localization of the respiration-related neurons is probably similar in newborns and adults of a given species (Bystrzycka et al., 1975; Farber and Lawson, Chapter 6, this volume). However, owing to immaturity of the neuronal connections, neurotransmitter and neuromodulator systems, age-dependent differences in the control of breathing (Duron and Marlot, 1980; Hansen and Chernick, 1983), and different ventilatory responses to a number of stimuli, including hypoxia, may exist.

III. Ventilatory Effects of Hypoxia in Newborn Animals

The respiratory effects of acute exposure to hypoxia have been investigated in newborn animals of several species. If the exposure to hypoxia persists beyond the time of respiratory excitation, a decrease in ventilation occurs; respiration may drop below the normoxic value. The initial ventilatory excitation is attributed to activation of peripheral chemoreceptors (Biscoe and Purves, 1967). Although several hypotheses have been proposed, the mechanisms for hypoxia-related fall in ventilation are still unknown. Various neurotransmitters and neuromodulators have been suggested as mediators of the decrease in breathing (see Lagercrantz, and Runold, Chapter 22, this volume). A decrease in metabolic rate (Cross et al., 1958), changes in respiratory mechanics (LaFramboise et al., 1983), alkalosis of the cerebrospinal fluid (CSF) (Aswal et al., 1980), and finally, a direct inhibitory effect of low oxygen on the central nervous system (Lawson and Long, 1983) have also been proposed. One other possibility is that central inhibitory effects of hypoxia are secondary to changes in the extracellular ion activities in the brain.

IV. Extracellular Ion Activities in the CNS of Mammals

Before ion-sensitive microelectrodes (ISMs) were introduced, extracellular ion activities were estimated from their concentrations in the plasma (Widdowson and McCance, 1956) and CSF (Cserr, 1965). Measurements made using ISMs indicated that during steady-state conditions, the extracellular ionic activities in the cerebral cortex of adult animals are indeed reflected by their compositions in the CSF (for review, see Hansen, 1985) and plasma (Hossman et al., 1977). However, during acute changes in ion activities, slow diffusion processes lead to a disequilibrium between these three spaces. Also, homeostatic mechanisms of the brain, low ion permeability in brain capillaries, and active ion transport mechanisms in the choroid plexus and brain endothelial cells all help maintain concentrations of cations in the ECS constant, even when their plasma levels increase (Cserr and Bundgaard, 1984). These mechanisms are especially efficient for ions such as K^+, Ca^{2+}, and Mg^{2+} (Davson and Welch, 1971). Therefore, owing to the characteristics of the blood-brain barrier and other homeostatic mechanisms, ion activities in the ECS may be controlled independently of their variations in plasma.

At rest, a constant ion gradient across neuronal membranes is maintained by active processes, namely, the ionic pumps that move ions against their concentration and electrical gradients. During neuronal excitation, passive movements of ions are triggered by voltage changes and influences of neurotransmitters and neuromodulators. Thus, the net movement of ions across the membrane is controlled by the active and passive transport systems. Changes in the levels of ions on both sides of the membrane may occur when passive ion transport exceeds the ion pump activity or when activity of the ion pumps is impaired. Several methods have been used to measure ion concentrations. However, the fast, stimulus-dependent changes in the extracellular ion activities are detectable only by ISMs. Using ISMs, excitation-dependent increase of extracellular potassium activity (Krnjevic and Morris, 1972) and decrease in extracellular calcium activity (Heinemann et al., 1977) were first described. These two ions, of primary importance for synaptic transmission and neuronal function, have been most frequently studied (for review, see Hansen, 1985).

Under "normal" steady-state conditions, i.e., when the oxygen supply meets the metabolic demands of brain tissues, resting level of extracellular potassium activity (aK^+_e) in the cerebral cortex of adult animals was reported as 2.5-3.5 mM (Morris, 1974; Hansen, 1985). Similar values were recorded in the medulla, spinal cord, and hippocampus (Richter et al., 1978; Krnjevic and Morris, 1972; Krnjevic et al., 1980). In these same areas, extracellular calcium activity (aCa^{2+}_e) was in the range of 1.2-1.5 mM (Krnjevic et al., 1980; Acker and Richter, 1985).

During physiological excitation, changes in ionic activities in the ECS are rather small. Within the portion of NTS containing inspiration-related neurons in adult cats, aK^+_e varies by no more than 0.5 mM during the respiratory cycle (Richter et al., 1978; Acker and Richter, 1985). During stimulation of afferent fibers or a direct neuronal stimulation, aK^+_e always increases but generally does not exceed 10-12 mM (Coles and Tsacopoulos, 1979; Heinemann and Lux, 1977; Lothman et al., 1975). The presence of the saturation level of aK^+_e has been attributed to the inhibitory effect of increased aK^+_e on synaptic transmission (Krnjevic and Morris, 1976), stimulation of reuptake mechanisms (Heinemann and Lux, 1977), and uptake by the glial cells (Coles and Tsacopoulos, 1979). It may also depend on a reduced K^+ conductance due to a drop in the interstitial pH developed during neuronal activation (Urbanics et al., 1978). The maximal value of aK^+_e of 12 mM is only exceeded under conditions of lack of energy substrate or during spreading depression (Hansen, 1977).

In the environment of respiration-related neurons in the NTS of adult cats, aCa^{2+}_e fluctuates within the respiratory cycle by about 0.4 mM (Acker and Richter, 1985). However, activity-dependent changes in aCa^{2+}_e are not as systematic as those seen in aK^+_e. During repetitive electrical stimulation of afferent fibers or direct stimulation of several regions of the mammalian CNS, aCa^{2+}_e may either decline (Heinemann et al., 1977; Krnjevic et al., 1980) or rise (Somjen, 1979, 1980) by 0.01-0.4 mM. During spreading depression, aCa^{2+}_e in the brain cortex decreases to extremely low values (Kraig and Nicholson, 1978).

We have only limited data on the extracellular ion activities in the CNS of newborn animals. The plasma concentration of K^+ is 10.2 mM in the fetus; it decreases to 8 mM in the newborn and to 4.9 mM in adult humans (Widdowson and McCance, 1956). During the first day of life, the plasma K^+ concentration is about 6 mM in kittens and 7 mM in rats; it decreases by the third week after birth to an adult value of 3.5 mM (Skoglund, 1967). Skoglund (1967) reports that in newborns of both species, the plasma Ca^{2+} has a tendency to decrease with age, but it varies only slightly above the adult value of 1.5 mM.

The developmental decrease in the plasma K^+ concentration was attributed to a fall in the extracellular fluid volume (Friis-Hansen, 1957) and therefore redistribution of K^+ from the extra- to the intracellular compartments (Widdowson and McCance, 1956). The decrease in aK^+_e during development could reflect an overall gradual increase in oxidative metabolism of the brain and activation of the ionic pumps. With increasing age, more Na^+ could be actively transferred out of the cells and more K^+ could be transported to the cells. These changes would lead to the elevated K^+ gradient between the inside and outside of the cell and consequently would be responsible for an increase in resting membrane potential. Increasing membrane potential with maturation would provide a good explanation for an increasing sensitivity to afferent inputs during postnatal development (Skoglund, 1967).

Despite the developmental changes in levels of K^+ in the plasma, studies using the ISMs indicated that the resting value of $aK+_e$ in the cerebral cortex did not change with age and varied from 3.0 to 3.7 mM in 5-19-day-old rats (Mutani et al., 1974; Mares et al., 1976). Resting values of $aK+_e$ in the respiration-related region of the NTS in rabbits during the first 3 weeks of life varied from 2.6 to 6.3 mM. The group mean of $aK+_e$ was 4.1 ± 1.1 mM (Trippenbach et al., 1987). In the same medullary area of adult cats, $aK+_e$ levels were more stable and varied only between 3 and 3.5 mM (Acker and Richter, 1985). Although $aK+_e$ had a tendency to decrease with maturation, the relationship between $aK+_e$ and the animal's age was not significant. This suggests that $aK+_e$ in the sensorimotor cortex and brainstem does not depend on the brain maturation. Possibly homeostatic mechanisms and the blood-brain barrier buffer the brain against changes in plasma K^+. On the other hand, lack of a clear developmental trend in aK^+_e in rabbits may have been due to different degrees of brain tissue damage, different sensitivity to ketamine and acepromazine (both used as the anesthetic), and differences in body temperature between newborns of different ages. It is clear that more investigations are necessary to describe developmental changes in the ionic environment of specific areas of the brain.

Activity-related fluctuations in aK^+_e have been recorded by us in newborns. In Figure 1, aK^+_e associated with medullary respiration-related neurons in a 4-day-old rabbit increases and decreases with the same periodicity as phrenic nerve output. The amplitude of aK^+_e fluctuations, ranging between 0.03 and 0.07 mM (Trippenbach et al., 1990), was lower than that recorded in the same area in adult cats (0.1-0.5 mM; Richter et al., 1978; Acker and Richter, 1985). These cyclic changes in aK^+_e probably depend on the pre- and postsynaptic excitation of the inspiration-related neurons. Diminished amplitude of the activity-related fluctuations early in life may be explained

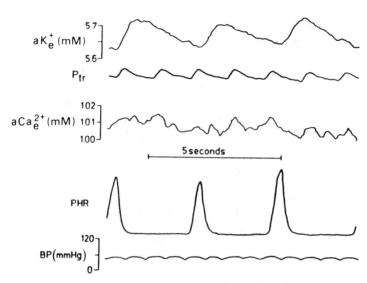

Figure 1   Extracellular potassium activity (aK^+_e) and extracellular calcium activity (aCa$^{2+}_e$) recorded in the NTS of 4-day-old rabbit during artificial ventilation with 50% O$_2$. Fluctuations in aK^+_e are related to integrated phrenic nerve activity (PHR). Delay between peak amplitude of PHR and aK^+_e is due to time response of ion-sensitive microelectrode. Variability in aCa$^{2+}_e$ is not related to respiratory rhythm. P$_{tr}$, Tracheal pressure; BP, arterial blood pressure. (Modified from Trippenbach et al., 1990.)

by less excitatory synaptic connections between neurons and the greater ECS in the immature than adult brain (Van Harreveld, 1972). This age difference may be further accentuated by the absence of the activity-related shrinkage of ECS in newborns. In an in vitro preparation, stimulation of an isolated optic nerve did not produce activity-related ECS shrinkage in animals younger than 2 weeks (Ransom et al., 1985). The appearance of activity-related shrinkage of the ECS coincided with the development of mature glial function (Skoff et al., 1976). This temporal relationship suggests that dynamic ECS shrinkage results from fluid and electrolyte shifts into the glial cells (Ransom et al., 1985). Spontaneous neuronal excitation would have a minimal effect on volume of the ECS even in adult animals. Obviously, this mechanism becomes more important during electrical stimulation of afferent pathways.

The group mean value of aCa$^{2+}_e$ of 1.3 $\pm$ 0.4 mM measured in newborn rabbits does not differ from values recorded in the brainstem (Acker and Richter, 1985) and other regions of the CNS (Krnjevic et al., 1980; Dingledine and Somjen, 1981) of adult animals and in the plasma of newborn rats and rabbits (Skoglund, 1967). Levels of aCa^+_e do not systematically vary with phrenic output (Fig. 1).

V. Extracellular Ion Activities During
Low Oxygen Supply

Effects of low oxygen supply on aK^+_e have been studied to better understand the mechanisms of cortical electrical silence, seizures, loss of consciousness, and spreading depression. In these investigations, long-term brain anoxia was produced by drastic methods. Only a few studies dealt with effects of short-term or smaller changes in the brain O_2 levels.

Large increases in aK^+_e of the cerebral cortex (to 60-100 mM) have been measured in association with long-term brain anoxia and spreading depression in adult animals (for review, see Hansen, 1985). Similarly, large changes in aK^+_e accompanying terminal anoxia were recorded in the cerebral cortex of newborn and young rats (Hansen, 1977; Mares et al., 1976). During hypoxia, the maximal increase in aK^+_e of the cat sensorimotor cortex varied between 5 and 40 mM (Kirshner et al., 1975). A transient increase in aK^+_e, by about 2 mM above its resting value, was recorded during brief anoxic episodes when brain surface PO_2 decreased by 5 mm Hg in the adult guinea-pig cortex. The aK^+_e increase was proportional to a decrease in the O_2 level (Morris, 1974). A less pronounced increase in aK^+_e was observed during 1-min exposures to hypoxia or anoxia when tissue PO_2 in the NTS of adult cats fell to 0 mm Hg. In the latter study, aK^+_e increased by 1-1.5 mM and aCa^{2+}_e decreased by 0.3-0.5 mM from its resting levels (Acker and Richter, 1985). On the other hand, elevation of tissue PO_2 seems to have little or no effect on aK^+_e, at least in the cortex of adult animals (Morris, 1974).

The study of Kirshner et al., (1975) pointed out the critical role played by blood pressure in determining the magnitude of the changes in aK^+_e during hypoxia in the cortex of adult cats. These authors found a strong inverse correlation between mean arterial blood pressure and aK^+_e levels at low arterial PO_2's. The threshold level of hypoxia that produced extracellular K^+ accumulation, provided normal blood pressure was maintained, corresponded to an arterial PO_2 of 20-23 mm Hg. Decreases in blood pressure induced by hypoxia invariably led to both increases in aK^+_e from 3.4 to 20 mM and a simultaneous slowing or flattening of the ECoG.

Although there are several studies describing the dependence of the cerebral cortex activity on aK^+_e levels, little is known about the relationship between activity of the specific control systems and the ionic environment. For example, Mares et al. (1976) reported that in one of their 19-day-old rats under pentobarbital anesthesia, aK^+_e of the sensorimotor cortex increased to 8.9 mM during irregular breathing and returned to the resting value of about 3 mM shortly after recovery of respiratory rhythmicity. On the basis of this observation, the authors concluded that the resting level of aK^+_e is a sensitive indicator of an inadequate respiration. Acker and Richter (1985) reported hypoxia-related prolongation of phrenic and inspiration-related neuron activities, sometimes followed by their block (i.e., central apnea, monitored by absence of phrenic nerve activity) simultaneously with an increase in aK^+_e by no more than 1.5 mM above control value in adult cats. Figure 2 illustrates two examples from our studies of how low oxygen tension affects central inspiratory activity (monitored by efferent vagal or phrenic activities), aK^+_e, and tissue PO_2 (PtO_2) in 4-h- and 21-day-old rabbits (Trippenbach et al., 1990). Ventilation with 100% N_2, provoked a decline in PtO_2 from control value of 14 mm Hg to 4.8 mm Hg in younger rabbits (Fig. 2A). In the older rabbit, ventilation with 10% O_2 resulted in the PtO_2 decrease from 24 mm Hg to 4 mm Hg. Comparably, low PtO_2 had different effects on the pattern of breathing in the two rabbits. In the 4-h-old animal, gasping and apnea were recorded at 4 min after the onset of anoxic exposure. Excitation of central inspiratory activity was not present during hypoxia but it occurred during the recovery period. On the other hand, in the older animal, respiratory excitation, although slightly diminished after the initial response, persisted during the total period of hypoxic exposure. The difference between these two rabbits in the respiratory effects of low PtO_2 in the medullary tissue implies that the respiratory neuronal network is more sensitive to lack of oxygen immediately after birth than later during development.

In the examples illustrated in Figure 2, and in all other newborns, 3-5-min exposures to hypoxia provoked three phase changes in aK^+_e. The initial slow increase (phase I) was followed by the secondary fast rate of rise (phase II) until a saturation level was reached (phase III). A three or even four-phase increase in aK^+_e was observed also in the cortex of newborn, young, and adult rats during brain anoxia. In the study on newborn rats, a transition from the first to the second phase appeared at aK^+_e of 6.5-8.6 mM (Mares et al., 1976). These values are similar to those observed in the NTS of newborn rabbits as a maximal response to a short-term hypoxic exposure. The difference between these observations may be due to the regional differences in the aK^+_e response to hypoxia. In adult rats during terminal anoxia, the rates of aK^+_e increase in the reticular formation, zona incerta, and hypothalamus were lower than in neocortex, hippocampus, thalamus, caudate nucleus, and amygdala (Bures and Buresova, 1981). Differences in glial cell composition from individual regions might also influence the K^+ response.

The initial slow aK^+_e response to low oxygen supply recorded in the respiration-related area of the NTS in newborn rabbits was probably due to a net efflux of K^+ from afferent terminals of peripheral chemoreceptors (Biscoe and Purves, 1967) and from the synaptically driven respiratory neurons. In agreement with this assumption, hypoxia-dependent excitation of PHR, if present, was recorded during phase I. This slow, but steady rise in aK^+_e suggests that already during early hypoxia, K^+ efflux exceeded capacity of

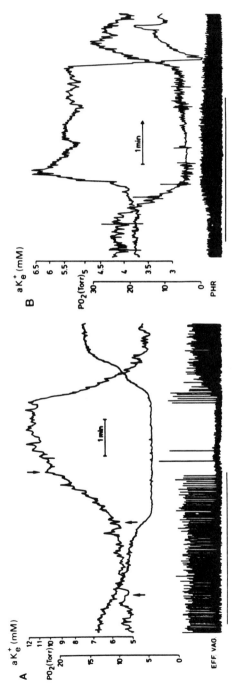

Figure 2 Extracellular potassium activity (aK^+_e) and tissue PO_2 (Pto_2) in 4-h-old (A) and 24-day-old (B) rabbits during control and hypoxic conditions. (A) Integrated efferent vagal activity (EFF VAG) and (B) integrated phrenic activity (PHR) were recorded as index of central inspiratory activity. For technical reasons, in younger animal phrenic activity could not be recorded, and it was replaced by rhythmical vagal output correlated with respiration before paralysis. Central apnea in 4-h-old and excitation of PHR in 24-day-old rabbits were recorded at the same Pto_2. During control, animals were ventilated with 50% O_2. Horizontal lines indicate duration of ventilation with 100% N_2 (A), and 10% O_2 (B). Arrows in A indicate beginning of phase I, II, and III of K^+_e increase. These phases can easily be distinguished in the older rabbit (B). Note overshoot in Pto_2 during recovery after hypoxia, suggesting a rapid increase in local blood flow. For further explanation, see text.

active transport mechanisms. The more rapid increase in aK^+_e during phase II implies a further, more pronounced, impairment of the medullary ionic homeostasis and therefore increased leakage of K^+ out of cells. The increased K^+ efflux may be associated with an inward movement of Na^+ during depolarization of the membrane and may be secondary to a decreased rate of neuronal metabolic processes due to low oxygen availability or increased release of excitatory amino acids during hypoxia (Hajos et al., 1986).

The presence of a saturation level in aK^+_e during low oxygen supply (Fig. 2) suggests that the mechanisms limiting the increase in aK^+_e above 12 mM in the adult brain (see above) are also active in newborns. The functional importance of such mechanisms probably increases with maturation since the highest aK^+_e values during hypoxia (11.8 mM, Fig. 2A) and in response to vagal stimulation during hypoxia (16 mM) were present in the youngest rabbits (Trippenbach et al., 1990). During hypoxia the maximal aK^+_e recorded in the NTS of 4-h to 28-day-old rabbits was higher by 1-5.95 mM (mean $aK^+_e = 6.8 \pm 2.3$ mM) than the mean control value. This hypoxia-related aK^+_e response seems to be more pronounced than the 1-1.5 mM increase from resting values recorded at tissue Po_2 of 0 mm Hg in the same area in adult cats (Acker and Richter, 1985). Despite the observation that the large increase in aK^+_e occurred in the youngest rabbits, age dependence of the maximal aK^+_e response to hypoxia was not obvious during postnatal development. It is possible that the difference in the maximal aK^+_e response to hypoxia in newborn rabbits and adult cats is related to the duration of the hypoxic exposure rather than to the animals' age (see above).

Although phrenic excitation was present during phase I of the increase in aK^+_e and apnea and/or gasp-like pattern of respiratory activity occurred always during phase II and phase III, our results also show that the control of the pattern of breathing did not depend only on the aK^+_e levels in the NTS of newborn rabbits. Note in Figure 2A that gasp-like inspiratory activity during hypoxia and respiratory excitation during recovery after hypoxia were recorded at the same aK^+_e. With the exception of the 21-day-old animal presented in Figure 2 (2B), two distinctly different patterns of PHR at the same aK^+_e during hypoxia and recovery were observed in all young rabbits tested. Therefore, it is likely that the aK^+_e increase is accompanied by activation of other mechanisms in the CNS during hypoxia, such as decrease in body metabolism or increased release of excitatory amino acids or of inhibitory neurotransmitters (see Section III). This is in agreement with the recent results of Michaelis et al. (1988), who indicated that the initial event in the neuromodulatory depressing action of adenosine is an increase in the membrane K^+ conductance leading to an increase in aK^+_e.

In contrast to the lack of observable developmental changes in the maximal response, the rate of increase in aK^+_e during hypoxia is greater in older animals. Durations of both phases decreased and their slopes increased between day 1 and day 28 (e.g., Fig. 2). The age dependence implies a better K^+ homeostasis in the immature medullary tissue than later in life. This conclusion is supported by age-dependent effects of hypoxia we observed on a K^+_e response to vagal stimulation (Trippenbach et al., 1990). The presence of aK^+_e response to vagal stimulation suggests that the ISM was located close to the inspiration-related neurons receiving vagal inputs (see Section II). Examples of this aK^+_e response in a 6-h-old rabbit are shown in Figure 3. In this and all other newborns, during control conditions (ventilation with 100% O_2) and hypoxia, aK^+_e rapidly increased at the beginning of stimulation and started to decline shortly after the maximal aK^+_e response was recorded. Group mean increases of 3.3 ± 1.7 mM and 2.8 ± 2.4 mM were recorded under control and hypoxic conditions, respectively. The maximal aK^+_e response was not affected by low oxygen levels, but the rate of decline (measured as a given decrease in aK^+_e for a unit of time) was lower during hypoxia (Fig. 3B) than control. The rate of aK^+_e decline was more affected in rabbits during their third week than on the first day of life. These results support the idea of greater resistance of the brain ionic homeostasis to lack of oxygen in the neonate.

It is well known that newborn mammals better tolerate hypoxia (Adolph, 1971) and that the immature brain tolerates anoxia longer than the adult (Fazekas et al., 1941). Slow changes in aK^+_e in the immature brain may be an expression of the high resistance to hypoxia early in postnatal life. The better ion homeostasis in the immature brain during hypoxia may reflect smaller demands for active ion transport and lower metabolic rate. The low metabolic rate of the brain tissue can be further reduced during hypoxia (Duffy et al., 1975). Moreover, the rate of K^+ release should be directly proportional to the surface of the somatodendritic membranes in relation to a unit volume of tissue. In the immature brain with small neurons, minimal arborization, and fewer synapses (Caley and Maxwell, 1968; Eayrs and Goodhead, 1959; Aghajanian and Bloom, 1967), the functional surface for ionic dissipation is smaller than in the adult. Finally, changes in aK^+_e may be minimized by a relatively large ECS during the perinatal period (Bondareff and Pysh, 1968; Van Harreveld, 1972). All these factors seem to compensate for the immaturity of the glial component early in life.

We observed inconsistent effects of hypoxia on aCa^{2+}_e in the respiration-related area of NTS in newborn rabbits.

In summary, levels of aK^+_e within the respiration-related areas in newborns are similar to or only slightly higher than those recorded in the same area in adult animals. On the other hand, the medullary aK^+_e values are lower than those reported in the plasma during the same period of development in several species. This suggests that the homeostatic mechanisms of

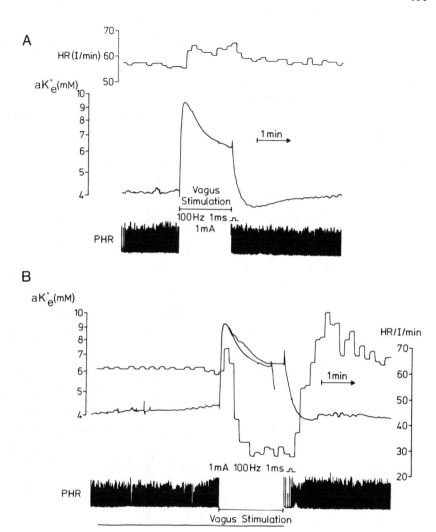

Figure 3 Effects of vagal stimulation on extracellular potassium activity (aK^+_e) recorded during control (A) and hypoxic (B) conditions in 6-h-old rabbit. PHR, Integrated phrenic activity; HR, heart rate. Duration and parameters of vagal stimulation are indicated in both A and B. Horizontal line indicates time of exposure to 10% O_2. (B) Profile of aK^+_e changes recorded during control conditions is superimposed on aK^+_e tracing during hypoxia. In this example, a 2-mM decrease from maximal response was achieved after 30 s under control conditions and after 50 s under hypoxic conditions. Note under both conditions undershoot of aK^+_e after discontinuation of vagal stimulation. (Modified from Trippenbach et al., 1990.)

the brain are already mature after birth, and together with a low permeability for cations in brain capillaries and active ion transport, the increased levels of K^+ in the plasma are not reflected in the ECS of the newborn brainstem. Slow changes in aK^+_e during hypoxia may contribute to the defense mechanisms protecting newborns against diminished availability of the energy substrate and may contribute to better adaptation to hypoxia of newborns than adults. Better ion homeostasis early in life may be explained by smaller demands for active ion transport in the immature brain.

References

Acker, H., and Richter, D. W. (1985). Changes in potassium activity in the extracellular space of inspiratory neurones within the NTS of cats. In *Neurogenesis of Central Respiratory Rhythm*. Edited by A. L. Bianchi and M. Denavit-Saubié. Hingham, MA, MTP Press, pp. 183-186.

Adolph, E. F. (1971). Physiological adaptation to hypoxia in newborn rats. *Am. J. Physiol.* 221:123-127.

Aghajanian, G. K., and Bloom, F. E. (1967). The formation of synaptic junctions in developing rat brain: A quantitative electron microscopic study. *Brain Res.* 6:716-727.

Aoki, M., Mori, S., Kawahara, K., Watanabe, H., and Ebata, N. (1980). Generation of spontaneous respiratory rhythm in high spinal cats. *Brain Res.* 202:51-63.

Aswal, S., Majcher, J. S., Vain, N., and Longo, L. L. (1980). Patterns of fetal lamb regional blood flow during and after prolonged hypoxia. *Pediatr. Res.* 14:1104-1110.

Baumgarten, R., and Kanzow, E. von (1958). The interaction of two types of inspiratory neurons in the region of the tractus solitarius of the cat. *Arch. Ital. Biol.* 96:361-373.

Berger, A. J. (1977). Dorsal respiratory group neurons in the medulla of cat: Spinal projections, responses to lung inflation and superior laryngeal nerve stimulation. *Brain Res.* 135:231-254.

Berger, A. J., Averill, D. B., and Cameron, W. E. (1984). Morphology of inspiratory neurons located in the ventrolateral nucleus of the tractus solitarius of the cat. *J. Comp. Neurol.* 224:60-70.

Berger, A. J., and Dick, T. E. (1987). Connectivity of slowly adapting pulmonary stretch receptors with dorsal medullary respiratory neurons. *J. Neurophysiol.* 58:1259-1274.

Bianchi, A. L., and Barillot, J. C. (1982). Respiratory neurons in the region of the retrofacial nucleus: Pontile, spinal and vagal projections. *Neurosci. Lett.* 31:277-282.

Biscoe, T. J., and Purves, M. J. (1967). Carotid body chemoreceptor activity in the newborn lamb. *J. Physiol. (London)* 190:443-454.

Bondareff, W., and Pysh, J. J. (1968). Distribution of extracellular space during postnatal maturation of rat cerebral cortex. *Anat. Rec.* 160: 773-780.

Bures, J., and Buresova, O. (1981). Cerebral [K$^+$] increase as an index of the differential susceptibility of brain structures to terminal anoxia and electroconvulsive shock. *J. Neurobiol.* 12:211-220.

Bystrzycka, E., Nail, B. S., and Purves, M. J. (1975). Central and peripheral neural respiratory activity in the mature sheep foetus and newborn lamb. *Resp. Physiol.* 25:199-215.

Caley, D. W., and Maxwell, D. S. (1968). An electron microscope study of neurons during postnatal development of the rat cerebral cortex. *J. Comp. Neurol.* 133:17-44.

Cartheuser, C. F. (1987). Progressive hypoxia until brain electrical silence: A useful model for studying protective interactions. *Can. J. Physiol. Pharmacol.* 66:1398-1406.

Coles, J. A., and Tsacopoulos, M. (1979). Potassium activity in photoreceptors, glial cells and extracellular space in the drone retina; changes during photostimulation. *J. Physiol. (London)* 290:525-549.

Cross, K. W., Tizard, J., and Trythall, D. (1958). The gaseous metabolism of the newborn infant breathing 15% oxygen. *Acta Paediatr. Scand.* 47:217-237.

Cserr, H. F. (1965). Potassium exchange between cerebrospinal fluid, plasma and brain. *Am. J. Physiol.* 209:1219-1226.

Cserr, H. F., and Bundgaard, M. (1984). Blood-brain interfaces in vertebrates: A comparative approach. *Am. J. Physiol.* 246 *(Regulatory Integrative Comp. Physiol.* 15):R277-R288.

Davson, H., and Welch, K. (1971). The permeation of several materials into fluids of the rabbit's brain. *J. Physiol. (London)* 218:337-351.

Dingledine, R., and Somjen, G. (1981). Calcium dependence of synaptic transmission in the hippocampal slice. *Brain Res.* 207:218-222.

Duffy, T. E., Kohle, S. J., and Vannucci, R. C. (1975). Carbohydrate and energy metabolism in perinatal rat brain: Relation to survival in anoxia. *J. Neurochem.* 24:271-276.

Duron, B., and Marlot, D. (1980). Nervous control of breathing during postnatal development in the kitten. *Sleep* 3:323-330.

Eayrs, J. T., and Goodhead, B. (1959). Postnatal development of the cerebral cortex in the rat. *J. Anat.* 93:385-402.

Euler, C. von, Hayward, J. N., Martilla, I., and Wyman, R. J. (1973). Respiratory neurones of the ventral nucleus of the solitary tract of the cat. Vagal input, spinal connections and morphological identification. *Brain Res.* **61**:1-22.

Fazekas, J. F., Alexander, F. A. D., and Himwich, H. E. (1941). Tolerance of the newborn to anoxia. *Am. J. Physiol.* **134**:281-287.

Friis-Hansen, B. (1957). Changes in body water compartments during growth. *Acta Paediatr.* **46**:Suppl. 110.

Hajos, F., Garthwaite, G., and Garthwaite, J. (1986). Reversible and irreversible neuronal damage caused by excitatory amino acid analogue in rat cerebellar slices. *Neuroscience* **18**:417-436.

Hansen, A. H., and Chernick, V. (1983). Development of respiratory control. *Physiol. Rev.* **63**:437-483.

Hansen, A. J. (1977). Extracellular potassium concentration in juvenile and adult rat brain cortex during anoxia. *Acta Physiol. Scand.* **99**:412-420.

Hansen, A. J. (1985). Effect of anoxia on ion distribution in the brain. *Physiol. Rev.* **65**:101-148.

Heinemann, U., and Lux, H. D. (1977). Ceiling of stimulation induced rises in extracellular potassium concentration in the cerebral cortex of cat. *Brain Res.* **120**:231-249.

Heinemann, U., Lux, H. D., and Gutnick, M. J. (1977). Extracellular free calcium and potassium during paroxysmal activity in the cerebral cortex of the cat. *Exp. Brain Res.* **27**:237-243.

Hossman, K. A., Sakaki, S., and Zimmermann, V. (1977). Cation activities in reversible ischemia of the cat brain. *Stroke* **8**:77-81.

Hubbard, J. I., and Willis, W. D. (1962). Reduction of transmitter output by depolarization. *Nature (London)* **193**:1294-1295.

Kalia, M. (1982). Anatomical organization of central respiratory neurones. *Annu. Rev. Physiol.* **43**:105-120.

Kalia, M., and Richter, D. W. (1985). Morphology of physiologically identified slowly adapting lung stretch receptor afferents stained with intra-axonal horseradish peroxidase in the nucleus of the tractus solitarius of the cat. I. A light microscopic analysis. *J. Comp. Neurol.* **241**:503-520.

Kirshner, H. S., Blank, W. F., Jr., and Myers, R. E. (1975). Brain extracellular potassium activity during hypoxia in cat. *Neurology* **25**:1001-1005.

Kraig, R. P., and Nicholson, G. (1978). Extracellular ionic variations during spreading depression. *Neuroscience* **3**:1045-1059.

Krnjevic, K., and Morris, M. E. (1972). Extracellular K^+ activity and slow potential changes in spinal cord and medulla. *Can. J. Physiol. Pharmacol.* **50**:1214-1217.

Krnjevic, K., and Morris, M. E. (1976). Input-output relation of transmission through cuneate nucleus. *J. Physiol. (London)* 257:791-815.

Krnjevic, K., Morris, M. E., and Reiffenstein, R. J. (1980). Changes in extracellular Ca^{2+} and K^+ activity accompanying hippocampal discharges. *Can. J. Physiol. Pharmacol.* 58:579-583.

Kuffler, S. W., Nicholls, J. G., and Orkand, R. K. (1966). Physiological properties of glial cells in the central nervous system in amphibia. *J. Neurophysiol.* 29:768-787.

LaFramboise, W. A., Guthrie, R. T., Standaert, T. A., and Woodrum, D. E. (1983). Pulmonary mechanics during the ventilatory response to hypoxemia in the newborn monkey. *J. Appl. Physiol. Respir. Environ. Exercise Physiol.* 49:319-325.

Lawson, E. E., and Long, W. A. (1983). The central origin of the biphasic breathing pattern during hypoxia in newborns. *J. Appl. Physiol.* 55: 483-488.

Lewis, D. V., and Schuette, W. H. (1976). NADH fluorescence, $[K^+]_o$ and oxygen consumption in cat cerebral cortex during direct cortical stimulation. *Brain Res.* 110:523-535.

Llinas, R. (1968). A possible mechanism for synaptic inhibition. In *Structure and Function of Inhibitory Neuronal Mechanisms.* Edited by C. von Euler B. Skoglund, and U. Söderberg. London, Pergamon Press, pp. 249-250.

Lothman, E., LaManna, J. Cordingley, G., Rosenthal, M., and Somjen, G. (1975). Responses of electrical potentials, potassium levels and oxidative metabolic activity of cerebral neocortex of cats. *Brain Res.* 88:15-36.

Mares, P., Kriz, N., Brozek, G., and Bures, J. (1976). Anoxic changes of extracellular potassium concentration in the cerebral cortex of young rats. *Exp. Neurol.* 53:12-20.

Merrill, E. G. (1981). Where are the real respiratory neurons? *Fed. Proc.* 40:2389-2394.

Michaelis, M. L., Johe, K. K., and Moghadam, B. (1988). Studies of the ionic mechanism for the neuromodulatory actions of adenosine in the brain. *Brain Res.* 473:249-260.

Morris, M. E. (1974). Hypoxia and extracellular potassium activity in the guinea-pig cortex. *Can. J. Physiol. Pharmacol.* 52:872-882.

Mutani, R., Futamachi, K. J., Prince, D. A., (1974). Potassium activity in immature cortex. *Brain Res.* 75:27-39.

Nicholson, C. (1980). Dynamics of the brain cell microenvironment. *Program Bull.* 18:177-322.

Ransom, R. R., Yamate, C. L., and Connors, B. W. (1985). Activity-dependent shrinkage of extracellular space in rat optic nerve: A developmental study. *J. Neurosci.* 5:532-535.

Richter, D. W., Camerer, H., and Sonnhof, U. (1978). Changes in extracellular potassium during the spontaneous activity of medullary respiratory neurons. *Pflügers Arch.* **376**:139-149.

Skoff, R. P., Price, D. L., and Stocks, A. (1976). Electron microscopic autoradiographic studies of gliogenesis in rat optic nerve. *J. Comp. Neurol.* **169**:313-323.

Skoglund, S. (1967). Plasma ion concentration as a basis for an hypothesis regarding neuronal excitability changes during development. *Acta Soc. Med. Upsal.* **72**:76-84.

Sokoloff, L., Reivich, M., Kennedy, C., Des Rosiers, M. H., Patlak, C. S., Pettigrew, K. D., Sakurada, O., and Shinohara, M. (1977). The [^{14}C] deoxyglucose method for the measurement of the local cerebral glucose utilization: Theory, procedure, and normal values in the conscious and anesthetized albino rat. *J. Neurochem.* **28**:897-916.

Somjen, G. (1979). Responses of extracellular calcium activity ($[Ca^{2+}]_o$) during afferent nerve stimulation and during seizures in spinal cord of cats. *Neurosci. Abstr.* **5**:730.

Somjen, G. (1980). Responses of extracellular calcium activity in spinal cord compared to those in the cerebral cortex. *J. Neurophysiol.* **44**: 617-632.

Trippenbach, T., Richter, D. W., and Acker, H. (1987). Potassium (aK^+_e) and calcium (aCa^{2+}_e) activities in extracellular space of the dorsal medullary respiratory neurons in newborn rabbits. *Fed. Proc.* **46**:1419 (Abstract).

Trippenbach, T., Richter, D. W., and Acker, H. (1988). Extracellular potassium (aK^+_e) and calcium (aCa^{2+}_e) activities during normoxia and hypoxia in newborn rabbits. In *Hypoxia: The Tolerable Limits.* Edited by J. R. Sutton, C. S. Houston, and G. Coates. Benchmark Press, Indianapolis (Abstract).

Trippenbach, T., Richter, D. W., and Acker, H. (1990). Hypoxia and ion activities within the brain stem of newborn rabbits. *J. Appl. Physiol.* **68**:2494-2503.

Urbanics, R., Leninger-Follert, E., and Lübbers, D. W. (1978). Time course changes of extracellular H^+ and K^+ activities during and after direct electrical stimulation of the brain cortex. *Pflügers Arch.* **378**: 47-53.

Van Harreveld, A. (1972). The extracellular space in the vertebrate central nervous system. In *The Structures and Function of Nervous Tissue.* Vol. 4. Edited by G. H. Bourne. New York, Academic Press, pp. 447-511.

Widdowson, E. M., and McCance, R. A. (1956). The effect of development on the composition of the serum extracellular fluids. *Clin. Sci.* **15**: 361-365.

22

Hypoxia and Neuropharmacology of Breathing

HUGO LAGERCRANTZ and MICHAEL RUNOLD

Karolinska Hospital and Karolinska Institute
Stockholm, Sweden

I. Introduction

The overall ventilatory response to hypoxia is the net result of summation of several antagonistic factors (Gautier, 1986). Hypoxia can exert its effect by both excitatory and inhibitory mechanisms to the respiratory controller. The net effect on minute ventilation depends on several factors, but of greatest importance are the PO_2 level and postconceptual age of the subject.

In adults, ventilation increases in response to hypoxia; i.e., in order to increase oxygen availability the organism reacts with enhanced respiratory motor output. In addition, a number of sympathoadrenal responses (Spyer, 1988), such as tachycardia, hypertension, increased muscle blood flow, sweating, and piloerection (the fight-and-flight response), accompany the hypoxic response. However, during more sustained and severe hypoxia other mechanisms are also brought into action. Thus, a decline in ventilation can be seen after the initial ventilatory increase (Vizek et al., 1987; Weiskopf and Gabel, 1975; Weil and Zwillich, 1976).

The mammalian fetus reacts to hypoxia quite differently than the adult, while the infant exhibits an intermediate reaction. The fetal response to

hypoxia is characterized by respiratory arrest, bradycardia, decreased muscle blood flow, and a reduction in fetal movements (Dawes, 1984). Metabolism is probably immediately reduced. This response, which to some extent can be regarded as a "paralysis reflex," might be purposeful to save
oxygen. A similar, but less pronounced response can be seen in some newborn animals, including preterm human infants, i.e., a marked biphasic
ventilatory response accompanied by a decrease in metabolism (Bonora
et al., 1984; Bryan et al., 1986; Cross and Oppé, 1952, Hill, 1959).

It has been clearly demonstrated that the peripheral chemoreceptors
mediate the initial ventilatory increase during hypoxia, in both adults and
infants (Watt et al., 1943; Cherniack et al., 1971; Schwieler, 1968). On the
other hand, the mechanisms causing the subsequent decrease in breathing
during prolonged hypoxia are not known.

Hypoxia influences the control of breathing at various levels. Hypoxia
has direct effects on central respiratory neurons (see Chapter 21, this volume)
and chemoreceptors (see Chapter 11, this volume). Hypoxia also has indirect
effects on breathing by affecting metabolism and temperature control and
cardiovascular mechanisms.

Four main hypotheses have been suggested: (1) failure of the peripheral chemoreflex either at the central or peripheral level (McCooke and Hanson, 1985; Bureau et al., 1984); (2) a change in pulmonary mechanics during hypoxia, which results in a reduced ventilation (Woodrum et al., 1981);
(3) reduced metabolism (Hill, 1959), which indirectly leads to a decrease in
respiration through a reduced oxygen consumption and CO_2 production;
(4) hypoxic depression of respiration-related areas in the medulla oblong-
ata-pons, either by a direct effect (Cross and Oppé, 1952) or via inhibitory
influences from structures above the medulla, presumably at a suprapontine
level (Dawes et al., 1983; Gluckman and Bennet, 1986).

All these proposed mechanisms may involve a large number of neuroactive agents, both excitatory and inhibitory, on the respiratory controller
during hypoxia. The substances can be released from nerve terminals and
cause an instantaneous switch on/switch off effect, like the classical *neurotransmitters.* Others may act as the so-called *neuromodulators* by changing,
for example, the effect of another neurotransmitter. A third category of
neuroactive agents, circulating *neurohormones,* act on target tissues at
some distance from the site of release.

II. Neuroactive Agents in the Regulation
of Breathing

To establish the role of a neuroactive agent in the control of ventilation
during hypoxia, certain research strategies can be employed (Lagercrantz,
1987; Moss and Inman, 1989).

A. Release of Neuroactive Agent in Respiratory-Related Structures During Hypoxia

Release of a neurotransmitter or neuromodulator during hypoxia can be studied by using the microdialysis technique (Ungerstedt, 1984). A dialysis probe is inserted stereotaxically into respiration-related nuclei. However, the resolution of the technique is hampered by the size of the probes. Presently, the smallest diameter is about 0.3 mm. Another way of measuring the release of neuroactive agents has recently been described (Duggan and Hendry, 1986; Hendry et al., 1988). A glass microelectrode is coated with fluorescence-labeled antibodies against specific neuropeptides such as substance P. One advantage of this technique is the smaller size compared to the microdialysis probe.

Besides more conventional biochemical and histochemical methods, the mRNA expression of a number of neuroactive agents can be studied, either by in situ hybridization or by Northern blot technique. These techniques may prove useful in studying the effect of sustained hypoxia on different neuromodulators, such as the neuropeptides.

B. Administration of Neuroactive Agents

Administration can be made locally using stereotaxically directed microsyringes or micropipettes (pressure injection or iontophoresis). Other routes of administration include the cerebroventricular system or systemic injection. It is important to consider the site of injection, since local and systemic injections can produce different results (Millhorn, 1986).

C. Blocking the Hypoxic Ventilatory Response with a Specific Antagonist

This classical pharmacological method requires the availability of very specific antagonists that do not interact with other systems. Presently, many neuromodulators lack specific antagonists.

D. Effects of Augmenting or Depleting Endogenous Concentrations of a Putative Neuroactive Agent on Respiratory Control

Herein lies the same problem of finding a specific drug that can stimulate or block the synthesis of a neuroactive agent.

III. Respiratory-Controlling Structures Affected by Hypoxia

A. Brainstem

The neurochemical anatomy of the brainstem structures controlling respiration during development is not known in detail (see Chapter 1, this volume).

Spontaneously firing neurons assumed to be respiratory rhythm generators have been identified in the rostral ventrolateral medulla in neonatal brainstem/ spinal cord preparations of rats (Onimaru et al., 1987; Onimaru and Homma, 1987). Respiratory rhythm activity recorded from the C_4 root disappeared after electrolytic lesion of this region. The area seems to correspond to the nucleus rostroventrolateralis in the adult rat, which contains adrenergic neurons. The effect of hypoxia on the neural activity of this region is not known.

Respiratory drive inputs from peripheral chemoreceptors and sensory afferents from the lung are assumed to be integrated in the dorsal respiratory group, the nucleus tractus solitarius (nTS). Chemical and nonchemical drive inputs seem to converge at the region of the nucleus paragigantocellularis lateralis (von Euler, 1986; Budzinska et al., 1985).

The immaturity of the respiratory network is probably the main cause of the instability of breathing of, for example, the preterm infant (Bryan et al., 1986). The maturity of the neuronal pathways as a cause of neonatal apnea is corroborated by the finding that the number of apneas is correlated to a slow conduction time in the brainstem as measured by evoked auditory responses (Henderson-Smart et al., 1983). It is possible that this labile control system during early development is more easily disturbed by hypoxia than in the adult. In this respect it should be pointed out that in order to elicit an action potential in phrenic motor neurons, a relatively high synaptic input is needed (Sears, 1977).

Instability of infant breathing may also be related to the inhibitory neurotransmitters/neuromodulators which seem to dominate in the fetal and neonatal central nervous system. Inhibitory amino acids such as GABA and taurine occur in relatively higher concentrations in the mammalian brain during the perinatal period than in the adult, while an excitatory amino acid like aspartate occurs in relatively low concentrations in infants (Man et al., 1987). The dominance of inhibitory neurotransmitters is further supported by the observation of more flat (mainly containing inhibitory neurotransmitters) than spherical (mainly excitatory) synaptic vesicles in the spinal cord of the kitten as compared with the adult cat (Arvidsson et al., 1987).

After birth there might be a major change in the balance of various neuroactive agents, possibly related to the sudden increase in PO_2. Using the Northern blot technique, we recently analyzed mRNA for substance P, cholecystokinin, and somatostatin, in discrete respiration-related areas of the rabbit pup (Lagercrantz et al., 1989). There was a sudden enhancement in the expression of these neuropeptides on the day of birth. This pattern was not seen in the striatum, which was used as a hybridizing control. It would be of considerable interest to see how short- and long-term hypoxia can affect the expression of various neuroactive agents in the respiratory-related areas.

B. Suprapontine Inputs to the Brainstem Respiratory Controller

The depressive effect of hypoxia on fetal respiration can be reversed by mid-collicular transection (Dawes et al., 1983). Further studies have revealed that there might be a ventrolateral area in the pons, which is responsible for this inhibition (Gluckman and Johnston, 1987). By lesioning the area in the fetal sheep, hypoxia elicited stimulation of breathing. Following the lesion, breathing was also more continuous. The neurotransmitters exerting this inhibitory effect have not been identified.

There is also a suprapontine stimulatory area enhancing ventilation during hypoxia. This effect of hypoxia is mainly seen in carotid-body denervated adult animals or during hyperoxic hypoxia, and it appears that the hypothalamus is important in eliciting this response. After a decrease in ventilation with hypoxia, carotid-sinus-denervated cats get tachypneic. This response could be modified by catecholamines and theophylline, which caused slowing of breathing (Gautier and Bonora, 1980, 1983). To what extent this mechanism is operative during hypoxia in the fetus and the neonate is not known.

C. Peripheral Chemoreceptors

The influence of the peripheral chemoreceptors on respiratory drive in the fetus is somewhat controversial (Blanco et al., 1984). The chemoreceptors of the fetal sheep have a low firing rate at PO_2 of 20-24 torr, i.e., the normal PO_2 of the fetus (Blanco et al., 1984). In the first few days after birth the sensitivity of the chemoreceptors is "reset" to a higher PO_2 level.

The neurochemical mechanism behind this resetting of the peripheral chemoreceptors is not clear; however, several lines of evidence indicate that substance P and dopamine might be of particular interest. Substance P has been suggested to mediate the hypoxic drive, at both a peripheral and central level. This is based on the findings that substance P receptor antagonists could block the hypoxia-induced increase in chemoreceptor discharge (Prabhakar et al., 1984), and that local application of these antagonists onto the IV ventricle could block the increase in ventilation (Yamamoto and Lagercrantz, 1985). The concentrations of substance P are very low in the peripheral chemoreceptors in the newborn kitten (Scheibner et al., 1988). A few weeks after birth substance P can be detected in significant concentrations in the glomus cells of the kitten and the human.

Dopamine is well known to modulate peripheral chemoreceptor input. In many species dopamine inhibits chemoreceptor discharge. We recently found relatively high dopamine levels in the carotid bodies of fetal and new-born rats. We also found a higher dopamine turnover in 0-6-h-old rat pups than in older ones. The emergence of the carotid chemoreflex was preceded by a rapid decrease in dopamine turnover in the carotid bodies (Hertzberg et et al., 1989).

D. The Sympathoadrenal System

During asphyxia high concentrations of catecholamines are released both in the fetal sheep (Jones and Robinson, 1975) and in the newborn infant (Lager-crantz and Bistoletti, 1977). These catecholamines originate to a large extent from the adrenal medulla and paraganglia (Jones et al., 1987). The sympa-thetic nervous system is probably not completely developed at birth, at least not in the rat and possibly not in the human (Lagercrantz and Slotkin, 1986). Noradrenaline is the predominant catecholamine released in the human in-fant during asphyxia (85%).

Although all types of adrenergic receptors are developed at an early stage, there seems to be some dominance of α_2-adrenergic receptors, possibly explaining parts of the inhibitory actions of catecholamines on, for example, lipolysis (Marcus, 1988).

E. Tissue Neuroregulatory Systems

Hypoxia can also affect control of breathing by releasing neuroactive agents or neurohormones from tissue in general, which also can influence the control of respiration. Particularly, the effects of adenosine and prostaglandins have been discussed in relation to perinatal asphyxia.

Adenosine is a purine metabolite that is present throughout the body. It can be formed by dephosphorylation of AMP, a reaction that is mainly catalyzed by 5-nucleotidase. Another potential source of adenosine is through the hydrolysis of *S*-adenosylhomocysteine, but the former mechanism is probably the most important during hypoxia. The formation and release of adenosine is due to an imbalance between metabolic activity and oxygen supply (Berne, 1986). Thus during hypoxia adenosine levels increase, for ex-ample, in the brain (Zetterström et al., 1982; Winn et al., 1981) and in plasma (Moss et al., 1987). An increase in metabolic activity or a decrease in sub-strate supply to the cell may also lead to an increase in adenosine formation. Adenosine is rapidly eliminated, with a half-life in plasma or less than 10 s (Ontyd and Schrader, 1984).

The adenosine concentration in the plasma of newborn infants is two to threefold higher than in resting adults (Irestedt et al., 1989). The adeno-sine metabolite hypoxanthine is inversely correlated to pH in the arterial blood of the newborn infant. In fact, hypoxanthine seems to be a good quantitative biochemical marker of perinatal asphyxia (Saugstad, 1975; Thiringer, 1982).

Prostanoids are usually not regarded as circulating neuroactive agents in the adult. However, relatively high amounts of prostaglandin E2 are formed in the placenta and transferred to the fetus, particularly during late gestation of the fetal sheep (Thorburn and Rice, 1987). This prostaglandin has been assumed to inhibit fetal breathing movements, an inhibition that is relieved immediately at birth (Kitterman et al., 1983). Whether hypoxia can increase plasma prostaglandin level is unclear. Asphyxia in neonatal guinea pig was found to increase brain levels of prostanoids (Allen et al., 1982). Acute hypoxemia caused a twofold increase in cerebral levels of PGE2 in beagle puppies (Ment et al., 1986).

IV. Fetal Hypoxia

The present knowledge of the fetal responses to hypoxia is nearly exclusively based on studies of the chronically instrumented fetal sheep. When we discuss the effects of hypoxia on the respiratory control in the fetus, we have to consider its prevailing oxygen levels. The arterial PO_2 is normally about 20-24 mm Hg in the fetal sheep, which would be considered severe hypoxemia if it occurred postnatally. Although the fetus is not hypoxic, it is possible that even a relatively mild hypoxic challenge may cause different neurochemical effects as compared with postnatal conditions.

GABA, or γ-aminobutyric acid, has been proposed to cause the hypoxia-induced ventilatory depression. Central administration of GABA and GABA analogs markedly inhibit fetal breathing. However, since hypoxia can cause inhibition of breathing also during infusion of the GABA antagonist picrotoxin (Johnston and Gluckman, 1983), it is unlikely that GABA is the main factor mediating the inhibition of fetal breathing during hypoxia.

Excitatory amino acids of the brain increase during hypoxia. With the microdialysis probe, the concentrations of glutamate and aspartate were shown to be substantially elevated during perinatal asphyxia (Hagberg et al., 1987). These amino acids could stimulate breathing (Foutz et al., 1988). However, the respiratory controller may be so inhibited when these substances are released that the depression cannot be overcome.

Endogenous opioids have been suggested to be involved in the respiratory depression of the fetal sheep via both μ- and δ-receptors (Moss and Scarpelli, 1984; Moss and Inman, 1989). β-Endorphin has been found to decrease fetal breathing movements (McMillen and Walker, 1986). Plasma β-endorphin concentrations increase during hypoxia in the fetal sheep (Stark et al., 1986). The effects of the opioid receptor antagonist naloxone on the ventilatory response are conflicting (Moss et al., 1986; Gluckman and Bennet, 1986). Furthermore, it can be noted that naloxone enhances catecholamine release during fetal hypoxia, which complicates the interpretation (Padbury et al., 1987).

Catecholamines are also released in high concentrations during fetal asphyxia. Although β-adrenergic stimulation has been reported to increase fetal breathing movements in sheep (Jansen et al., 1986), this effect does not seem to overcome the hypoxic depression. Adrenergic drugs produce continuous breathing, but do not affect the hypoxia-induced respiratory arrest (Joseph and Walker, 1989). It is possible that centrally released catecholamines inhibit breathing by acting on α_2-receptors. The α_2-receptor subtype dominates during early development and has major inhibitory actions (Marcus, 1988). The specific α_2-receptor agonist clonidine has been reported to depress fetal breathing (Bamford et al., 1986).

Adenosine is the neuroactive agent that fulfills most of the criteria to be responsible for the hypoxia-induced inhibition of breathing. It is formed during fetal asphyxia and it causes inhibition of fetal breathing (Szeto and Umans, 1984). Furthermore, adenosine has recently been found to affect fetal eye movement and electrocorticogram in a similar way as hypoxemia, elicited by administration of 9% O_2 to the ewe (Koos and Matsuda, 1990). However, if adenosine is responsible for the hypoxic inhibition, it is not clear how this can be reconciled with the observation that midcollicular transection removes the inhibition.

Prostaglandins. Prostaglandin E_2 inhibits fetal breathing (Kitterman et al., 1983) without altering sleep state. Administration of prostaglandin synthetase inhibitors, such as indomethacin or meclofenemate, was also found to stimulate fetal breathing (Kitterman et al., 1979), but there was no evidence that it affected the inhibition of breathing induced by hypoxia (see Gluckman and Bennet, 1986).

V. Infant Hypoxia

The transition at birth is associated with major changes of the expression and release of neuroactive agents. There is a surge of catecholamines and neuropeptides, as indicated by plasma analysis (Lagercrantz and Slotkin, 1986; Marchini et al., 1988), and a sudden expression of some neuropeptides in the brainstem (Lagercrantz et al., 1989)—events that might substantially affect respiratory control. There is also a rapid decrease in the synthesis of neurotransmitters, for example dopamine, in the peripheral chemoreceptors (Hertzberg et al., 1989). The neurochemical transition at birth is probably associated with the rise of the arterial P_{O_2}. Hypoxia in the neonate might therefore affect the control of breathing in different ways as compared to the fetus. This is reflected in the resetting of the peripheral chemoreceptors which occurs after a few of days in the lamb (Blanco et al., 1984) and in the human infant (Hertzberg and Lagercrantz, 1987).

During neonatal hypoxia there is a substantial release of a number of neuroactive agents that can affect respiratory control (Table 1). However, the PO_2 level at which a substantial release of neuroactive agents can be measured varies; i.e., different neuroactive agents appear to have different PO_2 thresholds.

Of the neuroactive amino acids *GABA* has been studied most extensively. The turnover of GABA is increased during neonatal asphyxia and higher concentrations of GABA were found in the cerebrospinal fluid of asphyxiated infants (Hedner et al., 1982). GABA has been found to decrease breathing in the newborn guinea pig (Wennergren and Wennergren, 1980). However, evidence that GABA antagonists can block the depressive phase of ventilation during hypoxia has not been published.

Table 1 Some Neuromodulators Released During Hypoxia Which Inhibit Respiration

Neuro-modulator	Release during hypoxia	Respiratory inhibition	Antagonist blocking the hypoxic depression	Ref.
Enkephalin	+ +	+ + +	Yes	Moss and Inman, 1989
Somatostatin	+ +	+ + +	No antagonist available	Yamamoto et al., 1988
GABA	+ +	+ +	No	Johnston and Gluckman, 1983; Wennergren and Wennergren, 1980
Prostaglandin E2	+ +	+ +	No	Kitterman et al., 1979; Ment et al., 1986
Adenosine	+ + +	+ + +	Yes	Runold et al., 1986, 1989; Zetterström et al., 1982

Met-enkephalin has been suggested to cause hypoxic ventilatory depression of the neonate. A slight increase of plasma enkephalin was observed in hypoxic piglets (Moss et al., 1987). Enkephalin (δ-receptor agonists) analogs markedly depress breathing, but they also decrease metabolism (Schaeffer and Haddad, 1985). A number of studies indicate that naloxone or naltrexone can block the ventilatory depression in rabbit pups (Grunstein et al., 1981) and piglets (Moss et al., 1987) and possibly also in the human baby (De Boeck et al., 1984). However, a relatively high concentration of naloxone had to be used since it is mainly a μ-receptor antagonist and the respiratory depression is to a considerable extent mediated by μ-receptors (1 mg/kg). Further, its interaction with other neuroactive agents like catecholamines has to be considered (see above).

Dopamine, which was found to stimulate fetal breathing, seems to be involved in the ventilatory response to hypoxia. After giving a combined D1 and D2 antagonist [*cis*(z)-flupentixol] to rabbit pups, we found that the initial stimulatory effect of hypoxia was markedly enhanced and the secondary depressive effect was abolished. Specific D2 antagonists (sulpiride and domperidone) were found to potentiate the first phase only, while a specific D1 antagonist (SCH 23390) also effectively blocked the hypoxic depression. Furthermore, the dopamine agonist apomorphine caused a biphasic hypoxic ventilatory response in decerebrate rabbit pups, which normally lack this latter drop in ventilation (Srinivasan et al., 1989). Preliminary results from our laboratory, using the microdialysis probe, have shown that hypoxia causes an increase of extracellular dopamine in the adult rabbit, an effect that was abolished by peripheral chemodenervation. Thus, hypoxia might activate suprapontine dopaminergic neurons via the peripheral chemoreceptors, which in turn inhibit medullary respiratory neurons. In the adult this depressive effect should normally be overcome by the stimulatory effect. This may not be the case in the newborn infant. The idea of some hypoxic inhibitory input via the peripheral chemoreceptors is further supported by the findings that respiratory depression could be acutely produced by activation of the peripheral chemoreceptors with sodium cyanide during sustained hypoxia in rabbit pups (Schramm and Grunstein, 1987).

Adenosine is probably released in significant amounts first during severe hypoxia (e.g., 6% O_2) in the piglet (Moss et al., 1987). The adenosine analog PIA decreased ventilatory minute volume mainly by affecting the frequency. This effect was more marked in younger than in older animals (Runold et al., 1986). The effect was also more pronounced when the animals were studied in cooler environments. The hypoxia-induced ventilatory inhibition could be blocked with theophylline (Runold et al., 1989). Further evidence for the role of adenosine at this PO_2 level is the finding that increased endogenous concentrations of adenosine caused by administration of dipyridamole produce a more severe ventilatory depression (Fig. 1).

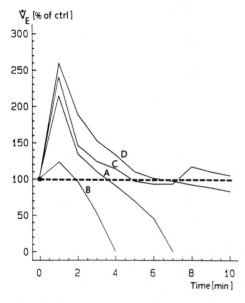

Figure 1 Ventilatory response to hypoxia (6% O_2 in N_2) in 4-day-old urethan-anesthetized rabbit pups. Curve A shows the "normal" response with initial stimulation of breathing and then a depressive phase. Curve B shows the effect after pretreatment with dipyridamole, which is supposed to inhibit the uptake of endogenously released adenosine. Curves C and D show the effect of aminophylline and 8-psδ theophylline, which are supposed to antagonize the released adenosine. Dotted line indicates the prehypoxia control level. (From Runold et al., 1989.)

Thus, adenosine fulfills all four criteria required (see above) to be involved in the ventilatory responses to hypoxia in the newborn, provided the asphyxia is severe enough. It is more doubtful that adenosine is responsible for the biphasic response during mild hypoxia.

The action of adenosine is complex. It seems to cause a direct inhibitory effect on respiratory neurons in the brainstem. Besides the direct respiratory inhibitory effects, adenosine causes a general depression (Berne, 1986), sedation, and lowering of the metabolic rate and body temperature in the newborn (Runold et al., 1986). This further supports the idea that adenosine might be involved in perinatal asphyxia, which is characterized by apnea, sedation, hypotonia, and hypothermia, in spite of the release of excitatory amino acids and catecholamine surge (Lagercrantz and Slotkin, 1986). However, the inhibitory effect on breathing in the newborn is not seen during adult life. Thus, in adult humans (Maxwell et al., 1986) and animals (Runold et al., 1988) adenosine stimulates respiration. This stimulatory effect is mediated both by peripheral chemoreceptors (McQueen and Ribeiro, 1983; Runold et al., 1990) and by vagal pulmonary afferents (Runold et al., 1988).

VI. Clinical Aspects

Some hints about the roles of various neuroactive agents in respiratory control during hypoxia have been obtained by clinical studies. Apneic infants have been reported to have higher concentrations of GABA in cerebrospinal fluid (Hedner et al., 1982). Higher dopamine concentrations have been found in the carotid bodies of victims of the sudden infant death syndrome (SIDS) (Perrin et al., 1984). A remarkably increased endorphin concentration was found in one infant who later died of hypoventilation (Brandt et al., 1980), but there was no significant difference in met-enkephalin content in various brain parts of SIDS and control infants (Bergström et al., 1984). On the other hand, significantly higher substance P concentrations were found in the medulla of the SIDS infants, possibly suggesting increased synthesis due to preceding chronic hypoxia or lower turnover.

Infants who succumbed to SIDS were also found to have increased concentrations of the adenosine metabolite hypoxanthine in the vitreous body of the eye, although the control group consisted mainly of adults (Rognum et al., 1988). These findings also suggest that SIDS is preceded by chronic hypoxia, which causes release of inhibitory neuroactive agents.

Neonatal apnea, particularly in preterm infants, can be treated by blocking the effect of inhibitory neuroactive agents. The most effective agents, which are also the most widely used, are the xanthine derivatives, such as theophylline and caffeine (Aranda and Turmen, 1979; Kuzemko and Paala, 1973), which block adenosine receptors (Fredholm, 1980). It is not clear whether endogenous adenosine causes apnea by direct effects on inhibitory A1 receptors on the central respiratory neurons or at a presynaptic level, inhibiting the release of excitatory neurotransmitters such as the excitatory amino acids (Fredholm and Dunwiddie, 1988). The latter effect seems likely and can perhaps explain why the xanthines produce such a general excitation of the infant, involving jitteriness and other effects.

VII. Conclusions

The net effect of hypoxia on breathing is the result of several antagonistic inputs. A number of excitatory and inhibitory neurotransmitter/neuromodulator systems are involved at various levels in the control of breathing. For example, during hypoxia neuroactive agents can affect breathing by direct effects and also indirectly by effects on metabolism and temperature regulation. The respiratory control system during early development is relatively labile. The immaturity of the organization of neuronal network and the apparent dominance of inhibitory synapses in the fetus and newborn make the respiratory control system less robust than in the adult. There-fore, it is possible that inhibitory neuroactive agents that are released during hypoxia, for example, can more easily produce inhibition of breathing dur-ing the perinatal period than later in life.

References

Allen, L. G., Louis, T. M., and Kopelman, A. E. (1982). Brain prostaglan-dins E2 and F2 following neonatal asphyxia in the guinea pig. *Biol. Neonate* **42**:8-14.

Aranda, J. V., and Turmen, T. (1979). Methylxanthines in apnea of pre-maturity. *Clin. Perinatol.* **6**:87-108.

Arvidsson, U., Svedlund, J., Lagerbäck, P-Å, and Cullheim, S. (1987). An ultrastructural study of the synaptology of γ-motoneurons during the postnatal development in the cat. *Dev. Brain Res.* **37**:303-312.

Bamford, O. S., Dawes, G. S., Denny, R., and Ward, R. A. (1986). Effects of the alpha 2-adrenergic agonist clonidine and its antagonist idazoxan on the fetal lamb. *J. Physiol. (London)* **381**:29-37.

Bergström, L., Lagercrantz, H., and Terenius, L. (1984). Post-mortem analyses of neuropeptides in brains from sudden infant death victims. *Brain Res.* **323**:279-285.

Berne, R. M. (1986). Adenosine: An important physiological regulator. *News In Physiological Sciences* **1**:130-134.

Blanco, C. E., Dawes, G. S., Hanson, M. A., and McCooke, H. B. (1984). The response to hypoxia of arterial chemoreceptors in fetal sheep and new-born lambs. *J. Physiol. (London)* **351**:25-37.

Bonora, M., Marlot, D., Gautier, H., and Duron, B. (1984). Effect of hy-poxia on ventilation during postnatal development in conscious kit-tens. *J. Appl. Physiol.* **56**:1464-1471.

Brandt, N. J., Terenius, L., Jacobsen, B. B., Klinken, L., Nordius, Å, Brandt, S., Blegvady, N., and Yssing, N. (1980). Hyperendorphin syndrome in a child with necrotizing encephalomyelopathy. *N. Engl. J. Med.* **303**: 914-916.

Bryan, A. C., Bowes, G., and Maloney, J. E. (1986). Control of breathing in the fetus and the newborn. In *Handbook of Physiology. The Respira-tory System.* Edited by N. S. Cherniack and J. G. Widdicombe. Bethesda, MD, American Physiological Society, pp. 621-647.

Budzinska, K., Euler, C. von, Kao, F. F., Pantaleo, T., and Yamamoto, Y. (1985). Effects of graded focal cold block in rostral areas of the med-ulla. *Acta Physiol. Scand.* **124**:329-340.

Bureau, M. A., Zinman, R., Foulon, P., and Begin, R. (1984) Diphasic ventil-atory response to hypoxia in newborn lambs. *J. Appl. Physiol.* **56**:84-90.

Cherniack, N. S., Edelman, N. H., and Lahiri, S. (1971). Hypoxia and hypercapnia as respiratory stimulants and depressants. *Respir. Physiol.* **11**:113-126.

Cross, K. W., and Oppé, T. E. (1952). The effect of inhalation of high and low concentrations of oxygen on the respiration of the premature infant. *J. Physiol. (London)* **117**:38-55.

Dawes, G. S. (1984). The central control of fetal breathing and skeletal muscle movements. *J. Physiol. (London)* **346**:1-18.

Dawes, G. S., Gardner, W. N., Johnston, B. M., and Walker, D. W. (1983). Breathing in fetal lambs: The effect of brain stem section. *J. Physiol. (London)* **335**:535-553.

De Boeck, C., Van Reempts, P., Rigatto, H., and Chernick, V. (1984). Naloxone reduces decrease in ventilation induced by hypoxia in newborn infants. *J. Appl. Physiol.* **56**:1507-1511.

Duggan, A. W., and Hendry, I. A. (1986). Laminar localization of the sites of release of immunoreactive substance P in the dorsal horn with antibody-coated microelectrodes. *Neurosci. Lett.* **68**:134-140.

Euler, C. von (1986). Brain-stem mechanisms for generation and control of the breathing pattern. In *Handbook of Physiology. The Respiratory System.* Edited by N. S. Cherniack and J. G. Widdicombe. Bethesda, MD, American Physiology Society, pp. 1-67.

Foutz, A. S., Champagnat, J., and Denavit-Saubié, M. (1988). N-methyl-D-aspartat (NMDA) receptors control respiratory off-switch in cat. *Neurosci. Lett.* **87**:221-226.

Fredholm, B. B. (1980). Are methylxanthine effects due to antagonism of endogenous adenosine? *Trends Pharmacol. Sci.* **1**:129-132.

Fredholm, B. B., and Dunwiddie, T. V. (1988). How does adenosine inhibit transmitter releases. *Trends Pharmacol. Sci.* **9**:130-134.

Gautier, H. (1986). Hypoxic enigma: Central effects of hypoxia on ventilatory and thermoregulatory control. In *Neurobiology of the Control of Breathing.* Edited by C. von Euler and H. Lagercrantz. New York, Raven Press, pp. 19-25.

Gautier, H., and Bonora, M. (1980). Possible alterations in brain monoamine metabolism during hypoxia-induced tachypnea in cats. *J. Appl. Physiol.* **49**:769-777.

Gautier, H., and Bonora, M. (1983). Ventilatory response of intact cats to carbon monoxide hypoxia. *J. Appl. Physiol.* **55**:1064-1071.

Gluckman, P. D., and Bennet, L. (1986). Neuropharmacology of fetal and neonatal breathing. In *Respiratory Control and Lung Development in the Fetus and Newborn. The Scientific Basis of Clinical Practice.* Reproductive and Perinatal Medicine. Vol. II. Edited by B. M. Johnston and P. D. Gluckman, Ithaca, NY, Perinatology Press, pp. 249-277.

Gluckman, P. D., and Johnston, B. M. (1987). Lesions in the upper lateral pons abolish the hypoxic depression of breathing in unanaesthetized fetal lambs in utero. *J. Physiol. (London)* 382:373-383.

Grunstein, M. M., Hazinski, T. A., and Schlueter, M. A. (1981). Respiratory control during hypoxia in newborn rabbits: Implied action of endorphins. *J. Appl. Physiol.* 51:122-130.

Hagberg, H., Andersson, P., Kjellmer, I., Thiringer, K., and Thordstein, M. (1987). Extracellular overflow of glutamate, aspartate, GABA and taurine in the cortex and basal ganglia of fetal lambs during hypoxia-ischemia. *Neurosci. Lett.* 78:311-317.

Hedner, T., Iversen, K., and Lundborg, P. (1982). GABA concentrations in the cerebrospinal fluid of newborn infants. *Early Hum. Dev.* 7:53-58.

Henderson-Smart, D. J., Pettigrew, A. G., and Campbell, D. J. (1983). Clinical apnea and brain-stem neural function in preterm infants. *N. Engl. J. Med.* 308:353-357.

Hendry, I. A., Morton, C. R., and Duggan, A. W. (1988). Analysis of antibody microprobe autoradiographs by computerized image processing. *J. Neurosci. Methods* 23:249-256.

Hertzberg, T., Hellström, S., Lagercrantz, H., and Pequignot, J. M. (1989). Development of the arterial chemoreflex and turnover of carotid body catecholamines in the newborn rat. *J. Physiol. (London)* 425:211-225.

Hertzberg, T., and Lagercrantz, H. (1987). Postnatal sensitivity of the peripheral chemoreceptors in newborn infants. *Arch. Dis. Child.* 62:1238-1241.

Hill, J. (1959). The oxygen consumption of new-born and adult mammals. Its dependence on the oxygen tension in the inspired air and on the environmental temperature. *J. Physiol. (London)* 149:346-373.

Irestedt, L., Dahlin, I., Hertzberg, T., Lagercrantz, H., and Sollevi, A. (1989). Adenosine concentration in umbilical cord blood of newborn infants after vaginal delivery and cesarean section. *Pediatr. Res.* 26:106-108.

Jansen, A. H., Ioffe, S., and Chernick, V. (1986). Stimulation of fetal breathing activity by beta-adrenergic mechanisms. *J. Appl. Physiol.* 60:1938-1945.

Johnston, B. M., and Gluckman, P. D. (1983). GABA-mediated inhibition of breathing in the late gestation sheep fetus. *J. Dev. Physiol.* 5:353-360.

Jones, C. T., and Robinson, R. O. (1975). Plasma catecholamines in foetal and adult sheep. *J. Physiol. (London)* 248:15-33.

Jones, C. T., Roebuck, M. M., Walker, D. W., Lagercrantz, H., and Johnston, B. M. (1987). Cardiovascular, metabolic and endocrine effects of chemical sympathectomy and of adrenal demedullation in fetal sheep. *J. Dev. Physiol.* 9:347-367.

Joseph, S. A., and Walker, D. (1989). Evidence of the effects of catecholamines on breathing movements and electrocortical activity in late gestation fetal sheep in utero. SSFP 16th Meeting, Reading, England, July. (Abstract), pp. G10.

Kitterman, J. A., Liggins, G. C., Clements, J. A., and Tooley, W. H. (1979). Stimulation of breathing movements in fetal sheep by inhibitors of prostaglandin synthesis. *J. Dev. Physiol.* **1**:453-466.

Kitterman, J. A., Liggins, G. C., Fewell, J. E., and Tooley, W. H. (1983). Inhibition of breathing movements in fetal sheep by prostaglandins. *J. Appl. Physiol.* **54**:687-692.

Koos, B. J., and Matsuda, K. (1990). Fetal breathing, sleep state and cardiovascular responses to adenosine in sheep. *J. Appl. Physiol.* **68**:489-495.

Kuzemko, J. A., and Paala, J. (1973). Apnoeic attacks in the newborn treated with aminophylline. *Arch. Dis. Child.* **48**:404-406.

Lagercrantz, H. (1987). Neuromodulators and respiratory control in the infant. *Clin. Perinatol.* **14**:683-695.

Lagercrantz, H., and Bistoletti, P. (1977). Catecholamine release in the newborn infant at birth. *Pediatr. Res.* **11**:889-893.

Lagercrantz, H., Persson, H., Srinivasan, M., and Yamamoto, Y. (1989). The developmental expression of some neuropeptide genes in respiration related areas of the rabbit brain. *J. Physiol. (London)* (Abstract) **417**:25P.

Lagercrantz, H., and Slotkin, T. A. (1986). The "stress" of being born. *Sci. Am.* **254**:100-107.

Man, E. H., Fisher, G. H., Payan, I. L. Cadilla-Perezrios, R., Garcia, N. M., Chemburkar, R., Arends, G., and Frey, W. H. (1987). D-aspartate in human brain. *J. Neurochem.* **48**:510-515.

Marchini, G., Lagercrantz, H., Winberg, J., and Uvnäs-Moberg, K. (1988). Fetal and maternal plasma levels of gastrin, somatostatin and oxytocin after vaginal delivery and elective cesarean section. *Early Hum. Dev.* **18**:73-79.

Marcus, C. (1988). Regulation of lipolysis in human adipocytes. A developmental study. Academic thesis. Karolinska Institute, Stockholm.

Maxwell, D. L., Fuller, R. W., Nolop, K. B., Dixon, C. M. S., and Hughes, J. M. B. (1986). Effects of adenosine on ventilatory response to hypoxia and hypercapnia in humans. *J. Appl. Physiol.* **61**:1762-1766.

McCooke, H. B., and Hanson, M. A. (1985). Respiration of conscious kittens in acute hypoxia and effect of almitrine bismesylate. *J. Appl. Physiol.* **59**:18-23.

McMillen, I. C., and Walker, D. W. (1986). Effect of beta-endorphin on fetal breathing movements in sheep. *J. Appl. Physiol.* **61**:1005-1011.

McQueen, D. S., and Ribeiro, J. A. (1983). On the specificity and type of receptor involved in carotid body chemoreceptor activation by adenosine. *Br. J. Pharmacol.* **80**:347-354.

Ment, L. R., Stewart, W. B., Duncan, C. C., Pitt, B. R., and Cole, J. S. (1986). Beagle puppy model of perinatal cerebral infarction: Regional cerebral prostaglandin changes during acute hypoxemia. *J. Neurosurg.* **65**:851-855.

Millhorn, D. E. (1986). Physiological significance of pharmacological studies of respiratory regulation. In *Neurobiology of the Control of Breathing.* Edited by C. von Euler and H. Lagercrantz. New York, Raven Press, pp. 89-95.

Moss, I. R., Denavit-Saubié, M., Eldridge, F. L., Gillis, R. A., Herkenham, M., and Lahiri, S. (1986). Neuromodulators and transmitters in respiratory control. *Fed. Proc.* **45**:2133-2147.

Moss, I. R., and Inman, J. G. (1989). Neurochemicals and respiratory control during development. *J. Appl. Physiol.* **67**:1-13.

Moss, I. R., Runold, M., Dahlin, I., Fredholm, B. B., Nyberg, F., and Lagercrantz, H. (1987). Respiratory and neuroendocrine responses of piglets to hypoxia during postnatal development. *Acta Physiol. Scand.* **131**:533-541.

Moss, I. R., and Scarpelli, E. M. (1984). CO_2 and naloxone modify sleep/ wake state and activate breathing in the acute fetal lamb preparation. *Respir. Physiol.* **55**:325-340.

Onimaru, H., Arata, A., and Homma, I. (1987). Localization of respiratory rhythm-generating neurons in the medulla of brainstem-spinal cord preparations from newborn rats. *Neurosci. Lett.* **78**:151-155.

Onimaru, H., and Homma, I. (1987). Respiratory rhythm generator neurons in medulla of brainstem-spinal cord preparation from newborn rat. *Brain Res.* **403**:380-384.

Ontyd, J., and Schrader, J. (1984). Measurement of adenosine, inosine, and hypoxanthine in human plasma. *J. Chromatogr.* **307**:404-409.

Padbury, J. F., Agata, Y., Polk, D. H., Wang, D. L., and Callegari, C. C. (1987). Neonatal adaptation: Naloxone increases the catecholamine surge at birth. *Pediatr. Res.* **21**:590-593.

Perrin, D. G., Cutz, E., Becker, L. E., Bryan, A. C., Madapallimatum, A., and Sole, M. J. (1984). Sudden infant death syndrome: Increased carotid-body dopamine and noradrenaline content. *Lancet* **2**:535-537.

Prabhakar, N. R., Runold, M., Yamamoto, Y., Lagercrantz, H., and Euler, C. von (1984). Effect of substance P antagonist on the hypoxia-induced carotid chemoreceptor activity. *Acta Physiol. Scand.* **121**:301-303.

Rognum, T. O., Saugstad, O. D., Oyasaeter, S., and Olaisen, B. (1988). Elevated levels of hypoxanthine in vitreous humor indicate prolonged cerebral hypoxia in victims of sudden infant death syndrome. *Pediatrics* **82**:615-618.

Runold, M., Lagercrantz, H., and Fredholm, B. B. (1986). Ventilatory effect of an adenosine analogue in unanesthetized rabbits during development. *J. Appl. Physiol.* **61**:255-259.

Runold, M., Lagercrantz, H., Prabhakar, N. R., and Fredholm, B. B. (1989). Role of adenosine in hypoxic ventilatory depression. *J. Appl. Physiol.* **67**:541-546.

Runold, M., Prabhaker, N. R., Mitra, J., and Cherniack, N. S. (1988a). Mechanisms of respiratory augmentation by adenosine: role of vagal pulmonary afferents. In *Respiratory Psychophysiology*. Edited by C. von Euler and M. Katz-Salamon. London, MacMillan Press, pp. 175-182.

Runold, M., Prabhakar, N. R., Mitra, J., Overholt, J., and Cherniack, N. S. (1990). Effect of adenosine on chemosensory activity of the cat aortic body. *Respiration Physiology* **80**:299-306.

Saugstad, O. D. (1975). Hypoxanthine as a measure of hypoxia. *Pediatr. Res.* **9**:158-161.

Schaeffer, J. I., and Haddad, G. G. (1985). Regulation of ventilation and oxygen consumption by δ- and μ-opioid receptor agonists. *J. Appl. Physiol.* **59**:959-968.

Scheibner, T., Read, D. J. C., and Sullivan, C. E. (1988). Substance P immunoreactivity in the developing kitten carotid body. *Brain Res.* **453**:72-78.

Schramm, C. M., and Grunstein, M. M. (1987). Respiratory influence of peripheral chemoreceptor stimulation in maturing rabbits. *J. Appl. Physiol.* **63**:1671-1680.

Schwieler, G. (1968). Respiratory regulation during postnatal development in cats and rabbits and some of its morphological substrate. *Acta Physiol. Scand.* (Suppl. 304): pp. 123.

Sears, T. A. (1977). The respiratory motor neurone and apneusis. *Fed. Proc.* **36**:2412-2420.

Spyer, K. M. (1988). Central nervous system control of the cardiovascular system. In *Autonomic Failure*. Edited by Sir R. Bannister. Oxford, Oxford University Press, pp. 56-79.

Srinivasan, M., Lagercrantz, H., and Yamamoto, Y. (1989). A possible dopaminergic pathway mediating hypoxic depression in neonatal rabbits. *J. Appl. Physiol.* **67**:1271-1276.

Stark, R. I., Wardlaw, S. L., and Daniel, S. S. (1986). Characterization of plasma beta-endorphin immunoactivity in the fetal lamb: Effects of gestational age and hypoxia. *Endocrinology* **119**:755-761.

Szeto, H. H., and Umans, J. G. (1984). The effects of a stable adenosine analog of fetal behaviour, respiration and cardiovascular functions. In *The Physiological Development of the Fetus and Newborn*. Edited by C. T. Jones and P. W. Nathanielsz. London, Academic Press, pp. 649-652.

Thiringer, K. (1982). Hypoxanthine as a measure of foetal asphyxia. Academic thesis. University of Gothenburg, Gothenburg.

Thorburn, G. D., and Rice, G. E. (1987). Fetal maturation and the timing of parturition: A comparative analysis. *Proc. Aust. Physiol. Pharmacol. Soc.* **18**:51-68.

Ungerstedt, U. (1984). Measurement of neurotransmitter release by intracranial dialysis. In *Measurement of Neurotransmitter Release in Vivo.* Edited by C. A. Marsden. Chictester, Wiley, pp. 81-105.

Vizek, M., Pickett, C. K., and Weil, J. V. (1987). Biphasic ventilatory response of adult cats to sustained hypoxia has central origin. *J. Appl. Physiol.* **63**:1658-1664.

Watt, J. G., Dumke, P. R., and Comroe, J. H. (1943). Effects of inhalation of 100 per cent and 14 per cent oxygen upon respiration of unanesthetized dogs before and after chemoreceptor denervation. *Am. J. Physiol.* **138**:610-617.

Weil, J. V., and Zwillich, C. W. (1976). Assessment of ventilatory response to hypoxia. *Chest* (Suppl. 70):124-128.

Weiskopf, R. B., and Gabel, R. A. (1975). Depression of ventilation during hypoxia in man. *J. Appl. Physiol.* **39**:911-915.

Wennergren, G., and Wennergren, M. (1980). Respiratory effects elicited in newborn animals via the central chemoreceptors. *Acta Physiol. Scand.* **108**:309-311.

Winn, H. R., Rubio, R., and Berne, R. M. (1981). Brain adenosine concentration during hypoxia in rats. *Am. J. Physiol.* **241**:H235-H242.

Woodrum, D. E., Standaert, T. A., Mayock, D. E., and Guthrie, R. D. (1981). Hypoxic ventilatory response in the newborn monkey. *Pediatr. Res.* **15**:367-370.

Yamamoto, Y., and Lagercrantz, H. (1985). Some effects of substance P on central respiratory control in rabbit pups. *Acta Physiol. Scand.* **124**:449-455.

Yamamoto, Y., Runold, R., Prabhakar, N., Pantaleo, T., and Lagercrantz, H. (1988). Somatostatin in the control of respiration. *Acta Physiol. Scand.* **134**:529-533.

Zetterström, T., Vernet, L., Ungerstedt, U., Tossman, U., Jonzon, B., and Fredholm, B. B. (1982). Purine levels in intact rat brain. Studies with an implanted perfused hollow fibre. *Neurosci. Lett.* **29**:111-115.

23

Control of Breathing During Asphyxia and Autoresuscitation

BRADLEY T. THACH and MARGARET S. JACOBI

Washington University School
of Medicine
St. Louis, Missouri

WILLIAM M. GERSHAN

Medical College of Wisconsin
Milwaukee, Wisconsin

I. Introduction

Renewed attention to an old problem in the field of respiratory control—the gasping pattern of respiration—comes from research on perinatal asphyxia and infantile apneic spells. Observations by several investigators (Chernick et al., 1975; Jansen and Chernick, 1974; Woodrum et al., 1972; Bystryzcka et al., 1975) suggest that an asphyxial gasp is frequently the mechanism of lung inflation at birth. This gasp, in effect, produces recovery from birth asphyxia. Other workers have been concerned with the prenatal diagnosis of asphyxia. Boddy and Dawes (1975) and Patrick et al. (1976) have made extensive study of fetal respiratory patterns. They have shown that a characteristic "gasping" pattern of respiration is a reliable predictor of fetal death in the sheep and human fetus. Attention to asphyxial gasps also comes from efforts to link infantile apneic spells to the sudden infant death syndrome (SIDS). Examining the hypothesis that "interruption of the apnea may be the phase relevant to SIDS." Guntheroth and Kawabori (1975) and French et al. (1972) have investigated respiratory activity during acute asphyxiation in the newborn and adult monkey. These studies have focused attention on an old unsolved problem: after a period of progressive asphyxiation, respiration ceases altogether, only to return again, seconds or minutes later, in the form of infrequent breaths or "gasps." In the past these hypoxic gasps have often been termed "agonal," implying that they represent disintegration of brain function. Only rarely have gasps been viewed as coordinated and functional respiratory activity. However, the gasp is obviously functional. When asphyxiated animals are given access to air, they are able to rapidly and spontaneously "autoresuscitate" by virtue of the infrequent gasps (Guntheroth, 1974; Guntheroth and Kawabori, 1975) (Fig. 1). Therefore, from the clinical perspective, death attributed to an apneic spell implies failure of the gasp mechanism to autoresuscitate.

II. The Pattern of Respiration During Acute Asphyxiation: Hyperpnea, Apnea, and Gasping

In its broadest sense "gasp" is used interchangeably with "big breath." Gasp has also been used in reference to the last respiratory pattern to appear prior to death in the course of asphyxiation. Dawes has categorized the sequence of respiratory patterns observed in progressively asphyxiated newborn rabbits and monkeys as primary hyperpnea, primary apnea, secondary gasping, and secondary apnea (Dawes, 1968). These patterns are illustrated by a tracing from one of our own studies in the newborn rabbit in Figure 2.

This sequence of patterns has been observed when asphyxiation is produced in a variety of ways. Thus tracheal occlusion (Davis, 1961; Guntheroth and Kawabori, 1975; Lawson and Thach, 1977), submersion in saline (Da-

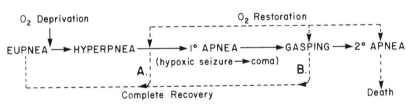

Figure 1 Schematic diagram showing the progressive changes in respiratory pattern and central nervous system activity that occur with acute anoxia. At onset of the primary apnea stage, a "hypoxic seizure" occurs, at which time the electroencephalograph becomes isoelectric. Muscle tone disappears and the animal appears lifeless (Dehaan and Field, 1959). This stage is best termed "hypoxic coma." Hypoxic gasping occurs after primary apnea. If oxygen is given prior to primary apnea, recovery is prompt (pathway A). Gasping can also restore blood oxygen, normal respiratory pattern, and neurological status ("autoresuscitation," pathway B). After hypoxic gasping, spontaneous recovery does not occur.

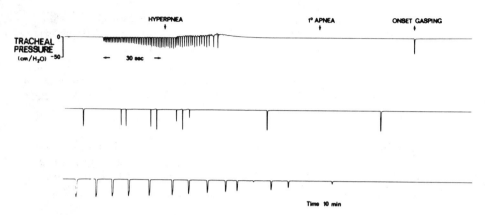

Figure 2 Tracheal pressure trace from a 3-day-old unanesthetized tracheotomized rabbit following airway occlusion at end expiration until cessation of respiratory activity. Three consecutive segments of the continuous trace are shown. Inspiratory activity is evidenced by negative deflections of the trace. Time from onset of asphyxia until last spontaneous gasp is 10 min. (From Lawson and Thach, 1977.)

vis, 1961; Godfrey, 1968a), and exsanguination (Dawes, 1968) all are said to result in a primary hyperpnea, followed by a "primary" or "hypoxic" apnea and secondary gasping. Acute hypoxemia produced by placing an animal in a 100% N_2 (Adolph, 1969; Campbell et al., 1971; Godfrey, 1968b; Holowach-Thurston et al., 1974; Jacobi and Thach, 1989) or a CO_2 (Peiper, 1963) environment is reported to give the same sequence of patterns. Other innovative studies point to the intrinsic central nervous system origin of these patterns. Since secondary (e.g., "hypoxic") gasping is manifested by coordinated activities of respiratory muscles and muscles depressing the mandible, Tschirgi and Gerard (1947) studied the "respiratory pattern" following decapitation. The same basic pattern sequence was observed in mandibular movements of the decapitated heads.

Although the duration of each respiratory phase is variable, this asphyxial sequence of patterns is observed in the fetus, newborn, and adult. Most observations on the hypoxic gasping phase came from studies of the newborn, in whom time to last gasp may last up to 1 h (Fazekas et al., 1941). In the adult rat (Guntheroth, 1974) and monkey (Guntheroth and Kawabori, 1975), Guntheroth has reported that the respiratory sequence of hyperpnea through gasping does not differ strikingly from that seen in the newborn animal with the exception that the gasping phase lasts only seconds in the adult and many minutes in the newborn. Our own studies in adult and newborn rabbits have confirmed these observations (Thach and Lawson, 1976, 1979). The control of respiratory movements in the fetus may be very different from that of the adult. However, when the fetal animal is asphyxiated while making rapid irregular respiratory movements, the same respiratory pattern sequence as occurs postnatally is reportedly observed (Bystryska et al., 1975; Dawes et al., 1972).

There do not appear to be marked species differences in the respiratory sequence following asphyxiation. Hyperpnea followed by apnea and hypoxic gasps has been observed in the mouse (Holowach-Thurston et al., 1974; Jacobi and Thach, 1989), rat (Adolph, 1969), guinea pig (Dawes et al., 1972), rabbit (Campbell et al., 1971; Davis, 1961; Godfrey, 1968a,b; Lawson and Thach, 1977), hamster, cat, and dog (Adolph, 1969; Guntheroth and Kawabori, 1975), sheep (Bystryzcka et al., 1975; Dawes et al., 1972), and rhesus monkey (Dawes et al., 1960; Guntheroth and Kawabori, 1975).

A. Evidence for a Medullary Gasp Center

On the basis of the sequential change in respiratory pattern observed during progressive asphyxiation, and the effects on this pattern of surgical transection of the brainstem, Lumsden (1922) proposed that the terminal hypoxic gasps result from activity of a medullary gasping center which becomes activated as the higher respiratory centers (e.g., pons) become depressed. This view was also held by Peiper (1963) on the basis of clinical and laboratory observations of the respiratory pattern during asphyxiation. In Lumsden's scheme, the gasp center is perceived as being inactive during eupnea and, in fact, is actually inhibited by the other respiratory centers. Barcroft (1934), in contrast, suggested that asphyxial gasps, as they occur in the newborn or fetus, represent a transformation of the normal activity of medullary respiratory centers which he perceived as the primary centers. In the Barcroft hypothesis the characteristic pattern recognizable as "gasping" occurs when the medullary centers are deprived of input from higher centers. The restriction of input from pontine and other higher centers is presumed to be caused by lack of oxygen. This hypothesis does not require a separate medullary gasp center. Lawson and Thach (1977) and, more recently, Davis et al. (1986) have provided support for this concept by showing that the hallmark features of gasps (short Ti, long Te, large amplitude) gradually appear during the hyperpneic phase. That is, the gasp pattern is the result of progressive, gradual changes in the output of the medullary centers prior to actual onset of the hypoxic gasping phase. This being the case, there is no need for a "gasp center" to explain the gasping pattern.

A somewhat different hypothesis from these two has been put forth by Jansen and Chernick (1975). It was initially observed that intravenous injections of cyanide elicit "gasps" from the apneic lamb fetus (Jansen and Chernick, 1974). In subsequent experiments it was found that local application of cyanide-soaked pledgets to the ventral surface of the medulla also elicits gasp-like efforts (Jansen and Chernick, 1975). Systemic effects and peripheral chemoreceptor influences were not required for this response. It was concluded that hypoxic depression of nearby medullary neurons was, in this preparation, a stimulus for "gasping." This was given as evidence for a medullary "gasp center" which is stimulated rather than depressed by hypoxia and therefore behaves fundamentally differently than the medullary "centers" involved in eupnea. A major problem in relating Jansen and Chernick's work to the more traditional theories of asphyxial gasp regulation stems from the fact that "gasping" as used by Jansen and Chernick is nonspecific. Since the phase of primary hyperpnea was not observed in their fetal preparation, one cannot necessarily assume that the respiratory efforts observed in response to local cyanide are akin to secondary gasping seen during progressive asphyxiation.

Recent studies by St. John and co-workers suggest that gasps originate from a medullary respiratory center anatomically separate from the main respiratory centers (St. John and Knuth, 1981; St. John et al., 1984, 1985). They have found that a localized area in the lateral tegmentum must be intact for gasping to take place, and moreover, that gasp-like inspiratory efforts occur when this area is stimulated. However, as with the Jansen and Chernick studies, the relation of gasping in St. John's studies to hypoxic gasps is not entirely clear. The problem is that gasping in this model is usually produced by surgical lesioning or cooling of the brainstem to interrupt connections between pons and medulla rather than by hypoxia in an intact nervous system.

B. Energy Metabolism as a Regulator of Respiratory Center Activity During Asphyxiation

It has long been appreciated that hypoxic gasping in the infant animal is associated with profound hypoxemia, circulatory failure, or both. Tschirgi and Gerard (1947), among others (Himwich et al., 1941), showed that such gasping in the neonatal animal takes place under nearly anaerobic conditions. The energy for gasping appears to be supplied by glycolysis since pretreatment of animals with fluoride inhibits glycolysis and prevents hypoxic gasping during asphyxia (Tschirgi and Gerard, 1947).

It is further known that activity of the central nervous system determines when gasping ceases altogether since electrical stimulation of phrenic nerves produces vigorous contraction of the diaphragm long after the last gasp (Dawes, 1968) Thus, neural energy metabolism may be a critical factor in determining activity of the gasping mechanism. This conclusion is further supported by the observation that infusions of glucose (Dawes et al., 1972; Holowach-Thurston et al., 1974) or bicarbonate (Dawes et al., 1963) prolong duration of hypoxic gasping in asphyxiated animals. Furthermore, cardiovascular function appears important for maintenance of gasping, since gasping duration in anoxia is positively correlated with cardiac glycogen (Shelley, 1961) and since elimination of cerebral perfusion reduces gasping duration (Hansen, 1977). On the basis of these observations, it could be speculated that the onset of hypoxic apnea and secondary gasping might closely reflect the availability of energy substrates for the medullary respiratory neurons. Against this hypothesis is the observation of Duffy et al. (1972) that high-energy phosphate substances are abundant in the respiratory regions of the medulla at onset of hypoxic apnea in the rat. Of particular interest was their observation that high-energy phosphate substrates are reduced in the forebrain and pons at onset of apnea—an observation that supports both Lumsden's and Barcroft's theories of hypoxic gasping.

Adolph's theory (1969) attributes respiratory frequency depression to the availability of O_2 in the primary pacemaker cells. He drew attention to the similar prolongation of cardiac and respiratory cycles with profound hypoxia and to the prompt recovery of both on administration of oxygen. Since this response is demonstrable in the isolated heart, it does not appear to be a reflex phenomenon. Adolph suggests that pacemaker cells both in the heart and in the respiratory centers spontaneously reduce their activity on the basis of a mechanism functioning at the cellular level that is sensitive to a critical level of hypoxia. Unlike the theory of Barcroft or Lumsden, Adolph's view pertains only to frequency and does not consider cycle phases or gasping force; it is thus incomplete. However, for the purpose of discussion it can be viewed as a separate hypothesis for the origin of hypoxic gasps. The Adolph hypothesis is different from either Lumsden's or Barcroft's in that it stresses the local effects of hypoxia on the cells that generate the gasp. It thereby differs from the others, which attribute gasps solely to activation of a normally inactive gasp center (Lumsden) or to alteration of the normal respiratory pattern by a mechanism of selective deafferentation of the "primary" medullary centers (Barcroft). In partial support for this view is the observation of a PaO_2 threshold for gasping in the range of 8-12 Torr (Lawson and Thach, 1977; Guntheroth and Kawabori, 1975).

C. Influence of Peripheral Chemoreceptor

The role of the peripheral chemoreceptor in regulation of respiratory pattern during asphyxiation is of critical importance. It is known that neural discharge from the carotid body is increased at the time of the last gasp and remains active for some time thereafter (Schwieler, 1968).

Furthermore, it is still widely held that the carotid body is essential for hypoxic gasping (Dawes, 1968). Tschirgi and Gerard (1947) showed that simultaneous hypoxia of the peripheral chemoreceptor (carotid body) and the brainstem of cross-perfused rats was required for secondary gasping to take place. Oxygenation of the carotid body, in a systemically asphyxiated animal, eliminated gasping. However, Tschirgi and Gerard's observations were limited to mandibular movements only. Activity of the primary respiratory muscles was not recorded. Therefore, the effect of chemoreceptor on the actual respiratory muscles during gasps is unknown.

Both Chernick et al. (1975) and Harned et al. (1967) have observed deep breaths, or "gasps," initiated by profound asphyxia or cyanide infusion in the lamb fetus. This activity was reported to be little changed by carotid body denervation and/or vagotomy. It is not, however, evident that these gasps are analogous to the hypoxic (i.e., "secondary") gasping of the air-breathing animal. It is, therefore, difficult to relate their observations to those of Tschirgi and Gerard. St. John and co-workers (St. John and Knuth, 1981; St. John et al., 1984) have shown that gasp-like breathing in normoxic, brainstem-lesioned cats is little affected by carotid body or sciatic nerve afferents which normally have potent effects on respiration. However, as mentioned above, the relation of this gasp-like breathing to hypoxic gasping is still not entirely clear.

D. Role of Cutaneous, Laryngeal, and Vagal Pulmonary Afferents

It is frequently said that cutaneous tactile (Barcroft, 1934) or painful (Dawes, 1968) stimuli elicit gasps in the apneic fetus or from the newborn during asphyxial primary apnea (Tizard, 1964). On the other hand, these studies have not used an experimental design that clearly distinguishes elicited from spontaneous gasps. In fact, much of the information on elicitation of "gasps" appears to be derived from observations on the well-oxygenated apneic fetus, and this further confuses the issue.

Vagal afferents appear to be important for periodic spontaneous deep breaths or sighs (Glogowska et al., 1972; Reynolds, 1962; Thach and Taeusch, 1976), which closely resemble hypoxic gasps with respect to inspiratory force and duration and the apparent stimulatory influence of asphyxia. However, vagal feedback does not appear to be required for the occurrence of hypoxic gasping since Godfrey (1968a) did not report gross differences in the sequence of respiratory patterns during progressive asphyxiation in vagotomized newborn rabbits.

Also, a number of studies have investigated the marked inhibitory effect on breathing when laryngeal receptors are stimulated with fluids (Johnson et al., 1975) or the superior laryngeal nerves are stimulated directly (Lawson, 1981; Donnelly and Haddad, 1986; Sanocka et al., 1988). It has been found that whereas the resultant apnea may be very prolonged in anesthetized preparations, the duration is much less marked as the level of anesthesia is reduced. Although apnea produced by this reflex may continue to the point of profound asphyxia, respiration is not abolished altogether, since hypoxic gasps occur once PaO_2 has fallen to the gasp threshold level (Johnson et al., 1975). In other words, the laryngeal chemoreceptor reflex appears to inhibit eupneic respiration but not gasping.

III. The Role of Gasping in Spontaneous Recovery from Hypoxic Apnea

A. Experimental Studies of "Autoresuscitation"

Although the ability of hypoxic gasps to produce spontaneous recovery from prolonged apnea in the newborn has been known for many years, comparatively few studies have examined this capability after the neonatal period. Adolph (1969) found that the newborn of several species are capable of spontaneous recovery from total anoxia following much longer anoxic exposure than the adult. He also found that the adult sometimes fails to self-resuscitate after one or more unsuccessful gasps. Guntheroth (1974) coined the term "autoresuscitation" and suggested that more mature animals lose the ability to autoresuscitate, although this was apparently not the case as far as adult rats were concerned. Likewise, Mathew et al. (1984) found that adult rabbits were proficient in autoresuscitation and that upper airway maintaining muscles were maximally active during gasping, suggesting that passive airway collapse might be an unlikely cause of autoresuscitation failure.

More recently Jacobi and Thach (1989) studied the effect of maturation on the ability to autoresuscitate in two strains of Swiss Webster mice (SWR and SW). Hypoxic (i.e., "primary") apnea was induced with 97% N_2, 3% CO_2. Air was given at onset of apnea (Fig. 3). The survival rates in SWR mice were 100% at 1-16 days, 19% at 17-23 days, and 90% in adults. The rates in these same age groups in SW mice were 100%, 70%, and 95%, respectively. The decrease in survival in 17-23-day-olds was statistically significant for both strains. The difference between strains is of interest because in subsequent studies we have found that this autoresuscitation defect is entirely lacking in weanling BALB/C mice (Gershan et al., 1990a). The fact that different breeds of mice vary so greatly in susceptibility to hypoxia suggests that genetic factors underlie the autoresuscitation defect in SWR weanling mice.

In Swiss Webster mice, seizure-like activity is frequently observed during hypoxia apnea and gasping, but this activity does not occur significantly more often in mice that fail to recover (Jacobi and Thach, 1989; Fig. 3). Simultaneous recording of movement and breathing pattern indicates that this activity does not interfere with the occurrence of gasping. Therefore, in adults and very young mice, autoresuscitation by gasping is a powerful mechanism for recovery from hypoxia, but at an intermediate age, in Swiss

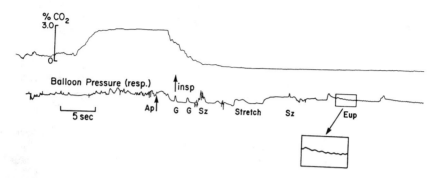

Figure 3 Example of successful autoresuscitation. Results of exposure to anoxia in an unanesthetized adult mouse showing primary apnea and autoresuscitation after exposure to air. The pressure signal is from a latex balloon wrapped around the thorax. The concentration of CO_2 is inversely proportional to the oxygen concentration: when it reaches a plateau at 3%, the gas composition in the chamber is 3% CO_2, 97% N_2. When air is flushed in, the CO_2 concentration returns to zero. The arrow shows the point at which the animal was seen to go still, coinciding with primary apnea. Air was then given. The times at which apnea (Ap), gasps (G), seizure-like activity (Sz), and stretching were seen are shown. Decerebrate posturing (e.g., "hypoxic seizures") just before primary apnea is seen in all animals. The seizure-like activity after primary apnea is influenced by maturation and does not occur in all animals. Gasps are followed by eupnea (Eup), which is of low amplitude relative to the gasps. Neurological and behavioral recovery is rapid and complete. (From Jacobi and Thach, 1989.)

Webster mice, at least, it is frequently unsuccessful. These studies, therefore, show an effective autoresuscitation mechanism in the adult which was somewhat unsuspected. They also indicate a vulnerable age (17-23 days) which was totally unexpected. One mechanism for failure to survive was clear—failure to gasp. By far the most common pattern, however, as gasping followed by death (Fig. 4).

The cause of the autoresuscitation failure in weanling mice is unclear. Several relevant points can be made, however. First, cardiovascular rather than respiratory function might underlie the failure to autoresuscitate. During successful autoresuscitation in mice there is recovery from bradycardia, with an abrupt increase in rate that occurs just prior to return of eupneic breathing (Gershan et al., 1990a). Significantly, this "cardiac resuscitation" does not occur in weanlings that fail to autoresuscitate, even though they have gasping activity (Jacobi et al., 1989a). Additionally, repeated exposure to anoxia induces susceptibility to autoresuscitation failure in BALB/C mice with a similar pattern (i.e., failure of cardiac resuscitation) to that of SWR weanlings (Gershan et al., 1990b). Repeated hypoxic exposure might cause depletion of substrates for anaerobic glycolysis in the circulatory

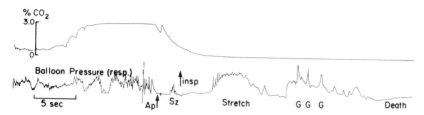

Figure 4   Example of failure to autoresuscitate. Exposure to anoxia in a 23-day-old mouse. Procedure and symbols are as explained in Figure 3 legend. After primary apnea (Ap), mouse gasps (G), but this is not effective in producing autoresuscitation. (From Jacobi and Thach, 1989.)

system, or buildup of toxic metabolites, that could be the basis for failed cardiac resuscitation. In support of this view are the observations of Swann and Brucer (1949), who noted that circulatory failure often precedes respiratory failure in the acutely anoxic dog. A potentially relevant observation is that pentobarbital anesthesia completely corrects the autoresuscitation defect in weanling SWR mice (Jacobi et al., 1989b). This result could be mediated through the known effect of pentobarbital in reducing metabolic rate and thereby prolonging anoxic survival.

B. Clinical Occurrence of Autoresuscitation

What causes the first breath following birth of the infant has long been a matter of debate. Dawes (1968) clearly stated the problem: "in the face of so many violent changes, it is a nice matter to decide which if any, is the essential stimulus." Nevertheless, it is clear from the observations of Chernick et al. (1975) and Woodrum et al. (1972) in the fetal sheep and Dawes et al. (1960) in the fetal monkey that, in the absence of other identifiable stimuli, acute and profound asphyxia alone is a stimulus for "gasps." These gasps are capable of inflating the lung and rapidly oxygenating the blood. Furthermore, they rapidly convert to eupneic respiration as asphyxia abates (Dawes, 1968; Godfrey, 1968b). More recently, Bystryzcka et al. (1975) have observed that umbilical cord occlusion does not usually result in a stage of primary hyperpnea in the fetal sheep. Instead, the first significant respiratory efforts often appear to correspond to the breaths of secondary gasping. Bystryzcka et al. maintain that, in the sheep at least, the hypoxic gasp mechanism functions to generate the first postnatal breath.

With only limited data allowing one to differentiate hypoxic gasps from eupneic or hyperpneic breaths, it is difficult to maintain that gasps function as the first postnatal breaths in humans. Several observations, however, support this point of view. Schmidt (1950) made pneumographic and motion picture recordings of normal deliveries. He reported that the first post-natal breath was usually an isolated "gasp" with a large inspired volume and short inspiratory duration. The gasp was preceded and followed by respiratory apnea of 3-5 s. In support of these direct observations is circumstantial evidence which suggests that the first breath in humans might be analogous to hypoxic gasping during asphyxiation. In order to inflate the fluid-filled lung, the first breath must be forceful. Karlberg (1960) has measured pleural pressure changes during the first breath in normal infants. Pressure swings of -50 to -70 cm H_2O were found to occur during the first breath. These pressure changes probably represent maximum respiratory output. The second and third breaths were found to be less forceful. The inspiratory duration of the first breath should also be a clue to the first breath's neural origin. This was noted by Karlberg to be "short" compared to subsequent breath and therefore resembles experimentally induced hypoxic gasps. It should be pointed out that the first breath of infants recorded by Mortola and co-workers (1982) in some ways more closely resembles eupneic or hyperpneic breaths than gasps. Hence, the issue of the origins of the first breath is still open to debate. In any case, the studies of Koch and Wendel (1968) and others (Avery and Fletcher, 1974) have shown that the first breath in human infants often takes place under conditions of rather marked hypoxemia, with O_2 saturations in the range of 18-25%. Therefore, the degree of hypoxemia at birth is comparable to that at which asphyxial gasps occur in postnatal animals.

In certain human infants, at least, it seems that the first breath is clearly analogous to the gasps of hypoxic (i.e., secondary) gasping since rather long periods of apnea, presumably analogous to primary (i.e., "hypoxic") apnea, are commonly observed before onset of the first breath in these "birth asphyxiated" infants (Barcroft, 1934; Smith, 1950; Tizard, 1964).

C. Neonatal Apneic Spells

Relatively more attention has been paid to hypoxic gasps during the immediate perinatal period than during later postnatal life. However, several noteworthy observations suggest that such gasping activity can be of clinical significance.

During the neonatal period intermittent spells of apnea lasting 20 s or longer and associated with cyanosis, bradycardia, and acute neurological depression are relatively common (Avery and Fletcher, 1974). A number of studies have been directed to determining factors that might precipitate spells. Few studies have investigated physiological mechanisms leading to spontaneous resolution of apnea. In the premature infant, the majority of apneic pauses lasting 15 s or longer resolve spontaneously (Daily et al., 1969). Most spells lasting 20 s or longer terminate when cutaneous stimulation is administered (Daily et al., 1969). Prior to the era of modern resuscitation techniques and during a time when handling was believed harmful to premature infants, Peiper (1963) made extensive observations of the natural course of apneic spells in premature infants. Many of these observations were made with motion picture documentation and with strain gage pneumography. Although many infants having repeated, frequent apneic spells died, Peiper observed that spontaneous recovery from the first or second prolonged spell was more frequent than not. Spontaneous recovery could be expected to occur even after 2 or 3 min of absent respirations. During prolonged apneic spells respiratory recovery was invariably initiated by a series of "gasps," which were "shorter and deeper than that of normal respiration" and accompanied by action of accessory muscles—"the lower jaw drops down and the mouth is opened . . . the head is jerked backward." Peiper regarded these resuscitative gasps in infants as phenomena identical to that of hypoxic gasping which he had studied in animals.

D. "Breath Holding" Apneic Spells of Older Infants

A good deal of attention has been focused on the interplay of pulmonary and cardiac reflexes that culminate in producing breath-holding spells or "infantile syncope" (Gastaut and Gastaut, 1968; Gauk et al., 1963; Lombroso and Lerman, 1967). Regardless of which factors initiate the spell, it is agreed that the end result is severe asphyxia of the central nervous system. Opisthotonic posturing, sometimes termed "hypoxic seizure," has been observed at the height of the spell, at which time the EEG is flat. The posturing is believed to be a phenomenon of "decerebration" analogous to that produced by cyanide injection (Ward and Wheatley, 1947; Wheatley et al., 1947), surgical decerebration (Ward, 1947), or acute anoxia in the laboratory animal.

When the decerebrate posturing occurs, the infant is usually apneic. Yet these infants all recover rapidly, frequently in the absence of resuscitative measures. Little attention has been devoted to the recovery period. Two observations are noteworthy. Peiper (1963) was able to observe the recovery period in an infant in whom breath-holding spells could be provoked at will. He observed that the terminal episode of apnea was interrupted by one or several isolated "gasps," similar to those observed during apneic spells in premature infants. Employing more elaborate techniques for physiological recording, Gauk et al. (1963) have confirmed Peiper's observations in two infants. Their patients lost consciousness and developed decerebrate posturing after 40 s of apneic breath holding. Within another 10 s, the apnea was terminated by an isolated "gasp." Within several seconds following the single gasp, arterial oxygen saturation, as measured by ear oximetry, rose from 40 to 65% in one instance and from less than 30 to 55%

in another. Following the gasp, a "eupneic" pattern of respiration was observed to return rapidly. Accordingly, the sequence of events in breath-holding spells closely resembles hypoxic apnea, gasping, and autoresuscitation in animal models.

E. Apneic Spells in Sleep and Apnea of Undetermined Etiology

Over the past 20 years it has repeatedly been proposed that some cases of sudden infant death syndrome (S.I.D.S) result from fatal episodes of sleep apnea. Numerous reports and studies have shown that periodic episodes of prolonged apnea (greater than 20 s) occur relatively commonly in infants 1 month to 1 year of age (Steinschneider, 1972; Guilleminault et al., 1975, 1976). These reports have provided little information on the pattern of respiration during recovery from prolonged apnea. Yet, judging from the frequency of spells and the high survival rate of infants, spontaneous recovery undoubtedly occurs often.

The mechanism of spontaneous recovery from apnea in these infants is conjectural, but a prominent theory is that recovery involves arousal from sleep. Other recovery mechanisms are likely also important. Recent studies of experimental asphyxia in animals suggest that during prolonged apnea in REM sleep, hypoxic apnea and loss of cortical electrical activity occur before there is an intervening period of arousal (Fewell and Baker, 1987). In this situation, autoresuscitation by gasping would be the only remaining recovery mechanism. Two clinical observations suggest that the sequence of events observed during the phase of recovery from asphyxia in breath-holding infants may occur in other apneic asphyxial states. Thus Moskowitz et al. (1976) reported a case of idiopathic sleep apnea in a 14-day-old boy. Recovery was preceded by hypoxic seizure and "gasping." During one observation period, after 60 s of apnea, a "myoclonic seizure" occurred presumed to be secondary to hypoxia. Following the seizure ventilation resumed, initiated by an isolated forceful "gasp," which in turn was followed by resumption of regular respirations. Similarly, an episode of prolonged obstructive sleep apnea progressing to the point of "hypoxic seizures" with failure of arousal has been reported in an adult (Sullivan et al., 1983).

Related observations in young infants are also noteworthy. Werne and Garrow (1953) reviewed a series of SIDS deaths in which the death was witnessed. They noted that "gasps" were seen in a number of infants at the time of death. In another attempt to reconstruct events immediately prior to death in SIDS, Stevens (1965) collected four cases in which sudden death occurred in infants who were also under observation at the time of death. Stevens believed these infants differed from other SIDS cases only in that they were hospitalized when death occurred. All infants were reported to be healthy until several hours prior to death, except that each had a mild upper respiratory infection. In each case, admission to the hospital was precipitated by one or more prolonged apneic spells. Between spells the infants appeared comparatively well. In each case, death occurred during an apneic spell similar in its onset and initial course to preceding spells. Steven's description of events during these spells is not unlike that of Gauk et al. for breath-holding infants. At onset of the spell, respiration ceased without warning. Within 40-60 s the infants developed cyanosis and then opisthotonic posturing. Some infants are known to have regurgitated at this time. After a few seconds the spasm terminated with muscular relaxation and respiration was noted to return with a series of deep "fish-like gasps." In several instances these gasps can appropriately be termed spontaneous since children were not being resuscitated when the gasps began. The infants appeared to be weakened by each successive apneic spell, and death occurred during a final apneic spell during which there was failure to resume spontaneous ventilation.

As in all other clinical cases of apnea discussed above, the initiating factors leading to apnea in these four infants are obscure. On the other hand, a reasonable hypothesis for the mechanism of recovery from individual spells can be offered on the basis of observations on breath-holding infants and animal studies. The hypothesis is that asphyxia resulting from the initial apnea led to hypoxic seizure. Following the seizure, the gasping mechanism became activated. Reoxygenation of the blood occurred as a consequence of one or more hypoxic gasps. Return of regular respirations then occurred as a consequence of reoxygenation. Having a somewhat different regulating mechanism than eupneic breaths, the gasps were not inhibited by whatever factors suppressed respiration at the onset of the spell. Finally, and most important, the conditions that precipitated the spell initially, whatever their source, were removed during the course of this rather violent sequence of events, and so return of regular respirations was not impeded. The ultimate failure to recovery by these children may be related to the deleterious effect of repeated hypoxic exposure on the autoresuscitation mechanism.

IV. Summary

The terminal respiratory pattern that appears during the course of progressive asphyxia has been termed "gasping." Gasping occurs when hypoxia produces coma and areflexia. The role played by the normal medullary respiratory centers, as opposed to specialized "gasping centers," in the generation of gasping pattern is unclear. Gasps are highly functional since they can readily initiate spontaneous recovery from anoxic coma ("autoresuscitation") with complete reversal of cardiovascular and central nervous system depression. The strength and limitations of the autoresuscitation mechanism are the focus of current research. Autoresuscitation by gasping has been well documented to occur during the course of prolonged apnea in infants. Death attributed to an apneic spell implies failure of the gasp mechanism to autoresuscitate.

References

Adolph, E. E. (1969). Regulations during survival without oxygen in infant mammals. *Respir. Physiol.* **7**:356-368.

Avery, M. E., and Fletcher, B. D. (1974). *The Lung and Its Disorders in the Newborn Infant.* Philadelphia, W. B. Sanders.

Barcroft, J. (1934). *Features in the Architecture of Physiologic Function,* Cambridge, Cambridge University Press.

Boddy, K., and Dawes, G. S. (1975). Fetal breathing. *Bri. Med. Bull.* **31**:3-7.

Bystryzcka, E., Nail, B. S., and Purves, M. J. (1975). Central and peripheral neural respiratory activity in the mature sheep foetus and newborn lamb. *Respir. Physiol.* **25**:119-215.

Campbell, A. C. M., Cross, K. W., Dawes, G. S., and Hyman, A. I. (1971). A comparison of air and O_2 in a hyperbaric chamber or by positive pressure ventilation in the resuscitation of newborn rabbits. *J. Pediatr.* **68**:133-158.

Chernick, V., Fariday, E. E., and Pagtakahn, R. D. (1975). Role of peripheral and central chemoreceptors in the initiation of fetal respiration. *J. Appl. Physiol.* **38**:407-410.

Daily, W. J., Klaus, M., and Meyer, H. B. (1969). Apnea in premature infants: Monitoring, incidence, heart rate changes and effects of environmental temperature. *Pediatrics* **43**:10-18.

Davis, J. A. (1961). The effect of anoxia in newborn rabbits. *J. Physiol.* **155**:56P.

Davis, P. J., Macefield, G., and Nail, B. S. (1986). Respiratory muscle activity during asphyxic apnoea and opisthotonos in the rabbit. *Respir. Physiol.* **6**:28-294.

Dawes, G. S. (1968). *Foetal and Neonatal Physiology.* Chicago, Year Book Medical Publishers.

Dawes, G. S., Fox, H. E., and Richards, R. T. (1972). Variations in asphyxial gasping with fetal age in lambs and guinea-pigs. *Q. J. Expl. Physiol.* **7**:131-138.

Dawes, G. S., Jacobson, H. N., Mott, J. C., and Shelley, J. H. (1960). Some observations in fetal and newborn rhesus monkeys. *J. Physiol.* **169**:167-184.

Dawes, G. S., Jacobson, H. N., Mott, J. C., Shelley, J. H., and Stafford, A. (1963). The treatment of asphyxiated, mature, foetal lambs and Rhesus monkeys with intravenous glucose and sodium carbonate. *J. Physiol.* **169**:167-184.

Dehaan, R., and Field, J. (1959). Anoxic endurance of cardiac and respiratory function in the adult and infant rat. *Am. J. Physiol.* **197**:445-448.

Donnelly, D. F., and Haddad, G. G. (1986). Respiratory changes induced by prolonged laryngeal stimulation in awake piglets. *J. Appl. Physiol.* **61**:1018-1024.

Duffy, T. E., Nelson, S. R., and Lowry, O. H. (1972). Cerebral carbohydrate metabolism during acute hypoxia and recovery. *J. Neurochem.* **19**:959-977.

Fazekas, J. F., Alexander, A. D., and Himwich, H. E. (1941). Tolerance of the newborn to anoxia. *Am. J. Physiol.* **134**:281-287.

Fewell, J. E., and Baker, S. B. (1987). Arousal from sleep during rapidly developing hypoxemia in lambs. *Pediatr. Res.* **22**:471-477.

French, J. W., Morgan, B. C., and Guntheroth, W. G. (1972). Infant monkeys—A model for crib death. *Am. J. Dis. Child.* **123**:480-484.

Gastaut, H., and Gastaut, Y. (1968). Electroencephalography and clinical study of anoxic convulsions in children: Their location within groups of infantile convulsions and their differentiation from epilepsy. *Electroencephalogr. Clin. Neurophysiol.* **10**:607-620.

Gauk, S. W., Kidd, L. and Prichard, J. S. Mechanisms of seizures associated with breath holding spells. *New. Eng. J. Med.* **268**:1436-1441.

Gershan, W. M., Jacobi, M. S., and Thach, B. T. (1990a). Maturation of cardiorespiratory interactions in spontaneous recovery from hypoxic apnea ("autoresuscitation"). *Pediatr. Res.*, **28**:87-93.

Gershan, W. M., Jacobi, M. S., and Thach, B. T. (1990b). Spontaneous recovery from hypoxic apnea: a comparison of adept and inept mice. *Am. Rev. Resp. Dis.* **141**:A812.

Glogowska, M. Richardson, P. S. Widdicombe, J. D., and Wining, A. J. (1972). The role of the vagus nerves, peripheral chemoreceptors and other afferent pathways in the genesis of augmented breaths in cats and rabbits. *Respir. Physiol.* **16**:179-196.

Godfrey, S. (1968a). Respiratory and cardiovascular changes during asphyxia and resuscitation of fetal and newborn rabbits. *Q. J. Exp. Physiol.* **3**:97-118.

Godfrey, S. (1968b). Blood gases during asphyxia and resuscitation of fetal and newborn rabbits. *Respir. Physiol.* **4**:309-321.

Guntheroth, W. G. (1974). Primary apnea, hypoxic apnea and gasping. In *S.I.D.S.* Edited by R. R. Robinson. Toronto, Canadian Foundation for the Study of Infant Deaths, pp. 261-267.

Guntheroth W. G., and Kawabori, I. (1975). Hypoxic apnea and gasping. *J. Clin. Invest.* **56**:1371-1377.

Guilleminault, C., Ariagno, R., Souquet, M., and Dement, W. C. (1976). Abnormal polygraphic findings in near-miss sudden infant death. *Lancet* **1**:1326-1327.

Guilleminault, C., Peraita, R., Souguet, M., and Dement, W. C. (1975). Apneas during sleep in infants: Possible relationship with sudden infant death syndrome. *Science* **190**:167-699.

Hansen, A. J. (1977). Extracellular potassium concentration in juvenile and adult rat brain cortex during anoxia. *Acta Physiol. Scand.* **99**:412-420.

Harned, H. S., Griffin, C. A., Berryhill, W. S., Mackinney, L. G., and Sugioka, K. (1967). Role of carotid chemoreceptors in the initiation of effective breathing of the lamb at term. *Pediatrics* **39**:329-336.

Himwich, H. E., Bernstein, A. O., Herrlich, H., Chester, A., and Fazekas, J. F. (1941). Mechanisms for the maintenance of life in the newborn during anoxia. *Am. J. Physiol.* **135**:387-391.

Holowach-Thurston, J., Hauhart, R. E., and Jones, E. M. (1974). Anoxia in mice: Reduced glucose in the brain with normal or elevated glucose in plasma and increased survival after glucose treatment. *Pediatr. Res.* **8**:238-243.

Jacobi, M. S., Gershan, W. M., and Thach, B. T. (1989a). Failure of recovery from hypoxic apnea associated with abnormality of gasping. *Pediatr. Res.* **25**:367A.

Jacobi, M. S., Gershan, W. M., and Thach, B. T. (1989b). Fascilitatory effect of phenobarbital on spontaneous recovery from hypoxic apnea in mice. *FASEB J.* **3**:855.

Jacobi, M. S., and Thach, B. T. (1989). Effect of maturation on spontaneous recovery from hypoxic apnea by gasping. *J. Appl. Physiol.* **66**:2384-2390.

Jansen, A. H., and Chernick, V. (1974). Respiratory response to cyanide after peripheral chemodenervation. *J. Appl. Physiol.* **36**:1-5.

Jansen, A. H., and Chernick, V. (1975). Site of central chemosensitivity in fetal sheep. *J. Appl. Physiol.* **39**:1-6.

Johnson, P., Salisbury, D. M., and Storey, A. T. (1975). Apnea induced by stimulation of sensory receptors in the larynx. In *Symposium on Development of Upper Respiratory Anatomy and Function.* Edited by J. F. Bosma and J. Showacre. Washington, DC, US Government Printing Office, pp. 160-183.

Karlberg, P. (1960). The adaptive changes in the immediate postnatal period, with particular reference to respiration. *J. Pediatr.* **6**:8-604.

Koch, G., and Wendel, H. (1968). Adjustment of arterial blood gases and acid balance in the normal newborn infant during the first week of life. *Biol. Neonate* **42**:136-161.

Lawson, E. E., and Thach, B. T. (1977). Respiratory patterns during progressive asphyxia in newborn rabbits. *J. Appl. Physiol.* **43**:468-474.

Lawson, E. E. (1981). Prolonged central respiratory inhibition following reflex apnea. *J. Appl. Physiol.* **50**:874-879.

Lombroso, C. T., and Lerman, P. (1967). Breathholding spells (cyanotic and pallid infantile syncope). *Pediatrics* **39**:563-581.

Lumsden, T. E. (1922). Observations on the respiratory centers in the cat. *J. Physiol.* (London) **57**:13-160.

Mathew, O. P., Thach, B. T., Abu-Osba, Y. K. Brouillette, R. T., and Roberts, J. L. (1984). Regulation of upper airway maintaining muscles during progressive asphyxia. *Pediatr. Res.* **18**:819-822.

Mortola, J. P., Fisher, J. T., Smith, B., Fox, G., Weeks, S., and Willis, D. (1982). Onset of respiration in infants delivered by cesarian section. *J. Appl. Physiol.* **52**:716-724.

Moskowitz, M. A., Fisher, J. N., Simpser, M. D., and Strieder, D. (1976). Periodic apnea, exercise hypoventilation and hypothalamic dysfunction. *Ann. Intern. Med.* **84**:171-173.

Patrick, J. E., Dalton, K. J., and Dawes, G. S. (1976). Breathing patterns before death in fetal lambs. *Am. J. Obstet. Gynecol.* **12**:733-78.

Peiper, A. (1963). *Cerebral Function in Infancy and Childhood.* New York, Consultants Bureau.

Reynolds, L. B. (1962). Characteristics of an inspiration—augmenting reflex in anesthetized cats. *J. Appl. Physiol.* **17**:683-688.

Sanocka, W. M., Donnelly, D. F., and Haddad, G. G. (1988). Cardiovascular and neurophysiologic changes during graded duration of apnea in piglets. *Pediatr. Res.* **23**:402-408.

Schmidt, L. (1950). *Der "erst" Atemzug* (The "first" breath). *Monatschr. Kinderhk.* **98**:213 (cited by Peiper, 1963).

Schwieler, G. H. (1968). Respiratory regulation during postnatal development in cats and rabbits and some of its morphological substrate. *Acta Physiol. Scand.* (Suppl. 304):1-123.

Shelley, H. J. (1961). Glycogen reserves and their changes at birth and in anoxia. *Br. Med. Bull.* **17**:137-143.

Smith, C. A. (1950). Physiologic background of fetal anoxia at birth and cyanosis in the newborn. *Am. J. Dis. Child.* **79**:1-9.

St. John, W., Bledsoe, T. A., and Sokol, H. W. (1984). Identification of medullary loci critical for neurogenesis of gasping. *J. Appl. Physiol.* **56**:1008-1019.

St. John, W. M., Bledsoe, T. A., and Tenney, S. M. (1985). Characterization by stimulation of medullary mechanisms underlying gasping neurogenesis. *J. Appl. Physiol.* **58**:121-128.

St. John, W., and Knuth, K. V. (1981). A characterization of the respiratory pattern of gasping. *J. Appl. Physiol.* **50**:984-993.

Steinschneider, A. (1972). Prolonged apnea and the sudden infant death syndrome: Clinical and laboratory observations. *Pediatrics* **50**:646-654.

Stevens, M. H. (1965). Sudden unexplained death in infancy. *Am. J. Dis. Child.* **110**:243-247.

Sullivan, C. E., Berthon-Jones, M., and Issa, F. G. (1983). Remission of severe obesity-hypoventilation syndrome after short-term treatment during sleep with nasal continuous positive airway pressure. *Am. Rev. Respir. Dis.* **128**:177-181.

Swann, H. G., and Brucer, M. (1949). The cardiorespiratory and biochemical events during rapid anoxic death. II. Acute anoxia. *Tex. Rep. Biol. Med.* **7**:539-552.

Thach, B. T., and Lawson, E. E. (1976). Hypoxic depression of respiration in the newborn and adult rabbit. *Pediatr. Res.* **13**:542 (Abstract).

Thach, B. T., and Lawson, E. E. (1979). Death from upper airway obstruction: decreased survival time in 33-day old rabbits. *Pediatr. Res.* **13**:542 (Abstract).

Thach, B. T., and Taeusch, H. W. (1976). Sighing in newborn human infants: the role of the inflation augmenting reflex. *J. Appl. Physiol.* **41**:502-509, 1976.

Tizard, J. P. M. (1964). Indications for oxygen therapy in the newborn. *Pediatrics* **34**:771-786.

Tschirgi, R. D., and Gerard, R. W. (1947). The carotid-mandibular reflex in acute respiratory failure. *Am. J. Physiol.* **150**:358-364.

Ward, A. A. (1947). Decerebrate rigidity. *J. Neurophysiol.* **10**:89-103.

Ward, A. A., and Wheatley, M. D. (1947). Sodium cyanide: sequence of changes of activity induced at various levels of the central nervous system. *J. Neuropathol. Exp. Neurol.* **6**:292-294.

Werne, J., and Garrow, I. (1953). Sudden, apparently unexplained death during infancy. Pathological findings in infants observed to die suddenly. *Am. J. Pathol.* **29**:817-827.

Wheatley, M. D., Lipton, B., and Ward, A. A. (1947). Repeated cyanide convulsions without central nervous pathology. *J. Neuropathol. Exp. Neurol.* **6**:408-411.

Woodrum, D. E., Paber, J. T., Wennberg, R. P., and Hodson, W. A. (1972). Chemoreceptor response on initiation of breathing in the fetal lamb. *J. Appl. Physiol.* **33**:120-125.

AUTHOR INDEX

Italic numbers give the page on which the complete reference is listed.

A

Abboud, E. F., 566, 574, *575, 577*
Abbrecht, P. H., 391, *425*
Abe, H., 635, *638*
Aberdeen, E., 630, *640*
Abrams, I. F., 560, 569, 573, *579, 586*
Abroms, I. F., 248, 262, *265, 268*
Abu-Osba, Y. K., 229, 233, *242,* 564, 587, 688, *698*
Acevedo-Duncan, M., 80, *97*
Acker, H., 291, 293, 296, 306, *312, 316,* 327, *336,* 647, 648, 649, 650, 651, 653, 654, *656, 660*
Adams, D. J., 119, *145,* 163, *172*
Adams, E. M., 354, *370*
Adams, F. H., 492, *511*
Adams, K. G., 502, *518*
Adams, P. R., 166, *174*
Adamson, T. M., 484, *507,* 555, 556, *575, 576, 583*
Aderem, A., 81, *109*
Adler, S. M., 248, 249, 252, *269,* 573, *579*

Adolph, E. E., 683, 684, 686, 688, *695*
Adolph, E. F., 654, *656*
Adrian, R. H., 438, *449*
Agata, Y., 668, *677*
Aghajanian, G. K., 654, *656*
Agostoni, E., 247, 248, 253, *263,* 274, *284, 473,* 560, *575*
Agranoff, B. W., 17, *68*
Agrawal, H. C., 623, *635*
Aguggini, G., 250, *264,* 495, *508*
Aguilar, J. S., 73, *102*
Aitkin, P. G., 596, *607*
Akabas, S. R., 457, 458, 459, 461, 462, 463, 468, 469, 470, *474, 475, 480, 481*
Akers, R. F., 33, *54,* 83, *95*
Akesson, T. R., 43, *54*
Akita, Y., 78, *105*
Akiyama, Y., 552, *585*
Alarie, Y., 224, *244*
Albani, M., 571, *579*
Albers, J. W., 391, *425*
Albers, R. W., 17, *68*
Albert, K. A., 81, *95, 108*

Albertini, M., 250, *264*, 495, *508*
Alcorn, D., 555, *575*
Alcorn, D. G., 484, *507*
Aldrich, R. W., 161, 168, *172*
Aldrich, T. K., 390, *416*
Alexander, F. A. D., 654, *658*
Alexander, J., 257, *266*
Alexander, S. C., 626, *636*
Alkon, D. L., 72, 83, *95, 105*
Allen, D. G., 439, *449*
Allen, J. C., 489, *516*
Allen, J. L., 491, *514*
Allen, L. G., 667, *673*
Alliod, C., 74, *105*
Alsberge, M., 324, 330, 331, 333, 334, *339*
Alter, C., 618, *639*
Altiere, R. J., 483, 484, 485, 486, 500, 501, *509, 517*
Altose, M. D., 257, 259, *266*, 456, 457, *476, 478*
Alving, B. O., 115, *145*
Amaral, D. G., 532, *546*
Amyot, R., 501, *513*
Anch, A. M., 521, *546*
Andersen, L. B., 468, *478*
Andersen, W. B., 79, *102*
Anderson, B. S., 631, 633, *635*
Anderson, J. P., 596, *612*
Anderson, J. V., Jr., 574, *575*
Anderson, J. W., 229, 230, *239, 242*
Anderson, M. L., 83, *105*, 633, *636*
Andersson, P., 667, *675*
Andersson, R., 503, *507*
Andersson, R. G. G., 483, 498, *507*
Andreasen, K. I., 81, *96*
Andreasen, T. J., 81, *96*
Andrews, W., 566, 570, *582*
Andronikou, S., 329, *339*
Andry, D., 125, 141, *145*
Anggard, A., 23, 24, *63*
Angqvist, K. A., 468, *475*
Aoki, M., 645, *656*
Apfeldorf, W., 72, *106*
Arad, I., 504, *517*
Araki, H., 633, *640*

Aranda, J. V., 672, *673*
Arata, A., 664, *677*
Archer, P. W., 360, *371*
Arends, G., 664, *676*
Arens, J., 248, 252, *268*
Ari, Y. B., 606, *607*
Arita, H., 346, 347, 355, 357, *361, 366*
Arita, H. N., 205, *213*
Arividsson, U., 664, *673*
Armstrong, C. J., 237, *241*, 492, 496, *510*
Armstrong, D. L., 82, *95*
Armstrong, R. B., 462, 464, 468, *477*
Arnold, A. P., 397, *422*
Arnould, P., *286*
·Aronson, P. A., 356, *365*
Ascher, P., 26, *61, 65*
Ase, K., 78, 80, *95, 99*
Aserinsky, E., 569, *575*
Ashford, M. L. J., 602, *613*
Ashkenazi, A., 357, *369*
Aswal, S., 646, *656*
Atweh, S., 49, *54*
Atwood, H. L., 414, *423*
Auld, P. A. M., 573, *582*
Auld, V., 74, *99*
Ausoni, S., 410, 411, 414, *425*
Averill, D. B., 644, *656*
Avery, M. E., 358, *366*, 573, *575*, 691, *695*
Aw, T. Y., 631, 633, *635*
Axel, R., 28, 29, *61*
Axelrod, J., 72, *95*
Azzaroni, A., 535, *546*

B

Bachofen, H., 257, 259, *263*
Bacon, J. P., 114, *145*
Baden, D. G., 606, *609*
Bader, C. R., 160, *172*
Badura, R. J., 472, 473, *479*
Baeuerle, P. A., 84, *95*
Bagchi, J. C., 80, *101*
Baghdoyan, H. A., 527, *545*

Bagust, J., 380, 386, 387, *416*
Bähler, M., 91, *95*
Bahr, R., 440, 444, *453*
Bailey, A., 557, *584*
Baines, A. J., 91, *95*
Bainton, C. R., 255, *268*, 279, *287*
Baitinger, C., 86, *107*
Bakalyar, H. A., *102*
Baker, D. G., 486, 497, 506, *507,
513*
Baker, J., 247, *269*
Baker, J. N., 625, *635*
Baker, P. F., 300, *312*
Baker, P. T., 595, *607*
Baker, S. B., 693, *696*
Baker, T. L., 534, *541*, 570, 572,
575, 594, *607*
Balasz, R., 309, *314*
Baldwin, K. M., 414, 415, *416*
Balestrino, M., 596, *607*
Balfour, H. H., 3, *7*
Balice-Gordon, R. J., 380, *416*
Ball-Burnett, M. E., 462, *477*
Ballantyne, D., 128, *153*, 206, 209,
210, *213, 215, 255, 268,* 277,
278, 279, *287*
Ballivet, M., 74, *105*
Baltimore, D., 84, *95*
Bambrick, L., 160, *172*
Bamford, O. S., 668, *673*
Bancalari, E., 248, 249, 250, *265,
266,* 473, *477,* 504, *510*
Bandler, R., 532, 533, *541*
Bang, R., 606, *609*
Baraban, J. M., 78, *110*
Barany, M., 414, *417*
Baranyi, A., 527, *548*
Barbeau, H., 23, *54*
Barcroft, J., 684, 687, 691, *695*
Barillot, J. C., 645, *656*
Barker, D. L., 124, *145, 151*
Barker, J. L., 115, *145*, 157, *175*,
310, *317*
Barman, S. M., 349, *362*
Barnard, D. G., 360, *371*
Barnard, E. A., 27, *64*

Barnard, R. J., 471, *476*
Barnes, P. J., 483, 484, 486, 498,
500, 505, *507, 514*
Baron, M., 295, *315*
Barrat, E., 486, *511*
Barrett, J. N., 157, *172*, 296, *312*
Barrett, P., 72, *106*
Barrington, K. J., 569, 570, *575*
Bartfai, T., 19, 23, *60, 61, 63*
Barthelemy, P., 229, *241*
Bartlett, D., 229, 230, *240*, 254, *263*
Bartlett, D., Jr., 155, *172*, 222, 225,
237, *240, 241,* 280, 281, *283,*
522, *546*, 562, 563, *576, 578*
Bartoli, A., 256, *263*
Bartoshik, L. M., 126, *153*
Basbaum, A. I., 180, *193*
Basbaum, C. B., 506, *507*
Bassai, M., 534, *541*
Battaglia, F. C., 626, *638*
Baue, A. E., 633, *635*
Baulieu, E. E., 39, *54*
Baumann, A., 74, 108
Baumgarten, R., 644, *656*
Baust, W., 522, *541*
Babb, M., 527, *542*
Baxter, J. D., 48, *67*
Baumgold, J., 160, *172*
Bawa, P., 434, *450*
Bayer, S. M., 623, *636*
Bayev, A. V., 224, *241*
Bayley, H., 32, *59*
Bayliss, D. A., 21, 30, 35, 36, 37,
39, 40, 41, 42, 43, 51, 53, *54,
57, 64, 69,* 351, *367*
Bayliss, D. B., 50, *68*
Baylor, D. A., 600, *607*
Bayol, A., 489, 490, *507*
Bayon, A., 48, *54*
Bazzi, M. D., 79, *95*
Bazzy, A. R., 155, 169, *174,* 457,
458, 459, 460, 461, 462, 463,
464, 466, 468, 469, 470, *474,
480, 481*
Beals, J. K., 177, *196*
Beaman-Hall, C. M., 86, *109*

Beardsmore, C. S., 567, *576*
Beaudet, A., 349, *367*
Becker, C. M., 74, *99*
Becker, L., 560, *584*
Becker, L. E., 322, *339*, 672, *677*
Beckley, S. C., 48, 49, *66*
Bedard, P., 23, *54*
Begin, R., 321, 322, 334, *336, 337,*
    360, *363*, 571, 572, *577*, 593,
    607, 608, 662, *673*
Beland, J., 485, 500, *515*
Belcastro, A. N., 414, 415, *417*
Belenky, D. A., 321, *336*, 572, *576*
Bell, R. M., 78, 79, 80, *99, 102*
Bellemare, F., 432, 433, 442, 444,
    448, *449*, 461, *475*
Bellemare, G., 461, *475*
Bellingham, M. C., 275, *283*
Belman, M. J., 391, 397, 404, *417,*
    426, 466, 467, *475, 480*
Belmonte, C., 308, *312*
Belsito, O., 524, 537, *547*
Beltz, B., 124, *145*
Ben-Air, Y., 603, *608, 610*
Bengtsson, F., 603, *613*
Benlabed, M., 259, *268*
Bennet, L., 662, 667, 668, *674*
Bennett, H. S., 13, *57*
Bennett, M., 379, 380, 389, *417*
Bennett, M. K., 84, 86, 87, *95, 101*
Bennett, V., 82, 91, *95*
Benoit, P., 470, *475*
Benovic, J. C., 73, *107*
Benovic, J. L., 29, *68*
Benowitz, L. I., 76, 81, *96*
Beneveniste, H., 309, *312*
Benveniste, J., 489, 490, *507*
Bereiter, D., 129, *145*
Berger, A. J., 129, 131, *145*, 271,
    275, 277, 279, *283, 285*, 355,
    362, 359, *362*, 382, 386, *417,*
    419, 422, 537, *544*, 644, *656*
Berger, K., 354, 355, *362*
Berger, W., 354, 355, *362*
Bergstrom, J., 462, 469, *475*
Bergstrom, L., 672, *673*

Berkowitz, B. A., 489, *509*
Berlucchi, G., 522, *541*
Bernard, C., 616, *636*
Berndt, J., 354, 355, *362*
Berne, R. M., 666, 671, *673, 679*
Bernstein, A. O., 685, *697*
Berridge, M. J., 31, *54, 55*, 76, *96,*
    305, *312, 319*
Berry, N., 80, *95*
Berryhill, W. S., 687, *697*
Berssenbrugge, A., 42, *68*, 539, *541*
Berthiaume, Y., 360, *363*
Berthon-Jones, M., 693, *699*
Berthoud, H. R., 129, *145*
Bertin, R., 502, *517*
Bertrand, D., 160, *172*
Bertrand, F., 276, *287*, 534, *541, 542*
Berwald-Netter, Y., 160, *172*
Besson, J. M., 532, *545*
Besterman, J., 84, *107*
Bettini, G., 457, 461, *480*
Betto, R., 411, *418*
Betz, H., 74, *99*
Betz, W. J., 379, 380, 386, *417, 424*
Beuerman, R. W., 220, *240*
Bhakthavathsalan, A., 626, 628, 629,
    638
Bhansali, P., 39, 42, *68*
Bhide, S., 323, 324, 330, 331, 333,
    334, *339*
Bhutani, V. K., 232, *240*, 495, *508*
Bialojan, C., 94, *96*
Bianchi, A. L., 279, *283*, 352, *362,*
    365, 534, *541*, 645, *656*
Biedenbach, M. A., 220, *240*
Bieger, D., 272, *283*
Biggs, W. H., III, 31, *69*
Bigland-Ritchie, B., 389, *417*, 432,
    433, 434, 436, 438, 440, 442,
    444, 445, 446, *450, 451, 452,*
    453
Bilezikjian, L. M., 31, 51, *65*
Birnbaumer, L., 29, *55*, 73, *106*
Bischoff, A., 201, 208, *214*
Biscoe, T. J., 272, *283*, 291, 292, 295,
    296, 299, 300, 302, 304, 305, 307,

[Biscoe, T. J.]
  308, 309, 310, *312, 313, 315, 319,*
  322, 324, 326, 334, *336, 337,* 645,
  651, *657*
Bishop, B., 257, 259, *263*
Bissonnette, J. M., 358, *366*
Bistoletti, P., 666, *676*
Bitter, N., 633, *636*
Bixby, J. L., 379, *417*
Bjorklund, A., 18, 19, 45, *55, 60,*
  180, *194*
Bjure, J., 274, *288*
Bjuro, T., 468, *475*
Black, A. M. S., 308, *313*
Blackman, J. G., 14, *55*
Blackshear, P. J., 31, *55,* 81, *99, 108*
Blager, F. B., 3, *7*
Blake, J. S., 490, *513*
Blanc, W. A., 3, *7*
Blanchard, P. W., 321, 322, 334,
  337, 593, *608*
Blanco, C. E., 322, 324, 325, 326,
  327, 330, 333, 334, *336,* 391,
  392, 395, 401, 404, 408, 413,
  417, 418, 426, 497, 508, 593,
  607, 665, 669, *673*
Bland, B. H., 540, *548*
Blank, W. F., Jr., 650, *658*
Blatz, A. L., 74, *96*
Blau, H. M., 380, 386, *426*
Blaustein, J. D., 37, 38, 43, 44, *55*
Blaustein, M. P., 14, *55,* 300, *312,*
  601, *607*
Bledsoe, S. W., 341, 355, *362*
Bledsoe, T. A., 685, 687, *698*
Bley, K. R., 74, 82, *109*
Bloom, F. E., 16, 24, 48, *54, 55,*
  654, *656*
Bloom, S. R., 487, 499, 502, *516*
Blum, F., 626, 629, *637*
Bochina, T. V., 633, *641*
Bockaert, J., 25, *68*
Bocking, A. D., 558, *580*
Boddy, K., 358, 360, *362,* 555, 557,
  576, 593, *607,* 681, *695*
Bodegard, G., 255, *263,* 570, *576*

Bodian, D., 13, *55*
Bodine-Fowler, S., 391, 392, *423*
Bodine, S. C., 378, *417*
Boggs, D. F., 155, *172,* 229, 230,
  240, 280, 281, *283*
Boileau, R., 485, *511*
Bolser, D. C., 261, *263, 268*
Bolton, D. P. G., 569, *576*
Bonora, M., 662, 665, *673, 674*
Bondareff, W., 654, *657*
Bonifacino, J. S., 72, *109*
Bonigk, W., 76, *101*
Boning, D., 462, *481*
Bonner, T. I., 28, *55,* 357, *362*
Bonora, M., 252, 255, *265,* 563, 571,
  572, *576,* 593, *607*
Booth, C. M., 380, *418*
Booth, F. W., 471, *477*
Borison, H. L., 343, 345, 354, *362*
Borison, R., 343, *362*
Boron, W. R., 300, *314*
Bosley, T. M., 309, *314*
Bosma, J., 564, *576*
Bots, R. S., 554, *584*
Bottevo, L., 78, *108*
Bouhuys, A., 486, 492, 493, 505, *516*
Bourke, R. S., 600, *607*
Boushey, H. A., 229, 230, *240,* 507,
  508
Bouveret, P., 410, *427*
Bouvier, M., 72, *110*
Bouws, P., 554, *584*
Bowden, J. A., 459, *481*
Bowery, N. G., 26, *55*
Bowes, G., *283,* 358, *362,* 522, *541,*
  552, 555, 556, 559, 573, *576,*
  577, 583, 584, 594, *607, 611,*
  662, 664, *673*
Bowker, R. M., 180, *193, 198*
Bowman, G. S., 626, 629, *637*
Boxer, P., 525, *545*
Boyarski, L. L., 343, 345, *367*
Boychuk, R. B., 252, *263,* 593, *612*
Bradley, G. W., 235, *240*
Bradley, K. H., 489, *508*
Bradley, M., 467, *478*

Brady, J. P., 321, *336*, 571, 572, 573, *576, 585*
Brady, S. T., 91, *104*
Brambilla-Sant'Ambrogio, F., 228, *243*
Bramble, D. M., 144, *145*
Brancatisano, T. P., 563, *576*
Brandt, N. J., 672, *673*
Brann, M. R., 28, *55*, 357, *362*
Breckenridge, C. G., 521, *544*
Breese, G., 308, *317*
Breese, G. R., 112, 125, 141, *145, 151, 152*
Bregestoviski, P., 26, *65*
Brennan, R. W., 626, 629, *637*
Brenner, R., 74, *98*
Brian, J., 485, *518*
Briggs, R. W., 618, *640*
Brink, C., 489, 490, 502, 503, *507, 508, 509*
Briskey, J. E., 392, *419*
Brockerhoff, H., 80, *96*
Brodecky, V., 358, *362*, 555, *583*
Brodin, E., 19, 34, *55, 60*, 224, *242*
Brodin, L., 125, *145, 146*, 169, *173*
Brody, J. S., 334, *338*, 594, *610*
Bromley, S. M., 573, *576*, 594, *607*
Brooke, M. H., 407, 414, *417*
Brooks, R. J., 605, *613*
Brooks, S. P. J., 605, *613*
Brouillette, R. T., 564, 573, *581, 587*, 688, *698*
Brown, A. C., 220, *240*
Brown, A. M., 354, *363*
Brown, D. L., 349, *363*
Brown, G., 113, *146*
Brown, J. C., 19, 20, 49, *65*
Brown, M. C., 379, 380, 386, *418*
Brown, T. H., 33, *55*
Brownstein, M. J., 18, *62*, 129, *152*
Brownstein, S., 18, 52, *66*
Brozanski, B. S., 159, 169, *172*, 259, *269*, 384, 385, *418*
Brozek, G., 648, 650, 651, *659*
Bruce, D. A., 633, *640*
Bruce, E., 257, *266*

Bruce, E. N., 262, *268*, 341, 343, 360, *363*, 560, 563, 566, 574, *575, 577, 586*
Brucer, M., 690, *699*
Brum, V. C., 39, *57*
Brumley, G., 358, *366*
Brundage, K. L., 492, 493, 494, 495, 500, *508*
Bryan, A. C., 248, 249, 250, 260, *265, 266, 267*, 322, *329*, 401, 404, 408, *423, 424*, 448, *451*, 457, 459, 464, 467, 469, 472, 473, *475, 478, 479*, 522, *548*, 552, 559, 560, 562, 563, 566, 567, 568, 570, 574, *577, 580, 581, 582, 583, 584, 586*, 594, *607*, 662, 664, 672, *673, 677*
Bryan, M. H., 248, 249, 250, 260, *265, 267*, 457, 459, 473, *479*, 495, 504, *518*, 559, 560, 562, 566, 567, 568, 570, *580, 582, 583, 584, 586*
Buchan, A., 19, 20, 49, *65*
Buchanan, J. T., 169, *173*
Buck, C. A., 75, *96*
Buckley, N. J., 28, *55*, 357, *362*
Budzinska, K., 205, *213*, 350, *363*, 664, *673*
Buerk, D. G., 306, *317*
Bugaisky, L. B., 410, 411, *418*
Buie, A. E., 606, *613*
Buller, A., 408, *418*
Bundgaard, M., 646, *657*
Bunge, M., 3, *7*
Bunn, J. C., 130, *146*
Burch, R. M., 72, *95*
Burdick, B., 489, *516*
Burdsall, J. A., 502, 506, *518*
Bureau, M. A., 246, *264*, 321, 322, 334, *336, 337*, 360, *363*, 571, 572, *577*, 593, *607, 608*, 662, *673*
Bures, J., 599, *613*, 648, 650, 651, *657, 659*
Buresova, O., 651, *657*
Burgess, S., *109*

Burgin, K., 85, *101*
Burgoyne, R. D., 74, *96*
Burke, D., 446, *451*
Burke, R. E., 258, *263*, 376, 377,
    378, 379, 381, 382, 383, 384,
    388, 392, 407, 413, *418, 427,*
    434, 444, *450*
Burnstock, G., 485, 505, *508*
Burri, P. H., 484, 498, *508*
Burrows, M., 114, 116, *151*
Burton, H., 129, *151*
Burton, R. F., 355, 356, *363*
Busse, C., 571, *586*
But, V. I., 224, *240*
Butler, A. L., 351, *366*
Butler-Browne, G. S., 410, 411, *418,*
    427
Butler, S., 617, 618, *639*
Buttner, N., 76, *96*
Bylund-Fellenius, A. C., 468, *475*
Byrne, J. H., 32, *67*, 134, *146*
Bystrzycka, E., 256, *263*, 645, *657,*
    681, 684, 690, *695*

C

Cabana, T., 178, 179, 180, 181, 183,
    189, 190, 191, 192, *193, 196*
Cabezas, G. A., 485, 486, 494, *508*
Caces, R., 570, 571, *583*
Cade, D., 401, *424*, 457, 473, *479,*
    560, *584*
Caddy, K. W. T., 295, 296, 304, 305,
    312, 315, 334, *337*
Cadilla-Perezrious, R., 664, *676*
Cahoon, R. L., 594, *611*
Cain, S. M., 592, *608*
Calabrese, R. L., 116, 128, *146, 152*
Calancie, B., 434, *450, 453*
Calaresu, F. R., 186, *195*
Caldwell, J. H., 379, 380, 386, *417,*
    418
Caldwell, P. C., 305, *314*
Caley, D. W., 654, *657*
Callahan, B., 569, *586*
Callahan, D., 456, *475*

Callanan, D., 247, 253, 255, 259,
    264
Callegari, C. C., 668, *677*
Camerer, H., 647, 648, *660*
Cameron, R., 90, *97*
Cameron, W. E., 159, 160, *172*, 384,
    385, 386, *418, 422*, 644, *656*
Campbell, A. C. M., 683, 684, *695*
Campbell, D. J., 571, *581*, 664, *675*
Campbell, E. J. M., 259, *263*
Campbell, K. P., 74, *98*
Campbell, P. J., 414, 415, *416*
Campochiaro, P., 46, *55*
Candelore, M. R., 28, *57, 69*
Cantini, M., 411, *418*
Capon, D. J., 357, *369*
Car, A., 272, *283*
Carafoli, E., 304, 305, *314*, 601, *608*
Carbone, E., 309, *316*
Cardaman, R. C., 389, *427*
Cardasis, C. A., 471, *475, 476*
Carlo, W. A., 248, 260, *267, 269,*
    563, 566, *577, 579, 582*
Carlson, V. R., 523, *542*
Carlton, S. M., 540, *545*
Caron, M. G., 28, 29, *57, 62, 68,* 72,
    73, *98, 107, 110*
Carpenter, D. O., 354, *363*, 381, *421*
Carpenter, G., 72, 75, *96*
Carraro, U., 411, *418*
Carrier, D. R., 144, *145*
Carrive, P., 532, *541*
Carroll, J. L., 321, 322, 334, *337,*
    593, 595, *611*
Carson, S., 248, 249, 252, *268*
Cartaud, J., *56*
Carter, A-J., 602, *608*
Carter, M. S., 47, *62*
Cartheuser, C. F., 644, *657*
Casey, P. J., 72, 74, *96, 98*
Cassens, R. G., 392, *419*
Cassini, P., 189, *194, 197*
Castagna, M., 77, *96*
Castellucci, V. F., 32, 33, *55, 59, 67,*
    83, *99*
Catani, C., 411, *418*

Cates, D., 358, *370*, 558, 560, 570, 571, 573, 574, *577, 583, 585,* 593, *612*

Caton, D. C., 5, *7*

Catterall, W. A., 74, 83, *96, 99*

Catterton, W. Z., 274, *285*

Caudry, I. H., 633, *635*

Cauna, N., 220, *240*

Caverson, M. M., 349, 350, *363*

Caviness, V. S., Jr., 192, *194*

Ceccarelli, B., 14, 19, *56*

Ceccatelli, S., 19, 20, *60*

Cederblad, G., 468, *475*

Cerrina, J., 489, 490, *507*

Ceruti, E., 321, *336*, 572, *576*

Cevolani, D., 535, *546*

Chad, J. E., 299, 310, *314, 315*

Chamberlain, S., 377, 378, *418*

Champagnat, J., 111, 133, 136, 141, *146*, 667, *674*

Chan, H. S., 387, *427*

Chan, K-F., J., 80, *100*

Chan, L., 38, *56*

Chan, S. Y., 79, *104*

Chan-Palay, V., 19, *56*

Chance, B., 617, 628, 631, 632, 633, 634, 635, *636, 637, 638, 639, 641*

Chandler, S. H., 525, 526, *542*

Chang, F-C. T., 135, *146*

Chang, K-J., 84, *107*

Chang, M-S., 75, *96*

Chang, S., 633, *639*

Changeux, J-P., 25, 30, 32, *56, 58, 63*, 470, *475*

Chapman, P. F., 33, *55*

Charbonneau, H., 75, 94, *108*

Charlton, C. G., 48, *56*

Charp, P. A., 80, *105*

Chase, M. H., 525, 526, 527, 539, *542, 545, 546, 548*

Chase, T. N., 343, *367*

Chasnoff, E. P., 605, 606, *612*

Chaudry, I. H., 390, *416, 636*

Chauhan, A., 80, *96*

Chauhan, V. P. S., 80, *96*

Chaussain, M., 567, *583*

Chen, E., 77, *97*

Chen, H-B., 73, *105*

Chen, K. S., 191, *194*

Chen, V., 467, *478*

Chen, Z. B., 351, *363*

Cherniack, N. S., 257, 259, *266*, 341, 343, 350, 351, 352, 354, *363, 366, 368, 370*, 662, 671, *674, 678*

Chernick, V., 200, 206, *214*, 358, 359, 360, *366, 367*, 487, *511*, 554, 555, 557, 573, *575, 581*, 668, 670, *674, 675*, 681, 684, 685, 687, 690, *695, 697*

Cherubini, E., 603, *608, 610*

Chester, A., 685, *697*

Cheung, T. S., 392, *418*

Chien, K., 603, *609, 634, 636*

Chih, C. P., 606, *608*

Chin, H., 75, *96*

Chinkers, M., 75, *96, 107*

Chirgwin, J. M., 47, *62*

Choksi, R. M., 411, *425*

Chon, W., 161, *175*

Christopher, K. L., 3, *7*

Chung, F-Z., 28, *56*

Churchill, T. A., 605, *613*

Cidlowski, J. A., 37, 39, 40, 41, 42, 43, *54*

Cimler, B. M., 81, *96*

Ciriello, J., 186, *195*, 349, 350, *363*

Citri, Y., 84, *102*

Citterio, G., 247, 248, 253, *263*, 274, *284*

Civelli, O., 48, *56*

Claiborne, B. J., 122, *146*

Clamann, P. H., 391, *418, 443, 450*

Clanton, T. L., 467, *476*

Clark, B. J., 617, *636*

Clark, F. J., 247, 248, 252, *263*

Clark, J. H., 38, *56*

Cleland, C. L., 128, *148*

Clement, M. G., 250, *264, 495, 508*

Clemente, C. D., 537, 538, *548*

Clements, A., 556, *582*

Clements, J. A., 668, 669, *676*

Clendeninn, N. J., 49, *56*
Clerici, C., 501, *509*
Clewlow, F., 554, 556, *577, 585*
Close, R. I., 408, *419*
Closter, J., 617, 618, *639*
Clyman, R. I., 556, *587*
Coburn, R. F., 485, 486, 487, *509*
Cockcroft, S., 74, *96*
Coghill, G. E., *194*
Cohen, A. H., 125, *146*
Cohen, C. A., 457, 459, *476*
Cohen, D. R., 92, *104*
Cohen, G., 566, *577*
Cohen, H. L., 329, 330, *337*
Cohen, M., 489, *509*
Cohen, M. I., 129, 132, 135, *146,
    147,* 258, *264,* 271, 274, 275,
    277, *284, 285*
Cohen, P., 93, 94, *97*
Cohen, P. J., 626, *636*
Cohen, P. T. W., 93, *97*
Cohlan, B. A., 260, *268,* 562, 567,
    568, *586*
Colasante, C., 472, *479, 480*
Colbran, R. J., 85, 88, 89, *97*
Cole, J. S., 667, 669, *677*
Cole, S., 322, *337*
Colebatch, H. J. H., 485, 486, 494,
    514
Coleridge, H. M., 231, *240, 259, 264,*
    489, 495, 496, 497, 499, *509*
Coleridge, J. C. G., 231, *240, 259,
    264,* 489, 495, 496, 497, 499,
    509
Coles, J. A., 647, *657*
Colin, A. A., 560, *581*
Colley, P. A., 33, *54,* 83, *95*
Collins, S., 62
Colrain, I. M., 568, *577*
Colvard, D. S., 38, *69*
Colwill, J. R., 626, *638*
Commissiong, J. W., 178, *194*
Comroe, J. H., Jr., 308, *313, 316,*
    662, *679*
Comtois, A., 461, *476*
Conn, P. J., 24, *56*

Connelly, B. J., 492, 493, 494, 495,
    499, 500, 505, *508, 518*
Connor, J. A., 118, *147,* 156, *172*
Connors, B. W., 600, *608,* 649, *659*
Conrad, H., 631, *636*
Conradi, S., 384, 387, 389, *419*
Conway, K., 85, *101*
Cook, D. L., 602, *613*
Cook, N. J., 76, *101*
Coon, R. L., 239, *241,* 246, *267*
Cooper, B. R., 125, 141, *145*
Cooper, C. C., 392, *419*
Cooper, D. R., 80, *97*
Cooper, J. A., 75, *100*
Cooper, R. G., 443, *451*
Coote, J. H., 180, *194*
Corda, M., 279, *286*
Cordingley, G., 647, *659*
Corke, B. C., 281, 282, *286*
Cornblath, M., 621, *639*
Cossu, G., 78, *108*
Costill, D. L., 468, *476*
Cote, A., 321, 322, 334, *337, 593,*
    596, *608*
Cotman, C. W., 25, *56,* 72, 74, 83,
    97, 606, 614
Cott, J. M., 125, 141, *145*
Coulter, J. D., 180, *193, 198*
Couraud, F., 160, *172*
Coussen, F., *102*
Coussens, L., 77, *97*
Couture, J., 501, *513*
Couturier, S., 74, *105*
Cowan, W. M., 615, *636*
Coyle, J. T., 46, 49, *55, 57*
Cozine, R. A., 342, 343, *363*
Cragg, P., 343, 355, *364*
Cranage, S., 555, *583*
Crance, J. P., *286*
Crandall, J. E., 192, *194*
Creogh, E. E., 42, *69*
Crepel, F., 160, *173*
Crill, W. E., 157, *172*
Crockett, J. L., 471, *476*
Crompton, M., 304, *314*
Cross, B. A., 309, *314*

Cross, K. W., 254, 255, *264*, 571, *577*, 646, *657*, 662, *674*, 683, 684, *695*

Crouch, T. H., 93, *102*

Crow, M. T., 414, *419*

Crozier, D. N., 467, *478*

Crutcher, K. A., 178, *194*

Crystal, R. G., 489, *508*

Cserr, H. F., 646, *657*

Cuatrecasas, P., 78, 79, 84, 87, 88, 89, 92, *102, 106, 107, 110*

Cuculis, J. J., 182, 183, *197*

Cuello, C., 19, 23, 34, *60, 61,* 498, *512*

Cuenoud, S., 410, 411, *418*

Culberson, J. L., 177, *196*

Cullen, J. H., 39, *57*

Cullheim, S., 384, 387, *419, 424,* 664, *673*

Cummins, T. R., 597, 601, *608*

Cunningham, D. J. C., 594, *608*

Curran, T., 92, *104*

Curzi-Dascalova, L., 570, *577*

Curzi, L., 259, *268*

Cutz, E., 322, *339*, 672, *677*

Czernik, A. J., 90, 91, *105, 108*

Czyzyk-Krzeska, M. F., 21, 30, 35, 36, *57, 64,* 211, *213*

D

Dabrowski, B., 467, 469, *478*

Dahlin, I., 666, 670, *675, 677*

Daikoku, S., 45, *57*

Daily, W. J., 691, 692, *695*

Dale, N., 26, 32, *57, 64*

Dalla Libera, L., 411, *418*

Dalle, D., 321, 322, 334, *337,* 593, *608*

Dalsgaard, C.-J., 186, *195*

Dalton, K. J., 681, *698*

Dampney, R. A. L., 349, *364,* 533, *541*

Daniel, E. E., 485, 492, *509*

Daniel, S. S., 667, *678*

Danieli Betto, D., 411, *418*

Daniels, J., 468, *476*

Dann, O., 49, *60*

Danselaar, Y., *452*

Darovec-Beckerman, C., 502, *518*

Da Silva, A. M. T., 351

Daubenspeck, J. A., 563, *578*

Davenport, P. W., 5, *7,* 248, 252, 256, *267*

Davi, M., 358, *370,* 558, 560, 574, *577, 585*

Davidson, N., 74, *99*

Davidson Ward, S. L., 3, *7*

Davies, A., 247, 253, 255, 257, 259, *264*

Davies, A. M., 274, 281, 282, *284*

Davies, C., 248, 252, *268*

Davies, J., 26, *57*

Davies, L. P., 46, *61*

Davies, N. W., 310, *317*

Davis, C., 485, 492, *509*

Davis, G. M., 246, *264*

Davis, H., 291, *314*

Davis, J. A., 682, *695*

Davis, J. M., 322, *337,* 623, *635*

Davis, P., 38, 44, *64,* 602, 603, *613*

Davis, P. J., 684, *695*

Davis, W. J., 126, *147*

Davson, H., 646, *657*

Dawes, G. S., *285,* 322, 324, 325, 326, 330, 333, 334, *336,* 358, 360, *362,* 497, *508, 509,* 554, 555, 557, *576, 578,* 593, *607,* 662, 665, 668, 669, *673,* 674, 681, 682, 683, 684, 685, 687, 690, *695, 696, 698*

Dawson, N. J., 442, *450*

Dawson, W. W., 224, *240*

Daynes, J., 279, *286*

Daze, M. A., 600, *607*

Dbaly, J., *508*

Deal, E. C., Jr., 354, *366*

Dean, J., 19, *60*

Dean, J. B., 21, 30, *64,* 348, 351, *364, 367*

De Boeck, C., 670, *674*

de Boer-van-Huizen, R., 191, *197*

Debono, M., 160, *173*
De Camilli, P., 90, *97, 100, 108*
De Castro, F., 291, *314*
Defendini, R. F., 3, *7*
De Gennaro, L. J., 90, *100*
De Graan, P. N. E., 81, 82, *97, 109*
De Graff, A. C., 322, *337*
deGroot, W. J., 521, *546*
Dekin, M. S., 5, *7*, 111, 112, 116,
  117, 118, 119, 120, 133, 134,
  135, 136, 137, 138, 139, 140,
  141, 142, 143, 144, *147, 150,*
  157, 161, 162, 163, 167, 171,
  172
Dekker, L. V., 82, *97*
De Laney, R. G., 253, *267,* 322, 323,
  325, 327, 332, *338*
Delapaz, R., 606, *613*
de la TorreVerduzco, R., 571, 572,
  573, *585*
Del Castillo, J., 14, *57*
Delcomyn, F., 113, *147*
De Lonzo, R. I., 87, *109*
De Lorenzo, G., 183, *196*
Delpiano, M. A., 293, 296, 306, 307,
  314, 316, 334, 337
Delpierre, S., 229, *241*
De Maille, J. G., 80, *107*
Demedts, M., 354, 357, *371*
Dement, W. C., 522, 534, *541, 546,*
  558, 563, 570, *575, 579, 584,*
  693, *697*
Dempsey, J., 539, *541*, 593, *608*
Dempsey, J. A., 35, 39, 42, *57, 68,*
  462, 467, *477*
Denavit-Saubie, M., 141, *146*, 276,
  287, 667, *674, 677*
De Neef, K. J., 231, 232, *240,* 486,
  509
Dennett, X., 387, *427*
Dennis, J., *56*, 569, *586*
Dennis, M. J., 379, 397, *419*
Denny, R., 668, *673*
Dermietzel, R., 343, 349, *364, 371*
De Robertis, E. D. P., 13, *57*
DeSalles, A. A. F., 540, *545*

Deshmukh, D. S., 80, *96*
Despos, P. J., *480*
Des Rosiers, M. H., 644, *660*
Detari, L., 537, *542*
Dev, N. B., 351, 356, 357, *364*
Devillers-Thiery, A., *56*
de Weer, P., 300, *314*
Dhoot, G. K., 410, 411, *419*
Diamond, J., 389, *419*
Diamond, L., 483, 484, 485, 486,
  500, 501, *509, 517*
Diaz Arnesto, L., 23, *61*
Diaz, E., 358, *364*
Dick, T. E., 382, 386, *419, 422,* 537,
  544, 644, *656*
Diemer, N. H., 309, *312*
DiFiore, J. M., 260, *267,* 566, *579*
Di Marco, A. F., 247, *264*
Dingledine, R., 649, *657*
Distel, H., 192, *194*
Di Tirro, F. J., 180, 183, 188, 189,
  194, 196
Dixon, C. M. S., 671, *676*
Dixon, G., 467, *476*
Dixon, M., 247, 253, 255, 259, *264*
Dixon, R., *57*
Dixon, R. A., 28, *69*
Dixon, R. A. F., 73, *97*
Dobbing, J., 168, *173*, 616, *636*
Dodd, D. S., 563, *576*
Doerschuk, C. F., 467, *479*
Dohlman, H. G., 28, *57*, 72, *98*
Doll, C. J., 606, *613*
Dom, R., 177, 179, *196*
Donahoe, P. K., 488, *513*
Donaldson, S. R., 349, *365*
Donlon, E., 618, *639*
Donnelly, D. F., 155, 157, *173, 174,*
  200, 201, 206, *215,* 274, 275,
  277, 280, 281, *284, 285,* 322,
  337, 459, *475,* 596, 597, 599,
  601, 603, 604, 605, *608, 609,*
  612, 687, *696, 698*
Donoghue, S., 129, 131, *148*, 323,
  334, *338*
DoPico, G. A., 593, *608*

Dostrovsky, J. O., 272, *286*
Doucet, J-P., 91, *107*
Douglass, J., 48, *56, 67*
Douglas, J. S., 484, 486, 488, 489,
490, 491, 492, 493, 502, 503,
505, *508, 509, 510, 515, 516,
518*
Douglas, N. J., 572, *578*
Dow, R. S., 485, *512*
Dowling, H., 358, *362*
Dowling, M., 555, 556, *576*
Downing, S. E., 274, 275, 282, *284,
286*, 492, *509*
Dragunow, M., 92, *98*
Drake, J., 467, *476*
Drejer, J., 309, *312*
Dretchen, K. L., 23, 34, *60*
Dreyfus-Brisac, C., 523, *545*, 552,
553, 554, *578*
Driscoll, J. M., 3, *7*
Drummond, A. H., 143, *148*
Dubinsky, J. M., 306, *315*
Du Bois, F. S., 222, *240*
Dubowitz, V., 392, *419*
Duchen, M., 295, 296, 304, 305, 310,
315
Duchen, M. R., 295, 296, 299, 300,
304, 305, 309, 310, *312, 313,
320, 334, 337*
Duffy, T. E., 620, 621, 625, 626,
627, 629, 630, 632, 634, *636,
637, 641,* 654, *657,* 686, *696*
Duggan, A. W., 186, *194,* 663, *674,
675*
Dumke P. R., 662, *679*
Duncan, C. C., 667, 669, *677*
Duncan, P. G., 484, 489, 490, 491,
502, *503, 509, 510*
Dunkley, P. R., 85, 86, *98, 106*
Dunmire, C. R., 222, 223, *243,* 280,
286
Dunn, R., 74, *94*
Dunwiddie, T. V., 672, *674*
Dupin, E., 160, *172*
Dupont, J. L., 160, *173*
Durand, M., 573, *585*

Durgesian, S., 86, *103*
Duron, B., 207, *213,* 222, 231, 232,
233, 234, 237, *242,* 250, 254,
255, 256, 260, *264, 267,* 384,
385, *419, 425,* 486, 494, *513,*
522, *542,* 559, 560, 571, 572,
576, 578, 593, *607,* 645, *657,*
662, *673*
Duthie, K. L., 494, *510*
Dutton, R. E., 573, *575*
Duxson, M. J., 397, *420, 425*
Dyme, I. Z., 574, *575*

E

Eayrs, J. T., 168, *173,* 654, *657*
Ebata, N., 645, *656*
Eberle, L. P., 329, *337*
Eccles, J., 408, *418*
Eccles, J. C., 12, *57*
Eccles, R., 408, *418*
Eckert, R., 299, 310, *314, 315*
Eckert, R. C., 3, *7*
Eddinger, T. J., 414, *420*
Edelman, G. M., 75, *98*
Edelman, N. H., 357, *370,* 596, *612,*
662, *674*
Eden, G. J., 327, 328, 329, *338*
Edgerton, V. R., 378, 391, 392, *417,
423, 426,* 471, *476*
Edström, L., 377, 391, 392, 407,
420, 464, *478*
Edwards, P., 566, *582*
Edwards, R. A., 606, *609*
Edwards, R. H. T., 389, *420,* 429,
436, 443, *450, 451,* 457, 458,
459, *476, 479*
Edwards, W. H., 360, *369*
Eerbeck, O., *452*
Effenberger, M., 349, *371*
Efmenko, G. P., 633, *641*
Eghboli, B., 554, *586*
Egizii, R., 272, *286*
Egreteau, L., 259, *268*
Eichhorn, W., 571, *586*
Eigen, H., 495, *511*

Einthoven, W., 483, *510*
Eisele, J. H., 248, *266*
Eisen, J. S., 122, 123, 124, *145, 148, 152*
Eisner, D. A., 295, 299, 300, 305, 310, *313, 320*
Ekholm, J., 261, *264*
Elde, R., 49, 50, *58, 61,* 112, *152*
Eldred, E., 378, 391, 392, *417, 423*
Eldridge, F. L., 19, 34, 35, *64,* 130, *148,* 341, 343, 350, 351, 353, 354, 355, 356, *364, 367, 368, 370,* 667, *677*
Elfvin, L. G., 19, *60*
Eliet-Flescher, J., 523, *545*
Ellenberger, H. H., 128, *148*
Ellis, L., 79, 81, *101*
Ellis, S. B., 74, *98*
Elsner, N., 114, 116, *151*
Emerson, P., 19, 53, *61*
Emson, P. C., 48, 49, *59, 61*
Enad, J. G., 391, 404, 407, 413, *420, 426*
Engel, L. A., 563, *576*
Engelhardt, J. K., 527, *546*
England, S. J., 235, 237, *240, 241,* 250, 256, *265,* 401, 404, 408, *423,* 448, *451,* 560, 562, 563, 564, 565, 568, 574, 577, 578, *580, 583*
Enomoto, S., 525, *542*
Epstein, A., 3, *7*
Epstein, M. A. F., 252, *266,* 569, 570, 574, *580*
Epstein, R. A., 3, *7,* 252, *266,* 569, 570, 574, *580*
Erb, M. J., 335, *338*
Erecinska, M., 617, *636*
Erickson, J. T., 21, 23, 30, 34, 35, *58, 64*
Erondu, N. E., 86, 87, *98, 101*
Ershler, W. B., 177, *197*
Eskay, R. L., 129, *152*
Euler, U. S., 297, *315*
Evanich, M. J., 465, *476*
Evans, R. M., 38, *58*

Evans, W., 468, *476*
Everitt, B., 19, *60*
Ewald, D., 299, *314*
Eyzaguirre, C., 291, 295, 308, *312, 315, 317*
Ezure, K., 201, *214*

F

Faden, J. S., 381, *427*
Fagenholz, S. A., *284,* 570, 574, *578*
Fagg, G. E., 25, 26, *58*
Fahey, J. T., 595, *609*
Fahim, M. A., 470, *476, 480*
Fahrenkrug, J., 23, 24, *63*
Fairbanks, M. K., 536, *545*
Fallert, M., 201, *214*
Fanaroff, A. A., 566, *577*
Fang, He., 160, *172*
Farber, J. L., 603, *609,* 634, *636*
Farber, J. P., 171, *173,* 192, *195,* 200, 203, 206, 207, *214,* 231, 232, 233, 234, 235, 237, *242,* 246, 250, 254, 255, 256, 257, 258, 259, 260, 261, *264, 265,* 484, 486, *512,* 562, 564, *578*
Farel, P. B., 177, *195*
Farese, R. V., 80, *97*
Fariday, E. E., 681, 687, 690, *695*
Farkas, G. A., 464, 467, 469, *476*
Fatt, P., 14, *58,* 309, *315*
Faulkner, J. A., 391, *425*
Fawcett, J. W., 615, *636*
Fazekas, J. F., 632, *637,* 654, *658,* 685, *697*
Fedorko, L., 279, *285, 286*
Fedorko, L. M., 155, *174,* 250, 256, *265*
Feinberg, I., 523, *542*
Feinmark, S. J., 83, *106*
Feinstein, P. G., *102*
Feldberg, W., 349, *364*
Felder, R. B., 129, 131, *148,* 323, 334, *338*
Feldman, H. A., 252, 253, *265*

Feldman, J. L., 128, 129, 130, 132, 135, 140, *146, 148, 153, 159, 175*, 201, *215*, 258, *264*, 271, 274, 277, 278, 279, *284, 285, 286*, 341, 351, 353, *364, 370*

Felig, P., 462, *476*

Feller, R., 504, *510*

Fencl, V., 342, 354, 355, *364, 369*

Feng, Z. C., 605, 606, *608, 609*

Fenner, A., 358, *366*, 569, *578*

Fenton, J., 443, *451*

Feramisco, J. R., 32, *67*

Ferguson, C. C., 486, *515*

Ferrari, G., 535, *546*

Fewell, J. E., 155, *174*, 558, *578*, 667, 668, *676*, 693, *696*

Fidone, S. J., 293, 295, 308, *315, 318*, 592, *609*

Fields, H. L., 180, *194*

Filtz, T. M., 349, *365*

Findley, L. J., 567, *578*

Fineberg, N., 571, *579*

Finer-Moore, J., 25, *58*

Finer, N. N., 248, *265*, 569, 570, *575, 579*

Fink, B. R., 568, *579*

Fink, J. S., 32, *67*

Fink, K., 38, *69*

Fink, W., 468, *476*

Finley, B. L., 615, 616, *637*

Fischer, E. H., 80, *107*

Fischer, J. A., 84, *103*

Fischer, R. L., 557, *576*

Fish, I., 168, *173*

Fisher, A. M., 189, *196*

Fishbein, P., 630, *640*

Fisher, G. H., 664, *676*

Fisher, J. N., 693, *698*

Fisher, J. T., 203, 223, 225, 226, 227, 229, 232, 233, 234, 235, 236, 237, *240, 241, 243, 244,* 246, 248, 249, 250, 254, 255, 258, 261, *265, 267,* 272, 274, *284, 287,* 473, *476,* 492, 493, 494, 495, 496, 497, 499, 500,

[Fisher, J. T.]
505, *508, 510,* 564, *579, 584,* 691, *698*

Fisher, R., 358, 360, *362,* 593, *607*

Fishman, A. P., 456, *478*

Fishman, M. C., 82, 108, 293, 295, 308, *315, 337*

Fishman, N. H., 253, *265*

Fisone, C., 19, *60*

Fitts, R. H., 467, *479*

Fitzgerald, J. W., 459, *481*

Fitzgerald, R., *481*

Fitzsimmons, R. B., 410, *421, 424*

Fladby, T., 386, 411, 413, *420*

Flamm, E. S., 309, *318*

Flamm, R., 124, *145*

Flamm, R. E., 115, 123, 124, 131, *148*

Fleetwood-Walker, S., 23, *65*, 180, *194*

Fleming, P. J., 260, *265*

Fleshman, J. W., 381, *420*

Fletcher, B. D., 691, *695*

Florez, J., 351, *366*

Flugge, G., 43, *58*

Foley, J. O., 222, *240*

Folgering, H., 308, *315*

Folts, J. D., 633, *636*

Fong, Y-L., 85, 88, 89, *97, 98*

Fontaine, B., 32, *56, 58*

Forehand, C. J., 177, *195*

Forn, J., 15, *58*

Fornaris, E., 486, *511*

Forssberg, H., 125, *148*

Forster, H. V., 35, *57,* 593, *608*

Forster, R. E., 631, 634, *639*

Fort, M. D., 483, *501*

Fossel, E. T., 618, *639*

Foster, A. C., 26, *58*

Foster, G. A., 45, 47, 49, *58*

Foster, G. H., 457, 461, *480*

Foster, R. E., 135, *146*

Foulon, P., 321, 322, 334, *337,* 571, *577,* 593, *608,* 662, *673*

Fournier, M., 378, 379, 382, 383, 388, 391, 392, 397, 404, 407, 413, *420, 426*

Foutz, A. S., 558, 570, *579*
Fox, A. P., 74, 82, *109,* 310, *318*
Fox, G., 223, 235, *243,* 691, *698*
Fox, G. S., 248, *265,* 564, 567, *579,* 584
Fox, H. E., 554, 555, *578,* 684, 685, *695*
Fox, R. E., 252, 253, *265*
Francke, U., 77, *97*
Francke, Y., 62
Frank, M., 126, *153*
Frank, N. R., 485, *518*
Franke, R., 73, *105*
Franklin, B. A., 467, *479*
Frantz, I. D., 248, 249, 252, *268*
Frantz, I. D., III, 248, 249, 252, 257, 260, 262, *268, 269,* 560, 562, 567, 568, 569, 570, 573, *579, 586, 587*
Fraser, C. M., 28, *56*
Fraser, G., 568, *577*
Fraser, I. G., 462, *477*
Fredholm, B. B., 666, 669, 670, 671, 672, *674, 677, 678, 679*
Freedman, J., 19, *60,* 489, *516*
Fregosi, R. F., 462, 467, *477*
Freid, G., 13, 24, *59*
Freis, E. D., 594, *609*
Freissmuth, M., 72, 74, *98*
French, A. S., 433, *452*
French, J. W., 275, 282, *284,* 681, *696*
Fresch, J. E., 78, *98*
Frey, W. H., 664, *676*
Fricks, C., 192, *197*
Fried, G., 24, *63*
Friedman, W. A., 381, *420*
Friedman, W. F., 492, *510*
Frielle, T., 62
Frierdich, G., 606, *610*
Friis-Hansen, B., 648, *658*
Froese, A. B., 401, *424,* 457, 473, *479,* 560, *584*
Frostig, R. D., 534, *542*
Frostig, Z., 534, *542*
Frumin, M. J., 128, *154*

Frydrychova, M., 255, *268*
Frye, G. D., 125, 141, *145*
Frysinger, R. C., 529, 530, 531, 532, *539, 542, 543, 549*
Fugii, T., 77, *105*
Fujisawa, A., 635, *638*
Fujisawa, H., 85, *108*
Fujiwara, N., 307, 310, *315,* 597, 600, *609*
Fukuda, K., 28, *62*
Fukuda, Y., 346, 348, 354, *364, 365*
Fukunaga, K., 86, *98*
Fuller, R. W., 671, *676*
Fulton, B. P., 159, 166, 169, *173,* 384, 387, 388, *420, 427*
Funk, G. D., 606, *613*
Furbush, F., 442, 444, 445, 446, *450, 453*
Furuichi, T., 76, *98*
Furutani, Y., 25, *65*
Futamachi, K. J., 648, *659*
Fuxe, K., 19, *61,* 112, 141, *150*

G

Gabel, R. A., 661, *679*
Gabella, G., 485, *510*
Gabriel, M., 571, *579*
Gadek, J. E., 467, *476*
Galioto, F. M., 322, *337*
Gallego, R., 387, *420*
Gallman, E. A., 21, 30, 37, 39, 43, *54, 64*
Galvan, M., 157, *173*
Gammeltoft, S. A., 39, *59*
Gandevia, S. C., 432, 446, *450, 451*
Gandhi, M. R., 155, 169, *174,* 200, 201, 206, *215,* 593, 594, *609*
Ganong, A. H., 26, *58,* 72, 74, 83, *97*
Ganten, D., 19, *61*
Garbers, D. L., 75, *96, 107*
Garcia, J. H., 633, *636*
Garcia, N. M., 664, *676*

Gardette, R., 160, *173*
Gardey, C., 502, *517*
Gardiner, P. F., 471, *477*
Gardner, K., 82, *95*
Gardner, W. N., 497, *509*, 555, 557, 578, 662, 665, *674*
Garfinkel, A., 539, *543*
Garner, S., 443, *451*
Garrow, I., 693, *699*
Garthwaite, G., 653, *658*
Garthwaite, J., 25, *59*, 653, *658*
Gastaut, H., 692, *696*
Gastaut, Y., 692, *696*
Gateau, O., 489, 490, *507*
Gates, E. P., 3, *7*
Gatti, P. J., 351, *365*
Gatto, C., 502, *510*
Gauda, E. B., 566, *579*
Gaudy, J. H., 252, 255, *265*
Gaultier, C., 255, 259, *265, 268, 475*
Gauthier, G. F., 410, *421*
Gauthier, P., 382, *421*
Gautier, H., 252, 255, *265*, 534, *542*, 562, 563, 571, 572, *576*, 593, *607*, 661, 662, 665, *673, 674*
Gazelius, B., 34, *55*
Gean, P. W., 157, *173*
Gebber, G. L., 349, *362*, 536, *546*
Geelhoed, G. C., 484, *512*
Geiss, W. P., 492, *510*
Geller, H. M., 161, *175*
Geller, M. M., 357, *370*
Gerard, R. W., 683, 685, 686, *699*
Gerhardt, T., 248, 249, 250, *265, 266*, 473, *477*, 504, *510*
German, K. J., 467, *479*
Gershan, W. M., 689, 690, *697*
Gershan, W. T., 688, 689, *696*
Gershon, M. D., 492, 499, 502, 505, *510*
Gertler, R. A., 470, *477*
Gesell, R., 356, *365*
Getting, P. A., 111, 112, 114, 116, 117, 118, 119, 120, 121, 122, 133, 134, 135, 136, 137, 140,

[Getting, P. A.]
141, 142, 143, *147, 149, 150, 151, 153*, 157, 159, 161, 162, 163, 165, 166, 167, 168, 171, *172, 173, 174*
Gettrup, E., 114, *154*
Ghandi, M. R., 570, 572, *580*
Ghosh, M., 618, *639*
Ghosh, N. C., 631, *640*
Giannini, D., 440, *452*
Gibson, G., 630, *636*
Gibson, H., 443, *451*
Giedde, A., 626, 629, *637*
Giese, W. P., 74, *108*
Gilbert, D., 618, *639*
Gilbert, R. F. T., 48, *59*
Gilbey, M. P., 180, *194*
Gill, D. L., 300, *315*
Gill, P. K., 354, *365*
Gillespie, J. R., 257, *266*
Gillespie, L. A., 540, *548*
Gillespie, M. N., 501, *517*
Gillis, R. A., 23, 34, *60*, 351, *365, 366, 371*, 667, *677*
Gilman, A. G., 29, *59*, 72, 74, *96, 98, 102*
Gilman, M. Z., 32, *67*
Ginsberg, A. D., 462, *481*
Ginsberg, M. D., 633, *636*
Ginsborg, B. L., 14, *55*
Girard, P. R., 78, 79, *98*
Giraudat, J., *56*
Gispen, W. H., 81, 82, *97, 109*
Giulian, G. G., 408, 414, *424*
Glass, H. G., 626, 627, 634, *637*
Glasser, R. L., 534, *548*
Gleason, M. L., 74, *104*
Glebovsky, V. D., 224, *241*, 387, 408, *421*
Glenn, L. L., 558, *579*
Glick, B. S., 74, *104*
Glogowska, M., 254, *266*, 687, *696*
Gluckman, P. D., 207, *214*, 360, *365*, 555, 557, 558, *579, 582*, 662, 665, 667, 668, 669, *674, 675*

Go, M., 77, *98*
Godfrey, S., 504, *517*, 683, 684, 687, 690, *696*
Goding, G. S., 272, 280, *284*
Goedel, D. V., 75, *96*
Goedert, M., 49, *59*
Goelet, P., 32, 33, *55, 59*, 83, *99*
Goh, A. S. F., 566, *579*
Goiny, M., 34, *55*
Goldberger, A., 38, *69*
Goldenring, J. R., 87, *109*
Goldin, A. L., 74, *99*
Goldman, M., 259, *266*
Goldman, M. D., 262, *268*, 560, *586*
Goldman, W. F., 291, *312*
Goldring, R. A., 90, *103*
Goldstein, A., 48, *56*
Goldstein, J. D., 555, *579*
Goldstein, M., 13, 19, 22, 24, 45, 47, 49, *56, 58, 59, 60*
Gollnick, P. D., 462, 464, 467, 468, *477, 479*
Goma, H., 635, *641*
Gomez-Del Rio, M., 504, *510*
Gommers, J., 554, *584*
Gomperts, B. D., 74, *96*
Gonsalves, S. F., 345, *362*
Gonzalez, C., 293, 295, 296, 305, 306, 308, *315, 316, 318, 320,* 334, *339, 592, 609*
Gonzalez, G. A., 31, *69*
Gonzalo-Ruiz, A., 540, *545*
Goodbourn, S., 84, *103*
Goode, G., 177, *196*
Goodhead, B., 168, *173*, 654, *657*
Goodman, H. M., 48, *67*
Goodman, N. W., 308, *316*
Goodman, R. H., 31, 32, 51, *65, 67*
Goodwin, C. W., 627, 630, 631, 633, *637, 638, 639*
Gootman, P. M., 129, *147*, 277, *284*, 329, 330, *337*
Gorcs, T., 23, 34, *58*, 351, *365*
Gorczyca, W., 461, *476*
Gordon, D., 74, *99*
Gordon, D. A., 379, 392, *421, 424*

Gordon-Weeks, P. R., 81, *99*
Gordon, R. D., 309, *314*
Gorelick, F. S., 89, 91, *99, 108*
Goritski, W., *369*
Gorza, L., 410, 411, 414, *425*
Goto, M., 45, *59*
Goujoux, M., 554, *586*
Gould, J. B., 569, 571, *579, 586*
Graf, P. D., 254, 256, *268*, 485, 486, 494, *508, 511*
Grafe, P., 157, *173*
Graff, J. M., 81, *99, 108*
Granata, A. R., 349, *365, 370*
Grant, D. K., 552, *585*
Grassino, A., 448, *449*, 459, 461, *475, 476, 477*
Gravel, D., 471, *477*
Gray, E. G., 13, *59*
Graybiel, A. M., 532, *543*
Greaser, M. L., 408, 414, *424*
Green, H., 404, 414, *426*, 462, *477*
Green, J. H., 259, *263*
Green, M., 457, 459, *479*, 496, *510*
Green, S., 484, 504, 505, *514*
Green, W. L., 293, 295, 308, *315*
Greenberg, M. E., 92, *99*
Greenberg, S. M., 32, *59*
Greene, L. A., 92, *99*
Greengard, P., 15, 30, 31, *58, 59, 60, 63, 65*, 81, 86, 87, 89, 90, 91, 92, *95, 97, 99, 100, 101, 102, 103, 104, 105, 107, 108, 109, 110*
Gregoire, N. M., 626, 629, *637*
Gregory, J. E., 231, 233, *241*
Gregson, N. A., 618, 621, *637*
Grelot, L., 279, *283*, 352, *362, 365*
Grenningloh, G., 74, *99*
Grierson, J. P., 161, *175*
Griffin, C. A., 687, *697*
Griffiths, G. B., 567, *579*
Griffiths, T., 310, *319*
Grillner, S., 114, 116, 125, *145, 146, 148, 149, 151*, 169, *173*
Grillo, F. G., 356, *365*
Grimaud, Ch., 229, *241*, 274, *285*

Grimby, G., 259, *266*
Grob, D., 390, *424*
Grodin, W. K., 252, *266*, 574, *580*
Grogaard, J., 274, *285*
Grogan, P., 159, *173*
Grollman, E. F., 300, *315*
Gross, D., 457, 459, *476, 477*
Gross, E. M., 322, *337*
Grunstein, M. M., 246, 251, 252,
    266, 322, *340,* 498, 499, *511,*
    670, *675, 678*
Grundstrom, N., 483, *498*
Gruner, J. A., 91, *104*
Grunstein, J. S., 498, 499, *511*
Grynkiewicz, G., 299, 300, *316*
Guazzi, M., 594, *609*
Guerriero, V., Jr., 80, *101*
Guertzenstein, P. G., 349, *364, 365*
Guilbaud, G., 532, *545*
Guillemin, R., 48, *54*
Guilleminault, C., 693, *697*
Gulston, G., 401, *424*, 457, 473, *479,*
    562, *580*, 560, *584*
Gundersen, K., 411, *425*
Gunds, S., 410, *423*
Gunn, T. R., 555, 558, *579, 582*
Gunst, S. J., 489, *511*
Guntheroth, W. G., 275, 282, *284,*
    681, 682, 683, 684, 686, 688,
    696
Guppy, M., 355, *366*
Guslits, B. G., 448, *451*
Gustafson, B., 157, 159, *173*
Guth, L., 408, 413, *421*
Guthrie, P. B., 603, *611*
Guthrie, R. D., 159, 160, 169, *172,*
    250, 256, 259, *267, 269,* 321,
    339, 360, *365*, 384, 385, 386,
    418, 422, 483, 501, *513,* 568,
    569, 571, 574, *579, 580, 582,*
    587, 593, *610, 614,* 662, *679*
Guthrie, R. T., 646, *659*
Gutmann, E., 408, *421*, 470, *481*
Gutnick, M. J., 169, *175*, 600, *608,*
    646, 647, *658*
Guyenet, P. A., 349, *363*

Guyenet, P. G., 349, *365*
Guyenet, P. G., 2349, *371*
Guz, A., 248, 255, 256, *263, 266*
Gyepes, M. T., 492, *511*
Gyulai, L., 617, 628, *637*

H

Hackett, J. T., 349, *371*
Haddad, G. G., 3, 5, 7, 155,
    157, 159, 161, 165, 166, 169,
    173, 174, 252, *266,* 274, 280,
    281, *284, 285,* 321, 322, *337,*
    338, 457, 459, 461, 462, 463,
    475, 480, 481, 498, 511, 553,
    569, 570, 572, 574, *580,* 593,
    594, 595, 596, 597, 599, 601,
    603, 604, 605, 606, *608, 609,*
    612, 670, *678, 687, 696, 698*
Haddock, M. P., 224, *244*
Haga, T., 28, *62*
Hagan, R., 562, *580*
Hagbarth, K., 446, *451*
Hagberg, H., 667, *675*
Hagiwara, M., 79, 84, *99*
Hahn, H. L., 494, *511*
Hainfeld, J., 25, *62*
Hairston, L. E., 525, *543*
Hajos, F., 653, *658*
Hale, P. T., 410, *421*
Hales, C. N., 602, *613*
Hall, A., 74, *99*
Hall, J., *479*
Halliwell, J. V., 166, *174*
Halsey, J. H., 633, *636*
Hamburger, F., 177, *195*
Hamill, O. P., 160, *174,* 295, *316*
Hamill, R. W., 186, *197*
Hamilton, N., 467, 469, *478*
Hamilton, P., 462, *481*
Hamilton, R. B., 272, *285*
Hamilton, R. D., 255, *266*
Hamm, T. M., 392, *421, 424*
Hammarberg, C., 380, 387, 388, *421*
Hamosch, P., 351, *365, 366, 371*
Han, V. K. M., 45, *62*

Hanamori, T., 272, *285*
Handa, N., 635, *638*
Hanks, E. C., 568, *579*
Hannun, Y. A., 79, 80, *99*
Hansen, A. H., 645, 646, 647, 650, *658*
Hansen, A. J., 310, *316*, 596, 599, 600, 602, 603, 604, *609, 610, 611, 612*, 686, *697*
Hansen, E. T., 356, *365*
Hanson, M., 357, *370*
Hanson, M. A., 321, 322, 324, 325, 326, 327, 328, 329, 330, 333, 334, 335, *336, 338*, 357, 360, *366*, 497, *508*, 593, *607*, 662, 665, 669, *673, 676*
Hanson, P., 86, *103*
Hara, Y., 48, 49, *68*
Harada, Y., 354, *366*
Harding, R., 229, 237, *241*, 555, 557, 558, 562, 563, 564, *580, 584*
Harding, R. P., 272, *285*
Hardy, P. H., 182, 183, *197*
Harf, A., 501, *509*
Harlan, R. E., 43, 44, *59, 68*
Harned, H. S., 687, *697*
Harper, C., 665, *678*
Harper, R. K., 539, *543*
Harper, R. M., 135, *146*, 386, *426*, 521, 522, 523, 524, 529, 530, 531, 532, 533, 534, 536, 539, *542, 543, 545, 546, 547, 548, 549*, 566, 569, 570, *581, 586*
Harpold, M. M., 74, *98*
Harris, A. J., 379, 397, *419, 420, 425*
Harris, S. M., Jr., 90, *97*
Harris-Warrick, R. M., 115, 123, 124, 131, *145, 148*
Hartline, D. K., 122, *153*
Hartman, C. G., 178, *195*
Hartree, E. F., 292, *316*
Hashimoto, K., 77, 80, *102*
Hashimoto, T., 78, *99*
Hashimoto, Y., 85, 89, 93, *97, 99, 100*

Hasizume, I., 39, *62*
Hasselbalch, K. A., 39, *59*
Hasselberger, F. X., 620, *638*
Hathorn, M. K. S., 569, 570, *580*
Hatward, J. M., 129, *148*
Hauhart, R. E., 683, 684, 685, *697*
Hauri, P. J., 524, 537, *547*
Havens, B., 523, *548*
Haxhiu, M. A., 354, *366*
Hayakawa, Y., 32, *60*
Hayashi, F., 39, *62*
Hayashi, S., 484, 489, 503, *511*
Hayashida, Y., 295, *315*
Hayes, R. L., 540, *545*
Hayward, J. N., 274, *284*
Haywood, J. N., 644, *657*
Hazinski, T. A., 670, *675*
Hazlett, J. C., 179, *196*
He, F., 384, 385, *418*
Headley, P. M., 186, *194*
Hedner, J., 112, 141, *149*, 351, *363*
Hedner, T., 112, 141, *149*, 351, *363*, 672, *675*
Hehre, D., 504, *510*
Heidmann, O., *56*
Heidmann, T., *56*
Heinemann, U., 647, *658*
Heisey, S. R., 342, 354, 355, *369*
Heisler, N., 605, *610*
Heitler, W. J., 121, *149*
Hejtmancik, J. F., 44, *67*
Held, D., 342, 354, 355, *369*
Heldt, G. P., 498, *517*, 560, *580*
Helke, C. J., 23, 34, 48, *56, 69*, 343, *367*
Hellerstein, H. K., 467, *479*
Hellstrom, S., 666, 668, *675*
Hellund, B., 23, *63*
Hemmer, B., 462, *481*
Hemmings, H., 15, 31, *60*
Hemmings, H. C., Jr., 92, *105*
Hempstead, J. L., 92, *104*
Henderson-Smart, D. J., 203, *215*, 235, *241*, 473, *477*, 559, 566, 567, 568, 570, 571, 572, 573, *577, 581*, 594, *610*, 664, *675*

Henderson, P., 45, *62*
Hendren, W. H., 488, *513*
Hendricks, C., 566, *581*
Hendricks, J. C., 524, *544*
Hendry, I. A., 663, *674, 675*
Henneman, E., 381, *421*
Herbert, A., 26, *65*
Herbert, D. A., 346, *368*, 497, 506, *507, 513, 515*
Herbert, E., 48, *56, 67*
Herkenham, M., 667, *677*
Herman, S., 569, *576*
Hermansen, L., 469, *477*
Hernandez, M. J., 626, 629, *637*
Herrick-Davis, K., 24, *63*
Herrlich, H., 685, *697*
Herschenson, M. B., 560, 574, *581*
Hershey, A. D., 47, *62*
Herschman, H. R., 75, *100*
Hertzberg, T., 274, *288*, 321, *338*, 666, 668, 669, *675*
Hescheler, J., 293, 296, 306, 307, *314, 316,* 334, *337*
Hess, G. L., 239, *241,* 246, *267*
Hess, P., 310, *316*
Heuser, J. E., 14, *60*
Hewick, R. M., 78, *102*
Heyman, M. A., 595, *612*
Heymans, C., 322, *338*
Hiatt, P., 495, *511*
Hibi, S., 523, *542*
Hickey, R. F., 254, 256, *268*
Hicks, J., 443, *451*
Hidaka, H., 79, 84, *99*
Higashi, H., 307, 310, *315*, 597, 600, *609*
Higgins, B. L., 303, *319*
Higgs, E. A., 80, *104*
High, J. A., 379, *423*
Hilaire, G., 129, *149*, 382, *421, 423*
Hilberman, M., 631, 634, *639*
Hill, J., 662, *675*
Hill, J. G., 186, *194*
Hill, J. R., 595, *610*
Hiller, E. J., 504, *515*
Hilton, S. M., 349, *366*

Himwich, H. E., 632, *637,* 654, *658,* 685, *697*
Himwich, W. A., 623, *635*
Hirano, M., 272, *288*
Hirose, T., 25, 28, *62, 65,* 76, *101*
Hislop, A. A., 558, *578*
Hitzig, B. M., 355, *366*
Ho, R. H., 180, 183, 185, 188, 189, *194, 196, 197*
Hobbs, A. W., 410, *421*
Hobbs, S., 321, 322, 334, *337,* 593, *608*
Hobson, J. A., 351, *370,* 536, *544*
Hochachka, P. W., 355, *366,* 602, 603, 606, *610, 613*
Hodgkin, A. L., 155, *174,* 300, *312,* 623, *637*
Hodgman, J., 523, *543,* 569, 570, *581*
Hodgman, J. E., 483, 501, *511*, 523, *548*
Hodgson, J. A., 383, 384, *427*
Hodson, W. A., 358, 360, *365, 366,* 568, 569, 574, *579, 580,* 681, 690, *699*
Hoenicke, H., 569, *578*
Hofer, M. A., 322, *338*
Hoff, H. E., 521, *544*
Hofmann, E., 569, 570, *581*
Hofmann, F., 82, *100*
Hogg, J. C., 485, *518*
Hoh, J. F., 410, *421*
Hohimer, A. R., 358, *366*
Hökfelt, T., 13, 18, 19, 20, 22, 23, 24, 32, 34, 45, 46, 48, 49, 50, 53, *55, 56, 58, 59, 60, 61, 63, 65, 68,* 112, 141, 150, 186, *195, 196,* 498, 500, *512*
Hokin, L. E., 305, *319*
Holets, V. R., 19, *60*
Hollander, H., 192, *194*
Hollinger, B., 488, *518*
Holloszy, J. O., 471, *477*
Holm, J., 468, *475*
Holmes, F. B., 94, *97*
Holmstedt, B., 24, *63*

Holowach-Thurston, J., 683, 684, 685, *697*
Holstege, G., 529, 532, 533, 534, 540, *544, 545*
Holtman, J. R., 23, 34, *60*
Holtman, J. R., Jr., 351, *366*, 537, *544*
Holtzman, D., 620, *637*
Holtzman, E., 25, *62*
Holtzman, M. J., 507, *508*
Holzbauer, M., 48, *61*
Holzer, P., 48, *61*
Homma, I., 350, *363*, 664, *677*
Honda, H., 633, *640*
Honda, Y., 39, *62*, 346, *364*
Honma, Y., 560, 574, *581*
Hooker, A. M., 414, 415, *416*
Hooper, D., 440, *452*
Hooper, N. K., 124, *145*
Hooper, S. L., 123, 124, 144, *145, 150, 152*
Hopkins, D. A., 272, *283*, 529, 532, *544*
Hopkins, W. G., 380, *418*
Hoppenbrouwers, T., 523, *543, 548,* 569, 570, *581*
Hoppin, F. G., Jr., 483, 492, *517*
Hori, T., 343, *366*
Horita, A., 125, 141, *145*
Horiuchi, A., 90, *108*
Horiuchi, M., 222, *243*
Hornbein, T. F., 341, 355, *362*
Horne, R. S. C., 359, *362*
Horner, R. L., 255, *266*
Horowitz, B., 358, *366*
Horowitz, J. G., 467, *479*
Horowitz, J. H., *637*
Horwitz, A. F., 75, *96*
Hoshimaru, M., 47, *62*
Hosking, G. P., *450*
Hossman, K. A., *637*, 646, *658*
Hounsgaard, J., 310, *316*, 600, *610*
House, C., 80, *100, 101*
Howe, P. E., 322, *337*
Howell, R. G., 563, *581*
Howell, S., *481*

Howell, T. W., 74, *96*
Hoyle, G., 116, 144, *154*
Hsieh, T-S., 84, *107*
Huang, F. L., 77, 78, 80, *100, 110*
Huang, K-P., 77, 78, 80, *100, 110*
Huang, Q., 201, 205, *214*
Hubbard, J. H., 354, *363*
Hubbard, J. I., 14, *61*, 644, *658*
Huber, F., 114, 116, *151*
Hudgel, D. W., 566, 572, *578, 581*
Huganir, R., 15, 31, *60*
Hugelin, A., 129, *150*, 276, *287*, 534, *541*
Hughes, J. M. B., 671, *676*
Hughes, S., 410, *421*
Hughson, R. L., 462, *477*
Huguenard, J. R., 160, *174*
Hui, A., 74, *98*
Huizar, P., 384, 387, 388, *420, 422*
Hultgren, H. N., 192, *195*
Hultman, E., 441, *451*, 462, 469, *475, 477*
Humbertson, A. O., 177, 178, 180, 183, *194, 196*
Hume, R. I., 114, 116, 117, 118, 119, 120, 121, 122, *149, 150, 151*
Humphrey, D. R., 354, *363*
Hunkapillar, M. W., 25, *67*
Hunt, J. R., 573, *581*
Hunt, S. P., 49, *59*
Hunter, T., 75, 94, *100*
Hurlbut, W. P., 14, *56*
Hurle, M. A., 351, *366*
Hursh, J. B., 231, *241*
Hurst, J. H., 354, *362*
Huszczuk, A., 247, 253, 255, 259, *264*
Hutchison, A. A., 5, *7, 563, 582*
Huttenlocker, P. R., 624, *637*
Huttner, W. B., 90, *97, 100*
Huxley, A. F., 155, *174*
Hwang, J. C., 222, 225, *241*
Hwang, Q., 352, *370*
Hyman, A. I., 683, 684, *695*
Hyman, C., 85, *100*

I

Ianuzzo, C. D., 401, 404, 408, *422*, 462, 464, 467, 469, 472, *478*
Iber, C., 42, *68*, 539, *541*
Ichigama, A., 28, *62*
Ichikawa, K., 346, 347, 355, 357, *361, 366*
Ichinose, M., 501, *511*
Igarashi, K., 77, *105*
Iida, H., 49, *68*
Ikeda, K., 114, 121, *154*
Ikonomidou, C., 606, *610*
Imajoh, S., 78, *105*
Inagaki, S., 48, 49, *68*
Inayama, S., 25, *65*
Imazato-Tanaka, C., 470, *481*
Ingebritsen, T. S., 89, *107*
Ingwall, J. S., 618, *639*
Inman, J. G., 5, *7*, 662, 667, 669, *677*
Inoue, H., 501, *511*
Inoue, M., 77, *100*
Ioffe, S., 200, 206, *214*, 358, *367*, 554, 555, 557, *581*, 668, *675*
Irestedt, L., 666, *675*
Irvin, C., 39, 42, *68*
Irvin, C. G., 485, 486, *511*
Irvine, R. F., 31, *55*, 305, *312, 319*
Isabel, J. B., 492, *511*
Isaza, G. D., 457, *478*
Iscoe, S., 271, 274, 279, *283, 285*, 352, *362, 365*
Issa, F., 360, *367*
Issa, F. G., 566, *579, 582*, 693, *699*
Ito, H., 78, *102*
Ito, S., 499, *511*
Itoh, N., 32, *60*
Itoh, T., 620, 623, *637*
Iverfeldt, K., 23, *61*
Iversen, K., 672, *675*
Iversen, L. L., 25, 48, *56, 61*
Ivy, G. O., 192, *195*
Iwamura, Y., 558, *582*
Iyengar, R., 29, *55*
Izumo, S., 410, *423, 424*

J

Jackson, D., 594, *611*
Jackson, D. C., 605, 606, *610, 613*
Jacobi, M. S., 659, 683, 684, 688, 690, *696, 697*
Jacobowitz, D., 492, *512*
Jacobs, J. A., 3, *7*
Jacobs, L., 308, *313, 316*
Jacobs, R. A., 3, *7*
Jacobsen, B. B., 672, *673*
Jacobson, H. N., 684, 685, *695, 696*
Jacquin, T., 111, 133, 136, *146*
Jaeger, C. B., 45, *61*
Jahn, H., 82, *100*
Jahn, R., 91, *107*
Jahnsen, H., 310, *316*, 600, *610*
Jain, S. K., 256, *263*
James, O., 571, *579*
Jammes, Y., 229, *241*, 274, *285*, 486, *511*
Jan, L. Y., 74, *101*, 157, *174*
Jan, Y. N., 74, *101*, 157, *174*
Jancso, G., 186, *195*
Jankowska, E., 114, 116, 125, *150, 151*
Janowsky, J. S., 615, 616, *637*
Jansen, A. H., 200, 206, *214*, 324, *338*, 358, 359, 360, *367*, 487, *511*, 554, 555, 557, *581*, 668, *675*, 681, 684, 685, *697*
Jansen, J. K. S., 379, 380, 386, 411, 413, *418, 420, 427*
Jansen, J. R. C., 231, 232, *240*, 486, *509*
Jarvie, P. E., 86, *98*
Jarvis, D., 501, *514*
Jasper, H. H., 537, *544*
Jean, A., 272, *283*
Jeanneret, Grosjean, A., 501, *513*
Jeanrenaud, B., 129, *145*
Jeffcoate, S. L., 19, 53, *61*, 112, 141, *150*
Jeffery, P., 237, *240*
Jeffrey, H. E., 573, *582*
Jellies, J., 125, 126, *150*

Jelsema, C. L., 72, *95*
Jeng, H. J., 553, 569, *580*
Jenkinson, D. H., 14, *61*
Jensen, K. F., 84, 92, *105*
Jessel, T. M., 28, 29, *61*
Jirikowski, G. F., 43, *61*
Jobsis, F. F., 291, *317*
Jodkowski, J. S., 160, *172*, 279, *285*, 382, 386, *422*
Joh, T. H., 48, 49, *66*, 349, *365*, *370*
Johannson, H., 626, *637*
Johansson, O., 19, 22, 49, 50, 53, *58*, *60, 61*, 112, 141, *150*, 186, *195*
Johansson, R., 442, 444, *450*
Johansson, R. S., 434, 436, 442, *450*, *451, 452, 453*
Johansson, S., 503, *507*
Johe, K. K., 653, *659*
Johnsen, D. C., 566, *579*
Johnson, B. E., *285*
Johnson, B. M., 555, 557, *578*
Johnson, D. E., 502, *510*
Johnson, J. D., 81, *99*
Johnson, J. W., 26, *61*
Johnson, M. G., 502, *510*
Johnson, P., 155, *174*, 229, *241, 244*, 272, 274, 275, 281, 282, *285*, 322, 329, 333, *336*, 497, *508*, 535, *544*, 555, 558, 563, *578*, *580*, 593, *607*, 687, 688, *697*
Johnson, R. L., Jr., 457, 461, *480*
Johnson, S., 161, 162, *172*
Johnson, S. M., 111, 133, 135, 136, 137, *147, 150*
Johnston, B. M., 207, *214*, 360, *365*, 497, *509*, 554, 555, 556, 557, 558, *577, 579, 582, 585*, 662, 665, 666, 667, 669, *674, 675*
Johnston, G. A. R., 46, *61*
Johnston, P. A., 90, *108*
Jolesz, F., 408, *422*
Jolivet, A., 38, *69*
Jonason, J., 112, 141, *149*
Jones, B. E., 349, *367, 532, 544*
Jones, C. T., 666, *675*

Jones, D. A., 436, 438, 442, *450*, *451*
Jones, D. P., 631, 633, *635*
Jones, E. M., 683, 684, 685, *697*
Jones, K. J., 38, 39, *64*
Jones, S. F., 14, *61*
Jones, S. P., 380, 386, 397, 411, 413, *422*
Jones, T. R., 492, *509*
Jongsma, H. W., 554, *584*
Jonsson, G., 19, *56*
Jonzon, A., 258, *266*
Jonzon, B., 666, 669, *679*
Jordan, C. L., 397, *422*
Jordan, D., 129, 131, *148*, 277, 278, 279, *287*, 323, 334, *338*, 487, *512*
Jordan, J., 200, *214*, 231, 232, *242*, 484, 486, *512*
Jordan, L. M., 125, *150, 154*
Jorfedlt, L., 468, *478*
Joseph, S. A., 668, *676*
Jouvet, M., 524, 527, 537, *544, 547*, 558, *582*
Ju, G., 19, *60*
Juan, G., *481*
Juel, C., 438, *451*
Jukes, M. G. M., 125, *150*
Julius, D., 28, 29, *61*

K

Kabat, H., 596, *612*
Kaczmarek, L. K., 14, *63*, 82, *101*
Kadefors, B., 458, 459, *478*
Kahng, M. W., 633, *639*
Kaibuchi, K., 77, *96*
Kaila, K., 305, *316*
Kairiss, E. W., 33, *55*
Kaiser, D. G., 39, *68*
Kaiser, K. K., 407, 414, *417*
Kalapesi, Z., 573, *585*
Kales, A., 552, *585*
Kalia, M., 19, *61*, 129, *150*, 237, *241*, 261, *266*, 487, 499, *512*, 645, *658*

Kalimo, H., 606, *614*
Kampine, J. P., 239, *241*, 246, *267*
Kanamori, N., 527, *547*
Kanaseki, T., 272, *288*
Kanazir, S. D., 90, *108*
Kanda, K., 470, *481*
Kandel, E. R., 32, 33, *55, 57, 59, 67,* 83, *99*
Kangawa, K., 28, *62,* 76, *101*
Kannan, M., 485, *509*
Kannan, M. S., 492, *509*
Kao, F. F., 205, *213,* 350, *363,* 664, *673*
Kao, H-T., 90, *108*
Kao, L. C., 504, *512*
Kaplan, J. K., 310, *317*
Kapp, B. S., 529, *546*
Karlberg, P., 691, *697*
Karlin, A., 25, *62*
Karlsson, J., 464, 468, *477, 478*
Karrmann, D., 444, *450*
Kasper, C. E., 408, 414, *424*
Kass, I. S., 601, 604, *610*
Kastella, K. G., 535, *544*
Kastenschmidt, L. L., 392, *419*
Katayama, Y., 540, *545*
Kater, S. B., 603, *611*
Kato, A. C., 160, *172*
Katona, P. G., 257, *267*
Kattwinkel, J., 258, *266*
Katz, B., 13, 14, *57, 58, 62,* 309, *315*
Katz, D. M., 327, 334, 335, *338*
Katz, F., 79, 81, *101*
Katzman, R., 17, *68*
Kauer, J. A., 83, 92, *103, 105*
Kaupp, U. B., 76, *101*
Kawabori, I., 681, 682, 683, 684, 686, *696*
Kawahara, K., 645, *656*
Kawai, Y., 48, 49, *68*
Kawano, H., 45, *57*
Kay, A. R., 143, *150*
Keeler, J. R., 343, *367*
Keenan, C. L., 33, *55*

Keens, T. G., 3, *7,* 401, 404, 408, *422,* 464, 467, 469, 472, *478,* 504, *512*
Keilin, D., 292, *316*
Keiner, M., 38, *66*
Kellerth, J. O., 380, 384, 387, 388, *421, 422*
Kelly, A. M., 397, 410, 411, *422, 424, 425*
Kelly, D. H., 322, *340*
Kelly, E. A., 490, *513*
Kelly, E. N., 250, 256, *265*
Kelly, G., 260, 261
Kelly, P. T., 83, 85, 86, 87, 89, 92, *101, 103, 107*
Kelly, S. S., 389, *422, 470, 480*
Kelsen, S. G., 257, 259, *266,* 390, *422,* 456, 457, 465, *478*
Kemilof, F. K., 633, *641*
Kemp, B. E., 80, *100, 101*
Kendergan, B. A., 467, *475*
Kennedy, C., 644, *660*
Kennedy, D., 144, *151*
Kennedy, M. B., 74, 83, 84, 86, 87, 88, 89, 90, 92, *95, 98, 101, 102*
Kennedy, T. E., 32, 33, *55,* 82, *108*
Kent, G., 562, 563, 564, 565, *578*
Kent, J., 617, *636*
Kerman, K. E. O., 300, *318*
Kernell, D., 389, *423,* 444, 445, *452*
Kety, S., 626, *638*
Keuhl, T. J., 401, 404, 408, *423*
Keynes, R. D., 623, *637*
Keynes, R. J., 380, *418*
Khatib, M., 382, *421, 423*
Khorana, H. G., 73, *105*
Kida, K., 502, *512*
Kien, J., 126, 127, *151*
Kikkawa, N., 77, *102*
Kikkawa, U., 77, 78, 79, 80, *95, 96, 98, 99, 102, 105, 107, 108*
Kikyotani, S., 25, *65*
Kiley, J. P., 130, *148,* 354, 355, 356, *364, 367*
Killackey, H. P., 192, *195*
Kim, Y. J., 459, *475*

Kimelberg, H. K., 600, *607*
Kimura, A., 499, *511*
Kimura, H., 39, *62*
Kimura, K., 635, *638*
King, D. G., 121, *153*
King, D. W., 468, *477*
King, J. S., 189, *198*
King, M. M., 89, 93, *99, 102*
King, R. A., 534, *548*
Kino, J. F., 80, *105*
Kirby, G. C., 295, 296, 304, 305, *312, 313, 315,* 334, *337*
Kirchner, J. A., 222, *244*
Kirkpatrick, C. T., 483, *512*
Kirkwood, P. A., 129, *151*
Kirshner, H. S., 650, *658*
Kishimoto, A., 77, 79, 80, *95, 100, 102, 105, 108*
Kitterman, J. A., 556, 557, 558, *578, 582, 584, 587,* 667, 668, 669, *676*
Kjellmer, I., 667, *675*
Klain, D. B., 573, *582*
Klarsfeld, A., 32, *56, 58*
Klaus, M., 254, 255, *264,* 691, 692, *695*
Klaus, M. H., 257, *267*
Klausen, K., 468, *478*
Klausner, R. D., 72, *109*
Klee, C. B., 93, *102*
Klein, J. P., 255, *268,* 279, *287*
Klein, W. L., 73, *102*
Kleineke, J., 623, *640*
Klesh, K. W., 483, 501, *513*
Klimova-Cherkasova, I. V., 224, *240*
Klingenberg, M., 617, *638*
Kluge, K. A., 523, *543*
Kniffki, K. D., 446, *452*
Knill, R., 248, *266,* 566, 570, *582*
Knopf, J. L., 78, *102*
Knox, C. K., 255, 256, 259, *266*
Knuth, K. V., 685, 687, *698*
Knuth, S. L., 563, *576*
Kobilka, B. K., *62*
Kobilka, T. S., 62
Koch, G., 691, *697*

Koch, W. J., 74, *98*
Koenig, J. S., 274, 281, 282, *284*
Koepchen, H. P., 144, *151*
Kogo, N., 205, *213,* 346, 347, 355, 357, *361*
Koh, E. T., 532, *547*
Kohle, S. J., 626, 627, 632, 634, *636,* 654, *657*
Kohn, L. D., 300, *315*
Kojima, M., 28, *62*
Koke, J. R., 633, *636*
Komisaruk, B. R., 540, *544*
Kong, F. J., 382, *419*
Konishi, A., 384, 387, *425*
Konuo, Y., 78, *105*
Koos, B. J., 360, *367,* 557, *582,* 668, *676*
Kopelman, A. E., 667, *673*
Kordeli, E., *56*
Koretsky, A. P., 440, *452*
Korn, G., 570, *577*
Korner, M., 84, *102*
Kosch, P. C., 248, 252, 253, 256, 260, *265, 267, 268,* 562, 563, 567, 568, *577, 581, 582, 586*
Kose, A., 78, *102*
Koshiya, N., 205, *213*
Koshland, D. E., Jr., 80, *104*
Koslo, R. J., *508*
Kostreva, D. R., 239, *241,* 246, *267*
Kotani, H., 47, *62*
Koulakoff, A., 160, *172*
Kovar, I., 274, *285*
Koyano, H., 295, *315*
Kozar, L. F., 237, *241,* 250, *268,* 274, *287,* 563, 564, 572, 573, 574, *576, 580, 585, 586,* 594, *607*
Kraemer, G. W., 125, 141, *151*
Kraft, A. S., 79, *102*
Krahenbuhl, G., 468, *476*
Kraig, R. P., 647, *658*
Kramis, R., 540, *548*
Krans, H. M. J., 73, *106*
Krastins, I. R. B., 467, *478*
Krause, J. E., 47, *62*

Krause, W. J., 254, *267*
Krauss, A. N., 562, 573, *582*
Krebs, E. G., 76, 80, *107, 109*
Krebs, H., 182, *195*
Kreisman, N. R., 606, *612*
Krettek, J. E., 532, *544*
Kreutter, D., 72, *106*
Krieger, A. J., 357, *371*
Krieger, D. T., 18, *62*
Krinks, M. H., 93, *102*
Kristan, W. B., 114, 116, *151*
Kriz, N., 599, *613*, 648, 650, 651, *659*
Kriz, R. W., 78, *102*
Krnjevic, K., 389, *423*, 441, *452*, 596, 597, 600, 603, *608, 610, 611,* 646, 647, 649, *658, 659*
Kroeger, E. A., 492, *517*
Kronauer, R. E., 569, 570, *587*
Krous, H. F., 200, *214*, 231, 232, *242*, 484, 486, *512*
Krukoff, T. L., 186, *195*
Krupinski, J., *102*
Kubin, L., 279, *286*
Kubo, T., 28, *62*
Kubota, K., 558, *582*
Kudo, N., 387, *420*
Kuehl, T. J., 472, *479*
Kuei, J. H., 390, *423*, 465, *478*
Kuffler, D. P., 380, *427*
Kuffler, S. W., 644, *659*
Kugelberg, E., 377, 391, 392, 407, *420*, 464, *478*
Kuhar, M., 18, 19, 45, 49, *54, 55*
Kuhn, L. J., 77, *105*
Kuipers, J. R. G., 595, *612*
Kukulka, C. G., 442, *450*
Kulik, T. J., 502, *510*
Kumar, P., 326, 330, 335, *338*
Kuna, S. T., 563, 566, *582*
Kunesch, E. J., 446, *451*
Kunis, D. M., 600, 606, *608, 613*
Kuno, K., 354, *365*
Kuno, M., 14, *62*, 354, *366*, 384, 387, 388, *420, 422*
Kuo, J. F., 31, *55*, 78, 79, *98, 110*

Kupferman, I., 114, *151*
Kuroda, T., 77, 80, *102*
Kurth, C. D., 5, *7*
Kushmerick, M. J., 414, *419*
Kushner, P. D., 124, *145, 151*
Kuwana, S., 355, 357, *361, 366*
Kuypers, H. G. J. M., 49, *60*, 177, 180, *195*
Kuzemko, J. A., 672, *676*
Kwanbunbumpen, S., 14, *61*
Kwiatkowski, A. P., 89, *102*
Kwiatkowski, K., 570, 571, *583*

L

Labat, C., 489, 490, *507*
Labruyere, J., 606, *610*
La Framboise, W. A., 250, 256, *267*, 459, 465, 473, *481*, 571, *582*, 593, *610*, 646, *659*
Lagerback, P. A., 664, *673*
Lagercrantz, H., 42, *64*, 112, *154*, 274, *288*, 321, *338*, 360, *367*, 662, 664, 665, 666, 668, 669, 670, 671, 672, *673, 675, 676, 677, 678, 679*
Lahiri, S., 322, 323, 325, 327, 328, 329, 332, 334, 335, *338, 339*, 594, *610*, 662, 667, *674, 677*
Lai, T. L., 553, 569, *580*
Lai, Y., 86, 89, *99, 104*
Lall, A., 308, *313*
Lam, R. L., 274, *286*
LaManna, J., 647, *659*
La Manna, J. C., 603, 605, 606, *612*
Lamarche, J., 321, 322, 334, *337*
Lamb, T. W., 346, *368*
Lambert, T. F., 555, *583*
Lamborghini, J. E., 160, *175*
Lamp, S. T., 602, *613*
Landau, L. I., 484, *512*
Lang, R., 19, *61*
Langercratz, H., 24, *63*
Langfitt, T. W., 633, *640*
Langston, C., 502, *512*
La Pointe, M., 471, *477*

Larimer, J. L., 125, 126, *150, 151*
Larnicol, N., 384, 385, *425*
Larsell, O., 485, *512*
Larson, J., 33, *63*
Lasher, R. S., 87, *109*
Laskowski, M. B., 379, *423*
Lau, C. K., 38, *69*
Lauder, J. M., 45, *62, 69*, 182, *195*
Laufer, R., 30, 32, *56, 63*
Laundau, L. I., 504, *517*
Lauzon, A. M., 498, *514*
Lavigne, C. M., 461, *475*
Lawff, D., 566, *582*
Lawing, W. L., 348, 351, *364, 367*
Lawson, E. E., 45, *64*, 201, 208, 211,
    213, 214, 274, 280, *285, 286,*
    330, 333, *339*, 573, *583*, 593,
    611, 646, *659*, 682, 683, 684,
    686, 687, *697, 699*
Layzer, R. B., 440, *452*
Lazdunski, M., 606, *607*
Leahy, F. N., 573, *585*
Leake, B., 523, *543*
LeBlond, J., 596, 597, 600, *610, 611*
LeBrun, F., 570, *577*
Leduc, B. M., 554, 555, *578*
Lee, A. F. S., 571, *579*
Lee, C. H., *556*, 557, *584, 587*
Lee, D., 358, *370*, 558, 570, 571,
    583, 585
Lee, J. A., 439, *449*
Lee, J. C., 274, 275, 282, *284*
Lee, L-Y., 230, *242*
Lee, M-H., 78, *102*
Lee, S. Y., *106*
Leeson, C. R., 254, *267*
Leff, A. R., 490, *513*
Lefkowitz, R. J., 28, 29, *57, 62,*
    68, 72, 73, *98, 107, 110*
Le Greves, P., 19, *60*
Lehninger, A. L., 304, *314*
Leibsten, A. G., 349, *371*
Leichnetz, G. R., 540, *545*
Leigh, J. S., 617, 628, 631, 634,
    637, 639
Leigh, J. S., Jr., 617, *636*

Leistner, H. L., 252, *266*, 569, 570,
    574, *580*
Leith, D. E., 467, *478*
Leitner, L. M., 304, *319*
Lende, R. A., 189, *195*
Leninger-Follert, E., 647, *660*
Lennard, P. R., 114, 116, 117, 118,
    119, 120, 121, 122, *149, 151*
Lenney, W., 504, *512*
Lent, R., 168, *174*
Lenzi, P., 535, *546*
Leonard, C. S., 15, *63*, 91, *103*
Leonhardt, S., 24, *63*
Lerman, P., 692, *698*
Leslie, R. A., 19, *63*
LeSoeuf, P. N., 401, 404, 408, *423,*
    484, *512*, 560, 568, *583*
Lester, H. A., 25, *63*
Leszczynski, L., 563, *581*
Letinski, M. S., 397, *422*
Leung, A. T., 74, *98*
Leusen, I. R., 341, *367*
Levetzki, A., 29, *68*
Levin, R. M., 492, *512*
Levine, D. N., 376, 379, 392, 407,
    413, *418*, 434, 444, *450*
Le Vine, H., 87, 88, 89, *102, 103*
Le Vine, H., III., 78, 84, 87, 92, *106,*
    107, 110
Levison, H., 401, 404, 408, *422*, 456,
    464, 467, 469, 472, *478*, 560,
    563, 568, *583, 586*
Levison, R. S., 456, *478*
Levitan, I. B., 82, *103*
Levitan, L., 14, *63*
Levitzki, A., 29, *63*
Lewis, D. M., 377, 378, 380, 386,
    387, *416, 418*
Lewis, D. V., 143, *151*, 644, *659*
Lewis, J. K., 125, 141, *151*
Lewis, M. I., 467, *475*
Lewis, R. E., 414, 415, *416*
Li, A., 351, *357*
Lichtman, J. W., 44, 45, *66*, 615, *640*
Liddell, E. G. T., 113, *151*, 376, 381,
    423

Lidov, H. G. W., 45, *63*
Liebeskind, J. C., 532, *545*
Lieble, G. J., 621, 631, *640*
Liftshits, R. L., 633, *641*
Liggins, G. C., 554, 555, 556, *578, 582*, 667, 668, 669, *676*
Liley, A. W., 14, *63*
Liljestrand, G., 297, *315*
Lin, C. R., 86, *103*
Lind, J., 564, *576*
Lindeman, R. C., 282, *287*
Linden, D. J., 33, *54*, 83, *95*
Lindenberg, L. B., 322, *337*
Linderoth, B., 34, 55
Lindh, B., 19, 20, *60*
Lindsey, B. G., 261, *263, 268*
Lindstrom, D. P., 274, *285*
Lindstrom, L., 458, 459, *478*
Ling, D. S. F., 161, *175*
Lingle, C. J., 122, *151*
Lione, A. D., 621, *639*
Lippold, O. C. J., 442, 444, *450*
Lipscombe, D., 74, 82, *109*
Lipscomb, W. T., 343, 345, *367*
Lipski, J., 275, 279, *283, 286*
Lipton, B., 692, *699*
Lipton, M. A., 125, 141, *145*
Lipton, P., 601, 604, *610*
Lisman, J. E., 90, *103*
Lister, G., 593, 595, *609, 611*
Liu, G., 159, *175*
Liu, M., 626, 628, 629, *638*
Livingstone, M. S., 74, *103*
Ljungdahl, A., 19, 48, *60, 63,* 112, *150*
Ljunggren, B., 600, *611*
Llados, F., 308, *320*
Llinas, R., 15, *63*, 91, *103, 104,* 155, 157, 169, *175*, 525, *545*, 644, *659*
Lobel, P., 25, *62*
Lock, J. E., 502, *510*
Loeschcke, H. H., 341, 342, 343, 344, 345, 346, 348, 351, 356, 357, *364, 365, 368, 370*
Loewy, A. D., 129, *151, 153,* 349, *368*, 532, *547*

Lofgren, O., 42, *64*
Logeat, F., 38, *69*
Loizzi, R., 222, 223, *243*
Lombroso, C. T., 692, *698*
Lomo, T., 410, 411, 414, *425*, 470, *479*
Long, W. A., 280, *286*, 330, 333, *339*, 360, *367*, 556, 573, *583*, 593, *611*, 646, *659*
Longo, L. L., 646, *656*
Loomis, C. R., 78, 79, *99, 102*
Lopata, M., 465, *476*
Lopes, J., 260, *267*
Lopes, J. M., 457, *479*, 559, 562, 563, 567, 568, *583*
Lopez-Barneo, J., 293, 296, 305, 306, *316, 320,* 334, *339*
Lopez, F., 527, *548*
López-López, J. R., 293, 296, 305, 306, *316, 320,* 334, *339*
Lorenco, R. V., 465, *476*
Lothman, E., 647, *659*
Lou, L. L., 85, *107*
Louis, T. M., 667, *673*
Loury, O. H., 686, *696*
Lovinger, D. M., 33, *54*, 83, *95*
Lowe, D. G., 75, *96*
Lowey, S., 410, *421*
Lowry, O. H., 620, *638*
Lübbers, D. W., 306, *312*, 327, *336*, 647, *660*
Lubka, R., 322, *337*
Lucier, G. E., 200, 211, *215*, 272, 275, 279, 280, 282, *287*
Lucke, S., 411, *418*
Lüdi, H., 304, *314*
Luff, A. R., 414, *423*
Lumsden, T. E., 684, *698*
Lund, S., 125, *150*
Lundbag, L., 238, *242*
Lundberg, A., 125, *150*
Lundberg, D., 112, 141, *149*
Lundberg, J., 19, 23, 24, *60, 63*
Lundberg, J. M., 112, *150*, 186, *195*, 224, 238, *242*, 485, 498, 500, *512, 513*

Lundborg, P., 672, *675*
Lundgren, G., 24, *63*
Lundholm, E., 503, *507*
Lundholm, K., 468, *475*
Lundholm, L., 503, *507*
Lunteren, E., 354, *366*
Lutz, P. L., 603, 605, 606, *608, 609, 612*
Lux, H. D., 143, *154*, 296, 309, 310, *316, 317*, 646, 647, *658*
Lydic, R., 527, 534, *545*, 563, *584*
Lunch, G., 33, *63*
Lynne-Davies, P., 459, *481*

Mc

Mc Allen, R. M., 345, 349, 351, 352, *364, 368*, 487, *513*
Mc Cance, R. A., 646, 647, 648, *660*
McCarley, R. W., 536, *544*
Mc Carter, J. M., 401, 404, 408, *423*
Mc Carter, R. J. M., 472, *479*
Mc Carthy, L. E., 343, 354, *362*
Mc Clelland, M. E., 229, *241*, 272, *285*, 555, 563, *580*
Mc Clelland, M. F., *285*
Mc Closkey, D. I., 308, *317*
Mc Comas, A. J., 443, *451*
Mc Connell, S. D., 489, *508*
Mc Cooke, H. B., 322, 324, 325, 326, 327, 330, 333, 334, *336, 338*, 497, *508*, 662, 665, 669, *673, 676*
Mc Cormick, D. A., 159, 160, 166, *175*
Mc Crady, E., 178, *196*
McCrady, E., Jr., 484, *513*
Mc Crimmon, D. R., 128, *148*, 277, 278, 279, *286*
McCulloh, K., 573, *581*
Mc Cully, K., 617, *636*
Mc Dermott, A. B., 26, 28, 29, 61, 64, 310, *317*
Mc Donald, D. M., 291, *317*
Mc Ewen, B. S., 38, 39, 44, *64*

Mc Ginty, D. J., 523, 536, 537, *545, 548*, 569, 570, 572, *575, 581, 586*, 594, *607*
Mc Guiness, T. L., 15, 31, *60, 63*, 86, 87, 91, *101, 103, 104, 105*
McIlroy, M. B., 560, *580*
Mc Ilwain, H., 132, *154*
Mc Kenna, E., 74, *98*
McKenzie, D. K., 446, *451*
Mc Kinney, W. T., 125, 141, *151*
Mc Laughlin, P. J., 186, *198*
Mc Murry, W. C., 623, *636*
Mc Namara, M. C., 45, *64*
McNicol, K. J., 484, 492, *515, 517*
Mc Phie, D. L., 83, *105*
Mc Queen, D. S., 308, *317*, 671, *676*

M

MacCullum, M., 560, 573, 574, *577, 585*
Mac Dermott, A. B., 160, *175*
Macefield, G., 446, *451*, 684, *695*
Macfabe, D. F., 537, *548*
MacFadyen, U. M., 567, *576*
Machida, C. M., 358, *366*
Machulskis, A., 502, 506, *518*
Mack, M. M., 490, *513*
Mackinney, L. G., 687, *697*
Macklem, P. T., 397, *425*, 457, 459, 461, 476, 477, *480, 481*, 485, 486, 511, *518*
Macklis, J. D., 351, *370*
Mac Leod, V. H., 180, *194*
Mac-Quin-Mavier, I., 501, *509*
Madapallimatum, A., 322, *339*, 672, *677*
Madison, D. V., 74, 82, 83, *103, 109*
Madison, R., 351, *370*
Maeda, A., 28, *62*
Maeda, N., 76, *98*
Maeyama, T., 224, *244*
Maggi, C. A., 492, 499, *513*
Maggio, J. E., 47, *64*
Magleby, K. L., 74, *96*, 296, *312*
Magnante, D., 246, *267*, 567, *584*

Magoun, H. W., 128, *153*, 537, *546*
Mahdavi, V., 410, *423*
Mahowald, M. W., 524, *545*
Mailman, R. B., 125, 141, *145*
Majcher, J. S., 646, *656*
Majcherczyk, S., 308, *318*
Makowski, E. L., 626, *638*
Malcomb, J. L., 343, *368*
Malenka, R. C., 83, 92, *103, 105*
Maley, B., 112, *152*
Malgren, L. T., 223, *243*
Malhotra, V., 74, *104*
Malinow, R., 83, 92, *103*
Malkowick, D., 492, *512*
Maloney, J. E., 358, *362*, 484, *507*, 552, 555, 556, 559, *575, 576, 577, 583*, 594, *607*, 662, 664, *673*
Mamalaki, C., 27, *64*
Mamelak, A. N., 351, *370*
Man, E. H., 664, *676*
Manabe, M., 201, *214*
Mandel, G., 31, 51, *65*
Maniatis, T., 84, *103*
Mann, G. L., 524, *544*
Mann, L. I., 626, 628, 629, *638*
Mannard, D., 433, *452*
Manton, M. A., 502, *518*
Mantyh, P. W., 49, *59*
Marangos, P. J., 487, 499, 502, *516*
Marchal, F., 274, 281, 282, *285, 286*
Marchetti, C., 309, *316*
Marchini, G., 668, *676*
Marcus, C., 666, 668, *676*
Marder, E., 122, 123, 124, *145, 148, 152*
Mares, P., 648, 650, 651, *659*
Margreth, A., 414, *425*
Marino, P. A., 569, 570, *580*
Marino, P. L., 346, *368*
Marks, J. D., 522, 529, *543, 545*
Marlot, D., 207, *213*, 222, 231, 232, 233, 234, 237, *242*, 250, 254, 255, 256, 260, 262, *264, 267, 269*, 385, *419*, 486, 494, *513*, 522, *542*, 559, 560, 571, 572, *576, 578*, 593, *607*, 645, *657*, 662, *673*

Marlow, T. A., 254, 260, *265*
Marshall, W. H., 354, *363*
Marsland, D. W., 569, *586*
Marson, L., 186, *195*, 349, *368*
Martilla, I., 644, *657*
Martin, C. B., 554, *584*
Martin, G. F., 177, 178, 179, 180, 183, 185, 188, 189, 190, 191, 192, *193, 194, 197*
Martin, J. B., 18, *62*
Martin-Moutot, N., 160, *172*
Martin, R. J., 257, 260, *267*, 485, 486, *511*, 563, 566, 568, 571, 572, 574, *575, 577, 578, 579, 583*
Martin, T. P., 391, 392, *423*
Martling, C. R., 498, *512*
Marttila, I., *284*, 534, *542*
Marttila, J., 129, *148*
Marty, A., 295, 296, *316, 319*, 300, *319*
Marzella, L., 633, *639*
Massion, W. H., 342, 343, 344, *368*
Masumura, M., 634, 635, *641*
Mathew, O. P., 224, 225, 226, 227, 228, 229, 230, 233, 237, *239, 242, 243, 244*, 272, 274, *287*, 566, *583, 585*, 688, *698*
Matsuda, K., 668, *676*
Matsui, H., 62
Matsumoto, M., 635, *638*
Matsuo, H., 28, *62*, 76, *101*
Matsuyama, T., 635, *638*
Mattsson, K., 305, *316*
Mattson, M. P., 603, *611*
Matus, A., 25, *58*
Mauk, M. D., 83, 92, *103*
Mauray, F. E., 556, *587*
Mauro, A., 14, *56*
Mauss, A., 48, *54*
Mauxion, F., 84, *102*
Max, S. R., 470, 471, *476, 481*
Maxwell, D. S., 654, *657*, 671, *676*
Maxwell, L. C., 401, 404, 408, *423*, 472, *479*
May, W. S., 78, *110*

Mayer, M. L., 310, *317*
Mayevsky, A., 617, *638*
Maynard, D. M., 122, *152*
Mayock, D. E., 321, *339*, 472, 473, *479*, 571, *587*, 593, *614*, 662, *679*
Mazza, N. M., 3, *7*
Mazzei, G. J., 79, *98*
Mead, J., 130, *146*, 257, 259, *266*, 574, *581*
Means, A. R., 38, *66*, 80, 88, *98, 101*
Mebel, P. E., 485, 486, 494, *514*
Mediavilla, A., 351, *366*
Meech, R. W., 115, 119, *152*, 296, *317*
Meesmann, M., 277, 278, 279, *287*
Mega, A., *369*
Megirian, D., 274, *287*, 522, 524, *547*, 563, *583, 586*
Mei, N., 229, *241*, 274, *285*, 486, *511*
Meiners, L., 529, 532, *544*
Meister, B., 19, 20, *60*
Mela, L., 627, 630, 631, 633, *636, 637, 638, 639, 640*
Melancon, P., 74, *104*
Melander, T., 19, 20, *60*
Meldrum, B. S., 310, *319*
Meli, A., 492, 499, *513*
Melichna, J., 408, *421*
Mellins, R. B., 3, *7*, 155, 169, *174*, 252, *266*, 321, *338*, 498, *511*, 553, 569, 570, 572, 574, *580*, 593, 594, *609, 612*
Mellström, A., 384, 387, *422, 423*
Mendell, L. M., 381, *421*
Mendoza, R. V., 472, *480*
Menger, F. M., 79, *110*
Menon, A. A., 566, *585*
Menon, A. P., 248, 249, *269*
Mense, S., 446, *452*
Ment, L. R., 667, 669, *677*
Merchenthaler, I., 43, *61*
Mergner, W. L., 633, *639*
Merrick, D., 74, *99*

Merrill, E. G., 279, *284, 286,* 644, 645, *659*
Merrill, S., 360, *371*
Merton, P. A., *452*
Mesulam, M. M., 129, *150*, 487, *512*
Metcalf, I., 248, 252, *268*
Mettler, F. A., 129, *152*
Metzger, J. M., 467, *479*
Meyer, H. B., 691, 692, *695*
Meyer, H. G., 182, 183, *197*
Mezey, E., 129, *152*
Micevych, P. E., 43, *54*
Michael, L., 489, *516*
Michaelis, M. L., 653, *659*
Michoud, M. C., 501, *513*
Middendorf, W. F., 485, *513*
Midzenski, M., 489, 490, 503, *508*
Mikami, A., 28, *62*
Mikawa, K., 77, 80, *102*
Mikoshiba, K., 76, *98*
Miledi, R., 14, *62*, 389, *419, 423,* 441, *452*
Milerad, J., 42, *64*, 274, *288*
Miles, R., 348, *368*
Miletic, V., 186, *197*
Milgrom, E., 38, *69*
Milic-Emili, J., 248, 249, 251, 252, 255, *266, 267, 268, 269,* 456, *476*, 567, *584*
Miller, A. J., 222, 223, *243*, 280, *286*
Miller, A. L., 621, 623, *639*
Miller, C., 74, *104*
Miller, H. C., 571, 573, *583*
Miller, J. B., 410, *427*
Miller, J. P., 116, 121, 122, 124, 125, 128, *152, 153*
Miller, L. D., 627, 630, 631, 633, *637, 638, 639*
Miller, M. J., 260, *267*, 566, *577, 579*
Miller, R. G., 440, *452*
Miller, R. J., 48, 49, *66*
Miller, S. G., 74, 86, 88, 89, *104*
Miller, T., 489, *516*
Miller, T. B., 342, 355, *364*

Millhorn, D. E., 19, 20, 21, 23, 30, 34, 35, 36, 37, 39, 40, 41, 42, 43, 46, 49, 50, 51, 53, *54, 55, 56, 57, 58, 60, 64, 65, 68, 69,* 130, *148,* 186, *196,* 341, 343, 348, 350, 351, 354, 355, 356, *364, 365, 367, 368,* 663, *677*
Mills, E., 291, 308, *317*
Mills, J. E., 259, 260, *267*
Mills, J. W., 343, 345, 351, 352, 357, *369*
Milner, A. D., 504, *512, 514, 515*
Milner-Brown, H. S., 440, *452*
Milner, T. A., 349, *368*
Milstein, J. M., 620, *639*
Minagawa, H., 49, *68*
Miserocchi, G., 252, *267*
Mishina, M., 28, *62*
Misra, U., 79, 84, *104, 109*
Mitchell, H. W., 484, 489, 490, 491, *516*
Mitchell, R., 23, *65*
Mitchell, R. A., 271, *283,* 342, 343, 344, 346, *368,* 497, 506, *507, 513, 515,* 521, *547*
Mitchell, R. W., 490, *513*
Mitchelson, F., 507, *513*
Mitra, J., 351, 352, 354, *368, 370,* 671, *678*
Mitsumasu, T., 272, *288*
Mittman, M. J., 467, *475*
Mittnacht, S., 603, *609,* 634, *636*
Miura, M., 501, *511*
Miyahara, L., 523, *543*
Miyake, H., 634, 635, *641*
Miyake, R., 77, *102*
Miyata, T., 25, *65,* 76, *101*
Miyata, Y., 384, 387, 388, *422*
Miyawaki, A., 76, *98*
Mizuno, N., 272, *286,* 384, 387, *425*
Mochly-Rosen, D., 80, *104*
Moerschbaecher, J. M., 351, *371*
Moghadam, B., 653, *659*
Mognoni, P., *473*
Mohammad, M. A., 489, *513*
Mohandas, T. K., 74, *99*

Mohl, B., 114, *145*
Mokashi, A., 329, 332, *339*
Moldofsky, H., 522, *548,* 559, *586*
Mollenholt, P., 19, *60*
Mollover, M. E., 45, *63*
Monaghan, D. T., 72, 74, 83, *97,* 606, *614*
Moncada, S., 80, *104*
Monod, N., 523, *545*
Monster, A. W., 389, *423,* 445, *452*
Monteau, R., 129, *149,* 382, *421, 423*
Montgomery, S. P., 345, *362*
Monti-Bloch, L., 295, 308, *315, 317*
Montminy, M. R., 31, 51, *65, 69*
Moore, C. L., 620, *637*
Moore, J. J., 502, 506, *518*
Moore, M., 358, *370,* 558, 573, 574, *583, 585*
Moore, P. J., 327, 328, 329, *338,* 360, *366*
Moore, R. L., 462, 467, *479*
Moore, T. J., 621, *639*
Moosavi, S. S. H., 567, *576*
Morad, M., 310, *317*
Morale, D., 411, *418*
Morales, F. R., 525, 526, 527, *542, 545, 546*
Moreau, G., 593, 595, *611*
Morgan, B. C., 275, 282, *284,* 681, *696*
Morgan, C. A., 498, *513,* 593, 595, *611*
Morgan, J. I., 92, *104*
Morgan, S. E. G., 484, *512*
Mori, S., 645, *656*
Mori, T., 77, 79, *108*
Moriette, G., 567, 573, 574, *583*
Morikawa, Y., 488, *513*
Morris, A. J. R., 457, 459, *479*
Morris, M. E., 646, 647, 649, 650, *658, 659*
Morris, P. L., 488, *515*
Morrison, A. R., 524, *544*
Morrison, D. E., 351, 353, *370*
Morrison, H., 125, 141, *151*
Morrison, S., 349, *369*

Morrison, S. F., 349, *368*, 536, *546*
Morrissey, M. A., 462, *477*
Morrow, J. S., 91, *106*
Mortola, J. P., 223, 232, 234, 235, *241, 243,* 246, 248, 249, 250, 255, *264, 265, 267,* 274, *284,* 385, *424,* 473, *476,* 492, 495, 498, *508, 513, 514, 515,* 564, 567, 571, *579, 584, 586,* 593, 595, *608, 611,* 691, *698*
Morton, C. R., 663, *675*
Moruzzi, G., 522, 537, *541, 546*
Moser, R., 304, *314*
Mosinger, J. L., 606, *610, 611*
Moskowitz, M. A., 693, *698*
Moss, I. R., 5, *7*, 360, *369*, 662, 666, 667, 669, *677*
Moss, M., 593, 595, *611*
Moss, R. L., 408, 414, *420, 424*
Motoyama, E. K., 334, *338*, 483, 501, *513*, 594, *610*
Mott, J. C., 684, 685, *695, 696*
Moulins, M., 144, *150*
Moxham, J., 457, 459, *479*
Mozes, R., 246, 249, 260, *269*
Mueller, R. A., 112, 125, 141, *145, 151, 152*
Mulle, C., *56*
Muller, D., 33, *63*
Muller, N., 401, *424*, 457, 459, 473, *479*
Muller, N. L., 260, *267*, 457, *479*, 559, 560, 562, 563, 567, 568, *583, 584, 586*
Muller, R. E., 602, *608*
Mulligan, E., 322, 323, 324, 327, 328, 329, 330, 331, 332, 333, 334, 335, *339*
Mullins, L. J., 300, *317, 319*
Mulloney, B., 121, *149*
Munoz, N. M., 490, *513*
Munson, J. B., 381, *420, 427, 428*
Murai, D. T., 556, 557, *584, 587*
Murakami, K., 79, *104*
Murakami, T., 525, *542*

Murphy, E., 250, *268*, 274, *287*, 572, 573, 574, *585, 586*
Murphy, R. A., 489, *516*
Murphy, T. M., 490, *513*
Mussini, I., 411, *418*
Mutani, R., 648, *659*
Mutch, W. A. C., 600, *611*
Mutich, R. L., 483, 501, *513*
Myers, R. E., 650, *658*
Myers, R. Z., *639*

N

Nadal-Ginard, B., 410, *423, 424*
Nadel, J. A., 253, 254, 256, *265, 268,* 483, 484, 485, 486, 494, 495, 496, 507, *508, 511, 514, 518*
Naeye, R. L., 322, *339*
Nagamine, Y., 31, *65*
Nagashima, H., 390, *416*
Nagata, T., 79, 84, *99*
Nail, B., 229, *241,* 274, *285*
Nail, B. S., 222, 225, *243,* 645, *657,* 681, 684, 690, *695*
Nair, P., 306, *320*
Nair, P. K., 306, *317*
Nairn, A. C., 81, 89, 91, 92, *95, 99, 105, 110*
Nakabayashi, H., 77, 78, 80, *100, 110*
Nakamura, Y., 525, 526, *542*
Nakanishi, S., 47, *62*
Nakayama, T., 73, *105*
Nakazawa, T., 631, *639*
Namba, T., 390, *424*
Naor, Z., 79, *105*
Narin, A. C., 15, 31, *55, 60*
Narusawa, M., 410, *424*
Nastainczyk, W., 82, *100*
Nathanielsz, P. W., 557, *584*
Nattie, E. E., 342, 343, 345, 351, 352, 356, 357, 360, *369, 370*
Navarrete, R., 388, *424*
Nawa, H., 47, *62*
Neer, E. J., 82, *108*
Nef, P., 74, *105*

Neher, E., 295, 296, *316*
Neil, E., 307, *318*, 322, 323, 333, *338*
Neil, J. J., *368*
Neilsen, M., 332, *339*
Nelsestuen, G. L., 79, *95*
Nelson, S. R., 686, *696*
Nemeth, P., 392, *424*
Nemeth, P. M., 392, 411, *421, 425*
Nestler, E. J., 65
Netick, A., 522, 534, *541, 546,* 563, 570, *575, 579, 584*
Neubauer, J. A., 348, 354, 357, 358, *369, 370*
Ngai, S. H., 128, *154*, 342, 343, *363*, 568, *579*
Nghiem, H., *56*
Ni, H., 529, 533, *546, 549*
Nicholls, D. G., 300, *318*
Nicholls, J. G., 600, *607*, 644, *659*
Nichols, R. A., 91, *105*
Nicholson, C., 600, *611*, 643, *659*
Nicholson, G., 647, *658*
Nicoll, R. A., 83, 92, *103*, 157, *175*, 186, *196*
Nicoll, R. H., 83, *105*
Niedel, J. E., 77, *105*
Niedergerke, R., 300, *318*
Nielsen, M., 354, *369*
Nienhuis, R., 527, *548*
Niidome, T., 76, *101*
Niimi, Y., 558, *582*
Nijhuis, J. G., 327, 328, 329, *338*, 360, *366*, 554, *584*
Nilsson, G., 19, 48, *60, 63*
Nilsson, K., 503, *507*
Nioka, S., 617, 618, 631, 632, 633, 634, 635, *636, 638, 639, 641*
Nishimura, M., 309, *318*
Nishino, T., 332, *339*
Nishizuka, Y., 31, *65*, 76, 77, 78, 79, 80, *95, 96, 98, 99, 100, 102, 105, 107, 108*
Nisimaru, N., 129, *151*
Noble, E. G., 267, 269, *278*
Noble, L. M., 260, *267*

Noble, M. I. M., 248, *266*
Nochomovitz, M. L., 390, *422,* 465, *478*
Noda, M., 25, *65*
Nogami, H., 501, *511*
Noguchi, A., 502, 506, *518*
Nolop, K. B., 671, *676*
Noma, A., 299, 310, 311, *318, 319*
Nomoto, S., 470, *481*
Nomura, H., 77, *98*
Nomura, S., 272, *286*
Norberg, K., 600, *611*
Nordesjo, L. O., 468, *478*
Nordin, M., 446, *451*
Norell, G., 186, *195*
Norgren, R., 272, *285*
Norman, W. P., 23, 34, *60*, 351, *365*
Norwood, C. R., 618, *639*
Norwood, W. I., 618, *639*
Noventa, D., 411, *418*
Nowak, L., 26, *65*
Noworaj, A., 567, *579, 584*
Nowycky, M. C., 310, *318*
Numa, S., 25, 28, *62, 65, 76, 101*
Nunokawa, T., 631, *639*
Nurse, C. A., 293, 308, *318, 319*, 334, *339*
Nyberg, F., 666, 670, *677*
Nye, P. C. G., 357, *366*
Nyström, B., 392, *424*

O

Obata, K. I., 32, *60*
Obesco, A., 293, 295, *318*
O'Brien, M. J., 559, 560, *585*
O'Brien, P., 467, *478*
O'Brien, R. A. D., 379, 380, *424*, 470, *479*
O'Callaghan, C., 504, *514*
Ochi, M., 635, *641*
O'Connell, K., 570, 574, *578*
O'Connell, S. M., 161, *175*
O'Day, T. L., 259, *269*
O'Donnell, M., 485, 500, *509*
Oertel, W. H., 19, 43, 46, *58, 65*

Oestreicher, A. B., 81, 82, *97, 109*
Ogilvie, M., 206, *215*
Ogita, K., 77, 79, *105, 107*
Ogura, J. H., 274, *286*
Ohga, A., 499, *511*
Ohmstede, C-A., 84, 92, *105*
Ohno, S., 78, *105*
Ohno-Shosaku, T., 602, *613*
Oishi, K., 80, *105*
Okado, N., 177, 191, *196*
Okamoto, H., 32, *60*
Okamura, Y., 45, *57*
Okken, A., 257, *267*, 568, 571, *583*
Okubo, S., 593, 595, *611*
Olaisen, B., 672, *677*
Olds, J. L., 83, *105*
O'Leary, D. D. M., 192, *196, 197*
O'Leary, D. M., 615, *636*
Olinsky, A., 248, 249, 250, 260, *267*
Oliveras, J. L., 532, *545*
Ollerenshaw, S., 501, *514*
Olney, J. W., 606, *610, 611*
Olsen, C. E., 600, *610*
Olsen, C. R., 485, 486, 494, *514*
Olson, E. B., 39, 42, *57*
Olson, L., 19, 48, 49, *60, 66*
Olster, D. H., 37, 38, 43, 44, *55*
O'Malley, B. W., 38, *56, 66*
O'Neill, S. C., 295, 299, 300, 305, 310, *313, 320*
Oneyser, C., 74, *105*
Onimaru, H., 664, *677*
Ono, Y., 77, *105*
Ontyd, J., 666, *677*
Oppe, T. E., 662, *674*
Oppenheim, R., 177, 191, *196*
Oprian, D., 73, *105*
Orci, L., 74, *104*
O'Regan, R. G., 307, 308, *318*, 323, 333, *339*
Orem, J., 206, *215*, 522, 523, 528, 534, *545, 546,* 559, 563, 570, *584*
Orenstein, D. M., 467, *479*
Orkand, P. K., 600, *611*
Orkand, R. K., 644, *659*

Orth, K., *102*
Orthner, F. H., 456, *479*
O'Shea, J. J., 72, *109*
Ostberg, A. J. C., 379, 380, *424,* 470, *479*
Oswaldo-Cruz, E., 185, *197*
Otis, A. B., 483, *514*
Otsuka, M., 16, *66*, 186, *197*
Otto, M. R., 351, 353, *370*
Ou, L. C., 343, 345, 351, 352, 357, *369*, 592, *613*
Ouimet, C. C., 87, *105*
Overholt, J. L., 351, 352, *368*, 671, *678*
Oxford, G. S., 306, *315*
Oyasaeter, S., 672, *677*
Ozand, P. T., 621, *639*
Ozawa, K., 633, *640*

P

Paala, J., 672, *676*
Paber, J. T., 681, 690, *699*
Pack, A. I., 206, *215*, 253, *267*, 539, *546*
Padbury, J. F., 668, *677*
Padykula, H. A., 471, *476*
Pagala, M. K. D., 390, *424*
Pagtakahn, R. D., 681, 687, 690, *695*
Paintal, A. S., 254, *267*, 292, 312, *318*
Pajot, N., 567, *583*
Palacek, F., 255, *268*
Palacios, L., 472, *480*
Palacios-Pru, E., 472, *479, 480*
Palade, G. E., 13, *66*
Palay, S. L., 13, 19, *56, 66*
Palecek, F., 224, *243*
Palkovits, M., 18, 19, 52, *66*, 129, *152*
Pallot, D. J., 291, 308, *318*
Pallotta, B. S., 296, *312*
Palmer, M. R., 48, 49, *66*
Pan, Y., 360, *371*
Pandya, K. H., 486, 503, 506, *514*
Pang, I-H., 72, *108*

Pang, L. M., 461, 462, 463, *475, 480*
Panitch, H. B., 491, *514*
Pantaleo, T., 205, *213,* 279, *286,* 350, *363,* 664, 669, *673, 679*
Pappenheimer, J. R., 342, 354, 355, *364, 369*
Papper, E. M., 568, *579*
Pardy, R. L., 466, 467, *480*
Park, D. H., 349, 350, *363, 365, 370*
Park, S., 72, *106*
Parker, P. J., 77, *97*
Parkes, M. J., 360, *366*
Parmalee, A. H., Jr., 553, *584*
Parmeggiani, P. L., 535, *546*
Parsons, B., 38, 44, *64*
Pascoe, J. P., 529, *546*
Patel, P., 467, *478*
Patlak, C. S., 644, *660*
Paton, J. Y., 3, *7*
Patrick, J. E., 681, *698*
Patterson, D. L., 295, 296, 304, 305, *312, 313, 315,* 334, *337*
Patterson, L., 343, 355, *364*
Patton, B. L., 74, 88, 89, *104*
Payan, I. L., 664, *676*
Peachy, L. D., 438, *449*
Pearson, K. G., 114, 116, *151*
Pearson, R. B., 80, *101*
Pecot-Dechavassine, M., 470, *482*
Peiper, A., 683, 692, *698*
Peirson, D. J., 42, *69*
Pelle, I., 468, *478*
Pelletier, G., 13, 24, *66*
Pepelko, W. E., 595, *611*
Pequignot, J. M., 666, 668, *675*
Peraita, R., 693, *697*
Peralta, E. G., 357, *369*
Pereda, A. E., 527, *546*
Periasamy, D., 410, *423*
Perin, M. S., 90, *108*
Perkel, D. J., 83, 92, *103*
Perkett, E. A., 274, *287*
Perkins, J. P., 76, *109*
Permutt, S., 573, *575*
Pernow, B., 16, 19, 48, 53, *61, 66*
Pernow, P., 19, *60*

Peroutka, S. J., 24, *66,* 73, *106*
Perrin, D. G., 322, *339,* 672, *677*
Perrino, B. A., 93, *100*
Perry, S. V., 410, 411, *419*
Perschka, M., 74, *108*
Pert, C. B., 49, *57*
Peschle, C., 78, *108*
Pessacq, T. P., 485, 488, *514*
Peterson, E. L., 116, 128, *146, 152*
Peterson, G. L., 357, *369*
Peterson, I., 458, 459, *478*
Petraitis, M., 49, *56*
Petroski, R. E., 161, *175*
Petrucci, T. C., 91, *106*
Pette, D., 392, 410, *424, 426*
Pettigrew, A. G., 379, 389, *417,* 571, *581,* 664, *675*
Pfaff, D. W., 38, 39, 43, 44, *64, 66, 67, 68*
Pfaffman, C., 126, *153*
Pfenninger, K. H., 79, 81, 83, 85, *100, 101, 106*
Philips, C. M., 333, *339*
Phillipson, E. A., 237, *241,* 250, 253, 254, 256, *265, 268,* 274, *283, 287,* 522, *541, 546,* 552, 563, 564, 569, 572, 573, 574, *576, 580, 584, 585, 586,* 594, *607, 611*
Phillis, J. W., 16, *66*
Pian, M. S., 498, *517*
Piazza, T., 498, *514*
Piccoli, S., 247, 248, 253, *263*
Pickel, V. M., 48, 49, *66*
Pickens, D. L., 249, *269,* 274. *287*
Pickett C. K., 572, *578,* 661, *679*
Piehl, K., 464, 468, *477*
Piercey, M. F., 129, *147,* 277, *284*
Pietruschka, F., 293, 296, 306, *316, 318,* 334, *340*
Pin, J. P., 25, *68*
Pindzola, R. R., 183, 185, *197*
Pinset-Härström, I., 410, *427*
Pinter, M. J., 159, *173*
Pinter, S., 358, 360, *362,* 557, *576,* 593, *607*

Piomelli, D., 83, *106*
Pitt, B. R., 667, 669, *677*
Pitton, G., 410, 414, *425*
Pitts, R., 128, *153*
Platika, D., 293, 295, 308, *315*
Platzker, A. C. G., 504, *512*
Pleschka, K., 382, *427*
Pockett, S., 472, *480*
Poenie, M., 299, 300, *316*
Pohl, S. L., 73, *106*
Pokorski, M., 346, *370*
Polacheck, J., 248, 252, *268*
Polak, J. M., 487, 499, 502, *516*
Polk, D. H., 668, *677*
Polk, S., 484, *517*
Pollet, R. J., 80, *97*
Pongs, O., 74, *108*
Ponte, J., 258, *268*, 295, 296, 304, 305, 308, *312, 313, 315,* 334, 335, *337, 340*
Poon, M., 125, *153*
Poore, E. R., 557, *584*
Posey-Daniel, V., 485, *509*
Posner, J. D., 457, *478*
Post, C., 19, *60*
Potter, P. C., 28, *56*
Potter, V. R., 621, 631, *640*
Power, G. G., 557, *582*
Pozzan, T., *319*
Prabhaker, N. R., 351, 352, 354, *368, 370,* 665, 669, 670, 671, *677, 678, 679*
Prange, A. J., 125, 141, *145, 151*
Praud, J. P., 259, *268*
Prechtl, H. F. R., 552, 553, 554, 559, 560, *585*
Prendiville, A., 484, 504, 505, *514, 516*
Price, D. L., 649, *660*
Price, J. L., 532, *544, 546*
Price, M. T., 606, *610*
Prince, D. A., 159, 160, 166, 169, *174, 175*
Priola, D. V., 492, *510*
Prochiantz, A., 26, *65*
Proske, U., 231, 233, *241*

Purves, D., 44, 45, *66*
Purves, M. J., 258, *268*, 302, 306, *312, 313,* 322, 324, 326, 327, 335, *336, 338, 340,* 343, 355, *364,* 645, 651, *657,* 681, 684, 690, *695*
Putnam, M. D., 522, *546*
Pygott, V., 472, *480*
Pysh, J. J., 619, *640,* 654, *657*

Q

Quastel, J. H., 620, 623, *637*
Quattrochi, H. A., 351, *370*
Quilligan, E. J., 556, *585*
Quinn, D. J., 625, *640*
Quinn, W. G., 74, *103*

R

Radda, G. K., 618, *640*
Radford, M., 504, *515*
Radvanyi-Bouvet, M. F., 567, *583*
Raffestin, B., 489, 490, *507*
Raftery, M. A., 25, *67*
Raines, P. L., 621, *639*
Rakic, P., 615, *640*
Ramachandran, J., 357, *369*
Ramon y Cajal, S., 11, *67*
Randic, M., 186, *197*
Rands, E., 28, *57, 69*
Raney, R. A., 3, *7*
Ransom, B. R., 600, 602, 603, *608, 613*
Ransom, R. R., 649, *659*
Ranson, S. W., 128, *153*
Rappaport, Z. H., 309, *318*
Rasmussen, H., 72, *95, 106*
Rattner, A., 84, *102*
Rawson, M. D., 624, *637*
Ray, C., 14, *55*
Raymond, P., 3, *7*
Raynor, R. L., 79, 80, *105, 110*
Read, D. J. C., 203, *215,* 235, *241,* 274, *287,* 456, *475,* 559, 567, 571, 572, 573, 574, *581, 582, 585,* 594, *610,* 665, *678*

Read, J. C., 473, *477*
Recasens, M., 25, *68*
Rechtschaffen, A., 552, *585*
Reddy, V. K., 189, *197*
Redet, W. V., 39, *57*
Redfern, P. A., 387, *424*
Reed, M., 502, *512*
Reed, R. R., *102*
Reed, W. R., 564, 566, *585*
Reese, T. S., 14, *60*
Reeves, R. B., 355, *370*
Regan, J. W., 62
Regen, D. M., 621, *639*
Register, R. B., 28, *57, 69*
Reich, E., 31, *65*
Reid, L. M., 483, 484, *515, 555*, *579*
Reid, S. A., 381, *428*
Reid, W. D., 466, 467, *480*
Reiffenstein, R. J., 647, 649, *659*
Reiley, J., 630, *640*
Reinecke, M., 472, *480*
Reinking, R., 392, *421*
Reinking, R. M., 392, *424*
Reis, D. H., 349, *365, 370*
Reis, D. J., 48, 49, *66*, 349, *368, 369*, *370*
Reiser, P. J., 408, 414, *424*
Reite, M., 594, *611*
Reivich, M., 626, 633, *636*, 644, *660*
Rejman, J. J., 38, *69*
Relier, J. P., 567, *583*
Remahl, S., 384, 387, *424*
Remmers, J. E., 128, *153*, 206, 210, *215*, 247, 255, 261, *268, 269*, 277, 279, *283, 287*, 521, 522, 534, *542, 546*, 562, 563, 566, *576, 583, 585*
Remmers, J. R., 352, 360, *362, 365*, *367*
Requena, J., 300, *317, 319*
Revah, T., *56*
Rexed, B., 182, *197*
Reynolds, H. C., 484, *515*
Reynolds, L. B., 687, *698*
Rhee, L., 77, *97*

Rhee, S. G., *106*
Riabowol, K. T., 32, *67*
Ribchester, R. R., 379, 380, 386, *417*
Ribeiro, J. A., 671, *676*
Ricardo, J. A., 532, *547*
Rice, G. E., 667, *679*
Rich, D. P., 85, 86, 89, *97, 98*
Richard, C. A., 539, *543*
Richards, R. T., 554, 555, *578*, 684, 685, *695*
Richardson, B. W., 343, *368*
Richardson, C. A., 497, *515*
Richardson, G. B., 157, 163, *173*
Richardson, J., 485, 500, *515*
Richardson, J. B., 483, 486, *515*
Richardson, M. A., 272, 280, *284*
Richardson, P. S., 229, 230, *240*, 254, *266*, 687, *696*
Richerson, B. S., 358, *366*
Richerson, G. B., 111, 112, 133, 136, 141, 142, 143, *147, 153*
Richter, D., 111, 133, 136, *146*
Richter, D. W., 128, 139, 140, *153*, 201, 206, 208, 209, 210, *213*, *214, 215*, 255, *268*, 277, 278, 279, *287*, 645, 647, 648, 649, 650, 651, 653, 654, *656, 658*, *660*
Ridderstrale, Y., 357, *370*
Ridge, R. M. A., 380, 386, 397, 411, 413, *418, 422, 424*
Rieger, G. E., 43, *61*
Ries, A. L., 567, *578*
Rifenberick, D. H., 470, *481*
Rigatto, E., 358, *370*, 558, *585*
Rigatto, H., 155, *175*, 252, *263*, 322, 329, 333, *336*, 358, *370*, 497, 498, *508, 515*, 558, 560, 570, 571, 572, 573, 574, *577, 583*, *585*, 593, *607, 612*, 670, *674*
Riley, D. A., 411, *427*, 470, *480*
Rink, T. J., *319*
Ritchie, A. K., 306, *319*
Ritchie, B. C., 358, *362*, 555, 556, *576, 583*

Rivington, R. N., *480*
Robbins, N., 470, *476, 477, 480*
Robbins, P. A., 594, *608*
Roberts, J. L., 44, 48, *67, 69,* 564, 566, *585,* 688, *698*
Robertson, C. H., Jr., 457, 461, *480*
Robertson, H. A., 92, *98*
Robin, E. D., 595, 606, *611, 612*
Robinson, A. J., 391, *418,* 443, *450*
Robinson, J. S., *285,* 358, 360, *362,* 555, 557, *576,* 593, *607*
Robinson, P. J., 80, *101*
Robinson, P. M., 484, *507,* 555, *575*
Robinson, R. O., 666, *675*
Robotham, J. L., 401, 404, 408, *423,* 472, *479*
Rocha-Miranda, C. E., 185, *197*
Rochester, D. F., 457, 461, *480*
Rochette, F., 354, 357, *371*
Rodbell, M., 73, *106*
Rodier, P. M., *640*
Roebuck, M. M., 666, *675*
Roehling, T., 569, *578*
Rogawski, M. A., 157, *175*
Rognum, T. O., 672, *677*
Rohmert, W., 448, *452*
Rohrkasten, A., 82, *100*
Romagnano, M. A., 186, *197*
Roman, C., 272, *283*
Romaniuk, J. R., 247, *264*
Ronnevi, L. O., 384, 387, 389, *419*
Rories, C., 38, *67, 69*
Rose, D., 384, 385, *425*
Rosen, C. L., 593, 595, *612*
Rosen, H., 48, *67*
Rosenberg, S. I., 223, *243*
Rosenfeld, M. G., 86, *103*
Rosenheimer, J. L., 470, 471, *480*
Rosenstein, R., 354, *362*
Rosenthal, J. L., 379, *425*
Rosenthal, M., 603, 605, 606, *608, 609, 612,* 647, *659*
Rosman, P., 602, *613*
Ross, B., 489, *516*
Ross, B. H., *453*
Ross, C., 349, *370*

Ross, C. A., 349, *370*
Ross, E. M., 74, *106*
Ross, J. J., 397, *420, 425*
Ross, W. R. D., 459, *477*
Rossen, R., 596, *612*
Rosser, B. W. C., 411, *425*
Rossi, A., 492, 495, *515*
Rostas, J. A. P., 85, 86, 87, *98, 106, 109*
Roth, G. I., 343, *366*
Roth, Z., 617, 628, *637*
Rothberg, K. G., 488, *515, 516*
Rothfeld, J. M., 44, *67*
Rothlein, J., 91, *107*
Rothman, J. E., 74, *104*
Rothwell, J. C., 432, *451*
Rottenberg, H., 304, *319*
Roumy, M., 304, *319*
Rouslin, W., 633, *640*
Roussos, C. S., 397, *425,* 457, 459, 461, 464, 467, 469, *476, 481*
Rovainen, C. M., 125, *146*
Rowlerson, A., 380, 386, 397, 413, *422*
Roy, R. R., 378, 391, 392, *417, 423*
Rubenstein, S. D., 232, *240,* 495, *508*
Rubin, D., 568, 571, *583*
Rubinstein, N. A., 410, *422, 424*
Rubio, R., 666, *679*
Ruckebusch, Y., 554, *586*
Rudolph, A. M., 595, *612*
Rudomin, P., 258, *263,* 278, *287*
Runold, M., 665, 666, 669, 670, 671, *677, 678, 679*
Ruppersberg, J. P., 74, *108*
Russchen, F. T., 532, *547*
Russell, D. F., 121, *153*
Russell, J. A., 483, 485, 486, 503, *513, 515*
Russo, A. F., 86, *103*
Russo, G., 78, *108*
Routtenberg, A., 33, *54,* 76, 79, 81, 83, *95, 96, 104*
Ruggiero, D. A., 349, *365, 370*
Russell, D. F., 122, 123, 124, *153*
Russell, S. J., 358, *367*

Rutter, N., 504, *515*
Ryan, J. P., 491, *514*
Ryser, M., 322, *339*

S

Sabol, S. L., *69*
Sacks, R. D., 391, 404, *426*
Sadoul, N., 457, 458, 459, *481*
Saetta, M., 246, *267*, 498, *515, 567, 571, 584, 586*
Saggin, L., 410, 411, 414, *425*
Sahyoun, N. E., 78, 79, 80, 82, 84, 85, 86, 87, 88, 89, 92, *102, 103, 104, 105, 106, 107, 108, 109, 110*
Saito, N., 78, *99, 102*
Saitoh, T., 89, *107*
Sakai, K., 524, 527, *547*
Sakaki, S., 646, *658*
Sakanaka, M., 48, 49, *68*
Sakmann, B., 74, *108*, 295, *316*
Sakmar, T., 73, *105*
Saleh, J., 606, *613*
Salisbury, D., 274, 275, 281, 282, *285*
Salisbury, D. M., 687, 688, *697*
Saltin, B., 462, 464, 468, 469, *476, 477, 478*
Salvador, H. S., 593, *613*
Samaha, F. J., 408, 413, *421*
Samalson, L. E., 72, *109*
Sameshima, H., 557, *582*
Sampson, S. R., 272, *283*, 302, 307, 308, 313, 319, 324, *336*
Samson, F. E., 625, *640*
Samuelsson, B., 19, *60*
Sandblom, J. P., 358, *364*
Sander, L., 571, *579*
Sandercock, T. G., 391, *425*
Sanders-Busch, E., 24, *56*
Sanders, M., 84, *107*
Sands, J., 168, *173, 616, 636*
Sanes, J. R., 379, *423*
Sankaran, K., 560, 574, *577,* 593, *612*

Sano, K., 77, *96*
Sano, Y., 45, *59*
Sanocka, W. M., 596, 605, *612,* 687, *698*
Sant'Ambrogio, F. B., 224, 225, 226, 227, 228, 229, 230, 237, *239, 241, 242, 243,* 257, *264, 272,* 274, *287,* 492, *510*
Sant'Ambrogio, G., 203, *214,* 224, 225, 226, 227, 228, 229, 230, 233, 234, 235, 237, *239, 240, 241, 242, 243, 244,* 246, 249, 257, 258, 261, *264, 265, 267,* 272, 274, *284, 287,* 492, 494, *510*
Santiago, T. V., 596, *612*
Santicioli, P., 492, 499, *513*
Saper, C. B., 129, *153,* 532, *547*
Sapru, H., 357, *371*
Sar, M., 38, 43, *67*
Saracino, F., *473*
Sarelius, I. H., 343, *368*
Sargeant, C. W., 3, *7*
Saria, A., 224, 238, *242,* 485, 498, *512, 513*
Sartore, S., 410, 414, *425*
Sarvey, J. M., 33, *69*
Sasaki, C. T., 222, *243*
Sastre, J. P., 524, 527, *547*
Sato, M., 384, 387, *425*
Saubert, C. W., IV., 462, 464, 468, *477*
Sauder, R. A., 484, 492, *515*
Sauerland, E. K., 521, 522, 524, 525, *543, 546, 547,* 566, *586*
Saugstad, O. D., 666, 672, *677, 678*
Savoca, J., 306, *320*
Sawada, T., 635, *641*
Sawamura, S., 78, *99*
Sawchenko, P. E., *197*
Sayeed, M. M., 633, *635*
Scarpa, A., 304, *319*
Scarpelli, E. M., 667, *677*
Scattergood, W., *57*
Schacher, S., 32, 33, *57, 59, 67,* 83, *99*

Schade, J. P., 600, *613*
Schaeffer, J. I., 155, 169, *174*, 670, *678*
Schaffner, A. E., 293, *315, 337*
Schalk, U., 569, *578*
Schalling, M., 19, 20, *60*
Schechtman, V. L., 523, *543*
Schecter, W. S., 593, 595, *612*
Schefft, G., 274, *287*
Schefft, G. L., 248, 249, *269*
Scheibner, T., 501, *514*, 665, *678*
Schenck, C. H., 524, *545*
Schenkel, E., 525, 527, *547*
Schersten, T., 468, *475*
Scheurmier, N., 235, *240*
Schiaffino, S., 410, 411, 414, *425*
Schiebler, W., 90, 91, *100, 107*
Schimerlik, M. I., 357, *369*
Schlaefke, M. E., 343, *370*
Schlueter, M. A., 670, *675*
Schmalbrunch, H., 397, *425*
Schmidt, B. J., 125, *154*
Schmidt, C. F., 626, *638*
Schmidt, L., 690, *698*
Schmidt, M., 354, 355, *362*
Schmidt, R., 446, *451*
Schmieden, V., 74, *99*
Schnader, J. Y., *481*
Schneider, J. M., 626, *638*
Schneider, T., 82, *100*
Schneider, W. C., 621, 631, *640*
Schofield, P. R., 73, 74, *99, 107*
Scholtz, K., 32, *67*
Scholz, W. K., 86, *107*
Schor, N., 635, *641*
Schousboe, A., 309, *312*
Schrader, J., 666, *677*
Schrader, W. T., 38, *56*
Schramm, C. M., 322, *340*, 670, *678*
Schroter, K. H., 74, *108*
Schuette, W. H., 644, *659*
Schulman, H., 84, 85, 86, 92, *103, 107*
Schulte, F. J., 571, *579, 586*
Schultz, D. W., 620, *638*
Schultz, E., 471, *481*

Schultz, H. D., 489, 495, 496, 497, 499, *509*
Schultzberg, M., 19, 45, 47, 49, *58, 60*
Schulz, I., 305, *319*
Schulz, S., 75, *107*
Schulzberg, M., 112, *150*
Schultz, H., 633, *640*
Schwartz, A., 74, *98*
Schwartz, J. H., 32, *59*, 83, 89, *106, 107*
Schwartz, K., 410, 411, *418, 427*
Schweitzer, T. W., 459, *481*
Schwieler, G. H., 233, 237, *244*, 250, 255, 256, *263, 268*, 329, 334, *340*, 484, 486, 492, 493, 505, *515, 516*
Schworer, C. M., 85, 88, 89, *97*
Scott, J. D., 80, *107*
Scott, T. R., 135, *146*
Sears, T. A., 114, 116, 129, *151*, 664, *678*
Sedin, G., 258, *266*
See, W. R., 343, *370*
Seeburg, P. H., 48, *67*, 73, 74, *99, 107*
Segal, M., 157, *175*
Seidel, C. L., 489, *516*
Seiger, A., 48, 49, *66*
Sejersted, O. M., 438, 440, 444, *453*
Sekar, C. M., 305, *319*
Sekiguchi, K., 77, 79, *98, 107*
Sell, S. M., 410, *427*
Sellick, H., 259, 260, *267*
Selstam, U., 274, *285*
Selverston, A. I., 116, 121, 122, 124, 125, 128, *146, 152, 153*
Sembrowich, W. L., 462, 464, *477*
Sen, R., 84, *102*
Senba, E., 48, 49, *68*
Seow, C. Y., 492, *516*
Serafini, T., 74, *104*
Serfözö, P., 23, *61*
Seroogy, K. B., 19, 20, 35, 36, 46, 50, *57, 60, 64, 65, 68*
Seshia, M. M., 252, *263*, 593, *612*

Sessle, B. J., 200, 211, *215*, 275, 278, 279, 280, 282, *286, 287*
Seta, K., 633, *640*
Seubert, P., 33, *63*
Sevarino, K. A., 31, 51, *65*
Severinghaus, J., 342, 343, 344, *368*
Severinghaus, J. W., 521, *547*
Sevilla, N., 29, *68*
Seybold, V., 502, *510*
Shadmehr, R., 390, *423*, 465, *478*
Shaffer, T. H., 232, *240*, 491, 495, *508, 514*
Shahid Salles, K., 606, *610*
Shamban, A., 621, 623, *639*
Shams, H., 354, *370*
Shander, A., 390, *416*
Shannon, D. C., 322, *340*, 569, 570, 574, *578, 586*
Shannon, R., 261, *263, 268*
Shapiro, E., 83, *106*
Shapovalov, A. I., 384, *425*
Sharman, D. F., 48, *61*
Sharp, A. H., 74, *98*
Sharratt, M., 404, 414, *426*
Shearman, M. S., 79, *105*
Sheller, J. R., 507, *508*
Shelley, J. H., 684, 686, *695, 698*
Shepherd, R. E., 462, 464, *477*
Sheppard, M. N., 487, 499, 502, *516*
Sherrey, J. H., 522, 524, *547*, 563, *583, 586*
Sherrington, C. S., 11, *68*, 113, *151, 153*, 376, 381, *423, 425*
Sherry, J. H., 274, *287*
Shields, S., 85, *101*
Shields, S. M., 89, *107*
Shiffman, D., 635, *641*
Shik, M. L., 125, *149*
Shimoji, K., 307, 310, *315*, 597, 600, *609*
Shin, T., 224, *244*
Shine, J., 48, *67*
Shinick-Gallagher, P., 157, *173*
Shiosaka, S., 48, 49, *68*
Shirahata, M., 329, *339*
Shiriaev, B. I., 384, *425*

Shivers, B. D., 43, *68*, 73, *107*
Shoemaker, W. J., 48, *54*
Shoubridge, E. A., 618, *640*
Shouse, M. N., 524, 537, *547*
Shrago, E. S., 633, *636*
Shug, A. L., 633, *636*
Shultz, C. W., 343, *367*
Sibley, D. R., 29, *68*, 72, 73, *107, 110*
Sica, A. L., 200, 201, 206, *215*, 275, 277, *284*, 329, 330, *337*
Sick, T. J., 603, 605, 606, *608, 609, 612*
Siddique, T., 74, *99*
Sidi, D., 595, *612*
Sidman, R. L., 615, *640*
Sieck, G. C., 378, 379, 382, 383, 386, 388, 390, 391, 392, 395, 397, 401, 404, 407, 408, 413, *417, 423, 425, 426, 428*, 465, *478*, 522, 534, *542, 543, 547*
Siegel, G., 17, *68*
Siegel, J. M., 525, 527, *547, 548*
Siegelbaum, S. A., 76, *96*
Siemkowicz, E., 596, *612*
Siesjo, B. K., 600, 603, *611, 613*, 626, 633, *637, 640*
Sigal, I. S., 28, *57, 69, 73, 97*
Sigger, J. N., 558, *580*
Siggins, G. R., 141, *146*
Sigworth, F. J., 295, *316*
Sihra, T. S., 81, 91, *105, 108, 109*
Silberstein, L., 380, 386, *426*
Silver, A., 349, *365*
Silver, I. A., 309, *314, 319*
Silverman, M., 484, 504, 505, *514, 516*
Silverstein, P. R., 633, *640*
Simon, E. J., 49, *56*
Simon, R. P., 310, *319*
Simpser, M. D., 693, *698*
Simpson, H., 567, *576*
Sinclair, J. D., 343, *368*
Singh, T. J., 80, *100*
Sinwell, T., 618, *639*
Siwicki, K., 123, *152*

Sjogaard, G., 438, *452*
Sjostrom, M., 468, *475*
Skagerberg, G., 180, *193*
Skatrud, J., 539, *541*
Skatrud, J. B., 39, 42, *57, 68*
Skirboll, L., 23, 34, *60,* 186, *195*
Skoff, R. P., 649, *660*
Skoglund, S., 233, *244*, 255, 261, *263, 268,* 384, 387, 389, *419, 422, 423,* 647, 648, 649, *660*
Skoogh, B. E., 486, 506, *516*
Skorupa, A., 73, *102*
Skovgaard, B. J., 125, *154*
Sladeczek, F., 25, *68*
Slater, C. R., 470, *479*
Slaughter, C., *102*
Slotkin, T. A., 308, *317*, 666, 668, 671, *676*
Smatresk, N. J., 322, 323, 328, 329, 335, *339*
Smejkal, V., 255, *268*
Smickley, J., 563, 566, *582*
Smith, B., 223, 235, *243*, 507, *584*, 691, *698*
Smith, C. A., 691, *698*
Smith, D., 404, 414, *426*, 617, 618, *639*
Smith, D. A., 532, *549*
Smith, D. H., 357, *369*
Smith, D. O., 389, *426,* 470, 471, *480*
Smith, D. S., 633, *636*
Smith, D. V., 272, *285*
Smith, E. L., 471, *481*
Smith, H., 332, *339*
Smith, J. B., 248, *265*, 303, *319*, 564, *579, 584*
Smith, J. C., 128, *148, 153,* 159, *175,* 351, 353, *370*
Smith, L., 303, *319*
Smith, M. K., 85, 88, 89, *97*
Smith, P. G., 308, *317*
Smith, P. R., 349, *366*
Smith, R. B., 485, 488, *517*
Smith, R. T., 471, *481*

Smith, S. J., 119, *145*, 163, *172, 175,* 310, *317*
Smith, T. C., 626, *636*
Smith, T. G., 115, *145*
Smull, N. W., 571, 573, *583*
Snell, T. C., 126, *153*
Snyder, D. H., 470, *481*
Snyder, F. F., 626, 627, 634, *637*
Snyder, S., 16, *68*
Snyder, S. H., 78, *110*
Soderberg, U., 342, 346, *364*
Soderling, T. R., 80, 85, 86, 88, 89, 93, *97, 98, 99, 100, 108*
Sodie, T., 308, *315*
Soejima, M., 311, *319*
Soileau, L. C., 380, 386, *426*
Soja, P. J., 526, 527, *542, 548*
Sokol, H. W., 685, 687, *698*
Sokoloff, L., 644, *660*
Solanki, L., 392, *421, 424*
Sole, M. J., 672, *677*
Soling, H. D., 623, *640*
Sollevi, A., 666, *675*
Solobodin, R. L., 633, *641*
Somjen, G., 381, *421*, 596, *607*, 647, 649, *657, 659, 660*
Somjen, G. G., 295, 296, 310, *315*
Sommer, D., 258, *264*
Sonnhof, U., 647, 648, *660*
Sorkness, R. L., 496, 498, *516*
Sotrel, A., 262, *268*, 560, *586*
Souguet, M., 693, *697*
Souhrada, J. F., 3, *7*
Souhrada, M., 488, *516*
Spalding, M. J., 462, *477, 478*
Sparrow, M. P., 484, 489, 490, 491, *513, 516*
Speck, D. F., 128, 129, *148, 153,* 201, *215,* 277, 278, 279, *286*
Spelsberg, T. C., 38, *67, 69*
Spindel, E., 19, 53, *61*
Spiratana, D., 387, *427*
Spiro, S. G., 457, 459, *479*
Spitzer, N. C., 160, *175*
Sposi, N. M., 78, *108*
Spriet, L., 441, *451*

Spurgeon, H. A., 535, *544*
Spyer, K. M., 129, 131, *148*, 277, 278, 279, *287*, 323, 334, *338*, 487, *512, 513,* 661, *678*
Sreter, F. A., 408, *422*
Srinivasan, M., 664, 668, 670, *676, 678*
Stahlman, M. T., 274, *285*
Staines, W., 19, *60*
Standaert, M. L., 80, *97*
Standaert, T. A., 321, *336, 339,* 360, *365,* 459, 465, 473, *481,* 568, 569, 571, 572, 574, *576, 579, 580, 582, 587,* 593, *610, 614,* 646, *659,* 662, *679*
Standant, T. A., 472, 473, *479*
Stanfield, B. B., 191, 192, *194, 197,* 615, *636*
Stanley, N. N., 456, *478*
Stanton, P. K., 33, *69*
Stark, A., 248, 252, *267*
Stark, A. R., 252, 253, 257, 260, *265, 268,* 560, 562, 563, 567, 568, 569, 570, 571, 574, *581, 582, 586, 587*
Stark, R. I., 667, *678*
Staron, R. S., 410, *426*
Staten, L. D., 83, *105*
Staub, N. C., 485, 486, 494, *514*
Stea, A., 293, *319*
Stebbins, C. L., 471, *481*
Stecenko, A. A., 484, 492, *515, 517*
Steele, A. M., 200, 201, 206, *215,* 329, 330, *337*
Steer, M. L., 29, *68*
Steeves, J. D., 125, *154*
Stehbens, W. E., 291, *313*
Stein, P. S. G., 144, *154*
Stein, R. B., 433, *452*
Steinberg, G. K., 606, *613*
Steinberg, R., 600, *611*
Steinbusch, H. W. M., 13, 19, 24, 53, *60, 61, 66*
Steiner, J. P., 82, *95*
Steinhardt, R. A., 300, *312*
Steinschneider, A., 693, *698*

Stelzner, D. J., 177, *197*
Stensaas, L. J., 308, *317*
Stent, G. S., 114, 116, *151*
Stephens, J. A., 434, *452*
Stephens, N. L., 483, 492, *516, 517*
Stephenson, F. A., 27, *64,* 72, 74, *108*
Sterbenz, G. C., 357, *370*
Sterling, G. M., 222, 225, *243*
Sterman, M. B., 523, 524, 537, 538, *543, 547, 548,* 569, 570, *581*
Stern, E., 553, *584*
Stern, R. C., 467, *479*
Sternberger, L. A., 182, 183, *197*
Sternweis, P. C., 72, 108
Stevens, C. F., 74, *108,* 118, *147,* 156, *172*
Stevens, M. H., 693, *698*
Stevenson, J. H., 621, *639*
Stevenson, M., 570, *586*
Stewart, D. J., 537, *548, 549*
Stewart, W. B., 667, 669, *677*
St. John, W., 201, 205, *214*
St. John, W. M., 222, 225, *241,* 343, 345, 351, 352, 357, *369, 370,* 534, *548,* 685, 687, *698*
Stockdale, F. E., 410, *427*
Stocker, M., 74, *108*
Stockle, H., 600, *611*
Stocks, A., 649, *660*
Stogryn, H. A. F., 235, *240,* 401, 404, 408, *423,* 560, 562, 563, 564, 565, *578, 583*
Stokes, M. J., 443, *451*
Stoll, B., 282, *286*
Stone, J. A., 462, *481*
Storey, A. T., 200, 211, *215,* 229, *244,* 272, 274, 275, 280, 281, 282, *285, 286, 287,* 687, 688, *697*
Storey, K. B., 605, *613*
Storm, D. R., 81, *96*
Story, J. M., 605, *613*
Stover, H., 470, *482*
Straddling, J. R., 563, 564, *580*
Strader, C. D., 25, 28, *57, 67, 69,* 73, *97*

Stradling, J. R., 237, *241*
Streb, H., 305, *319*
Strehler, E. E., 410, *423*
Strehler, M. A., 410, *423*
Strieder, D., 693, *698*
Strittmater, S. M., 82, *108*
Strohl, K. P., 562, 563, 566, *577,
  587*
Strong, R., 248, 252, *268*
Stropp, J. Q., 489, *511*
Strott, C. A., 38, *66*
Stroud, R. M., 25, *58*
Strumwasser, F., 296, *317*
Stuart, D. G., 392, *421*
Stühmer, W., 74, 76, *101, 108*
Stumpf, W. E., 38, 43, *61, 67*
Stumpo, D. J., 81, *108*
Sturgess, N. C., 602, *613*
Stys, P. K., 602, 603, *613*
Suarez, R. K., 606, *613*
Subramanian, H. V., 617, 618, 631,
  632, 634, *638, 639*
Sudhoff, T. C., 90, *108*
Sugimori, M., 15, *63*, 91, *103, 104*
Sugioka, K., 687, *697*
Suh, P. G., *106*
Sullivan, C., 501, 504, *514, 517*, 522,
  541, 546, 566, 572, 573, 574,
  576, 579, 582, 585, 586
Sullivan, C. E., 250, *268*, 274, *283,
  287*, 665, *678*, 693, *699*
Sullivan, J., 73, *102*
Sullivan, M. C., 180, *193*
Sultzman, L. A., 78, *102*
Sumal, K. K., 48, 49, *66*
Sun, M. K., 349, *371*
Sundaram, K., 357, *371*
Sundell, H., 274, 281, 282, *285, 286*
Sundell, H. W., 274, *285*
Sundler, F., 220, *244*
Sunyer, T., 86, *108*
Suthers, G. K., 203, *215*
Sutton, D., 282, *287*
Sutton, F. D., 42, *69*
Sutton, L. A., 411, *427*
Suzue, T., 128, *154*

Suzuki, K., 78, *105*
Suzuki, M., 222, *243*
Svedlund, J., 664, *673*
Swaminathan, S., 3, *7*
Swan, J. H., 310, *319*
Swandulla, D., 143, *154*
Swann, H. G., 690, *699*
Swanson, L. W., *197*
Swarman, K. F., 620, *639*
Swynghedauw, B., 410, *427*
Sypert, G. W., 381, 420, *427, 428*
Syrovy, I., 408, *421*
Szerb, J. C., 537, *548*
Szeto, E., 566, *582*
Szeto, H. H., 626, 628, 629, *638,
  668, 678*
Sziber, P. P., 74, *103*
Szymeczek, C. L., 21, 30, 50, 51, 53,
  64, 68, 69
Szymusiak, R., 537, *548*

T

Tabachnik, E., 560, 563, 568, *583,
  586*
Taeusch, H. W., 569, 573, *579*, 687,
  699
Taeusch, H. W., Jr., 248, 249, 252,
  265, 268, 269
Taghert, P. H., 117, *154*
Taira, M., 525, *542*
Takagi, H., 49, *68*
Takahashi, H., 25, 28, *62, 65*
Takahashi, T., 16, *66*
Takai, A., 94, *96*
Takai, Y., 77, 79, *96, 100, 102, 108*
Takashima, T., 501, *511*
Takatsuki, K., 48, 49, *68*
Takei, K., 90, *108*
Takuwa, N., 72, *106*
Takuwa, Y., 72, *106*
Tamarova, Z. A., 384, *425*
Tan, E. D., 324, 327, *336, 338*
Tan, K., 529, 532, *544*
Tanabe, T., 25, *65*, 76, *101*
Tanaka, C., 78, *99, 102*

Tanaka, D., 246, *266*
Tanaka, D. T., 498, 499, *511*
Tanaka, K., 635, *638*
Tanaka, S.-I., 77, 80, *102*
Tanaka, Y., 77, *102*, 272, *288*
Tang, W.-J., *102*
Taraskevich, P. S., 379, *425*
Tarpley, H. L., 621, *639*
Tate, K. A., 387, *427*
Tatemoto, K., 86, *103*
Tatemoto, T., 23, *63*
Tatooles, G. J., 492, *510*
Tatton, W. G., 192, *198*
Taussig, L. M., 504, *517*
Taylor, A., 309, *313, 434,*
    452
Taylor, E. M., 282, *287*
Taylor, I. M., 485, 488, *517*
Taylor, W. L., 88, *98*
Teager, H., 571, *579*
Teague, W. G., 498, *517*
Ten Bruggencate, G., 600, *611*
ten Donkelaar, H. J., 191, *197*
Tenney, S. M., 192, *195*, 592, *613*,
    685, *698*
Teppema, L. J., 354, 357, *371*
Tepper, R. S., 492, 493, 495, *511,*
    517
Terada, S., 76, *101*
Terenius, L., 13, 19, 20, 24, 49, *58,*
    59, 60, 61, 65, 186, *196*, 672,
    673
Terreberry, R. R., 539, *543*
Terzuolo, C. A., 525, *545*
Tessier, J., 537, *544*
Testa, U., 78, *108*
Thach, B. T., 682, 683, 684, 687,
    688, 689, 690, *696, 697, 698,*
    699
Thatch, B. T., 229, 233, *242*, 248,
    249, 254, 262, *268, 269*, 274,
    281, 282, *284, 287*, 560, 564,
    566, 571, 573, *579, 585, 586,*
    587
Theriault, E., 192, *198*
Thiringer, K., 666, 667, *675, 679*

Thomas, C. K., 433, 434, 442, 444,
    446, *450, 451, 452, 453*
Thomas, S. F., 467, *475*
Thompson, E. B., 492, 499, 502, *505*
Thompson, H. K., 354, *363*
Thompson, J., 567, *576*
Thompson, S. H., 119, *145*, 161,
    163, *172, 175*
Thompson, W., 380, *427*
Thompson, W. J., 380, 386, 411,
    416, 426, 427
Thomson, J., 404, 414, *426*
Thomson, J. M., 462, *481*
Thor, K. B., 23, 34, *69*
Thorburn, G. D., 557, *584*, 667, *679*
Thordstein, M., 667, *675*
Thurlbeck, W. M., 502, *512, 517*
Tibes, U., 462, *481*
Tildon, J. T., 621, *639*
Tisi, G. M., 567, *578*
Titler, M., 24, *63*
Tizard, J., 646, *657*
Tizard, J. P. M., 687, 691, *699*
Tobimatsu, T., 85, *108*
Toda, N., 484, 489, 503, *511*
Todres, I. D., 569, *586*
Tohyama, M., 48, 49, *68*
Tokuzen, M., 45, *57*
Tomanek, R. J., 392, *427*
Tomaszewski, K. S., 527, *548*
Tominaga, M., 77, 80, *102*
Tomita, T., 485, *509*
Tonks, N. K., 75, 94, *108*
Tooley, W. H., 254, 255, *264*, 498,
    517, 556, *582*, 667, 668, 669,
    676
Tordet, C., 502, *517*
Torrance, R. W., 357, *366*
Torri, G., *473*
Torvik, A., 129, *154*
Tossman, U., 666, 669, *679*
Towell, M. E., 593, *613*
Towers, B., 492, *511*
Towle, A. C., 45, *62*, 112, 141, *152*
Townsend, E. R., 573, *576*, 594, *607*
Toyosato, M., 25, *65*

Trachy, R. E., 272, 280, *284*
Trautmann, A., 300, *319*
Trelease, R. B., 386, *426*, 522, 529, *543, 547*
Tremblay, J., 485, *511*
Trenchard, D., 248, 256, *263, 266*
Trinder, J., 568, *577*
Trippenbach, T., 238, *244*, 246, 249, 257, 260, 261, 262, *269*, 499, *517*, 534, *542*, 648, 651, 653, 654, *660*
Trivedi, R. D., 354, *366*
Trouth, C. O., 360, *371*
Trube, G., 602, *613*
Trump, B. F., 633, *639*
Trythall, D., 646, *657*
Trzebski, A., 351, *367*
Tsacopoulos, M., 647, *657*
Tsairis, P., 376, 379, 392, 407, 413, *418*, 434, 444, *450*
Tschirgi, R. D., 683, 685, 686, *699*
Tsien, R. W., 74, 82, 83, 92, *103, 109*, 310, *316, 318*
Tsien, R. Y., 299, 300, *316, 319*
Tsous, N. G., 626, *638*
Tsubone, H., 224, 225, 237, *243*
Tsukitani, Y., 94, *97*
Tsukuda, M., 79, *107*
Tsuruo, Y., 19, 20, *60*
Tuffery, A. R., 470, 471, *481*
Turcotte, J. C., *55*
Turino, G. M., 3, *7*
Turmen, T., 672, *673*
Turner, D. J., 484, *512*
Tusiewicz, K., 522, *548*, 559, *586*
Tyc-Dumont, S., 159, *173*
Tyler, J. M., 39, *69*

U

Uchida, C., 79, 84, *99*
Uddman, R., 220, *244*
Ueda, T., 90, *109*
Ulfhake, B., 384, 387, *419, 424*
Ullrich, A., 72, 75, 77, *97*
Ulrich, C. E., 224, *244*

Ultsch, G. R., 605, *613*
Umans, J. G., 668, *678*
Ungerstedt, U., 34, *55*, 663, 666, 669, *679*
Unwin, N., 74, *109*
Urbanics, R., 647, *660*
Urena, J., 293, 296, 305, 306, *316, 320*, 334, *339*
Usuda, N., 79, 84, *99*
Uvnas-Moberg, K., 668, *676*

V

Vaillancourt, P., 255, *269*
Vain, N., 646, *656*
Valdeolmillos, M., 295, 299, 300, 305, 310, *313, 320*
Valdman, B. M., 633, *641*
Valenzuela, D., 82, *108*
Vallano, M. L., 86, 87, *109*
Vallbo, A., 434, 436, *451*
Van den Abbeele, A., 274, *285*
Vandenbark, G. R., 77, *105*
Vanderwolf, C. H., 537, 539, 540, *542, 548, 549*
Van Essen, D., 379, 380, 386, *418*
Van Essen, D. C., 379, *417, 421*
VanEykern, L. A., 559, 560, *585*
VanHarreveld, A., 600, *613*, 643, 649, 654, *660*
Van Hoff, C. O. M., 81, *109*
van Lunteren, E., 562, *587*
Vannucci, R. C., 620, 621, 625, 626, 627, 629, 632, 634, *636, 637, 641*, 654, *657*
VanReempts, P., 573, 574, *583*, 670, *674*
Van Rossum, G. D. V., 623, *641*
Vapalahti, M., 633, *640*
Vaughan, R. L., 274, *287*
Veech, R. L., 631, 634, *639*
Velasquez, T. M., 334, *338, 594, 610*
Venter, J. C., 28, *56*
Verdoorn, T., 45, *62*
Verhofstad, A. A. J., 13, 19, 24, 46, 53, *60, 61, 65, 66*, 186, *196*

Vernet, L., 666, 669, *679*
Vernon, P., 85, *101*
Versprille, A., 231, 232, *240*, 486, *509*
Versteeg, D. H. G., 82, *97*
Viana, F., 382, 386, *422*
Vibert, J. F., 276, *287*
Vidruk, E. H., 308, *319*, 496, 498, *516*
Villar, M., 19, *60*
Virgona, V., 498, *513*, 593, 595, *611*
Visser, T. J., 23, 34, *60*
Vizek, M., 661, *679*
Voipio, J., 305, *316*
Volgyesi, G., 459, *479*, 560, *584*
Volk, E. A., 169, *172*
Vollestad, N. K., 438, 440, 444, *453*
Volterra, A., 76, *96*
Von, S., 297, *315*
von Euler, C., 1, 7, 129, *148*, 201, 205, 210, *213, 214,* 247, 248, 252, *263, 264, 284,* 342, 346, 350, *363, 364,* 534, *542,* 644, *657,* 664, 665, *673, 674, 677*
von Kanzow, E., 644, *656*
Voss, M. D., 275, *283*
Vrbova, G., 379, 380, 388, 392, 410, *424,* 470, *479*
Vyskocil, F., 470, *481,* 599, *613*

W

Wada, K., 76, *98*
Wada, S., 224, *244*
Wadepuhl, M., 124, 125, *153*
Waggener, T. B., 260, *268,* 562, 567, 568, 569, 570, *586, 587*
Wagner, J. A., 31, 51, *65*
Wagner, M. A., 90, *108*
Wagner, P. D., 567, *578*
Walaas, S. I., 81, 89, *109, 110*
Waldman, S., 573, *582*
Waldron, M. A., 237, *241, 244,* 492, 493, 494, 495, 496, 499, 500, 505, *508, 510, 518*

Waldrop, T. G., 34, 35, *64*, 130, *148*, 350, 354, *368*
Walker, A. M., 359, *362*
Walker, D. W., 360, *371,* 497, *509,* 518, 554, 555, 556, 557, *577, 578, 585,* 662, 665, 666, 667, 668, *674, 675, 676*
Walker, J. J., 189, *198*
Wall, J., 25, *62*
Wallace, J. A., 45, *69*
Wallen, L. D., 556, 557, *584, 587*
Wallen, P., 125, *146,* 169, *173*
Wallin, E. U., 446, *451*
Walmsley, B., 383, 384, *427*
Walsh, D. A., 32, *67,* 76, *109*
Walther, C., 472, *480*
Walti, H., 567, *583*
Walton, K., 159, 166, 169, *173,* 384, 387, 388, *420, 427*
Waltzer, R., 183, *196*
Wand, C. D., 28, *56*
Wang, D. L., 668, *677*
Wang, J. K. T., 81, 89, 91, *99, 108, 109*
Wang, S. C., 128, *154*
Wang, Y. Z., 354, *366*
Wannamaker, E. M., 467, *478*
Warburton, D., 504, *512*
Ward, A. A., 692, *699*
Ward, R. A., 668, *673*
Wardlaw, S. L., 667, *678*
Warembourg, M., 38, *69*
Warner, P., 571, *577*
Washio, H., 470, *481*
Wassmeyer, B., 201, *214*
Wasterlain, 634, *641*
Watanabe, H., 645, *656*
Watanabe, S., 39, *62*
Watchko, J. F., 259, *269,* 459, 465, 472, 473, *479, 481*
Waterfield, M. D., 77, *97*
Watkins, J. C., 26, *57*
Watson, J. E., 80, *97*
Watson, L. A., 494, *510*
Watt, J. G., 662, *679*
Waxham, M. N., 83, 92, *103*

Waxman, S. G., 602, 603, *;613*
Ways, N. R., 74, *98*
Webber, C. L., Jr., 382, *427*
Weber, E. D., 177, *197*
Webster, E., 626, 627, 634, *637*
Weeks, S., 223, 235, *243,* 248, *265,*
   564, 567, *579, 584,* 691, *698*
Weibel, E. R., 485, 486, 494, *508*
Weidman, P. J., 74, *104*
Weil, J. V., 42, *69,* 572, *578,* 594,
   611, 661, *679*
Wein, A. J., 492, *512*
Weinberger, R. P., 85, 86, 87, *106,*
   109
Weiner, M. W., 440, *452*
Weiskopf, R. B., 661, *679*
Weiss, G. K., 535, *544*
Weiss, J. N., 602, *613*
Weiss, K. R., 114, *151*
Weiss, S., 25, *68*
Weisser, K., 254, 255, *264*
Weissman, A. M., 72, *109*
Welch, K., 646, *657*
Weller, K. L., 532, *549*
Wen, J., 200, *214,* 231, 232, *242,*
   484, 486, *512*
Wendenburg, A., 569, *578*
Wennberg, R. P., 681, 690, *699*
Wennergren, G., 274, *288,* 359, *371,*
   669, *679*
Wennergren, M., 359, *371*
Werne, J., 693, *699*
Wernig, A., 470, *482*
Wesche, J., 440, *453*
Wessberg, P., 112, 141, *149*
West, T. G., 606, *613*
Westbrook, G. L., 160, *175,* 310, *317*
Westerberg, E., 606, *614*
Westerblad, H., 439, *449*
Westerman, R. A., 380, 386, 387,
   416, 427
Westling, G., 434, *451, 452, 453*
Westlund, K. N., 180, *194, 198*
Whalen, P., 322, *339*
Whalen, R. G., 410, 411, *418, 427*
Whalen, W. J., 306, *317, 320*

Wharton, R., 630, *640*
Wheatley, M. D., 692, *699*
Whitaker, V. P., 14, *69*
Whitcomb, J. M., 192, *194*
White, D. P., 572, *578*
White, J. G., 620, *639*
White, N., 19, 53, *61,* 112, 141, *150*
White, S. R., 186, *196*
Whitsett, J. A., 488, 502, 506, *518*
Whittaker, J. A., 360, *371*
Whittam, R., *641*
Whitton, J., 401, *424,* 457, 473, *479,*
   560, *584*
Widdicombe, J., 228, *244*
Widdicombe, J. D., 687, *696*
Widdicombe, J. G., 222, 2;25, 229,
   230, *240, 243,* 247, 253, 254,
   255, 259, 260, *264, 266, 267,*
   269, 483, 485, 494, 496, *510,*
   514, 518
Widdowson, E. M., 646, 647, 648,
   660
Wiebe, H., 593, *612*
Wieizouk, S., 410, *423*
Wieloch, T., 310, *320*
Wieloch, T. W., 606, *614*
Wiersma, C. A. G., 114, 121, *154*
Wigender, E., 84, *109*
Wight, D., 461, *475*
Wigston, D. J., 470, *482*
Wigstrom, H., 157, *173*
Wikberg, K., 503, *507*
Wilcox, J. N., 44, *69*
Wilkes, D., 560, 574, *581*
Wilkie, R. A., 495, 504, *518*
Wilkins, J. F., 665, *678*
Wilkinson, A. R., *285*
Wilkinson, M. H., 358, *362,* 555,
   556, 569, 570, *575, 576, 583*
Willenberg, I. M., 349, *371*
Williams, C., *69*
Williams, D. L., 618, 621, 622, *637,*
   641
Williams, H., 462, *478*
Williams, M. E., 74, *98*
Williams, S., 635, *641*

Williamson, E., 407, 414, *417*
Willis, D., 564, *584*
Willis, W. D., 644, *658*
Willows, A. O. D., 116, 117, 144, *154*
Wills, M., 488, 489, 490, 491, *518*
Wilson, D. F., 389, *427*, 617, *636*
Wilson, D. M., 113, 114
Wilson, G. V., 568, *577*
Wilson, J. E., 622, *641*
Wilson, N. J., 485, *518*
Wilson, P., 539, *541*
Wilson, S. L., 564, *587*
Wimpress, S. P., 567, *576*
Winberg, J., 668, *676*
Winick, M., 168, *173*, 620, *641*
Wining, A. J., 687, *696*
Winn, H. R., 666, *679*
Winning, A. J., 254, 255, *266*
Winslow, J. W., 357, *369*
Winters, R. W., 3, *7*
Winterstein, H., 354, *371*
Wise, J. C. M., 229, 230, 237, *240*, 247, 253, 255, 259, *264*
Wiseman, G., 38, *69*
Wistrand, P. J., 358, *364*
Withington-Wray, D. J., 487, *512*
Witte, M. K., 248, *269*
Wohl, M. E. B., 560, *581*
Wolf, M., 78, 79, 80, 82, 84, *107, 109, 110*
Wolff, C. B., 594, *608*
Wolfson, M. R., 491, *514*
Wollman, H., 626, *636*
Wollner, J. C., 155, *174*
Wolotsky, P., 129, *147*, 277, *284*
Wong, K. S., 143, *150*
Woo, P., 223, *243*
Wood, J., *369*
Wood, J. G., 78, 79, *98*
Wood, L. E., 25, *67*
Wood, L. M., 487, *512*
Wood, M. G., Jr., 79, *110*
Wood, R. P., II., 3, *7*
Woodhams, P. L., 309, *314*
Woodrum, D., 646, *659*

Woodrum, D. E., 250, 256, *267*, 321, *336, 339,* 360, *365,* 459, 465, 472, 473, *479, 481,* 568, 569, 571, 572, *576, 579, 580, 582, 587,* 593, *610, 614,* 662, *679,* 681, 690, *699*
Woods, J. J., 432, 433, 436, 440, 442, 444, 445, 446, *449, 450, 453*
Woolcock, A., 485, 501, *514, 518*
Woolf, G. M., 522, *541*, 573, *576*
Worley, P. F., 78, *110*
Wozniak, J. A., 248, 252, 256, *267,* 563, *582*
Wrobel-Kuhl, K., 633, *636*
Wu, J-Y., 19, 46, *65*
Wu, P. H., 16, *66*
Wu, W. C-S., 81, *95, 110*
Wurth, M. A., 633, *635*
Wuttke, W., 43, *58*
Wyke, B. D., 222, *244*
Wyman, 114, *154*
Wyman, R. J., 129, *148*, 644, *657*
Wyszogrodski, I., 249, *269*
Wyzinski, P. W., 536, *544*

X

Xia, Y., 606, *609*
Xu, Z. S., 47, *62*

Y

Yafuso, N., 501, *511*
Yakushev, V. S., 633, *641*
Yamate, C. L., 649, *659*
Yamada, H., 201, *214*
Yamada, K. A., 343, 351, *371*
Yamamoto, K. R., 31, 38, 43, 44, *69*
Yamamoto, W. S., 343, *366*, 456, *479*
Yamamoto, Y., 34, *55*, 112, 132, *154,* 205, *213,* 247, *264,* 350, *363,* 664, 665, 668, 669, *673, 676, 677, 678, 679*
Yamuy, J., 527, *546*
Yanagisawa, M., 186, *197*

Yanaihara, N., 32, *60*
Yang-Feng, T., 62
Yang-Feng, T. L., 77, *97*
Yang, T. Z., 532, *544*
Yarden, Y., 72, 75, *110*
Yasuda, I., 77, 80, *102*
Yemm, R., 433, *452*
Yip, R., 85, *101*
Yodh, N., 25, *62*
Yoneda, S., 635, *638*
Yoshida, A., 39, *62*
Yoshida, Y., 78, *100, 272, 288*
Yoshikawa, K., *69*
Yoshikawa, S., 76, *98*
Yoshimasa, T., 72, *110*
Yoshimura, M., 307, 310, *315,* 597, 600, *609*
Yoshioka, H., 634, 635, *641*
Yoshizaki, K., 293, 308, *315*
Younes, M., 248, 251, 252, *266, 268*
Younes, M. K., 247, 255, *269*
Young, A. C., 28, *55,* 357, *362*
Young, T. N., 81, *99*
Young, W., 309, *318*
Young, W. S., III., 78, *100*
Yu, P., 495, *511*
Yunis, K., 593, *608*

Z

Zacks, S. I., 397, *422*
Zagelbaum, G., 457, 459, *476*

Zagon, I. S., 186, *198*
Zajac, F. E., 376, 379, 381, 392, 407, 413, *418, 427,* 434, 444, *450*
Zaman, A., 634, 635, *641*
Zapata, P., 295, 308, *315, 320*
Zawalich, W., 72, *106*
Zengel, J. E., 381, *428*
Zetterstrom, R., 255, *263*
Zetterstrom, T., 666, 669, *679*
Zeuthen, T., 600, *610*
Zhan, W-Z., 397, *426, 428*
Zhang, H., 275, 277, *284*
Zhang, J., 529, 530, 531, 533, 536, *542, 546, 549*
Zhou, D., 352, *370*
Zhou, Q., 79, *110*
Ziccone, S. P., 387, *427*
Ziegler, M. G., 180, *198*
Ziff, E. B., 92, *99*
Zimmerman, I., 160, *172*
Zimmermann, V., 646, *658*
Zinkin, P., 552, *585*
Zinman, R., 246, 249, 260, *269,* 322, 334, *337,* 571, *577,* 593, *608*
Ziskind-Conhaim, L., 159, *175,* 384, 387, 389, 397, *419, 428*
Zotterman, Y., 225, *244*
Zuperku, E. J., 239, *241,* 246, *267*
Zwillich, C. W., 42, *69,* 527, *545,* 572, *578,* 661, *679*

SUBJECT INDEX

A

Acetylcholine (ACh), 13-15
Acute asphyxiation, patterns of
  respiration during, 682-688
  energy metabolism as regulator
    of respiratory center ac-
    tivity, 685-686
  evidence for a medullary gasp
    center, 684-685
  influence of peripheral chemo-
    receptor, 686-687
  role of cutaneous, laryngeal,
    and vagal pulmonary af-
    ferents, 687-688
Adducin, 82
Adenosine, 5, 668, 669, 670-671
Adrenergic/sympathetic control
  of airway smooth muscle,
  502-506
  β-adrenergic function, 502-505
  development of α-adrenergic
    receptors, 505-506
Adult diaphragm, motor unit
  properties in, 378-379
Afferent innervation of the upper
  and lower airways, 220-
  224, 230-231
Afferent input to respiration, 4-5
Airway and pulmonary receptors,
  219-244
  lower airway receptors, 230-
  239

[Airway and pulmonary receptors]
  afferent innervation, 230-231
  functional characteristics, 231-239
  upper airway receptors, 220-230
  afferent innervation, 220-224
  functional characteristics, 224-230
Airway smooth muscle control in the
  newborn, 483-518
  adrenergic/sympathetic control of
    smooth muscle, 502-506
  autonomic airway innervation in
    the adult, 485-487
  species differences, 484-485
  vagal control of airway smooth
    muscles, 487-501
  cholinergic excitatory innervation,
    487-498
  nonadrenergic noncholinergic in-
    hibitory innervation, 499-501
  noncholinergic excitatory innerva-
    tion, 498-499
  summary of vagal airway innerva-
    tion, 501
Amino acids, excitatory, 667
γ-Aminobutyric acid (GABA), see
  GABA
Apnea, 2
  apneic spells in sleep, 693-694
  "breath holding" apneic spells in
    older infants, 692-693
  neonatal, 672, 691-692
  of undetermined etiology, 693-
    694

753

Arnold-Chiari malformation, 3
Arterial chemoreceptors, workings of,
    289-320
  cellular elements and identity of
      the transducer, 291-293
  centrifugal pathway, 307-308
  effects of hypoxia on neurons,
      309-311
  intracellular recordings from Type
      I cells, 295-299
    action of cyanide, 296-299
  measurements of intracellular cal-
      cium, 299-305
    effects of hypoxia, 300-304, 304-
        305
    response to cyanide, 299-300,
        304-305
  other recent investigations, 305-
      307
  prospects for the future, 311-312
  single-cell studies, 293-295
  the transmitters, 308-309
Asphyxia, control of breathing dur-
    ing, 681-699
  pattern of respiration during acute
      asphyxiation, 682-688
    energy metabolism as regulator
        of respiratory center activity,
        685-686
    evidence for a medullary gasp
        center, 684-685
    influence of peripheral chemo-
        receptor, 686-687
  role of cutaneous, laryngeal, and
      vagal pulmonary afferents,
      687-688
  role of gasping in spontaneous
      recovery from hypoxic apnea,
      688-694
    apnea of undetermined etiology,
        693-694
    apneic spells in sleep, 693-694
    "breath holding" apneic spells
        of older infants, 692-693
    clinical occurrence of autore-
        suscitation, 690-691

[Asphyxia]
    experimental studies of "auto-
        resuscitation," 688-690
    neonatal apneic spells, 691-
        692
Aspiration, 2
Autoresuscitation, 688-690
  clinical occurrence of, 690-
      691

B

Brain development, energy metab-
    olism in, 615-641
  energy metabolism in early life,
      616-626
    biochemical changes in the mito-
        chondria, 619-620
    development of carbohydrate,
        fat, and amino acid metabol-
        ism, 620-623
    development of circulation-energy
        metabolism coupling, 625-626
    development of Na-K ATPase,
        623-625
    ionic homeostasis in developing
        brain, 623
    mitochondrial alterations in the
        developing brain, 618-619
    principle of energy metabolism
        during development, 616-618
  hypoxic tolerance in the developing
      brain, 626-635
    age-dependent mitochondrial
        function during hypoxia, 631-
        632
    age-dependent response to ische-
        mia, 634-635
    developing oxygen gradient in
        vasculatures and in brain,
        628-630
    fetal hemoglobin during brain
        tissue hypoxia, 627-628
    hypoxia response in the in vivo
        brain and in isolated mito-
        chondria, 630-631

[Brain development]
  ischemic responses and energy
    metabolism in developing
    brain, 633-634
  mitochondrial responses during
    oxygen deprivation, 630
Brain protein kinase C, 78
Brainstem, effect of hypoxia on,
  663-664
Brainstem neurons in early life, 155-
  175
  active membrane properties and
    ionic currents, 160-168
  action potential waveform, 160-
    161
  membrane potential and cellular
    properties, 161-168
  implications for future research,
    171
  importance to breathing, 169-171
  morphology and connectivity, 168-
    169
  passive electrophysiological prop-
    erties, 157-160
  repetitive firing properties, 157
  synaptic input versus membrane
    properties, 156-157
"Breath holding" apneic spells of
  older infants, 692-693
Breathing behavior, 1-2

C

Calcitonin gene-related peptide
  (CGRP), 32
Ca^{2+}/calmodulin-dependent protein
  kinase (CaMKinase) type II,
  84-93
  developmental expression, 85-86
  genetic modulation, 92-93
  mechanism of activation, 87-90
  molecular properties, 84-85
  regional and subcellular distribu-
    tion, 86-87
  regulation of synaptic communica-
    tion, 90-92

Carbon dioxide (CO_2) response of
  newborns, 573-574
Carotid bodies, discharge properties
  of, 321-340
  carotid chemoreceptor afferent
    nerve responses
    in the fetus, 324-325
    in the neonate, 325-335
  carotid chemoreceptor recordings,
    322-324
    recording from the petrosal gang-
      lion, 323-324
    recording from sinus nerve
      strands, 322-323
  development of carotid body che-
    moreceptors, 321-322
Catecholaminergic projections to the
  spinal cord, development of,
  183-185
Catecholamines, 5, 665, 668
Central pattern generator (CPG),
  113-114, 115-125
  cellular building blocks, 116-121
  oscillation and pattern formation
    in, 128-129
  plasticity, 121-125
Central respiratory chemoreceptors,
  341-371
  central chemoreception during de-
    velopment, 358-360
  evidence for the existence of cen-
    tral chemoreception, 341-342
  expanded role for neurons within
    the VLM, 348-354
  location of, 342-348
  mechanism of, 354-358
C-fiber afferents, 237-239
Chemical messenger, 15-16
Chemical synaptic transmission, 11-69
  cellular and molecular mechanisms
    of synaptic transmission, 12-31
    postsynaptic neuron, 24-31
    presynaptic neuron, 13-24
  ontogenesis of neurotransmitter
    and peptide systems in the
    brain, 44-53

[Chemical synaptic transmission]
  gene expression for peptides dur-
    ing development, 50-53
  ontogeny of classical neurotrans-
    mitters, 45-47
  ontogeny of neuropeptides, 47-49
  regulation of gene expression by
    synaptic and hormonal signals,
    31-44
  activity-dependent regulation of
    gene expression, 32-36
  ligand-dependent regulation of
    gene expression, 36-43
  regulation of gene expression for
    transmitter enzymes and neuro-
    peptides by steroid hormones,
    43-44
Chest wall reflexes influencing breathing, *see* Pulmonary and chest
  wall reflexes influencing breathing
Classical transmitters:
  coexistence, in mammalian CNS,
    of peptides and, 20
  ontogeny of, 45-47
Complexity of respiratory control, 2-3
Consciousness, state of, importance
  to respiratory control of, 3
Corticospinal projections in the spinal
  cord, development of, 189-191
Creatine kinase equilibrium system, 618
Cutaneous afferents in the regulation
  of respiratory pattern during
  asphyxiation, 687-688

                          D

Descending spinal pathways, development of, 177-198
  brainstem-spinal projections, devel-
    opment of, 178-189
  monoaminergic projections in the
    spinal cord, 180-185
  origin of brainstem-spinal path-
    ways at different stages of de-
    velopment, 178-180

[Descending spinal pathways]
  selected peptidergic projections
    to the spinal cord, 185-189
  corticospinal projections, develop-
    ment of, 189-191
Diaphragmatic fatigue, metabolic
  basis of, 461-464
  blood flow, 461-462
  tissue substrates, 462-464
Diaphragm muscle cells, 375-428
  diaphragm fiber type proportions
    during postnatal development,
    407-415
  correlation between diaphragm
    contractile properties and fi-
    ber type proportions, 413-414
  fiber type composition of muscle
    units, 411-413
  histochemical classification of
    fiber types in the neonate,
    407-409
  histochemical classification of
    muscle fiber types, 407
  immunohistochemical classifica-
    tion of muscle fiber types,
    409-410
  influence of activity on fiber
    type differentiation, 410-411
  rate of myofibrillar ATPase ac-
    tivity in diaphragm muscle
    fibers, 414-415
  factors contributing to diaphragm
    muscle fatigue, 389-407
  capillary density and diaphragm
    fatigue resistance, 404
  central versus peripheral fatigue,
    389
  fatigue resistance of the neo-
    natal diaphragm and muscle
    fiber oxidative capacity, 397-
    404
  fiber cross-sectional area and
    diaphragm fatigue resistance,
    404-407
  motor unit fatigue resistance and
    oxidative capacity, 391-392

[Diaphragm muscle cells]
neurotransmission failure and
diaphragm fatigue, 390-391
SDH variability among diaphragm
muscle fibers, 392-397
motor unit physiological properties,
375-381
classification of motor units,
375-376
diaphragm innervation and synapse elimination, 379-380
motor unit properties in the
adult diaphragm, 378-379
motor unit properties during
postnatal development, 380-381
motor unit tension, 377-378
neural control of muscle tension,
381-389
diaphragm motor unit recruitment, 382-384
frequency coding of motor units
during postnatal development,
387-389
frequency coding of motor unit
tension, 386-387
intrinsic motoneuron properties
during postnatal development,
384-385
intrinsic properties of motoneurons, 381-382
motor unit recruitment during
postnatal development, 385-386
Discharge properties of carotid bodies, 321-340
carotid chemoreceptor afferent
nerve responses
in the fetus, 324-325
in the neonate, 325-335
carotid chemoreceptor recordings,
322-324
recordings from the petrosal
ganglion, 323-324
recordings from sinus nerve
strands, 322-323
development of carotid body chemoreceptors, 321-322

Dopamine, 665-666, 670
Dorsal respiratory group (DRG), 128-
129
Dorsal respiratory group (DRG) neurons, 132-143, 644-645
cellular organization, 133-136
effect of laryngeal reflexes on,
277-278
intrinsic membrane properties, 136-
139
pattern formation in the DRG,
139-143

E

87 kDal MARCKS protein, 81
Electrophysiological changes in the
CNS activity, 521-549
ascending projection systems, 535-
538
control of motoneurons at the membrane level during sleep, 525-
528
descending forebrain systems and
respiratory control, 528-535
descending "arousal" system to
respiratory musculature, 528-534
respiratory patterns from descending temperature influences, 534-
535
electrophysiologic aspects of REM
sleep, 522, 523-525
relationship of slow-wave activity
to respiratory patterning, 538-
540
Emotion, importance to respiratory
control of, 3
End-expiratory volume, breathing
with increased in, 257-259
Endogenous opioids, 667-668
Energy metabolism in the developing
brain, 615-641
energy metabolism in early life,
616-626
biochemical changes in the mitochondria, 619-620

[Energy metabolism in the developing brain]
  development of carbohydrate, fat, and amino acid metabolism, 620-623
  development of circulation-energy metabolism coupling, 625-626
  development of Na-K ATPase, 623-625
  ionic homeostasis in developing brain, 623
  mitochondrial alterations in the developing brain, 618-619
  principle of energy metabolism during development, 616-618
  hypoxic tolerance in the developing brain, 626-635
    age-dependent mitochondrial function during hypoxia, 631-632
    age-dependent response to ischemia, 634-635
    developing oxygen gradient in vasculatures and in brain, 628-630
    fetal hemoglobin during brain tissue hypoxia, 627-628
    hypoxia response in the in vivo brain and in isolated mitochondria, 630-631
    ischemic responses and energy metabolism in developing brain, 633-634
    mitochondrial responses during oxygen deprivation, 630
Energy metabolism regulation of respiratory center activity during asphyxiation, 685-686
Enkephalin, 669
Enkephalin (ENK)-LI axons, 188-189
Met-Enkephalin, 670
Excitatory amino acids, 667
Excitatory postsynaptic potentials (EPSPs), REM sleep and, 527-528

F

Fatigue processes in human voluntary contractions, 439-444
  role of CNS motor drive, 444
  role of excitation/contraction coupling and metabolic changes, 439-441
  role of impaired action potential transmission, 441-444
Fatigue resistance of the neonatal diaphragm and muscle fiber oxidative capacity, 397-404
Fetal breathing movements, 554-558
  metabolic control, 557-558
  upper airway activity, 558
Fetal hemoglobin during brain tissue hypoxia, 627-628
Fetal hypoxia, neurochemical effects on breathing caused by, 667-668
Fetus:
  carotid chemoreceptor afferent nerve responses in, 324-325
  response to hypercarbia of, 557-558
  response to hypoxemia of, 557
Functional abnormalities of respiratory control, 3-4

G

GABA (γ-aminobutyric acid), 5, 667, 669
GAP43 protein, 81-82
Gasping pattern of respiration, *see* Asphyxia, control of breathing during
Gene expression, regulation by synaptic and hormonal signals of, 31-44
  activity-dependent regulation of gene expression, 32-36
  ligand-dependent regulation of gene expression, 36-43
  regulation of gene expression for transmitter enzymes and neuropeptides by steroid hormones, 43-44

Glossopharyngeal afferents, 225
L-Glutamate receptors, 25-27
Gonadal steroid hormones, behavioral effects of, 38-39
Growth factors, 5
GTP-binding proteins, 73-74

H

Hemoglobin, fetal, during brain tissue hypoxia, 627-628
Hormones, 5
Human muscles and their CNS interactions, 429-453
  contractile properties of human motor units, 433-436
  fatigue processes in human voluntary contractions, 439-444
    role of CNS motor drive, 444
    role of excitation/contraction coupling and metabolic changes, 439-441
    role of impaired action potential transmission, 441-444
  fatigue of respiratory muscles, 446-448
  implications for neonatal respiratory muscle failure, 448
  neuromuscular fatigue defined, 429-430
  potential causes of fatigue, 436-439
    excitation/contraction coupling and metabolic changes, 438-439
    influence of stimulus frequency, 436-438
  problems in fatigue studies, 431-433
    interspecies variations, 432-433
    multiple causes, 431-432
    stimulated versus voluntary contractions, 432
  relation between CNS motor drive and muscle contractile properties, 444-446
    fatigue-induced reflex inhibition, 445-446
    motor unit firing rates, 444

[Human muscles and their CNS interactions]
  sites of fatigue and methods of investigation, 430-431
Hydrocephalus, 3
Hypercapnia, 2
Hypercarbia, fetal response to, 557-558
Hypertension, 661
Hypoglossal (HYP) neurons, 159
Hypoxemia, fetal response to, 557
Hypoxia, 2, 571-573
  effects on arterial chemoreceptors of, 300-304
  effects on neurons of, 309-311
  neuropharmacology of breathing and, 661-679
    clinical aspects, 672
    fetal hypoxia, 667-668
    infant hypoxia, 668-671
    neuroactive agents in regulation of breathing, 662-663
    neuromodulators released during hypoxia which inhibit respiration, 669
    respiratory-controlling structures affected by hypoxia, 663-667
  response of newborn to, 571-573
  tolerance in the developing brain to, 626-635
    age-dependent mitochondrial function during hypoxia, 631-632
    age-dependent response to ischemia, 634-635
    developing oxygen gradient in vasculatures and in brain, 628-630
    fetal hemoglobin during brain tissue hypoxia, 627-628
    hypoxia response in the in vivo brain and in isolated mitochondria, 630-631
    ischemic responses and energy metabolism in developing brain, 633-634

[Hypoxia]
  mitochondrial responses during
    oxygen deprivation, 630
  ventilatory effects of, in newborn
    animals, 645-646
  ventilatory response to, 6, 591-
    594
  cellular and molecular response,
    595-606
  integrated response, 592-595
Hypoxic apnea, role of gasping in
    spontaneous recovery from,
    688-694

                    I

Infants:
  apneic spells and SIDS in, 681-682,
    693-694
  "breath holding" apneic spells of,
    692-693
    neurochemical effects on breath-
      ing caused by hypoxia in, 668-
      671
  See also Neonates; Newborn
Invertebrate motor networks, 111-154
  implications for control of breath-
    ing, 127-143
    dorsal respiratory group neurons,
      132-143
    oscillation and pattern formation
      in the respiratory CPG, 128-
      129
    respiratory drive, 129-131
  motor organization of invertebrates,
    112-127
    central pattern generators, 115-
      125
    redefining the command func-
      tion, 125-127
  summary, 144
Ion channels, 82-83
Ionic environment in the CNS, 643-
  660
  anatomical organization of respira-
    tory-related neurons, 644-645

[Ionic environment in the CNS]
  extracellular ion activities in the
    CNS of mammals, 646-649
  extracellular ion activities during
    low oxygen supply, 650-656
  ventilatory effects of hypoxia in
    newborn animals, 645-646
Ionic homeostasis in developing brain,
  623
Ion-sensitive microelectrodes (ISMs),
  measurements of extracellular
  ion activities in the CNS of
  mammals by, 646-649
Ischemia, age-dependent response to,
  633-635

                    L

Laryngeal afferents, 225-230
  in regulation of respiratory pattern
    during asphyxiation, 687-688
Laryngeal reflexes, 271-288
  effect and mechanism of action,
    274-280
    effect on DRG neurons, 277-278
    effect on respiratory muscle drive,
      275-277
    effect on VRG neurons, 278-280
    respiratory and autonomic re-
      sponses, 274-275
  future directions, 282-283
  laryngeal nerve fibers, 272
    fiber trajectory, 272
  maturational changes, 280-282
    nerve fibers, 280
    respiratory and autonomic re-
      sponses, 280-282
Lower airway receptors, 230-239
  afferent innervation of, 230-231
  functional characteristics, 231-
    239
    C-fiber afferents, 237-239
    rapidly adapting receptors, 237
    slowly adapting receptors, 233-
      237
    sympathetic afferents, 239

M

Maturation, rates of, 6
Medullary gasp center, 684-685
Metabolic basis of diaphragmatic
    fatigue, 461-464
  blood flow, 461-462
  tissue substrates, 462-464
Metabolic control of fetal breathing
    movements, 557-558
Miniature end-plate potentials
    (MEPPs), 14
Mitochondria:
  age-dependent mitochondrial func-
      tion during hypoxia, 631-632
  biochemical changes in, 619-620
  in the developing brain, altera-
      tions in, 618-619
  isolated, hypoxia responses in, 630-
      631
  responses during oxygen depriva-
      tion, 630
Monoaminergic projections to the
    spinal cord, development of,
    180-185
  catecholaminergic projections, 183-
      185
  serotoninergic projections, 182-
      183

N

Neonatal apnea, 672
Neonatal apneic spells, 691-692
Neonate:
  carotid chemoreceptor afferent
      nerve responses in, 325-335
    the kitten, 327-329
    the lamb, 325-327
    the piglet, 329-335
  See also Infants; Newborn; Post-
      natal development
Neural control of muscle tension,
    381-389
  diaphragm motor unit recruitment,
      382-384

[Neural control of muscle tension]
  frequency coding of motor units
      during postnatal development,
      387-389
  frequency coding of motor unit
      tension, 386-387
  intrinsic motoneuron properties
      during postnatal development,
      384-385
  intrinsic properties of motoneu-
      rons, 381-382
  motor unit recruitment during post-
      natal development, 385-386
Neuroactive peptides found in mam-
    malian nervous tissue, 16, 17
Neurochemicals, 5
Neurohormones, 662
Neuromodulators, 662
Neuromuscular fatigue, 429-453
  contractile properties of human
      motor units, 433-436
  fatigue processes in human volun-
      tary contractions, 439-444
    role of CNS motor drive, 444
    role of excitation/contraction
        coupling and metabolic changes,
        439-441
    role of impaired action potential
        transmission, 441-444
  fatigue of respiratory muscles, 446-
      448
  implications for neonatal respira-
      tory muscle failure, 448
  neuromuscular fatigue defined,
      429-430
  potential causes of fatigue, 436-439
    excitation/contraction coupling
        and metabolic changes, 438-439
    influence of stimulus frequency,
        436-438
  problems in fatigue studies, 431-
      433
    interspecies variations, 432-433
    multiple causes, 431-432
    stimulated versus voluntary con-
        tractions, 432

[Neuromuscular fatigue]
  relation between CNS motor drive
      and muscle contractile proper-
      ties, 444-446
    fatigue-induced reflex inhibition,
      445-446
    motor unit firing rates, 444
  sites of fatigue and methods of
      investigation, 430-431
Neuronal firing, three different pat-
      terns of, 157, 158
Neuronal signaling, 71-110
  CaMKinase II, 84-93
    developmental expression, 85-86
    genetic modulation, 92-93
    mechanism of activation, 87-90
    molecular properties, 84-85
    regional and subcellular distribu-
      tion, 86-87
    regulation of synaptic communi-
      cation, 90-92
  families of signaling molecules,
      73-76
  protein kinase C, 76-84
    cell-specific expression, 78
    compartmentalization, 78-79
    developmental expression, 78
    genetic modulation, 83-84
    long-term potentiation, 83
    mechanism of activation, 79-
      80
    molecular properties, 77-78
    substrates, 80-83
  signal termination by phosphopro-
      tein phosphatase, 93-94
  signaling pathways, 71-73
Neuropharmacology of breathing,
      hypoxia and, 661-679
  clinical aspects, 672
  fetal hypoxia, 667-668
  infant hypoxia, 668-671
  neuroactive agents in regulation
      of breathing, 662-663
  neuromodulators released during
      hypoxia which inhibit respira-
      tion, 669

[Neuropharmacology of breathing]
  respiratory-controlling structures
      affected by hypoxia, 663-667
    brain stem, 663-664
    peripheral chemoreceptors, 665-
      666
    suprapontine inputs to the brain-
      stem respiratory controller,
      665
    sympathoadrenal system, 666
    tissue neuroregulatory systems,
      666-667
Neurophysiological organization of
      respiratory neurons in early
      life, 199-215
  future directions/problems, 213
  studies using extracellular record-
      ing and related techniques,
      200-207
    discharge properties of expira-
      tory neurons in developing
      opossum, 204-206
    discharge properties of inspira-
      tory neurons in developing
      opossum, 201-204
    effects of excitatory amino acids
      in the rostral pons of opos-
      sums, 207
    interpretation of results with re-
      spect to models of respiratory
      rhythmogenesis, 206
    location of RRUs in opossums,
      200-201
    medullary RRUs in other imma-
      ture mammals, 206
  studies using intracellular recording
      techniques, 207-212
    effect of activation of laryngeal
      reflexes on bulbar respiratory
      neurons, 211-212
    triphasic pattern of membrane
      potential oscillation during
      the respiratory cycle, 208-
      211
Neurotransmitters, 662
  biosynthesis of, 16-18

Newborn:
  airway smooth muscle control in, 483-518
    adrenergic/sympathetic control of airway smooth muscle, 502-506
    autonomic airway innervation in the adult, 485-487
    species differences, 484-485
    vagal control of airway smooth muscles, 487-501
  respiratory muscles of, 558-567
    diaphragm and intercostal muscles, 559-562
    pharynx and nose, 564-567
    upper airway muscles, 562-567
  sleep state determination in, 553
Nicotinic acetylcholine receptor, 25
NonREM sleep in the newborn, 553
Nucleus of the solitary tract (NTS), 18-19

O

Opioids, 5
  endogenous, 667-668
Oxygen gradient development in vasculatures and in the brain, 628-630

P

Peptidergic projections to the spinal cord, development of, 185-189
Peptides:
  biosynthesis of, 16-18
  coexistence, in mammalian CNS, of classical transmitters and, 20
  neuroactive, found in mammalian nervous tissue, 16, 17
  ontogeny of, 47-49
Peripheral chemoreceptors:
  effect of hypoxia on, 665-666
  in regulation of respiration pattern during asphyxiation, 686-687

Phosphoprotein phosphatases, neuronal signal termination by, 93-94
Phosphorylation potential, 617-618
Piloerection (fight-or-flight response), 661
Postnatal development, diaphragm fiber type proportions during, 407-415
  correlation between diaphragm contractile properties and fiber type proportions, 413-414
  fiber type composition of muscle units, 411-413
  histochemical classification of fiber types in the neonate, 407-409
  histochemical classification of muscle fiber types, 407
  immunohistochemical classification of muscle fiber types, 409-410
  influence of activity on fiber type differentiation, 410-411
  rate of myofibrillar ATPase activity in diaphragm muscle fibers, 414-415
Postsynaptic neuron, 24-31
  G protein-linked receptors and signal transduction molecules, 27-31
  ligand- or transmitter-gated channels (ionophores), 24-27
    L-glutamate receptors, 25-27
    nicotinic acetylcholine receptor, 25
Presynaptic neuron, 13-24
  biosynthesis of neurotransmitters and peptides, 16-18
  coexistence of multiple chemical messengers in neurons, 18-24
  identification of chemical messengers, 15-16
  transmitter release, 13-15
Progesterone, 5
  facilitation of respiration and, 39-42
Prostaglandins, 5, 668, 669

Protein kinase C, 76-84
  cell-specific expression, 78
  compartmentalization, 78-79
  developmental expression, 78
  genetic modulation, 83-84
  long-term potentiation, 83
  mechanism of activation, 79-80
  molecular properties, 77-78
  substrates, 80-83
G Protein-linked receptors, 27-31
Pulmonary and airway receptors, 219-
    244
  lower airway receptors, 230-239
    afferent innervation, 230-231
    functional characteristics, 231-
      239
  upper airway receptors, 220-230
    afferent innervation, 220-224
    functional characteristics, 224-
      230
Pulmonary and chest wall reflexes in-
    fluencing breathing, 245-269
  implications for future research,
    262-263
  reflex responses associated with
    vagal airway receptors, 247-261
    effects during expiration, 254-
      259
    effects during inspiration, 247-
      254
    reflex responses with important
      inspiratory and expiratory
      components, 259-261
  reflexes associated with chest wall
    receptors, 261-262
  specific considerations with respect
    to development, 246-247

R

Rapid eye movement (REM) sleep:
  electrophysiologic aspects of, 522,
    523-535
  excitatory postsynaptic potentials
    (EPSPs) and, 527-528
  in the newborn, 553

Rapidly adapting receptors (RARs),
    237
Redundancy of respiratory control,
    2-3
Reflex responses with important in-
    spiratory and expiratory con-
    trol, 259-261
Respiration-related unit activity
    (RRUs) in the medulla, studies
    of, 200-201, 206
Respiratory arrhythmia, 5
Respiratory control, 1-7
  afferent input to respiration, 4-5
  breathing behavior and, 1-2
  complexity of, 2-3
  control network in early life, 199-
    215
    future directions/problems, 213
    studies using extracellular record-
      ing and related techniques,
      200-207
    studies using intracellular record-
      ing techniques, 207-212
  growth factors and, 5
  hormones and, 5
  importance of respiratory timing, 4
  neurochemicals and, 4
  rates of maturation, 6
  redundancy of, 2-3
  relation of emotion and state to, 3
  respiratory arrhythmia and, 5
  structural versus functional abnor-
    malities, 3-4
Respiratory motor output, 551-587
  determination of state, 552-554
    state definition in the fetus, 554
    state definition in the infant,
      552-554
  fetal breathing movements, 554-558
    metabolic control, 557-558
    upper airway activity, 558
  metabolic control, 571-574
    CO_2 response, 573-574
    hypoxia, 571-573
  newborn period, 558-567
    respiratory muscles, 558-567

[Respiratory motor output]
respiratory muscle output, 567-571
lung volume, 567-568
pattern of breathing, 569-571
ventilation, 568-569
Respiratory muscle fatigue, 6
Respiratory muscles' response to load, 455-482
development of respiratory muscles, 472-473
how respiratory muscle training modulates the response to acute loads, 466-472
response to acute loads, 456-466
diaphragmatic function, 457-461
metabolic basis of diaphragmatic fatigue, 461-464
neural output and neuromuscular transmission in response to load, 465-466
ventilatory response, 456-457
Respiratory timing, importance of, 4

S

Serotonin, 5
Serotoninergic projections to the spinal cord, development of, 182-183
Signaling molecules, families of, 73-76
Sleep:
apneic spells in, 693-694
control of motoneurons at the membrane levels during, 525-528
importance to respiratory control of, 3
sleep state determination in the newborn, 553
Slowly adapting receptors (SARs), 233-237
Somatostatin, 669
Somatostatin (SOM) immunoreactive neurons, 50-53

Structural abnormalities or respiratory control, 3-4
Substance P, 5
Succinate dehydrogenase (SDH), 391-392
variability among diaphragm muscle fibers, 392-397
Sudden death syndrome (SIDS):
hypoxia and, 672
infantile apneic spells and, 681-682, 693-694
Superior laryngeal nerve (SLN) afferent fibers, 272, 273
Suprapontine inputs to the brainstem, effect of hypoxia on, 665
Sweating, 661
Sympathetic/adrenergic control of airway smooth muscle, 502-506
β-adrenergic function, 502-505
development of α-adrenergic receptors, 505-506
Sympathetic afferent fibers, 239
Sympathoadrenal system, effect of hypoxia on, 666
Synapsin I, 15
CaMKinase II phosphorylation of, regulation of synaptic communication by, 90-92

T

Tachycardia, 661
Theophylline, 665
Thyrotropic-releasing hormone (TRH), 5
effect on ventral NTS cells of, 141-143
Thyrotropin-releasing hormone (TRH) mRNA, 52-53
Time, respiratory, importance of, 4
Tissue neuroregulatory system, effect of hypoxia on, 666-667
Tonic vagal input, 256-257
Transmitters, *see* Classical transmitters; Neurotransmitters
Trigeminal afferents, 224-225

U

Upper airway receptors, 220-230
  afferent innervation of, 220-224
  functional characteristics, 224-230
    glossopharyngeal afferents, 225
    laryngeal afferents, 225-230
    trigeminal afferents, 224-225

V

Vagal afferents in the regulation of respiratory pattern during asphyxiation, 687-688
Vagal airway receptors, reflex responses associated with, 247-261
  effects during expiration, 254-259
  effects during inspiration, 247-254
  reflex responses with important inspiratory and expiratory components, 259-261
Vagal control of airway smooth muscles in the newborn, 487-501
  cholinergic excitatory innervation, 487-498
  nonadrenergic noncholinergic inhibitory innervation, 499-501
  noncholinergic excitatory innervation, 498-499
  summary of vagal airway innervation, 501
Ventilatory effects of hypoxia in newborn animals, 645-646

Ventilatory response to acute respiratory load, 456-457
Ventilatory response to hypoxia, 6, 591-614
  cellular and molecular response, 595-606
    changes in membrane potential during hypoxia, 596-597
    comparative hypoxia studies, 603-606
    hypoxia and CNS function, 596
    intracellular metabolism, 601-602
    membrane ionic fluxes: channels and pumps, 597-601
    protective mechanisms, 606
    reversibility of neuronal hypoxic changes, 602-603
  integrated response, 592-595
    metabolic response, 595
    ventilatory response, 592-594
Ventral region of the tractus solitarius (v-NTS) neurons, 161-168
Ventral respiratory group (VRG), 128-129
Ventral respiratory group (VRG) neurons, 644-645
  effect of laryngeal reflexes on, 278-280
Ventrolateral medulla (VLM), 341
  expanded role for neurons within, 348-354
  location of central chemoreceptors on, 342-348
  mechanisms of central chemoreceptors on, 354-358